Advanced Technologies and Societal Change

This series covers monographs, both authored and edited, conference proceedings and novel engineering literature related to technology enabled solutions in the area of Humanitarian and Philanthropic empowerment. The series includes sustainable humanitarian research outcomes, engineering innovations, material related to sustainable and lasting impact on health related challenges, technology enabled solutions to fight disasters, improve quality of life and underserved community solutions broadly. Impactful solutions fit to be scaled, research socially fit to be adopted and focused communities with rehabilitation related technological outcomes get a place in this series. The series also publishes proceedings from reputed engineering and technology conferences related to solar, water, electricity, green energy, social technological implications and agricultural solutions apart from humanitarian technology and human centric community based solutions.

Major areas of submission/contribution into this series include, but not limited to: Humanitarian solutions enabled by green technologies, medical technology, photonics technology, artificial intelligence and machine learning approaches, IOT based solutions, smart manufacturing solutions, smart industrial electronics, smart hospitals, robotics enabled engineering solutions, spectroscopy based solutions and sensor technology, smart villages, smart agriculture, any other technology fulfilling Humanitarian cause and low cost solutions to improve quality of life.

Soney C. George · Benjamin Tawiah
Editors

Nanotechnology in Societal Development

Springer

Editors
Soney C. George
Centre for Nanoscience and Technology
Amal Jyothi College of Engineering
Kanjirappally, Kerala, India

Benjamin Tawiah
Department of Industrial Art
College of Art and Built Environment (CABE)
Kwame Nkrumah University of Science and Technology (KNUST)
Kumasi, Ghana

ISSN 2191-6853 ISSN 2191-6861 (electronic)
Advanced Technologies and Societal Change
ISBN 978-981-97-6183-8 ISBN 978-981-97-6184-5 (eBook)
https://doi.org/10.1007/978-981-97-6184-5

This Springer imprint is published by the registered company Springer Nature Singapore Pte Ltd.
The registered company address is: 152 Beach Road, #21-01/04 Gateway East, Singapore 189721, Singapore

Dedicated to the brilliant authors whose contributions have enriched the pages of this book, and to the trailblazing pioneers of nanotechnology shaping our future, and to the innovative leaders driving progress in the industry.

—Prof. Soney C. George and
Dr. Benjamin Tawiah

Foreword

It is my distinct honor and privilege to write the foreword for the enlightening and thought-provoking book, *Nanotechnology in Societal Development*. As we stand at the forefront of a new era defined by remarkable advancements in science and technology, the profound impact of nanotechnology on societal progress has never been more palpable than before.

This book represents a seminal contribution to the discourse on nanotechnology, offering a comprehensive exploration of its multifaceted implications for societal development. Through a rich tapestry of insights from esteemed experts, it navigates the complex intersections of nanoscience, engineering, and societal dynamics, shedding light on the transformative potential and ethical considerations inherent in the widespread adoption of nanotechnology.

Nanotechnology, with its ability to manipulate matter at the nanoscale, has transcended disciplinary boundaries, permeating diverse fields such as medicine, energy, environment, and communication. The chapters within this book encapsulate the collective wisdom of scholars and practitioners who have delved into the frontiers of nanotechnology, unraveling its promise in revolutionizing healthcare, enabling sustainable solutions, and shaping the landscape of information technology.

As I perused the pages of this book, I was struck by the depth of insight and the breadth of perspectives encapsulated within its chapters. From the ethical dimensions of nanotechnology to the potential risks and rewards it presents, this publication serves as a compass, guiding readers through the complex terrain of nanotechnology's societal implications.

I commend the contributors for their erudition and dedication in elucidating the intricate dynamics of nanotechnology and its ramifications for society. Their collective expertise has endowed this book with a richness of knowledge that will undoubtedly inspire scholars, researchers, policymakers, and enthusiasts to engage in critical discourse and innovative endeavors in this burgeoning field.

I am confident that *Nanotechnology in Societal Development* will serve as a beacon, illuminating the path toward responsible and equitable integration of nanotechnology into the fabric of our society. It is my fervent hope that this book will ignite dialogue, foster collaboration, and catalyze ethical and sustainable advancements in nanotechnology for the betterment of humanity.

In closing, I extend my gratitude to the author and the contributors for their scholarly contributions, and to the readers for their curiosity and commitment to advancing knowledge. May this book ignite a fervor for exploration and ethical stewardship in the realm of nanotechnology, propelling us toward a future where the societal benefits of this transformative field are realized to their fullest potential.

Prof. (Dr). Sabu Thomas
Professor of Polymer Engineering
and Nanotechnology
Mahatma Gandhi University
Kottayam, Kerala, India

Preface

It is with great pleasure and enthusiasm that we (the editors) present this book, *Nanotechnology in Societal Development*, to the readers. Nanotechnology, the manipulation of matter at the nanoscale, has emerged as a transformative force with the potential to revolutionize various aspects of our lives. As the field of nanotechnology continues to advance at a rapid pace, its societal impact becomes increasingly significant, influencing fields such as medicine, energy, environment, and communication. This book aims to provide a comprehensive overview of the role of nanotechnology in societal development, exploring its applications, implications, and ethical considerations. Through a multidisciplinary approach, this book delves into the potential benefits and challenges associated with the widespread integration of nanotechnology into our daily lives. From enhancing medical treatments and enabling sustainable energy solutions to addressing environmental concerns and shaping the future of information technology, nanotechnology holds immense promise for societal progress.

In these pages, readers will find a collection of insightful contributions from leading experts in the field, offering diverse perspectives on the intersection of nanotechnology and societal development. The chapters within this book cover a wide range of topics, including nanomedicine, nanoengineering, nanoelectronics, and nanotoxicology, cultural, policy and regulatory perspectives on nanotechnology, among others, providing a holistic understanding of the multifaceted implications of nanotechnology.

We are deeply grateful to the contributors who have shared their expertise and insights, enriching this book with valuable knowledge and perspectives. Their dedication and commitment to advancing the discourse on nanotechnology have been instrumental in shaping this publication. We would also like to express my appreciation to the readers, whose interest in this rapidly evolving field motivates the continual exploration and dissemination of knowledge. It is our sincere hope that this book will serve as a valuable resource for researchers, students, policymakers, and anyone seeking a deeper understanding of the profound impact of nanotechnology on societal development.

As we stand on the cusp of a new era defined by the convergence of nanotechnology and societal progress, this book endeavors to illuminate the opportunities and challenges that lie ahead. May it inspire thoughtful dialogue, foster innovation, and contribute to the responsible advancement of nanotechnology for the betterment of society.

Kanjirappally, India — Prof. Soney C. George
Kumasi, Ghana — Dr. Benjamin Tawiah

Acknowledgments

The writing of this book, *Nanotechnology in Societal Development*, has been a collaborative effort that would not have been possible without the invaluable contributions and support of numerous individuals and organizations. We extend our heartfelt gratitude to all those who have played a role in bringing this project to fruition.

First and foremost, we are deeply thankful to the esteemed contributors whose expertise and dedication have enriched the content of this book. Their insightful perspectives and scholarly contributions have been instrumental in shaping a comprehensive and multidimensional exploration of nanotechnology's impact on societal development. We are indebted to the researchers, scientists, and practitioners whose pioneering work in the field of nanotechnology continues to drive innovation and progress. Their groundbreaking discoveries and tireless commitment to advancing the frontiers of nanoscience have laid the foundation for the transformative potential of nanotechnology in addressing societal challenges.

We would like to express my appreciation to the editorial and production team whose meticulous efforts have been essential in bringing this book to fruition. Their professionalism, attention to detail, and commitment to excellence have been indispensable in ensuring the quality and coherence of this publication.

Furthermore, we are grateful to the academic and research institutions that have provided resources, facilities, and support for the advancement of nanotechnology. Their commitment to fostering a conducive environment for scientific inquiry and technological innovation has been pivotal in driving the progress of this field.

We also extend our thanks to the readers and scholars whose interest and engagement in the subject of nanotechnology have inspired the continual exploration and dissemination of knowledge. It is our sincere hope that this book will serve as a valuable resource for furthering understanding and dialogue in this dynamic and rapidly evolving field.

Lastly, we would like to acknowledge the unwavering support of our management; Amal Jyothi College of Engineering, India, and the Kwame Nkrumah University of Science and Technology, Kumasi, Ghana, head of Centre for Nanoscience and Technology, AJCE, and the head of Department of Industrial Art, KNUST, colleagues, our families, and friends, for their immense support throughout the process of conceptualizing and developing this book. Their encouragement, understanding, and patience have been a source of strength and motivation.

It is with deep appreciation and humility that we acknowledge the collective efforts of all those who have contributed to the realization of this publication. Their involvement has been indispensable in shaping this book into a meaningful and substantive contribution to the discourse on nanotechnology and its societal implications.

Prof. Soney C. George
Dr. Benjamin Tawiah

About This Book

The book *Nanotechnology in Societal Development* is an important contribution to the field of nanotechnology and its impact on society. It provides a comprehensive perspective on the global scientific, technological, and societal implications of nanotechnology and explores future opportunities and research directions.

The book covers a range of topics related to nanotechnology and its societal impact, including its potential benefits and challenges. It offers insights from renowned nanotechnology experts and provides a long-term perspective on future opportunities in this field. The authors evaluate the impact of nanoscience and nanotechnology research and provide recommendations to decision-makers in academia, industry, and government for productive and responsible paths forward for nanotechnology research and development.

In addition to discussing the scientific and technological aspects of nanotechnology, the book also addresses the ethical, legal, social, economic, and educational issues that arise from its implementation. It highlights the need for innovative and responsible governance of nanotechnology to ensure maximum benefit for humanity while minimizing any potential negative impacts.

Furthermore, the book emphasizes the importance of public perception and engagement in shaping the development and acceptance of nanotechnology. It acknowledges that societal response to nanotechnology can be influenced by various factors such as trust, risk perception, and perceived benefits. Understanding these factors is crucial for effective governance and responsible implementation of nanotechnology.

In conclusion, *Nanotechnology in Societal Development* is a valuable resource for anyone interested in the field of nanotechnology and its implications for society. It provides a comprehensive overview of the global impact of nanotechnology, highlights future research directions, and addresses the ethical and societal challenges associated with its implementation. The book aims to guide decision-makers and researchers toward productive and responsible paths for nanotechnology research and development.

Contents

1 **Nanotechnology in Societal Development** 1
Benjamin Tawiah, Emmanuel Abankwah Ofori, and Soney C. George

2 **Nanotechnology: Social Acceptance, Cultural Impact and Privacy** 65
Peramjeet Singh, Priya Gupta, Vernica Verma, Neetu Yadav, and Narendra Kumar Pandey

3 **The Promise of Nanotechnology in Healthcare** 95
Raphael Kanyire Seidu, Anna Doersam, Emmanuel Abankwah Ofori, and Shouxiang Jiang

4 **The Role of Nanotechnology in Agriculture and Food Security** 149
Leticia Donkor and Eric Banan-Mwine Daliri

5 **Nanotechnology and the Environment: Opportunities and Challenges** 191
Bismark Sarkodie, Benjamin Tawiah, and Bin Fei

6 **Nanotechnology, Global Development: Bridging the Digital Divide** 229
Sujata Deb

7 **Nanotechnology and Education: Preparing the Next Generation of Future Leaders** 247
Benjamin Tawiah, Raphael Kanyire Seidu, George Kwame Fobiri, and Cynthia A. Chichi

8 **The Intersection of Nanotechnology and Biotechnology Implications for Human Health** 271
Akash A. Podutwar, Purva U. Chandorkar, Anuruddha R. Chabukswar, Satish A. Polshettiwar, and Swati C. Jagdale

9 Nanotechnology, Artificial Intelligence, and the Future of Manufacturing 307
Promila Sharma, Varinder Singh, Anita Thakur, and Ashok Kumar Malik

10 Nanotechnology and Water: Emerging Solutions for a Global Crisis 333
Benjamin Tawiah, Anna M. Abraham, Justin J. Kappen, Nikhi Maria Raju, C. Surej Rajan, and Soney C. George

11 Nanotechnology and Materials Science: Innovations for Industry and Infrastructure 375
Yuyang Zhou, Jingjing Wang, Chunguang Liang, and Feiyang Zheng

12 The Role of Nanotechnology in Space Exploration and Colonization 395
Peeyush Phogat, Shreya, Ranjana Jha, and Sukhvir Singh

13 Nanotechnology in Defense and Social Justice: Opportunities and Risks 439
S. K. Suja and S. Mathiya

14 Nanotechnology and Governance: Regulatory Framework for Responsible Innovation 481
Chinnaiah Amutha, Ardhra Gopan, Indhiran Pushbalatatha, Mariappan Ragavi, and Joseph Antony Reneese

15 Considerations in the Development and Deployment of Nanotechnology 505
Vernica Verma, Priya Gupta, Peramjeet Singh, and Narendra Kumar Pandey

Contributors

Anna M. Abraham Center for Nanoscience and Technology, Amal Jyothi College of Engineering Kanjirappally, Koovappally, Kerala, India

Chinnaiah Amutha Department of Animal Behaviour and Physiology, School of Biological Sciences, Madurai Kamaraj University, Madurai, India

Anuruddha R. Chabukswar School of Health Sciences and Technology, Department of Pharmaceutical Sciences, Dr. Vishwanath Karad MIT World Peace University, Kothrud, Pune, India

Purva U. Chandorkar School of Health Sciences and Technology, Department of Pharmaceutical Sciences, Dr. Vishwanath Karad MIT World Peace University, Kothrud, Pune, India

Cynthia A. Chichi Department of Industrial Art (Fashion and Textiles), Kwame Nkrumah University of Science and Technology, Kumasi, Ghana

Eric Banan-Mwine Daliri Department of Biological Models, Institute of Biochemistry, Life Sciences Center, Vilnius University, Vilnius, Lithuania

Sujata Deb Department of Physics, The Assam Royal Global University, Guwahati, Assam, India

Anna Doersam The Henry Royce Institute, Manchester, UK;
Department of Materials, The University of Manchester, Oxford Rd, Manchester, UK

Leticia Donkor Department of Food Process Engineering, School of Engineering Sciences, College of Basic and Applied Sciences, University of Ghana, Legon, Accra, Ghana

Bin Fei School of Fashion and Textiles (SFT), Hong Kong Polytechnic University, Hong Kong SAR, China

George Kwame Fobiri Department of Fashion Design and Textiles Studies, Kumasi Technical University, Kumasi, Ghana

Soney C. George Center for Nanoscience and Technology, Amal Jyothi College of Engineering Kanjirappally, Koovappally, Kerala, India;
Centre for Nanoscience and Technology, Amal Jyothi College of Engineering, Koovappally, Kottayam, India

Ardhra Gopan School of Biological Sciences, Madurai Kamaraj University, Madurai, India

Priya Gupta Department of Physics, University of Lucknow, Lucknow, Uttar Pradesh, India;
Electronics Department, Manyawar Kanshiram Government Polytechnic, Tirwa, Kannauj, UP, India

Swati C. Jagdale School of Health Sciences and Technology, Department of Pharmaceutical Sciences, Dr. Vishwanath Karad MIT World Peace University, Kothrud, Pune, India

Ranjana Jha Department of Physics, Netaji Subhas University of Technology, New Delhi, India

Shouxiang Jiang School of Fashion and Textiles, The Hong Kong Polytechnic University, Kowloon, Hong Kong SAR, China;
Research Institute for Intelligent Wearable Systems, The Hong Kong Polytechnic University, Kowloon, Hong Kong SAR, China

Justin J. Kappen Center for Nanoscience and Technology, Amal Jyothi College of Engineering Kanjirappally, Koovappally, Kerala, India

Chunguang Liang National Engineering Laboratory for Modern Silk, College of Textile and Clothing Engineering, Soochow University, Suzhou, China

Ashok Kumar Malik Department of Chemistry, Punjabi University, Patiala, Punjab, India

S. Mathiya Department of Chemistry, Lady Doak College, Tamil Nadu, Madurai, India

Emmanuel Abankwah Ofori Department of Materials Science and Engineering, Hubei University of Automotive Technology, Shiyan, Hubei, China

Narendra Kumar Pandey Department of Physics, University of Lucknow, Lucknow, UP, India

Peeyush Phogat Department of Physics, Netaji Subhas University of Technology, New Delhi, India

Akash A. Podutwar School of Health Sciences and Technology, Department of Pharmaceutical Sciences, Dr. Vishwanath Karad MIT World Peace University, Kothrud, Pune, India

Satish A. Polshettiwar School of Health Sciences and Technology, Department of Pharmaceutical Sciences, Dr. Vishwanath Karad MIT World Peace University, Kothrud, Pune, India

Indhiran Pushbalatatha School of Biological Sciences, Madurai Kamaraj University, Madurai, India

Mariappan Ragavi School of Biological Sciences, Madurai Kamaraj University, Madurai, India

C. Surej Rajan Center for Nanoscience and Technology, Amal Jyothi College of Engineering Kanjirappally, Koovappally, Kerala, India;
Department of Chemical Engineering, Amal Jyothi College of Engineering Kanjirappally, Koovappally, Kerala, India

Nikhi Maria Raju Center for Nanoscience and Technology, Amal Jyothi College of Engineering Kanjirappally, Koovappally, Kerala, India;
Department of Chemical Engineering, Amal Jyothi College of Engineering Kanjirappally, Koovappally, Kerala, India

Joseph Antony Reneese School of Biological Sciences, Madurai Kamaraj University, Madurai, India

Bismark Sarkodie School of Textile and Clothing, Anhui Polytechnic University, Anhui Province, Wuhu, China

Raphael Kanyire Seidu School of Fashion and Textiles, The Hong Kong Polytechnic University, Hung Hom, Kowloon, Hong Kong SAR, China

Promila Sharma Department of Chemistry, RIMT University, Mandi, Gobindgargh, Punjab, India;
Department of Chemistry, Punjabi University, Patiala, Punjab, India

Shreya Department of Physics, Netaji Subhas University of Technology, New Delhi, India

Peramjeet Singh Department of Physics, University of Lucknow, Lucknow, UP, India

Sukhvir Singh Department of Physics, Netaji Subhas University of Technology, New Delhi, India

Varinder Singh Department of Chemistry, RIMT University, Mandi, Gobindgargh, Punjab, India

S. K. Suja Department of Chemistry, Lady Doak College, Tamil Nadu, Madurai, India

Benjamin Tawiah Department of Industrial Art (Textiles), Kwame Nkrumah University of Science and Technology, Kumasi, Ghana;
School of Fashion and Textiles (SFT), Hong Kong Polytechnic University, Hong Kong SAR, China;

School of Fashion and Textiles, The Hong Kong Polytechnic University, Hung Hom, Kowloon, Hong Kong SAR, China;
Department of Industrial Art (Fashion and Textiles), Kwame Nkrumah University of Science and Technology, Kumasi, Ghana

Anita Thakur Department of Chemistry, RIMT University, Mandi, Gobindgargh, Punjab, India

Vernica Verma Department of Physics, University of Lucknow, Lucknow, UP, India

Jingjing Wang Suzhou Mingwang New Material Technology Co. Ltd, Suzhou, China

Neetu Yadav Department of Physics, University of Lucknow, Lucknow, Uttar Pradesh, India

Feiyang Zheng National Engineering Laboratory for Modern Silk, College of Textile and Clothing Engineering, Soochow University, Suzhou, China

Yuyang Zhou National Engineering Laboratory for Modern Silk, College of Textile and Clothing Engineering, Soochow University, Suzhou, China

Chapter 1
Nanotechnology in Societal Development

Benjamin Tawiah, Emmanuel Abankwah Ofori, and Soney C. George

Introduction to Nanotechnology: Foundations and Evolution

Over millennia, scientists have predicted the advancement and transformation of technology and instruments made possible by nanotechnology, which manipulates matter at the atomic level and inspires hope for advancement [1]. The history of nanotechnology is a subject of significant debate; nonetheless, some scientists date the earliest glimpses of these discoveries to the early 1800s, while others describe them in the late 1980s [2]. Maxwell's Demon, a novel theory by James Clerk Maxwell regarding sub-microscopic entities that may control individual molecules and atoms, was first proposed in 1867. This was one of the first times that nanotechnology was mentioned as a notion [3]. In a similar vein, other theories contend that the earliest known evidence of nanotechnology dates to 1959, or possibly even earlier, when physicist Richard Feynman delivered his well-known lecture "There is plenty of room at the bottom," emphasising the usage of nanomaterials in technology, allowing for the precise manipulation of atomic-level micrometre instrumentation and production tools [2] and coined the term nano dentistry [4].

B. Tawiah (✉)
School of Fashion and Textiles (SFT), The Hong Kong Polytechnic University, Hung Hom, Kowloon, Hong Kong SAR, China
e-mail: benjamin.tawiah@knust.edu.gh

Department of Industrial Art (Fashion and Textiles), Kwame Nkrumah University of Science and Technology, Kumasi, Ghana

E. A. Ofori
Department of Materials Science and Engineering, Hubei University of Automotive Technology, Hubei, China

S. C. George (✉)
Centre for Nanoscience and Technology, Amal Jyothi College of Engineering, Koovappally, Kottayam 686518, India
e-mail: soneycgeorge@amaljyothi.ac.in

S. C. George and B. Tawiah (eds.), *Nanotechnology in Societal Development*, Advanced Technologies and Societal Change, https://doi.org/10.1007/978-981-97-6184-5_1

Nano is said to have Greek origins, meaning "very small or dwarf". Generally, nanotechnology is a term used when referring to materials that are 1–100 nm in size [5, 6] (see Fig. 1.1). Thus, a nanometer is one billionth of a metre, smaller than the wavelengths of visible light and about the size of a few atoms lined up next to each other [7]. Although the word "technology" refers to an improvement in tools, "nanotechnology" was coined by the Japanese researcher Norio Taniguchi [8, 9] of the Tokyo University of Science in 1974 to describe machining with tolerances less than a micron. Unfortunately, the field of nanotechnology did not come into being until 1981, when Nobel Prize winners Gerd Binning and Heinrich Rohrer created the scanning tunnelling microscope (STM), which allowed researchers to observe and manipulate individual atoms [7, 10]. This kickstarted the direct manipulation techniques envisioned by Feynman and sparked intense research interest in the nanoworld. In 1985, Richard Smalley, Curl, and Kroto had discovered spherical carbon nanoparticles called fullerenes (buckyballs) [11] and by 1986, Prof. Kerie Eric Drexler, a researcher and writer published the first book on nanotechnology [12].

The year 1991 saw the discovery of carbon nanotubes by Sumio Iijima, which are microscopic cylinders composed of rolled-up graphene sheets with exceptional strength, conductivity, and semiconducting qualities [7, 14]. Carbon nanoparticles thus provided an incredible new family of materials for applications ranging from targeted medication delivery to supercapacitors. For instance, Doxil, the first licenced nanoparticle-based cancer medication of composite nanoparticles. It demonstrated an improved efficacy in treating Kaposi's sarcoma in 1995 [15]. The advancement of nanoscience and nanotechnology was the explicit aim of several prominent government and company endeavours initiated during the 1990s. In the early 2000s, this prompted the establishment of the first National Nanotechnology Initiative (NNI) to build quantum mirrors and corrals [7]. In the years that followed, several more recent breakthroughs were made, such as nanometer-scale electronic devices and DNA

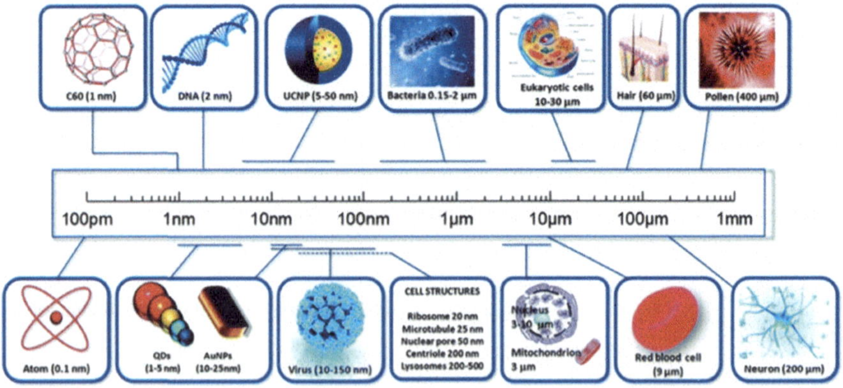

Fig. 1.1 A comparison of sizes of nanomaterials [13]. Reprinted with permission from the Royal Society of Chemistry

self-assembly of sophisticated structures. In 2011, nanotechnology evolved quickly in producing new nanomaterials and advancements including speedier computer chips. Richard Zsigmondy introduced the idea of nanomaterials early in the twentieth century [4]. Nonetheless, nanotechnology is very young and opens up new areas for scientific study [16]. Tyre producers have used carbon black to enhance tyre performance since the 1920s, long before they were aware of nanoparticles [17]. The modern world is greatly impacted by the numerous applications of nanotechnology in both society and the environment. Their many features, including the large surface area, durability, and conductivity, have led to the replacement of many bulk materials with nanotechnology, which has applications in practically every industry [18]. For example, nanomaterials such as carbon nanotubes, graphene, metal oxides, nano clays, and polymers are employed in a variety of applications, including beauty items [19–21], pharmaceuticals [22, 23], information and communication technology [24], defence, automobiles [25, 26], sports [27, 28], chemical synthesis [29], water purification [30, 31], magnetic and electrical applications [32, 33], food industry [34, 35], biotechnology [36], etc. The distinctive and erratic differences in characteristics between the qualities at the macro and nanoscales are mostly caused by surface and quantum effects, which are more common at the nanoscale level [37]. This permits exact property tweaking through specialised nanofabrication techniques and synthesis pathways, as well as the very arrangement and orientation of atoms and molecules. For example, aluminium that is stable at the macroscale turns flammable at the nanoscale, and gold nanoparticles look red rather than golden at the nanoscale [38]. These artificial or organic nanoparticles can be applied in a variety of ways, such as thin film coating, polymers, connecting materials, reinforcement or filler materials, and diagnostic tools. To enhance performance in the human and social dynamics priority area, the National Science Foundation has integrated convergent nanoscale technologies in the Nanoscale Science and Engineering priority area (since FY 2003) [39]. Nonetheless, ethical, environmental, legal, and social concerns must coexist with responsible development. The United Nations, realising that ending poverty and other forms of deprivation is intrinsically linked to improving the environment, the economy, and public health, issued an urgent call to action for all countries in 2017 with its Sustainable Development Goals (SDGs) to end inequality worldwide [9]. Despite the progress and contribution technology has made to societal development, its downside from the perspective of many has overshadowed the positives with few documentations highlighting the positives to the negatives. As a hotspot worth discussing, this introductory chapter emphases the role of nanotechnology in the realm of Societal development, highlighting its economic implications, ethical dimensions, environmental impact and sustainability, risk assessment, cultural representation and societal imagination, and envisioning its future scenarios in facilitating societal progress.

Economic Implications of Nanotechnology

A globally sustainable economy depends on manufacturing. It is the prime force behind invention and higher value employment creation in both industrialised and emerging nations [40]. Nanotechnology is poised to transform economies and industries globally through the unprecedented control, precision, and manufacturing capabilities it enables at the atomic scale [24]. It promises better materials, technologies, more productive industrial processes, and new chances for sector-wide growth [9]. However, ethical, legal, environmental, health and social ramifications must be balanced with economic gains to unlock its full potential. The global nanotechnology industry is expected to reach $74.1 billion by 2032 at a cumulative annual growth rate (CAGR) of 34.5%, from a value of about $3.78 billion in 2020 [41]. Significantly, the National Science Foundation predicts that by 2025, the global nanotechnology sector will need roughly 6 million jobs. Additionally, significant expansions are expected in semiconductor manufacturing, energy, drug delivery, accessible drinking water, life sciences, textiles, food technology, and consumer-based nanoparticle-enhanced goods [7]. The economically most significant aspects of nanotechnology today and most likely in the near future are the use and usage of nanomaterials in optical and magnetic components, electronic and mechanical devices, quantum computing, tissue engineering, and other biotechnologies [7]. In addition to promoting energy efficiency in solar cells, battery catalysts, and filtering systems, lighter, stronger composites and environmentally friendly production techniques reduce fuel costs in the transportation industry [42]. Food shelf life and safety are improved via antimicrobial nano-packaging [43, 44] and medical nanotechnology aims to lessen healthcare costs through early diagnosis, targeted therapy, and rapid healing applications [45]. Since nanoscale product sales are likely to rise and their full promise depends on science and conscience, these translate to direct and indirect economic rewards. Therefore, the unit discusses the economic implications of nanotechnology under the subtopics below.

Emergence of New Industries and Markets Driven by Nanotechnology

Nanotechnology is enabling the emergence of entirely new industries and transforming existing markets [24, 46]. At the nanoscale, unique optical, electrical, and other qualities enable advancements in electronics, transportation, materials, agricultural and environmental science, energy, medicine, consumer products, etc. [36, 47] (see Fig. 1.2). With massive nanotechnology product research and development, more practical materials with unique applications could emerge. As of 2015, nano-related products and services were expected to be worth over $1 trillion [41]. As a result, it was predicted that nano businesses would develop at the highest rate in history, surpassing both the information technology and telecommunication sectors

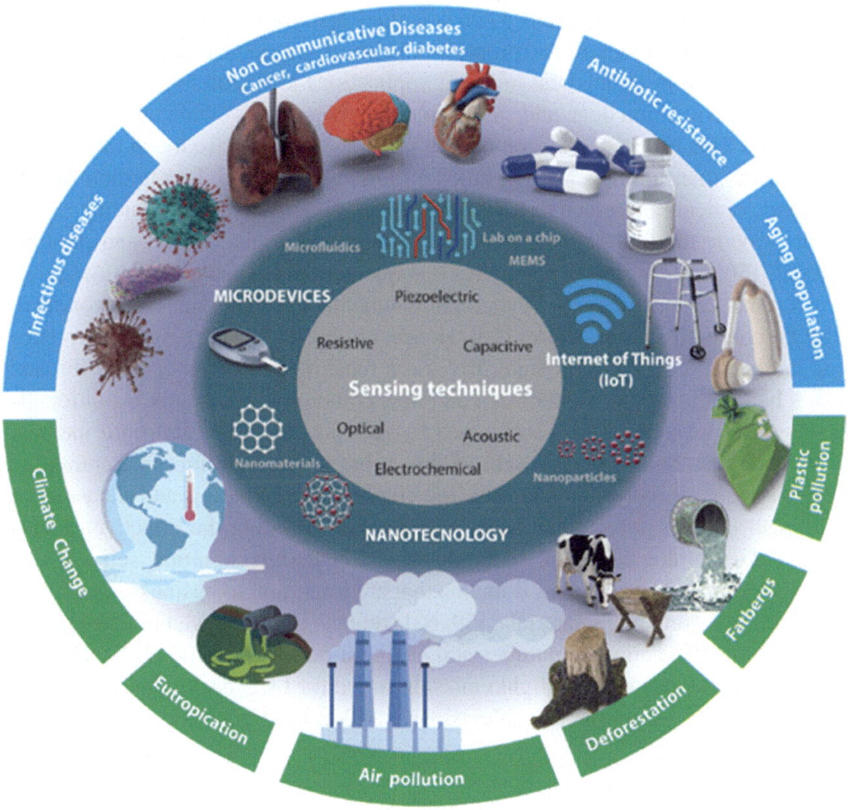

Fig. 1.2 Diagram demonstrating the application of micro/nanotechnology [50]. Open access by MDPI

combined [7]. Importantly, commercialising nanotechnology research results has great potential to benefit society and is profitable. [47]. In short, a wide range of daily products—including computers, cell phones, paints, medications, and antibacterial materials [48], to name a few—are utilising nanotechnologies. Therefore, any industry interested in gaining a competitive edge via the mastery of cutting-edge technology would find nanotechnology intriguing [49]. Importantly, the booming entry of nanomaterials and nanoparticles has foreseen application in numerous areas of society as illustrated in Fig. 1.2.

Electronics and Computing

Nanoelectronics is offering continued miniaturisation and performance improvements for electronics and computing. Nanoparticles (NPs) with at least one dimension less than 100 nm make up these compounds [51, 52]. The significance of these materials became apparent when scientists discovered that a substance's size may affect its physio-chemical properties, such as its optical quality, and therefore, can be used in biomedical imaging applications [53]. Among them are metal nanoparticles, quantum dots, and carbon-based nanoparticles (also known as carbon nanotubes). This is because carbon nanotubes and graphene have much higher electron mobility than silicon, enabling quicker transistors [54, 55]. Due to their distinct structural, optical, and electrical properties, one-dimensional semiconductors and metals provide the foundation for a new class of electronic, sensor, and photonic materials [56, 57]. Similarly, nanowires and quantum dots enable dense, low-power memory and logic devices for beyond complementary metal–oxide–semiconductor (CMOS) computing architectures [58]. Hence, products like faster computer processors and higher-density solid-state drives are reaching consumers, while research pushes boundaries on transistor sizes and new computing paradigms. Additionally, displays and optics are also transformed by nanophotonics research into metamaterials and plasmonics [58]. In light of this, a brain-like neuromorphic computing system is an especially alluring area for combining computing and data storage on a small chip. It provides an ideal platform for imitating the effectiveness and features of the biological brain, including speech, pattern recognition, self-learning, and logic computing, as well as domains that are important to business, like AI and deep learning, using nanomaterials. [59–62] (refer to Fig. 1.3). Manufacturing with nanomaterials like carbon nanotubes and quantum dots on demand enables flexible displays, precise narrow-band light emission suited for LCD backlights, and greater resolution for screens in phones, televisions, and VR headsets [63].

Energy

Over the years, combustion has historically been a significant source of energy conversion for humans but has exhibited adverse environmental effects. Recently, the world has witnessed an increase in primary energy consumption by 86% as of mid-2010 [64]. Because fossil fuels are not renewable, it has been predicted that the world would face constraints and a shortage of them in the years to come [36]. Consequently, scientists have modified their study approaches to produce renewable energy from easily accessible resources at minimal expense, and nanoparticles have been determined as reliable materials. By 2011, it was estimated by Lux Research that the energy storage nanotechnology industry reached approximately US$3.7 billion [65]. Importantly, nanomaterials aim to improve energy generation, storage, and transmission efficiency. Solar cells based on quantum dots, perovskites, and nanoparticle inks promise low-cost, high-efficiency photovoltaics [66]. For example, photoelectrochemical (PEC) and electrochemical water splitting, piezoelectricity creation, and reduction to fuel precursors are all common uses for nanoparticles (NPs) and carbon nanomaterials (CNMs) [67, 68]. In a novel approach to energy generation, nanogenerators are currently being developed that can use piezoelectric technology to convert mechanical energy into electricity [69]. Some energy generating devices that use nanomaterials are depicted in Fig. 1.4. Importantly, graphene and carbon nanotube

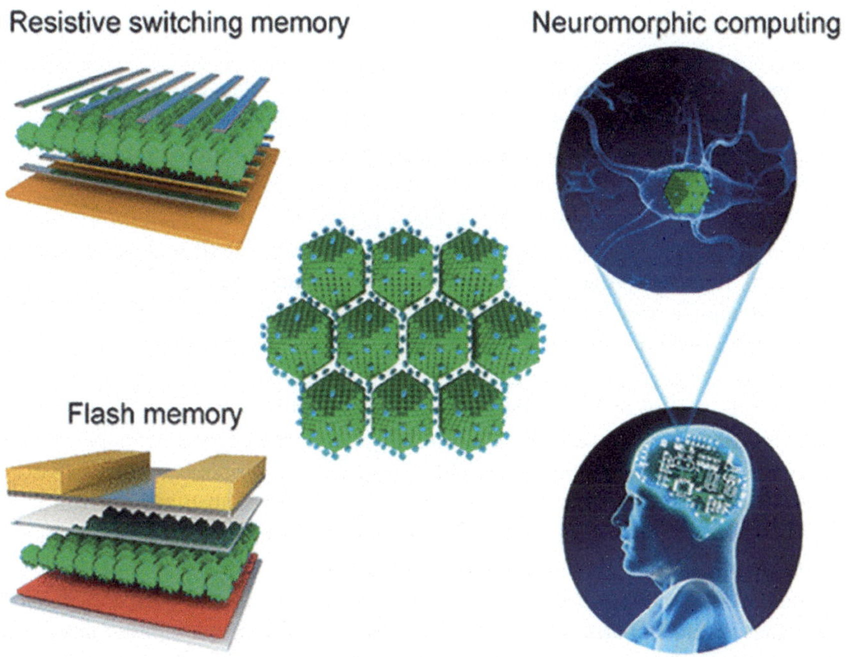

Fig. 1.3 Nanomaterials in electronics and computing [58]. Reprinted with permission from ACS publications

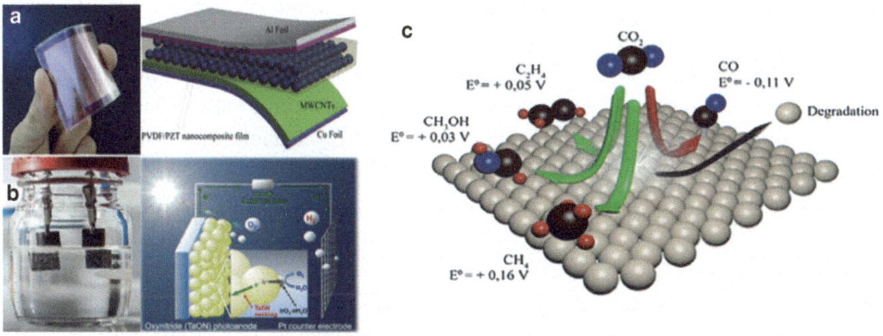

Fig. 1.4 Energy generation approaches from **a** Piezoelectrics actuators **b** Water splitting **c** CO_2 reduction [36]. Open access article by Elsevier B.V.

anodes store more lithium for better batteries, while nano grids and aerogels provide insulation for more efficient heating, ventilation, and air conditioning (HVAC) in buildings. This emphasises the importance of better-performing infrastructure in saving tens of billions of dollars in annual electricity bills.

Medicine

Nanomedicine is already saving lives through targeted drug delivery mechanisms and diagnostic biosensors enabled by nanoparticles and nanotubes [70]. Their attractiveness is related to properties like their high surface area-to-volume ratio, quantum effects, and other unusual occurrences in situations where accuracy, sensitivity, and selectivity are crucial [71]. From an economic perspective, the market for liposomal doxorubicin, which includes the brand-name medication Doxil and other generic versions, is now valued at over $1 billion in the United States. Based on an 8% compound annual growth rate, the market is expected to reach $2.1 billion globally in 2028 [72]. Interestingly, worldwide demand for liposomal medications is growing at an average yearly growth rate of 12.8%. Nanomaterials like dendrimers, liposomes, and polymers can encapsulate medications and deliver them specifically to cancer spots, avoiding systemic damage [73]. Similarly, nanoparticle contrasts also enhance MRI resolution to spot tumours early [74]. For magnetic resonance imaging (MRI), iron oxide nanoparticles (IONPs) have become a prospective substitute for traditional contrast agents (CAs) because of their better magnetic properties, strong biocompatibility, and ease of functionalization [75]. Consequently, it has been proposed that tumours of the breast, stomach, colon, kidney, liver, or brain can be identified using nanomaterials functionalized with epithelial growth factor receptor antibodies, short peptides like RGD, or aptamers, among other things. [75, 76]. Implanted biosensors that monitor blood glucose, organ function, and other parameters can wirelessly communicate data to smartphones and the cloud [77]. Fundamentally, nanofibers made from recyclable polymeric materials have properties such as high surface area, tunable pore size, and exceptional tissue regeneration results [78], whereas nanopore DNA sequencing allows for faster and less expensive genomes. Antimicrobial nanoparticles, such as nano silver, used in surfaces, textiles, and filters, help to keep hazardous germs at bay in public settings. The impregnated textile's capacity to reduce microbial growth or eliminate microbes is valuable in wound dressings, medical staff uniforms, bedsheets, and other applications [79].

Consumer Goods

Nanomaterials are improving consumer items by enhancing specialty fabrics, anti-ageing cosmetics, scratch-resistant coatings, and more durable sporting goods. Cosmetics and dermatological products [80] based on nanomaterials are utilised for therapies, skin diseases, and diagnostic purposes. Dermatological products have a significant therapeutic impact on the skin and other areas of the human body when it comes to medications and formulations for the treatment of disorders including wrinkles, photoaging, dry skin, dark spots, hair damage, and hyperpigmentation. [81]. Other nanoparticles utilised in the nanocosmetics sector include fullerenes, liposomes, aquasomes, niosomes, dendrimers, and polymeric or metallic nanoparticles. These nanoparticles provide particular properties like increased absorption, enhanced aesthetic value, and increased shelf life [82]. More particularly, the development of sunblock lotion creams along with other sun-protective cosmetics is encouraged by the use of nanomaterials. Titanium dioxide nanosuspensions, which provide sun protection in cosmetics, and cinnamates extracted from carnauba wax are the secret ingredients [83]. Likewise, solutions of nanoparticles are being used as nanostructured lipid carriers (NLCs) for pharmacological and dermatological uses [84]. They demonstrate the capabilities of direct drug targeting, occlusion, improved skin penetration and absorption, and regulated drug-carrying and realising qualities. Additionally, these carrier nanoemulsions exhibit excellent resistance to challenging physiological and environmental conditions since the lipid nanocarriers have been studied and shown to be healthy for both medical and cosmetic uses [85].

Environmental Remediation

Nanotechnology also offers more sustainable solutions for pressing environmental issues from water purification as membrane technologies to CO_2 capture [86] (see Fig. 1.5a). Utilizing mono-, bi-, tri-, and tetra metallic oxide nano adsorbents with increased surface energy, porosity, surface area, and size-dependent surface morphology, highly active adsorption sites have been generated for the adsorption of pollutants [42]. At the hybridised carbon–metal interfaces, for instance, carbon-based nanomaterials (CNMs) can reduce contaminant reaction potentials or preconcentrate pollutants [87, 88]. This facilitates the ensuing catalytic/redox processes [89]. Magnetic nanoparticles and nano filters in water remove chemicals (Fig. 1.5b), while metal oxides capture flue gas CO_2 emissions. The University of Stanford's Jennifer Dionne research group has revealed phase-changing nanoparticles and estimated that nanomaterials might absorb approximately a gigaton of CO_2 per year at coal plants, bringing hope for ecosystem restoration. Carbon nanomaterial-based heavy metal sensors (CMNHs) are capable of detecting biomacromolecules, phenolics, pesticides, heavy metals, and antibiotics. The performance of sensing can be enhanced by many surface functional groups, excellent electron and electrochemical charge transfer, high sensitivity, excellent photostability, and biocompatibility [90]. For instance, the graphene-Bi framework can identify heavy metal ions in situ at concentrations of 0.02–4 μg/L, 40–300 μg/L of Zn(II), and 1–120 μg/L of Pb(II) and Cd(II). It is claimed that the massive SSA, fast mass and electron transmission, and tunable nanoarchitecture of the nanohybrids are exclusive [91]. Additionally, as absorption sensing materials, Raghav [42] notes that dye-doped polymeric and silica nanoparticles have higher luminous intensity depending on the molecule. Since ozone and chlorination have substantially improved human health, nanoparticles of zinc, silver, cerium, titanium oxides, fullerenes, and carbon nanotubes (CNTs) exhibit strong antibacterial qualities. When dangerous metal ions (such as nC60, CNTs, and Ce oxide NPs) leach from the NMs, they either directly penetrate the cell membrane of the microorganism or produce reactive oxygen species (ROS, amino fullerene, and TiO_2) with less DBPs [92, 93].

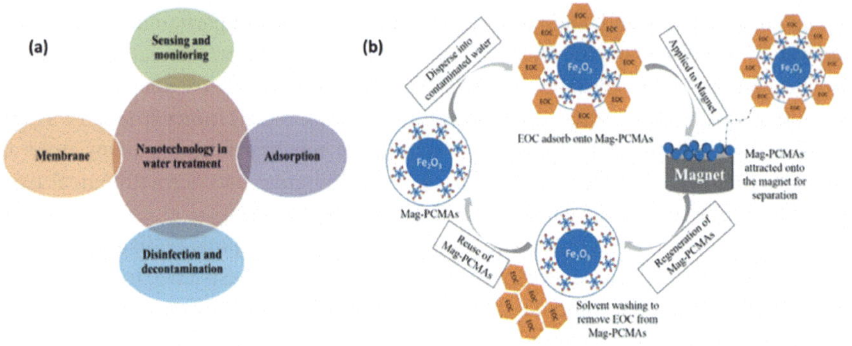

Fig. 1.5 **a** Nanotechnology applications for water treatment [42], **b** Adsorbents with magnetic nanoparticles for organic pollutants [94]

Job Creation and Shifts in Employment Patterns

The commercialization of nanotechnologies is having a substantial impact on jobs and employment trends across sectors [95]. The consequences of developing technologies, such green nanotechnologies, for economic growth, including effects on employment and wages, particularly concern policymakers [65]. Although forecasting the number of new occupations is impossible, it is estimated that about 6 million jobs consisting of entirely new positions centred on nanoscience research, applied nanoengineering, and integration of nanomaterials into industrial processes, linked sectors, and services are emerging. Importantly, when traditional products are phased out, existing workers may transition into green nanotechnology operations [96]. For example, industries ranging from energy to health are already experiencing shifts in skill demands that will only increase. Estimates of the number of jobs in nanotechnology in the United States as of 2015 varied from tens of thousands to about one million [96]. Depending on how many jobs are classified as green, a sizable chunk of the nanotechnology workforce will be working in green jobs. Depending on the criteria used, this proportion may also include some positions in other green sector industries [65, 97]. Over time, costs and benefits usually materialise in different ways. For instance, research and development (R&D) expenses are frequently incurred early in a new technology's life cycle; in spite of this, R&D expenses may continue as long as the technology is regularly upgraded and modified. Interestingly, public R&D in nanotechnology started to significantly rise in the 2000s, with the field starting to take off in the mid-1990s [98]. Different advantages might materialise at different times, as Tassey [99] noted. Early adopters in academia and business are witnessing a surge in labour requirements for R&D and specialised manufacturing. Intermediate results are being attained through patents, prototyping, and luring venture capital, which can lead to the development of new products, processes, and business expansion. Long-term benefits to the industry, economy, and society also materialise [65]. Furthermore, nanoscience and engineering universities are increasing degree programmes to meet expected talent shortages. According to Nanowerk, over 162 academic institutions worldwide presently offer a total of 271 degree programmes in nanotechnology subjects. Growth will also spur indirect economic activity and job creation in related businesses, ranging from equipment sellers to software providers who enable computer simulations and modelling at nanoscales [100]. As nanotechnology becomes more entrenched in products and services, IT consultants will become more involved in system integration tasks. Within specific industries, such as medicine, new therapy programmes may arise, shifting skill needs towards diagnostic expertise. Oncologists educated in AI-assisted biopsy analysis may have a competitive advantage over generalists since focused nanoparticle drug delivery allows for more personalised therapies [101]. Such assistance, when combined with effective immigration reform for highly talented workers and entrepreneurs, can secure national innovation advantages while contributing to economic growth.

Strategic Competitiveness of Countries

As already discussed, nanotechnology is positively reshaping economies and strengthening competitiveness in a variety of areas, including catalysing industry growth, raising productivity across sectors, and providing countries with strategic competitive advantages [102]. Thus, nanotechnological innovation and commercialization are driving substantial industry growth worldwide [24]. When the techniques for research and production are perfected, and the foundational infrastructure and education are put in place, it has the potential to eventually equal or surpass the digital revolution in terms of economic significance and societal influence [103]. The worldwide nanotechnology industry, according to Spherical Insights, was estimated to be worth more than $3.78 billion in 2020 and is projected to rise at a compound annual growth rate (CAGR) of 34.5% to reach $74.1 billion by 2032. Currently, China, the USA, and Japan together represent over 70% of the global nanotechnology market share, as these countries have made major public and private investments [104]. China, for example, outpaces all countries in its commitment to nanotechnology, which is supported by the country's purchasing power. In 2012, the Chinese Academy of Sciences (CAS) spearheaded the nation's strategic pioneering initiative on nanotechnology, which was funded with one billion yuan (US$152 million) and spread over five years. Because of this sustained dedication, China has become a major player in the field of nanotechnology, expanding our understanding of the challenges and possibilities that lie ahead as well as the possible effects of nanoparticles on public health [105]. Likewise, the National Institute of Standards and Technology (US) calculates that there are about 1,700 nanotechnology businesses globally. Currently, the United States spends around $2 billion annually on large industries, which funds over half of nanotechnology research and development [106]. The total amount of funding for nanotechnology research in Japan increased from around 82.5 billion yen in 2002 to about 90.4 billion yen in 2003 [107]. As a result, Japan's financing for nanotechnology is now nearly equal to the US federal budget. Likewise, Germany, South Korea, and Taiwan are expanding national manufacturing bases and exports. Moving forward, developing countries will profit more from nanotech commercialization. Beyond market growth, nanotechnology enables businesses and industries to increase productivity by developing unique optical, electrical, mechanical, thermal, and other properties of engineered nanoscale materials and devices to facilitate the development of more efficient manufacturing processes and advanced products [24, 108]. For example, nanoparticles and nanotubes are strengthening plastics, metals, and concrete used widely in construction and infrastructure [109, 110]. Importantly, the use of biodegradable green nanoparticles that enable lighter weighing enhances sustainability, reuse, and recycling, hence reducing waste. Furthermore, nano coatings that resist wear and corrosion have made machines and equipment more durable and long-lasting in industries such as energy, healthcare, and transportation [111], resulting in lower fuel and maintenance costs. Through these productivity enhancements, earlier lab-to-market translation, and reduced costs, nanotechnology contributes significantly to business competitiveness [112]. Building national research excellence and high-tech industry capabilities in nanotechnology provides strategic economic benefits. According to a survey published in Asia Pacific Nanotech Weekly, also as confirmed from data from the Scopus database, the number of publications on nanotechnology has increased exponentially over the last 20 years as shown in Fig. 1.6a with China second only to the United States [107, 113] (Fig. 1.6b).

This implies that nanotech R&D generates knowledge, intellectual property, publications, highly skilled talent, and technology breakthroughs that serve as competitive assets [112, 114]. Priorities that have benefited from the R&D department include semiconductors, advanced nanomaterials design, nanoelectronics, nanomedicine, green nanotech, nanosensors, nanophotonics, as already discussed [115].

Opportunities for Inclusive Growth and Potential Disparities

It has been noted that nanotechnology shares traits with scientific advancements that have the potential to provide technological opportunities that might revolutionise current sectors while spawning new ones [116]. For instance, nanotechnology is acknowledged for creating value-added materials that are essential to the production of improved consumer goods, such as the transparent electrodes formed of carbon nanotube networks or silver nanowires that are used to fabricate touchscreens [117]. Additionally, value-added components derived from nanotechnology include functionalized ink dispersions, raw nanoparticles used to create thin films, and roll-to-roll nanomanufacturing or printing techniques [118]. According to Morse [118], nanomanufacturing is a framework for transitioning simple innovations to better consumer products. Realising the potential of technology necessitates worldwide

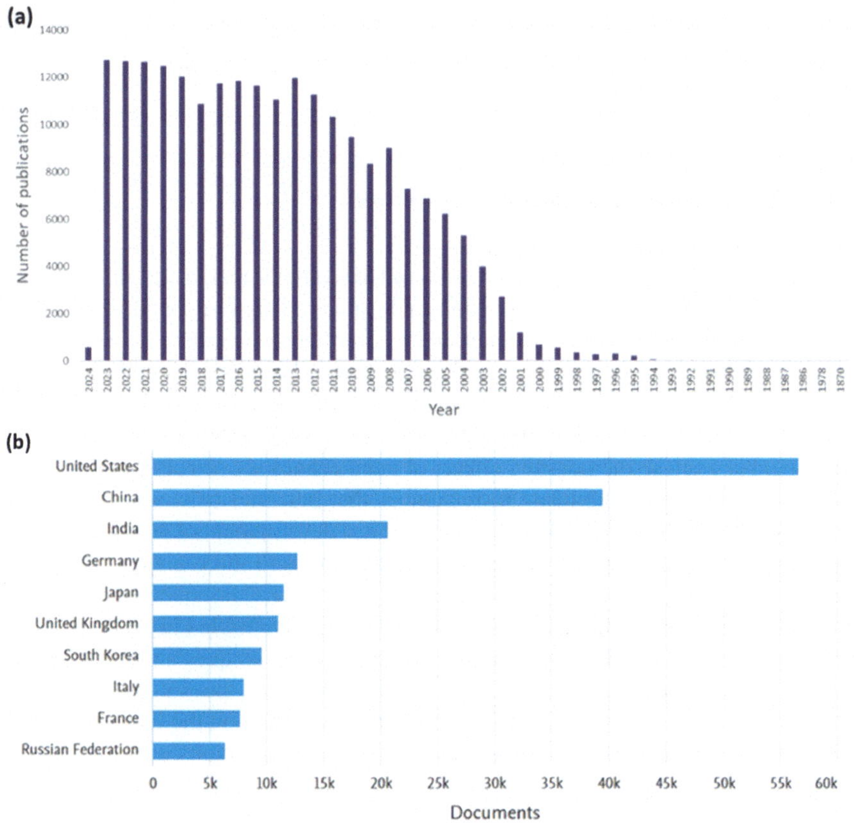

Fig. 1.6 **a** Rate of publication on nanotechnology over the years **b** top ten countries in terms of publication on nanotechnology (Data from Scopus database as of January 9, 2024)

collaboration in priority-setting for innovation aimed at long-term human development [9]. Collaborative investment in applications such as economical point-of-care diagnostics, sensors monitoring environmental toxicants, or sustainable energy solutions directly benefits marginalised communities while building the groundwork for equitable leapfrogging [107]. Investments in local skill-building, technology transfers, and infrastructure development help countries contextualise innovations like pollution remediation, water filters, or precision agriculture to specific demands affordably [112]. Additionally, mandating transparency and participation in standard-setting bodies averts regulatory disconnects between emerging scientific insights and policies governing assimilation [119]. Therefore, transnational R&D ecosystems anchored to public good foster opportunities inclusively. Currently, nanotechnology remains concentrated in countries with the majority of world communities awaiting trickle-down benefits. Such nano-divides stem from trajectories shaped by concentrated capital chasing affluent markets rather than accessing goals aligned to development metrics around poverty alleviation or capability building. As a result, progress is made more equitably by maintaining the public good as the primary orientation, harmonizing the pace and breadth of nano-innovations with society's preparedness to assimilate responsibly.

Ethical Dimensions of Nanotechnology

As humanity faces significant challenges, new technologies that tackle almost every element of energy, the environment, health, and climate change are sought after, either directly or indirectly. But advancement must nevertheless be in line with the interests of the public, highlighting the significance of ethical concern [120]. With current knowledge, there may be inherent risks associated with nanoparticles and their uses. To find examples of nonconformity with safety norms and regulations, it is necessary to thoroughly analyse the possible effects of nanotechnology principles, processes, and applications. It is significant to remember that potential risks will depend on the particular nanomaterials and uses. Nano-ethical development of nanotechnology necessitates addressing numerous challenges that fall into three major categories: nano-toxicity, environmental protection, and privacy [121]. Consequently, upholding ethics encourages technological development that is responsible [17]. The primary task at hand is now the development and validation of techniques and tools for the identification, characterisation, and analysis of nanomaterials, the evaluation of nanomaterial exposure, and the creation of thorough information on the dangers related to nanomaterials and nanotechnologies [122]. Due to the ambiguity and multiplicity of ethical issues surrounding nanotechnology, as Schummer pointed out [123], it is appropriate to categorize them into specific and general issues. According to him, some ethical issues might arise from particular technological applications, goods, and research procedures, as well as how such things are seen in a larger scientific and societal framework. The possible hazards that nanoparticles pose to human health and the environment seem to be the most urgent ethical issues facing nanotechnology. But the social ramifications of nanotechnology are far-reaching, encompassing everything from consumer goods to privacy, security, the environment, and ethical issues like human-nanotechnology interactions with medical devices and applications [123]. The development of nanotechnology, which employs minuscule sensors and gadgets to facilitate the monitoring of individuals and locations, is responsible for a great deal of the technological privacy concerns [124]. If appropriate safeguards are in place to direct innovation, nanotechnologies can significantly improve people's quality of life. Discussions about how to deploy ethical nanotechnology applications must keep up with the rapid pace of technological advances in this field. Thus, key insights are given below.

Ethical Considerations in Manipulating Materials at the Nanoscale

The ability to engineer materials and devices at the nanometer scale poses monumental opportunities and risks for societal development [125]. Without a doubt, the ability to modify materials at the nanoscale, as well as the inclusion of nanostructures and nano processes into technological advances, will enable the production of low-cost, higher-performing goods [126]. As nanoscience quickly innovates across sectors such as healthcare, manufacturing, robotics, and information technology, ethical norms and governmental interventions are urgently required to direct these seismic powers towards equitable progress over instability. Important core issues such as permission, transparency, access distribution, and environmental implications require concerted attention before exponential technological change alienates populations without suitable integration frameworks [127]. The United Kingdom's Royal Society and Royal Academy of Engineers collaborated to issue a study on nanotechnology in 2004 following a thorough investigation. The statement says that "consideration of the ethical and social implications of advanced technologies should be included in the formal training of all research students and staff working in these areas" [128]. Importantly, radical nanomedical interventions such as targeted nanoparticle drug delivery, early disease detection via nanosensor bodily scans, and even reversing ageing via cell repair nanobots raise concerns about privacy, consent, and access allocation, which have been a focus of study for most researchers. Another strategy is to highlight the larger environmental effects of resource extraction and nanomaterial life cycles to aid with sustainable transition initiatives [129]. As a result, the fabrication of 2D and 3D nanoproducts often has tougher purity criteria. Due to their poor process yields and thus low material efficiency, many of the procedures required to produce them result in significant waste. Additives, water, and solvents are also often used in these processes. There are two ethical frameworks that are observed at the manipulation size Environmental Assessment (EA) and Life Cycle Risk Assessment (LCRA) at the nanoscale [130]. This serves the notion that resource flows and manufacturing could dramatically shift from globalised networks toward hyper localised automation via advanced nanotechnology [131]. Such fluxes necessitate consciously equitable policies given the sector's potential to either conquer or deeply worsen ecological challenges like climate change [132]. With such a massive responsibility, the imperatives now include bringing together society's brightest brains to drive this change fairly through greener designs and manufacturing processes, as well as proactive engagement. As a result, Michael Berger [133] of Nanowerk LLC offers a possible remedy by incorporating the following green chemistry principles—described by the US Environmental Protection Agency (EPA) on their "Green Chemistry" webpage—into the processes involved in nanomanufacturing:

- Adopt chemical syntheses practices that minimise waste and create safer substances and goods.
- Develop chemical syntheses techniques that minimizes danger to flora and fauna.
- Make use of renewable raw materials and feedstocks.

- Stay away from chemical derivatives.
- Optimise the atom economy.
- Make use of safer solvents and reaction parameters.
- Boost energy efficiency.
- Evaluate immediately to stop pollution.
- Reduce the possibility of mishaps.

Implications of Nanotechnology on Human Identity and Ethics

Nanotechnology's exponential development trajectory signals potential impacts on human identity and ethics that could profoundly shape societal futures [134]. As the ability to engineer matter at microscopic scales emerges, from revolutionising medicine to reinventing manufacturing, proactive examination of the accompanying moral and social consequences has become critical. If done correctly, deliberate integration initiatives that emphasise consent and access parity have the potential to break long-held human restrictions in remarkably life-affirming ways. Without meaningful public input driving innovative paths, unchecked progress risks sidelining ethical review and further entrenching existing imbalances [135]. Radical nanoparticle therapies and enhancements gather much speculation about transforming health and cognition [136]. Cancer interventions using targeted drug delivery appear to hold remarkable potential [137, 138], but normalising molecular biology and re-engineering raises difficult problems about how we define human enhancement versus therapy and how they establish boundaries across cultural viewpoints [139]. Nanotechnology-enabled neuro-interfaces promise seamless memory enhancement or artificially enhanced intelligence [140]. However, deploying cerebral prosthetics before equal distribution of existing healthcare risks focusing resources solely on enabling elites unless intentionally democratised. To frame proper answers, ethical supervision must grow in lockstep with exponential technological advancement [141]. Thus, social scientific and economic research can assist producers and governments in making the best judgements when deploying new technology, maximising its usefulness to humans. Alternatively, democratising access to nano factories could enable broadly decentralised, localised manufacturing.

Privacy, Security, and Control in the Nanotechnology Era

Information and communication technologies (ICT) saw tremendous growth and demand in the early twenty-first century, which in turn enabled numerous industries, including manufacturing, transportation, healthcare, and agriculture, to reach their full potential. As a result, many cutting-edge technological platforms emerged, including cloud computing, artificial intelligence (AI), and the Internet of Things (IoT).[142], which facilitated machine-to-machine communication [142]. The

Internet of Things (IoT) often refers to sensors, actuators, smart phones, and any other common digital device that can be connected to the internet that generates and exchanges large amounts of data [143]. As a result, international communication and information exchange have significantly improved. IoT on a nanoscale, also known as the Internet of Nano-Things, or IoNT, has unquestionably revolutionised how people utilise the Internet and device-to-device communication [144]. Technologies include nanorobots, nanosensors, and nanomachines that can exchange data, communicate, and interact with bigger systems comprise the network of interconnected nanoscale devices [144, 145]. For example, IoNT devices monitor temperature, humidity, water quality, air pollutants, and more in smart industrial settings. [146]. Technologies include nanorobots, nanosensors, and nanomachines that can exchange data, communicate, and interact with bigger systems comprise the network of interconnected nanoscale devices [147, 148]. Nano-sensors implanted in the body, for example, monitor biomarkers and transmit health data, whereas magnetic nanoparticles, due to their small size, provide highly sensitive imaging techniques for early cancer detection [77]. However, the increasing ability to engineer materials at microscopic scales also creates risks around privacy, security, and control that could challenge norms of consent and autonomy unless proactively addressed [149, 150]. Because of the constant possibility of personal data theft and misuse, some people view the privacy risks associated with nanotechnology as being similar to those associated with older communication technologies in the past. Some contend that the widespread use of invisible sensors brought about by nanotechnology would make privacy protection obsolete [121]. The largest concern regarding nanotechnology and privacy is its potential to greatly increase surveillance through the use of minuscule devices that are imperceptible to humans. In his article "On Nano-Panopticism: A Sociological Perspective," Michael Mehta [151] examines the social effects of the advancement of microsensor technology. Hiroshima University and Nippon Hoso Kyokai (NHK) researchers have discovered the photoconductive capabilities of silicon nano-crystal film. These films should allow for more control over the crystal grain size, which will enable the creation of tiny, highly sensitive video cameras through the use of charge-coupled devices. Implanted medical diagnostic devices can, in general, be used to determine physiological health and malfunctioning self-monitoring systems. By implanting nanodevices controlled by microchips within the human body [152], these devices could communicate with devices outside the body to report results, like early signals that a tumour, heart damage, or infection is developing. They could also give patients with Alzheimer's immediate access to heart-directed drugs and assistive cognitive equipment. They could also give patients with Alzheimer's immediate access to heart-directed drugs and assistive cognitive equipment [153]. The light-sensing photothermal reaction (LSPR), which targets tumour tissues in a heat-specific manner and kills cancer cells, is used in one study to expose gold nanoparticles to the right wavelength of light [154, 155]. Similarly, Yezdani et al. [156] emphasised a report by Professor Jennifer of Rice University where gold nanoparticle-coated nanoshells with a diameter of 120 nm were used to eradicate cancerous tumours in mice by conjugating antibodies or peptides on the surface of the nanoshells, forming bonds with the malignant cells. Using an infrared laser to target the tumour, sufficiently heating the gold eliminates the cancer cells. Thus, nanotechnology helps with surveillance on multiple levels, including the development of Radio Frequency Identity Chips (RFIDs) [152, 157], as well as invisible tags and nanosensors that may record and trace individuals' movements [121]. Thus, healthcare and security may require full traceability of persons' motions and physiological characteristics, which becomes attainable with digital convergence, i.e. the combination of nanotechnology, digital technology, and biotechnology. Another major security risk to the IoNT is the possibility of hostile actors gaining access to and manipulating nanodevices [149, 158]. These devices are usually integrated into or linked to larger systems, thus a breach of one device might have a significant impact and jeopardise the security and privacy of those larger systems. Privacy and civil freedoms may be seriously affected by this, particularly if governments or other powerful institutions use the devices. As a result, considerably improved nanotechnology activity has altered encryption and authentication processes, preventing unauthorised access to nanodevices. The Medizinproduktegesetz (Medical Devices Act) in Germany, for instance, deals with the national application of European Directives 90/385/EWG for active implantable medical devices. This includes the use of multi-factor authentication and public key infrastructure (PKI) as well as secure identification and authentication mechanisms for nanodevices [142]. Consumer data gathering already sparks issues about informed permission, but nanotechnology exponentially multiplies the data available to collect about purchasing habits via embedded sensors. With responsible implementation guided by ethical evolutionary social contracts, however, nanotechnology can catalyse sustainable abundance benefiting all of society.

Equitable Access and Ethical Distribution of Nanotech Innovations

Emerging technologies come with social and ethical debates of which nanotechnology is no exception [159]. Nanotechnology's immense potential for bettering lives hinges on equitable distribution across countries and communities, particularly for applications satisfying basic needs [160]. Importantly, nanotechnology is an enabling finding that is represented in a wide range of applications and advancements, from novel textile, packaging, and food materials to sophisticated medical procedures and therapies [161]. For example, after magnetic clearance of nanoparticles, in situ fluorescence technology can be used to identify SARS-CoV-2 more accurately [144, 162]. While the use of nano goods indicates the rapid pace of scientific progress, it also highlights the unknown risks and uncertainties associated with nanotechnological products [159]. Concerns about uneven access to nanotechnology have been raised in the scientific community. Healthcare providers, from physicians to biomedical firms, are under pressure to meet patient demand for innovative treatments, diagnostics, and mobile information access. At the same time, the expenses associated with developing new medical goods are increasing. The goal is to enhance life quality at a fair cost by identifying diseases early and providing prompt, efficient treatment for them [163]. Currently, access discrepancies continue between rich regions rapidly turning nano-innovations into commercial products and most of the world's countries waiting for advantages to trickle down. As a result, nano-enabled products, including those at any stage of pharmaceutical production, will be unaffordable in developing countries [159, 164]. Evans [165] makes the case that everyone should benefit from scientific advancements and that global action should be taken to guarantee equitable access. According to Schroeder et al. [166], it makes sense for the group that gains from a patent regime to share that benefit with the group that loses out. Similar to other nanomedical technologies, the COVID-19 nano vaccines were developed and put into use in specific social and cultural settings, which meant that complex moral issues went beyond the technological aspect of these accomplishments [167]. Accepting the idea that justice is essential to global health and ought to serve as the cornerstone of initiatives aimed at raising people's standard of living is crucial [168]. A 'nano-divide' between affluent and poor populations and nations might arise by misusing or limiting access to medical nanotechnologies, exacerbating already-existing disparities in health outcomes and power and resource imbalances [169]. Importantly, in achieving equitable distribution for nano-innovations that benefit the public needs emphasises comprehensive governmental measures to connect market mechanisms with social welfare agendas. Hence, multilateral engagement and capacity building, reinforced by intellectual property, provide opportunities to the most marginalised communities [170]. Therefore, proactive governance fosters assimilating nanotechnology's benefits justly rather than furthering disparities.

Environmental Impact and Sustainability

Sustainability and future studies are inextricably intertwined, albeit the periods involved may differ depending on whether the stakeholders are futurists or environmentalists [171]. However, a growing environmental movement has made it necessary for businesses and nations to begin changing their practices to secure the planet's existence [172]. The three fundamental pillars of sustainability, according to the UN Brundtland Commission, are the environment, the economy, and society. Therefore, sustainable development is that which satisfies current demands without jeopardising those of future generations. To preserve the earth and its resources for future generations, society must adhere to or adopt the Sustainable Development Goals (SDGs), a collection of 17 main goals (see Fig. 1.7). [173]. However, according to the contributions of numerous scientists, nanotechnology is positively associated with the significant contributions to sustainability, expressing their agreement as a substantial contributor. Notwithstanding the many advantages of applying nanotechnology, the globe is currently confronted with the challenge of rising demand for basic resources including food, energy, water, and commodities as well as services [42]. Currently, providing greater living standards to all people while minimising the

Fig. 1.7 Sustainable development goals (SDGs). (Image credit: https://www.un.org/sustainabledevelopment/blog/2015/12/sustainable-development-goals-kick-off-with-start-of-new-year/)

effects of human activity on the climate and environment is a significant problem. Nanotechnology is expected to deliver more sustainable answers to global concerns affecting humanity. A review of corporate sustainability reports found that analysis focused mainly on the health and environmental safety impacts.

Environmental Risks Associated with Nanomaterials

Nanotechnology is thought to be a more environmentally friendly manufacturing process. While practically everything may be dangerous in high concentrations, the more important question to consider is how poisonous nanomaterials are in different levels that could be used? [174]. The fundamental material, size, shape, and coatings all affect how poisonous a nanomaterial could be. The identification and comprehension of the detrimental effects of nanomaterials necessitate accurate methodology and data interpretation that takes assumptions into account. Various research organisations employ various cell lines, culture settings, and incubation durations in nanoparticle toxicity investigations. According to Patra et al. [175] research, 33 nm citrate-capped gold nanospheres were non-cytotoxic to human hepatocellular liver cancer cells or baby hamster kidney cells, however they were cytotoxic to a human carcinoma lung cell line at various doses. Agricultural pollution and water filtering are two significant problems that are attributed to nanotechnology. Additionally, nanotechnology can produce end products that are so costly that they cannot be employed in industry [18]. Consequently, there is an increasing number of national

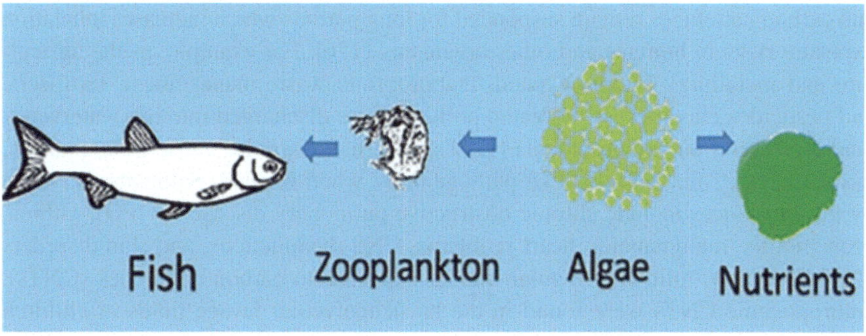

Fig. 1.8 Algae-zooplankton-fish food chain [176]. Reprinted with Permission from Springer Nature

and international concerns regarding the environmental hazards linked to nanomaterials, raising doubts about the overall environmental potential of nanotechnology and providing justification for an immediate investigation by environmental and health researchers in addition to some nano researchers. A noteworthy concern is that animals at the base of food chains may acquire nanomaterials through different contaminants that include potentially toxic substances [176, 177]. Most studies have focused on the effects of nanoparticles on model food chain systems, which include fish, amphibians, semiaquatic animals, earthworms, invertebrates, algae, and cells of mammals and amphibians [178, 179]. Nanomaterials may be absorbed by simple species like plankton, and as a result, they go up the food chain and provide food for fish and birds, among other species. For instance, Moller et al. [178] use an experimental model based on an algal-zooplankton food chain to study the toxic impact, bioaccumulation, and contamination of AuNPs (Fig. 1.8). Ankistrodesmus falcatus, a zooplankton grazer, was discovered to have two different sources of AuNPs in its body: contaminated phytoplankton diet and direct water intake.

Bozich et al. [180] also investigated the negative impacts of negatively and positively charged ~4–5 nm gold NPs (AuNPs) functionalized with various chemical molecules against Daphnia Magna during their whole life cycle. It was discovered that the toxicity of the modified AuNPs was also impacted by their aggregation state in extracellular and intracellular media, with the positively charged AuNPs being more toxic than the negatively charged AuNPs. Another worry is the enormous surface area of nanomaterials per unit of volume which substantially influences their accumulation states in the environment [181]. The surface area of NPs is thought to have a major influence on their toxicity [182, 183]. This enables them to bind to and transport other environmental toxins such as pesticides and flame retardants, resulting in increased bioavailability and combined toxicity effects of nanomaterials and other contaminants in organisms [184]. It is feasible to inhale airborne particles containing nanoparticles with diameters ranging from a few nanometers to several micrometres due to the very tiny size of produced nanomaterials [174]. When released into the air, some nanomaterials like titanium dioxide, silicon dioxide, and carbon nanotubes remain suspended for long periods which increases inhalation exposure risks in humans and other organisms [176]. For example, as the agriculture and metallurgy sectors expand, inappropriate waste management, fertilisers, and pesticides classified as inorganic pollutants are discharged into our waterways, land (soil), and air (environment) [181]. Some nanoparticles cause inflammation, tissue damage, and asbestos-like pathogenicity when inhaled. Ailments produced by the substances include chronic obstructive pulmonary disease (COPD), asthma, bronchiolitis, malignancies, heart problems, CNS dysfunctions, and skin disorders [185]. Recently, following regular human exposure to carbon nanotubes (CNTs), anthropogenic CNTs were found in the bronchoalveolar lavage fluids of children with asthma in Paris [186]. Importantly, the small size makes environmental pollution identification and remediation extremely challenging due to limited analytical tools capable of quantifying nanomaterial concentrations in complicated environmental matrices. As a result, while nanomaterials provide scientific improvements, releasing them into the environment, whether through unintentional spills during production/transport or end-of-life disposal must be done with caution due to their high mobility and reactivity.

Strategies for Sustainable Nanomaterial Production and Disposal

Over 90% of waste is burned or dumped in an open manner in low-income countries, where the United Nations Environment Programme estimates that each year, close to 11.2 billion tonnes of solid waste are produced. This is a major cause of environmental degradation and adverse health effects [187]. This predicament has prompted the development, implementation, and reinforcement of many policy initiatives. Nanomaterial synthesis, or nanofabrication, is not a difficult task, but examining nanoparticle form, size, and dimensions is. Thus, nanofabrication refers to the process of producing, assembling, or forming tiny structures of nanomaterials [188]. Numerous novel techniques, such hydrothermal (time-consuming and not cost-effective), sol–gel (long time for nanomaterial synthesis), co-precipitation (possibility of pollutants addition), and others, have been utilised for the manufacturing and modification of nanomaterials [126]. As shown in Fig. 1.9a, there are two ways to approach nanosynthesis: either through the top-down or bottom-up process. It is essential to have a comprehensive grasp of these systems' physicochemical properties, identification, mode of release into the environment, and toxicity to living things in order to comprehend the environmental effect [189]. The inherent potential toxicity of nanomaterials is mostly determined by their ability to reach and penetrate different environmental compartments, such as soil, water, and air. The number of nanomaterials released into the biosphere has a direct correlation with this effect. Detailed investigation of the nanomaterial's complete life cycle is necessary to evaluate and quantify the amount emitted [190], taking into consideration how they are

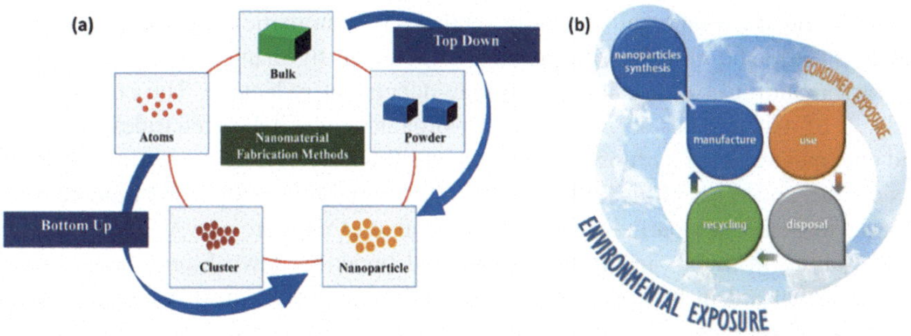

Fig. 1.9 **a** Top-down and bottom-up approaches for fabricating nanomaterials [126] Reprinted with permission from Elsevier. **b** The lifespan of goods incorporating nanoparticles [127]. Open access article by MDPI

utilised and integrated into finished goods, starting with the processes involved in the creation of nanomaterials and concluding with recycling and disposal practices (refer to Fig. 1.9b). While releases of nanomaterials can happen at any point in the life cycle, they become a significant issue when those nanoparticles are utilised in the finished product (e.g., washing garments or fabrics or using aerosols containing nanoparticles). This characteristic makes effect evaluation considerably more difficult because it is the most difficult stage of the life cycle because it is impossible to establish procedures and involves many unpredictable factors, such as climatic variations [191].

As structural modifications to the NP-containing material occur, such as drilling, cutting, or other operations requiring high energy and temperature to create or change the form, the emission rises throughout the final product production stage [192]. Thus, incorporating sustainability concepts in a greener synthesis method during manufacturing and disposal is critical for responsible nanotechnology development [193]. In contrast, green synthesis involves substituting toxic chemical products with safer alternatives and creating innovative methods to reduce or eliminate harmful substances to the environment and human health in order to produce environmentally friendly nanoparticles [194]. In order to provide a unique platform for the aseptic manufacturing of nanomaterials, the need for green technology has spurred study into the safety of nanoscience on living things. Technologies for biological or biomimetic green synthesis based on plant extracts are being researched to replace non-sustainable processes since they use less energy and less dangerous chemicals [195]. Banana peel extract (BPE) was used by Ibrahim [196] as a capping and reducing agent in the synthesis of silver nanoparticles, which is an economical and environmentally beneficial method. Under a variety of conditions, the silver nanoparticles showed 1.16–1.32 fold antibacterial activity against common bacteria and yeast pathogens. In a similar vein, Vaibhav et al..[197] describe the synthesis of silicon dioxide nanoparticles using bamboo leaf extract. Additional significant agricultural wastes, including grass, leaves, cereal bran, sugarcane bagasse, rice husk, fruit peels, and microorganisms including bacteria, fungus, algae, actinomyces, and lichens, are crucial to the environmentally friendly synthesis of nanoparticles utilising nanotechnology [198]. The bioreduction process is mediated by many enzymes present in fungus, including hydrogenase and nitrate-dependent reductase. The composition and size of nanocrystals are influenced by organic acids, polysaccharides, proteins, and enzymes secreted by fungus [199]. Aspergillus niger, Fusarium solani, Coriolus vesicolous, and Aspergillus fumigatus are a few fungi that mediate the bioreduction process to make nanoparticles like gold and silver. Fungi are very good at binding walls, taking in metals inside their cells, and breaking down foreign substances. Enzymes that aid in hydrolysing the bulk substance into nanostructures can also be produced by them. The purification process is considerably easier because the produced nanoparticles are extracellular [194]. Another important strategy is the deliberate engineering of nanomaterials to reduce hazards while retaining function [200]. For example, creating nanomaterials while minimising harmful component leaching or lowering reactivity profiles. Many important barriers stand in the way of safer bio-design, including the paucity of information in toxicology, differences between studied and real-world nanomaterial exposures, and trade-offs between usefulness and safety. In light of this, several approaches have been put up to lessen (aspects of) the danger associated with nanomaterials. These include coating, size management, doping, grafting, loading, controlling form and crystallinity, lowering the persistence of nanomaterials, and replacement [200]. The separation of nanomaterial production and processing, in situ synthesis of nanomaterials, improved integrity and durability of nanocomposites, product design for disassembly, and effective material recycling can all lead to a reduction in the number of nanoparticles released into the environment. Most importantly, changing flow patterns and minute fluid volumes can be achieved by using cleaner manufacturing processes to boost production efficiency [201]. Concerning the disposal of nanomaterials, researchers emphasise that following the 3Rs (reduce, reuse, recycle) is an effective method. In a sustainable approach, typical consumer products containing nanomaterials such as electronics, clothing, sunscreens, and tennis rackets are reused across society before finally recycling the raw materials at end-of-life, as this maintains utility and prevents immediate waste disposal. The creation of NPs and NP-enabled technologies is made possible, crucially, by the twelve principles of green chemistry, which offer a workable framework. These guidelines aim to direct the creation of chemical pathways that provide more sustainable results [202, 203].

Role of Nanotechnology in Addressing Environmental Challenges

Environmental degradation and climate change are enormous, interconnected concerns that civilization faces. Climate change is currently the most serious environmental danger. In the last 40–50 years, the surface temperature has risen by 1.5–2 °C. The earth's temperature will rise in the next 50–100 years, creating severe living circumstances with disastrous effects [204]. However, the rise of nanotechnology gives chances to promote sustainability by enabling cleaner industrial processes, decreasing greenhouse gases, remediating existing pollutants, and monitoring ecosystems [205] (refer to Fig. 1.10). A wide range of issues in fields including agriculture, the environment, energy, food, medicine, catalysis, and material science can be resolved by using nanoscale devices thanks to nanomaterials [206]. Because of their small size, engineered nanomaterials have unique chemical, optical, and structural capabilities for specific applications [207], and therefore in dealing with the challenges of environmental concerns. Nanomaterials such as titanium dioxide nanoparticles, nanoclays, and carbon nanotubes, for example, assist in enhancing production efficiency by lowering energy, waste, and costs [108, 208]. A photocatalytic copper tungsten oxide nanogrid was developed by researchers at State University of New York (SUNY) Stony Brook in the wake of the 2010 Deepwater Horizon oil spill. Through the use of a solar-powered nanotechnology solution, these nanoparticles convert oil into biodegradable compounds when exposed to sunlight and eliminate any pollutants [209]. According to estimates from the World Health Organization (WHO), millions more people's health and well-being are harmed and around 3.4 million people die annually as a result of polluted water and inadequate sanitation, the majority of whom are children. Many illnesses and conditions, including cholera, typhoid, hepatitis A, and dysentery, are brought on by water contamination. When contaminants in waste streams are effectively filtered using nano-enabled membranes, sectors including oil, gas, chemicals, and pharmaceuticals benefit from reduced environmental impact [210]. Absorbent materials at the nanoscale have a greater capability for pollutant absorption due to the notable high surface area of nanoparticles.

Moreover, the increased surface area per unit volume compared to their macro-counterpart offers an added advantage of better material interaction, energy conveyance, and adsorption of greenhouse gases [211]. Modern agricultural technologies like nanopesticides, nanofertilisers, and precision farming technology have

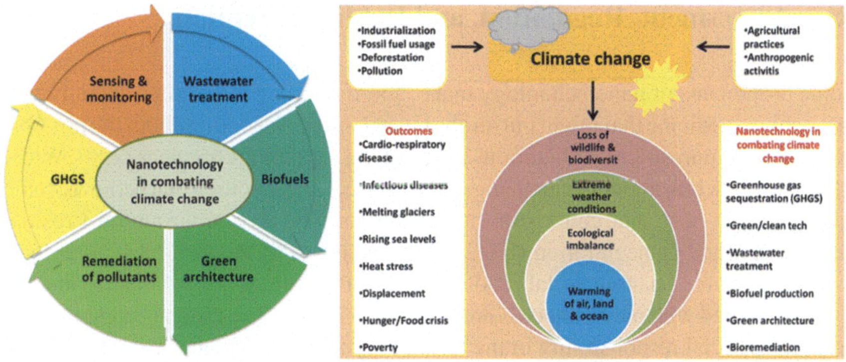

Fig. 1.10 Ways nanotechnology is used in fight climate change [204]. Open access article by Elsevier B.V.

been made possible by nanotechnology, which helps to balance ecosystems and increase production [212]. As a major player, nanosensors have several advantages for agricultural farms. They make it possible to monitor environmental parameters, pesticides, herbicides, fertilisers, and other elements necessary to increase food crop productivity. They also enable the detection of pathogens, contaminants (such as drugs and heavy metals), and related toxins in food packaging [213] [214]. Crucially, nanotechnology offers a number of ways to cut greenhouse gas emissions and limit global warming by the terms of the Paris Agreement. These include the use of nanotechnology in energy-efficient buildings with insulation materials, lighter cars with alloys strengthened by nanotechnology to save fuel, and nanocatalysts to store intermittent renewable energy and produce lower emissions as well as nanocarbon membranes to capture carbon [211, 215]. Nanotechnology breaks down dyes and other colourless contaminants in wastewater streams through environmental remediation [216]. This reduces minute particulate matter emissions from companies and cars, while also hastening the adoption of renewable energy and electric mobility. Nanotechnology also contributes to the advancement of hydrogen fuel cells, biofuel catalysts, and smarter grids. The high surface area-to-volume ratio of nanoparticles like nano-iron allows for quicker removal of heavy metal pollutants like lead, arsenic, and mercury than bulk approaches [217]. Nanomaterials' unique semiconductor, optical, and conductive properties make them ideal for improving solar cell efficiency while carbon nanotubes and graphene promote charge transfers [218]. Similar characteristics boost battery electrode storage capacity and durability by reducing lithium dendrite growth. Furthermore, tiny nanosensors and spectroscopic tags provide detailed, real-time environmental tracking capabilities via a network of wireless nano-enabled sensors in remotely reporting hyperlocal pollution levels, soil conditions, ecosystem alterations, and animal protection [219]. These help create spatially distributed historical databases to understand anthropogenic changes relative to ecological baselines.

Risk Assessment, Regulation, and Public Perception

Some proponents of nanotechnology make "pie-in-the-sky" assertions about breakthroughs in medicine that might put an abrupt end to any discussion about the possible drawbacks of unrestricted, ubiquitous usage of the technology are near [220]. While a lot of experts talk about the amazing things that nanotechnology can achieve, some talk about the risks that come with nanomaterials. The Action Group on Erosion, Technology, and Concentration (ETC Group) activist Pat Mooney has called for a halt to commercial nanomaterial production until regulations are established and the risks associated with nanomaterials are better understood. The group is committed to promoting social responsibility in the technology industry. Mooney draws attention to the worry that innocuous particles can result in potent chemical reactions and biological harm when they are created at the nanoscale and then inhaled. It's crucial

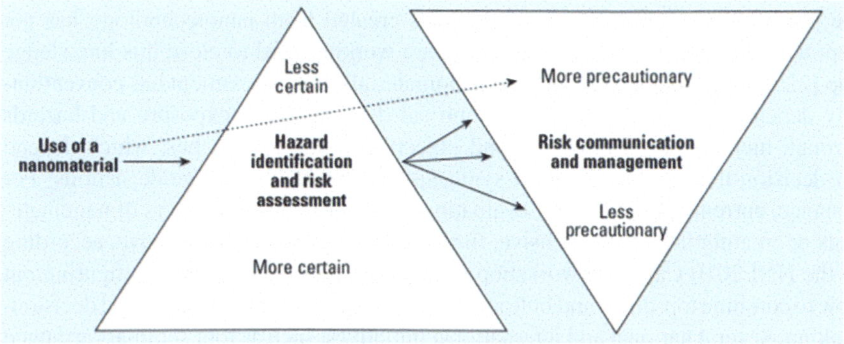

Fig. 1.11 Risk management decision-making for nanoparticles with the appropriate level of controls [150]. An open-access article by EHP publishing

to remember that risk evaluations are often highly political and partially subjective. Since production, time, values, and policies are all specific, all risk predictions are value-laden [221]. Risk assessment, as a regulatory framework, has served as the foundation for evaluating product and environmental concerns, including those associated with nanotechnology [222]. To prevent or mitigate recognised possible hazards, risk management is necessary to analyse and characterise the risk as well as the reasons creating or impacting the risk (see Fig. 1.11). Numerous global organisations are engaged in risk management related to nanotechnology [18]. Assessments of the present state of nanotechnology have been provided by national institutions, international organisations (such as the Asia Pacific Nanotechnology Forum 2005), and European Union governmental bodies (European Commission), with an emphasis on the significance of meticulous risk analysis [18]. The European Council highlights how critical it is to closely monitor any possible risks across the whole life cycle of products based on nanotechnology. Although nanotechnology seems to have many advantages, the broad use of poorly understood nanomaterials may come with unexpected health and environmental risks [223].

Challenges in Assessing and Managing Nanotechnology-Related Risks

Today, the world faces significant social and environmental challenges because of environmental issues related to high resource consumption in industrialised countries, as well as numerous economic, social, and environmental promises made about high technologies such as nano, bio, and ICT. A portion of the problems arise from an oversimplified and linear understanding of the expectations around these technologies, which are commonly framed in legislative attempts as either market pull or technical push [224]. Nevertheless, the risk assessment necessary to understand the potentially harmful effects of products created from nanotechnology has not kept up with their growth, and scientists are working hard to close this knowledge gap [225]. Regarding chemicals and nanomaterials, risk assessment has conventionally depended on comprehensive, empirical data regarding exposure and hazards through methods like bottom-up and objective-driven approaches, which depend on decision-makers' information synthesis and acquisition to guide actions. For instance, current approaches for evaluating the environmental dangers of nanomaterials or creating laws are expensive, time-consuming, and data-intensive, according to the NNI 2010 Capstone workshop [222]. Still, a major challenge is figuring out how to combine top-down and bottom-up methods to support risk-informed decision-making. Several national and international initiatives, such as four seminars arranged by the National Nanotechnology Initiative in 2009–2010, have concentrated on how important policymakers may use this information [226]. The absence of toxicity information on the many nanomaterials being developed, however, is a significant obstacle [227]. Dose–response toxicity studies in humans, animals, and cells are the primary focus of traditional chemical risk assessment. Setting legislation or exposure limits and identifying hazards are complicated by the lack of data. One review, for instance, financed by the European Commission found that there is still not enough data to finish the comprehensive risk assessments needed to guide the regulatory decision-making process regarding the safety of NMs [228]. Similar to this, Hristozov et al. [229] highlight the several gaps in exposure and toxicity data that prevent a reliable risk evaluation of produced nanomaterials. Additionally, there is a fundamental deficiency in the methods and instruments needed to identify, quantify, characterise, and hence keep an eye on nanomaterials across a range of media. Techniques to identify and describe nanoparticles in complex matrices, such as final formulations, food, cosmetics, and environmental systems, are still being prototyped and developed [230]. The lack of multifunctional, multimedia nano-detection methodologies prevents workers, employers, product makers, and regulators from conducting routine monitoring for known and unknown nanoparticles in real-world circumstances. The lack of basic scientific and technological knowledge has made it extremely difficult to set standards for nanotechnology; yet, politics has historically influenced innovation [231]. Institutional regulatory duties have a significant impact on assessment-related concerns. Because of their special properties and potential for new hazards, the Royal Society and Royal Academy of Engineers in the United Kingdom suggested in 2004 that nanoparticles be categorised as novel chemicals for regulatory purposes. A substance's bulk and nano forms are identical, and one tonne remains the threshold for requiring registration and, as a result, risk information for the import or fabrication of a nanomaterial. In spite of this, the registration, evaluation, authorization, and restriction of chemicals (REACH) framework theoretically governs nanoparticles in Europe [232]. Risk management requires balancing the costs and advantages of using nanotechnology against any possible hazards. To proactively acquire data, build creative models appropriate for nanoscale phenomena, apply robust risk management frameworks, and communicate responsibly about this developing technology, a coordinated ecosystem of diverse academics, corporations, regulators, and society is required. Similarly, stringent governance that promotes regularly updated safety standards and tracked breakthroughs in nanotechnology research would be necessary to overcome assessment roadblocks.

International Regulatory Frameworks and Standards

The advancement of nanotechnology in different areas has resulted in an increase in risk management in terms of biological and environmental interactions, as well as exposure levels. Similarly, internationally harmonised frameworks for standardised testing and responsible development are also growing. Prominent intergovernmental efforts are currently underway to promote responsible nanotechnology growth through a combination of science-based safety assessment methodologies, risk mitigation recommendations, ethical guidelines, capacity-building modules, and adaptive governance policies tailored to dynamic lifecycles. In public policymaking, three sorts of instruments are used: regulatory, economic financial, and soft instruments. The first is frequently referred to as "sticks," the second as "carrots," and the third as the "sermon" [233]. Several international organisations, including the United Nations Educational, Scientific, and Cultural Organization (UNESCO), the International Council of National Science Academies, the International Organization for Standardization (ISO), and the Organization for Economic Cooperation and Development (OECD) have started multi-stakeholder initiatives to address these issues. The ISO encompasses 161 national standard organisations worldwide [234]. Three advisory groups operating under ISO/TC 229—ISO/TC 229/TG 02, ISO/TC 229/TG 03, and ISO/TC 229 SG—formed the Technical Committee on Nanotechnologies, which was established in 2005 with the goal of developing standards and technical reports on terminology and nomenclature, metrology and instrumentation, calibration, reference materials, test methodologies, modelling, and simulations for nanomaterial characterisation [235]. For instance, ISO/TS 80,004 outlines procedures for evaluating important physicochemical characteristics for toxicity evaluation, including particle size distribution, zeta potential, water solubility, and degree of agglomeration or aggregation [235]. There were 35 participating nations and 16 observer countries in ISO/TC 229 as of April 2018. Comparably, the Organization for Economic Cooperation and Development (OECD) formed the Working Party on Manufactured Nanomaterials in 2006, and since then, it has released guidelines that include protocols for safety assessment that are supported by science and include impacts on biotic systems, degradation, accumulation, and genotoxicity/carcinogenicity [236]. For instance, the suitability of current techniques for nanomaterial physicochemical characterisation and eco/toxicity testing was assessed in its Preliminary Review of OECD Test Guidelines for their Applicability to Manufactured Nanomaterials [237]. Eleven nanomaterials were tested and evaluated for about fifty-nine (59) suitable end points in their most current testing and assessment. A regional standardisation organisation made up of thirty-four (34) European nations is called the European Committee for Standardization (CEN). Three major EU standardization groups, CEN, CENELEC, and ETSI, established a technical committee on nanotechnology in 2005. CEN provides regulatory guidelines for the following fields: commercial and stakeholder issues; test methodologies; modelling and simulation; nanotechnology products and processes; health, safety, and environmental regulations; and classification, terminology, nomenclature, metrology, measurement, and characterisation (including calibration procedures) [234]. Working with associations, the European Union is developing REACH laws on nanotechnology that would impact the production, import, and marketing of nanomaterials. The World Commission on the Ethics of Scientific Knowledge and Technology was established by UNESCO to provide moral guidelines for evaluating potentially disruptive technologies such as nanotechnology. It preserves human dignity while seeking for morally acceptable compromises between nations [238]. Globally, the US, UK, EU, China, Japan, and Korea have all passed national laws pertaining to nanotechnology-related safety, health, and environmental concerns [234, 239]. They comprise worker protective equipment, allowable exposure levels, disposal and recycling processes, and mandated reporting criteria prior to the commercialization of goods in order to expedite and permit balanced factory monitoring.

Public Perception, Awareness, and Education About Nanotechnology

The way society views research, development, prevention, and control of possible nanomaterial risks may have an impact on how nanotechnology is portrayed [150]. A lot of social acceptance concerns hovers around the difficulty of dealing with ambiguity. Disapproval of inventions by society has frequently hurt commercialization [240]. With the introduction of the first wave of nanoproducts into the market, including medications, cosmetics, nutritional supplements, and pesticides, society is naturally worried about the potential effects these materials may have on human health, the environment, and safety. For example, in Germany, the first product recall due to health concerns over nanoparticles was recorded [241]. Existing studies on people's attitudes towards nanotechnology and adoption of nanotechnology applications have generally studied the impact of factors at the person or context levels [242]. As nanotechnology advances and becomes more integrated into society, public perception, awareness, and education become increasingly crucial. Despite the field's rising impact on items such as sunscreens, stain-resistant clothes, and even food, public awareness of nanotechnology is currently low [243]. Increased communication and education between scientists, government, industry leaders, and the general public can help build appropriate understanding and informed policies. For example, surveys demonstrate gaps and scepticism around nanotechnology in the public consciousness. In a study conducted by Scheufele and Lewenstein [244], just sixteen percent (16% of respondents considered themselves to be at least "somewhat informed" in the field of nanotechnology. In another study, eighty percent (80%) of the respondents to Cobb and Macoubrie's study [245] stated they had heard little to nothing. However, the respondents expressed optimism regarding nanotechnology as opposed to anxiety, believing that its advantages will exceed its drawbacks. "New and better ways to detect and treat human diseases" is their preferred possible benefit of nanotechnology, while "losing personal privacy to tiny new surveillance devices" is the most important potential risk to avoid. After two years, Macoubrie [246] revealed a still lower degree of awareness, with 39% saying they had only heard a little and 56% saying they had heard nothing at all. Hart Research Associates reports that 42% of Americans have heard nothing at all, and 27% of Americans have heard very little. Even among the populations most likely to use items containing nanoparticles, they found surprisingly little comprehension. The public is mostly ignorant of nanotechnology and so ill-prepared to discuss research priorities or laws, which is something that scientists, regulators, industry, and scientific communicators anticipate from these results. According to Miller [247], we don't know much about the public's concerns about particular features of nanotechnology or what kind of reaction they would like to see. A lack of public knowledge is sometimes used by those who oppose public engagement with science to imply that it limits the public engagement's capacity to access popular opinion or the "wisdom of the crowd." In contrast, Gehrke [243] conducted a study in which more than half of the respondents (59%) showed at least a basic understanding of the scalar definition of nanotechnology, which is something extremely tiny but potentially very important. Gehrke believes that statistical data misses the entire behaviour and perception of society members. Merely 9% of participants showcased a high level of proficiency, including several applications, mostly in the field of information technology, while 32% showed minimal to no comprehension. Museums, the internet, TV, and newspapers might all communicate advancements, issues, and realities to provide a balance of viewpoints. Communication, on the other hand, should avoid oversimplification and explicitly differentiate between hypothetical and current capabilities. Progress based on informed consent and consensus has the best opportunity of ethically progressing nanotechnology for greater social benefit. Clear, accurate public awareness and perception allow communities to thoughtfully weigh risks versus rewards instead of blanket calls to ban advancement.

Role of Government Agencies and International Bodies in Oversight

Government agencies and international organisations play an essential role in realising the promise of nanotechnology while monitoring and regulating the associated hazards by social interests. The US Food and Drug Administration (FDA), Environmental Protection Agency (EPA), National Nanotechnology Initiative (NNI), European Chemicals Agency, China's Ministry of Science and Technology (MOST), Japan's Nanotechnology Manufacturing Science and Technology Centre (NMSTC), and India's Department of Science and Technology (DST) are key government bodies involved in nano oversight within major countries driving large investments [248, 249]. The FDA formulates guidance on nanomaterial characterisation, toxicity testing, and risk assessment strategies required for pharmaceutical and biomedical applications before approving clinical trials or commercial launches. The EPA examines environmental fate, exposure levels, and ecological impact data for engineered nanomaterials across production, usage, and disposal lifecycle phases while evaluating registration and restriction requirements [250]. Funding bodies like NNI and MOST enable risk-focused research from a fundamental level to a translational scale [251, 252]. International organisations such as the OECD, ISO, and WHO develop frameworks for harmonised standards for nanomaterial physicochemical characterisation, effects monitoring, and risk estimation that individual countries can contextualise across regional legislation. Beyond scientific elements, social aspects such as promoting public awareness, enabling responsible innovation ecosystems, and resolving unintended repercussions are critical for governance [253]. For example, the European Commission fosters talks between enterprises, politicians, researchers, and civil society groups to align nanotechnology evolution with social values and environmental goals specified in the European Green Deal [254]. Policy discourse in emerging nations focuses on adjusting global paradigms to domestic infrastructural capacities and population requirements while limiting overexposure. The requirement for transparency in governance fosters confidence and allows for the reconciliation of innovative opportunities, responsible norms, and access priorities for nano-enabled transformation that benefits the public at large.

Cultural Representations and Societal Imagination

Emerging technologies are dramatically reshaping human capabilities to engineer, manipulate, and rebuild the fabric of the natural world at the molecular level. However, even before such extreme manipulation was possible in laboratories, the cultural imagination was ablaze with imaginative images of nano-enabled futures for society. Fiction frequently provides the creative space to imagine how developing technologies might affect mankind before they become entrenched in socio-political institutions, allowing a priori meditation on ideal futures rather than retrospective redressal alone [255]. Some argued that people's quality of life will be endangered by uncontrollably self-replicating systems, miniature weapons of mass destruction, covert surveillance methods, and unidentified impacts of nanotubes [256]. Popular novels, such as Michael Crichton's novel Prey and Neal Stephenson's The Diamond Age, illustrate the disastrous consequences of self-replicating nanobots destroying human society while depicting technological monopolies [157, 257]. Thus, futurist cautionary tales demand protection and control. More balanced visions in Star Trek, Ender's Game, and Big Hero 6 use radical nanotechnology as motifs to envisage advanced societies where human desires and intent line more with collective well-being [258]. Such speculative visions enable collective reflection on the long-term aspirations and risks of technologies reshaping reality at fundamental levels, even before applications grow from theoretical possibility to commercial systems. Speculative fiction sparks discussion about existential hazards, ethics in the face of unknowns, and shared responsibility among stakeholders [157]. Along with literary tales and film, informational diagrams, scientific artwork, and infographics deciphering via visual metaphors impact public perception and attitudes towards acceptability [259]. Thus, cultural translations through the visual landscapes of science fiction, info-art, and experiential displays serve as a dynamic record of collective hopes and worries about converging technology and society. Understanding public mood by analysing cultural artefacts as socio-historical records of multivocal futurism provides critical insights alongside technical roadmaps for responsibly realising nanotechnology's benefits. This lays the stage for exploring the different facets of public imagination regarding nanotechnology as both cautionary tales highlight risks and more aspirational narratives imagining how responsibly guided technological advancement may produce better futures.

Depictions of Nanotechnology in Art, Media, and Popular Culture

Numerous science fiction and popular books depict dystopian science fiction worlds where human life and culture are destroyed and boundaries between nature and culture collapse due to fearsome hybrid forms of existence caused by nanotechnology [260]. This has inspired both enthusiasm and anxiety because of the huge consequences for humanity. These different opinions have found cultural expression in art, media portrayals, and popular narratives that impact public attitudes towards this emerging science and technology. It is everyone's responsibility to preserve and pass on cultural identities and legacy to future generations [261]. This type of culture is typical of the twenty-first century and modern cultures all around the world [157]. Culture is valued through the preservation of history through digital technologies and public interaction. As a result, creating novel conservation approaches that are easily available, economical, and simple to implement is critical to maintaining access [262], and nanoscience has been shown worthy of use. Artworks and experiential media also have a visceral appeal for reflective appraisal. In Florence, Italy, wall paintings at the Brancacci Chapel were cleaned in the late 1980s, marking the beginning of the use of nanoscience in art conservation [263]. When restoring the Brancacci Chapel paintings, an oil-in-water microemulsion containing dodecane nanodroplets stabilised in water by a surfactant was used to remove wax stains from the surface. The surface of the wall murals in Venice's San Salvador church was coated with extensive layers of old acrylic coatings that were removed using a hydrophobically modified hydroxyethyl cellulose (HMHEC) dispersion including an oil-in-water microemulsion (see Fig. 1.12a, b). Similar to this, near the close of the twentieth century, nanostructured calcium hydroxide ($Ca(OH)_2$) became known as a very useful restoration material due to its ability to both neutralise acidity,

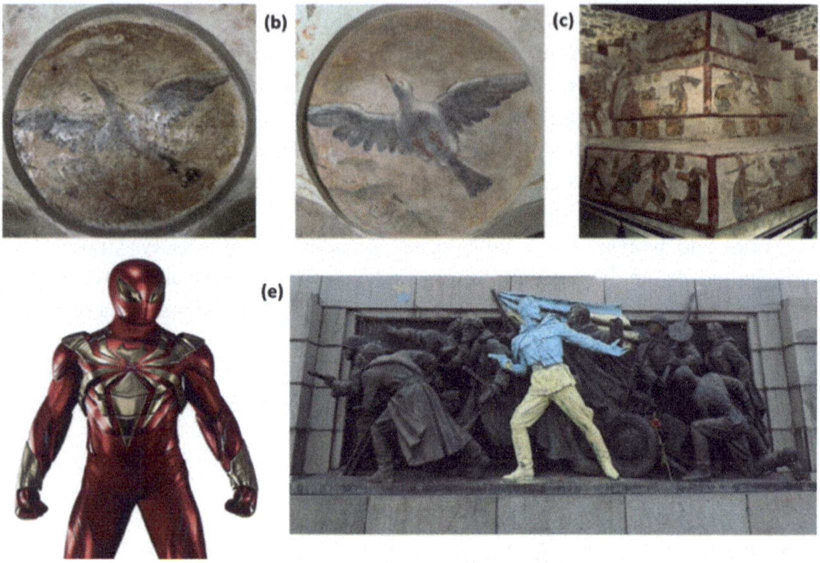

Fig. 1.12 **a, b** Old acrylic coatings removed from the wall murals at the Venice, Italy, sacristy of the San Salvador church [262] **c** Maya wall paintings in the United Nations Calakmul, Mexico is home to the Educational, Scientific, and Cultural Organization World Heritage Site [267] **d** Spider man's iron suit [268] **e** Monument of the Bulgarian Soviet Army [269] Open access article by article Atlas Obscura

which breaks down cellulose in wood and paper artefacts, and reinforce fragile wall paintings and limestones [264, 265]. One notable case study for this approach is the consolidation of Maya wall murals in Calakmul, Mexico (Fig. 1.12c). Giorgi describes the Santa Maria Novella church in Florence's nano-wall painting [266]. Another product sector that may benefit from "intelligent" nanotechnology is paint. For instance, researchers at the University of Strathclyde in Glasgow, Scotland, have created "smart paint," which is paint with tiny sensors implanted in it.

Aspirations and worries are framed by various pop culture forms. Following them provides a cultural pulse on envisioned futures, risk perceptions, and governance ethical arguments. Importantly, they enable public articulation and ideas for inclusive policy to shape the direction of this shift. Drexler's concept of the 'assembler,' the self-producing nanomachine, inspired popular culture and science fiction [260]. One example is the novel "Prey" by Crichton, which tells the story of a group of sentient nanorobots that spectacularly escape the control of the engineering team that created them. Nanotechnology in the form of nanites has been used in a variety of media, film, and television shows, including Avengers: Infinity War, Iron Man's Mark L armour suit, Spider Man's Iron Spider suit created by Stark (see Fig. 1.11d), and Black Panther, to name a few [268]. Similarly, on 17 June 2011, Sofia [269] reported on Bulgaria's capital city's famed Soviet monument, which was transformed overnight into a pop culture stage for some of the world's most well-known icons, including the Joker, Wolverine from the X-Men, Superman, Captain America, Santa Claus, and

Ronald McDonald. The Nano Art international competition also includes attractively designed molecular structures for public participation. Such sensory-based explanation of scientific concepts strives for deeper involvement and informed assessment of cultural assimilation issues.

Influence of Cultural Narratives on Public Perception and Acceptance

As interest in nanotechnologies grows, so do study patterns that have emerged in tandem with an explosion of media and public opinion analyses across national boundaries [270]. Art, fiction, cinema, experiential media, and popular representations all play an important part in developing public imagination and attitudes towards acceptance or opposition [271]. The value notion underpins a method for measuring cultural differences. This strategy is based on the notion that fundamental values influence people's attitudes and behaviours, and that the relevance of different values varies between cultures [272]. Discrimination based on cultural assumptions and misconceptions can impede inclusive and ethical governance. Politics is fundamentally concerned with how varied people coexist peacefully, allocate resources, and plan shared destinies [273]. Understanding the processes by which the public makes sense of developing technologies and develops reactions to them is widely acknowledged as crucial for the design and management of instinctive platforms for public engagement and involvement [274]. The acceptability of new technology is therefore influenced by cultural narratives around public attitudes and acceptance levels based on media frames, ideological predispositions, or religious convictions. The results of a UK study based on the DEEPEN research on nanotechnology indicate that people's willingness to adopt new technologies relies on how they weigh the benefits and hazards of each technology, or if they assume that people are either "for" or "against" a particular technology [274]. According to Felt et al. [275], technological adoption is focused on maximising the benefits of technology while minimising its downsides. Others acknowledge that the concept of rising technology can 'mess with nature,' and it is based on the ancient conception of nature having sacred aspects that establish rules or order in the human world. Reacting to nanotechnology's promising potential to increase control over the natural world and human race, the 'messing with nature' narrative was widely accepted. This was shown in programmes that relied on what Bensaude-Vincent refers to as a completely 'artificialist vision of nature.' [276]. Nonetheless, polls in the food sector found that the benefits of nanotechnology outweigh the dangers for lay people, but a sizable minority of respondents did not answer the risk-versus-benefit question. In other words, the participants were unable to weigh the pros and hazards of nanotechnology. This research implies that people's perceptions about nanotechnology are still rather flexible [272]. Hence, public perceptions are primarily fuelled by public fears [277], and the odds between good and bad.

Shaping of Societal Aspirations and Anxieties Through Cultural Lenses

According to the Oxford Dictionary, an aspiration is "a hope or ambition of achieving something". Aspirations therefore are not expectations: we may aim or hope to achieve something, yet not necessarily realistically expect to achieve it. The notion of aspiration, in its broadest sense, allows an individual to entertain ambitions that may or may not be within the range of outcomes that the individual's situation allows her to achieve. According to Ferrara [278], an aspiration is a goal or desire to accomplish something. Consequently, goals are not the same as expectations. In its broadest definition, it allows for the pursuit of goals that may or may not be within the realm of possibilities afforded by the individual's circumstances. Assumptions, symbols, rhetorics, biases, and blindspots mediate public perception of the effects of emergent advances, influencing reactionary hopes or worries [279]. However, social conceptions and deliberate interventions in governance narratives assist shape collective responses to technology along moral axes that align with ideals, given the fluidity of cultural perspectives. Risk expectations, especially for nanomaterials, have been brought to the attention of policymakers with the help of activists' counterclaims by critical Environmental Non-governmental Organizations (ENGOs) [280]. The NNI and Observatory Nano have established guiding principles for the 2020 timeframe, despite their emphasis on commerce. Future manufacturing facility frameworks, the use of nanotechnology to address sustainability issues like energy, innovative uses of biosystems and medicine, information technologies, novel materials and associated systems and procedures, and adherence to international regulations are a few of these [281]. Through genome sequencing, the information codes used by living systems have been gradually revealed in their incredible diversity, but at a faster pace and at a cost that is far lower and dropping more quickly [282]. Human genetic mutations and the roles of heredity and environment in these changes have been studied by researchers. The recent field of DNA-based data storage exemplifies proactive efforts in matching cultural perspectives and governance mechanisms from the start to responsibly shape goals before commercialization [281]. Recognising the societal concerns over genetic alteration, researchers created both technical and legal safeguards for encoding synthetic DNA while thoroughly describing the possibilities and restrictions to avoid misdirected aspirations or scaremongering. As a result, governmental structures and cultural rhetorics symbiotically influence social imagination in generally ethical directions [283]. The need for openness about the use of synthetic nanoparticle substances in commercialized items was brought up in the Grenelle de l'environnement round table discussions on sustainable development with the growing environmental concerns [281]. However, explicit regulatory rules for defining boundary instances and promoting transparency helped to consistently realign the cultural narrative closer to realistic evaluations. Requirements such as openly revealing synthesis techniques and the usage of selecting agents create faith in complying norms [284], pushing the promises of nanomaterials forward. Meanwhile, literary descriptions eventually normalise nano-consumer items, relieving lingering worries through familiarity. As a result, reasonable aspirations prevailed as technologies such as nanomedicines, durable commodities, and quality sensors responsibly entered general use, increasing the everyday quality of life based on priorities [285]. According to a study conducted by McGrail [280] on the people's perception of nanotechnology, participants emphasised that they would rather that some of the organisations voicing concerns also acknowledged some of the toxicity studies conducted about the evidence for potential negative effects of nanoscience, especially in the area of health. Additionally, Australian research may be used to increase knowledge-intensive jobs, provide businesses a competitive edge, and increase exports. It could also aid in the development of better climate change technology. This encourages growth through safe design, which serves as the central quandary [280].

Impacts of Cultural Representations on Technological Adoption

Emerging technologies spark both excitement and concern given their disruptive implications for restructuring reality [286]. Cultural expressions through art, literature, and media have a significant impact on shaping public perception, aspirations, and risk fears, which influence the adoption and integration trajectories of inventions with broad impact potential [287, 288]. Speculative visions enable social contemplation of potential futures deriving from scientific revolutions and extraordinary capabilities before technical stabilisation. Science fiction has imagined dystopias caused by failed technologies, as well as bold answers to humanity's major dilemmas through responsible innovation [289]. Complex tensions, paradoxes, and priorities are distilled into cultural motifs and tale archetypes that encode warnings, morality, and hope. These appear emotionally before becoming formalised into ethics, legislation, and conventions [290]. As a result, targeted governance measures that promote balanced and inclusive cultural representations can prevent such technological fallacies from influencing societal outcomes. For example, cultural discourses surrounding emergent biotechnologies such as GMOs and synthetic biology are still intertwined with Frankenstein's implications of "tampering with nature" [291]. These alarmist stereotypes, accentuated by dramatic media coverage, tend to establish negative impressions of ethics. They neglect balanced scientific assessments of biosafety methods that minimise dangers for low-risk research. As a result, public funding for beneficial projects is hampered. Similar inequities hampered nanotechnology at first until concerted communication efforts addressed its cultural framing. Because of the association with invisible threats, metaphors likening nanoparticles to undetectable "magic bullets" heightened anxiety and terror [292, 293]. Campaigns in academia, government, and business offering realistic assessments of rigorous safety evaluations and preventative supervision control helped shift balanced perceptions [294].

Hence, nanotechnology has seen relatively easy adoption in commercial applications like targeted medicine delivery, high-quality sensors, and strong yet lightweight building materials [295, 296]. The trajectory of automation and AI is also dependent on cultural representations appropriately aligning with complicated reality before reaching dissemination stages [297]. In essence, cultural representations that resound loudly throughout media, literature, and social networks risk informing generalised public perceptions of developing technology more viscerally than specialised expert assessments [298]. However, proactive governance initiatives can correct technology flaws before incorrect impressions become established. Such cultural governance tools, when combined, create good adoption paths for technical breakthroughs rather than reactively attempting to correct misperceptions after spread [299].

Envisioning Future Scenarios: Nanotechnology Integration

Emerging technologies promise radical shifts but also carry uncertainties regarding widespread impact and envisioning plausible futures helps contemplate opportunities and challenges for responsible innovation [300]. As nanotechnology progresses from lab proofs-of-concept to industrial applications, it holds enormous potential but also significant concerns [47]. In healthcare, for example, nano-robots could deliver medications precisely targeted to cancer cells while minimising negative effects and enabling regenerative tissue repairs through cellular manipulations [301]. Integrating self-sensing and self-repairing nano-engineered materials in environmental areas enables resilient, flexible architecture, prevents disasters, and improves the quality of air, water, and agricultural waste [50, 302]. Access to benefits and opportunities in R&D is provided by significant improvements across industries, which necessitates recognising imbalances exacerbated by technical disparities [303]. Speculative diligence addressing diverse ramifications, particularly social embedding and access, helps emergent innovations to be guided towards responsible innovation benefiting the general public good. Thus, the different envisioned application areas of nanotechnology are described.

Nanotechnology in Energy Production and Storage

A global understanding of the need for green transitional aims has led to the daily emergence of new renewable energy technologies in a variety of designs and forms [304]. In recent years, the progress of nanostructured materials has considerably increased the efficiency of energy storage, conversion, and utilisation. Their nanoscale size and unique optical, electrical, and magnetic capabilities can transcend the limits of traditional energy systems [305]. The anticipated high energy-conversion efficiencies at a lower footprint illustrate nanotechnology's great power for furthering renewable energy and electrified mobility goals provided translation constraints can be properly handled through continued materials research and responsible development [306]. Unique types of nanoparticles, such as thiolate group-stabilised nanoclusters of gold, silver, and other metal atoms, have provided a visual representation of how nanotechnology may function in this field in the future [307]. Importantly, nanomaterials like quantum dots, nanowires, graphene, and metal oxides are applied in solar cells to manipulate photon absorption and electron transport to improve solar cell performance beyond traditional boundaries. For example, when the characteristic dimensions of a semiconductor are reduced to a specific size, quantum confinement results in an increased bandgap. The material's size may be easily altered to modify the bandgap, which enables the optical absorption and emission spectra to be precisely tailored to meet the requirements of the intended applications [305]. They serve as electrode and electrolyte membranes in batteries and supercapacitors, allowing for enhanced charge conductivity, voltage, and capacitor storage density for high performance with faster charge/discharge [308]. Unlike conventional devices, this novel structure is tipped to have good electrochemical properties as well as an appropriate material density and structural design. This may be engineered so that, during multicycle charging and discharging, the internal structure of the ion exchange pores lessens the impact of volumetric shrinkage. The nanoscale catalyst particles in fuel cells maximise surface area for higher chemical reaction efficiency during hydrogen fuel synthesis and oxidation, resulting in increased power density and lower costs [309, 310]. They can also function as nanoscale coating materials as thin films to reduce reflection losses from solar surfaces, increase the heat transfer efficiency of steam turbines, or improve overall energy conversion ratios [311].

Nanotechnology-Enabled Education and Communication

Nanotechnology has the potential to alter education and communication systems by exploiting nanoscale phenomena for benefits such as interaction, personalisation, and quick information transfer across networks [312]. It is anticipated that the use of nanotechnology in education and communication will be critical, as augmented textbooks with integrated printed electronics will be able to dynamically display adaptive content and multimedia beyond static pages in classrooms [313]. Wearable devices with sensing brain signals may enable immersive educational experiences by assessing cognitive load, engagement, and stress signs to provide personalised feedback [314]. Bio digital measures can supplement anatomical structures by interacting with historical events through avatars or building prototypes in engineering simulators via hyper-realistic virtual reality settings powered by haptics [315]. Importantly, nanotechnology is prepared to advance electronics power usage by allowing them to work on flexible, printable substrates rather than rigid silicon boards, allowing for widespread integration with common objects and surfaces [316]. Enhancing real settings and embodied cognition processes rather than just virtual screens represents a paradigm change in learning experience design. These interactive tools also democratise information exchange for improved livelihoods outside of formal contexts. Similarly, wireless nano-enabled surfaces can allow public access to digital connectivity in remote locations that now lack infrastructure, thereby empowering education throughout communities. Rapid information dissemination via flexible graphene-based antenna grids that may be easily deployed allows for efficient humanitarian response coordination [317]. Importantly, integrating nanotechnology alongside pedagogical and social innovations can make learning more accessible, engaging, and powerful by eliminating barriers to equal development across demographic and geographic boundaries.

Smart Cities, Urban Planning, and Infrastructure Innovations

Nanotechnology is set to provide next-generation solutions in developing future smart cities with intelligent, connected, and responsive infrastructure for urban planning. Nano-engineered construction materials like self-sensing concretes containing conductive nanotubes for damage detection [318], and self-healing composites with microcapsules that can fill cracks autonomously [319] using nanocellulose and graphene with unprecedented strength-to-weight ratios can enable resilient, adaptive buildings and infrastructure. These systems can also handle harsh heat and cold, as well as degradation, indicating that metropolitan environments will be considerably more sustainable. Integrating nanosensors into systems can assist in monitoring structural health and avoid disasters through early action [320]. Importantly, nano-coated glass not only keeps buildings naturally light but also temperature-managed for effective HVAC by optimising heat flux regulations [321]. Nanomaterials-based solar cells, fuel additives, and batteries will provide locally generated, cost-effective, and renewable energy. Piezoelectric nanogenerators that gather vibration energy could deliver trickle charges and improve environmental friendliness. Distributed communication networks spanning infrastructure can be established using integrated nanosensors, transmitters, and data storage devices [322]. For example, nano-enabled millimetre wave technology created by NASA engineers for 5G can use radio airwaves to transfer data without the need for significant infrastructure and damage to our cities. Pollution control and remediation solutions can use nano-filters for water purifiers, catalytic converters to reduce car emissions, and nano-absorbents for toxic chemical cleanup to improve environmental quality and public health [323]. Data is already being stored by traffic signs, which will improve security and safety for residents of future smart cities. In order to track traffic patterns throughout the nation, the Chinese are currently adding nano latex inks to their road signs. This has the potential to significantly enhance the environment, security, and traffic flow.

Synergies with Other Emerging Technologies (AI, Biotech, etc.)

Integrating nanotechnology with other developing technological advances, such as artificial intelligence and biotechnological tools, offers the potential to expedite scientific discoveries across several domains. New applications might be made possible by the confluence of these fields. Every modern digital gadget that can be connected to the internet and create and interchange vast amounts of data, including sensors, actuators, smart mobile devices, and more, is referred to as the Internet of Things (IoT) [143, 324]. The development of nanosensors supplemented with AI processing could provide a previously unheard-of ability to sense, analyse, and respond to chemical events. Nanomaterials have aided the advancement of precision medicine at all phases of treatment. New omics collection methods, such as single-molecule nanopore sequencing, allow for quick and sensitive single-molecule detection as well as extended sequence read length, preserving genetic context [325]. Nanoparticle medication delivery techniques guided by AI pathfinding models will provide precision targeting with minimal systemic side effects, while nanomachine-bio interfaces enable regenerative neuron-by-neuron repairs [326]. AI systems may then interpret and extract insights from these massive nanosensor datasets to optimise system performance, enable predictive maintenance, and accelerate medical diagnosis [327]. Nanosensors embedded in smart infrastructure systems, industrial equipment, or even the human body might offer huge volumes of real-time data on component stresses, microfractures, temperature variations, and biological indicators [50, 328]. Exciting prospects for environmentally friendly food production are presented by the integration of agricultural precision and artificial intelligence (AI), in which farmers use both technologies to react in real time to changes in crop development [329]. To improve targeting, absorption, distribution, nutrient capture, and long-term impacts on soil microorganisms, current models for crop yield and nutrient cycle can be combined with nanoinformatics techniques to create nanoscale agricultural compounds with the optimal safety and functionality profiles [329].

Responsible Innovation and Ethical Governance

The US National Science Foundation has estimated that the competition to become the world's top innovator in technology will be worth trillions of dollars over the next ten years. Like with biotechnology, there are also demands for the general population to be involved in shaping the future growth of this novel field of science and technology, as well as for urgent attention to the social, ethical, legal, and environmental challenges related to nanotechnology [256]. Realising nanotechnology's transformative potential for societal flourishing necessitates proactive efforts for responsible and ethical governance throughout its innovation lifecycle [330], as these aspects necessitate the development of adaptive governance systems that balance economic competitiveness with ethical responsibility goals [331]. A responsible framework for guiding innovation routes comprises proactively assessing and minimising risks, fairly distributing gains, and supporting long-term development objectives. This requires aligning R&D priorities with socially acceptable and legal applications for emerging nanosystems. Furthermore, ethical governance must broaden access to the benefits of nanotechnology across society and globally. Public funds can stimulate innovation for public goods such as clean air, water, decarbonisation, infrastructure enhancements, and inexpensive diagnostics and medications [107]. Multi-stakeholder involvement in road mapping activities helps align innovation routes with ethical ideals and priorities while delivering information that builds public trust. Adopting ethical responsibility principles such as universal access, trust, safety, transparency, and justice can thus guide nanotechnology towards socially desirable and sustainable innovation pathways, and governance policies can thus balance intellectual property monopolies with equitable access imperatives.

Balancing Innovation with Ethical Considerations

The enormous potential of nanotechnology to serve humanity necessitates a careful balance of disruptive innovation and ethical deliberation, hence giving routes for nanoscale alterations into various product forms. However, the same qualities that distinguish nanoparticles pose concerns about their safety, equity, and societal impact [332]. These ethical considerations necessitate complex governance that improves innovation routes to match moral goals such as sustainability, accountability, and justice [333]. The unique physical, chemical, and biological behaviours may have negative ecosystem and health effects due to enhanced toxicity, bioaccumulation, and biomagnification compared to conventional forms [334]. As a result, governance rules must balance the rate of advancement with proper oversight, and regulatory definitions must expressly include nanoforms rather than relying on standard bulk material categorisations [335]. Environmental health and safety (EHS) testing as a criterion for market approval embraces the precautionary principle while allowing low-risk applications [336]. Adapting limits in response to new risk information enables governance iteration as scientific understanding grows over time. To balance innovation and ethics, governments must analyse unanticipated secondary impacts beyond direct toxicity, such as economic, social, and ethical ramifications, while incorporating value-sensitive design societal expectations into R&D phases [337]. Importantly, funding priorities provide another tool for ethically influencing research by matching nanotechnology development with environmental goals [9]. Good governance thus necessitates a participatory mechanism involving diverse stakeholders in decision-making since perceptions of risks and benefits vary across groups. Public engagement through deliberative exercises provides inputs for policy priorities calibration and multi-stakeholder dialogue in road-mapping, foresight, and anticipation research also aligns innovation with societal values [338]. Such integrated governance harmonising precaution and progress positively unleashes nanotechnology's revolutionary potential for society ethically by design.

Ensuring Responsible Development and Deployment of Nanotechnology

More research on the environmental consequences of nanomaterials from various backgrounds is being requested because of recent breakthroughs in the disciplines of material science and nanotechnology. These days, economic processes depend on information in all its forms. Nations that manage and develop their intellectual property well can function more effectively, creating jobs and income [339]. The effects of nanotechnology on the environment may be divided into two categories: the first addresses how new developments in the field may help the environment, while the second focuses on the potential for new kinds of pollution to arise from the discharge of nanomaterials into the surroundings [340]. As a result, responsible innovation frameworks that could provide advice so that nanotechnology advances ethically towards promoting human well-being rather than exacerbating existing disparities is necessary [341]. The adoption of new technology is not a simple, self-evident process. Rather, it is ingrained in social interactions and requires political support [342]. On the one hand, they can evolve spontaneously via the combined efforts of institutional and economic players [343]. Government planning should gather the appropriate enabling players in the decision-making due to the intricate nature of the ecosystems of innovation (EoI) rollout process. These players from academia, business, and state agencies must then collaborate as the main group steering the creation of an EoI for nanotechnology. Thus, setting R&D objectives that are connected to scale-up and product commercialization goals that are socially, legally, and morally acceptable via the long-term obligations needed to implement it is the first step in sustainably directing the advancement of nanotechnology [344]. Furthermore, enough levels of human capital must be trained and maintained. Despite a notable surge in private investment in nanotechnology over the past ten years, these commitments continue to pose a considerable barrier to entry for enterprises and venture investors [345]. The necessity for an open innovation strategy in research is another significant aspect of the deployment of nanotechnology due to its multidisciplinarity [148]. Many observers predicted in the early 1990s that nanotechnology would be concentrated in a few countries due to the massive investments necessary for its adoption. This theory was proven false, and today's nanotechnology enjoys the advantages of a global collaborative environment where anybody may access and contribute to the discovery process. In fact, the great majority of nanoscience institutes worldwide are connected, and the knowledge-sharing network is the real advantage of this design [344]. Regulatory policies must be updated to instil appropriate development practices. For example, identifying nanoforms in existing chemical inventories, testing requirements, and risk management rules eliminates unintentional oversight gaps that are simply dependent on bulk material categorisation [346]. Thus, regulating toxicity testing and environmental effect assessments for engineered nanomaterials makes prudence a crucial component of creating responsibility. Importantly, standardisation bodies encourage responsible development by developing testing methods, characterisation techniques, reference materials, and toxicological markers to improve scientific understanding of novel features and to assess hazards more correctly [347, 348].

Promoting Transparency, Accountability, and Inclusivity in Nanotech Advancements

The realisation of the immense potential of nanotechnology for societal upliftment requires governance mechanisms promoting transparency, accountability, and inclusivity across its innovation pipeline. Ensuring transparency in nanotechnology development pathways is vital for evidence-based public debate and trust [119]. Therefore, innovation and technology must ensure that existing social orders are not undermined while still being socially acceptable, morally sound, and harmless—that is, free from prejudice and dishonesty. Marketable goods and services must be restricted by an inherently safe design and an appropriate control plan in order to guarantee that they do not endanger people or the environment [349]. A greater need and chance for stakeholder and community engagement to enhance decision legitimacy and sustainability arises from a growing scepticism of government and business strategies for nanotechnology development [350]. One way that public education and participation overlaps with governance is the development of a science and technology governance framework that better includes the public. Enhancing bidirectional communication between manufacturers and the public is vital, as is including risk and benefit evaluations into the conversation [351]. Surveys evaluating acceptance and concerns across demographic criteria such as gender, ethnicity, age, and geography indicate priority issues for governance systems and construct responsible advancement pathways reflecting varied requirements [352]. Researchers should proactively disseminate data from safety studies, risk assessment models, and uncertainty characterisation in open-access repositories. Furthermore, manufacturers giving fair knowledge must set realistic public expectations by addressing residual uncertainties, as regulatory systems require transparency to earn public trust and fight corporate lobbying effects [351]. In addition to participatory planning, downstream oversight necessitates diversity and accountability. As a result, nanotechnology industries should form ethics advisory boards comprised of stakeholders such as workers, academia, public interest groups, and local community representatives to assist with concerns such as labour displacement [150]. Allowing public interest litigation strengthens external accountability through judicial review of administrative actions and policy gaps. As a result, requiring traceability infrastructure and supply chain disclosures on nanomaterial flows holds manufacturers accountable for safe environmental and health stewardship when life cycles end [353].

Collaborative Efforts Between Academia, Industry, and Policymakers

Nanotechnology is garnering increased investment from governments and private sector organisations around the world. The EU alone invested around €896 million in nanotechnology-related research between 2007 and 2011. The global investment in nanotechnology is expected to be in the trillions of dollars, with the United States committing up to $2 billion [47]. The efforts of academics, businesses, and non-governmental organisations to determine and create regulations relevant to the environmental, health, and safety (EHS) aspects of engineered nanomaterials demonstrate a shared interest in the responsible development and application of nanotechnologies [354]. An increase in worldwide financing has affected the quantity of academic publications about nanotechnology. National nanotechnology initiatives have been started in more than 60 countries as of 2016 [112]. The governmental and private sectors are expected to support nanotechnology advancements more, allowing them to continue growing. For example, the National Nanotechnology Initiative (NNI) has prompted numerous countries to launch extensive research and development (R&D) programmes in nanotechnology, and the US federal government has continuously invested in nanotechnology innovation to advance its advancement, from President Obama's 2011 introduction of the Advanced Manufacturing Partnership (AMP) to former President Bill Clinton's efforts [107]. Additionally, the likelihood of a successful product fit for the industry has grown with the inclusion of fresh approaches in the research plan. The Royal Society and the Royal Academy of Engineering in the United Kingdom released the first widely read publication evaluating the benefits and drawbacks of nanotechnology in 2004. This publication emphasised the importance of formal training for all research students and personnel working in these domains [128]. One year after the first EU (European Union) action plan on nanosciences and nanotechnologies (covering the period 2005–2009) was adopted, the Scientific Committee on Emerging and Newly Identified Health Risks adopted a report on the environmental and health aspects of nanotechnology in 2006 [354]. Research-driven innovation uses scientific or technological breakthroughs from laboratories as a source of creative inspiration [9] for the benefit of society. However, because there aren't enough human-centred methods for introducing these technological advancements, consumers may reject such improvements. Therefore, the Centers for Nanotechnology in Society at Arizona State University and the University of California, Santa Barbara investigated the ethical and legal concerns surrounding this technology to ease public feeling and comprehend the social dynamics of nanotechnology [355]. Similarly, Yasri and Wiwanitkit [356] and Allon et al. [357] have conducted extensive studies on the recent ethical issues associated with nanomedicine and nanomedical technology in a bid allay public fears because it has the potential to significantly boost innovation and value generation in the private sector. As current chemicals are examined and prioritised based on a variety of characteristics, including danger, persistence, bioaccumulation, and exposure, research groups might partner with appropriate organisations to produce cutting-edge solutions through data-driven innovations [354, 358]. Based on current customer gaps or desires, this data is then strategically leveraged to construct or enhance products and services. One can wonder which approach is superior. Each method has advantages and disadvantages. While either can lead to innovation, combining the two is preferred [47]. Importantly, university degrees in nanotechnology, research initiatives, and deliverables serve as a suitable benchmark for assessing the worth of academic work. Therefore, it is important to consider how investments in nanotechnology would affect export-oriented industries, lower manufacturing costs, create jobs, develop new goods and services, and establish new businesses [17] which together, will help push the scope of nanotechnology forward.

Conclusion: Charting a Path Forward

The immense promise of nanotechnology ushered in an era of dramatic advancements across industries ranging from communications to medicine is dependent on continuous research and responsible development. In exploiting the maximum societal benefits of nanotechnology, an interdisciplinary approach must be taken. Thus, progress will be dependent on continued financing and institutional support for fundamental research and development (R&D) to unleash nanotechnologies potential, as well as translational research to take breakthroughs from the lab to real-world implementations. Policy frameworks should thus, encourage ethical scientific research and commercialisation that is aligned with public interests. Regulatory organisations must take the lead in defining safety standards and processes for new nanomaterials as well as future nano systems integrated with biotechnology and artificial intelligence. Collaborative projects across academics, government, and industry can address concerns associated with runaway technology while sustaining pace. Most crucially, the potential and the risk of nanotechnology is ultimately tied to how humanity employs these evolving capabilities. Setting an ethical compass for nanotech R&D and enrolment ensures that societal outcomes improve the quality of life for all. Prioritising investment in applications that address critical issues such as sickness, environmental degradation, and equal prosperity might help push innovation in positive ways. Charting a thoughtful, conscientious course for nanotechnology development will enable unlocking its full potential for society while aligning scientific progress to shared human values. With sound policy and leadership, people stand to harness nanotechnology breakthroughs to create a safer, healthier, and more just world for generations to come.

Summary of Key Insights and Findings from the Chapter

The primary conclusions from all the data analysed so far in this chapter show that the development of nanotechnology dates back to the 1800s and is credited to renowned physicist Richard Feynman's "There is plenty of room at the bottom" lecture, which emphasises the use of nanomaterials in technology. As a result, a plethora of technological tools, including nanomachines, nanorobots, and nanodevices, have emerged. These are used in the production of tiny instruments and production tools with precise control at the atomic level in a variety of industries, including environmental remediation, health, agriculture, communication, and cosmetics. Importantly, nanotechnology is poised to transform the economy enabling more efficient industrial processes, superior materials, disruptive technologies, and new growth opportunities across sectors, not forgetting its quest to provide jobs to the youth. Therefore, depending on the percentage of nanotechnology that is classified as green and the definitions applied, a sizeable portion of the workforce will be employed in green nanotechnology occupations. However, balancing economic gains with ethical, legal, environmental, health, and social implications is vital towards achieving the UN's SDG Agenda 2030 goals which are geared towards sustainability. This calls on key players such as policymakers, regulatory bodies, the general public, and the government to join forces towards achieving a greener production or synthesis process, distribution and deployment, at the nanoscale level, ensuring privacy in manufactured nanoproducts and towards an even access across countries and continents. One instance is the precise identification of SARS-CoV-2 with the use of nanoparticles in fluorescence technology to achieve magnetic clearing. Though many are of a negative view regarding nanotechnology, it has transformed the health sector especially. Understanding the risks involved, the necessary risk management approaches, and following standard regulations have yielded positive results, thanks to interventions from the NNI, EPA, ISO, OECD, UNESCO, and the Royal Society and Royal Academy of Engineers. As nanotechnology continues to advance and integrate into more areas of society, public perception, awareness, and education become increasingly important. Currently, public knowledge about nanotechnology is fairly low despite the field's growing impact on products like sunscreens, stain-resistant clothing, and even food, according to literature. The clear, accurate public awareness and perception allow communities to thoughtfully weigh risks versus rewards instead of blanket calls to ban advancement. The study again highlights that public perceptions are rooted in societal imagination, cultural representations, and art and media. For instance, nanotechnology has influenced numerous dystopian science fiction stories in both popular and science fiction, as well as in historical artefacts like the Brancacci Chapel's wall paintings and the Maya wall paintings at Calakmul,

Mexico. The excitement and concern help re-evaluate the implications of nanotechnology for future developments in areas including fuel cells, education, smart cities, and the synergistic employment in other technologies including artificial technology. With funding from the government and private sectors and the call for cooperation between industries and academia, via responsible design and ethical considerations, nanotechnology is set to transform the globe and enhance the transition of society towards a sustainable and much-improved development.

Emphasis on the Intertwined Nature of Nanotechnology and Societal Development

Nanotechnology has evolved as a revolutionary force that intricately intertwines with social growth, presenting unparalleled possibilities and discoveries across numerous areas. Nanotechnology has an impact from medicine and electronics to environmental sustainability and economic prosperity. It is of necessity to recognise the interconnectivity of various domains and the responsibilities that come with managing the dynamic terrain of nanotechnology and its discourse in societal development. This complex interaction necessitates careful examination of ethical, environmental, and social ramifications, emphasising the importance of fostering a harmonious and equitable coexistence of nanotechnology with societal advancement. Nanotechnology has reshaped diagnostic capacities, drug delivery systems, and imaging technologies in the field of healthcare. Importantly, the nanoscale instruments' precision enables targeted medicines and personalised medicine, suggesting significant shifts in the norms of healthcare. Via the advances in treatment and medical procedures, the societal ramifications of improved health outcomes and increased longevity cannot be understated. Similarly, nanotechnology has been critical in tackling environmental issues. The innovative manufacturing contributes to effective water purification, pollution management, and sustainable energy solutions by utilising nanoscale materials as catalysts and sensors. This emphasises its impact on environmental sustainability by portraying the possibility of a positive contribution to the well-being of the earth and its inhabitants. Economically, the incorporation of nanotechnology into many industries pushes growth and progress. Improved consumer goods, more efficient manufacturing processes, job possibilities, and higher productivity all contribute to global economic development. The interdependence of nanotechnology and economic progress highlights the importance of ethical and inclusive methods that ensure the advantages of nanotechnological advancements are spread equitably. A call to action comes as we navigate the interwoven landscapes of nanotechnology and societal development. This is an appeal for collaboration, transparency, and ethical responsibility to encourage the adoption of nanotechnology at all levels towards the contribution to societal growth.

Fostering Harmonious and Equitable Relationship Between Nanotechnology and Societal Progress

The world is currently at a crossroads in human history, where rapid breakthroughs in nanotechnology have the potential to transform every area of our civilization in unexpected ways, bringing up new opportunities while also posing unknown threats if handled carelessly. The path forward depends profoundly upon the responsible ethical minds driving the exciting new frontier. This necessitates the need for scholars, innovators, officials, business executives, civil groups, and every citizen to pay close attention to the ripple effects of emerging nanotechnologies, not just for progress but for the actual improvement of humanity's shared quality of life. The grandest possibilities glimmering on the horizon of nanotechnologies can only be realised responsibly if society undertakes two profoundly interconnected efforts at the same time: enabling nanoscience's progression through continuous resource support and open knowledge exchange while upholding rigorous safety standards, and facilitating thoughtful dialogue, debate, and community participation around implementation to steer innovations towards equitable access and immunisation. Adopting a precautionary approach guarantees that progress is guided by ethical ideals and a commitment to minimising potential harm. Thus, rigorous regulatory frameworks, transparent communication, and a coordinated effort involving manufacturers, policymakers, industry stakeholders, and the public are all required. Additionally, a careful evaluation of the potential impacts on the environment and human health is necessary. Toxicological investigations and risk assessments are critical for understanding the potential risks connected with nanoparticles. It is critical to establish criteria that govern the safe use of nanotechnology and its products while prioritising human and environmental well-being. Nonetheless, efforts should be made to guarantee that the benefits of nanotechnology are available to different communities and do not worsen existing societal imbalances. This necessitates a commitment to resolving concerns of price, distribution, and ethics in the development and deployment of nanotechnological solutions. Without caution, tomorrow's omnipresent nanotechnologies may enrich lives in unimaginable ways, while unforeseen consequences could exacerbate inequality or damage natural systems. Thus, the meeting of science and humanity evolves hand in hand, which can dramatically transform the future for the better.

References

1. Visciano, S.: Chapitre 1: nanotechnologies, bioethics and human dignity. J. Int. Bioéthique = Int. J. Bioethics **22**, 17–36, 206 (2011)
2. Shetty, N.J., Swati, P., David, K.: Nanorobots: future in dentistry. Saudi Dental J. **25**(2), 49–52 (2013)
3. Nistor, M.T., Rusu, A.G.: Chapter 3—nanorobots with applications in medicine. In: Vasile, C. (ed.) Polymeric Nanomaterials in Nanotherapeutics, pp. 123–149. Elsevier (2019)
4. Verma, S.K., Chauhan, R.: Nanorobotics in dentistry—a review. Indian J. Dentist. **5**, 62–70 (2014)

5. Mazumder, S., Biswas, G.R., Majee, S.B.: Applications of nanorobots in medical techniques. IJPSR **11**, 3150 (2020)
6. Nehru, S., Misra, R., Bhaswant, M.: Multifaceted engineered biomimetic nanorobots toward cancer management. ACS Biomater. Sci. Eng. **8**(2), 444–459 (2022)
7. Madkour, L.H.: Introduction to nanotechnology (NT) and nanomaterials (NMs). In: Madkour, L.H. (ed.) Nanoelectronic Materials: Fundamentals and Applications, pp. 1–47. Springer International Publishing, Cham (2019)
8. Khan, A.U., et al.: Selected nanotechnologies and nanostructures for drug delivery, nanomedicine and cure. Bioprocess Biosyst. Eng. **43**(8), 1339–1357 (2020)
9. Pokrajac, L., et al.: Nanotechnology for a sustainable future: addressing global challenges with the international network for sustainable nanotechnology. ACS Nano **15**(12), 18608–18623 (2021)
10. Anirban, A.: 40 years of scanning tunnelling microscopy. Nat. Rev. Phys. **4**(5), 291–291 (2022)
11. Bayda, S., et al.: The history of nanoscience and nanotechnology: from chemical-physical applications to nanomedicine. Molecules **25**(1) (2019)
12. Mani, S., et al.: Nano-robotics: the future of health and dental care. IP Int. J. Periodontol. Implantol. **6**, 6–10 (2021)
13. Gnach, A., et al.: Upconverting nanoparticles: assessing the toxicity. Chem. Soc. Rev. **44**(6), 1561–1584 (2015)
14. A brief history of some landmark papers. Nat. Nanotechnol. **5**(4), 237–237 (2010)
15. Krausz, A.E., et al.: Nanoparticle-encapsulated doxorubicin demonstrates superior tumor cell kill in triple negative breast cancer subtypes intrinsically resistant to doxorubicin. Precis. Nanomed. **1**(3), 173–182 (2018)
16. Freitas, R.A.: What is nanomedicine? Nanomed. Nanotechnol. Biol. Med. **1**(1), 2–9 (2005)
17. Babatunde, D., et al.: Environmental and societal impact of nanotechnology. IEEE Access, pp. 1–1 (2019)
18. Purohit, R., et al.: Social, environmental and ethical impacts of nanotechnology. Mater. Today Proc. **4**(Part D), 5461–5467
19. Raj, S., et al.: Nanotechnology in cosmetics: opportunities and challenges. J. Pharm. Bioallied Sci. **4**(3), 186–193 (2012)
20. Abbasi, B.H., et al.: Chapter 5—nanomaterials for cosmeceuticals: nanomaterials-induced advancement in cosmetics, challenges, and opportunities. In: Nanda, A., et al. (eds.) Nanocosmetics, pp. 79–108. Elsevier (2020)
21. Fytianos, G., Rahdar, A., Kyzas, G.: Nanomaterials in cosmetics: recent updates. Nanomaterials **10**, 979 (2020)
22. Mazayen, Z.M., et al.: Pharmaceutical nanotechnology: from the bench to the market. Futur. J. Pharm. Sci. **8**(1), 12 (2022)
23. Ojha, S., Saikia, D., Bora, U.: Nanopharmaceuticals: synthesis, characterization, and challenges. In: Yata, V.K., et al. (eds.) Nanopharmaceuticals: Principles and Applications, Vol. 3, pp. 81–138. Springer International Publishing, Cham (2020)
24. Malik, S., Muhammad, K., Waheed, Y.: Nanotechnology: a revolution in modern industry. Molecules **28**(2) (2023)
25. Rathore, D.: Chapter 1—nanocomposites: an introduction. In: Song, H., et al. (eds.) Nanotechnology in the automotive industry, pp. 3–14. Elsevier (2022)
26. Shafique, M., Luo, X.: Nanotechnology in transportation vehicles: an overview of its applications, environmental, health and safety concerns. Materials (Basel) **12**(15) (2019)
27. Ćibo, M., et al.: Application and impact of nanotechnology in sport. In: 30th Scientific-Experts Conference of Agriculture and Food Industry. Springer International Publishing, Cham (2020)
28. Tang, M., Yang, L., Zhou, H.: Applications and safety of nanotechnology and nanomaterials in sports. In: Informatics and Management Science IV. Springer, London (2013)
29. Khalil, M., Kadja, G.T.M., Ilmi, M.M.: Advanced nanomaterials for catalysis: current progress in fine chemical synthesis, hydrocarbon processing, and renewable energy. J. Ind. Eng. Chem. **93**, 78–100 (2021)

30. Saleem, H., Zaidi, S.J.: Developments in the application of nanomaterials for water treatment and their impact on the environment. Nanomaterials **10** (2020). https://doi.org/10.3390/nano10091764
31. Nagar, A., Pradeep, T.: Clean water through nanotechnology: needs, gaps, and fulfillment. ACS Nano **14**(6), 6420–6435 (2020)
32. Lah, N.A.C., Zubir, M.N.M., Samykano, M.A.L.: Chapter 20—engineered nanomaterial in electronics and electrical industries. In: Mustansar Hussain, C. (ed.) Handbook of Nanomaterials for Industrial Applications, pp. 324–364. Elsevier (2018)
33. Thangadurai, T.D., et al.: Nanomaterials, properties and applications. In: Thangadurai, T.D., et al. (eds.) Nanostructured Materials, pp. 11–28. Springer International Publishing, Cham (2020)
34. Shafiq, M., et al.: An overview of the applications of nanomaterials and nanodevices in the food industry. Foods **9** (2020). https://doi.org/10.3390/foods9020148
35. Nile, S.H., et al.: Nanotechnologies in food science: applications, recent trends, and future perspectives. Nanomicro Lett. **12**(1), 45 (2020)
36. Khan, I., Saeed, K., Khan, I.: Nanoparticles: properties, applications and toxicities. Arab. J. Chem. **12**(7), 908–931 (2019)
37. Moon, G.D., et al.: Chemical transformations of nanostructured materials. Nano Today **6**(2), 186–203 (2011)
38. Pluchery, O., Remita, H., Schaming, D.: Demonstrative experiments about gold nanoparticles and nanofilms: an introduction to nanoscience. Gold Bull. **46**(4), 319–327 (2013)
39. Roco, M.C.: Science and technology integration for increased human potential and societal outcomes. Ann. N. Y. Acad. Sci. **1013**, 1–16 (2004)
40. Diallo, M.S., Fromer, N.A., Jhon, M.S.: Nanotechnology for sustainable development: retrospective and outlook. J. Nanopart. Res. **15**(11), 2044 (2013)
41. LLP, S.I.: Global nanotechnology market size to exceed USD 74.1 Billion by 2032. 2023, Spherical Insights & Consulting, 210p
42. Raghav, S., Yadav, P.K., Kumar, D.: Chapter 19—nanotechnology for a sustainable future. In: Hussain, C.M. (ed) Handbook of Nanomaterials for Manufacturing Applications, pp. 465–492. Elsevier (2020)
43. Ahmad, A., Qurashi, A., Sheehan, D.: Nano packaging—progress and future perspectives for food safety, and sustainability. Food Packag. Shelf Life **35**, 100997 (2023)
44. Kuswandi, B.: Nanotechnology in Food Packaging, pp. 151–183 (2016)
45. Malik, S., Muhammad, K., Waheed, Y.: Emerging applications of nanotechnology in healthcare and medicine. Molecules **28** (2023). https://doi.org/10.3390/molecules28186624
46. Haleema, et al.: Chapter 10—Recycling of nanomaterials by solvent evaporation and extraction techniques. In: Rai, M., Nguyen, T.A. (eds.) Nanomaterials Recycling, pp. 209–222. Elsevier (2022)
47. Rambaran, T., Schirhagl, R.: Nanotechnology from lab to industry—a look at current trends. Nanoscale Adv. **4**(18), 3664–3675 (2022)
48. Contado, C.: Nanomaterials in consumer products: a challenging analytical problem. Front. Chem. **3** (2015)
49. Łojkowski, W., Fecht, H.J., Swiderska Środa, A.: Quo Vadis nanotechnology? In: The Nano-Micro Interface, pp. 79–94 (2015)
50. Tovar-Lopez, F.J.: Recent progress in micro- and nanotechnology-enabled sensors for biomedical and environmental challenges. Sensors **23** (2023). https://doi.org/10.3390/s23125406
51. Laurent, S., et al.: Magnetic iron oxide nanoparticles: synthesis, stabilization, vectorization, physicochemical characterizations, and biological applications. Chem. Rev. **110**(4), 2574–2574 (2010)
52. Tiwari, J.N., Tiwari, R.N., Kim, K.S.: Zero-dimensional, one-dimensional, two-dimensional and three-dimensional nanostructured materials for advanced electrochemical energy devices. Prog. Mater. Sci. **57**(4), 724–803 (2012)

53. Dreaden, E.C., et al.: The golden age: gold nanoparticles for biomedicine. Chem. Soc. Rev. **41**(7), 2740–2779 (2012)
54. Astefanei, A., Núñez, O., Galceran, M.T.: Characterisation and determination of fullerenes: a critical review. Anal. Chim. Acta **882**, 1–21 (2015)
55. Aqel, A., et al.: Carbon nanotubes, science and technology part (I) structure, synthesis and characterisation. Arab. J. Chem. **5**(1), 1–23 (2012)
56. Holzinger, M., Le Goff, A., Cosnier, S.: Nanomaterials for biosensing applications: a review. Front. Chem. **2** (2014)
57. Shaalan, M., et al.: Recent progress in applications of nanoparticles in fish medicine: A review. Nanomed. Nanotechnol. Biol. Med. **12**(3), 701–710 (2016)
58. Lv, Z., et al.: Semiconductor quantum dots for memories and neuromorphic computing systems. Chem. Rev. **120**(9), 3941–4006 (2020)
59. Sun, Z., et al.: Logic computing with stateful neural networks of resistive switches. Adv. Mater. **30**(38), e1802554 (2018)
60. Xiong, W., et al.: Bilayered oxide-based cognitive memristor with brain-inspired learning activities. Adv. Electr. Mater. **5**(8), 1900439 (2019)
61. Park, Y., Lee, J.-S.: Artificial synapses with short- and long-term memory for spiking neural networks based on renewable materials. ACS Nano **11**(9), 8962–8969 (2017)
62. Yao, P., et al.: Face classification using electronic synapses. Nat. Commun. **8**(1), 15199 (2017)
63. Xiong, J., et al.: Augmented reality and virtual reality displays: emerging technologies and future perspectives. Light Sci. Appl. **10**(1), 216 (2021)
64. Chen, G., et al.: Smart textiles for electricity generation. Chem. Rev. **120**(8), 3668–3720 (2020)
65. Shapira, P., Youtie, J.: The economic contributions of nanotechnology to green and sustainable growth. In: Basiuk, V.A., Basiuk, E.V. (eds.) Green Processes for Nanotechnology: From Inorganic to Bioinspired Nanomaterials, pp. 409–434. Springer International Publishing, Cham (2015)
66. Hasan, M.A.M., et al.: 2D Nanomaterials for effective energy scavenging. Nano-Micro Lett. **13**(1), 82 (2021)
67. Ning, F., et al.: TiO_2/graphene/NiFe-layered double hydroxide nanorod array photoanodes for efficient photoelectrochemical water splitting. Energy Environ. Sci. **9**(8), 2633–2643 (2016)
68. Avasare, V., et al.: Room-temperature synthesis of TiO_2 nanospheres and their solar driven photoelectrochemical hydrogen production. Int. J. Energy Res. **39**(12), 1714–1719 (2015)
69. Wang, Z., et al.: Piezoelectric nanowires in energy harvesting applications. Adv. Mater. Sci. Eng. **2015**, 165631 (2015)
70. Yetisgin, A.A., et al.: Therapeutic nanoparticles and their targeted delivery applications. Molecules **25**(9) (2020)
71. Shah, S.S., et al.: Present status and future prospects of jute in nanotechnology: a review. Chem. Rec. **21**(7), 1631–1665 (2021)
72. Bhatia, S.N., et al.: Cancer nanomedicine. Nat. Rev. Cancer **22**(10), 550–556 (2022)
73. Nikzamir, M., et al.: Applications of dendrimers in nanomedicine and drug delivery: a review. J. Inorg. Organomet. Polym. Mater. **31**(6), 2246–2261 (2021)
74. Estelrich, J., Sánchez-Martín, M.J., Busquets, M.A.: Nanoparticles in magnetic resonance imaging: from simple to dual contrast agents. Int. J. Nanomed. **10**, 1727–1741 (2015)
75. Avasthi, A., et al.: Magnetic nanoparticles as MRI contrast agents. Top. Curr. Chem. **378**(3), 40 (2020)
76. Chavan, N., et al.: Magnetic nanoparticles—a new era in nanotechnology. J. Drug Deliv. Sci. Technol. **77**, 103899 (2022)
77. Ramburrun, P., Khan, R.A., Choonara, Y.E.: Design, preparation, and functionalization of nanobiomaterials for enhanced efficacy in current and future biomedical applications **11**(1), 1802–1826 (2022)
78. Blaker, J., Lee, K.-Y., Bismarck, A.: Hierarchical composites made entirely from renewable resources. J. Biobased Mater. Bioenergy **5**, 1–16 (2011)

79. Ballottin, D., et al.: Antimicrobial textiles: biogenic silver nanoparticles against Candida and Xanthomonas. Mater. Sci. Eng. C **75**, 582–589 (2017)
80. Kulkarni, D., et al.: Chapter 14—nanomaterials in cosmetics and dermatology. In: Husen, A., Siddiqi, K.S. (eds.) Advances in Smart Nanomaterials and their Applications, pp. 297–323. Elsevier (2023)
81. Das, S., Das, S., Das, M.K.: Chapter 2—nanocosmeceuticals: concept, opportunities, and challenges. In: Das, M.K. (ed.) Nanocosmeceuticals, pp. 31–69. Academic Press (2022)
82. Dubey, S.K., et al.: Emerging trends of nanotechnology in advanced cosmetics. Colloids Surf. B Biointerfaces **214**, 112440 (2022)
83. Raj, S., et al.: Nanotechnology in cosmetics: opportunities and challenges. J. Pharm. Bioallied Sci. **4**(3) (2012)
84. Singh, S., Pandey, S.K., Vishwakarma, N.: Chapter 22—functional nanomaterials for the cosmetics industry. In: C. Mustansar Hussain (ed.) Handbook of Functionalized Nanomaterials for Industrial Applications, pp. 717–730. Elsevier (2020)
85. Katz, L.M., Dewan, K., Bronaugh, R.L.: Nanotechnology in cosmetics. Food Chem. Toxicol. **85**, 127–137 (2015)
86. Wang, D., et al.: Next-generation multifunctional carbon-metal nanohybrids for energy and environmental applications. Environ. Sci. Technol. **53**(13), 7265–7287 (2019)
87. Khan, M., et al.: Graphene based metal and metal oxide nanocomposites: synthesis, properties and their applications. J. Mater. Chem. A **3**(37), 18753–18808 (2015)
88. Anwer, H., Park, J.-W.: Synthesis and characterization of a heterojunction $rGO/ZrO_2/Ag_3PO_4$ nanocomposite for degradation of organic contaminants. J. Hazard. Mater. **358**, 416–426 (2018)
89. Hua, S., et al.: Highly efficient p-type Cu_3P/n-type g-C_3N_4 photocatalyst through Z-scheme charge transfer route. Appl. Catal. B **240**, 253–261 (2019)
90. Yang, C., et al.: Recent trends in carbon nanomaterial-based electrochemical sensors for biomolecules: a review. Anal. Chim. Acta **887**, 17–37 (2015)
91. Shi, L., et al.: Facile fabrication of a novel 3D graphene framework/Bi nanoparticle film for ultrasensitive electrochemical assays of heavy metal ions. Anal. Chim. Acta **968**, 21–29 (2017)
92. Li, Q., et al.: Antimicrobial nanomaterials for water disinfection and microbial control: potential applications and implications. Water Res. **42**(18), 4591–4602 (2008)
93. Lee, J., et al.: Photosensitized oxidation of emerging organic pollutants by tetrakis C60 aminofullerene-derivatized silica under visible light irradiation. Environ. Sci. Technol. **45**(24), 10598–10604 (2011)
94. Huang, Y., Keller, A.A.: Magnetic nanoparticle adsorbents for emerging organic contaminants. ACS Sustain. Chem. Eng. **1**(7), 731–736 (2013)
95. Tsang, M.P., et al.: Evaluating nanotechnology opportunities and risks through integration of life-cycle and risk assessment. Nat. Nanotechnol. **12**(8), 734–739 (2017)
96. Invernizzi, N.: Nanotechnology between the lab and the shop floor: what are the effects on labor? J. Nanopart. Res. **13**(6), 2249–2268 (2011)
97. Chapple, K.: Defining the green economy: a primer on green economic development author cover photo clockwise from left: the author is grateful to (2008)
98. Shapira, P., Wang, J.: Follow the money. Nature **468**(7324), 627–628 (2010)
99. Tassey, G.: Methods for assessing the economic impacts of government R&D (2003)
100. Berger, M.: Nanotechnology education around the world. In: Nanoetchnology (2021)
101. Soltani, M., et al.: Enhancing clinical translation of cancer using nanoinformatics. Cancers (Basel) **13**(10) (2021)
102. Liu, P.P.: Nanotechnology (2023)
103. Roco, M.C.: Overview: affirmation of nanotechnology between 2000 and 2030 (2017)
104. Clunan, A.L., Hsueh, R., McManus, I.: Nanotechnology in a Globalized World. Strategic Assessments of an Emerging Technology. Northeastern University, the Naval Postgraduate, School Center on Contemporary Conflict, and the Defense Threat Reduction Agency (2014)

105. Qiu, J.: Nanotechnology development in China: challenges and opportunities. Natl. Sci. Rev. **3**, 148–152 (2016)
106. Foster, B.: 1.11 The future. In: Nano-Safety: What We Need to Know to Protect Workers, p. 38 (2023)
107. Jia, L.: Global governmental investment in nanotechnologies. Curr. Nanosci. **1**(3), 263–266 (2005)
108. Saleh, H.M., Hassan, A.I.: Synthesis and characterization of nanomaterials for application in cost-effective electrochemical devices. Sustainability **15** (2023). https://doi.org/10.3390/su151410891
109. Jalali Mosallam, S., et al.: The effect of carbon nanotubes on mechanical properties of structural lightweight concrete using LECA aggregates. Structures **35**, 1204–1218 (2022)
110. Mohajerani, A., et al.: Nanoparticles in construction materials and other applications, and implications of nanoparticle use. Materials (Basel) **12**(19) (2020)
111. Gu, Y., et al.: Technical characteristics and wear-resistant mechanism of nano coatings: a review. Coatings **10** (2020). https://doi.org/10.3390/coatings10030233
112. Rambaran, T., Schirhagl, R.: Nanotechnology from lab to industry—a look at current trends. Nanoscale Adv **4**(18), 3664–3675 (2022)
113. Gao, Y., et al.: China and the United States-global partners, competitors and collaborators in nanotechnology development. Nanomedicine **12**(1), 13–19 (2016)
114. Munawar, A., et al.: Nanosensors for diagnosis with optical, electric and mechanical transducers. RSC Adv. **9**(12), 6793–6803 (2019)
115. Sahu, M.K., Yadav, R., Tiwari, S.P.: Recent advances in nanotechnology. Int. J. Nanomater. Nanotechnol. Nanomed. **9**(1), 015–023 (2023)
116. Malik, S., Muhammad, K., Waheed, Y.: Nanotechnology: a revolution in modern industry. Molecules **28** (2023). https://doi.org/10.3390/molecules28020661
117. Mohamed, K.: 2.16 Nanoimprint lithography for nanomanufacturing. In: Andrews, D.L., Lipson, R.H., Nann, T. (eds.) Comprehensive Nanoscience and Nanotechnology, 2nd edn, pp. 357–386. Academic Press, Oxford (2019)
118. Morse, J.: Assessing the economic impact of nanotechnology: the role of nanomanufacturing (2012)
119. Stirling, A.: Keep it complex. Nature **468**(7327), 1029–1031 (2010)
120. Kagan, C., et al.: Nano Day: celebrating the next decade of nanoscience and nanotechnology. ACS Nano **10** (2016)
121. Ganascia, J.-G.: The new ethical trilemma: security, privacy and transparency. C. R. Phys. **12**(7), 684–692 (2011)
122. Baran, A.: Nanotechnology: legal and ethical issues. Ekonomia i Zarzadzanie **8** (2016)
123. Schummer, J.: Identifying ethical issues of nanotechnologies. Philippiniana Sacra **57** (2012)
124. Hoven, J.V., Vermaas, P.E.: Nano-technology and privacy: on continuous surveillance outside the panopticon. J. Med. Philos. **32**(3), 283–297 (2007)
125. Singh, K.K.: 7—legal issues in nanotechnology. In: Kumar, A., Singh, A.K., Choudhary, K.K. (eds.) Role of Plant Growth Promoting Microorganisms in Sustainable Agriculture and Nanotechnology, pp. 107–120. Woodhead Publishing (2019)
126. Ahire, S.A., et al.: The augmentation of nanotechnology era: a concise review on fundamental concepts of nanotechnology and applications in material science and technology. Results Chem. **4**, 100633 (2022)
127. Martínez, G., et al: Environmental impact of nanoparticles' application as an emerging technology: a review. Materials **14** (2021). https://doi.org/10.3390/ma14010166
128. McGinn, R.E.: What's different, ethically, about nanotechnology? Foundational questions and answers. NanoEthics **4**(2), 115–128 (2010)
129. Meyer, D., Upadhyayula, V.K.: The use of life cycle tools to support decision making for sustainable nanotechnologies. Clean Technol. Environ. Policy **16**, 757–772 (2013)
130. Shatkin, J.A.: Nanotechnology: Health and Environmental Risks. CRC Press (2017)
131. Rai, S., Rai, A.: Nanotechnology-the secret of fifth industrial revolution and the future of next generation. J. Nasional **7**(2), 61–66 (2017)

132. Som, C., et al.: The importance of life cycle concepts for the development of safe nanoproducts. Toxicology **269**(2), 160–169 (2010)
133. Berger, M.: Nanotechnology—not that green? (2008)
134. Gökçay, B., Arda, B.: Nanotechnology, nanomedicine; ethical aspects. Rev. Rom. Bioet. **13**(3) (2015)
135. Jia, H.: Research ethics: a safeguard for advanced technologies. Natl. Sci. Rev. **7**(11), 1787–1792 (2020)
136. Gerken, L.R.H., et al.: Catalytic activity imperative for nanoparticle dose enhancement in photon and proton therapy. Nat. Commun. **13**(1), 3248 (2022)
137. Veselov, V.V., et al.: Targeted delivery methods for anticancer drugs. Cancers (Basel) **14**(3) (2022)
138. Yu, H., et al.: Sequential drug delivery in targeted cancer therapy. Pharmaceutics **14** (2022). https://doi.org/10.3390/pharmaceutics14030573
139. Almeida, M., Ranisch, R.: Beyond safety: mapping the ethical debate on heritable genome editing interventions. Humanit. Soc. Sci. Commun. **9**(1), 139 (2022)
140. Acarón Ledesma, H., et al.: An atlas of nano-enabled neural interfaces. Nat. Nanotechnol. **14**(7), 645–657 (2019)
141. Bonizzato, M.: Neuroprosthetics: an outlook on active challenges toward clinical adoption. J. Neurophysiol. **125**(1), 105–109 (2020)
142. Alabdulatif, A., et al.: Internet of Nano-Things (IoNT): a comprehensive review from architecture to security and privacy challenges. Sensors **23** (2023). https://doi.org/10.3390/s23052807
143. Ali, N.A., Aleyadeh, W., AbuElkhair, M.: Internet of nano-things network models and medical applications. In: 2016 International Wireless Communications and Mobile Computing Conference (IWCMC) (2016)
144. Pramanik, P.K.D., et al.: Advancing modern healthcare with nanotechnology, nanobiosensors, and internet of nano things: taxonomies, applications, architecture, and challenges. IEEE Access **8**, 65230–65266 (2020)
145. Dong, B., et al.: Recent progress in nanoplasmonics-based integrated optical micro/nano-systems. J. Phys. D Appl. Phys. **53**(21), 213001 (2020)
146. Kulakowski, P., Solarczyk, K., Wojcik, K.: Routing in FRET-based nanonetworks. IEEE Commun. Mag. **55**(9), 218–224 (2017)
147. Malhotra, V.: Biomedical nanotechnology: current status and future prospects in disease diagnosis and therapy. Biomed. Syst. Emerg. Technol. **10**(3), 168 (2023)
148. Anjum, S., et al.: Emerging applications of nanotechnology in healthcare systems: grand challenges and perspectives. Pharmaceuticals (Basel) **14**(8) (2021)
149. Sadhu, P.K., Yanambaka, V.P., Abdelgawad, A.: Internet of things: security and solutions survey. Sensors **22** (2022). https://doi.org/10.3390/s22197433
150. Schulte, P.A., Salamanca-Buentello, F.: Ethical and scientific issues of nanotechnology in the workplace. Environ. Health Perspect. **115**(1), 5–12 (2007)
151. Mehta, M., On nano-panopticism: a sociological perspective. Interfaces Can. Chem. News (2002)
152. Aubert, H.: RFID technology for human implant devices. C. R. Phys. **12**(7), 675–683 (2011)
153. Roco, M.C., Bainbridge, W.S.: Overview converging technologies for improving human performance, in converging technologies for improving human performance. In: Roco, M.C., Bainbridge, W.S. (eds.) Nanotechnology, Biotechnology, Information Technology and Cognitive Science, pp. 1–27. Springer Netherlands, Dordrecht (2003)
154. Abdellatif, A.A.H., Tawfeek, H.M.: Development and evaluation of fluorescent gold nanoparticles. Drug Dev. Ind. Pharm. **44**(10), 1679–1684 (2018)
155. Singh, P., et al.: Gold nanoparticles in diagnostics and therapeutics for human cancer. Int. J. Mol. Sci. **19**(7) (2018)
156. Yezdani, U., et al.: Nanotechnology in diagnosis and treatment of various diseases and its future advances in medicine. World J. Pharm. Pharm. Sci. **7**, 1612–1633 (2018)

157. Vargolici, S.: The Influence of Popular Culture and Science Fiction on the Public Perception of Emerging Technologies—Focus: Nanotechnology. University of Maastricht, Netherlands (2011)
158. Akyildiz, I.F., Jornet, J.M.: The Internet of nano-things. IEEE Wirel. Commun. **17**(6), 58–63 (2010)
159. Sahoo, S.: Socio-ethical issues and nanotechnology development: perspectives from India. In: 2010 10th IEEE Conference on Nanotechnology, NANO 2010, pp. 1205–1210 (2010)
160. Maclurcan, D.: Nanotechnology and the Hope for a More Equitable World: A Mixed Methods Study (2010)
161. Wasti, S., et al.: Ethical and legal challenges in nanomedical innovations: a scoping review. Front. Genet. **14** (2023)
162. Cheong, J., et al.: Fast detection of SARS-CoV-2 RNA via the integration of plasmonic thermocycling and fluorescence detection in a portable device. Nat. Biomed. Eng. **4**(12), 1159–1167 (2020)
163. Horton, M., Khan, A., Maddison, S.: Delivering nanotechnology to the healthcare, IT and environmental sectors—a perspective from the 'London Centre for Nanotechnology.' BT Technol. J. **24**, 175–183 (2006)
164. Sharma, D., Patel, P., Shah, M.: A comprehensive study on Industry 4.0 in the pharmaceutical industry for sustainable development. Environ. Sci. Pollut. Res. **30**(39), 90088–90098 (2023)
165. Evans, D.: Ethics, nanotechnology and Health (2007)
166. Schroeder, D., et al.: Responsible inclusive innovation and the nano-divide. NanoEthics **10**, 177–188 (2016)
167. Salamanca-Buentello, F., Daar, A.S.: Nanotechnology, equity and global health. Nat. Nanotechnol. **16**(4), 358–361 (2021)
168. Eyawo, O., Viens, A.M.: Rethinking the central role of equity in the global governance of pandemic response. J. Bioethical Inquiry **17**(4), 549–553 (2020)
169. Maclurcan, D.: Nanotechnology and Global Equality. CRC Press (2012)
170. Cozzens, S., Wetmore, J.: Nanotechnology and the Challenges of Equity, Equality and Development (2011)
171. Tonn, B.E.: Futures sustainability. Futures **39**(9), 1097–1116 (2007)
172. Mulder, H.: Sustainable development and climate change: a business perspective. United Nations (2007)
173. Gautam, P., Kumar, S., Lokhandwala, S.: Chapter 11—energy-aware intelligence in megacities. In: Kumar, S., Kumar, R., Pandey, A. (eds.) Current Developments in Biotechnology and Bioengineering, pp. 211–238. Elsevier (2019)
174. Ray, P.C., Yu, H., Fu, P.P.: Toxicity and environmental risks of nanomaterials: challenges and future needs. J. Environ. Sci. Health C Environ. Carcinog. Ecotoxicol. Rev. **27**(1), 1–35 (2009)
175. Patra, H.K., et al.: Cell selective response to gold nanoparticles. Nanomedicine **3**(2), 111–119 (2007)
176. Maharramov, A.M., et al.: The engineered nanoparticles in food chain: potential toxicity and effects. SN Appl. Sci. **1**(11), 1362 (2019)
177. Jakubiak, M., et al.: Bioaccumulation and biosorption of inorganic nanoparticles: factors affecting the efficiency of nanoparticle mycoextraction by liquid-grown mycelia of pleurotus eryngii and trametes versicolor. Mycol. Prog. **13**(3), 525–532 (2014)
178. Møller, P., et al.: Role of oxidative damage in toxicity of particulates. Free Radic. Res. **44**(1), 1–46 (2010)
179. Safari, J., Zarnegar, Z.: Advanced drug delivery systems: nanotechnology of health design a review. J. Saudi Chem. Soc. **18**(2), 85–99 (2014)
180. Bozich, J.S., et al.: Surface chemistry, charge and ligand type impact the toxicity of gold nanoparticles to Daphnia magna. Environ. Sci. Nano **1**(3), 260–270 (2014)
181. Roy, A., et al.: Nanomaterials for remediation of environmental pollutants. Bioinorg. Chem. Appl. **2021**, 1764647 (2021)

182. Fröhlich, E.: The role of surface charge in cellular uptake and cytotoxicity of medical nanoparticles. Int. J. Nanomed. **7**, 5577–5591 (2012)
183. Schaeublin, N.M., et al.: Surface charge of gold nanoparticles mediates mechanism of toxicity. Nanoscale **3**(2), 410–420 (2011)
184. El-Kalliny, A.S., et al.: Nanomaterials: a review of emerging contaminants with potential health or environmental impact. Discov. Nano **18**(1), 68 (2023)
185. Manisalidis, I., et al.: Environmental and health impacts of air pollution: a review. Front. Public Health **8** (2020)
186. Jeevanandam, J., et al.: Review on nanoparticles and nanostructured materials: history, sources, toxicity and regulations. Beilstein J. Nanotechnol. **9**, 1050–1074 (2018)
187. Abdelbasir, S.M., et al.: Waste-derived nanoparticles: synthesis approaches, environmental applications, and sustainability considerations. Front. Chem. **8** (2020)
188. Virji, M., Stefaniak, A.: A Review of Engineered Nanomaterial Manufacturing Processes and Associated Exposure, pp. 1–3=125 (2014)
189. Dong, H., et al.: Aging of zero-valent iron-based nanoparticles in aqueous environment and the consequent effects on their reactivity and toxicity. Water Environ. Res. **92**(5), 646–661 (2020)
190. Nowack, B.: Evaluation of environmental exposure models for engineered nanomaterials in a regulatory context. NanoImpact **8**, 38–47 (2017)
191. Sengul, A.B., Asmatulu, E.: Toxicity of metal and metal oxide nanoparticles: a review. Environ. Chem. Lett. **18**(5), 1659–1683 (2020)
192. Ahirwar, H., et al.: Materials for orthopedic bioimplants: modulating degradation and surface modification using integrated nanomaterials. Coatings **10** (2020). https://doi.org/10.3390/coatings10030264
193. Mahmoud, A.E.D.: Nanomaterials: green synthesis for water applications. In: Kharissova, O.V., Martínez, L.M.T., Kharisov, B.I. (eds.) Handbook of Nanomaterials and Nanocomposites for Energy and Environmental Applications, pp. 1–21. Springer International Publishing, Cham (2020)
194. Aarthye, P., Sureshkumar, M.: Green synthesis of nanomaterials: an overview. Mater. Today Proc. **47**, 907–913 (2021)
195. Abd-Elsalam, K.A., Rajeshkumar, S., Periakaruppan, R.: Chapter 1—sustainable strategies for producing large-scale nanomaterials: a note from the editors. In: Abd-Elsalam, K.A., Periakaruppan, R., Rajeshkumar, S. (eds.) Agri-Waste and Microbes for Production of Sustainable Nanomaterials, pp. 1–13. Elsevier (2022)
196. Ibrahim, H.M.M.: Green synthesis and characterization of silver nanoparticles using banana peel extract and their antimicrobial activity against representative microorganisms. J. Radiat. Res. Appl. Sci. **8**(3), 265–275 (2015)
197. Vaibhav, V., Vijayalakshmi, U., Roopan, S.M.: Agricultural waste as a source for the production of silica nanoparticles. Spectrochim. Acta Part A Mol. Biomol. Spectrosc. **139**, 515–520 (2015)
198. Devatha, C.P., Thalla, A.K.: Chapter 7—green synthesis of nanomaterials. In: Mohan Bhagyaraj, S., et al. (eds.) Synthesis of Inorganic Nanomaterials, pp. 169–184. Woodhead Publishing (2018)
199. Bakshi, M., Mahanty, S., Chaudhuri, P.: Fungi mediated biosynthesis of nanoparticles and application in metal sequestration, pp. 423–434 (2017)
200. Reijnders, L.: Chapter 10—safer-by-design for nanomaterials. In: Rajendran, S., et al. (eds.) Nanotoxicity, pp. 215–237. Elsevier (2020)
201. Chen, T., Yin, S., Wu, J.: Nanomaterials meet microfluidics: improved analytical methods and high-throughput synthetic approaches. TrAC Trends Anal. Chem. **142**, 116309 (2021)
202. Gilbertson, L.M., et al.: Designing nanomaterials to maximize performance and minimize undesirable implications guided by the principles of green chemistry. Chem. Soc. Rev. **44**(16), 5758–5777 (2015)
203. Benelli, G.: Green synthesis of nanomaterials. Nanomaterials (Basel) **9**(9) (2019)

204. Chausali, N., Saxena, J., Prasad, R.: Nanotechnology as a sustainable approach for combating the environmental effects of climate change. J. Agric. Food Res. **12**, 100541 (2023)
205. Ray, U.: The environmental impact of nanotechnology. In: Moore, S. (ed.) AZoNano.com (2019)
206. Chausali, N., Saxena, J., Prasad, R.: Nanobiochar and biochar based nanocomposites: advances and applications. J. Agric. Food Res. **5**, 100191 (2021)
207. Gubala, V., et al.: Engineered nanomaterials and human health: part 1 preparation, functionalization and characterization (IUPAC Technical Report) **90**(8), 1283–1324 (2018)
208. Kumar, S.: Smart and innovative nanotechnology applications for water purification. Hybrid Adv. **3**, 100044 (2023)
209. Taylor-Smith, K.: How could nanotechnology address problems in the environment? In: Craig, M. (ed.) AZoNano.com (2020)
210. Tripathy, D.B., Gupta, A.: Nanomembranes-affiliated water remediation: chronology, properties, classification, challenges and future prospects. Membranes **13** (2023). https://doi.org/10.3390/membranes13080713
211. Subramanian, K.S., et al.: Nanotechnology for mitigation of global warming impacts. In: Venkatramanan, V., Shah, S., Prasad, R. (eds.) Global Climate Change: Resilient and Smart Agriculture, pp. 315–336. Springer Singapore, Singapore (2020)
212. Prasad, R., Bhattacharyya, A., Nguyen, Q.D.: Nanotechnology in sustainable agriculture: recent developments, challenges, and perspectives. Front. Microbiol. **8**, 1014 (2017)
213. Tripathi, M., et al.: Agro-nanotechnology: a future technology for sustainable agriculture. Int. J. Curr. Microbiol. Appl. Sci. **7**, 196–200 (2018)
214. Chausali, N., Saxena, J., Prasad, R.: Recent trends in nanotechnology applications of bio-based packaging. J. Agric. Food Res. **7**, 100257 (2022)
215. Rezaei, D.: Nanotechnology, a step toward sustainable architecture. J. Eng. Appl. Sci. **13**(16), 6692–6699 (2018)
216. Yousefi, R.: The environmental crisis and nanotechnology. Micro Nanosyst. **14**(3), 188–190 (2022)
217. Ali, Q., et al.: Nanoremediation for heavy metal contamination: a review. Hybrid Adv. **4**, 100091 (2023)
218. Moore, K., Wei, W.: Applications of carbon nanomaterials in perovskite solar cells for solar energy conversion. Nano Mater. Sci. **3**(3), 276–290 (2021)
219. Poddar, A.: Nanotechnology and biodiversity: a case study on nano-sensors for wildlife monitoring, vol. 12, pp. 209–218 (2023)
220. Gutierrez, E.: Privacy implications of nanotechnology. In: Electronic Privacy Information Center (2023)
221. Eduljee, G.H.: Trends in risk assessment and risk management. Sci. Total. Environ. **249**(1), 13–23 (2000)
222. Fadel, T.R., et al.: The challenges of nanotechnology risk management. Nano Today **10**(1), 6–10 (2015)
223. Upreti, G., et al.: Life cycle assessment of nanomaterials. In: Basiuk, V.A., Basiuk, E.V. (eds.) Green Processes for Nanotechnology: From Inorganic to Bioinspired Nanomaterials, pp. 393–408. Springer International Publishing, Cham (2015)
224. Jørgensen, M.S., Jørgensen, U., Clausen, C.: The social shaping approach to technology foresight. Futures **41**(2), 80–86 (2009)
225. Wilson, N.: Nanoparticles: environmental problems or problem solvers? Bioscience **68**(4), 241–246 (2018)
226. Bergeson, L.L.: Nanotechnology now—NNI releases reports from EHS workshops (2011)
227. Nasrollahzadeh, M., Sajadi, S.M.: Chapter 7—risks of nanotechnology to human life. In: Nasrollahzadeh, M., et al. (eds.) Interface Science and Technology, pp. 323–336. Elsevier (2019)
228. Johnston, H., et al.: Engineered nanomaterial risk: lessons learnt from completed nanotoxicology studies: potential solutions to current and future challenges. Crit. Rev. Toxicol. **43**(1), 1–20 (2013)

229. Hristozov, D.R., et al.: Risk assessment of engineered nanomaterials: a review of available data and approaches from a regulatory perspective. Nanotoxicology **6**(8), 880–898 (2012)
230. Hartmann, N., et al.: Environmental fate and behaviour of nanomaterials: new knowledge on important transformation processes (2014)
231. Hodge, G., Bowman, D.M., Maynard, A.: International Handbook on Regulating Nanotechnologies (2010)
232. Eisenberger, I., et al.: Nano regulation in the European Union (NanoTrust Dossier No. 017en—November 2010) (2010)
233. Borrás, S., Edquist, C.: The choice of innovation policy instruments. Technol. Forecast. Soc. Chang. **80**(8), 1513–1522 (2013)
234. Soltani, A.M., Pouypouy, H.: Chapter 19—standardization and regulations of nanotechnology and recent government policies across the world on nanomaterials. In: Ghorbanpour, M., Wani, S.H. (eds.) Advances in Phytonanotechnology, pp. 419–446. Academic Press (2019)
235. Benko, H.: ISO technical committee 229 nanotechnologies. In: Metrology and Standardization of Nanotechnology, pp. 259–268 (2017)
236. Rasmussen, K., et al.: Review of achievements of the OECD working party on manufactured nanomaterials' testing and assessment programme: from exploratory testing to test guidelines. Regul. Toxicol. Pharmacol. **74**, 147–160 (2016)
237. Kühnel, D., Nickel, C.: The OECD expert meeting on ecotoxicology and environmental fate—towards the development of improved OECD guidelines for the testing of nanomaterials. Sci. Total. Environ. **472**, 347–353 (2014)
238. UNESCO: World commission on the ethics of scientific knowledge and technology (COMEST). In UNESCO.org. (2023)
239. Dong, H., et al.: The nanotechnology race between China and the United States. Nano Today **11**(1), 7–12 (2016)
240. Oluwoye, J.O.: Age as a predictor of social acceptance of nanotechnology and nano-based food: a conceptual framework (2015)
241. Valverde, J.L.J.: Nanotechnology: risk assessment and risk management perspective **8**, 25–47 (2011)
242. Ganesh Pillai, R., Bezbaruah, A.N.: Perceptions and attitude effects on nanotechnology acceptance: an exploratory framework. J. Nanopart. Res. **19**(2), 41
243. Gehrke, P.J.: Public understanding of nanotechnology: how publics know. In: Gehrke, P.J. (ed.) Nano-Publics : Communicating Nanotechnology Applications, Risks, and Regulations, pp. 21–37. Springer International Publishing, Cham (2018)
244. Scheufele, D.A., Lewenstein, B.V.: The public and nanotechnology: how citizens make sense of emerging technologies. J. Nanopart. Res. **7**(6), 659–667 (2005)
245. Cobb, M.D., Macoubrie, J.: Public perceptions about nanotechnology: risks, benefits and trust. J. Nanopart. Res. **6**(4), 395–405 (2004)
246. Macoubrie, J.: Nanotechnology: public concerns, reasoning and trust in government. Public Underst. Sci. **15**, 221–241 (2006)
247. Miller, G.: Nanotechnology and the public interest: repeating the mistakes of GM foods? Int. J. Technol. Transfer Commer. **7** (2008)
248. Roco, M.C.: National Nanotechnology Investment in the FY 2011 Budget. American Society of Mechanical Engineers (2009)
249. Pavlicek, A., et al.: A European nano-registry as a reliable database for quantitative risk assessment of nanomaterials? A comparison of national approaches. NanoImpact **21**, 100276 (2021)
250. EPA: Research on evaluating risk across the life cycle of manufactured chemicals, materials, & products (2023)
251. Qiu, J.: Nanotechnology development in China: challenges and opportunities. Natl. Sci. Rev. **3**(1), 148–152 (2016)
252. Roco, M.C.: The long view of nanotechnology development: the national nanotechnology initiative at 10 years. J. Nanopart. Res. **13**(2), 427–445 (2011)

253. Renn, O., Roco, M.: Nanotechnology and the need for risk governance. J. Nanopart. Res. **8**, 153–191 (2006)
254. Bernstein, M., et al.: The European commission's green deal is an opportunity to rethink harmful practices of research and innovation policy. Ambio **52**, 1–10 (2022)
255. Marshall, H., Wilkins, K., Bennett, L.: Story thinking for technology foresight. Futures **146**, 103098 (2023)
256. Gaskell, G., et al.: Imagining nanotechnology: cultural support for technological innovation in Europe and the United States. Public Underst. Sci. **14**(1), 81–90 (2005)
257. Güzel, F.: Science fiction literature as thought experiment: an ethical analysis of michael crichton's prey. J. Am. Stud. Turkey, 27–51 (2022)
258. Koepsell, D.: Innovation and Nanotechnology: Converging Technologies and the End of Intellectual Property. Bloomsbury Academic (2011)
259. Vega, M.F.R.: Understanding society: metaphors in infographics, a tool to catch the eye?. In: Humanities. Tilburg University (2018)
260. Boholm, Å.: Nanotechnology, anthropomorphic matter and human machinery: an exploration of the longue durée of technological vision. J. Assoc. Soc. Anthropol. (2014)
261. Hosseini, H.: Popular culture in Asia: globalization, regionalization, and localization. Procedia Soc. Behav. Sci. **2**(5), 7356–7358 (2010)
262. Baglioni, P., Carretti, E., Chelazzi, D.: Nanomaterials in art conservation. Nat. Nanotechnol. **10**(4), 287–290 (2015)
263. Baglioni, P., Chelazzi, D., Giorgi, R.: Consolidation of wall paintings and stone. In: Baglioni, P., Chelazzi, D., Giorgi, R. (eds.) Nanotechnologies in the Conservation of Cultural Heritage: A Compendium of Materials and Techniques, pp. 15–59. Springer Netherlands: Dordrecht (2015)
264. Ambrosi, M., et al.: Stable dispersions of Ca(0H)2 in aliphatic alcohols: properties and application in cultural heritage conservation. Prog. Colloid Polym. Sci. **118**, 68–72 (2001)
265. Giorgi, R., et al.: Nanoparticles of calcium hydroxide for wood deacidification: decreasing the emissions of organic acid vapors in church organ environments. J. Cult. Herit. **10**, 206–213 (2009)
266. Giorgi, R., et al.: Nanotechnologies for conservation of cultural heritage: paper and canvas deacidification. Langmuir **18**, 8198–8203 (2002)
267. Royal Society of C. and C. Royal Society of: Nanoscience for the Conservation of Works of Art. The Royal Society of Chemistry (2013)
268. Roy, H., et al.: Nanotechnology in pop culture (2021)
269. Sofia, Monument to the Soviet army. In: Atlas Obscura (2014)
270. Magaudda, P.: Nanotechnologies and emerging cultural spaces for the public communication of science and technologies: an introduction. J. Sci. Commun. **11** (2012)
271. Azimi, M.: Cultural diversity: an overview. Int. J. Asian Soc. Sci. **2013**, 2147–2151 (2013)
272. Siegrist, M., Hartmann, C.: Consumer acceptance of novel food technologies. Nat. Food **1**(6), 343–350 (2020)
273. Touseef, M., et al.: The influence of cultural values and norms on political attitudes and behavior: a sociological study. Adv. Qual. Res. **1**, 40–50 (2023)
274. Macnaghten, P., Davies, S.R., Kearnes, M.: Understanding public responses to emerging technologies: a narrative approach. J. Environ. Plann. Policy Manage. **21**(5), 504–518 (2019)
275. Felt, U., Schumann, S., Schwarz, C.G.: (Re)assembling natures, cultures, and (nano)technologies in public engagement. Sci. Culture **24**(4), 458–483 (2015)
276. Macnaghten, P.: Researching technoscientific concerns in the making: narrative structures, public responses, and emerging nanotechnologies. Environ. Plann. A Econ. Space **42**(1), 23–37 (2010)
277. Rathore, A., Mahesh, G.: Public perception of nanotechnology: a contrast between developed and developing countries. Technol. Soc. **67**, 101751 (2021)
278. Ferrara, E.: Aspirations, social norms, and development. J. Eur. Econ. Assoc. **17** (2019)
279. De Witt, A.: Worldviews and their significance for the global sustainable development debate. Environ. Ethics **35**, 133–162 (2013)

280. McGrail, S.D.: Promising, Contesting and Abandoning Nanotechnology. The University of Melbourne (2011)
281. Roure, F.: Societal approach to nanoscience and nanotechnology: when technology reflects and shapes society. In: Nanosciences and Nanotechnology: Evolution or Revolution?, pp. 357–404. Springer (2015)
282. Feero, W.G., Guttmacher, A.E.: Genomics, personalized medicine, and pediatrics. Acad. Pediatr. **14**(1), 14–22 (2014)
283. Viseu, A.: Caring for nanotechnology? Being an integrated social scientist. Soc. Stud. Sci. **45**(5), 642–664 (2015)
284. Chehelgerdi, M., et al.: Progressing nanotechnology to improve targeted cancer treatment: overcoming hurdles in its clinical implementation. Mol. Cancer **22**(1), 169 (2023)
285. Saha, M.: Nanomedicine: promising tiny machine for the healthcare in future-a review. Oman Med. J. **24**(4), 242–247 (2009)
286. Taylor, N.B.: Emerging technologies, risk, peace, and conflict. In: Taylor, N.B. (ed.) Existential Risks in Peace and Conflict Studies, pp. 139–190. Springer International Publishing, Cham (2023)
287. Zaidel, D.W.: Chapter 2—Culture and art: Importance of art practice, not aesthetics, to early human culture. In: Christensen, J.F., Gomila, A. (eds.) Progress in Brain Research, pp. 25–40. Elsevier (2018)
288. Mohamed, H.B., Al-Suwaidi, K.: Cultural influence on art: understanding the connection (2023)
289. Reina-Rozo, J.D.: Science fiction and engineering: between dystopias, (e)utopias, and uchronias. In: Fritzsche, A., Santa-María, A. (eds.) Rethinking Technology and Engineering: Dialogues Across Disciplines and Geographies, pp. 225–238. Springer International Publishing, Cham (2023)
290. Keller, J., Wen Chen, E., Leung, A.K.Y.: How national culture influences individuals' subjective experience with paradoxical tensions. Cross Cult. Strateg. Manag. **25**(3), 443–467 (2018)
291. Trump, B.D.: Synthetic biology, GMO, and risk: what is new, and what is different? In: Trump, B.D., et al. (eds.) Synthetic Biology 2020: Frontiers in Risk Analysis and Governance, pp. 85–105. Springer International Publishing, Cham (2020)
292. Okkeh, M., et al.: Gold nanoparticles: can they be the next magic bullet for multidrug-resistant bacteria? Nanomaterials (Basel) **11**(2) (2021)
293. Tewabe, A., et al.: Targeted drug delivery—from magic bullet to nanomedicine: principles, challenges, and future perspectives. J. Multidiscip. Healthc. **14**, 1711–1724 (2021)
294. Klintman, M., et al.: Academia and society in collaborative knowledge production towards urban sustainability: several schemes—three common crossroads. Environ. Dev. Sustain. (2022)
295. Mitchell, M.J., et al.: Engineering precision nanoparticles for drug delivery. Nat. Rev. Drug Discov. **20**(2), 101–124 (2021)
296. Prajapati, S., et al.: Chapter 11—nanotechnology-based sensors. In: Pal, K., et al. (eds.) Biopolymer-Based Formulations, pp. 237–262. Elsevier (2020)
297. Schulz, J., Robinson, L., Wellman, B.: Digital Automation and AI: Trajectories and Cultures in and Outside the Workplace, pp. 1667–1671. SAGE Publications Sage CA, Los Angeles, CA (2023)
298. Saliu, H.: The role of media in intercultural communication in the age of globalization, pp. 325–335 (2015)
299. Majumdar, S., Guha, S., Parekh, N.: Technology and innovation for social change: an introduction, pp. 1–3 (2015)
300. Thiele, L.P.: Integrating political and technological uncertainty into robust climate policy. Clim. Change **163**(1), 521–538 (2020)
301. Kong, X., et al.: Advances of medical nanorobots for future cancer treatments. J. Hematol. Oncol. **16**(1), 74 (2023)

302. Saritha, G.N.G., Anju, T., Kumar, A.: Nanotechnology—big impact: How nanotechnology is changing the future of agriculture? J. Agric. Food Res. **10**, 100457 (2022)
303. Callaghan, C.W., Callaghan, N.C., Jogee, R.: Inequality in healthcare R&D outcomes: a model of process disruption. Dev. South. Afr. **36**(6), 874–888 (2019)
304. Wang, L., et al.: A holistic and state-of-the-art review of nanotechnology in solar cells. Sustain. Energy Technol. Assess. **54**, 102864 (2022)
305. Zhang, Q., et al.: Nanomaterials for energy conversion and storage. Chem. Soc. Rev. **42** (2013)
306. Abdalla, A., et al.: Nanotechnology utilization in energy conversion, storage and efficiency: a perspective review. Adv. Energy Convers. Mater. **1**, 38 (2020)
307. Xuan, J., et al.: Nanotechnology for energy storage and efficiency. In: Li, Z., Zheng, J., Ye, E. (eds.) Sustainable Nanotechnology. The Royal Society of Chemistry (2022)
308. Kumar, N., et al.: Recent advanced supercapacitor: a review of storage mechanisms, electrode materials, modification, and perspectives. Nanomaterials (Basel) **12**(20) (2022)
309. Omeiza, L.A., et al.: Nanostructured electrocatalysts for advanced applications in fuel cells. Energies **16** (2023). https://doi.org/10.3390/en16041876
310. Sun, Y., et al.: Advancements in cathode catalyst and cathode layer design for proton exchange membrane fuel cells. Nat. Commun. **12**(1), 5984 (2021)
311. Xu, K., et al.: A review of high-temperature selective absorbing coatings for solar thermal applications. J. Materiomics **6**(1), 167–182 (2020)
312. Hamza, E., Nafea, S.: Nanotechnology application for wireless communication system, pp. 115–130 (2022)
313. Hung, Y.H., Chen, C.H., Huang, S.W.: Applying augmented reality to enhance learning: a study of different teaching materials. J. Comput. Assist. Learn. **33** (2016)
314. Bustos-López, M., et al.: Wearables for engagement detection in learning environments: a review. Biosensors **12** (2022). https://doi.org/10.3390/bios12070509
315. Wang, D., et al.: Haptic display for virtual reality: progress and challenges. Virtual Reality Intell. Hardw. **1**(2), 136–162 (2019)
316. Luo, Y., et al.: Technology roadmap for flexible sensors. ACS Nano **17**(6), 5211–5295 (2023)
317. Riaz, A., Khan, S., Arslan, T.: Design and modelling of graphene-based flexible 5G antenna for next-generation wearable head imaging systems. Micromachines **14** (2023). https://doi.org/10.3390/mi14030610
318. Dinesh, A., Sudharsan, S.T., Haribala, S.: Self-sensing cement-based sensor with carbon nanotube: fabrication and properties—a review. Mater. Today Proc. **46**, 5801–5807 (2021)
319. Wang, Y., Pham, D., Ji, C.: Nanocomposites for extrinsic self-healing polymer materials, pp. 243–279 (2017)
320. Han, D., et al.: A comprehensive review and analysis of nanosensors for structural health monitoring in bridge maintenance: innovations, challenges, and future perspectives. Appl. Sci. **13** (2023). https://doi.org/10.3390/app132011149
321. Gangisetty, G., Zevenhoven, R.: A review of nanoparticle material coatings in passive radiative cooling systems including skylights. Energies **16** (2023). https://doi.org/10.3390/en16041975
322. Akyildiz, I., Jornet, J.: Electromagnetic wireless nanosensor networks. Nano Commun. Netw., 3–19 (2010)
323. Guerra, F.D., et al.: Nanotechnology for environmental remediation: materials and applications. Molecules **23** (2018). https://doi.org/10.3390/molecules23071760
324. Ezz El-Din, H., Manjaiah, D.H.: Internet of nano things and industrial internet of things. In: Acharjya, D.P., Geetha, M.K. (eds.) Internet of Things: Novel Advances and Envisioned Applications, pp. 109–123. Springer International Publishing, Cham (2017)
325. Gupta, P.K.: Single-molecule DNA sequencing technologies for future genomics research. Trends Biotechnol. **26**(11), 602–611 (2008)
326. Das, K.P.: Nanoparticles and convergence of artificial intelligence for targeted drug delivery for cancer therapy: current progress and challenges. Front. Med. Technol. **4**, 1067144 (2022)
327. Ghaffar Nia, N., Kaplanoglu, E., Nasab, A.: Evaluation of artificial intelligence techniques in disease diagnosis and prediction, 5 p (2023) (Electronic: 2731-0809)

328. Al-Atawi, A.A., et al.: Stress monitoring using machine learning, IoT and wearable sensors. Sensors (Basel) **23**(21) (2023)
329. Zhang, P., et al.: Nanotechnology and artificial intelligence to enable sustainable and precision agriculture. Nat. Plants **7**(7), 864–876 (2021)
330. Sandler, R.: Nanotechnology: The Social and Ethical Issues (2009)
331. Harthorn, B.H.: Innovative and responsible governance of nanotechnology for societal development. J. Nanopart. Res (2011)
332. Gottardo, S., et al.: Towards safe and sustainable innovation in nanotechnology: state-of-play for smart nanomaterials. NanoImpact **21**, 100297 (2021)
333. Daradkeh, M.: Navigating the complexity of entrepreneurial ethics: a systematic review and future research agenda. Sustainability **15** (2023). https://doi.org/10.3390/su151411099
334. Gupta, R., Xie, H.: Nanoparticles in daily life: applications, toxicity and regulations. J. Environ. Pathol. Toxicol. Oncol. **37**(3), 209–230 (2018)
335. Boverhof, D.R., et al.: Comparative assessment of nanomaterial definitions and safety evaluation considerations. Regul. Toxicol. Pharmacol. **73**(1), 137–150 (2015)
336. Sargent, J.F.: Nanotechnology and Environmental, Health, and Safety: Issues for Consideration (2008)
337. Wasti, S., et al.: Ethical and legal challenges in nanomedical innovations: a scoping review. Front. Genet. **14**, 1163392 (2023)
338. Gupta, P., et al.: Assessing the interactions of people and policy-makers in social participation for health: an inventory of participatory governance measures from a rapid systematic literature review. Int. J. Equity Health **22**(1), 240 (2023)
339. OECD, OSLO Manual: Guidelines for Collecting and Interpreting Innovation Data (2005)
340. Rajpoot, S., Ijmtst, E.: Impact of nanotechnology on environment—a review. Int. J. Mod. Trends Sci. Technol. **7**, 159–164 (2021)
341. Jotterand, F.: Emerging Conceptual, Ethical andPolicy Issues in Bionanotechnology. Berlin (2008)
342. Falk, A., Hartl, S., Sinner, F.: Regulation and safety implementation of nanotechnology for chemical enterprises in the Central Europe Space. J. Phys. Conf. Ser. **429**(1), 012068 (2013)
343. Stirling, D.A.: The Nanotechnology Revolution: A Global Bibliographic Perspective. CRC Press (2018)
344. Matteucci, F., et al.: Deployment and exploitation of nanotechnology nanomaterials and nanomedicine, vol. 1990, 020001pp (2018)
345. Pearsall, N.: Commercialization of Nanotechnology, LES Insights. Online J. Licensing Executives Soc. (2014)
346. Hansen, S.F., Baun, A.: European regulation affecting nanomaterials—review of limitations and future recommendations. Dose Response **10**(3), 364–383 (2012)
347. Forsberg, E.:The role of ISO in the governance of nanotechnology (2010)
348. Allan, J., et al.: Regulatory landscape of nanotechnology and nanoplastics from a global perspective. Regul. Toxicol. Pharmacol. **122**, 104885 (2021)
349. Li, W., et al.: The making of responsible innovation and technology: an overview and framework. Smart Cities **6**, 1996–2034 (2023). https://doi.org/10.3390/smartcities6040093
350. Russell, A.W.: Improving legitimacy in nanotechnology policy development through stakeholder and community engagement: forging new pathways. Rev. Policy Res. **30**(5), 566–587 (2013)
351. Roco, M.C., Mirkin, C.A., Hersam, M.C.: Nanotechnology research directions for societal needs in 2020: summary of international study. J. Nanopart. Res. **13**, 897–919 (2011)
352. Arabeyyat, Z.H., Jamaliah, M.M., Khalaf, M.A.: Public awareness of nanotechnology and its implications for health in Jordan. Sustainability **14** (2022). https://doi.org/10.3390/su14105786
353. Schmidt Charles, W.: Nanotechnology-related environment, health, and safety research: examining the national strategy. Environ. Health Perspect. **117**(4), A158–A161 (2009)
354. Morris, J., et al.: Science policy considerations for responsible nanotechnology decisions. Nat. Nanotechnol. **6**(2), 73–77 (2011)

355. Guston, D.H., Cameron, N., Mitchell, M.: The center for nanotechnology in society at Arizona state university and the prospects for anticipatory governance. In: Nanoscale: Issues and Perspectives for the Nano Century, pp. 377–92 (2007)
356. Yasri, S., Wiwanitkit, V.: Important ethical issues for nanomedicine. J. Res. Med. Sci. **22**(1) (2017)
357. Allon, I., et al.: Ethical issues in nanomedicine: tempest in a teapot? Med. Health Care Philos. **20**(1), 3–11 (2017)
358. Bozeman, B., Fay, D., Slade, C.P.: Research collaboration in universities and academic entrepreneurship: the-state-of-the-art. J. Technol. Transf. **38**(1), 1–67 (2013)

Benjamin Tawiah is a lecturer at Kwame Nkrumah University of Science and Technology. He obtained his Ph.D. from the Hong Kong Polytechnic University where he worked as postdoctoral Fellow and at the same time served as a research associate in the City University of Hong Kong. Dr. Tawiah has been engaged in basic and applied research in many areas including nanomaterial synthesis and applications, polymer nanocomposites, flame retardant polymer composites and second-generation batteries for energy storage in grid systems where he has published extensively.

Emmanuel Abankwah Ofori is a postgraduate researcher at Hubei University of Automotive Technology. Prior to that he had a BSc. Textile Design and Technology from the Industrial Art Department of Kwame Nkrumah University of Science and Technology, Kumasi. His research interest focuses on sustainable energy materials, nanomaterials and composites, textile science, polymers, and energy efficiency. Emmanuel has actively participated in several research projects, working with renowned authors to explore various facets of environmental sustainability, adaptation to climate change, green dyeing, and sustainable renewable energy consumption.

Soney C. George is the Dean Research and the Director of Centre for Nanoscience and Technology, Amal Jyothi College of Engineering, Kanjirappally, Kerala, India. He received his Ph.D from Mahatma Gandhi University and Post doctoral Studies from University of Blasie Pascal, France and Inha University, South Korea. His h.Index is 34 with a total citations of 4791. He also received National Technical Teachers award 2022 from All India Council for Technical Education, New Delhi, India.

Chapter 2
Nanotechnology: Social Acceptance, Cultural Impact and Privacy

Peramjeet Singh, Priya Gupta, Vernica Verma, Neetu Yadav, and Narendra Kumar Pandey

Introduction

Nanotechnology has become an emerging technology that plays an important role in civilization and has enormous positive effects on society. Several developed as well as developing countries, like the United States, Japan, Bharat, South Korea, and China, categorise nanotechnology as an important asset and a priority in their national plans based on science and technology. The socioeconomic ramifications of nanotechnology, as well as its possible hazards to human health and the environment, are also of concern. The incorporation of technology based on nanomaterials into national plans is a significant concern in each of these nations. It was observed from the collected data from 2000 to 2020 that passive nanostructures like polymers, ceramics, coatings, nanostructured metals, and nanoparticles showed higher risk and uncertainty as compared to other integrated nanosystems, molecular nanosystems, and converging technologies like 3D networking and robotics, atomic design, and cognitive technologies, respectively [1]. The nanoparticles can be synthesised by two processes: bottom-up which includes sol–gel, vapour deposition, chemical, bioreduction and atomic or molecular condensation and top which includes thermal process, optical process, mechanical process, sputtering, and chemical etching [2]. The feature of nanomaterials seems different from the visible world. For example, nanoparticles of gold show a red colour instead of golden and aluminium nanoparticles start producing flames at the nano levels. Several effects such as quantum and surface effects predominate at the nano levels than macro levels which makes them different. The proportionality of the atoms is more and continuously increases as the materials reach the nanoscale. In the coming days, appliances based on nanomaterials

P. Singh (✉) · P. Gupta · V. Verma · N. Yadav · N. K. Pandey
Department of Physics, University of Lucknow, Lucknow, Uttar Pradesh, India
e-mail: ps751994lko.uni@gmail.com

P. Gupta
Manyawar Kanshiram Government Polytechnic, Tirwa, Kannauj, Uttar Pradesh, India

S. C. George and B. Tawiah (eds.), *Nanotechnology in Societal Development*, Advanced Technologies and Societal Change, https://doi.org/10.1007/978-981-97-6184-5_2

will be in high demand due to their easy availability and lower cost. Nanoparticles are currently employed in semiconductors, laptops, computers, supercapacitors, and video players, and they are going to play a major role in the further advancements of electronic components and storage devices. It was found that nanotechnology plays an important role in various fields such as chemical science, physical science, material science, biological science, engineering fields, and information technology [3]. M. Albrecht et al. studied the biomedical applications of magnetite particles using atomic force microscopy (AFM) and magnetic force microscopy (MFM) and reported that the nanoparticle consists of single isolated magnetosomes having a size of 40 nm [4].

Drug consumption as well as therapy expenses are reduced by customised medications. Since the ecosystem is the primary beneficiary in the race to modify nanotechnology, it is crucial to environmentally friendly design technologies. In the process of fabricating new materials, it is essential to reduce contamination in the environment. Nowadays, solar cells are highly attractive for generating electricity. Researchers and academics are continuously making efforts to fabricate solar cells with minimal cost. The application of nanotechnology to transportation has greatly improved the comfort levels of human beings, resulting in the development of lightweight bicycles, fuel-efficient cars, and aeroplane wings that can withstand lightning strikes. The first generation of beneficial innovations is probably going to be in the identification and monitoring field. While identification surpasses response competence, as it typically attempts to, moral and rule quandaries will naturally surface. Consequently, nanoparticles will not replace existing techniques; instead, they will continue to operate along with them for a very long time. This might allow them more time to consider the possible moral and societal ramifications of nanotechnology. As a result, cultural and moral issues must be covered in the new education curriculum. In 1971, to foster new technology and facilitate collaboration between universities and advanced research institutes, the Indian government constituted the "Department of Science and Technology (DST)" as an independent agency. DST launched a programme called "Science for Equity, Empowerment, and Development (SEED)" to advance nanotechnology and motivate researchers to concentrate on initiatives that aim to improve living standards through socioeconomic development. The government has started various initiatives and projects to empower citizens by using various inventions of nanotechnology to fulfil the requirements and aspirations of citizens.

Social Acceptance and Challenges of Nanotechnology

It is simple to detect but challenging to quantify or estimate how emerging innovations will affect society. In addition to its consequences for employment and the relationship between people and administrations, nanotechnology is going to have major societal implications in the fields of defence and property rights. To understand the challenges in social influence of a particular technology over the course of the whole revolution, one has to recall the number of experts who applauded the emergence of paperless offices after the development of e-mails [5]. One field of research where nanotechnology is projected to see some of its most significant and rapid benefits is medicine. Recent developments include novel pharmaceutical delivery methods, methods for diagnosis, and surgical devices [6]. The practical use of nanomaterials in the automotive sector has resulted in several advantages, including the improvement of the strength and longevity of vehicles [7]. Various harmful chemicals, waste products, and pollutants released from the advancement of nanoparticles are among the causes of environmental concern. Particularly with regard to consumers, assessment monitoring is currently beginning to take an interest in protection products and beauty products that use nanoparticles. Nanoparticles are found to be harmful to human beings and other living organisms with respect to reproduction, respiratory, nervous, digestive, and immune systems [8].

Technology and recent advancements have historically been a major factor in generating assets, and the rapidly developing nano industry has the power to revolutionise every aspect of the economy [9]. Assuming that nanotechnologies and the techniques that come through them will transform the globe, we must begin with the fundamental belief that new discoveries and advancements at the nanoscale will have a noticeable and substantial influence on production. The nano market will be driven by the commercialization of technological advancements in the nanosphere and research findings at the cutting edge of science. It is obvious that responsive approaches are needed in research and development for growing sectors like nanotechnologies, where novel discoveries and advances are made frequently. In order to efficiently develop a significant concept for future generations, scientists demand immediate and unbureaucratic assistance, and entrepreneurs need to establish solid relationships with academic institutions to find the most suitable collaborators for a speedy translation of findings from studies into goods.

Environmental Concerns

Concerns about the environment are being brought up by nanotechnology, with its great potential in a variety of industries, like appliances, electricity, and medicine. These worries consist of the following.

Nano waste

The main risk of nanoparticles rises after their disposal in the environment like soil, water and air. The various importance of nanowaste in areas like solar panels, biosensors, electrical appliances, LEDs, water purifiers, etc. led to several challenges to living organisms and their habitats [10]. The emissions of airborne pollutants and liquid or solid collections of products (waste) during the fabrication of nanomaterials from factories into the environment harm the living standards of living organisms. Eventually, disseminated nanoparticles will accumulate on the outermost layers of the earth and oceans. These waste materials from nanomaterials that flow to water bodies through contaminated soil can damage aquatic life. They can also harm human

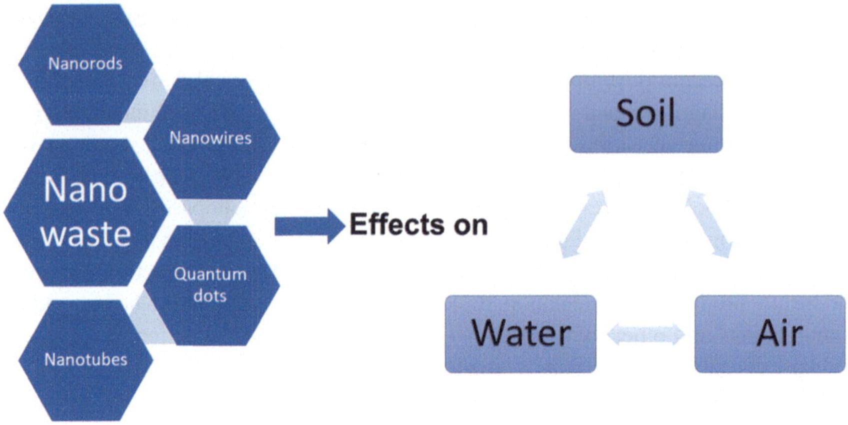

Fig. 2.1 Effects of nano waste on the environment like soil, water, and air

beings by reaching the surface of the earth through rainwater. The nanoparticles of carbon, silver, silicon, titanium oxide, and gold released during fabrication will affect the central nervous system, granulomas, skin, and pulmonary inflammation [11] (Fig. 2.1).

Energy and Resource Consumption

The fabrication of nanoparticles frequently consumes many resources and excess energy, which releases enormous amounts of carbon and harmful gases into the environment [12]. Several techniques for the synthesis of nanoparticles require the utilisation of chemical substances, other natural resources, and valuable metals. The mining and exploitation of these natural assets may worsen the damage to the environment by triggering habitat loss and harming diversity [13]. Environmental issues might arise from improper management of materials used in the manufacturing of nanomaterials, including chemicals, stabilising substances, and reducing agents. These substances can pollute water as well as soil if they are not disposed of properly, endangering both living organisms and their ecosystems [14]. A significant amount of water is essential for the synthesis of some nanoparticles for solution use, washing, and the cooling process. Water-intensive production techniques have an increased adverse effect on the environment in areas where there is a shortage of water [15]. Scientists, engineers, and industrialists should concentrate more on establishing environmentally friendly methods for fabricating nanoparticles to solve these challenges. This involves investigating more eco-friendly fabrication methods with organic components, renewable energy sources, and chemicals that don't harm the environment. Intending to reduce the adverse ecological consequences of producing nanoparticles, it is also essential to improve activities to decrease waste creation, consume less energy, and save resources.

Health concerns

The primary medical hazards associated with nanotechnology are those that may arise from contact with nanoparticles. Even though nanomaterials have many advantages, including improvements in the medical field, there may also be health hazards that should be taken into account [16]. The human eye cannot see nanoparticles. It is extremely difficult for scientists and researchers to recognise the presence of nanoparticles around their bodies or workplaces. These particles might be brought inside the house if they are present. There are a lot of products that have already made it to the consumer market, even though very little is known about how nanoparticles behave in the ecosystem and in living organisms. Currently, the general public uses several products made of nanoparticles, such as sunscreens and cosmetics, mattresses, stain-resistant clothes and fabrics, sports equipment like tennis and bowling balls, cooking oil, and chewing gum [17] (Fig. 2.2)

Ethical concerns

Since nanotechnology has such enormous potential risks, we really need to consider what might happen. Although it's possible that nanotechnology will never reach

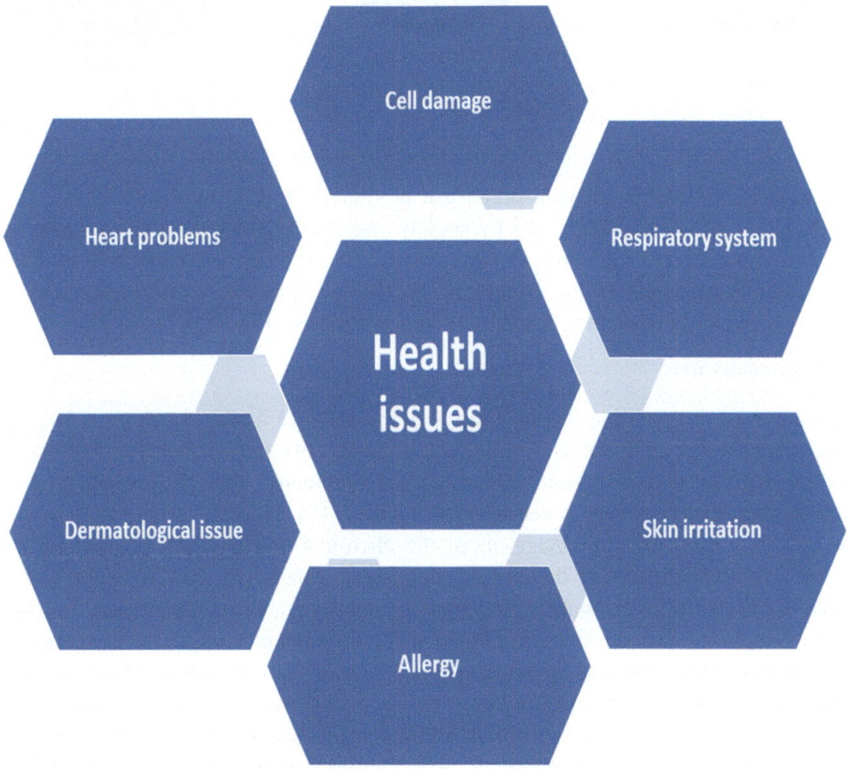

Fig. 2.2 Impact of nanotechnology on health

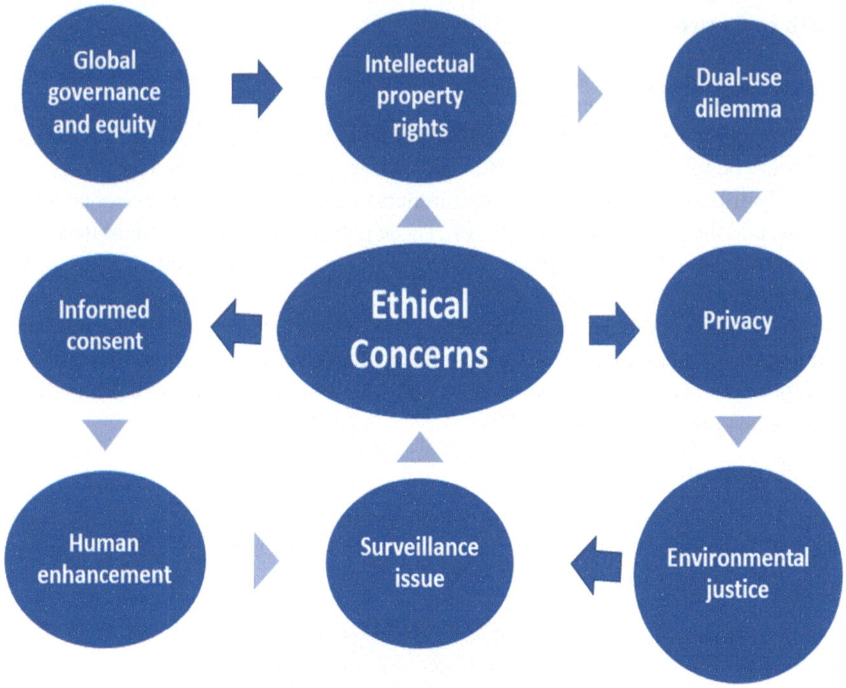

Fig. 2.3 Various ethical issues of nanotechnology

the heights of power and widespread use predicted by its proponents, before the science is permanently embraced by society, we should work through the process of identifying ethical dilemmas that may arise. To eradicate or significantly reduce nanotechnology's detrimental effects on society, we must consider the ethics of its development and establish laws that will support it [18] (Fig. 2.3).

Risk assessments

Evaluation of the possible threats and related hazards of nanotechnology is largely dependent on risk evaluations. Finding the precise traits and attributes of nanotechnology that can endanger the health of living organisms, their surroundings, and various biological systems is necessary to do this [19]. Potential risks can be identified with the aid of knowledge about the chemical and physical characteristics of nanoparticles. Following the characterization of a particular risk, the subsequent stage of any threat investigation is evaluating the danger of nanomaterials. Factual information for rational risk management will be obtained through risk assessment. Analysing chemical threats is frequently pertinent to nanoparticles. There are two things to discuss here. The material must first be identified, and any possible hazards must be investigated under controlled circumstances. Measurement and quantification of exposure are the second needs. A particular risk can be described based on the previous two phases, and a choice on the risk's height for every potential exposure condition must be made [20]. It is necessary to evaluate the methods by which nanoparticles may be exposed to living organisms or environmental conditions. This includes assessing the methods of exposure, such as swallowing, inhaling, or skin contact, during the different phases of the synthesis, application, and storage of nanoparticles. Researching the ability of nanoparticles to harm organisms, tissues, and cells, as well as any cumulative or long-term impacts, is part of this. Analysing data from epidemiology, doing experiments on animals, and conducting in vitro research are all possible components of the evaluation of any hazard [21]. To guarantee the continuous safeguarding of the natural world and public health, risk evaluations must be reviewed and updated on an ongoing basis in light of fresh research and technological developments. Participants can enhance the advantages while avoiding potential damages by methodically examining the risks connected with nanomaterials and making educated decisions. Scientists, policymakers, businesses, and other interested parties must work together to perform thorough and efficient assessments of risk in nanotechnology (Fig. 2.4).

Transparency

As in many other fields, transparency is a basic element of nanomaterials management and legislation. It entails transparency, responsibility, and unrestricted access

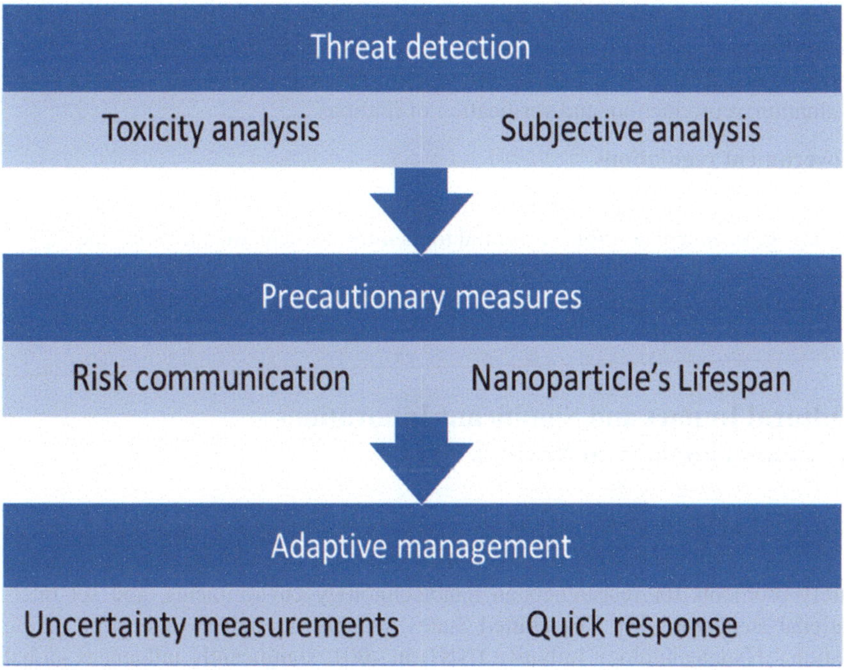

Fig. 2.4 Various threat assessments and actions of nanotechnology

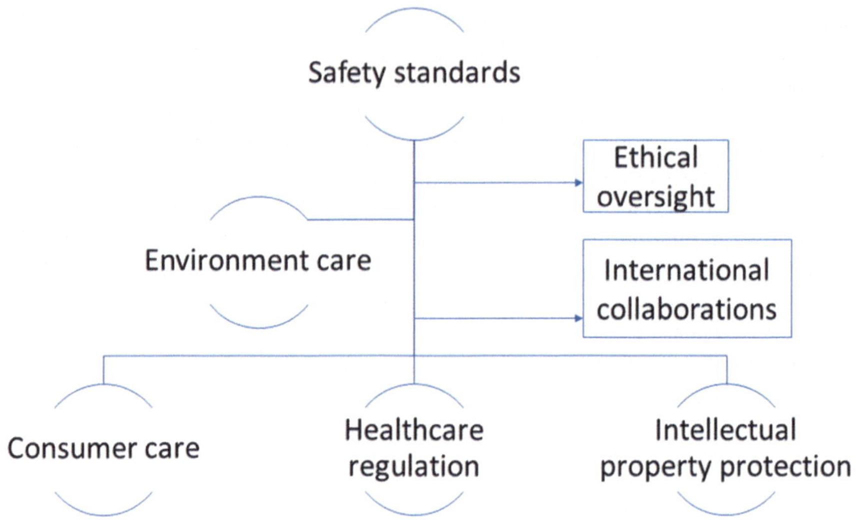

Fig. 2.5 Various aspects of government regulation on nanotechnology

to information for all parties involved. Transparency has various significant functions in the field of nanotechnology [22]. Technological advancement and innovation in nanotechnology are fostered by the open dissemination of results from studies, methodologies, and data. The scientific profession values authority and confidence because providing open access to research articles and information facilitates peer evaluation, reproduction, and verification of findings.

Government regulations

To maintain security, ethical behaviour, and the advantages of nanotechnology for society, government control is essential to its improvement, manufacture, and application. The government's oversight of nanomaterials includes the following aspects (Fig. 2.5).

Cultural Impact and Significant Innovations of Nanotechnology in Various Areas

The global adoption of nanotechnology has prompted concerns regarding its various implications. Efforts are underway worldwide to create or revise regulations, improve safety protocols for researchers in nanotechnology environments, and set international standards [23]. The United States, notably through the inception of the National Nanotechnology Initiative (NNI) in 2001, significantly influences global nanotechnology policies and strategies [24]. Nanotechnology has the potential to transform multiple sectors such as medicine, agriculture, consumer goods, environmental solutions, and electronics [25]. This has prompted heightened investments from both public and private entities in nanotechnology research. Nevertheless, the notable progress in nanotechnology poses challenges in comprehending how individuals perceive, form attitudes towards, and embrace nanotechnology applications. Researchers have diligently delved into uncovering the factors that shape people's attitudes and acceptance of nanotechnology, with a specific emphasis on understanding risk and benefit perceptions [26]. People who have a positive outlook on science and technology typically view nanotechnology favourably. Greater involvement and interest in science are associated with perceiving increased benefits of nanotechnology, attributed to a deeper understanding and decreased uncertainty [27]. Additionally, individuals who value the positive aspects of science and comprehend its capabilities are more likely to have a positive stance toward nanotechnology applications [28]. Engaging in comprehensive information processing indicates a higher likelihood of supporting nanotechnology. This correlation is supported by consumer behaviour studies, highlighting the impact of motivation, cognitive ability, and informational opportunities on the rational assessment and endorsement of emerging technologies.

Nanotechnology in Bioengineering

Nanotechnology can find applications in bioengineering like sorting and gene identification, rapid plant genome sequencing, plant diagnostics and therapy, growth enhancers and seed invigoration, explant sterilization, plant gene transformation, and advanced nanobionic plants. One well-known property of silver ions is their antibacterial properties when extracted from their bulk form. Silver nanoparticles are more effective because of their small size, which allows them to penetrate explant membranes at a lower rate [29]. The different features of nanoparticles, like their electrical, surface, particle size, optical, crystallite size, and magnetic characteristics, reveal novel instruments that are used for the identification of various harmful diseases. The assessment of the adverse and hazardous effects of nanotechnology is becoming more and more crucial with the development of nanotechnology in medicine. When nanomaterials dissolve with materials, they release harmful components, which can cause stress and damage genes in organisms [30].

Nanotechnology in Defense Sector

The possibilities provided by nanostructures and nanotechnology are realized by integrating nanoparticles with other relevant innovations, making them a catalyst for innovation that overcomes all domains. Within this framework, innovations in nanomaterials might assist in tackling challenges to national safety by facilitating better production of textiles and substances, automation, privacy, weapons, automobiles, and nanomedicine to enhance soldier's wellness. The chemical vapour deposition method has allowed nanoscientists and engineers to develop nano-teflon, which is used to make army combat jackets that are impenetrable and waterproof [31]. Sensing response, responsibility, response and recovery time, and the percentage of instrument error are the most important parameters, particularly when it comes to the military profession. Due to their unique characteristics, nanomaterials have proved beneficial in advanced defense. Microscopic frameworks, quantum capability, a larger surface-to-volume ratio, and small size constitute a few of the characteristics of nanomaterials. Modern warfare is developing quickly due to advancements in technology [32].

Nanotechnology in Agricultural Sectors

Agriculture may take advantage of the utilization of nanotechnology. For example, nanocapsules are capable of being employed for effectively dispensing nutrients, herbicides, and agricultural products [33]; nanoparticles can be used to determine sanitiser and diagnose animal and plant infections [34]. Nutritional absorption, bioavailability, and dissemination can all be enhanced by using nanocapsules. In addition, alimentary pathogens can be determined using nanoparticles, which may be further employed as antifungi and antimicrobe [35]. Researchers cannot ignore the appalling conditions of the agricultural and food sectors with respect to the significance of nanotechnology in agriculture. The challenge is getting worse due to soil degradation, shortages of food, starvation, and declining wildlife. In addition to their contribution to the food sector, sensors based on nanotechnology, when combined with nanoparticle-based inks, could allow quick identification of products [36].

Nanotechnology in Medical Industry

The medical sector, global pharmaceutical firms, advanced medical entrepreneurs, shareholders, and patients themselves all became more involved in the ongoing debates about nanotechnology as they were incorporated into pharmaceutical research, research studies, and contemporary healthcare [37]. Millions of patients worldwide suffer from health issues that could be addressed by using nanoparticles in the medical industry to investigate biological mechanisms at a resolution that has never been possible at the nanoscale [38]. Azamat Ali et al. reported that nanotechnology is innovative in fields of medicine like imaging, antibiotics, vaccinations, and drug distribution. Their examination indicates that pharma and drug organisations are major players in both public and private research centres. A connection analysis revealed that companies, scientific centres, and medical facilities have established global partnerships as well as connections with one another. India faces many challenges to commercialising studies on nanotechnology, one of which is a lack of financing for facilitating technological advancements [39].

Nanotechnology in Electrical Appliances

Researchers are continuously focused on consuming less power while transporting electrical gadgets by reducing their size and weight. Nanotechnology is widely applied in several devices, like desktop and laptop computers, television screens, smartphones and telecommunication equipment, musical instruments, LCDs, gaming devices, and media. Quantum dots play a crucial role in the application of nanotechnology in electrical appliances due to the quantum confinement effect, which shows a discrete energy state. It increases the energy band gap, which makes nanocrystals suitable for semiconductors [40]. C. Sumereder et al. studied the application of nanotechnology in electronics and reported that the top layer of ceramic material was coated with nanotechnology, and its durability was evaluated in both ambient and controlled surroundings in a humidity chamber. The angle of contact was measured objectively by optical means at specified times in order to establish an average for durability over time [41]. According to a research article by J. E. Contreras et al. on the electrical characteristics of nanotechnology, nanoparticles have revolutionised a variety of scientific and technical fields by providing consumers with valuable possibilities and innovative approaches. The growth of numerous profitable nanomaterials has resulted from the growing significance of nanomaterial concepts in modern times [42].

Nanotechnology as Sensors

The well-being and security of individuals need to identify pollutants like hazardous exhaust and chemical-based weapons. The last century has seen a rapid advancement that has released thousands of hazardous chemicals that adversely affect the environment and threaten human health too [43]. As a result, protecting the environment, especially the air we breathe, is very important. Many hazardous and dangerous gases, like carbon monoxide, nitrogen dioxide, ammonia, carbon dioxide, and hydrogen sulphide, are released by automobiles, factories, and other sources, and citizens are frequently in contact with them in metropolitan regions. Among the five senses that humans have been endowed with, the nose has an exceptionally versatile and responsive sense of smell. Human beings should have to immediately recognize it and, if possible, try to reduce it. As a result, it is crucial to fabricate highly responsive and sensitive sensors with high sensibility, minimum hysteresis, long stability, quick response, and recovery time because their results could help consumers effectively. One-dimensional metal oxide and metal-doped oxide nanoparticles that possess specific electrical, optical, surface, chemical, and physical characteristics are found to be suitable candidates for sensor fabrication. It is possible to synthesize sensors based on 1-D metal oxide nanoparticles such as MoO_3, ZnO, Bi_2O_3, V_2O_5, WO_3, Cu_2O, and SnO.

Various Issues and Concerns of Nanotechnology in Society

Progress in nanotechnology shows immense potential for improving many areas of medical services, with an increase in new concerns. Initial investigation demonstrates that specifically developed nanomaterials may exhibit inappropriate hazardous characteristics, posing issues regarding possible threats to the health and safety of humans and the environment. The use of nanomaterials offers novel possibilities for advancements in science to deal with unidentified concerns [44]. Divergent opinions exist regarding the appropriateness of regulating nanoparticles: either freely chosen or under authority control based on preventions [45]. It is critical to comprehend surface chemistry and how it relates to toxicity risks because 1 nm nanoparticles absorb almost 100% of surface atoms [46]. The fabrication of nanomaterials through different routes releases hazardous components into the environment, raising certain ecological and safety issues. The ecological mechanisms of nanotechnology and the harmful effects that result from their exposure are still insufficiently understood. It is yet unknown which nanoparticles pose a specific and severe risk to the environment's and public's health [47]. Nanoparticles play a crucial role in various biological, medical, and pharmaceutical applications by acting as efficient drug delivery systems that enhance drug effectiveness, reduce side effects, and target specific tissues and organs [48]. Ongoing research is focused on improving the design and functionality of nanocarriers for drug delivery, enhancing targeting abilities, and refining nanoparticle preparation methods to enhance drug delivery efficiency, clinical applicability, and industrial scalability [49]. Additionally, nanoparticles not only serve as drug carriers but also exhibit active properties, like the ability of metal-containing nanoparticles to selectively heat and destroy tumour cells when directed to tumours [50].

Furthermore, nanoparticles such as titanium dioxide and fullerenes demonstrate antimicrobial properties, making them valuable in the development of antimicrobial products like wound dressings [51]. The integration of nanotechnology in healthcare shows great potential for revolutionizing drug delivery methods and enhancing therapeutic outcomes with increased efficacy and precision. Nanotechnology plays a vital role in the transportation sector by providing structures and materials at the atomic and molecular levels that can perform tasks beyond the capabilities of conventional macroscopic materials [52]. It is crucial for the development, operation, and maintenance of infrastructure that facilitates vehicle movement, upkeep of transportation routes, and vehicle management. Research in transportation aims to bridge the gap between nanotechnology and transportation systems, with a particular focus on enhancing safety through advancements in surface properties, durability, and the interaction between vehicles and infrastructure [53]. The paper battery acts as both a high-capacity battery and a supercapacitor, blending two traditionally separate elements in standard electronics [54]. Crafted from nanocomposite paper, a combination of cellulose, it fuses the energy storage capabilities of a battery with the high-energy density of a supercapacitor, enabling it to deliver potent surges of power [55]. Various types of carbon nanotubes, differing in structures, thicknesses, lengths, and layers, are integrated into these devices, encompassing single-walled, double-walled, triple-walled, and multi-walled carbon nanotubes [56]. By incorporating aligned carbon nanotubes into cellulose, each around one-millionth of a centimetre thick, the paper battery can provide long-lasting energy, bursts of power, and reliable performance. Paper batteries hold the promise of powering upcoming medical devices, electronics, and hybrid vehicles, offering flexibility in diverse applications without sacrificing efficiency [57]. Functioning within a temperature span of −75 to 150 degrees Celsius, paper batteries operate by immersing paper in an ion-based liquid solution to trigger a chemical reaction between the electrodes and liquid, generating electricity as electrons travel from the cathode to the anode [58]. During recharging, the paper electrode swiftly stores energy, with ions moving rapidly through the thin electrode in just seconds. Improving battery performance can be accomplished by stacking multiple paper batteries together.

Environmental remediation, human activities, industrial processes, and agricultural practices release chemicals and harmful substances into the environment, causing disturbances in ecosystems and adverse effects on organisms. To combat environmental pollution, a range of remediation technologies have been developed, with bioremediation standing out for its eco-friendly principles. Bioremediation utilizes biological agents to tackle environmental challenges, either on the site of contamination or off-site. Despite its advantages compared to traditional methods, bioremediation faces obstacles such as slow processes and complexities with specific compounds [59]. Nanotechnology offers avenues to improve traditional bioremediation strategies by leveraging the distinctive characteristics of materials at the nanoscale. The growing utilization of nanoparticles and nanomaterials in bioremediation is due to their ability to infiltrate contamination areas inaccessible to larger particles and their enhanced reactivity towards pollutants [60]. The applications of nanotechnology in environmental clean-up include denitrification, dichlorination, degradation of organic pollutants, elimination of heavy metals, and anodic remediation. For instance, the successful breakdown of hazardous substances like carbon tetrachloride using iron nanoparticles coated with oxides demonstrates the potential of nanotechnology in remediation efforts [61]. Studies have confirmed the effectiveness of nanoscale iron particles in transforming and detoxifying various environmental contaminants, emphasizing the promising role of nanotechnology in addressing environmental challenges [62]. Water purification and access to clean and safe water is crucial for sustaining life, promoting health, conserving the environment, and fostering economic development. As the global population grows, the need for water is escalating, especially in regions experiencing water scarcity [63]. This has prompted the exploration of unconventional water sources like seawater and wastewater. However, traditional water treatment methods such as filtration, reverse osmosis, chlorination, and ozonation are proving insufficient due to the continuous discharge of new industrial pollutants into the environment annually. To address these evolving challenges, it is essential to develop novel water purification technologies that can effectively remove contaminants and are cost-effective, environmentally friendly, and energy-efficient.

Various Developments in the World in the Last Two Decades

The advancement of nanotechnology over the past few decades has been essential to all areas of science and research. The rapid development of nanotechnology offers possibilities for implementation in a variety of fields, including electronic devices, health and wellness, the environment, and medicines [64]. The ability to tailor nanomaterials rapidly improved with time, contributing to the development of a wide range of semiconductors [65]. From graphene to borophene, the emerging phase of material for semiconductors enables the large-scale manufacturing of powerful semiconducting devices. The advancement of nanotechnology in medicine includes advances in drug delivery, blood coagulation, cancer, surgical equipment, and robotics [66]. Nanoparticles are a low-cost and effective way of delivering medication to specific parts of the body. Nanotechnology shows great potential for transforming agriculture by utilizing nanostructured materials that correspond to biological processes at the nano-scale [67]. The use of nanotechnology in agriculture aims to exploit nanoscale particles with distinct characteristics to improve the productivity of crops and livestock. The levels of soil nutrients are vital in determining the quality and quantity of crops, and when nutrients are depleted, they need to be replenished for optimal harvests [68]. Conventional methods of fertilization often result in inefficiencies and environmental damage, such as water pollution. In contrast, nanotechnology allows for the precise release of nutrients using encapsulated fertilizers that react to specific cues like environmental factors and biological requirements [69]. This precise delivery system for nutrients enhances the effectiveness of fertilizers while reducing negative environmental consequences.

Nanotechnology is emerging as a promising solution to enhance water purification processes by refining current techniques and introducing innovative methods. Nanotechnology-based water purification approaches offer the flexibility to customize treatments for specific pollutants or target multiple contaminants simultaneously. These strategies utilize nanoparticles for various purposes, including antimicrobial agents for disinfection, catalysts and adsorbents for eliminating pesticides and organic substances, and filters for extracting heavy metals and inorganic compounds [70]. The utilization of nanotechnology in water purification is expected to achieve higher remediation rates while generating fewer harmful by-products, attributed to the increased surface area and reactivity of nanoparticles compared to larger particles.

Conventional water treatment processes encompass various stages such as pretreatment, coagulation, flocculation, sedimentation, disinfection, aeration, and filtration [71]. Nonetheless, a significant drawback of these methods is their incapacity to efficiently eliminate dissolved salts and specific soluble substances. Nanofiltration membranes present a viable solution by selectively sieving substances at low pressure, thereby preserving essential minerals in the water [72]. Nanofiltration has demonstrated effectiveness in removing salts, organic pollutants, biological impurities, nitrates, arsenic, and cadmium from a range of water sources. Furthermore, nanoceramic filters, carbon nanotubes, and nano-sized zero-valent iron integrated into nanofiltration membranes have exhibited the potential to eradicate bacteria, heavy metals, and organic compounds from polluted water. Despite concerns regarding the safety and potential impacts of nanoparticles, the integration of nanotechnology is poised to transform water purification practices, offering more sustainable alternatives to energy-intensive conventional methods [73]. Nanotechnology, known as nanomedicine in the healthcare field, offers a wide range of applications including disease prevention, diagnosis, treatment, and monitoring, as well as drug delivery, therapy, and disinfection. By utilizing molecular genomics and proteomics to track diseases at the molecular level, nanotechnology enables early disease detection and allows for treatment initiation before symptoms appear [74]. Nanomedicine addresses challenges faced by traditional drug delivery methods, such as issues with specificity, toxicity of medications, and limited bioavailability.

In transportation, nanotechnology is applied to enhance the strength, durability, skid resistance, and water resistance of materials, as well as to improve visibility and signage for safer transportation facilities. The use of nanocatalysts, fuel additives, tribological coatings, and carbon nanotube-reinforced composites contributes to the sustainability of transportation systems [75]. Innovations in tyre manufacturing, such as the integration of nanoscale rubber particles and nano clays, have resulted in improved performance, reduced fuel consumption, and enhanced wet skid resistance [76]. Nanotechnology plays a crucial role in enhancing the sustainability of the energy sector by providing cleaner and more efficient energy resources and utilization. It reduces dependence on traditional energy sources like electricity, natural gas, and fossil fuels, contributing significantly to various areas within the energy industry such as lighting, heating, renewable energy, energy storage, fuel cells, and hydrogen power generation and storage [77]. LED-based lighting systems, utilizing nanoscale semiconductors, are emerging as a sustainable option for environmentally friendly and energy-efficient lighting [78]. LEDs are known for their durability, minimal heat production, and high electrical efficiency, offering a greener alternative to conventional incandescent and fluorescent lighting. Foreseen advancements in nanotechnology for lighting are expected to lead to a reduction of over 10% in global energy consumption. Furthermore, ongoing research on nanofluids for improved heat transfer applications is enhancing the effectiveness of central heating systems [79]. Additionally, the integration of nanotechnology in utilizing renewable energy sources like solar, wind, and geothermal energy can help mitigate the environmental impact linked with traditional energy production methods. Innovations in nanotechnology aimed at lowering the cost of solar panels aim to increase the accessibility and competitiveness of solar-generated electricity compared to standard energy sources [80]. In modern warfare, there is an increasing demand for precise delivery of force while minimizing unintended damage. This can be accomplished through the strategic utilization of sensors, information technology, and advanced materials. Stronger and lighter structural components, in conjunction with dependable explosives and propellants that provide higher energy output, are crucial in meeting these requirements. Incorporating less susceptible corrosive materials in the construction of satellites enhances their longevity. Nanostructured materials offer significant potential for various structural uses, with nanocomposites already illustrating enhancements in weight and strength across diverse sectors. Additionally, nanotechnology facilitates the downsizing of components, which is vital for the development of the next era of unmanned aerial vehicles and combat aerial vehicles [81]. In the realm of space technology, nanomaterials like nanocrystalline tungsten-titanium diboride-copper composites exhibit promise in improving the effectiveness and lifespan of satellite ignitors. Moreover, ongoing initiatives aim to decrease the weight of solar cells utilized in satellites by investigating alternative nanomaterials such as aerogels, which also have applications in the manufacturing of spacecraft and defence gear like lightweight jackets. The Defence and Development has identified nanotechnology as a key research area, implementing a structured programme with seven Programme Component Areas in alignment with the US National Nanotechnology Initiative [82]. Roughly half of the defence and development nanotechnology funding is allocated to the Defense Advanced Research Projects Agency and the remaining portion to the armed forces, including the establishment of the 'Institute for Soldier Nanotechnologies at the Massachusetts Institute of Technology [83]. The defence and development investments in nanotechnology aim to enhance both offensive and defensive military capabilities, focusing on areas such as nanoelectronics, nanomaterials, nanosensors, and soldier protection equipment. The US Government has increased its financial support for nanotechnology, while countries like Germany, France, the United Kingdom, and Russia are also exploring nanotechnology-based materials and systems for military applications. However, most Asian and European countries do not have specific programs dedicated to defense-related nanotechnology. As the years passed the modification in nanotechnology grew fast which helped in the creation of various semiconductors and electronic devices. The new era of semiconductor materials changes from graphene to borophene which provides a large-scale formation of advanced semiconductor devices [84]. Nano transistors are minute electronic components that operate on a nanoscale, with gate electrodes as small as 70 nm and gate oxides separating them from current-carrying channels by just 1 nm [85]. The semiconductor industry is moving towards logic chips containing around half a billion nanoscale MOSFETs, translating to roughly 5–10 nano transistors per square micrometre [86]. Shrinking transistor sizes result in variations in atomic-scale dimensions that impact performance and circuit reliability. The emphasis is on developing innovative tools and methods for designing nanoscale transistors and circuits to produce cost-efficient, reliable, and high-yield silicon chips despite fluctuations in device characteristics [87]. Leading semiconductor manufacturers have already accomplished the fabrication of microchips with transistors smaller than 30 nm, significantly thinner than human hair [88].

Furthermore, nanotechnology is employed to strengthen construction materials, engine components, sensors, and avionics systems, and to create functional coatings

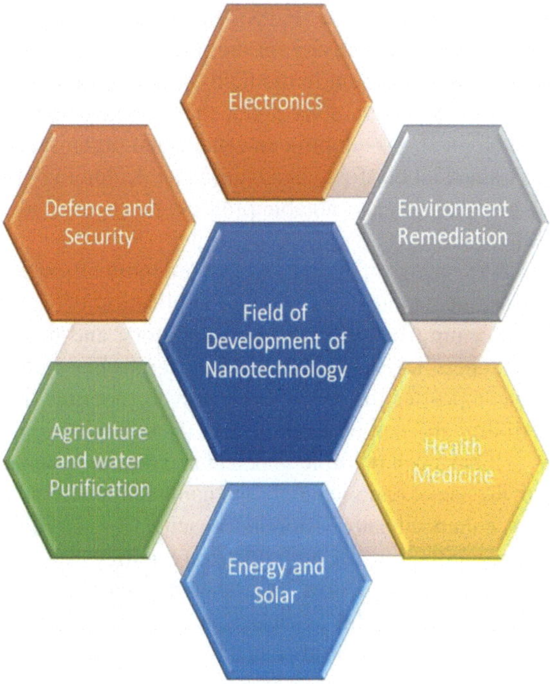

Fig. 2.6 Development of nanotechnology in different sectors

that enhance durability and sustainability. Research efforts focus on substituting traditional materials with lighter composites, incorporating nanoparticles to enhance material properties, and using nano-sized additives in concrete to enhance strength and resistance to abrasion. Nanotechnological advancements also target improving fuel combustion efficiency, reducing emissions, and enhancing the performance of fuel cells in automotive applications (Fig. 2.6).

India: The most prominent developing nation in the discipline of nanotechnology is India. In the previous 20 years, India has made significant progress in the fabrication of novel nanomaterials. The government of India has started several schemes in order to increase the contribution of researchers and scientists to innovations in the field of nanotechnology. Although resources such as petroleum and pristine water are necessary for life, their utilization is decreasing because of pollutants. Carbon nanotube filters, which extract large amounts of hydrocarbons from petroleum and impurities from water, were used in the cleansing process of the resources. Researchers at IIT Delhi fabricated a novel device called nanosensors to estimate the amount of mercury in water. India has also contributed to the invention of smart hydrogel nanoparticles, which serve as coatings for water-soluble drugs, cardiac pacemakers, and a few drugs synthesised from nanomaterials to treat psoriasis and tuberculosis.

Indian scientists have developed nanoparticles and nanocomposites that have the potential to increase the capacity of solar power cells and batteries as well as their ability to store energy. DST established a new programme named "Interdisciplinary

Cyber Physics Systems (ICPS)" that utilises computer-based techniques to monitor and regulate user activity on the internet. The government of India has shown an increase in investment in the field of nanotechnology since 2001 [89]. The projects related to nanotechnology divide their funding into two distinct areas: scientific and technological. Developing an established organisational foundation, infrastructure, and trained labour force to advance nanotechnology has been the primary priority of the government, and notable progress has been made in accomplishing these goals [90].

USA: Electrical appliances that are more efficient and compact are now possible because of nanotechnology. The application of nanoscale materials like graphene and quantum dots in batteries, transistors, and memory storage has been studied. Researchers have looked into nanomaterials like carbon nanotubes (CNTs) and some nanocomposites that show exceptional properties like durability, conductivity, and flexibility and can be employed in the automotive, aerospace, and building sectors [91]. It has been possible to detect and diagnose some rare diseases, such as cancer, with the use of nanomaterials and nanomedicine. Energy storage, solar energy generation, enhanced battery storage capacity, fuel cell modification, and novel renewable energy production all involve nanotechnology. According to A.K. Akhnoukh, carbon nanotube nanostructures have shown potential for being sustainable and durable building materials for construction endeavours. They stated that the manufacturing sector of America has advanced its use of nanotechnology and that ongoing studies and commercial initiatives involving the introduction of nanoparticles into building materials like cement have resulted in benefits like increased durability, reduced flexibility, and high tensile strength [92].

Britain: An important breakthrough in the fields of technology and medicine has been the production of remarkable nanomaterials such as graphene, which are widely recognised for their exceptional qualities. The first nanoparticles of graphene were initially fabricated at the University of Manchester (US). In addition to improvements in personalised therapeutics and targeted drug administration, nanoelectronics advancements like quantum computing and nanoscale transistors have led to smaller, quicker, more effective, and more reliable electronic equipment. To increase sustainability, the potential of nanotechnology in environmental applications has also been investigated. These applications include water purification and renewable energy.

Europe: Scientists are making efforts to create novel nanomaterials that are robust, effective catalysts, and have a wide range of uses in sensors. The development of nanoelectronics indicates the way towards the development of gadgets that are extremely rapid, small, and energy-efficient [93]. The advancement of nanotechnology has yielded significant advantages for the healthcare industry, including the delivery of drugs, personalised medicine, cancer therapy, wearable technology, imaging apparatus, and transplantation. Eco-friendly manufacturing was developed in Europe with the introduction of environmental safety legislation. Applications of nanotechnology have proliferated in the fields of robotics, energy storage, automotive, and aerospace.

Development in nanotechnology provided a major benefit for personalised treatment according to the symptoms of each patient, and RNA-based therapies were discovered to treat diseases or disorders associated with genes [94]. The most dangerous cancer disease needs to be identified in its early stages. Immunotherapies were introduced to fight against cancer cells by boosting the body's immunity. Nanoparticles were explored to improve the efficacy and stability of vaccines. Europe introduced safety regulations for the environment to develop eco-friendly manufacturing. Nanotechnology applications expanded across automotive, in which lightweight but strong nanomaterials were used for the construction of vehicles that led to minimum emissions and maximum efficiency of fuels. Nano-coatings were introduced to boost longevity and create scratch-free surfaces. In aerospace, the sensors were discovered for precise monitoring of the aircraft. Nanotechnology was used to improve the performance of batteries and energy storage devices and increase the efficiency of fuel cells [95]. Robotic equipment is the biggest example of the development of nanotechnology because of its sensing and responsive capability.

Japan: Japan has been deeply involved in nanotechnology research and development, making significant contributions across electronics, medicine, materials science, and other fields. Japanese advancements include smaller and more efficient electronic devices, nanomedicine for precise healthcare applications, novel nanomaterials, government-backed initiatives supporting research, and active international collaboration in the field. These efforts underline Japan's leadership and innovation in the realm of nanotechnology. They focused on contributing to enhancing sustainability and energy efficiency by developing new materials with unique properties. Nanotechnology has a great impact on nanomedicine and precise healthcare applications for the regeneration of medicine, RNA (mRNA and siRNA) therapies, and the treatment of neurological disorders by neurological therapies. Novel nanomaterials, government-backed initiatives supporting research, and active international collaboration in the field. These efforts have a crucial effect on Japan's leadership and innovation in the realm of nanotechnology.

Russia: Russia has actively engaged in nanotechnology research and development for a considerable period. The country has established various initiatives and institutions aimed at advancing this field. RUSNANO, established in 2007, was a pivotal entity in promoting nanotechnology by supporting research and development projects across multiple sectors like medicine, energy, electronics, and materials science. In healthcare, the nanoparticles were used for specific drug delivery, therapies to treat cancer without any side effects on healthy tissues, and image technologies for sensitive and more accurate medical imaging. In electronics, space, and satellite technologies, Russia has a very strong impact on the world. Russia has actively contributed to the field of renewable energy and environmental issues with reduced pollution, waste management, conservation, and biodiversity.

China: China is also one of the developed nations that has made significant advancements in the realm of nanotechnology. It has shown remarkable progress for two decades in various sectors, such as electronics, medicine, and energy, where nanotechnology played a pivotal role. China has emerged as a global leader in nanotechnology, driving innovation and economic development. In nanoelectronics, the development of 5G, an advanced communication system, works globally. In nanotechnology, China has worked heavily on nanomedicines such as regenerative

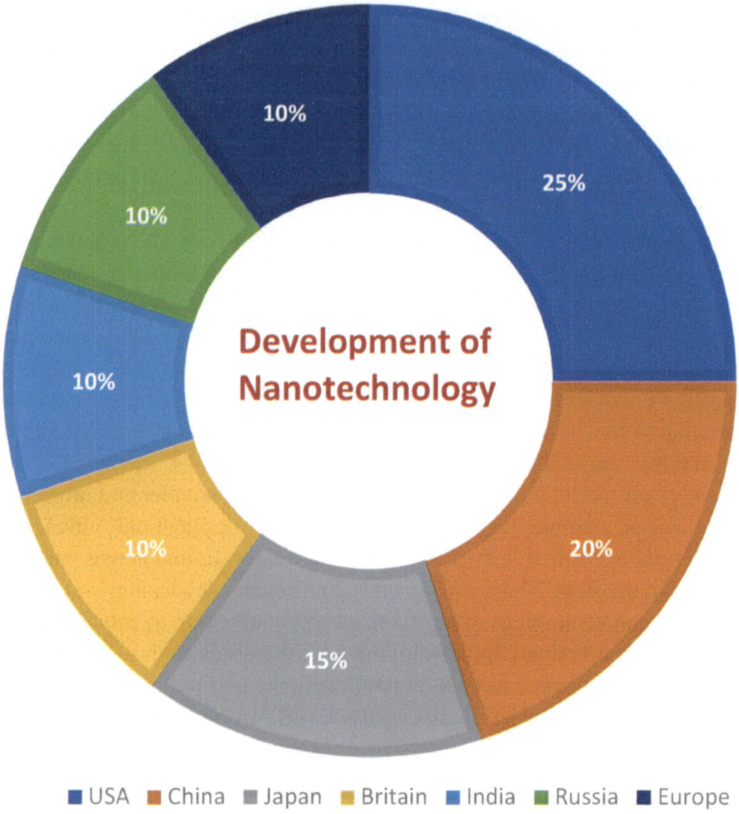

Fig. 2.7 The development of nanotechnology over two decades in these countries (% approximately) can be represented by a pie chart

medicine, genomics, precision medicine, and Traditional Chinese Medicine (TMC). The vaccine development for very infectious diseases like COVID-19 was invented in public health (Fig. 2.7).

Essence of Privacy

Privacy in nanotechnology refers to the protection of an individual's personal information and the ethical use of nanotechnological applications that may have privacy implications. Nanotechnology involves manipulating materials and devices at the nanoscale, often with dimensions on the order of a billionth of a metre. Nanotechnology has the potential to significantly transform our understanding of individual privacy. As nanotechnology makes computing capabilities smaller and more efficient, the collection, storage, sharing, and processing of large amounts of information become easier and more affordable [96]. This advancement also enhances surveillance devices and may lead to an increase in incentives for private companies producing security nanotechnology. Examining the potential privacy risks of nanotechnology requires an understanding of the technical and legal barriers that will shape its use [97]. In the past, Congress has preemptively addressed potential risks by passing legislation before the introduction of new products. For instance, anticipating two-way cable, Congress passed the Cable Communications Policy Act in 1984 to protect individual data with strong opt-in provisions [98]. Similarly, in 1991, the Telephone Consumer Protection Act prohibited telemarketing on cellular phones before it became widespread [99]. To address potential threats to the environment, health, safety, public welfare, and privacy posed by nanotechnology, Congress should enact legislation in advance of its adoption [100]. Acting early allows for the prevention of problems, as post-implementation corrections may be limited with this technology. Historically, privacy-protective legislation aims not to limit technology use but to ensure fair, transparent, and law-abiding data collection. This principle should apply to new nanotech devices designed for surveillance purposes, ensuring they adhere to legal standards. Nanotechnology introduces new surveillance devices that may invade individual privacy. Legislation should not necessarily prohibit surveillance but ensure it aligns with the Fourth Amendment. The Kyllo v. United States case sets a precedent by deeming thermal imaging of a home without a warrant a Fourth Amendment violation [101]. This principle should apply to potential nanotech surveillance devices, preserving established notions of privacy. One significant privacy risk related to nanotechnology is the potential implantation of microchips into humans [102]. While these chips offer benefits such as drug dispensing and assisting Alzheimer's patients, there's a risk of misuse.

Legislation should be in place to ensure responsible use, preventing potential applications in monitoring incarcerated individuals, those on parole, public assistance recipients, school-aged children, employees, and spouses. The main privacy concern with nanotechnology is its potential for intrusive surveillance, termed "nano-panopticism". Researchers propose establishing regulatory agencies for nanotechnology issues, rewriting industry ethical codes to include privacy measures, and increasing scientists' awareness of societal impacts. To protect individual privacy from emerging nanotech inventions, educating legislators on potential intrusions is crucial. Regulations instituted before privacy invasions can incorporate ethical considerations into decision-making processes. Including social scientists and ethicists in discussions on nanotechnology, especially in publicly supported research programmes, can ensure a more comprehensive evaluation of ethical implications.

In "Minding the Gap: Science and Ethics of Nanotechnology," the authors suggest managing the privacy and social implications of nanotechnology by dedicating a percentage of research funding to studying ethical, legal, and social implications. Strengthening research capacity at all levels and encouraging public involvement in discussions about nanotechnology's ethical and social implications are also recommended. In conclusion, nanotechnology's potential impact on privacy necessitates proactive legislative measures, ethical considerations, and public awareness initiatives. Balancing technological advancements with privacy protection is essential to ensuring the responsible and ethical development and use of nanotechnology in society.

Future Aspects

The present situation in the field of research and development in laboratories, academic institutions, industrial sectors, and commerce will undoubtedly yield a plethora of innovative goods. The fact that more and more significant scientific investigations have been sponsored in developing as well as developed countries indicates how much interest there is in this field on a global scale. Every developed nation focuses on highly skilled researchers and technologists with effective education who can suggest the importance and applications of nanomaterials to their governments and institutions. For the sake of humanity as a whole, nanotechnology possesses the ability to transform current social frameworks. The resources, drinking water, electricity, and healthcare sectors are among the areas where it can improve human efficiency and promote sustainable growth. Based on the demands of society, nanotechnology research and development (R&D) is predicted to contribute to novel and scientific innovations by 2030. Several current issues, like sustainable development, power conservation and generation, preservation, and transformation, will drive the transformation of nanotechnology from laboratories and institutes to commercial applications. The coming time will see the development of new innovations and new technology as a result of application-driven research that delivers fresh results from science and economic efficiency. Though new methods for responsible, proactive, collaborative leadership and real-time technological evaluation will be needed, these transformations will be advantageous to civilization.

Conclusion

In this chapter, we primarily focused on the developments and innovations of nanotechnology, which have transformed the electronic industry; however, it also poses several challenges. With several concerns like safety, ethical values, culture, and privacy, researchers should have to explore nanotechnology in more sectors. Even more effective, environmentally friendly, and smoothly integrated electronic devices are promised for the future as a result of ongoing studies and advancements in nanotechnology. The purpose of this section is to study and fabricate novel techniques that may considerably improve the discipline of nanotechnology. Recent advancements in nanoparticle technology have opened up exciting possibilities for designing nanosensors and other devices that are beneficial for human beings. A summary of nanotechnology, privacy, cultural value, and its many applications is presented in this chapter. The systematic use of different nanotechnologies can greatly enhance the quality of life for humans.

Acknowledgments The authors acknowledge their gratitude to the Physics department, at the University of Lucknow, for their qualitative support.

Author Contribution The initial and original content of this chapter was prepared and written by Dr. Peramjeet Singh, Dr. Priya Gupta, and Ms. Neetu Yadav. Prof. Narendra Kumar Pandey and Dr. Vernica Verma help with editing, reviewing, and rewriting.

Funding No specific grants from governmental, commercial, or nonprofit funding organizations were given for this research.

Conflict of Interest The authors declare that they have no known financial conflicts of interest or significant personal connections that could have influenced the findings of the study discussed in this chapter.

References

1. Roco, M.C.: The long view of nanotechnology development: the National Nanotechnology Initiative at 10 years. J. Nanopart. Res. **13**, 427–445 (2011). https://doi.org/10.1007/s11051-010-0192-z
2. Bayda, S., Adeel, M., Tuccinardi, T., Cordani, M., Rizzolio, F.: The history of nanoscience and nanotechnology: from chemical-physical applications to nanomedicine. Molecules **25**(1), 112 (2020). https://doi.org/10.3390/molecules25010112
3. Khan, K., Rehman, S.: Wonders of nanotechnology. Nanomater. Fascin. Attrib. **1**, 3 (2016)
4. Albrecht, M., Janke, V., Sievers, S., Siegner, U., Schuler, D., Heyen, U.: Scanning force microspy study of biogenic nanoparticles for medical applications. J. Magn. Magn. Mater. 290, 269–271 (2005). https://doi.org/10.1016/j.jmmm.2004.11.206
5. Wellner, G.: A Postphenomenological Inquiry of Cell Phones: Genealogies, Meanings, and Becoming. Lexington Books (2015)
6. Pathan, S.A., Iqbal, Z., Zaidi, S., Talegaonkar, S., Vohra, D., Jain, G.K., Azeem, A., Jain, N., Lalani, J.R., Khar, R.K., Ahmad, F.J.: CNS drug delivery systems: novel approaches. Recent Pat. Drug Deliv. Formulation **3**(1), 71–89 (2009)
7. Mohseni, M., Ramezanzadeh, B., Yari, H., Gudarzi, M.M.: The role of nanotechnology in automotive industries. In: New Advances in Vehicular Technology and Automotive Engineering, pp. 3–54 (2012)
8. Takemura, M.: Activities on Social Acceptance of Nanotechnology. NISTEP Science & Technology Foresight Center (2008)
9. Malik, S., Muhammad, K., Waheed, Y.: Nanotechnology: a revolution in modern industry. Molecules **28**(2), 661 (2023)
10. Zahra, Z., Habib, Z., Hyun, S., Sajid, M.: Nanowaste: Another future waste, its sources, release mechanism, and removal strategies in the environment. Sustainability **14**(4), 2041 (2022). https://doi.org/10.3390/su14042041
11. Srivastava, V., Gusain, D., Sharma, Y.C.: Critical review on the toxicity of some widely used engineered nanoparticles. Ind. Eng. Chem. Res. **54**(24), 6209–6233 (2015). https://doi.org/10.1021/acs.iecr.5b01610
12. Bystrzejewska-Piotrowska, G., Golimowski, J., Urban, P.L.: Nanoparticles: their potential toxicity, waste and environmental management. Waste Manage. **29**(9), 2587–2595 (2009). https://doi.org/10.1016/j.wasman.2009.04.001

13. Mondal, S., Palit, D.: Challenges in natural resource management for ecological sustainability. In: Natural Resources Conservation and Advances for Sustainability, pp. 29–59 (2022). https://doi.org/10.1016/B978-0-12-822976-7.00004-1
14. Briffa, J., Sinagra, E., Blundell, R.: Heavy metal pollution in the environment and their toxicological effects on humans. Heliyon **6**(9), e04691 (2020). https://doi.org/10.1016/j.heliyon.2020
15. Ge, Q., Su, J., Chung, T.S., Amy, G.: Hydrophilic superparamagnetic nanoparticles: synthesis, characterization, and performance in forward osmosis processes. Ind. Eng. Chem. Res. **50**(1), 382–388 (2011). https://doi.org/10.1021/ie101013w
16. Seaton, A., Tran, L., Aitken, R., Donaldson, K.: Nanoparticles, human health hazard and regulation. J. R. Soc. Interface **7**(suppl_1), S119–S129 (2010). https://doi.org/10.1098/rsif.2009.0252.focus
17. Hoyt, V.W., Mason, E.: Nanotechnology: emerging health issues **15**(2), 10–15 (2008). https://doi.org/10.1016/j.jchas.2007.07.015
18. Mnyusiwalla, A., Daar, A.S., Singer, P.A.: 'Mind the gap': science and ethics in nanotechnology. Nanotechnology **14**(3), R9 (2003)
19. Ray, P.C., Yu, H., Fu, P.P.: Toxicity, and environmental risks of nanomaterials: challenges and future needs. J. Environ. Sci. Health C **27**(1), 1–35 (2009)
20. Owen, R., Handy, R.: Formulating the problems for environmental risk assessment of nanomaterials, 5582–5588 (2007)
21. Chrishtop, V.V., Prilepskii, A.Y., Nikonorova, V.G., Mironov, V.A.: Nanosafety vs. nanotoxicology: adequate animal models for testing in vivo toxicity of nanoparticles. Toxicology **462**, 152952 (2021)
22. Snir, R.: Governance by disclosure: transnational convergence in the field of nanotechnology. Trans. Environ. Law **2**(1), 69–94 (2013)
23. Allan, J., Belz, S., Hoeveler, A., Hugas, M., Okuda, H., Patri, A., Rauscher, H., Silva, P., Slikker, W., Sokull-Kluettgen, B., Tong, W.: Regulatory landscape of nanotechnology and nanoplastics from a global perspective. Regul. Toxicol. Pharmacol. **122**, 104885 (2021)
24. Rathore, A., Mahesh, G.: Public perception of nanotechnology: A contrast between developed and developing countries. Technol. Soc. **67**, 101751 (2021)
25. Nasrollahzadeh, M., Sajadi, S.M., Sajjadi, M., Issaabadi, Z.: Applications of nanotechnology in daily life. Interface Sci. Technol. **28**, 113–143 (2019)
26. Alzoubi, L., Aljabali, A.A., Tambuwala, M.M.: Empowering precision medicine: the impact of 3d printing on personalized therapeutic. AAPS PharmSciTech **24**(8), 228 (2023)
27. Retzbach, A., Marschall, J., Rahnke, M., Otto, L., Maier, M.: Public understanding of science and the perception of nanotechnology: the roles of interest in science, methodological knowledge, epistemological beliefs, and beliefs about science. J. Nanopart. Res. **13**, 6231–6244 (2011)
28. Todaro, N.M., Gusmerotti, N.M., Daddi, T., Frey, M.: Do environmental attitudes affect public acceptance of key enabling technologies? Assessing the influence of environmental awareness and trust on public perceptions about nanotechnology. J. Clean. Prod. **387**, 135964 (2023)
29. Kalia, A.: Nanotechnology in bioengineering: transmogrifying plant biotechnology, In: Omics Technologies and Bio-Engineering, pp. 211–229. Academic Press (2018). https://doi.org/10.1016/B978-0-12-815870-8.00012-7
30. Ray, S.S., Bandyopadhyay, J.: Nanotechnology-enabled biomedical engineering: current trends, future scopes, and perspectives. Nanotechnol. Rev. **10**(1), 728–743 (2021). https://doi.org/10.1515/ntrev-2021-0052
31. Mohammed, I.A., Bankole, M.T., Abdulkareem, A.S., Afolabi, A.S., Kariim, I., Abubakre, O.K.: Nanotechnology applications in national defence: a review
32. Agarwal, S., Mahto, S., Agarwal, R.C.: Strengthening the growth of Indian defence by harnessing nanotechnology-a prospective. Defence Sci. J. **63**(1), 46 (2013)
33. Kumar, S., Nehra, M., Dilbaghi, N., Marrazza, G., Hassan, A.A., Kim, K.H.: Nano-based smart pesticide formulations: emerging opportunities for agriculture. J. Control. Release **294**, 131–153 (2019)

34. Verma, D.K., Sharma, A., Awasthi, L., Singh, H., Kumar, P., Rajput, P., Sinha, A., Chaubey, K.K., Kumar, A., Rai, N., Bachheti, R.K.: Recent development and importance of nanoparticles in disinfection and pathogen control. In: Nanomaterials for Environmental and Agricultural Sectors, pp. 83–106. Springer Nature, Singapore (2023)
35. Peters, R., Brandhoff, P., Weigel, S., Marvin, H., Bouwmeester, H., Aschberger, K., Rauscher, H., et al.: Inventory of Nanotechnology applications in the agricultural, feed and food sector. EFSA Support. Publ. **11**(7), 621E (2014)
36. Wasilewska, A., Bielicka, M., Klekotka, U., Kalska-Szostko, B.: Nanoparticle applications in food–a review. Food Funct. (2023)
37. Ganau, M., Paris, M., Syrmos, N., Ganau, L., Ligarotti, G.K.I., Moghaddamjou, A., Prisco, L., Ambu, R., Chibbaro, S.: How nanotechnology and biomedical engineering are supporting the identification of predictive biomarkers in neuro-oncology. Medicines **5**(1), 23 (2018). https://doi.org/10.3390/medicines5010023
38. Zhao, W., Karp, J.M., Ferrari, M., Serda, R.: Bioengineering nanotechnology: towards the clinic. Nanotechnology **22**(49) (2011). https://doi.org/10.1088/0957-4484/22/49/490201
39. Ali, A., Sinha, K.:Prospects of nanotechnology development in the health sector in India. Int. J. Health Sci. **2**(2), 109–125 (2014)
40. Kandil, M.: The Role of Nanotechnology in Electronic Properties of Materials. ENRRA, Cairo (2016)
41. Sumereder, C., Muhr, M.: The prospects of nanotechnology in electrical power engineering. In: 19th International Conference on Electrical Distribution, pp. 1–4 (2007)
42. Contreras, J.E., Rodriguez, E.A., Taha-Tijerina, J.: Nanotechnology applications for electrical transformers—a review. Electr. Power Syst. Res. **143**, 573–584 (2017). https://doi.org/10.1016/j.epsr.2016.10.058
43. Myers, S.S., Patz, J.A.: Emerging threats to human health from global environmental change. Annu. Rev. Environ. Resour. **34**, 223–252 (2009)
44. Kim, D.Y., Kadam, A., Shinde, S., Saratale, R.G., Patra, J., Ghodake, G.: Recent developments in nanotechnology transforming the agricultural sector: a transition replete with opportunities. J. Sci. Food Agric. **98**(3), 849–864 (2018)
45. Baranowska, N.N.: Public authority liability and the regulation of nanotechnology: a european perspective. Can. J. Law Technol. **16**(1), 3 (2018)
46. Mishra, M., Dashora, K., Srivastava, A., Fasake, V.D., Nag, R.H.: Prospects, challenges and need for regulation of nanotechnology with special reference to India. Ecotoxicol. Environ. Saf. **171**, 677–682 (2019). https://doi.org/10.1016/j.ecoenv.2018.12.085
47. O'Brien, N., Cummins, E.: Recent developments in nanotechnology and risk assessment strategies for addressing public and environmental health concerns. Hum. Ecol. Risk Assess. Int. J. **14**(3), 568–592 (2008). https://doi.org/10.1080/10807030802074261
48. Sahu, T., Ratre, Y.K., Chauhan, S., Bhaskar, L.V.K.S., Nair, M.P., Verma, H.K.: Nanotechnology based drug delivery system: current strategies and emerging therapeutic potential for medical science. J. Drug Deliv. Sci. Technol. **63**, 102487 (2021)
49. Alshawwa, S.Z., Kassem, A.A., Farid, R.M., Mostafa, S.K., Labib, G.S.: Nanocarrier drug delivery systems: characterization, limitations, future perspectives and implementation of artificial intelligence. Pharmaceutics **14**(4), 883 (2022)
50. Chandrakala, V., Aruna, V., Angajala, G.: Review on metal nanoparticles as nanocarriers: current challenges and perspectives in drug delivery systems. Emergent Mater. **5**(6), 1593–1615 (2022)
51. Pormohammad, A., Monych, N.K., Ghosh, S., Turner, D.L., Turner, R.J.: Nanomaterials in wound healing and infection control. Antibiotics **10**(5), 473 (2021)
52. Nasrollahzadeh, M., Sajadi, S.M., Sajjadi, M., Issaabadi, Z.: An introduction to nanotechnology. In: Interface science and technology, vol. 28, pp. 1–27. Elsevier (2019)
53. Deng, Z., Li, W., Dong, W., Sun, Z., Kodikara, J., Sheng, D.:Multifunctional asphalt concrete pavement toward smart transport infrastructure: design, performance and perspective. Compos. Part B Eng., 110937 (2023)

54. Fu, W., Turcheniuk, K., Naumov, O., Mysyk, R., Wang, F., Liu, M., Kim, D., Ren, X., Magasinski, A., Yu, M., Feng, X.: Materials and technologies for multifunctional, flexible or integrated supercapacitors and batteries. Mater. Today **48**, 176–197 (2021)
55. Khan, J., Khan, A., Rubab, B., Jamshaid, F., Al-Kahtani, A.A., Dahshan, A.: Exploring the progression of energy storage toward flexibility: metal-organic framework and conducting polymer aspects. Appl. Mater. Today **34**, 101906 (2023)
56. Cai, Y., Wang, Y., Cheng, L., Guo, S., Liu, T., Hu, Z., Yu, H., Chen, D., Li, Y., Yuan, H.: Structure design and assembly mode of carbon nanotube-based flexible electrode materials and flexible supercapacitors. J. Energy Storage **73**, 109179 (2023)
57. Yu, M., Peng, Y., Wang, X., Ran, F.: Emerging design strategies toward developing next-generation implantable batteries and supercapacitors. Adv. Funct. Mater., 2301877 (2023)
58. Abbott, A., Aldous, L., Borisenko, N., Coles, S., Fontaine, O., Garcia, J.D.G., Gardas, R., Hammond, O., Hardwick, L.J., Haumesser, P.H., Hausen, F.: Electrochemistry: general discussion. Faraday Discuss. **206**, 405–426 (2018)
59. Landa-Acuna, D., Acosta, R.A.S., Hualpa Cutipa, E., Vargas de la Cruz, C., Luis Alaya, B.: Bioremediation: a low-cost and clean-green technology for environmental management. Microb. Bioremediat. Biodegradation, 153–171 (2020)
60. Marcon, L., Oliveras, J., Puntes, V.F.: In situ nanoremediation of soils and groundwaters from the nanoparticle's standpoint: a review. Sci. Total. Environ. **791**, 148324 (2021)
61. Karn, B., Kuiken, T., Otto, M.: Nanotechnology and in situ remediation: a review of the benefits and potential risks. Environ. Health Perspect. **117**(12), 1813–1831 (2009)
62. Perez, H., Quintero Garcia, O.J., Amezcua-Allieri, M.A., Rodriguez Vazquez, R.: Nanotechnology as an efficient and effective alternative for wastewater treatment: an overview. Water Sci. Technol. **87**(12), 2971–3001 (2023)
63. LeRoy, P.: Troubled waters: population and water scarcity. Colo. J. Int'l Envtl. L. Pol'y **6**, 299 (1995)
64. Malik, S., Muhammad, K., Waheed, Y.: Emerging applications of nanotechnology in healthcare and medicine. Molecules **28**(18), 6624 (2023)
65. Jing, L., Zhou, W., Tian, G., Fu, H.: Surface tuning for oxide-based nanomaterials as efficient photocatalysts. Chem. Soc. Rev. **42**(24), 9509–9549 (2013)
66. Haleem, A., Javaid, M., Singh, R.P., Rab, S., Suman, R.: Applications of nanotechnology in medical field. Global Health J. (2023)
67. Katel, S., Upadhyay, K., Mandal, H.R., Yadav, S.P.S., Kharel, A., Rijan, R.: Nanotechnology for agricultural transformation: a review. Fundam. Appl. Agric. **6**(4), 403–414 (2021)
68. Nungula, E.Z., Mugwe, J., Nasar, J., Massawe, B.H., Karuma, A.N., Maitra, S., Seleiman, M.F., Dindaroglu, T., Khan, N., Gitari, H.I.: Land degradation unmasked as the key constraint in sunflower (Helianthus annus) production: role of GIS in revitalizing this vital sector. Cogent Food Agric. **9**(2), 2267863 (2023)
69. Haydar, M.S., Ghosh, D., Roy, S.: Slow and controlled release nanofertilizers as an efficient tool for sustainable agriculture: recent understanding and concerns. Plant Nano Biol., 100058 (2024)
70. Naseem, T., Durrani, T.: The role of some important metal oxide nanoparticles for wastewater and antibacterial applications: a review. Environ. Chem. Ecotoxicol. **3**, 59–75 (2021)
71. Gedda, G., Balakrishnan, K., Devi, R.U., Shah, K.J., Gandhi, V., Gandhi, V., Shah, K.L.: Introduction to conventional wastewater treatment technologies: limitations and recent advances. Mater. Res. Found **91**, 1–36 (2021)
72. Maroufi, N., Hajilary, N.: Nanofiltration membranes types and application in water treatment: a review. Sustain. Water Resour. Manag. **9**(5), 142 (2023)
73. Corsi, I., Venditti, I., Trotta, F., Punta, C.: Environmental safety of nanotechnologies: the eco-design of manufactured nanomaterials for environmental remediation. Sci. Total Environ. **864**, 161181 (2023)
74. Shrestha, B., Tang, L., Hood, R.L.: Nanotechnology for personalized medicine. In: Nanomedicine, pp. 555–603. Springer Nature, Singapore (2023)

75. Erol, M., Mustafov, S.D., Akalın, S.A., Uzunbayır, B., Ozdemir, E.T., Coskun, B., Ertekin, Z.: Carbon nanostructures for automotive and aerospace applications. In: Handbook of Functionalized Carbon Nanostructures: From Synthesis Methods to Applications, pp. 1–29. Springer International Publishing, Cham (2023)
76. Prajitha, V., Jibin, K.P., Abitha, V.K., Sisanth, K.S., Huskic, M., Meera, A.P., George, J.S., Thomas, S.: Advancing mechanical performance in sustainable engineering: Synergistic effects of graphene oxide/nano-silica hybrid nanofiller via latex coagulation in natural rubber composites. Polym. Compos. (2024)
77. Kabeyi, M.J.B., Olanrewaju, O.A.: Sustainable energy transition for renewable and low carbon grid electricity generation and supply. Front. Energy Res. **9**, 1032 (2022)
78. Mishra, R.K., Chianella, I., Goel, S., Nezhad, H.Y.: Nanomaterial integration in micro led technology: enhancing efficiency and applications
79. Singh, B.P., Sharma, K., Tyagi, S., Gautam, D., Chaudhary, M., Kumar, A., Vikal, S., Gautam, Y.K.: Green and sustainable technology for clean energy production: applications. In Handbook of Green and Sustainable Nanotechnology: Fundamentals, Developments and Applications, pp. 1–23. Springer International Publishing, Cham (2023)
80. Fernandez-Gonzalez, D.:A state-of-the-art review on materials production and processing using solar energy. Miner. Process. Extr. Metall. Rev., 1–43 (2023)
81. Kabir, M.S.N., Reza, M.N., Chowdhury, M., Ali, M., Samsuzzaman, Ali, M.R., Lee, K.Y., Chung, S.O.: Technological trends and engineering issues on vertical farms: a review. Horticulturae **9**(11), 1229 (2023)
82. Lu, B.: Fundamental Research on Nanomanufacturing. Springer Nature (2023)
83. Altmann, J.: Military uses of nanotechnology: perspectives and concerns. Secur. Dialogue **35**(1), 61–79 (2004)
84. Hou, C., Tai, G., Liu, Y., Wu, Z., Liang, X., Liu, X.: Borophene-based materials for energy, sensors and information storage applications. Nano Res. Energy 2(2), e9120051 (2023)
85. Greenwood, G., Kim, J.M., Nahid, S.M., Lee, Y., Hajarian, A., Nam, S., Espinosa-Marzal, R.M.: Dynamically tuning friction at the graphene interface using the field effect. Nat. Commun. **14**(1), 5801 (2023)
86. Plummer, J.D., Griffin, P.B.: Integrated Circuit Fabrication: Science and Technology. Cambridge University Press (2023)
87. Gong, S., Lu, Y., Yin, J., Levin, A., Cheng, W.: Materials-driven soft wearable bioelectronics for connected healthcare. Chem. Rev. **124**(2), 455–553 (2024)
88. Kimura, T., Uemura, T., Shibafuji, Y., Suyama, T., Ueno, H., Sekitani, T.: Biosignal amplifiers based on low-noise organic transistors with printed electrodes. Adv. Electr. Mater., 2201279 (2023)
89. Saikanth, D.R.K., Gawande, V., Chaithanya, G., Mishra, S., Swami, S., Singh, B.V., MP, H.K.: Development and current state of nanotechnology in the agricultural and food sectors: a review. Int. J. Environ. Clim. Change **13**(9), 1602–1612 (2023)
90. Beumer, K., Bhattacharya, S.: Emerging technologies in India: developments, debates, and silences about nanotechnology. Sci. Public Policy **40**(5), 628–643 (2013). https://doi.org/10.1093/scipol/sct016
91. Khan, I., Saeed, K., Khan, I.: Nanoparticles: properties, applications, and toxicities. Arab. J. Chem. **12**(7), 908–931 (2019)
92. Akhnoukh, A.K.: Overview of nanotechnology applications in construction industry in the United States. Micro Nanosyst. **5**(2), 147–153 (2013)
93. Gargini, P., Balestra, F., Hayashi, Y.: Roadmapping of nanoelectronics for the new electronics industry. Appl. Sci. **12**(1), 308 (2021)
94. Lima, E.S., Santos, D.D., Souza, A.L., Macedo, M.E., Bandeira, M.E., Junior, S.S.S., Fiuza, B.S.D., Rocha, V.P.C., dos Santos Fonseca, L.M., Nunes, D.D.G., Hodel, K.V.S.: RNA combined with nanoformulation to advance therapeutic technologies. Pharmaceuticals **16**(12), 1634 (2023)
95. Mahmud, M.Z.A.: A concise review of nanoparticles utilized energy storage and conservation. J. Nanomater. (2023)

96. Taha, T.B., Barzinjy, A.A., Hussain, F.H.S., Nurtayeva, T.: Nanotechnology and computer science: trends and advances. Memories Mater. Devices Circuits Syst. **2**, 100011 (2022)
97. Zukersteinova, A.: Skill needs in emerging technologies: nanotechnology. Cedefop Thessaloniki GR, 65 (2007)
98. Meyerson, M.I.: The Cable Communications Policy Act of 1984: a balancing act on the coaxial wires. Ga. L. Rev. **19**, 543 (1984)
99. Waller, S.W., Heidtke, D.B., Stewart, J.: The Telephone Consumer Protection Act of 1991: adapting consumer protection to changing technology. Loy. Consum. L. Rev. **26**, 343 (2013)
100. Sargent, J.F.: The National Nanotechnology Initiative: Overview, Reauthorization, and Appropriations Issues. Congressional Research Service (2010)
101. Basha, R.M.: Kyllo v United States: the fourth amendment triumphs over technology. Brandeis LJ **41**, 939 (2002)
102. Almansoori, L.M: A comparative study on microchip implants in humans and wearable devices (2023)

Dr. Peramjeet Singh has completed his Ph.D. from the Department of Physics, University of Lucknow, Uttar Pradesh, India, in the field of material science. He did his M.Sc. from Barkatullah Vishwavidyalaya, Bhopal, and his B.Sc. from GDC, Udhampur, University of Jammu, India, in 2017 and 2015, respectively. His area of research is nanotechnology, gas sensors, and the characterization of materials.

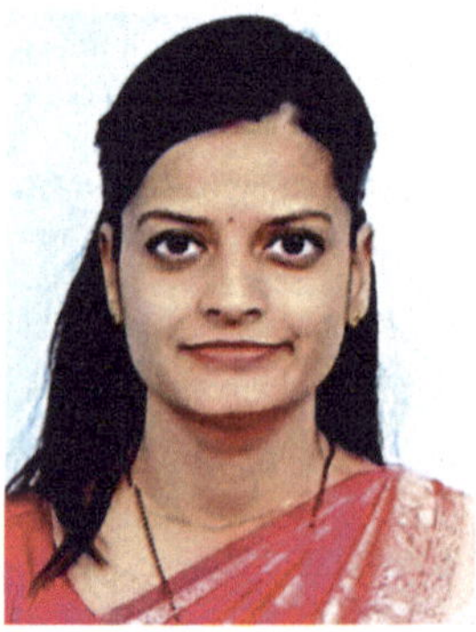

Dr. Priya Gupta has been working as a lecturer at Manyawar Kanshiram Government Polytechnic, Tirwa, Kannauj, since January 2022. She has completed a Ph.D. from Integral University, Lucknow, in collaboration with Lucknow University, Lucknow, in the year 2021. She did her M. Tech. from Madan Mohan Malviya Engineering College, Gorakhpur, in 2014, and her B.Tech. from Integral University, Lucknow.

Dr. Vernica Verma was a former assistant professor at the Department of Physics, BBD, Lucknow, and completed her Ph.D. in Physics from the Department of Physics, University of Lucknow. She did her B.Sc. and M.Sc. in Physics from the University of Lucknow in 2005 and 2007, respectively. Her specialisation in research is fabrication, characterization like XRD, SEM, AFM, UV-VIS, FT-IR, Raman, gamma irradiation, etc., and humidity and gas sensors.

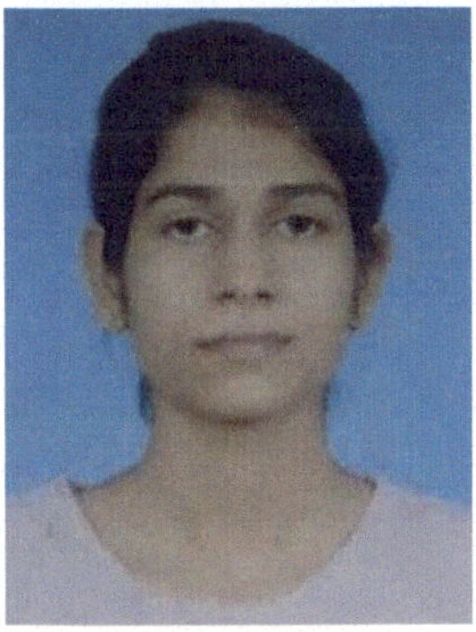

Ms. Neetu Yadav is currently pursuing Ph. D. in Material Science, Department of Physics, University of Lucknow. She did her M.Sc. in Physics from University of Lucknow in 2019 and B. Sc. from Kumaun University in 2017. Her specialisation in research is Fabrication, Characterization like XRD, SEM, AFM, UV-VIS, FT-IR, Raman, Gamma irradiation etc., and humidity and Gas Sensors.

Prof. Narendra Kumar Pandey received his degree (hons.) in Physics from St. Xavier's College, Ranchi, India, in 1985, and the M.Sc. degree in Physics and the Ph.D. degree from the Indian Institute of Technology Delhi, Delhi,India, in 1987 and 1995, respectively. He is currently a professor and head of department with the Department of Physics, University of Lucknow, Lucknow, India. His areas of interest include humidity and gas sensing, nanotechnology, and optical fibre, sensors, and fibre Bragg grating sensors.

Chapter 3
The Promise of Nanotechnology in Healthcare

Raphael Kanyire Seidu, Anna Doersam, Emmanuel Abankwah Ofori, and Shouxiang Jiang

Introduction

The conventional healthcare delivery system provides several challenges such as slow diagnosis, invasive monitoring procedures, non-targeted drug delivery methods, and lack of continuous provision of real-time monitoring data to detect diseases. This has led to several reported errors, late diagnosis, and treatments, a situation that affects the healthcare of patients. To overcome this, advanced technologies have been developed to contribute effectively to the delivery of affordable, good quality, and real-time data for immediate healthcare for people. This technological progress brings rapid diagnosis, prevention, and treatment of diseases. One such significant technological advancement for the healthcare sector is nanotechnology, which in recent years has provided great assistance [1, 2]. Nanotechnology is the overarching field that focuses on manipulating materials at the nanoscale. Hobson [3] reported that nanotechnology as a scientific domain highlights the fabrication of functional

R. K. Seidu (✉) · S. Jiang (✉)
School of Fashion and Textiles, The Hong Kong Polytechnic University, Hung Hom, Kowloon, Hong Kong SAR, China
e-mail: raphael-kanyire.seidu@connect.polyu.hk

S. Jiang
e-mail: kinor.j@polyu.edu.hk

A. Doersam
The Henry Royce Institute, Royce Hub Building, Manchester M13 9PL, UK

Department of Materials, The University of Manchester, Oxford Rd, Manchester M13 9PL, UK

E. A. Ofori
Department of Materials Science and Engineering, Hubei University of Automotive Technology, Shiyan, Hubei, China

S. Jiang
Research Institute for Intelligent Wearable Systems, The Hong Kong Polytechnic University, Kowloon, Hong Kong SAR, China

S. C. George and B. Tawiah (eds.), *Nanotechnology in Societal Development*, Advanced Technologies and Societal Change, https://doi.org/10.1007/978-981-97-6184-5_3

structures at nanoscale, using both chemical and physical techniques. The chemical, biological, surface, physical, mechanical, and electrical properties of nanomaterials have promoted the application of nanotechnology in different aspects of healthcare [4].

These nanomaterials are composed of 1–100 nm scale size for providing a unique building block to engineer several products for effective diagnosis and monitoring of vitals in the human body [5]. These nanomaterials are categorized as 0D (e.g. nanoparticles), 1D (e.g. nanorods, nanotubes), 2D (e.g. graphene), and 3D (e.g. nano prisms) [6] or alternatively in five categories based on their potential toxicity, pore size, origin, structural configuration, and dimension or size as reported in [7]. Further details on the significant role of nanostructures in advancing technology have been reported extensively in [8]. Premise on this, different fabrication methods for electrodes such as screen printing, transfer printing, dip coating, spray coating, inkjet printing [9–13], and nanomaterials (which includes reduced graphene oxide (RGO), carbon nanotubes (CNTs), graphene, graphite among others) [14–16] are fabricated on substrates most especially the surface of textile-based materials for enhanced sensitivity and performance.

With such unique scale and properties of nanomaterials, nanotechnology can provide a significant impact on healthcare such as rapid and accurate diagnosis, non-invasive monitoring procedures, targeted drug delivery methods, and continuous provision of real-time monitoring data to detect diseases [17]. For example, nanotechnology has widened the development of wearable sensing devices for motion detection, electrophysical and pressure sensing, drug delivery, and others in the medical field [18]. The products are used by both the young and the aged for effective diagnosis and treatment of any diseases. It is thus necessary to advance research and provide cost-effective, portable, and non-invasive wearable devices with real-time data for good healthcare delivery to individuals most especially the aged. This is because, as reported in [19], there will be an estimated increase of about 5% of people aged 65 and a 23% increase above age 65 by 2035. These non-invasive wearable devices based on nanotechnology could be used by these aged people without compromising their comfort and movement. However, the rapid rise in the destructive behavior of individuals has resulted in varying health problems and diseases. These have to be monitored and detected in time for effective treatments, a situation nanotechnology through wearable devices has sought to solve.

Wearable based on nanotechnology provides potential significance concerning tracking, recording, and transferring vital health data for healthcare providers for immediate treatment. As a result, diseases are detected at their early stages to help limit recurring deaths. These advancements have been addressed and reported by studies to adopt the appropriate nanomaterials and techniques to fabricate devices for healthcare applications. Herein, this chapter highlights the promise of nanotechnology in healthcare with specific references to nanomedicine, nanobots, nanofibers, and nanotech-based wearables.

Application Areas of Nanotechnology in Healthcare

Nanotechnology holds great promise for revolutionizing healthcare and medicine [20]. Through the use of nanomaterials, products with improved functionality and drastically lower costs are produced in large quantities to improve healthcare and lessen environmental impact [21, 22]. The Ostwald–Freundlich and Noyes–Whiney equations describe the exceptional saturation solubility of those that are less than 100 nm in particular [23]. Importantly, materials with special qualities at the nanoscale level are used in a variety of medical applications, such as medication delivery, therapies, the development of regenerative medicine, and enhanced medical diagnostics [24, 25]. In situations where precision, sensitivity, and selectivity are crucial, their attractiveness is attributed to characteristics including their high surface area-to-volume ratio, quantum effects, and other unusual phenomena [26]. Therefore, in the realm of healthcare, nanoparticles, nanosensors, and nanodevices have demonstrated amazing capabilities. Nanoparticles, for instance, are designed to contain medications and release them at particular bodily locations [22, 27]. This minimizes exposure to healthy tissues while allowing larger medication concentrations to build up in sick tissues. Moreover, efforts have been made to create nanocarriers that can transport medicines for neurological disorders across extremely difficult barriers, such as the blood–brain barrier [28], and in the field of tissue engineering and regenerative medicine, extracellular matrix mimicking nano-scaffolds offer structures on which cells proliferate. Consequently, growth factors can be released through the use of nanofibers and nanoparticles to promote tissue regeneration. Gagliardi et al. [29] claim that to optimize and promote tissue and cell interaction, factors such as charge, size, pattern of nanoscale medical molecules, and shape are modulated. However, nanotechnology has the potential to completely change healthcare in the twenty-first century and beyond with further study and advancement. Therefore, the different healthcare application areas are handled independently in the next section.

Nanomedicine

The field of nanotechnology and nanodrugs has yielded a huge array of novel applications. Remarkable advancements in nanomedicine have raised the bar for medication and improved patient outcomes, particularly in treating inflammatory and degenerative nervous system disorders [30]. As a subfield of nanotechnology, nanomedicine deals with highly targeted molecular medical interventions to treat diseases or restore damaged tissues like muscles, nerves, or bones. This is because when nanoparticles are delivered to the body, they do not trigger an immune response since they are made in a way that makes them non-immunogenic, or compatible with the immune system. On the other hand, nanoparticles work as a medication transporter that can precisely target damaged brain subregions, boosting the effectiveness of treatment [22]. Thus, applying nanotechnology to the discipline of nanomedicine, which focuses on medical diagnostics, can potentially lead to improved healthcare. Nanosensors implanted in the body monitor biomarkers and send health data, while magnetic nanoparticles allow for highly sensitive imaging techniques for early cancer detection [28]. Additionally, biomolecules are precisely analyzed in small quantities in labs-on-a-chip thanks to nanoscale components [31]. Thus, these point-of-care nano diagnostic devices potentially provide rapid test results to improve medical decision-making.

One of the most important and delicate parts of the human body is the nervous system, which is shielded by the blood–brain barrier (BBB), composed of impermeable endothelial cells with tight junctions, enzymatic activity, and active efflux transport systems that regulate homeostasis. Nowadays, breaking through the blood–brain barrier (BBB) to increase medicine efficacy is a major problem in drug development. Examples of substances that have demonstrated the ability to deliver a self-transporting medication into the brain include polysorbate 80/LDL, lactoferrin, and poly (butylcyanoacrylate)-based nanoparticles which are capable of transcytosis via receptors and deliver doxorubicin, hexapeptide, and dalargin, as well as melano transfer. Thus, nanomedicine offers remarkable treatments for a variety of deadly illnesses, including cancer, blood, lung, diabetes, inflammatory and infectious diseases, Parkinson's or Alzheimer's disease, neurological disorders (particularly neurodegenerative diseases), and orthopedic problems [32]. Analyses of numerous unique classes of nanotherapeutics are currently in progress. Similarly, therapeutic items based on nanotechnology have been reviewed in comparison to previously preferred medications [33]. While still an emerging field, nanomedicine offers remarkable opportunities to treat disease, repair tissue damage, and diagnose conditions at extremely early stages.

Drug Delivery

In the healthcare sector, various nanostructures with distinct physiochemical characteristics have been employed to enhance the effectiveness of medication administration to particular targets [22]. The concept of employing "magic bullets" to efficiently target specific sites in medicine dates back a century, to the German scientist Paul Ehrlich. Significantly, scientists have demonstrated that target organs or tissues designated as drug delivery systems may be selectively accumulated using nanoparticles, which is appropriate and practicable. Regarding the use of nanomaterials in drug delivery, the selection of the nanoparticle is based on the physicochemical features of drugs [34]. To synthesize the first sustained-release drug medication systems, drug-containing cores were coated with polymers of different thicknesses in 1952 [35]. However, by altering them, nanoparticles can now efficiently transport medications to specific areas of the human body [22, 36]. The very first generation of nanoparticle-based therapy included lipid systems like liposomes and micelles, which are now FDA-approved [37]. To ensure safe medication release and reduce cytotoxicity, it is imperative to assess a nanoparticle's biocompatibility [36]. This has made it possible to address and cure difficult illnesses, particularly cancers. As a result of the well-known increased penetration and retention (EPR) effect, the drug delivery system utilizing nanoparticles in this instance accumulates more at the tumor than its control formulation.

In addition, PEGylation, the surface well-known increased penetration and retention (EPR) effect, the drug delivery system utilizing nanoparticles in this instance accumulates more at the tumor than its control formulation. coating of poly(ethylene glycol) (PEG), extends the nanoparticles' blood circulation time and enhances their accumulation at the intended site. For example, the biodistribution of paclitaxel in tumor-bearing mice was investigated using paclitaxel nanocrystals containing both tritium-labeled paclitaxel and fluorescent probe molecules, as reported by Hollis et al. [38]. Approximately 1% of the total paclitaxel delivered was observed to have accumulated at the tumor site following intravenous injection via the tail vein. Taxols, a solution formulation, and nanocrystals both displayed comparable concentrations of paclitaxel at the tumor with comparable action. In addition, trigger-responsive structure gadget uses X-ray-generated and magnetic field-activated drug delivery methods [39, 40]. Therapeutic medications can be delivered to a specific region and medication release can be achieved as needed using a variety of stimuli-responsive techniques that include sensitive fragments. Because it increases the effectiveness of cellular compartment absorption and penetration, the properties and topologies of trigger-responsive nanoparticles can be changed by endogenous or external triggers. One way to monitor the release of medication at the location of pain is by using sonosensitizers that are triggered by ultrasound. Notably, drugs such as megestrol, paliperidone, fenofibrate, sirolimus, aprepitant, and other clinically effective medications have been developed using drug nanocrystal technology [41]. To avoid the need to remove the microparticles at the end of their useful lives, long-term microparticle formulations are made of biodegradable polymers. This form of green chemistry route of designing nanoparticles loaded with drugs is widely encouraged as it minimizes the hazardous constituents in the biosynthetic process, lessening the side effects of the medications [42].

Drug Design and Pharmaceuticals

Many therapeutic approaches have been proposed and studied as a result of the advancement of nanomedicine, drug discovery/design, and drug delivery systems. These approaches have benefits such as the ability to alter properties like solubility, drug release profiles, diffusivity, bioavailability, and immunogenicity to improve drug specificity and diagnostic accuracy [43]. Recently, almost every pharmaceutical company has emphasized standard procedures like expressing and cloning human receptors and enzymes to find novel medications. These procedures enable high throughput, automated screening, and the use of combinatorial chemistries [44]. Pharmaceutical nanotechnology is essentially concerned with the creation and growth of small structures, such as molecules, atoms, or compounds, that range in size

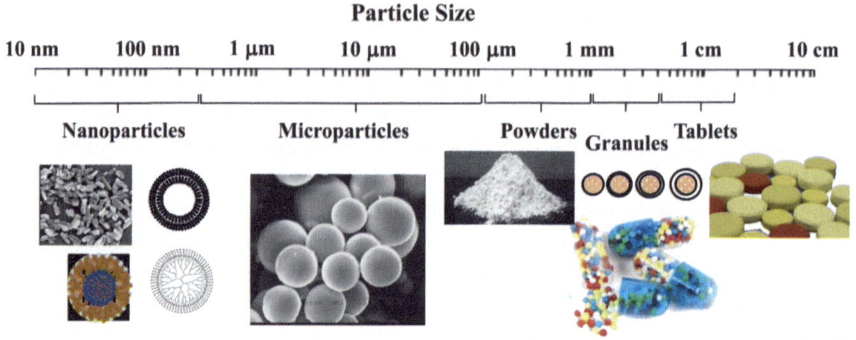

Fig. 3.1 Relative sizes of various pharmaceutical particles ranging from nanoparticles to tablets. Adapted with permission from Ref. [46] Copyright 2015 Elsevier

from 0.1 to 100 nm [45]. These drug particles find application in diverse fields, and their sizes and shapes span from conventional tablets and granules to microparticles and nanoparticles (Fig. 3.1) [46]. Importantly, pharmaceutical nanotechnology helps combat multiple diseases by detecting associated antigens and microbes and virus-causative diseases.

However, many nanoforms have been utilized as drug delivery methods to boost the therapeutic efficacy of drugs, including liposomes, solid lipid nanoparticles, dendrimers, and solid metal-containing nanoparticles [47]. For instance, liposomic medications are used to treat fungal infections, ovarian cancer, Kaposi's sarcoma, breast cancer, and lymphomatous meningitis, in that order. Examples of these medications are amphotericin B, doxorubicin, cytarabine, daunorubicin, camptothecin, and vancomycin [48]. In addition to polymers like drug conjugates, which extend the solubility and shelf life of medications, these particles are also made up of non-polymeric substances including carbon nanotubes, silica nanoparticles, and quantum dots. Most often, polymerics of polyethylene-glycol-based drug delivery systems employ biodegradable nanoparticles, whereas other materials, including carbon fibers, can handle fluid environments [49]. Over the past few decades, the FDA has approved some nano-pharmaceuticals for use in clinical settings. Oral, intravenous, and transdermal administration were used to administer authorized nanodrugs. For example, [50] study summarizes a list of nano-approved medicines in the pharmaceutical industries as presented in Table 3.1. Equally, Table 3.2 highlights several potential treatments that combine nanotechnology and pharmacology [51].

Cancer Treatment

Cancer accounts for one out of every six deaths worldwide in terms of health-related issues [52, 53]. Thus, among all causes of death, cancer is the second most common worldwide [54]. Malignant tumors, on the other hand, are thought to be the

Table 3.1 Summarized studies of nanotechnology in drug design and pharmaceutics [50]

Trade name	Material description	Advantages	Indications	Approved year
Estrasorb™	Micellar estradiol	Controlled drug delivery	Menopausal therapy	2003
Marqibo®	Liposomal vincristine	Increased delivery to tumor tissue, decreased systemic toxicity	Acute lymphoblastic leukemia	2012
Onivyde®	Liposomal irinotecan	Toxicity resulting from side effects	Pancreatic cancer	2015
BNT162b2 vaccine (developed by BioNTech and Pfizer)	Nucleoside-modified mRNA encoding the viral spike glycoprotein of SARS-CoV-2, encapsulated in lipid nanoparticles	Protection of the non-replicating RNA from degradation and allows it to be delivered into host cells after intramuscular injection	Prevention of coronavirus disease	2020
mRNA-1273 vaccine (developed by Moderna)	Lipid nanoparticle-encapsulated mRNA-based vaccine, which encodes the spike protein (S protein) of SARS-CoV-2	Lipid nanoparticles play a role in protecting and transporting the mRNA effectively to the right place in cells	Prevention of coronavirus disease	2020

result of a malfunctioning process that regulates cell growth and proliferation [55], causing damaged cells to divide and multiply uncontrollably. Given that conventional cancer medications, including chemotherapy, have issues with off-target and on-target effects on healthy cells, innovative strategies are therefore required to enhance cell-selective targeting [52, 54, 56]. Due to scientific advancements, it has become clear that administering drugs via nanoparticles kills tumor cells more efficiently and with fewer adverse effects. This has led to research on the use of nanomaterials as chemical biology tools in cancer therapeutics [57] (Fig. 3.2). Through the use of the Enhanced Permeability and Retention (EPR) effect, nanoparticles can be used in treatment to selectively aggregate in tumor tissues. This provides a platform for more effective therapeutic delivery and tailored drug delivery. Similarly, some of the most studied medications for liposome-based drug formulations are anticancer drugs, such as vincristine, paclitaxel, doxorubicin, and daunorubicin; liposomal forms of these drugs are presently being used in cancer therapy thanks to nanomaterials [58]. These compounds have been applied to decrease toxicity to healthy cells, enhance drug transport to cancer cells, and enhance patient outcomes. For example, liposomes exhibit stable, prolonged release of doxorubicin in acidic settings via a pH gradient mechanism. In neutral pH environments, polymeric nanoparticles provide regulated

Table 3.2 Treatment areas where nanotechnology is combined with pharmacology [51]

Treatment	Nanotechnology used	Pharmacology used
Cancer chemotherapy	Drugs can be delivered to cancer cells directly using nanoparticles	Chemotherapy drugs
Cancer	Ultrasound therapy and nanotechnology in medication delivery	Ultrasound-sensitive nanomaterials used in drug delivery
Chronic pain management	Nanoparticles to deliver pain medication directly to affected areas	Nonsteroidal anti-inflammatory medications (NSAIDs) and opioids
Rheumatoid arthritis	Antirheumatic medications that change disease (DMARDs) and novel approaches to drug delivery	Targeted biological and synthetic DMARDSs aim to prevent joint degradation and minimize inflammation. Liposomes, dendrimers, and nanoparticles with enhanced stability and bioavailability for targeted drug delivery
Diabetes management	Using nanoparticles to administer insulin or other drugs that lower blood sugar	Metformin, sulfonylureas, and insulin
Breast cancer	pH-sensitive acetylated dextran (AcDEX)-based nanoparticles with no harm to living organisms	Breast cancer imaging and medication-assisted treatment. Combination therapy with siRNAs targeted to BV6, STAT3, and CMD-conjugated trimethyl NIK
Cardiovascular disease	Using nanoparticles to administer medications that reduce blood pressure or cholesterol	Beta-blockers, ACE inhibitors, and statins
Neurological disorders	Medicines delivered by nanoparticles to treat neurological disorders	Medications that treat anxiety, depression, and seizures
Ophthalmic disorders	Utilizing nanoparticles to administer medication for eye disorders	VEGF (vascular endothelial growth factor) inhibitors, anti-inflammatory medications, and antibiotics

(continued)

Table 3.2 (continued)

Treatment	Nanotechnology used	Pharmacology used
Wound healing	Delivery of growth factors or other compounds to aid in the healing of wounds using nanoparticles	Platelet-derived growth factor (PDGF), fibroblast growth factor (FGF), and other growth factors
Osteoarthritis	Nanoparticles and nanocarriers intended for skin application	3D cell culture models using microfluidic technology. Hydrogels that utilize microemulsions for the topical delivery of medicinal substances, liposome and transfer a few different delivery mechanisms. Cubosome-based medication delivery

diffusion of paclitaxel while maintaining stability during the procedure. However, dendrimers are unstable and under certain basic conditions release methotrexate in a pulsatile way through swelling [59]. Importantly, the potential of nanoparticles to carry large pharmacological payloads, their extended half-lives and decreased toxicity, and their improved targeting efficiency provide several benefits over free medicines [60]. Additionally, their small size makes them the ideal candidates to provide excellent images of tumor locations in the lymphatic drainage system. Researchers highlight the diagnosis and treatment of cancer using gold nanoparticles (AuNPs) which have been documented. Because gold is inert, its cytotoxicity is reduced, and it has special optical and localized surface plasmon resonance (LSPR).

In a study, gold nanoparticles are exposed to light at the right wavelength through the light-sensing photothermal reaction (LSPR), which targets tumor tissues in a way that is heat-specific and kills cancer cells [61, 62]. Similarly, Yezdani et al. [47] highlight a study by Rice University Professor Jennifer using 120 nm-diameter gold nanoparticle-coated nanoshells to remove malignant tumors from mice. Bonds with the malignant cells are formed by conjugating antibodies or peptides on the nanoshell surface. Hence, via an infrared laser in targeting the tumor, sufficiently heating the gold destroys the cancer cells. In a different study, a nanoscale delivery device known as a nano cell which consisted of a core PLGA nanoparticle enclosed within a PEG-linked lipid envelop was used in a unique technique proposed by Banerjee [60] to achieve the temporal targeting of cancer cells and tumor vasculature. Within the lipid envelope on the outside of this nanostructure, the antiangiogenic substance combretastatin was trapped while the chemotherapeutic medication DOX was covalently bonded to the inner PLGA core. Significant regression of several tumors, including melanoma, was observed following the release of combretastatin, which caused vascular collapse and intratumoral trapping of the nanoparticles which released the chemotherapeutic drug in response to local hypoxia. Nevertheless, QD-based biosensors have been applied in recent studies to identify cancer biomarkers [63]. To identify circulating tumor cells (CTCs), a wide range of nanomaterials have been investigated. These include graphene oxide, magnetic nanoparticles (MNPs),

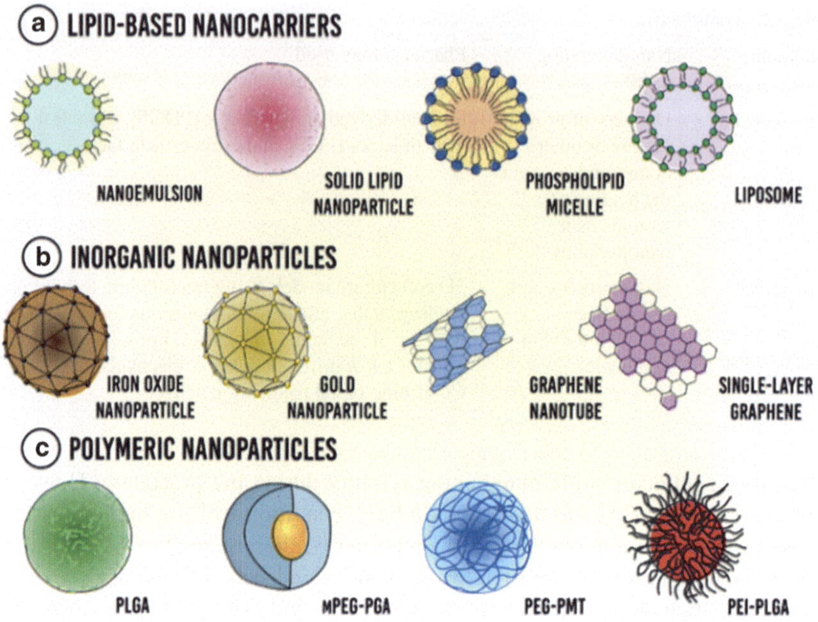

Fig. 3.2 Types of nanocarriers used for drug delivery in cancer therapy. **a** Lipid-based nanocarriers; **b** Inorganic nanoparticles; **c** Polymeric nanoparticles [57]

polymers, dendrimers, nanowires, nanopillars, silicon nanopillars, nanowires, and AuNPs. These nanoparticles help in cancer detection and prognosis since research indicates that they increase the sensitivity and specificity of CTC collection tools. A list of other studies conducted by scholars on the use of nanotechnology in cancer treatment is summarized in Table 3.3, together with their evaluated results.

Diabetes Treatment

Research from the International Diabetes Federation (IDF) reports that in 2015, 415 million persons lived with diabetes, with a projected increase to 552 million by 2030 [74, 75]. In the treatment of type 1 diabetes mellitus, regular insulin delivery is essential for glycaemic control [76]. Nonetheless, invasive subcutaneous or intravenous injection is the current standard of care, requiring daily injections and causing pain [77]. However, non-invasive insulin delivery methods have been developed using polylactide-co-glycoside (PLGA)-based MPs and NPs. This is because the medicine can be released to the target tissue, organ, or cell via PLGA MPs/NPs owing to the small dimensions of the nanomaterials, which allows them to bypass physiological barriers [78]. According to a preliminary study, PLGA nanospheres, which have a mean diameter of 400 nm, are easily inhaled through a respirator and could

Table 3.3 Summarized studies of nanomaterials in cancer treatment

Nanomaterial	Cancer type	Key findings	Refs.
Alginate folic-acid-modified chitosan (FCANA)	Colorectal cancer	For the purpose of early cancer diagnosis, colorectal cancer cells can easily absorb FCANA through folate receptor-mediated endocytosis	[64]
Paclitaxel	Non-small cell lung cancer	Good toxicity with a decent resection rate and disease response	[65]
Doxorubicin	Platinum-sensitive ovarian cancer	Safe and efficient treatment plans for individuals with a good risk–benefit ratio	[66]
Polyamidoamine dendrimer (PAMAM) doxorubicin (DOX)	Colon cancer cells	Reduced tumor growth due to effective encapsulation with a higher anti-tumor effect	[67]
Polyamidoamine dendrimer (PAMAM)	Liver cancer cells	Increased cancer cell death	[68]
Graphene quantum dots (doxorubicin)	Blood cancer cells (K562)	Enhanced water vapor permeability, pH-sensitive drug administration, and in vitro swelling qualities; little toxicity against blood cancer cells	[69]
Dendrimers	Breast cancer cell	Dendrimers have the potential to be utilized in vitro for carrier delivery in breast cancer treatment	[70]
5-aminolevulinic acid (ALA) dendrimers	Bladder cancer	An innovative in vivo method for bladder cancer detection using fluorescence	[71]
Ag-Pt nanoparticles	Glioblastoma and melanoma cancer cells	No harmful impact on healthy cells	[72]
Fumed silica nanoparticles	Cancer	For more accurate cancer detection, nanoparticles can bind to multi-site phosphorylated peptides	[73]

be used as delivery systems for insulin into the lungs [79]. This led to an instant reduction in the blood sugar level and a prolongation of hypoglycemia in guinea pigs to more than 48 h. Sheng et al. [80] used insulin-loaded N-trimethyl chitosan chloride (TMC)-coated polylactide-co-glycoside (PLGA) nanoparticles (Ins TMC-PLGA NPs) on diabetic rats to overcome several obstacles in insulin absorption. While given orally, it had a more potent hypoglycemia impact while treating diabetes as a drug delivery vehicle for macromolecular therapies. Parallel to this, the electrostatic self-assembly process was used for manufacturing poly(lactide-co-glycoside) (PLGA) and folic-acid-modified chitosan (FA-CS) composite nanocarriers (PLGA/FA-CS) that improve insulin bioavailability and overcome several obstacles for oral insulin delivery (Fig. 3.3) [81]. The resulting composite nanocarriers showed good stability in a protein solution and minimal cytotoxicity when tested on HT-29

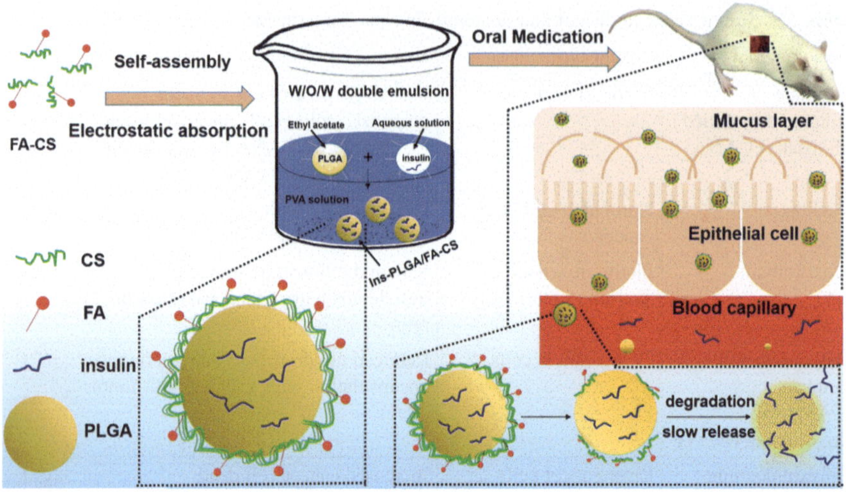

Fig. 3.3 Schematic representation for preparation of Ins-PLGA/FA-CS nanocarriers. Adapted with permission from Ref. [81] Copyright 2017 Elsevier

cells. In the presence of pepsin- and pancreatin-containing simulated gastric fluids (SGF and SIF, respectively), the chemical stability of loaded insulin was evaluated against digestive enzymes. An efficient hypoglycemic effect was observed after oral administration of insulin to diabetic rats, as opposed to subcutaneous insulin injection.

Cardiovascular Diseases

Cardiovascular diseases (CVD) are another field where the properties of nanoparticles may be leveraged [82]. The world's top cause of mortality, cardiovascular illnesses are worryingly on the rise as a result of sedentary lifestyles. Despite this, the disease is frequently ignored because it rarely causes symptoms. Heart-related disorders, including stroke, high blood pressure, diabetes mellitus, excessive alcohol intake, tobacco use, high cholesterol, low-density lipoprotein (LDL) levels, truncal obesity, and insufficient physical activity, are prevalent ailments that impact numerous individuals [83]. Estimates for 2015 alone highlight the number of CVD patients at 422.7 million, with 17.9 million fatalities [83], which is prevalent in the United States.

Nonetheless, new therapeutic and diagnostic approaches for the treatment of cardiovascular illnesses are made possible by nanotechnologies [84]. Reduced production of nitric oxide (NO) endothelial cells, which is believed to be the first stage of atherosclerosis, is associated with most cardiovascular risk factors, including smoking, high blood pressure, high cholesterol, homocystinuria, diabetes mellitus, and hypertension [82]. Because cardiovascular illnesses have limited NO bioavailability, gold and silica nanoparticles have been manufactured to improve NO delivery. Crucially, to overcome biological obstacles, nanoscale carriers with multiple functional components as therapeutic payload, a targeting moiety, and a nanoparticle core allow for better delivery to the heart and circulatory system. Gupta et al. [85], for instance, address the application of nanotechnology as tailored therapeutic delivery agents and diagnostic instruments for cardiovascular disorders. The application of nanocarriers, such as liposomes, which are nanometric vesicles composed of concentric lipid bilayers is emphasized by [86]. These vesicles can bind to receptor cysteine-rich protein-2 in cardiovascular illness, thereby acting as "heart-homing" ligands through CRPPR conjugation. It was discovered that both healthy and sick hearts had CRPPR-liposome accumulation and the cargo of liposomes quickly collected in the tissue and was still visible a day later. Notably, when loaded with stem cells, proteins, and small interference RNA (siRNAs), copolymers such as poly(lactic-co-glycolic acid) (PLGA), poly(cyclohexane-1,4-diylacetone ketone) (PCADK), and poly(ethyleneimine) (PEI) [87] have all demonstrated application toward treating cardiovascular inflammatory disease (Fig. 3.4). SiRNA is a kind of double-stranded RNA with 20–25 base pairs that can bind to complementary mRNA and cause mRNA to be destroyed inside cells to create RNA-induced silencing complexes (RISCs) [88]. Patil and Panyam [89] used PEI into PLGA NPs to encapsulate siRNA for gene silencing in vitro. Although it was anticipated that the cationic PEI would boost siRNA's loading capacity, the PLGA-PEI blending also demonstrated a notable reduction in PEI-induced cytotoxicity and an increase in particle absorption into the cell, making the hybrid blending a more successful way to deliver siRNA.

In a similar vein, the systematic administration of nanoparticle carriers to injured regions is crucial for their selective targeting and accumulation in inflammatory cells as vascular target carriers. Cyrus et al. [90] examine the effects of intramural delivery of rapamycin nanoparticles targeted at alpha(v)beta(3)-integrin on endothelial repair responses and vascular stenosis formation in a rabbit femoral artery model of vascular injury. Results show that MR molecular imaging verified the intramural binding of

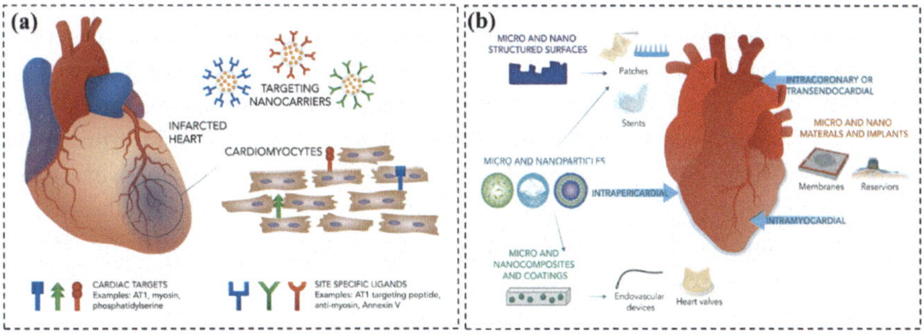

Fig. 3.4 **a** Schematic representation of potential cardiac targets for functionalized nanocarriers in an infarcted heart; **b** Schematic areas of nanotechnological inclusion in cardiovascular disease Adapted with permission from Ref. [87] Copyright 2017 Elsevier

paramagnetic nanoparticles targeted by integrins. Furthermore, two weeks later, the targeted rapamycin-treated segments of the MR angiograms exhibited much less luminal plaque than the contralateral control vessels, although at baseline the MR angiograms could not distinguish between the targeted and control arteries. This demonstrates how tailored nanoparticles can deliver therapy and form intramural bonds in vascular sites and lesions. Therapeutic payloads were delivered using nano-sized PEGylated liposomes functionalized with an AT1 targeting peptide, against angiotensin II type1. According to the study findings, ligand nanoparticles conjugate to specifically target cardiac cells in vitro and infarcted hearts after intravenous injection in vivo to deliver medicinal drugs to the heart [91]. Identifying and authenticating cardiac targets that are overexpressed on the surface of heart cells following injury is vital because the MI region lacks blood flow, making it impossible for nonspecific carriers to concentrate. Hence, [92] identifies an approach for the tuneable release of cardiovascular medicines via implantable nanochannel membrane devices for timed treatments. However, outside the primary domain of payload distribution from the target nanocarrier, nanotechnology shows innovation to support cardiovascular care.

Ethical and Social Implication of Nanomedicine in Healthcare

The advancement of nanotechnology for nanomedicine has shown great prospects for healthcare monitoring [93]. Nanomedicine offers a personalized and targeted approach for precise diagnosis, treatment, and prevention of diseases, scaffolds with nanoparticles used in tissue engineering, implant surfaces, and smart device materials [94, 95]. With such great future prospects, there remain several concerns with social and ethical implications worth considering with regard to privacy, access and equity, and regulatory concerns.

Nanomedicine provides the potential for easy diagnosis and detection of diseases in an individual, a situation which raises some serious ethical questions on the right to and impact of the information [96] and data protection. The right of the patient to know or not to know their health information detected by diagnosis devices raises serious questions [94]. Any misinterpretation of such information could affect the psychological well-being of the individual on some potential diseases they may or may not carry. The remote detection of potential diseases by nanoscale devices pushes for a bulk of surveillance and infrastructures capable of relaying vital information to healthcare providers. Such phenomena as opined by [93] give rise to the involvement of third parties to help ensure easy access to and communication between the healthcare providers and the devices used by individuals or patients, hence raising serious privacy and data protection issues. This poses potential risks where vital information on the individual's health could be misused and exploited against their wish [97].

Other ethical issues surround the use of targeted nano-encapsulated medicine for drug delivery. Herein, specific drugs are designed to deliver medicine to a targeted organ in the human body. However, how these targeted nano-encapsulated medicines can carry and deliver the drug without affecting other organs in the human body remains a growing concern in the health sector [93]. The uncertainties of the effects of these nanoscale products in the human body in the short to long term raise serious ethical questions on the techniques to be used for drug delivery. A wrong delivery of the drug to the targeted area or a wrong passage of the nanoscale product through a vessel in the human body could cause serious health problems. Aside from this, other ethical issues draw on how nanomedicine is used in personalized and regenerative medicine [94].

Advancement in nanomedicine requires several clinical trials to show the effectiveness of the innovation for healthcare delivery. This procedure raises some ethical and consent issues concerning the individuals involved in the clinical research trials. Participants in such research may not have a full understanding of the side effects of using such devices or how their bodies will react to these nanomedicine [98]. This is because the risks associated with the use of such nanoscale devices are under investigation, hence they are not properly informed or communicated with the full information [99]. This therefore calls for effective communication to inform participants of all side effects after they interact with the nanoscale products. Access to nanomedicine at a low cost by further has some social implications. Individuals in developing or underdeveloped countries in the World should be able to have access to this new technology for full benefits. The high-cost nature of these technologies could limit their use by individuals for healthcare benefits.

To ensure these nanoscale devices used for nanomedicine are safe and easily accessible for many to benefit, the issue of regulation and precautionary principles arises. These regulations provide strict guidelines on appropriate techniques during manufacturing and application in clinical trials for the safety of the human body and the environment. A situation that calls for collaboration by all stakeholders in nanotechnology and healthcare delivery. Similarly, precautionary principles as opined by [100] draw on strict approaches accepted by all within the field to regulate new innovation and technology, whose harmful effects are not fully understood. This is imperative since the harmful effects of nanoscale products used in nanomedicine are still under investigation, hence delays in clinical trials are necessary to limit any potential risks of this technology to the environment and public health [101].

Addressing these ethical issues will go a long way to ensure strict regulations and compliance with the use of nanomedicine for effective healthcare delivery and limit their harm to the human body.

Nanobots

The field of clinical medicine has benefited greatly from the advancements in medical robotics during the past few decades [102]. For upper limb rehabilitation, for example, robotic devices have been created to help individuals with movement problems regain their usefulness [103]. With the development of soft robotics, which combines biocompatibility and biomimicry with soft materials, new biomedical applications where a soft interaction with a patient is desired are now feasible [104]. Thus, technological advancements have led to the creation of nanobots, which are little robots on the nanoscale that can interact with DNA and cells [105]. Mostly utilized in nanomedicine, nanorobots enable access to difficult-to-reach bodily parts and carry out a range of medicinal functions [106, 107]. Importantly, nanorobots can be externally or independently operated and are used for precise activities such as molecular tissue repair, medication delivery, and cell manipulation [108, 109]. Compared to conventional chemotherapy, which also kills healthy cells and may be useful in diagnostics, nanobots considerably lessen side effects. Crucially, they are employed in the cellular-level capture of pictures or sensory signals from within the body. Thus, diseases like cancer can be diagnosed earlier and with more accuracy thanks to this new advent of technology [110, 111]. Additionally, biosensor-equipped nanobots potentially track biomarkers that reflect the course of a disease or an individual's health, providing real-time personalized health data to help guide treatment choices. According to Herrmann et al. [112], nanobots can conduct nanosurgery to remove plaque build-up in arteries or clear blood vessel blockages and also have therapeutic capabilities of their own. This function stems from the fact that minuscule particulates can enter the body in an unprecedented way, potentially revolutionizing the fields of prevention, diagnosis, and treatment medicine. These breakthroughs are thus discussed under major themes.

Drug Delivery

In contrast to conventional treatments, nanocarriers are designed to deliver therapeutics specifically to the site of joint inflammation, minimizing adverse and systemic consequences [108]. These delivery techniques rupture epithelial barriers to liberate the drugs, which are then carried to the site of action by bioactive molecules across many biological barriers [113]. This minimizes the unintended systemic effects of the drugs, including chemotherapeutic treatments. Motile and swimmable micro/nanorobots could precisely administer a therapeutic payload by swimming right into target areas [114] with common happenings in pharmaceuticals, biologics, living cells, and inorganic therapies. Comparing medical nanorobotics to passive diffusion technologies, the latter have less precision and speed in medication delivery. Generally speaking, in vivo applications have been driven in this direction [115]. For instance, fluorouracil medicine was delivered to a mouse model with tumor growth using magnetically guided nanorobots. Because the medication release was initiated externally, the nanorobotic platform was able to disperse a large quantity of the therapeutic agent in a specific region of the tumor [116]. Similarly, [117] designed and constructed a multi-component magnetic nanorobot using doxorubicin (DOX) and multi-walled carbon nanotubes (CNT) loaded with anticancer antibodies. This magnetic nanorobot could be driven by an external magnetic field to push itself through complicated biological fluids. Anticancer medication payloads within three-dimensional (3D) spheroidal tumors may be released upon stimulation by intracellular H_2O_2 or local pH fluctuations within the tumor microenvironments. The nanorobot, consisting of chemically linked magnetic Fe_3O_4 nanoparticles, opened a gate on the surface of the Fe_3O_4 nanoparticles to release DOX selectively into the intracellular lysosomal compartment of human colorectal cancer (HCT116) cells. Similarly, a hybrid magnetoelectric nanorobot with tailored drug delivery capabilities was fabricated by [118]. The drug release was initiated by an external magnetic field and by revolving magnetic fields, the device is precisely guided toward a desired spot wirelessly and executes magnetoelectrically assisted medication release on-demand for the elimination of cancer cells.

Wound Treatment

Injuries cost healthcare systems in the US alone about $25 billion annually, placing a heavy financial strain globally [119]. The complicated process of wound healing necessitates the proper extracellular milieu and extensive cooperation among numerous cell types [120]. Overwhelming inflammation, hypoxia, chronic infection, and elevated protease activity are common problems associated with chronic wounds. Crucially, the human body recognizes wounds and initiates the healing process using a variety of biochemical signals and mechanisms. Nonetheless, in situations where the lesion is excessively bleeding or with insufficient localized coagulant factors in the affected area, these biological mechanisms fail [120]. In addition, the incidence of peripheral vascular disease, metabolic syndrome, and type 2 diabetes has increased dramatically, leading to a rise in chronic wounds [121]. On the other hand, medical nanorobotics simulates such systems in this direction by employing active delivery for quick and efficient wound healing [122]. Innovatively, biocompatible materials, including proteins or DNA, can be used to create nanobots, which can blend in with biological systems and tissues without doing any harm [123]. Their actions within wounds can be programmed and guided to include debriding dead and damaged tissues, combating infection, supplying oxygen and nutrients, and attracting host stem cells to the injury site to regenerate skin and tissues [124]. The surface proteins on nanobots are designed to recognize and bind to necrotic tissue and detritus, which is useful for debridement [125] to degrade tissue remains enzymatically or manually remove pieces of dead tissue. Moreover, nanobots have the potential for precise therapeutic drug delivery in intricate wound settings by selectively transporting medications, growth hormones, and oxygen to the right places without endangering healthy tissues [126]. Thus, nanobots speed up and enhance body's inherent healing processes through these diverse roles in healthcare.

Medical Diagnostics

Novel receptor-functionalized micro and nanorobotic motors have transformed the binding and transportation capabilities of robotic devices. Through the application of a distinct sensing approach, nanorobots have exhibited the capacity to carry out accurate illness diagnosis through the isolation and identification of a broad spectrum of biological targets, including proteins, nucleic acids, and cancer cells [127, 128]. Recent research has, for instance, produced customized nanorobots that can subcutaneously interact with and even diffuse at cellular levels [125]. Crucially, early cancer identification is vital because tumor-killing nanorobots can significantly improve patient survival rates [129, 130]. According to Peng et al. [131], a logic gate nanomachine-based 3D DNA nanorobot was created to identify bi-specifically overexpressed cancer cell biomarkers. In addition, the DNA nanorobot can perform Boolean logic operations on the cancer cell membrane, which gives it a significant degree of therapeutic potential for treating malignancies. Yang et al. [132] investigated on Au-BP7@SP Janus nanohybrids and reports an improved photothermal impact and active motion thermophoresis. Through the conversion of mechanical energy into thermal energy and the subsequent death of malignant cells, the compelling momentum of these Janus particles efficiently boosts the temperature of the cells exposed to NIR light. Hence, the particles can locate the cancer accurately without harming the adjacent healthy tissue by greatly enhancing the therapeutic benefit and lowering the risk of adverse side effects. Such multifunctional nanodevices promise to provide rapid diagnosis and personalized treatment all from a single nanoparticle agent. Additionally, the development of nano-biomaterials for precision medicine is a highly customizable process. This careful design approach enables adjustments of the therapeutics' pharmacokinetics to match requirements for solubility, administration, or biodistribution and has seen success [133]. Table 3.4 illustrates a list of research findings on nanotechnology in medical diagnosis.

Table 3.4 Summarized studies of nanotechnology in medical diagnosis

Nanomaterials/technology	Target for detection	Applications	Refs.
Computed tomography (CT)	Tumor detection of hypervascular hepatoma	Medical imaging	[134]
Superparamagnetic iron oxide (SPIO)	Targeting of lung tumors by MRI	Medical visualization	[135]
Quantum dots	Detection of biomarkers	Protein optical imaging and tumor visualization	[136]
Luminescent nanomaterials	Early-stage identification of breast cancer via detection	Optical imagining of cells	[137]
Micelles in MRI	Detection of unhealthy cells	Medical imaging	[138]
Perfluorocarbon	Targeting blood cloth, cancer metastasis, and angiogenesis	Medical imaging	[139]

Enhanced Precision Medical Instruments

Medical instruments like surgical tools, implants, diagnostic systems, and assistive devices play a pivotal role in healthcare delivery for accurate diagnosis and effective treatment of diseases [140]. The goal of precision medicine is to improve therapeutic outcomes by developing patient-specific treatments in a clinical context, thereby addressing the numerous drawbacks of traditional one-size-fits-all techniques [141]. With its applications in the identification, diagnosis, and treatment of cancer, this has produced a number of encouraging outcomes. These include molecular imaging, targeted therapy, biomarker mapping, drug delivery, gene therapy, and drug carriage [142]. Crucially, compared to current approaches that necessitate extensive sample processing, the use of magnetic NPs or AuNPs, which are easy to employ, in biomarker detection tests saves time and money [133, 143]. Furthermore, minuscule tissues or individual cells can be handled and worked on with negligible damage thanks to incredibly delicate nano-tweezers and scalpels. This facilitates microsurgeries such as neurosurgeries, ophthalmic surgeries, and neonatal procedures. Nanoprobe technologies improve the imaging distinction of tumor versus healthy tissues, enabling more precise targeted tumor resections during cancer surgery [144]. Moreover, nano-enhanced robotics improves control and precision in minimally invasive robotic surgery systems [145], resulting in less tissue damage and a quicker return of the patient to regular activities. In a similar vein, highly sensitive sensors based on nanomaterials allow for the unparalleled real-time detection of biological events with speed and precision, enhancing implantable device monitoring [146]. For instance, iron oxide-based magnetic nanoparticles have also been employed in glucose sensors. These particles can be utilized independently or in conjunction with other systems like CNTs [147, 148]. Similarly, colorimetric assays using gold nanoparticles linked with antibodies can detect analytes down to even a single molecule because of their great sensitivity, affordability, and simplicity of use [149]. In the medical field, superparamagnetic iron oxide (SPION) MRI can offer reduced sensitivity but greater anatomical information and detailed visualization [150, 151].

Gene Therapy

Gene therapy is revolutionized by targeted gene delivery nanoparticles, which replace or repair damaged genes to potentially heal genetic diseases [152]. Gene therapy refers to the treatment or prevention of disease by introducing, altering, or silencing genetic material within a patient's cells. Its three facets are immunodeficiencies; mental diseases; cancer; heart disease; diabetes; HIV; and gene replacement, in which the target gene is delivered by plasmids or viral vectors. All of these conditions are active areas of gene therapy research and offer hope as a treatment when conventional drugs and procedures fail [153]. From a practical standpoint, gene therapy attempts to directly modify the genome, impact gene activity, or replace defective genes that cause disease. In other words, it introduces new genes that help treat or cure an illness

[154]. Additionally, due to its capacity to fabricate unique nanoscale objects with targeted, invasive, and intracellular effector functions, DNA origami holds promise for more flexible gene delivery applications [155]. Over time, researchers continue to study several approaches to gene therapy delivery and treatment. Recently, medical nanorobots readily treat genetic diseases by comparing the molecular structures of both DNA and proteins found in the cell to known or desired reference structures [152]. A minimally invasive technique combined with state-of-the-art nanodevices allows for the modification or editing of anomalies in DNA and protein structures relative to normal ones. In this regard, nanorobots may be a driving factor behind several medical advances, such as the ability to reverse the effects of Alzheimer's disease and assist paraplegics to walk again. Nonetheless, biological tools or "wet" nanotechnology can now be used to completely design organic living systems thanks to advances in biotechnology [156]. In some cases, chromosome replacement therapy can occasionally be more effective than repair [152]. Thus, niche genetic algorithm (NGA) technology is used in a nanorobot-assisted multifocal cancer detection procedure (MCDP) that was proposed by [157]. The NGA-inspired nanorobots, which swim in the high-risk tissue region to detect tumors, can be thought of as an auto-searching process that provides an optimal solution to an objective function within a parameter space based on certain limitations. This can effectively find tumor sites by resolving the multimodal optimization (MMO) problem, taking into account realistic in vivo propagation and nanorobot control.

Additionally, immuno-gene therapy is one possible therapeutic strategy for p53-deficient tumors [158]. Crucially, intracellular siRNA delivery for gene therapy was achieved by engineering a rolled amplified DNA strand wrapped in gold nanowire that can hybridize with siRNA [159]. Once the pressure gradient has been created, the scissor-like scissors of siRNA split the target mRNA, to achieve 94% silencing efficiency in a matter of processing minutes, enabling the nanorobot to enter the cancer cells with effectiveness. In a different investigation, [160] reported the direct killing of leukemia cancer cells by immobilizing asparaginase, which effectively kills asparagine throughout blood circulation and with its faster motor action, facilitates the enzymatic connection with the malignant cells (Fig. 3.5). Realizing routine clinical implementation still requires optimizing nanobot biodistribution, targeting, maximizing transfection efficiency, and demonstrating long-term safety which remains an ongoing research endeavor; thus, gene therapy aided by the remarkable capabilities of nanobots portends a revolutionary treatment option for genetic diseases.

Ethical and Social Implication of Nanobots in Healthcare

The use of nanobots ensures easy detection of diseases for effective drug delivery and safer surgeries [161]. This is made possible due to the tiny nature of the robots. With such great future prospects, there are several concerns with social and

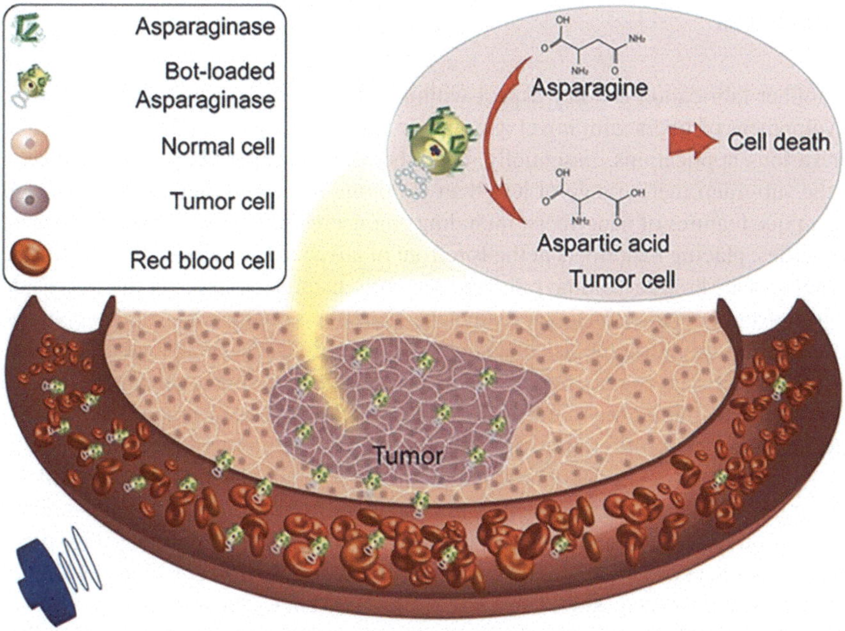

Fig. 3.5 Graphical representation of the anti-tumor activity of US-propelled asparaginase-modified nanobot engines Adapted with permission from Ref. [160] Copyright 2022 Elsevier

ethical implications worth considering with regard to privacy, patient consent, safety, biocompatibility, and regulatory concerns.

The consent of patients for any clinical trials is very important. The use of nanobots in clinical trials draws attention to consent issues where the patient is required to be told details of the procedure. This will help inform them for a better understanding and decision-making to consent willingly to the use of the nanobots for diagnosis and treatment. This further ties in with the issue of privacy. Herein, these tiny devices are used to gather and process vital health information of patients, a situation which exposes patient's health data to third parties if not protected. This requires confidentiality and protection of data by the healthcare provider and the manufacturer, a breach of this could raise serious legal issues and mistrust in the use of the technology [161].

Despite the great potential nanobots offer for diagnosis and treatment, the issue of safety and biocompatibility remains a great concern influencing its full use in clinicals [162]. The design of these nanobots utilizes different material components, and how these are compatible without causing harm when used in the human body remains a great question. Furthermore, effective regulatory frameworks for nanobots will provide and improve medical use aimed at addressing the issue of toxicity and biocompatibility [162].

Nanofibers

Nanofiber fabrication is a key aspect within this research area that investigates the development of fibers with novel and unique properties, opening up great potential for various applications. Essentially, nanofibers are ultrafine fibers with diameters in the sub-micrometer range of less than 1000nm. Several studies have recognized the unique features of nanofibers, including engineered biochemical and mechanical properties, placing nanofibers at the forefront of advanced materials [163–169]. The higher surface area-to-volume ratio and porosity distinguish fibers in the nanoscale from conventional fibers in larger diameters. Nanofibers engineered from biocompatible materials minimize the risk of adverse body reactions, thus making them suitable for implants, tissue engineering scaffolds, and wound dressing applications. Due to their ability to mimic the extracellular matrix of human tissues, nanofibers facilitate cell adhesion, proliferation, and differentiation. The porous structure of nanofibrous materials provides an optimal environment for cell attachment and metabolic exchange [170]. For drug delivery and tissue engineering applications, nanofiber degradation rates can be tailored to match specific drug release or tissue growth timelines. Nanofibers can be functionalized as electrically conductive, which is useful in nerve tissue engineering and for creating biosensors or designed to be optically active or inert, which is beneficial for applications in imaging and diagnostics. The fiber surface can be chemically modified to introduce specific functionalities, enhancing their interaction with biological systems or targeting specific cells or tissues [28]. In wound dressings, nanofibers can provide a protective barrier against microbial infection while maintaining breathability and moisture balance at the wound site. Nanofibers can be fabricated from a variety of natural and synthetic polymers and functionalized to obtain specific physical and chemical properties [171–173]. These findings indicate the versatility and potential of polymer nanofibers in a wide range of applications, demonstrating their significance as advanced materials for various biomedical fields.

Materials and Processing Techniques

The material selection involves considering various factors such as the intended application, desired properties of the nanofibers, and the fabrication technique to be employed [172]. Natural-source biopolymers have characteristics that are comparable to macromolecular compounds found in the human body, thus rendering them suitable for interactions in biological environments [174]. They are characterized by their biocompatibility and biodegradability and thus exhibit minimal toxicity and immune response [175]. A major advantage of biodegradable materials lies in their ability to eliminate the need for surgical removal of implanted devices, such as sutures [176, 177]. The degradation rate of these materials is crucial in selecting the appropriate material for specific applications.

Natural biopolymers include a diverse range of materials, including collagen, elastin, cellulose, chitosan, gelatin, and silk fibroin, derived from plant or animal sources. These materials have been extensively researched for diverse biomedical applications due to their favorable biological properties [172, 175]. For example, a substantial amount of research has been conducted on materials such as cellulose, silk fibroin, chitosan, alginate, hyaluronic acid, and collagen for their utilization in tissue engineering, wound healing applications, and drug delivery mechanism [174, 175]. Cellulose is a widely used material in nanofiber fabrication and is often processed via electrospinning due to its abundance, renewability, biocompatibility, and biodegradability [171, 174]. Silk fibroin is particularly chosen for its mechanical strength and biocompatibility, and it is utilized in nanofiber fabrication through techniques like electrospinning, self-assembly, and phase separation [174]. Chitosan, derived from chitin, offers biocompatibility, antimicrobial properties, and biodegradability, used in electrospinning and blow spinning for diverse biomedical and healthcare applications [172]. Synthetic biopolymers on the other hand are designed and produced through controlled synthesis and modification processes, resulting in customized biodegradable materials with specific mechanical characteristics and functional groups. These can be tailored to meet particular requirements, including controlled drug delivery, structural integrity, and specific degradation profiles [177]. The diversity of synthetic biopolymers, including PCL (poly-ε-caprolactone), PLA (polylactic acid) and its copolymers, e.g. PLGA (poly(lactic-co-glycolic acid)), and polyvinyl alcohol (PVA), highlight their significant impact on nanofiber engineering for various biomedical applications [172, 177–179]. It is important to highlight that not every synthetic biopolymer is suitable for nanofiber engineering, e.g. highly crystalline or insoluble biopolymers may pose challenges. PCL and its composites are extensively investigated in tissue engineering for its slow degradation rate, providing prolonged support for tissue formation [176, 178, 180]. Similarly, PLA and its copolymers are being researched for developing nanofibers with controlled drug delivery profiles due to their tuneable degradation rate and ability to encapsulate and release pharmaceuticals over time [177].

Fabrication methods in the field of nanofibers have advanced considerably in recent years, exploring a variety of processing techniques including electrospinning, solution blow spinning, microfluidic spinning, and force spinning, such as centrifugal spinning and draw spinning. These techniques have enabled the production of continuous nonwoven nanofiber mats, with electrospinning being the most widely used method [163, 171, 172, 177, 181–183]. Electrospinning, known for its simplicity and versatility in fiber collection, uses electrostatic force to extrude nanofibers from a viscoelastic solution, yielding fibers with diameters that span from micrometers to nanometers [184]. The setup typically comprises a high-voltage power supply, a spinneret (commonly a syringe and needle), and a grounded collector [184, 185]. By applying a high electric field to the polymer solution, a cone-shaped jet emerges, referred to as a Taylor cone. As the electrostatic repulsion surpasses the surface tension of the polymer solution, a jet is ejected from the tip of the cone and gradually thins due to elongation and solvent evaporation, eventually forming randomly oriented nanofibers [184, 186].

Current research has also focused on optimizing the electrospinning processes, aiming to enhance the efficiency, scalability, and functionality of nanofiber fabrication and reproducibility [187, 188]. There are several excellent reviews on advances in electrospinning reviewing research efforts on the influence of spinning and spinning solution parameters on fiber control, e.g. diameter and morphology [189], materials and composites as well as surface functionalization for advanced applications [190], and electrospinning setup, such as using rotating collectors or multi-needles to produce aligned or patterned nanofiber structures. Nanofiber yarn, an emerging electrospinning technique, can be manufactured using a rotating mandrel to manufacture yarns with aligned fibers and adjustable twists, influencing the structure of the collected nanofibers [185, 191, 192].

Solution blow spinning, a gas-assisted fiber fabrication method, has been recognized as a technique for producing fibers from materials characterized by low dielectric constants and/or electrical conductivities [182]. This technique extrudes fibers through high-pressure compressed air or another gas stream that stretches the polymer droplet at the tip of the nozzle. The droplet's surface is shaped into a cone that resembles Taylor's cone in electrospinning as the compressed gas leaves the nozzle. As the aerodynamic forces applied to the solution overcome the surface tension, the polymer solution is stretched forming a jet that moves from the needles toward the collector. As the jet accelerates, the solvent evaporates forming fibers. The fibers are collected on a variety of surfaces, such as rotating drums, flat surfaces, and liquids [182]. Parameters controlling the morphology and properties of the resulting fibers among others include the flow rate of the polymer solution, gas pressure, which is crucial for stretching the droplet and ultimately influence fiber diameter, and polymer concentration as well as the solvent system, which determines the viscosity of the solution. The technique offers advantages such as production efficiency and adaptability to a broad spectrum of polymers and offers considerable potential for fiber applications [182].

Microfluid fiber spinning begins with the design of a microfluidic device, which typically consists of channels molded into materials like PDMS (polydimethylsiloxane) or glass. Two different fluids are introduced into the device through separate inlets. The primary fluid acts as the core, which usually contains the polymer solution intended to form the fiber. The other fluid, known as the sheath fluid, surrounds the core fluid and can serve multiple purposes: it can act as a lubricant to facilitate smooth fiber extrusion, and in some cases, it also acts as a cross-linking agent to solidify the polymer [182]. Fiber formation occurs by solidifying the core fluid through photopolymerization, chemical or ionic cross-linking, or solvent exchange. Self-assembly represents another method, offering the advantage of the spontaneous organization of molecules or polymers into nanofibrous structures, but it might be limited by the need for precise control over assembly conditions. Additionally, phase separation is recognized for its ability to enable the processing of polymeric biomaterials into nanofibers, providing control over fiber composition and porosity using a relatively simple experimental setup. Each method's suitability varies based on the targeted healthcare applications, for in vivo and ex vivo. For example, bio-textiles refers to a broad category of medical devices and materials with specific nanofiber architecture and design used in biological environments, including engineered textile scaffolds, implants, drug delivery carries, hygiene textiles for wound care, wearable bioelectronic devices, and more [185, 193, 194]. These textile structures and membranes offer enhanced biocompatibility, mechanical properties, and controlled release of bioactive agents over conventional healthcare devices [174, 175].

Wound Treatment

Traditional wound dressings, like gauzes and bandages, offer basic protection yet often fall short in actively promoting healing or effectively combating infection when it occurs. The limited effectiveness of existing treatment methods for skin wounds contributes to considerable morbidity and mortality, primarily due to the risks of infection and dehydration. Chronic wounds, in particular, require advanced wound dressings that can create a moist environment, promote cell growth and proliferation, prevent microbial infiltration, and effectively treat infections. Nanofiber-based wound treatments offer promising solutions due to their unique properties, including high surface area, micro-porous structure, and drug or biomolecule loading capacity [164, 172, 175]. Nanofibrous wound treatment applications aim to promote scar-free tissue regeneration [195]. Additionally, incorporating antibacterial properties into nanofibers can help prevent infections and improve the overall effectiveness of wound treatments. Drug-loaded nanofibers have shown promise in wound and burn treatments, ensuring localized and controlled delivery of therapeutic agents [196]. With advancing research, nanofibrous treatments create multifunctional devices that not only promote healing and effectively prevent or treat infections but also monitor wound conditions. John et al. [197] highlighted the promise of nanofiber dressings to stop bleeding, reduce pain, and deliver therapeutics for enhanced wound healing.

A range of studies have explored the potential of nanofiber wound dressings in promoting wound healing. Studies have explored electrospun nanofibers for their applicability in wound care and as drug delivery systems, emphasizing their role in enhancing cell–material interaction [195]. Several polymers, including carboxyethyl chitosan, chitosan/arginine-chitosan, CS/SF, CS/PVA, collagen, and gelatin/PU, have been studied for their efficacy as wound dressing materials [164, 195]. Li et al. [198] highlights the benefits of gelatin-based nanofiber dressings fabricated by electrospinning, including their high porosity, air and moisture permeability, and ability to resemble the extracellular matrix and [166] found that gelatin nanofibers accelerated wound closure and promoted cell adhesion and proliferation. Kang et al. [199] researched chitosan-coated PVA nanofibers as an effective wound dressing, with the chitosan coating improving water resistance and tensile properties. Homaeigohar and Boccaccini [200] discusses the use of antibacterial biohybrid nanofibers, which can deliver antibiotics and antibacterial agents to control infection. Xie et al. [201] developed a dual growth factor-releasing nanofiber system that accelerated wound healing by promoting angiogenesis and tissue regeneration. Rath et al. [202] demonstrated that collagen nanofibers containing silver nanoparticles accelerated wound healing by providing an aseptic environment and promoting re-epithelization and collagen production. These studies collectively suggest that nanofiber wound dressings have the potential to significantly improve wound healing outcomes. Moreover, these nanofiber networks have shown promising results in promoting hemostasis and oxygen permeation, absorbing wound exudate, improving the delivery of drugs and antimicrobials to the dermis, and providing support for three-dimensional tissue formations.

Drug Delivery

The advancement of drug delivery systems in healthcare plays a pivotal role in enhancing therapeutic effectiveness and patient treatment. These systems are designed for precise delivery and controlled release of therapeutic agents. Nanofiber-based drug delivery systems, in particular, have seen substantial progress in recent research. Their ease of fabrication, diverse material compatibility, surface-to-area ratio, and pore architecture allow efficient drug encapsulation, low release, and high stability, making them highly applicable in various delivery applications [167, 203–205]. Vojoudi and Babaloo [206] provide an overview of the incorporation of biomolecules and drugs into electrospun nanofiber scaffolds, highlighting their potential in a range of biomedical applications, e.g. cancer treatment, wound dressing, growth factor delivery scaffolds, and carriers for stem cell delivery systems. Joshi et al. [207] discusses the latest developments in the topical delivery of drugs, with a focus on nanofibrous drug delivery systems and their applications in wound dressing and anticancer treatment. Goyal et al. [167] reviewed topical medication delivery, using nanoparticles and nanofibers, emphasizing their ability to deliver both hydrophobic and hydrophilic drugs. [173] further explore the potential of nanofibers in drug delivery in tissue engineering scaffolds, particularly in the delivery of proteins for tissue regeneration, antibiotics, and anticancer drugs. Kamble et al. [208] summarize the use of nanofibers in local drug delivery on skin tissue for wound healing or cosmetic use. In the area of oral drug delivery, nanofibers can be used in diverse formats, such as scaffolds, electro-sponge nanofibers, and multi-layered nanofiber-loaded meshes, with different polymers and formulation methods [204].

Various drug loading techniques have been explored, such as active loading, encapsulation during electrospinning, and the preparation of composite nanofibers, for sustained drug release [209]. Electrospun core–shell nanofibers and composite nanofibrous mats have shown potential for sustained release of proteins through active loading and encapsulation methods [205]. Active loading is the chemical binding or attachment of drug molecules to the nanofiber polymer matrix, often through specific interactions or conjugation techniques. Encapsulation, on the other hand, refers to the process of enclosing drug molecules within the physical structure of nanofibers without forming chemical bonds. This encapsulation approach allows the incorporation of a wide array of pharmaceuticals, including hydrophilic and hydrophobic compounds, within the nanofiber structure [205]. Drug encapsulation allows for a broader range of drug types, including hydrophilic and hydrophobic compounds, whereas active loading offers the advantage of precise and targeted drug placement within the nanofibers, which may be beneficial for specific drug delivery applications [205].

The loading of drugs, such as curcumin, diclofenac, and vitamin B12, into hybrid electrospun chitosan/phospholipid nanofibers has demonstrated successful transdermal delivery with controlled release behavior [208]. For instance, sustained release of curcumin was achieved over seven days, indicating the potential of nanofibers to control drug release over extended periods [210]. [211] developed a transdermal patch for the localized delivery of the antibiotic ciprofloxacin, encapsulated within hydrophilic, biodegradable nanofibers made from a composite of PVA (Polyvinyl Alcohol) and Sodium Alginate (NaAlg) through electrospinning. The study showcased the practicality and efficiency of embedding ciprofloxacin in the nanofibers for targeted and localized delivery.

The significant potential of nanofibers as medication systems is further demonstrated through the investigation of cellulose acetate/polycaprolactone nanofibers for prolonged drug delivery, presenting a significant advancement in achieving sustained-release profiles for improved therapeutic outcomes [187]. Non-biodegradable material-based systems release drugs through slow diffusion, whereas biodegradable materials offer more complex but controlled and gradual release patterns. Natural and synthetic composites are often preferable in drug delivery applications, combining the advantageous properties of both material types. The manipulation and degradation rate of biodegradable materials can be achieved through methods such as modifying the composition and crystallinity of polymer blends or copolymers [209]. For example, the methyl group in PLA makes it more hydrophobic and thus degrades slower than PGA [209]. Nanofiber structures, made from a variety of materials, exhibit efficient drug encapsulation, controlled slow-release properties, and robust stability, rendering them ideal for localized drug release management. [205]. Strategies such as blending and co-axial electrospinning have been explored to improve drug payload, particularly for hydrophilic drugs [212].

Tissue Engineering

Tissue engineering (TE), a multidisciplinary field intersecting engineering, chemistry, biology, and materials science, aims to regenerate damaged tissues or organs [170, 174]. Tissue engineering uses the body's innate ability to heal itself through tissue regeneration. TE is the process of creating functional tissues by the use of biological agents, cells, and biomaterials to restore, maintain, or improve tissue function. Due to their unique characteristics, nanofibers show great advancements in TE [213]. The use of nanofiber scaffolding aims to create a favorable microenvironment that supports cell attachment, migration, proliferation, and tissue growth, mirroring the characteristics of the ECM found in human tissues [170, 174].

Nanofiber scaffolds can be prepared using various techniques such as solution blow spinning, centrifugal spinning, and electrospinning which allows for the production of complex, biomimicking native tissue structures. Electrospinning, in particular, has shown promising results and is widely studied for TE applications due to its ability to create and replicate the ECM with controlled fiber diameter, fiber alignment, and a vast variety of polymers. Studies highlight the impact of scaffold topography on cytomorphology and alignment, outlining the regulatory effects and mechanisms of cell–scaffold interaction [165]. [214] investigated the impact of scaffold topography, such as average diameter and alignment of fibers, on orientation, protein expression, and attachment of endothelia cells. The findings show the different effects of micro- and nano-topography cellular response. Fiber alignment creates guidance, with nano-sized aligned fibers significantly enhancing cellular proliferation over a longer cell culture period. [189] electrospun PCL membranes and investigated the effect of fiber diameter on porosity, mechanical property, and cellular response. Results indicated a change in surface roughness correlating with increasing fiber diameter, which in turn facilitated improved cell attachment and growth. The study emphasized a positive correlation between surface roughness and improved cell behavior.

With advances in electrospinning, the bundling of nanofibers to form long-length nanofibrous yarns, which can be further processed, is increasingly explored [185, 192, 215, 216]. Scaffold engineering using textile technologies—knitting, weaving, and braiding—creates more complex structures that potentially better mimic the architecture and anisotropic characteristics of human tissues compared to nanofibers membranes [185, 193, 217, 218]. Nanofiber yarn scaffolds exhibit enhanced tensile strength and suture retention, making them applicable for load-bearing scaffolds such as tendon TE. These bio-textiles have shown promising results in engineering woven or knitted scaffolds for tissue regeneration that closely mimic the anisotropic and hierarchical characteristics of native tissues and organs [192]. The use of nanofiber yarn ß has been explored for the fabrication of soft tissue scaffolds for their anisotropic properties, with potential applications in smooth muscle and osteogenic differentiation [215]. Recent findings have highlighted the promising use of textile-based scaffolds for vascular implants. In particular, research groups are focusing on advancing tendon reconstruction techniques and innovative tubular textile structures tailored for large and small-diameter blood vessels, such as arteries and capillaries [194, 218].

Ethical and Social Implication of Nanofiber Applications in Healthcare

The use of nanotechnology in healthcare, particularly in the form of nanofibers, presents complex ethical implications and social challenges. Discussions in this area stem from multifaceted challenges, encompassing concerns about the long-term safety of micro-sized materials, the need for comprehensive safety assessments to identify and mitigate potential risks in health, industry, and environment as well as social acceptance and equity. Health hazards, although, biocompatible polymers are engineered to not elicit adverse reactions when in contact with body tissues or fluids, are once FDA-approved generally non-toxic and considered safe, concerns may arise to fully understand the interaction of nanofibers with biological systems that can vary significantly from traditional materials. Degradable biopolymers are engineered to degrade into byproducts that can be metabolized or excreted by the body [219]. However, the degradation process may yield intermediate or byproducts that could pose potential toxicity risks if they accumulate within the body. Unintended bioaccumulation and toxicity where nanofibers accumulate in organs or tissues, potentially causing harm, require extensive long-term studies to ensure safety. Moreover, additives or impurities introduced during production processes or inhalation exposure particularly for healthcare workers and patients posing respiratory issues further complicate the safety profile of these materials. Mitigating such risks is crucial to guarantee the overall safety of nanofiber-based healthcare products.

At the cellular level, studies focus on how nanofibers interact with cell membranes, potentially leading to cellular uptake, cytotoxicity, or inflammatory responses. Key concerns include the induction of oxidative stress, DNA damage, and apoptosis. Systemically, the biodistribution, pharmacokinetics, and long-term effects of nanofibers are of interest, particularly their potential to accumulate in organs, leading to systemic toxicity or immune responses [220, 221]. These evaluations are critical, considering the diverse routes through which nanofibers can interact with biological systems, including inhalation, dermal exposure, and direct implantation. The FDA approval process for nanomaterials used in healthcare devices and treatments involves thorough evaluations of safety and efficacy. However, the novel properties of nanoscale materials can complicate these assessments, leading to ethical concerns about the adequacy of current regulatory frameworks to address the unique challenges posed by nanotechnology.

Current market applications, products like the Spincare system (https://www.regenmedical.co.uk/products/nanomedic/), developed an electrospinning system that applies a nanofiber membrane directly to wounds and burns without physical contact, exemplify the practical utilization of nanofibers in enhancing healing processes while minimizing patient discomfort. Similarly, Nanofiber Solution development of 3D nanofiber scaffolds underscores the application of nanofibers in fostering environments conducive to tissue repair and regeneration (https://nanofibersolutions.com/regenerative-medicine/). The advent of nanofibers in healthcare allows for the development of treatments that are tailored to individual patient needs, heralding a new era of personalized medicine. While this personalized approach has the potential to enhance treatment efficacy significantly, it introduces several ethical considerations, such as personalized risk assessments, consent, and patient autonomy but also equitable distribution. The advanced nature and associated high costs of such technologies raise issues regarding availability to economically disadvantaged populations, possibly exacerbating existing healthcare disparities [222]. It is ethically imperative to address and mitigate these disparities to ensure equitable access to nanotechnology treatments across different socioeconomic groups.

Industrial and environmental implications of nanofiber products encompass the entire lifecycle of these materials from production to disposal that need careful consideration. Manufacturing processes have significant energy demands and may involve the use of potentially hazardous chemicals, raising concerns about industrial emissions and waste management [223, 224]. Production, usage, and disposal call for a rigorous examination of their ecological footprint, advocating for the inception of sustainable practices to mitigate potential environmental risks. Furthermore, the manufacturing and handling of nanofibers necessitate stringent occupational safety measures to prevent exposure risks, underscoring the importance of ensuring safe working conditions in nanofiber-related industries [225].

Addressing these ethical challenges requires a multidisciplinary approach, involving not only advancements in nanotechnology and healthcare but also robust ethical guidelines, and regulatory frameworks, to navigate these complexities.

Nanotech-Based Wearables

These are wearable devices that combine the unique quality of nanotechnology and sensors to monitor the health of an individual and hence have significant potential in healthcare. Advances in these wearables such as smart clothing, electronic-textiles (e-textiles), and smart watches among others [226] have evolved rapidly to combine unique electronics, actuators, sensors, and generation sources. These materials are closely worn on the human body to measure and gather real-time data, for effective analysis. This provides an insight into the health of the individual by detecting and quantifying signals from the skin [227]. For example, these wearables could advance healthcare monitoring of patients remotely using nanosensors that are effectively embedded into cloth. This helps provide an alert system for medical professionals and the wearer on changes in their body system for immediate healthcare. Several nanotech-based wearables have been fabricated for different applications like heart rate, sweat, and blood pressure monitoring.

Heart Beat Monitoring

The human heart is an essential organ, where monitoring of its functionality is imperative to help ensure the circulation of blood through the blood vessels. The effective monitoring and measuring of the heart beat is done using the heart rate, which according to [228] is the measure of the "number of contractions of the heart per minute (bpm)". Herein, the human skin is attached with an electrode, where the electrocardiogram (ECG) can be analyzed from the heart rate [229]. The monitoring of these vitals in the body is considered a unique indicator of the health status or well-being of the individual [226]. The discovery of carbon nanotubes (CNT) [230] and graphene [231] have advanced the possibilities of using nanostructures for sensing applications most especially piezoresistive sensors [232]. Furthermore, 2D materials such as transition metal dichalcogenides, hexagonal boron nitride, black phosphorus, metal phosphorus chalcogenide, and transition metal carbides (MXenes) have also been experimented on for sensing applications due to varying functionality and performance [233, 234].

Premise on this, the design, development, and combination of intelligent sensors or devices and nanotechnology creates an effective tool to facilitate real monitoring of the heartbeat to understand the psychological reaction, breathing, and heart conditions of the individual. For example, these are vital indicators when monitoring an athlete's training performance [235]. The use of nanofibers for wearables according to [236, 237] effectively captures accurate electrical signals of the heart which can be digitally processed. To evaluate the physiological signals of the human heart of an athlete, a recent study conducted by [237] fabricated a nanoparticle doped-graphene film that was embedded in clothing as a heart rate sensor module. The fabrication process largely employed graphene due to their unique mechanical properties which makes them good materials for strain sensors [238]. Results show that the porous graphene film (PGF-2) with 20 percent of nanoparticles as its pore template exhibited better sensitivity to small pressures and better stability. These properties make it suitable for capturing the signals in real time. Still on the use of graphene, a previous study by [239] reported on the fabrication of a graphene-based textile used for ECG monitoring (Fig. 3.6a, b). Here, vacuum filtration was used to reduce graphene oxide for the flexible graphene textile. Results show high sensitivity, good impedance of 116.2 KΩ to 24.7 KΩ, effective electrical performance, and superior long-term monitoring even at wet state. In further research conducted by [240], a porous flexible nano-graphene was fabricated on a substrate using a low-power laser diode for electrodes capable of ECG monitoring. The use of porous graphene according to [241] is largely due to their high electrical conductivity which can be induced using a photothermal process on carbon-rich material surfaces. In the experimental work, the substrate of the flexible polyimide is composed of an electrophysical electrode, where electrodes of negative and positive terminals are respectively positioned at the right and left wrist. This together with a clustering algorithm was tested to monitor and record the heart rate signals. Results show that the interface skin electrodes (Fig. 3.6c, d) have a specific contact area due to the material porosity. Additionally, out of 70 R-peaks, the electrodes were able to detect 66 R-peaks, resulting in a 94.3% accuracy for heart rate monitoring (Fig. 3.6e).

Aside from the use of graphene as an effective material for fabricating electrodes, carbon nanotubes (CNT) are also some promising materials used for nanostructures for sensing applications. For effective monitoring of electrocardiograms (EKGs) on stretchable clothing, [242] fabricated a carbon nanotube (CNT) fiber or threads by dip coating in CNT ink. The fabricated CNT threads are non-toxic and have excellent contact impedance with skin and it is relatively electrically conductive and strong [243, 244]. These CNT threads are sewn onto stretch clothing using zigzag stitches. Experimental tests revealed the ability of the threads to withstand greater flex fatigue, have high tensile strength, low interfacial impedance, and electrical conductivity, to detect and transmit signals from the human body. In a different approach, CNT

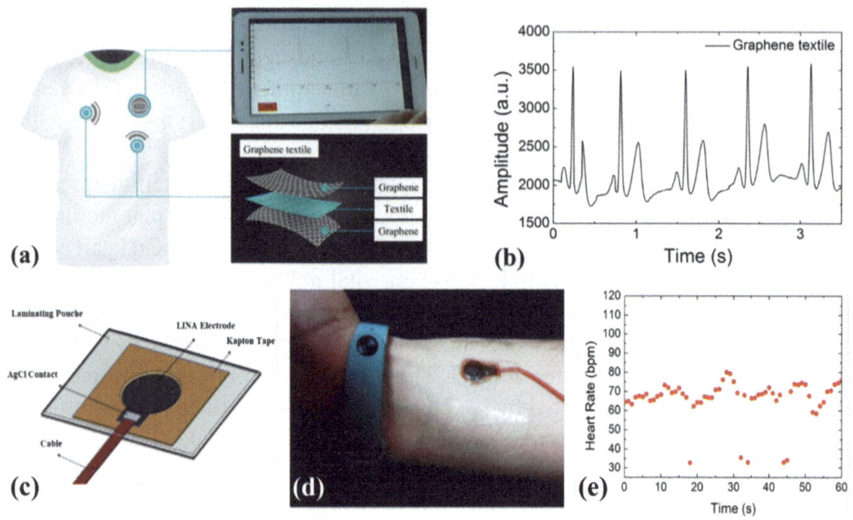

Fig. 3.6 **a** Design and application of the graphene textile. **b** ECG signals of the graphene textile [239]. **c** Flexible electrode schematic. **d** Image of a forearm attached with a flexible electrode. **e** RS-peaks frequency shows the heart rate over time [240]

(with excellent mechanical, thermal, and electrical properties) and polydimethylsiloxane PDMS, due to its optical transparency, flexibility, and elastic properties, were combined for use as a dry electrode. Herein, a study by [245], fabricated a dry electrode for ECG monitoring using CNT/PDMS composites in a two-step dispersal method (where under high shear flow conditions, highly concentrated CNTs are first dispersed and later diluted to the desired concentration). It was evident from the findings that the quality of signal from the composite electrodes is dependent on the size and composition. Further results revealed that there were no side effects when in contact with the skin with the electrodes being flexible and efficient for long-term ECG signal measurements. This advanced the production of high-performance and robust CNT/PDMS composite electrodes that could function under sweat and motion conditions for effective healthcare monitoring.

Sweat Monitoring

Sweat is typically a biological fluid produced from the pores of the human body. These are normally monitored to check the health condition of the human hence providing the electrolyte balance, hydration, and protein levels among others. The individual's physical and mental health can be determined by using the biochemical information from sweat from the human body [246]. This implies that valuable information on the potential issues and well-being of the individual can be obtained with different monitoring devices. For example, sweat from diabetic patients can be used to determine glucose levels [247]. The advancement of wearable devices with electrodes made from nanoparticles has improved the sensitivity of sweat sensors. This is as a result of the large specific surface area of the structures as compared to traditional conductive materials [248]. These wearables are capable of being close to the skin for non-invasive, painless, non-destructive, and continuous detection and monitoring [249]. Materials such as carbon-based and metal oxide nanomaterials are widely used for electrochemical sensors for sweat metabolites due to their good chemical and electrical behavior, good sensitivity, and conductivity [250].

Studies have been conducted using different nanomaterials for wearables that have direct interfacing with the skin surface in sweat monitoring. [251] fabricated a smart textile biosensor from modified graphene-based nanocomposites. With a 3.06% RSD, the developed sensor can detect lactate from sweat with 1.3mM limits of quantification LOQ and limits of detection of 0.4mM. The sensor produced still maintained good reproducibility and consistency in measuring lactate levels even after being cleaned with a washing machine. Glucose was detected in sweat using a flexible biosensor fabricated from reduced graphene oxide nanocomposites [252]. In the experimental process, electrostatic interaction was used to fabricate graphene oxide on silicon dioxide (SiO_2) nanospheres surface, which was later reduced to electroactive rGO/SiO_2 nanocomposites using tannic acid. The sensor exhibited a 60.8 $\mu A\ mM^{-1}\ cm^{-2}$ sensitivity and a 0.1–9 mM range of linear detection for glucose in sweat. CNT electrodes were used to fabricate a sweat wearable sensor [253]. Herein, a solid-state ion-selective electrode that is responsive to Na+ ions was fabricated by applying a coating of plasticized poly(vinyl chloride) infused with ionophore and ion exchanger onto the CNT electrodes. The sensor has a sensitivity of 58 mV/decade and a response time of 57s due to its flexibility, reliability, and stability. For uric acid detection in sweat, a flexible carbon nanofiber (CNF)-based sensor was fabricated with good electrical conductivity and sensitivity [254]. This therefore shows the potential of using CNF membrane for personalized non-invasive diagnosis and monitoring.

Aside from graphene and CNT, gold nanowire electrodes can be fabricated for glucose detection in sweat. A *Enokitake* mushroom-like gold nanowire electrodes were developed by modifying the electrodes with Prussian blue nanoparticles and glucose oxidase [255]. In an artificial sweat, glucose was detected with a sensitivity of 4.55 $\mu A{\cdot}mM^{-1}{\cdot}cm^{-2}$. This proves the potential of continuous biological signal monitoring. A related study by [256] reported on a highly sensitive biosensor to detect the concentration of ethanol in sweat, since an unusual level of ethanol in a person's biofluids can be attributed to health issues from internal diseases. Here, gold nanowire aerogel was used in fabricating the sensor using a substrate-assisted growth method. With 0.01 and 0.5 M range of ethanol concentration, the fabricated bioelectrode demonstrated rapid and efficient linear current reactions.

Blood Pressure Monitoring

Monitoring of one's blood pressure (BP) is important since it helps in the early detection of high blood pressure (hypertension) and is a contributory factor for death-causing cardiovascular diseases in the world [257]. Alternatively, hypertension has resulted in the development of wearable electrochemical nanosensors which according to [258] have exhibited a high surface area-to-volume ratio, high sensitivity, and quick response rate. These properties help in the detection of hypertension biomarkers such as uric acid, cortisol, nitric oxide, galectin-3, ghrelin, leptin, adiponectin, and nitric oxide among others [258]. Approximately 1.4 billion people are suffering from arterial hypertension; hence, rapid advancement of technology for effective and timely detection is important. According to [259], arterial catheterization, despite being the most dependable technique for continuous blood pressure tracking, is invasive and generally reserved for critically ill patients. Furthermore, sphygmomanometers offer a less invasive way to monitor blood pressure, but they come with certain limitations such as being non-portable and providing only intermittent readings [259]. Due to the properties of nanomaterials, studies have reported on the fabrication of wearable sensors, which are portable, non-invasive, and provide real-time data for monitoring blood pressure in the human body.

A highly shaped and sensitive flexible pressure sensor was fabricated by [260] using reduced graphene oxide which is introduced in two thin flexible polydimethylsiloxane (PDMS) sheets. Results show that, after 20,000 cycles of loading and unloading, these sensors maintained exceptional stability. They further demonstrated a high sensitivity of 50.9 per kilopascal within the range of 3 to 1000 Pascals and a rapid response time of 50 ms for effective monitoring of blood pressure. The pulse arrival time determined by the electrocardiogram (ECG) and blood pulse wave can facilitate continuous monitoring of blood pressure. As such, [261] fabricated a paired printed stretchable sensor with an accelerometer and ECG electrodes to monitor blood pressure. Herein, ink from silver nanoparticles was fabricated using aerosol spraying and direct printing methods. The accelerometer and ECG electrodes are attached to the right wrist and the chest, respectively. Results show that the accelerometer was used to measure the pulse waves and the blood pressure and ECG signal was measured with the electrode. Here, [262] reported on a wearable graphene bioimpedance electronic tattoo for continuous and effective arterial blood pressure monitoring. Results revealed that, with an accuracy of 0.2 ± 5.8 mm Hg and 0.2 ± 4.5 mm Hg for systolic and diastolic pressures, respectively, the fabricated bioimpedance platform can measure blood pressure in a non-invasive and continuous manner.

Transduction mechanisms such as triboelectric, piezoresistive, capacitive, and piezoelectric can be fabricated for use as flexible sensors to obtain electrical signals from the human skin for monitoring BP [259]. For example, piezoresistive sensors operate when there is a resistance change when activated. A wearable flexible piezoresistive sensor (FPS) patch system was fabricated for cuffless measurement of blood pressure using epidermal electrocardiogram and piezoresistive sensors [263]. In this work, three epidermal ECG electrodes were combined with the FPS made of a carbon black decorated fabric. Results show that the fabricated sensor has the potential to monitor subtle variations of physiological signals with an ultralow power consumption of 3 nW. The FPS can capture precise pulse waveform due to its fast response time, high sensitivity, and 0–35 Kpa linear region. A highly flexible piezoresistive pressure sensor was fabricated using rough conductive electrodes made of highly elastic and structured CNT thin films [264]. The sensor exhibited a response time of less than 20 ms and a pressure sensitivity of 12 800 folds.

Ethical and Social Implication of Nanotech-Based Wearables in Healthcare

The use of wearables based on nanotechnology has presented an effective opportunity to replace rigid devices and sensors to detect, measure, and quantify the signals of the human body for effective personal monitoring. These wearables have great potential for revamping the way the health of an individual is monitored for immediate response systems. This will help limit the threat of late detection of medical conditions that put the life of the individual at risk. The conception or perception of the public about nanotechnology is at its early stages [265]. However, there are several significant social and ethical implications worth considering to further advance their application. These issues are drawn on the security, privacy and protection of data, data quality, access to data, and regulatory concerns.

The most significant ethical issue in the use of wearable-based nanotechnology draws on the issue of data protection or security and privacy. Nanotech-based wearables are commonly worn close to the skin to effectively detect, monitor, record, and transfer vital information about the wearer. These real-time data on the individual's health are transferred via wireless connectivity for prompt decisions by their healthcare providers [266–268]. The collection of such large volumes of health information raises serious privacy concerns, data sharing, and security issues [269]. There are great concerns on the effective protection of health data to help limit its misuse. Alternatively, the act of data breaches by third parties puts the data of the individual's health at great risk. This thus calls for great efforts to safeguard the health information of patients by the manufacturers and health providers. A clear situation that improves the trust of patients concerning the security of their sensitive information. Aside from this, any mistrust from patients and the public will lead to resistance, which could further affect the utilization of these nanotech-based wearables for healthcare monitoring and delivery of early treatment.

Data from these nanotech-based wearables raises the issue of data quality since these wearables record and transfer vital health information. Hence, the quality, reliability, and accuracy of these scientific data for use by healthcare providers is of great importance. It is regarded as a fundamental value that helps individuals to benefit immensely from treatments provided at hospitals or clinics [270]. Such data quality is compromised with the issue of data variability obtained from different parts of the individual's body. These could be attributed to the varying sensors used to measure and record signals which adopt different techniques and parameters for processing [269, 271]. For example, varying data are collected for the measurement of saturation of oxygen from different parts of the human body using different sensor types [272]. Alternatively, poor data for a patient with heart-related problems could lead to unnecessary panic that could seriously affect their health [273]. This could further lead to mistrust of individuals concerning nanotech-based wearables for significant health monitoring. This raises concerns about common standards to be applied during data collection. It is thus vital for setting international standards to help validate the quality of the data obtained for the needed healthcare. The standardization of procedures used for the experiments coupled with the appropriate techniques and sensors used are important to validate the trustworthiness, reliability, accuracy, and quality of data collected from the individual irrespective of the part of the human body.

Access to vital information about one's health by the wearers or patients is of great concern with regard to the use of nanotech-based wearables for healthcare. It is worth noting that market-ready wearables are developed by large corporations who can easily collect health information of wearers for various advertising and marketing schemes [269]. Furthermore, device manufacturers in most cases do not share these data with users or wearers, hence making it different to effectively analyze and understand how their health conditions are detected. This is largely due to the non-sharing and non-access of data by the wearer to fully understand and benefit from it [274]. Easy access to these data by users according to [269] could improve the quality and trustworthiness of the data provided by these wearables and counter the political and economic powers held by tech developers and corporations [275].

Lastly, regulatory concerns are significant to help improve the innovation and utilization of nanotech-based wearables. Others include ensuring the safety of data through strict policy regulations by the Government and relevant stakeholders to help limit any misuse by third parties and mistrust by wearers. The setting of the regulations will provide some standards for the innovation of quality products and the subsequent security of data.

To successfully address these challenges, there is a need to conduct ethical and social research studies to understand user's perceptions on the utilization of these nanotech-based wearables for healthcare. Key findings from such studies will guide stakeholders, policymakers, and manufacturers to develop effective frameworks and policies or regulations for data security and access and innovative products for data quality.

Summary of Ethical and Social Acceptance Issues of Nanotechnology in Healthcare

Research on nanotechnology has prioritized healthcare, where advances in human health and quality of life can be achieved through the convergence of nanoscience, medicine, molecular biology, and cell biology. In essence, the trajectory of nanotechnology's future development and application is likely to be influenced by consumer acceptance of its various applications [276]. According to earlier studies, professionals involved in the development and regulation of nanotechnology believe that users will embrace applications they deem good and useful and reject those they perceive dangerous [277]. Despite the great potential of nanotechnology, its safety in humans, animals, and plants and its effects on the environment are of concern [278]. The legal, social, and ethical implications of nanomedical technology's wider therapeutic applications must be carefully considered, while also being aware of the associated risks, given the field's rapid advancement [279]. Several products based on nanotechnology were previously released; however, they were subsequently taken off the market due to public reports of negative consequences. For instance, the first product recall owing to health concerns about nanoparticles was documented in Germany [280]. However, the advancement of nanotechnology in an ethical manner requires tackling a number of issues that can be divided into three main categories: privacy, environmental protection, and nanotoxicity [281].

Clinical study participants draw attention to the possible hazards associated with nanochemical medications [282], which pose a barrier and create issues with obtaining participants' fully informed consent. Similarly, society believes that patients are in a potentially vulnerable position as a result of technological advancements and their integration with other technologies, like as AI and ML, where personal data could be utilized against patient interest [279]. Likewise, medical personnel handling nanomaterials in a healthcare context may also be in danger of ingesting, inhaling, or coming into touch with them through their skin. Unequal access is another issue that has been raised by society. Creating more economical and effective nanocarriers, guaranteeing the safety of nanocarriers in humans, and advancing knowledge of tumor biology are some other obstacles [59]. Nonetheless, the aforementioned issues empower regulatory organizations and international collaboration to establish standards that unify methods for developing, assessing, and testing nanotechnological products and devices. For example, study on the ethical and legal aspects of nanotechnology is the focus of the Centers for Nanotechnology in Society at Arizona State University and the University of California, Santa Barbara. Additionally, since 2014, the FDA's Center for Drug Evaluation and Research (CDER) has provided expertise to the Nanotechnology Risk Assessment Working Group in evaluating the possible impact of nanotechnology on pharmaceuticals [283]. In order to use nanotechnology in clinical settings and to address significant regulatory issues with products based on nanotechnology, FDA permission is required [44]. The Food and Drug Administration (FDA) has approved more than twenty medicines based on nanotechnology for use in clinical settings to date, and additional products are undergoing clinical studies [128]. For example, a prospective FDA-approved nanotechnology-based vaccination has demonstrated its significant utility in global prophylaxis against the COVID-19 pandemic, while gold nanoparticles and nanorods have demonstrated promise in photothermal therapy and imaging applications.

Ethically, numerous procedures have been taken to guarantee the sustainable and safe use of nanomaterials in the medical field. For example, thorough risk assessments are carried out to determine the possible risks and hazards associated with particular nanomaterials prior to their deployment in medical applications. Comparably, sufficient regulatory frameworks are established to guarantee the safe manufacture and application of nanomaterials as well as to reduce occupational exposure to nanomaterials [284, 285]. Furthermore, testing protocols are established to evaluate the effectiveness and safety of nanomaterials used in medicine. This includes using engineering controls, toxicity testing, characterization, quality control, standard operating procedures, personal protective equipment (PPE), and staff training initiatives [284]. Furthermore, building trust among stakeholders, including patients, healthcare providers, and the general public, requires open communication about the possible hazards and advantages of nanomaterials [286]. Since the ability to control when and how one's data is shared with others and used is becoming an increasingly important aspect of one's right to privacy, protecting privacy could limit the potential harm that could result from information misuse and expand the use of nanotechnology in healthcare.

Conclusion and Future Perspectives

Nanotechnology applications in nanomedicine and nanobots hold significant promise for transforming healthcare and medicine. Exciting advancements in the use of nanoparticles for more focused drug delivery to cancerous tumors, as well as methods for regenerative medicine and overcoming biological barriers like the blood–brain barrier for more accurate illness diagnostics, have already been designed. Furthermore, tiny nanorobots are used in biomedical applications to carry out sensing, actuation, and computational tasks inside the body. Therefore, the nanoplatform presents a fascinating chance to acquire multifunctionality and targeted techniques. In a more speculative scenario, shape-shifting nanobots might even be able to physically destroy or mend certain sick cells. Additionally, wearable cardiac monitor feedback mechanisms based on enhanced sensing due to physiochemical properties of nanoparticles, such as nanoscale gaps capable of label-free, single-molecule sensing and detection, plasmonic nanomaterials, core–shell nanostructures, nanoarrays, and single quantum dot-based nanosensors, could be used to detect physiologic changes and start the right treatment response. Thus, using nanorobots to monitor several physiological parameters in real time can help with early disease identification and diagnosis. Another potential application is in vivo diagnostics where wearables and smartphone apps are already being used by patients to record their

ECG and employ artificial intelligence (AI) algorithms to detect high heart rates and arrhythmias and deliver notifications. As a result, possible advancements might involve using nanotechnology to improve treatment methods, identify the source, and categorize severity. Furthermore, the integration of genomic profiling of patients with nanotechnology platforms and genome editing tools such as Clustered Regularly Interspaced Short Palindromic Repeats (CRISPR) can enhance precision and broaden the scope of personalized medicine. Advanced nanomedicine technologies may eventually contribute to radical anti-aging and life extension through tissue engineering, organ regeneration, and cellular repair. Medical nanorobot development has considerable promise for a range of innovative biomedical uses, such as tissue repair, immunological system control, cancer treatment, diagnostics, and microsurgery. Importantly, drug delivery applications have seen success using conventional nanoparticles. While still largely in the research stage, advancements in nanotechnology, nanomachines, and materials science are bringing more sophisticated medical nanobots closer to reality. Their tiny size allows access to the body in unprecedented ways, carrying the potential to transform medicine across prevention, diagnosis, and treatment. Continued research and development, clinical translation, and commercialization efforts in nanomedicine and medical nanobots will help drive better patient outcomes and expand treatment options for diseases in the coming decades.

References

1. Prabhu, S., Poulose, E.K.: Silver nanoparticles: mechanism of antimicrobial action, synthesis, medical applications, and toxicity effects. Int. Nano Lett. **2**, 32 (2012)
2. Wang, A.Z., Langer, R., Farokhzad, O.C.: Nanoparticle delivery of cancer drugs. Annu. Rev. Med. **63**, 185–198 (2012)
3. Hobson, D.W.: The commercialization of medical nanotechnology for medical applications. In: Prokop, A., Weissig, V. (eds.) Intracellular Delivery III: Market Entry Barriers of Nanomedicines, pp. 405–449. Springer International Publishing, Cham (2016)
4. Sindhwani, S., Chan, W.C.W.: Nanotechnology for modern medicine: next step towards clinical translation. J. Intern. Med. **290**, 486–498 (2021)
5. Kim, B.Y.S., Rutka, J.T., Chan, W.C.W.: Nanomedicine. N. Engl. J. Med. **363**, 2434–2443 (2010)
6. Chouhan, R.S., Jerman, I., Heath, D., Bohm, S., Gandhi, S., Sadhu, V., et al.: Emerging tri-s-triazine-based graphitic carbon nitride: a potential signal-transducing nanostructured material for sensor applications. Nano Select. **2**, 712–743 (2021)
7. Mekuye, B., Abera, B.: Nanomaterials: an overview of synthesis, classification, characterization, and applications. Nano Select. **4**, 486–501 (2023)
8. Nasrollahzadeh, M., Issaabadi, Z., Sajjadi, M., Sajadi, S.M., Atarod, M.: Chapter 2—types of nanostructures. In: Nasrollahzadeh, M., Sajadi, S.M., Sajjadi, M., Issaabadi, Z., Atarod, M. (eds.) Interface Science and Technology, pp. 29–80. Elsevier (2019)
9. Jeong, H., Lim, S.: A stretchable radio-frequency strain sensor using screen printing technology. Sensors. **16**, 1839 (2016)
10. Fan, X., Wang, N., Yan, F., Wang, J., Song, W., Ge, Z.: A transfer-printed, stretchable, and reliable strain sensor using PEDOT:PSS/Ag NW hybrid films embedded into elastomers. Adv. Mater. Technol. **3**, 1800030 (2018)
11. Liu, H., Jiang, H., Du, F., Zhang, D., Li, Z., Zhou, H.: Flexible and degradable paper-based strain sensor with low cost. ACS Sustain. Chem. Eng. **5**, 10538–10543 (2017)

12. Oh, J., Yang, J.C., Kim, J.-O., Park, H., Kwon, S.Y., Lee, S., et al.: Pressure insensitive strain sensor with facile solution-based process for tactile sensing applications. ACS Nano **12**, 7546–7553 (2018)
13. Kao, H.-L., Cho, C.-L., Chang, L.-C., Chen, C.-B., Chung, W.-H., Tsai, Y.-C.: A fully inkjet-printed strain sensor based on carbon nanotubes. Coatings **10**, 792 (2020)
14. Vu, C.C., Kim, J.: Human motion recognition by textile sensors based on machine learning algorithms. Sensors. **18**, 3109 (2018)
15. Liao, X., Zhang, Z., Kang, Z., Gao, F., Liao, Q., Zhang, Y.: Ultrasensitive and stretchable resistive strain sensors designed for wearable electronics. Mater. Horiz. **4**, 502–510 (2017)
16. Khan, A., Haque, M.N., Kabiraz, D.C., Yeasin, A., Al Rashid, H., Sarker, A.C., Hossain, G.: A review on advanced nanocomposites materials based smart textile biosensor for healthcare monitoring from human sweat. Sens. Actuators A **350**, 114093 (2023)
17. Crawford, M.: 10 Ways Nanotechnology Impacts Our Lives. Chemical and Pharmaceutical Manufacturing. The American Society of Mechanical Engineers, United States (2016)
18. Nag, A., Mukhopadhyay, S.C.: Nanoparticles-based flexible wearable sensors for health monitoring applications. In: Kumar, C.S.S.R. (ed.) Nanotechnology Characterization Tools for Environment, Health, and Safety, pp. 245–284. Springer, Berlin, Heidelberg (2019)
19. Bennett, J., Rokas, O., Chen, L.: Healthcare in the smart home: a study of past, present and future. Sustainability **9**, 840 (2017)
20. Haleem, A., Javaid, M., Singh, R.P., Rab, S., Suman, R.: Applications of nanotechnology in medical field: a brief review. Global Health J. **7**, 70–77 (2023)
21. Mashaghi, S., Jadidi, T., Koenderink, G., Mashaghi, A.: Lipid nanotechnology. Int. J. Mol. Sci. **14**, 4242–4282 (2013)
22. Khan, A.U., Khan, M., Cho, M.H., Khan, M.M.: Selected nanotechnologies and nanostructures for drug delivery, nanomedicine and cure. Bioprocess Biosyst. Eng. **43**, 1339–1357 (2020)
23. Abdellatif, A.A.H., Alsowinea, A.F.: Approved and marketed nanoparticles for disease targeting and applications in COVID-19 **10**, 1941–1977 (2021)
24. Malhotra, V.: Biomedical nanotechnology: current status and future prospects in disease diagnosis and therapy. Biomed. Syst. Emerg. Technol. **10**, 168 (2023)
25. Anjum, S., Ishaque, S., Fatima, H., Farooq, W., Hano, C., Abbasi, B.H., Anjum, I.: Emerging applications of nanotechnology in healthcare systems: grand challenges and perspectives. Pharmaceuticals (Basel) **14** (2021)
26. Shah, S.S., Shaikh, M.N., Khan, M.Y., Alfasane, M.A., Rahman, M.M., Aziz, M.A.: Present status and future prospects of jute in nanotechnology: a review. Chem. Rec. **21**, 1631–1665 (2021)
27. Yusuf, A., Almotairy, A.R.Z., Henidi, H., Alshehri, O.Y., Aldughaim, M.S.: Nanoparticles as drug delivery systems: a review of the implication of nanoparticles' physicochemical properties on responses in biological systems. Polymers (Basel) **15** (2023)
28. Ramburrun, P., Khan, R.A., Choonara, Y.E.: Design, preparation, and functionalization of nanobiomaterials for enhanced efficacy in current and future biomedical applications **11**, 1802–1826 (2022)
29. Gagliardi, A., Giuliano, E., Venkateswararao, E., Fresta, M., Bulotta, S., Awasthi, V., Cosco, D.: Biodegradable polymeric nanoparticles for drug delivery to solid tumors. Front. Pharmacol. **12**, 601626 (2021)
30. Rhaman, M.M., Islam, M.R., Akash, S., Mim, M., Noor alam, M., Nepovimova, E., et al.: Exploring the role of nanomedicines for the therapeutic approach of central nervous system dysfunction: at a glance. Front. Cell Dev. Biol. **10** (2022)
31. Boisseau, P., Loubaton, B.: Nanomedicine, nanotechnology in medicine. Comptes rendus hebdomadaires des séances de l'Académie des sciences (2011)
32. Javaid, A., Oloketuyi, S.F., Khan, M.M., Khan, F.: Diversity of bacterial synthesis of silver nanoparticles. BioNanoScience. **8**, 43–59 (2018)
33. Ebrahimi, E., Akbarzadeh, A., Abbasi, E., Khandaghi, A.A., Abasalizadeh, F., Davaran, S.: Retracted article: novel drug delivery system based on doxorubicin-encapsulated magnetic

nanoparticles modified with PLGA-PEG1000 copolymer. Artif. Cells Nanomed. Biotechnol. **44**, 290–297 (2016)
34. Patra, J.K., Das, G., Fraceto, L.F., Campos, E.V.R., Rodriguez-Torres, M.P., Acosta-Torres, L.S., et al.: Nano based drug delivery systems: recent developments and future prospects. J. Nanobiotechnol. **16**, 1–33 (2018)
35. Dokoumetzidis, A., Macheras, P.: A century of dissolution research: from noyes and whitney to the biopharmaceutics classification system. Int. J. Pharm. **321**, 1–11 (2006)
36. Naahidi, S., Jafari, M., Edalat, F., Raymond, K., Khademhosseini, A., Chen, P.: Biocompatibility of engineered nanoparticles for drug delivery. J. Control. Release **166**, 182–194 (2013)
37. Shi, X., Sun, K., Baker, J.R., Jr.: Spontaneous formation of functionalized dendrimer-stabilized gold nanoparticles. J. Phys. Chem. C **112**, 8251–8258 (2008)
38. Hollis, C.P., Weiss, H.L., Leggas, M., Evers, B.M., Gemeinhart, R.A., Li, T.: Biodistribution and bioimaging studies of hybrid paclitaxel nanocrystals: lessons learned of the EPR effect and image-guided drug delivery. J. Control. Release **172**, 12–21 (2013)
39. Salvatore, A., Montis, C., Berti, D., Baglioni, P.: Multifunctional magnetoliposomes for sequential controlled release. ACS Nano **10**, 7749–7760 (2016)
40. Yao, C., Li, J., Cao, X., Gunn, J.R., Wu, M., Jiang, S., Pogue, B.W.: X-ray-induced cherenkov optical triggering of caged doxorubicin released to the nucleus for chemoradiation activation. ACS Appl. Mater. Interfaces **12**, 44383–44392 (2020)
41. Peltonen, L., Hirvonen, J.: Pharmaceutical nanocrystals by nanomilling: critical process parameters, particle fracturing and stabilization methods. J. Pharm. Pharmacol. **62**, 1569–1579 (2010)
42. Lam, P.L., Wong, W.Y., Bian, Z., Chui, C.H., Gambari, R.: Recent advances in green nanoparticulate systems for drug delivery: efficient delivery and safety concern. Nanomedicine (Lond.) **12**, 357–385 (2017)
43. Patra, J.K., Das, G., Fraceto, L.F., Campos, E.V.R., Rodriguez-Torres, Md.P., Acosta-Torres, L.S., et al.: Nano based drug delivery systems: recent developments and future prospects. J. Nanobiotechnol. **16**, 71 (2018)
44. Deepak Kumar, D., Rajni Kant, P., Anil Kumar, S., Vaibhav, T.: Role of nanobiotechnology in drug discovery, development and molecular diagnostic. In: Margarita, S., Roumen, Z. (eds.) Applications of Nanobiotechnology, p. Ch. 2. IntechOpen, Rijeka (2020)
45. Goday, S.: Application of nanotechnology in drug development. Austin Pharm. Pharm. (2023)
46. Lee, B.K., Yun, Y.H., Park, K.: Smart nanoparticles for drug delivery: boundaries and opportunities. Chem. Eng. Sci. **125**, 158–164 (2015)
47. Yezdani, U., Khan, M., Kushwah, N., Verma, A., Khan, F., Khan, M.: Nanotechnology in diagnosis and treatment of various diseases and its future advances in medicine. World J. Pharm. Pharm. Sci. **7**, 1612–1633 (2018)
48. Witharana, C., Wanigasekara, J.: Applications of nanotechnology in drug delivery and design—an insight. Curr. Trends Biotechnol. Pharm. **10**, 78–91 (2016)
49. De Jong, W.H., Borm, P.J.A.: Drug delivery and nanoparticles: applications and hazards. Int. J. Nanomed. **3**, 133–149 (2008)
50. Bobo, D., Robinson, K.J., Islam, J., Thurecht, K.J., Corrie, S.R.: Nanoparticle-based medicines: a review of FDA-approved materials and clinical trials to date. Pharm. Res. **33**, 2373–2387 (2016)
51. Díaz, J.E.M.: Nanotechnology in pharmacology: advances and applications in drug delivery. Pharmacol. Clin. Res. **9**, 1–8 (2023)
52. Kaur, R., Bhardwaj, A., Gupta, S.: Cancer treatment therapies: traditional to modern approaches to combat cancers. Mol. Biol. Rep. **50**, 9663–9676 (2023)
53. Debela, D.T., Muzazu, S.G., Heraro, K.D., Ndalama, M.T., Mesele, B.W., Haile, D.C., et al.: New approaches and procedures for cancer treatment: current perspectives. SAGE Open Med. **9**, 20503121211034370 (2021)
54. Kashyap, B.K., Singh, V.V., Solanki, M.K., Kumar, A., Ruokolainen, J., Kesari, K.K.: Smart nanomaterials in cancer theranostics: challenges and opportunities. ACS Omega **8**, 14290–14320 (2023)

55. He, W., Li, Q., Lu, Y., Ju, D., Gu, Y., Zhao, K., Dong, C.: Cancer treatment evolution from traditional methods to stem cells and gene therapy. Curr. Gene Ther. **22**, 368–385 (2022)
56. Gavas, S., Quazi, S., Karpiński, T.M.: Nanoparticles for cancer therapy: current progress and challenges. Nanoscale Res. Lett. **16**, 173 (2021)
57. Lôbo, G.C.N.B., Paiva, K.L.R., Silva, A.L.G., Simões, M.M., Radicchi, M.A., Báo, S.N.: Nanocarriers used in drug delivery to enhance immune system in cancer therapy. Pharmaceutics (2021)
58. Lombardo, D., Kiselev, M.A., Caccamo, M.T.: Smart nanoparticles for drug delivery application: development of versatile nanocarrier platforms in biotechnology and nanomedicine. J. Nanomater. **2019**, 3702518 (2019)
59. Chehelgerdi, M., Chehelgerdi, M., Allela, O.Q.B., Pecho, R.D.C., Jayasankar, N., Rao, D.P., et al.: Progressing nanotechnology to improve targeted cancer treatment: overcoming hurdles in its clinical implementation. Mol. Cancer **22**, 169 (2023)
60. Banerjee, D., Sengupta, S.: Nanoparticles in cancer chemotherapy. Prog. Mol. Biol. Transl. Sci. **104**, 489–507 (2011)
61. Abdellatif, A.A.H., Tawfeek, H.M.: Development and evaluation of fluorescent gold nanoparticles. Drug Dev. Ind. Pharm. **44**, 1679–1684 (2018)
62. Singh, P., Pandit, S., Mokkapati, V., Garg, A., Ravikumar, V., Mijakovic, I.: Gold nanoparticles in diagnostics and therapeutics for human cancer. Int. J. Mol. Sci. **19** (2018)
63. Tabish, T.A., Hayat, H., Abbas, A., Narayan, R.J.: Graphene quantum dot–based electrochemical biosensing for early cancer detection. Curr. Opin. Electrochem. **30**, 100786 (2021)
64. Yang, S.-J., Lin, F.-H., Tsai, H.-M., Lin, C.-F., Chin, H.-C., Wong, J.-M., Shieh, M.-J.: Alginate-folic acid-modified chitosan nanoparticles for photodynamic detection of intestinal neoplasms. Biomaterials **32**, 2174–2182 (2011)
65. Lu, B., Sun, L., Yan, X., Ai, Z., Xu, J.: Intratumoral chemotherapy with paclitaxel liposome combined with systemic chemotherapy: a new method of neoadjuvant chemotherapy for stage III unresectable non-small cell lung cancer. Med. Oncol. **32**, 345 (2015)
66. Mahner, S., Meier, W., du Bois, A., Brown, C., Lorusso, D., Dell'Anna, T., et al.: Carboplatin and pegylated liposomal doxorubicin versus carboplatin and paclitaxel in very platinum-sensitive ovarian cancer patients: results from a subset analysis of the CALYPSO phase III trial. Eur. J. Cancer **51**, 352–358 (2015)
67. Pishavar, E., Ramezani, M., Hashemi, M.: Co-delivery of doxorubicin and TRAIL plasmid by modified PAMAM dendrimer in colon cancer cells, in vitro and in vivo evaluation. Drug Dev. Ind. Pharm. **45**, 1931–1939 (2019)
68. Jędrzak, A., Grześkowiak, B.F., Coy, E., Wojnarowicz, J., Szutkowski, K., Jurga, S., et al.: Dendrimer based theranostic nanostructures for combined chemo- and photothermal therapy of liver cancer cells in vitro. Colloids Surf. B Biointerfaces **173**, 698–708 (2019)
69. Javanbakht, S., Namazi, H.: Doxorubicin loaded carboxymethyl cellulose/graphene quantum dot nanocomposite hydrogel films as a potential anticancer drug delivery system. Mater. Sci. Eng. C **87**, 50–59 (2018)
70. Bulbake, U., Kommineni, N., Ionov, M., Bryszewska, M., Khan, W.: Comparison of cationic liposome and PAMAM dendrimer for delivery of anti-Plk1 siRNA in breast cancer treatment. Pharm. Dev. Technol. **25**, 9–19 (2020)
71. Singla, R., Sharma, C., Shukla, A.K., Acharya, A.: Toxicity concerns of therapeutic nanomaterials. J. Nanosci. Nanotechnol. **19**, 1889–1907 (2019)
72. Wang, L., Song, X., Yu, M., Niu, L., Zhao, Y., Tang, Y., et al.: Serum exosomal miR-377-3p and miR-381-3p as diagnostic biomarkers in colorectal cancer. Future Oncol. **18**, 793–805 (2021)
73. Liu, Y., Peng, N., Yao, Y., Zhang, X., Peng, X., Zhao, L., et al.: Breaking the nanoparticle's dispersible limit via rotatable surface ligands. Nat. Commun. **13**, 3581 (2022)
74. Whiting, D.R., Guariguata, L., Weil, C., Shaw, J.: IDF diabetes atlas: global estimates of the prevalence of diabetes for 2011 and 2030. Diabetes Res. Clin. Pract. **94**, 311–321 (2011)

75. Li, L., Jiang, G., Du, X., Chen, H., Liu, Y., Huang, Q., et al.: Preparation of glucose-responsive and fluorescent micelles via a combination of RAFT polymerization and chemoenzymatic transesterification for controlled release of insulin. RSC Adv. **5**, 75766–75772 (2015)
76. Ding, D., Zhu, Q.: Recent advances of PLGA micro/nanoparticles for the delivery of biomacromolecular therapeutics. Mater. Sci. Eng. C **92**, 1041–1060 (2018)
77. Arbit, E.: The physiological rationale for oral insulin administration. Diabetes Technol. Ther. **6**, 510–517 (2004)
78. Mundargi, R.C., Babu, V.R., Rangaswamy, V., Patel, P., Aminabhavi, T.M.: Nano/micro technologies for delivering macromolecular therapeutics using poly(d, l-lactide-co-glycolide) and its derivatives. J. Control. Release **125**, 193–209 (2008)
79. Kawashima, Y., Yamamoto, H., Takeuchi, H., Fujioka, S., Hino, T.: Pulmonary delivery of insulin with nebulized dl-lactide/glycolide copolymer (PLGA) nanospheres to prolong hypoglycemic effect. J. Control. Release **62**, 279–287 (1999)
80. Sheng, J., Han, L., Qin, J., Ru, G., Li, R., Wu, L., et al.: N-Trimethyl chitosan chloride-coated PLGA nanoparticles overcoming multiple barriers to oral insulin absorption. ACS Appl. Mater. Interfaces **7**, 15430–15441 (2015)
81. Xu, B., Jiang, G., Yu, W., Liu, D., Liu, Y., Kong, X., Yao, J.: Preparation of poly(lactic-co-glycolic acid) and chitosan composite nanocarriers via electrostatic self assembly for oral delivery of insulin. Mater. Sci. Eng. C **78**, 420–428 (2017)
82. Sim, S., Wong, N.K.: Nanotechnology and its use in imaging and drug delivery (review). Biomed. Rep. **14**, 42 (2021)
83. Iqbal, J., Iqbal, A., Mukhtar, H., Jahangir, K., Mashkoor, Y., Zeeshan, M.H., et al.: Cardioprotective effects of nanoparticles in cardiovascular diseases: a state-of-the-art review. Curr. Probl. Cardiol. **48**, 101713 (2023)
84. Sahu, T., Ratre, Y.K., Chauhan, S., Bhaskar, L.V.K.S., Nair, M.P., Verma, H.K.: Nanotechnology based drug delivery system: current strategies and emerging therapeutic potential for medical science. J. Drug Deliv. Sci. Technol. **63**, 102487 (2021)
85. Gupta, P., Garcia, E., Sarkar, A., Kapoor, S., Rafiq, K., Chand, S.H., Jayant, D.R.: Nanoparticle based treatment for cardiovascular diseases. Cardiovasc. Hematol. Disord. Drug Targets **19**, 33–44 (2019)
86. Zhang, H., Li, N., Sirish, P., Mahakian, L., Ingham, E., Curry, F.-R., et al.: The cargo of CRPPR-conjugated liposomes crosses the intact murine cardiac endothelium. J. Control. Release **163**, 10–17 (2012)
87. Wang, D.K., Rahimi, M., Filgueira, C.S.: Nanotechnology applications for cardiovascular disease treatment: current and future perspectives. Nanomed. Nanotechnol. Biol. Med. **34**, 102387 (2021)
88. Carthew, R.W., Sontheimer, E.J.: Origins and mechanisms of miRNAs and siRNAs. Cell **136**, 642–655 (2009)
89. Patil, Y., Panyam, J.: Polymeric nanoparticles for siRNA delivery and gene silencing. Int. J. Pharm. **367**, 195–203 (2009)
90. Cyrus, T., Zhang, H., Allen, J.S., Williams, T.A., Hu, G., Caruthers, S.D., et al.: Intramural delivery of rapamycin with alphavbeta3-targeted paramagnetic nanoparticles inhibits stenosis after balloon injury. Arterioscler. Thromb. Vasc. Biol. **28**, 820–826 (2008)
91. Dvir, T., Bauer, M., Schroeder, A., Tsui, J.H., Anderson, D.G., Langer, R., et al.: Nanoparticles targeting the infarcted heart. Nano Lett. **11**, 4411–4414 (2011)
92. Chen, C., Lumsden, A.B., Hanson, S.R.: Local infusion of heparin reduces anastomotic neointimal hyperplasia in aortoiliac expanded polytetrafluoroethylene bypass grafts in baboons. J. Vasc. Surg. **31**, 354–363 (2000)
93. Bruce, D.: Ethical implications of nanomedicine: implications of imagining new futures for medicine. In: Ge, Y., Li, S., Wang, S., Moore, R. (eds.) Nanomedicine: Principles and Perspectives, pp. 251–269. Springer, New York (2014)
94. Allon, I., Ben-Yehudah, A., Dekel, R., Solbakk, J.-H., Weltring, K.-M., Siegal, G.: Ethical issues in nanomedicine: tempest in a teapot? Med. Health Care Philos. **20**, 3–11 (2017)

95. Kubik, T., Bogunia-Kubik, K., Sugisaka, M.: Nanotechnology on duty in medical applications. Curr. Pharm. Biotechnol. **6**, 17–33 (2005)
96. Desai, N.: Challenges in development of nanoparticle-based therapeutics. AAPS J. **14**, 282–295 (2012)
97. Fisher, E., Boenink, M., van der Burg, S., Woodbury, N.: Responsible healthcare innovation: anticipatory governance of nanodiagnostics for theranostics medicine. Expert Rev. Mol. Diagn. **12**, 857–870 (2012)
98. Nel, A., Grainger, D., Alvarez, P.J., Badesha, S., Castranova, V., Ferrari, M., et al.: Nanotechnology environmental, health, and safety issues. In: Nanotechnology Research Directions for Societal Needs in 2020: Retrospective and Outlook, pp. 159–220. Springer Netherlands, Dordrecht (2011)
99. Resnik, D.B., Tinkle, S.S.: Ethics in nanomedicine. Nanomedicine **2**, 345–350 (2007)
100. Stebbing, M.: Avoiding the trust deficit: public engagement, values, the precautionary principle and the future of nanotechnology. J. Bioethical Inquiry **6**, 37–48 (2009)
101. Marchant, G.E., Sylvester, D.J., Abbott, K.W.: Risk management principles for nanotechnology. NanoEthics **2**, 43–60 (2008)
102. Li, M., Xi, N., Wang, Y., Liu, L.: Progress in Nanorobotics for advancing biomedicine. IEEE Trans. Biomed. Eng. **68**, 130–147 (2021)
103. Maciejasz, P., Eschweiler, J., Gerlach-Hahn, K., Jansen-Troy, A., Leonhardt, S.: A survey on robotic devices for upper limb rehabilitation. J. Neuroeng. Rehabil. **11**, 3 (2014)
104. Laschi, C., Mazzolai, B., Cianchetti, M.: Soft robotics: technologies and systems pushing the boundaries of robot abilities. Sci. Robot. **1**, eaah3690 (2016)
105. Durairaj, R.B., Shanker, J., Sivasankar, M.: Nano robots in Bio medical application. In: IEEE-International Conference on Advances in Engineering, Science and Management, ICAESM-2012 (2012)
106. Agrahari, V., Agrahari, V., Chou, M.L., Chew, C.H., Noll, J., Burnouf, T.: Intelligent micro-/nanorobots as drug and cell carrier devices for biomedical therapeutic advancement: promising development opportunities and translational challenges. Biomaterials **260**, 120163 (2020)
107. Medina-Sánchez, M., Schmidt, O.G.: Medical microbots need better imaging and control. Nature **545**, 406–408 (2017)
108. Nasra, S., Bhatia, D., Kumar, A.: Recent advances in nanoparticle-based drug delivery systems for rheumatoid arthritis treatment. Nanoscale Adv. **4**, 3479–3494 (2022)
109. Li, J., Esteban-Fernández de Ávila, B., Gao, W., Zhang, L., Wang, J.: Micro/Nanorobots for biomedicine: delivery, surgery, sensing, and detoxification. Sci. Robot. **2** (2017)
110. Collins, F.S., Varmus, H.: A new initiative on precision medicine. N. Engl. J. Med. **372**, 793–795 (2015)
111. Anderson, R.L., Balasas, T., Callaghan, J., Coombes, R.C., Evans, J., Hall, J.A., et al.: A framework for the development of effective anti-metastatic agents. Nat. Rev. Clin. Oncol. **16**, 185–204 (2019)
112. Herrmann, I.K., Urner, M., Koehler, F.M., Hasler, M., Roth-Z'Graggen, B., Grass, R.N., et al.: Blood purification using functionalized core/shell nanomagnets. Small **6**, 1388–1392 (2010)
113. Jain, K.K.: Drug delivery systems—an overview. Methods Mol. Biol. **437**, 1–50 (2008)
114. Soto, F., Wang, J., Ahmed, R., Demirci, U.: Medical micro/nanorobots in precision medicine. Adv. Sci. **7**, 2002203 (2020)
115. Erkoc, P., Yasa, I.C., Ceylan, H., Yasa, O., Alapan, Y., Sitti, M.: Mobile Microrobots for active therapeutic delivery. Adv. Ther. **2**, 1800064 (2019)
116. Hoop, M., Ribeiro, A.S., Rösch, D., Weinand, P., Mendes, N., Mushtaq, F., et al.: Mobile magnetic nanocatalysts for bioorthogonal targeted cancer therapy. Adv. Func. Mater. **28**, 1705920 (2018)
117. Andhari, S.S., Wavhale, R.D., Dhobale, K.D., Tawade, B.V., Chate, G.P., Patil, Y.N., et al.: Self-propelling targeted magneto-nanobots for deep tumor penetration and pH-responsive intracellular drug delivery. Sci. Rep. **10**, 4703 (2020)
118. Chen, X.Z., Hoop, M., Shamsudhin, N., Huang, T., Özkale, B., Li, Q., et al.: Hybrid magnetoelectric nanowires for nanorobotic applications: fabrication, magnetoelectric coupling, and magnetically assisted in vitro targeted drug delivery. Adv. Mater. **29** (2017)

119. Sen, C.K., Gordillo, G.M., Roy, S., Kirsner, R., Lambert, L., Hunt, T.K., et al.: Human skin wounds: a major and snowballing threat to public health and the economy. Wound Repair Regen. **17**, 763–771 (2009)
120. Das, S., Baker, A.B.: Biomaterials and nanotherapeutics for enhancing skin wound healing. Front. Bioeng. Biotechnol. **4**, 82 (2016)
121. Bader, A., Lorenz, K., Richter, A., Scheffler, K., Kern, L., Ebert, S., et al.: Interactive role of trauma cytokines and erythropoietin and their therapeutic potential for acute and chronic wounds. Rejuvenation Res. **14**, 57–66 (2011)
122. Soto, F., Chrostowski, R.: Frontiers of medical micro/nanorobotics: in vivo applications and commercialization perspectives toward clinical uses. Front. Bioeng. Biotechnol. **6**, 170 (2018)
123. Hamdi, M., Ferreira, A.: Methodology of design and characterization of bionano and nanorobotic devices. In: Hamdi, M., Ferreira, A. (eds.) Design, Modeling and Characterization of Bio-Nanorobotic Systems, pp. 41–73. Springer, Netherlands, Dordrecht (2011)
124. Nistor, M.T., Rusu, A.G.: Chapter 3—nanorobots with applications in medicine. In: Vasile, C. (ed.) Polymeric Nanomaterials in Nanotherapeutics, pp. 123–49. Elsevier (2019)
125. Chattha, G.M., Arshad, S., Kamal, Y., Chattha, M.A., Asim, M.H., Raza, S.A., et al.: Nanorobots: an innovative approach for DNA-based cancer treatment. J. Drug Deliv. Sci. Technol. **80**, 104173 (2023)
126. Hu, M., Ge, X., Chen, X., Mao, W., Qian, X., Yuan, W.E.: Micro/nanorobot: a promising targeted drug delivery system. Pharmaceutics **12** (2020)
127. McGuire, S.: Geneva, Switzerland: world health organization, international agency for research on Cancer, WHO Press. Adv. Nutr. **7**, 418–419 (2016)
128. Shi, J., Votruba, A.R., Farokhzad, O.C., Langer, R.: Nanotechnology in drug delivery and tissue engineering: from discovery to applications. Nano Lett. **10**, 3223–3230 (2010)
129. Bohunicky, B., Mousa, S.A.: Biosensors: the new wave in cancer diagnosis. Nanotechnol. Sci. Appl. **4**, 1–10 (2010)
130. Nehru, S., Misra, R., Bhaswant, M.: Multifaceted engineered biomimetic nanorobots toward cancer management. ACS Biomater. Sci. Eng. **8**, 444–459 (2022)
131. Peng, R., Zheng, X., Lyu, Y., Xu, L., Zhang, X., Ke, G., et al.: Engineering a 3D DNA-logic gate nanomachine for bispecific recognition and computing on target cell surfaces. J. Am. Chem. Soc. **140**, 9793–9796 (2018)
132. Yang, P.-P., Zhai, Y.-G., Qi, G.-B., Lin, Y.-X., Luo, Q., Yang, Y., et al.: NIR Light propulsive janus-like nanohybrids for enhanced photothermal tumor therapy. Small **12**, 5423–5430 (2016)
133. Mitchell, M.J., Billingsley, M.M., Haley, R.M., Wechsler, M.E., Peppas, N.A., Langer, R.: Engineering precision nanoparticles for drug delivery. Nat. Rev. Drug Discov. **20**, 101–124 (2021)
134. Cuccurullo, V., Di Stasio, G.D., Mazzarella, G., Cascini, G.L.: Microvascular invasion in HCC: the molecular imaging perspective. Contrast Media Mol. Imaging **2018**, 9487938 (2018)
135. Wang, Z., Qiao, R., Tang, N., Lu, Z., Wang, H., Zhang, Z., et al.: Active targeting theranostic iron oxide nanoparticles for MRI and magnetic resonance-guided focused ultrasound ablation of lung cancer. Biomaterials **127**, 25–35 (2017)
136. Thwala, L.N., Ndlovu, S.C., Mpofu, K.T., Lugongolo, M.Y., Mthunzi-Kufa, P.: Nanotechnology-based diagnostics for diseases prevalent in developing countries: current advances in point-of-care tests. Nanomaterials (Basel) **13** (2023)
137. Grebenik, E.A., Nadort, A., Generalova, A.N., Nechaev, A.V., Sreenivasan, V.K., Khaydukov, E.V., et al.: Feasibility study of the optical imaging of a breast cancer lesion labeled with upconversion nanoparticle biocomplexes. J. Biomed. Opt. **18**, 76004 (2013)
138. Vargo, K.B., Zaki, A.A., Warden-Rothman, R., Tsourkas, A., Hammer, D.A.: Superparamagnetic iron oxide nanoparticle micelles stabilized by recombinant oleosin for targeted magnetic resonance imaging. Small **11**, 1409–1413 (2015)
139. Prencipe, G., Tabakman, S.M., Welsher, K., Liu, Z., Goodwin, A.P., Zhang, L., et al.: PEG branched polymer for functionalization of nanomaterials with ultralong blood circulation. J. Am. Chem. Soc. **131**, 4783–4787 (2009)

140. Bhatia, R.: Emerging health technologies and how they can transform healthcare delivery. J. Health Manag. **23**, 63–73 (2021)
141. Aguado, B.A., Grim, J.C., Rosales, A.M., Watson-Capps, J.J., Anseth, K.S.: Engineering precision biomaterials for personalized medicine. Sci. Transl. Med. **10**, eaam8645 (2018)
142. Jin, C., Wang, K., Oppong-Gyebi, A., Hu, J.: Application of nanotechnology in cancer diagnosis and therapy—a mini-review. Int. J. Med. Sci. **17**, 2964–2973 (2020)
143. Papi, M., Palmieri, V., Digiacomo, L., Giulimondi, F., Palchetti, S., Ciasca, G., et al.: Converting the personalized biomolecular corona of graphene oxide nanoflakes into a high-throughput diagnostic test for early cancer detection. Nanoscale **11** (2019)
144. Zhang, Q., Hou, K., Chen, H., Zeng, N., Wu, Y.: Nanotech probes: a revolution in cancer diagnosis. Front. Oncol. **12**, 933125 (2022)
145. Salehahmadi, Z., Hajiliasgari, F.: Nanotechnology tolls the bell for plastic surgeons. World J. Plast. Surg. **2**, 71 (2013)
146. Cash, K.J., Clark, H.A.: Nanosensors and nanomaterials for monitoring glucose in diabetes. Trends Mol. Med. **16**, 584–593 (2010)
147. Baby, T.T., Ramaprabhu, S.: SiO_2 coated Fe_3O_4 magnetic nanoparticle dispersed multiwalled carbon nanotubes based amperometric glucose biosensor. Talanta **80**, 2016–2022 (2010)
148. Luo, L., Li, Q., Xu, Y., Ding, Y., Wang, X., Deng, D., Xu, Y.: Amperometric glucose biosensor based on $NiFe_2O_4$ nanoparticles and chitosan. Sens. Actuators B Chem. **145**, 293–298 (2010)
149. Yang, J., Wang, X., Sun, Y., Chen, B., Hu, F., Guo, C., Yang, T.: Recent advances in colorimetric sensors based on gold nanoparticles for pathogen detection. Biosensors (2023)
150. Long, C.M., van Laarhoven, H.W., Bulte, J.W., Levitsky, H.I.: Magnetovaccination as a novel method to assess and quantify dendritic cell tumor antigen capture and delivery to lymph nodes. Cancer Res. **69**, 3180–3187 (2009)
151. Kobukai, S., Baheza, R., Cobb, J.G., Virostko, J., Xie, J., Gillman, A., et al.: Magnetic nanoparticles for imaging dendritic cells. Magn. Reson. Med. **63**, 1383–1390 (2010)
152. Upadhyay, V., Sonawat, M., Singh, S., Merugu, R.: Nano robots in medicine: a review. Int. J. Eng. Technol. Manag. Res. **4**, 27–37 (2020)
153. Sayed, N., Allawadhi, P., Khurana, A., Singh, V., Navik, U., Pasumarthi, S.K., et al.: Gene therapy: comprehensive overview and therapeutic applications. Life Sci. **294**, 120375 (2022)
154. Landhuis, E.: The gene fix. Nature (2021)
155. Kretzmann, J.A., Liedl, A., Monferrer, A., Mykhailiuk, V., Beerkens, S., Dietz, H.: Gene-encoding DNA origami for mammalian cell expression. Nat. Commun. **14**, 1017 (2023)
156. Mazumder, S., Bhargava, S.S.: Full length review article nanorobots: current state and future perspectives (2014)
157. Shi, S., Chen, Y., Yao, X.: NGA-inspired nanorobots-assisted detection of multifocal cancer. IEEE Trans Cybern. **52**, 2787–2797 (2022)
158. Tristán-Manzano, M., Justicia-Lirio, P., Maldonado-Pérez, N., Cortijo-Gutiérrez, M., Benabdellah, K., Martin, F.: Externally-controlled systems for immunotherapy: from bench to bedside. Front. Immunol. **11**, 2044 (2020)
159. Esteban-Fernández de Ávila, B., Angell, C., Soto, F., Lopez-Ramirez, M.A., Báez, D.F., Xie, S., et al.: Acoustically propelled nanomotors for intracellular siRNA delivery. ACS Nano 10, 4997–5005 (2016)
160. Meisami, A.H., Abbasi, M., Mosleh-Shirazi, S., Azari, A., Amani, A.M., Vaez, A., Golchin, A.: Self-propelled micro/nanobots: a new insight into precisely targeting cancerous cells through intelligent and deep cancer penetration. Eur. J. Pharmacol. **926**, 175011 (2022)
161. Burak, A.: The Promise and Challenges of Using Nanobots in Medicine. Relevant Software LLC, Tech Label (2023)
162. Datta, B.S.: Medical Nanorobots: The Biocompatibility Constraints of Clinical Translation, pp. 228–232. Springer International Publishing, Cham (2022)
163. Barhoum, A., Pal, K., Rahier, H., Uludag, H., Kim, I.S., Bechelany, M.: Nanofibers as new-generation materials: from spinning and nano-spinning fabrication techniques to emerging applications. Appl. Mater. Today **17**, 1–35 (2019)

164. Abrigo, M., Mcarthur, S.L., Kingshott, P.: Electrospun nanofibers as dressings for chronic wound care: advances, challenges, and future prospects. Macromol. Biosci. **14**(6), 772–792 (2014)
165. Chandrasekaran, A.R., Venugopal, J.R., Sundarrajan, S., Ramakrishna, S.: Fabrication of a nanofibrous scaffold with improved bioactivity for culture of human dermal fibroblasts for skin regeneration. Biomed. Mater. **6**, 015001 (2011)
166. Dubský, M., Kubínová, Š, Sirc, J., Voska, L., Zajíček, R., Zajícová, A., et al.: Nanofibers prepared by needleless electrospinning technology as scaffolds for wound healing. J. Mater. Sci. Mater. Med. **23**, 931–941 (2012)
167. Goyal, R.C., Macri, L.K., Kaplan, H.M., Kohn, J.: Nanoparticles and nanofibers for topical drug delivery. J. Control. Release Official J. Control. Release Soc. **240**, 77–92 (2016)
168. Homaeigohar, S., Boccaccini, A.R.: Antibacterial biohybrid nanofibers for wound dressings. Acta Biomaterialia (2020)
169. Vashaghian, M., Zandieh-Doulabi, B., Roovers, J.P., Smit, T.H.: Electrospun matrices for pelvic floor repair: effect of fiber diameter on mechanical properties and cell behavior. Tissue Eng. Part A. **22**, 1305–1316 (2016)
170. Barnes, C.P., Sell, S.A., Boland, E.D., Simpson, D.G., Bowlin, G.L.: Nanofiber technology: designing the next generation of tissue engineering scaffolds. Adv. Drug Deliv. Rev. **59**, 1413–1433 (2007)
171. Lim, C.T.: Nanofiber technology: current status and emerging developments. Prog. Polym. Sci. **70**, 1–17 (2017)
172. Rasouli, R., Barhoum, A., Bechelany, M., Dufresne, A.: Nanofibers for biomedical and healthcare applications. Macromol. Biosci. **19**, 1800256 (2019)
173. Haidar, M.K., Eroglu, H.: Nanofibers: new insights for drug delivery and tissue engineering. Curr. Top. Med. Chem. **17**, 1564–1579 (2017)
174. Vasita, R., Katti, D.S.: Nanofibers and their applications in tissue engineering. Int. J. Nanomed. **1**, 15–30 (2006)
175. Leung, V., Ko, F.: Biomedical applications of nanofibers. Polym. Adv. Technol. **22**, 350–365 (2011)
176. Xu, L., Liu, Y., Zhou, W., Yu, D.: Electrospun medical sutures for wound healing: a review. Polymers **14**, 1637 (2022)
177. Syed, M.H., Khan, M.M.R., Zahari, M.A.K.M., Beg, M.D.H., Abdullah, N.: A review on current trends and future prospectives of electrospun biopolymeric nanofibers for biomedical applications. Eur. Polym. J. **197**, 112352 (2023)
178. Malikmammadov, E., Tanir, T.E., Kiziltay, A., Hasirci, V., Hasirci, N.: PCL and PCL-based materials in biomedical applications. J. Biomater. Sci. Polym. Ed. **29**, 863–893 (2018)
179. Ekrami, E., Khodabandeh Shahraky, M., Mahmoudifard, M., Mirtaleb, M.S., Shariati, P.: Biomedical applications of electrospun nanofibers in industrial world: a review. Int. J. Polym. Mater. Polym. Biomater. **72**, 561–575 (2023)
180. Azari, A., Golchin, A., Mahmoodinia Maymand, M., Mansouri, F., Ardeshirylajimi, A.: Electrospun polycaprolactone nanofibers: current research and applications in biomedical application. Adv. Pharm. Bull. **12**, 658–672 (2022)
181. Barhoum, A., Rasouli, R., Yousefzadeh, M., Rahier, H., Bechelany, M.: Nanofiber technologies: history and development. In: Barhoum, A., Bechelany, M., Makhlouf, A.S.H. (eds.) Handbook of Nanofibers, pp. 3–43. Springer International Publishing, Cham (2019)
182. dos Santos, D.M., Correa, D.S., Medeiros, E.S., Oliveira, J.E., Mattoso, L.H.C.: Advances in functional polymer nanofibers: from spinning fabrication techniques to recent biomedical applications. ACS Appl. Mater. Interfaces **12**, 45673–45701 (2020)
183. Ogueri, K.S., Laurencin, C.T.: Nanofiber technology for regenerative engineering. ACS Nano **14**, 9347–9363 (2020)
184. Xue, J., Wu, T., Dai, Y., Xia, Y.: Electrospinning and electrospun nanofibers: methods, materials, and applications. Chem. Rev. **119**, 5298–5415 (2019)
185. Wu, S., Dong, T., Li, Y., Sun, M., Qi, Y., Liu, J., et al.: State-of-the-art review of advanced electrospun nanofiber yarn-based textiles for biomedical applications. Appl. Mater. Today **27**, 101473 (2022)

186. Bhattarai, R.S., Bachu, R.D., Boddu, S.H.S., Bhaduri, S.: Biomedical applications of electrospun nanofibers: drug and nanoparticle delivery. Pharmaceutics (2019)
187. Haider, A., Haider, S., Kang, I.-K.: A comprehensive review summarizing the effect of electrospinning parameters and potential applications of nanofibers in biomedical and biotechnology. Arab. J. Chem. **11**, 1165–1188 (2018)
188. Ibrahim, H.M., Klingner, A.: A review on electrospun polymeric nanofibers: production parameters and potential applications. Polym. Testing **90**, 106647 (2020)
189. Kim, H.H., Kim, M.J., Ryu, S.J., Ki, C.S., Park, Y.H.: Effect of fiber diameter on surface morphology, mechanical property, and cell behavior of electrospun poly(ε-caprolactone) mat. Fibers Polym. **17**, 1033–1042 (2016)
190. Xue, J., Xie, J., Liu, W., Xia, Y.: Electrospun nanofibers: new concepts, materials, and applications. Acc. Chem. Res. **50**, 1976–1987 (2017)
191. Qi, Y., Wang, C., Wang, Q., Zhou, F., Li, T., Wang, B., et al.: A simple, quick, and cost-effective strategy to fabricate polycaprolactone/silk fibroin nanofiber yarns for biotextile-based tissue scaffold application. Eur. Polymer J. **186**, 111863 (2023)
192. Wu, S., Wang, Y., Streubel, P.N., Duan, B.: Living nanofiber yarn-based woven biotextiles for tendon tissue engineering using cell tri-culture and mechanical stimulation. Acta Biomater. **62**, 102–115 (2017)
193. Jiang, C., Wang, K., Liu, Y., Zhang, C., Wang, B.: Application of textile technology in tissue engineering: A review. Acta Biomater. **128**, 60–76 (2021)
194. Singh, C., Wong, C.S., Wang, X.: Medical textiles as vascular implants and their success to mimic natural arteries. J. Funct. Biomater. **6**, 500–525 (2015)
195. Miguel, S.P., Figueira, D.R., Simões, D., Ribeiro, M.P., Coutinho, P., Ferreira, P., Correia, I.J.: Electrospun polymeric nanofibres as wound dressings: a review. Colloids Surf. B **169**, 60–71 (2018)
196. Rasouli, R., Barhoum, A.: Advances in nanofibers for antimicrobial drug delivery. In: Barhoum, A., Bechelany, M., Makhlouf, A.S.H. (eds.) Handbook of Nanofibers, pp. 733–774. Springer International Publishing, Cham (2019)
197. John, J.V., McCarthy, A., Karan, A., Xie, J.: Electrospun nanofibers for wound management. ChemNanomat. Chem. Nanomater. Energy. Biol. More. **8**, 7 (2021)
198. Li, T., Sun, M., Wu, S.: State-of-the-art review of electrospun gelatin-based nanofiber dressings for wound healing applications. Nanomaterials (Basel) **12** (2022)
199. Kang, Y.O., Yoon, I.S., Lee, S.Y., Kim, D.D., Lee, S.J., Park, W.H., Hudson, S.M.: Chitosan-coated poly(vinyl alcohol) nanofibers for wound dressings. J. Biomed. Mater. Res. B Appl. Biomater. **92**, 568–576 (2010)
200. Homaeigohar, S., Boccaccini, A.R.: Antibacterial biohybrid nanofibers for wound dressings
201. Xie, Z., Paras, C.F., Weng, H., Punnakitikashem, P., Su, L.C., Vu, K., Tang, L., et al.: Dual growth factor releasing multi-functional nanofibers for wound healing
202. Rath, G., Hussain, T., Chauhan, G., Garg, T., Goyal, A.K.: Collagen nanofiber containing silver nanoparticles for improved wound-healing applications
203. Singh, B., Kim, K., Park, M.-H.: On-demand drug delivery systems using nanofibers. Nanomaterials **11**, 3411 (2021)
204. Kapahi, H., Khan, N.M., Bhardwaj, A., Mishra, N.K.: Implication of nanofibers in oral drug delivery. Curr. Pharm. Des. **21**(15), 2021–2036 (2015)
205. Patel, G.C., Yadav, B.K.N.: Polymeric nanofibers for controlled drug delivery applications (2018)
206. Vojoudi, E., Babaloo, H.: Application of electrospun nanofiber as drug delivery systems: a review. Pharm. Nanotechnol. (2022)
207. Joshi, M., Butola, B.S., Saha, K.: Advances in topical drug delivery system: micro to nanofibrous structures. J. Nanosci. Nanotechnol. **14**(1), 853–867 (2014)
208. Kamble, P., Sadarani, B., Majumdar, A., Bhullar, S.: Nanofiber based drug delivery systems for skin: a promising therapeutic approach. J. Drug Deliv. Sci. Technol. **41**, 124–133 (2017)
209. Monfared, M.H.A., Taghizadeh, S., Zare-Hoseinabadi, A., Mousavi, S.M., Hashemi, S.A., Ranjbar, S., Amani, A.M.: Emerging frontiers in drug release control by core–shell nanofibers: a review. Drug Metab. Rev. **51**, 589–611 (2019)

210. Mendes, A.C., Gorzelanny, C., Halter, N., Schneider, S.W., Chronakis, I.S.: Hybrid electrospun chitosan-phospholipids nanofibers for transdermal drug delivery. Int. J. Pharm. **510**, 48–56 (2016)
211. Kataria, K., Gupta, A., Rath, G., Mathur, R.B., Dhakate, S.R.: In vivo wound healing performance of drug loaded electrospun composite nanofibers transdermal patch. Int. J. Pharm. **469**, 102–110 (2014)
212. Singh, A., Rath, G., Singh, R., Goyal, A.K.: Nanofibers: an effective tool for controlled and sustained drug delivery
213. Babitha, S., Rachita, L., Karthikeyan, K., Shoba, E., Janani, I., Poornima, B., Purna, S.K.: Electrospun protein nanofibers in healthcare: A review. Int. J. Pharm. **523**, 52–90 (2017)
214. Li, X., Wang, X., Yao, D., Jiang, J., Guo, X., Gao, Y., et al.: Effects of aligned and random fibers with different diameter on cell behaviors. Colloids Surf. B **171**, 461–467 (2018)
215. Wu, S., Duan, B., Liu, P., Zhang, C., Qin, X., Butcher, J.T.: Fabrication of aligned nanofiber polymer yarn networks for anisotropic soft tissue scaffolds. ACS Appl. Mater. Interfaces **8**(26), 16950–16960 (2016)
216. Chikelu, C.W., Berns, M., Conover, D., Habas, R., Han, L., Street, R.M., Schauer, C.L.: Collagen nanoyarns: hierarchical three-dimensional biomaterial constructs. Biomacromol **24**, 1155–1163 (2023)
217. Jiao, Y., Li, C., Liu, L., Wang, F., Liu, X., Mao, J., Wang, L.: Construction and application of textile-based tissue engineering scaffolds: a review. Biomater. Sci. **8**, 3574–3600 (2020)
218. Doersam, A., Tsigkou, O., Jones, C.: A review: textile technologies for single and multi-layer tubular soft tissue engineering. Adv. Mater. Technol. **7**, 2101720 (2022)
219. Nel, A., Xia, T., Mädler, L., Li, N.: Toxic potential of materials at the nanolevel. Science **311**, 622–627 (2006)
220. Salvati, A., Åberg, C., dos Santos, T., Varela, J., Pinto, P., Lynch, I., Dawson, K.A.: Experimental and theoretical comparison of intracellular import of polymeric nanoparticles and small molecules: toward models of uptake kinetics. Nanomed. Nanotechnol. Biol. Med. **7**, 818–826 (2011)
221. Alghamdi, M.A., Fallica, A.N., Virzì, N., Kesharwani, P., Pittalà, V., Greish, K.: The promise of nanotechnology in personalized medicine. J. Pers. Med. **12**, 673 (2022)
222. Tutton, R.: Personalizing medicine: futures present and past. Soc Sci Med **75**, 1721–1728 (2012)
223. Roco, M.C., Mirkin, C.A., Hersam, M.C.: Nanotechnology research directions for societal needs in 2020: summary of international study. J. Nanopart. Res. **13**, 897–919 (2011)
224. Gottschalk, F., Sonderer, T., Scholz, R.W., Nowack, B.: Modeled environmental concentrations of engineered nanomaterials (TiO_2, ZnO, Ag, CNT, Fullerenes) for different regions. Environ. Sci. Technol. **43**, 9216–9222 (2009)
225. Maynard, A.D., Kuempel, E.D.: Airborne nanostructured particles and occupational health. J. Nanopart. Res. **7**, 587–614 (2005)
226. Berger, M.: How Nanotechnology Enables Wearable Electronics. Nanowerk LLC (2019)
227. Xie, L., Zhang, Z., Wu, Q., Gao, Z., Mi, G., Wang, R., et al.: Intelligent wearable devices based on nanomaterials and nanostructures for healthcare. Nanoscale **15**, 405–433 (2023)
228. Rajendra Acharya, U., Paul Joseph, K., Kannathal, N., Lim, C.M., Suri, J.S.: Heart rate variability: a review. Med. Biol. Eng. Comp. **44**, 1031–1051 (2006)
229. Prieto-Avalos, G., Cruz-Ramos, N.A., Alor-Hernández, G., Sánchez-Cervantes, J.L., Rodríguez-Mazahua, L., Guarneros-Nolasco, L.R.: Wearable devices for physical monitoring of heart: a review. Biosensors **12**, 292 (2022)
230. Iijima, S.: Helical microtubules of graphitic carbon. Nature **354**, 56–58 (1991)
231. Novoselov, K.S., Geim, A.K., Morozov, S.V., Jiang, D., Zhang, Y., Dubonos, S.V., et al.: Electric field effect in atomically thin carbon films. Science **306**, 666–669 (2004)
232. Sengupta, D., Kottapalli, A.G.P.: Nanomaterials-based bioinspired next generation wearable sensors: a state-of-the-art review. Adv. Electron. Mater., 2300436
233. Garg, M., Gupta, A., Sharma, A.L., Singh, S.: Advancements in 2D materials based biosensors for oxidative stress biomarkers. ACS Appl. Bio Mater. **4**, 5944–5960 (2021)

234. Rohaizad, N., Mayorga-Martinez, C.C., Fojtů, M., Latiff, N.M., Pumera, M.: Two-dimensional materials in biomedical, biosensing and sensing applications. Chem. Soc. Rev. **50**, 619–657 (2021)
235. Kim, K.O., Kim, G.J., Kim, J.H.: A cellulose/β-cyclodextrin nanofiber patch as a wearable epidermal glucose sensor. RSC Adv. **9**, 22790–22794 (2019)
236. Lei, Y., Wu, Y., Chowdhury, A.J.K.: Multi-target tracking algorithm in intelligent transportation based on wireless sensor network. Open Phys. **16**, 1000–1008 (2018)
237. Tang, X., Yang, A., Li, L.: Optimization of nanofiber wearable heart rate sensor module for human motion detection. Comput. Math. Methods Med. **2022**, 1747822 (2022)
238. Yu, G., Ma, H., Witarsyah, D.: Optimal path selection algorithm for mobile beacons in sensor network under non-dense distribution. Open Phys. **16**, 1066–1075 (2018)
239. Lou, C., Li, R., Li, Z., Liang, T., Wei, Z., Run, M., et al.: Flexible graphene electrodes for prolonged dynamic ECG monitoring. Sensors **16**, 1833 (2016)
240. Romero, F.J., Castillo, E., Rivadeneyra, A., Toral-Lopez, A., Becherer, M., Ruiz, F.G., et al.: Inexpensive and flexible nanographene-based electrodes for ubiquitous electrocardiogram monitoring. npj Flex. Electron. **3**, 12 (2019)
241. Jiao, L., Chua, Z.Y., Moon, S.K., Song, J., Bi, G., Zheng, H., et al.: Laser-induced graphene on additive manufacturing parts. Nanomaterials **9**, 90 (2019)
242. Taylor, L.W., Williams, S.M., Yan, J.S., Dewey, O.S., Vitale, F., Pasquali, M.: Washable, sewable, all-carbon electrodes and signal wires for electronic clothing. Nano Lett. **21**, 7093–7099 (2021)
243. Taylor, L.W., Dewey, O.S., Headrick, R.J., Komatsu, N., Peraca, N.M., Wehmeyer, G., et al.: Improved properties, increased production, and the path to broad adoption of carbon nanotube fibers. Carbon **171**, 689–694 (2021)
244. McCauley, M.D., Vitale, F., Yan, J.S., Young, C.C., Greet, B., Orecchioni, M., et al.: In vivo restoration of myocardial conduction with carbon nanotube fibers. Circ. Arrhythmia Electrophysiol. **12**, e007256 (2019)
245. Jung, H.C., Moon, J.H., Baek, D.H., Lee, J.H., Choi, Y.Y., Hong, J.S., Lee, S.H.: CNT/PDMS composite flexible dry electrodesfor long-term ECG monitoring. IEEE Trans. Biomed. Eng. **59**, 1472–1479 (2012)
246. Zhong, B., Jiang, K., Wang, L., Shen, G.: Wearable sweat loss measuring devices: from the role of sweat loss to advanced mechanisms and designs. Adv. Sci. **9**, 2103257 (2022)
247. Lee, H., Choi, T.K., Lee, Y.B., Cho, H.R., Ghaffari, R., Wang, L., et al.: A graphene-based electrochemical device with thermoresponsive microneedles for diabetes monitoring and therapy. Nat. Nanotechnol. **11**, 566–572 (2016)
248. Yin, J., Li, J., Reddy, V.S., Ji, D., Ramakrishna, S., Xu, L.: Flexible textile-based sweat sensors for wearable applications. Biosensors **13**, 127 (2023)
249. Hauke, A., Simmers, P., Ojha, Y.R., Cameron, B.D., Ballweg, R., Zhang, T., et al.: Complete validation of a continuous and blood-correlated sweat biosensing device with integrated sweat stimulation. Lab Chip **18**, 3750–3759 (2018)
250. Das, R., Nag, S., Banerjee, P.: Electrochemical nanosensors for sensitization of sweat metabolites: from concept mapping to personalized health monitoring. Molecules (2023)
251. Khan, A., Winder, M., Hossain, G.: Modified graphene-based nanocomposite material for smart textile biosensor to detect lactate from human sweat. Biosens. Bioelectron. X. **10**, 100103 (2022)
252. Xu, M., Zhu, Y., Gao, S., Zhang, Z., Gu, Y., Liu, X.: Reduced graphene oxide-coated silica nanospheres as flexible enzymatic biosensors for detection of glucose in sweat. ACS Appl Nano Mat. **4**, 12442–12452 (2021)
253. Roy, S., David-Pur, M., Hanein, Y.: Carbon nanotube-based ion selective sensors for wearable applications. ACS Appl. Mater. Interfaces **9**, 35169–35177 (2017)
254. Wei, X., Zhu, M., Li, J., Liu, L., Yu, J., Li, Z., Ding, B.: Wearable biosensor for sensitive detection of uric acid in artificial sweat enabled by a fiber structured sensing interface. Nano Energy **85**, 106031 (2021)

255. Zhai, Q., Gong, S., Wang, Y., Lyu, Q., Liu, Y., Ling, Y., et al.: Enokitake mushroom-like standing gold nanowires toward wearable noninvasive bimodal glucose and strain sensing. ACS Appl. Mater. Interfaces **11**, 9724–9729 (2019)
256. Guan, S., Xu, B., Yang, Y., Zhu, X., Chen, R., Ye, D., Liao, Q.: Gold nanowire aerogel-based biosensor for highly sensitive ethanol detection in simulated sweat. ACS Appl. Nano Mat. **5**, 11091–11099 (2022)
257. World Health Organization: Guideline for the Pharmacological Treatment of Hypertension in Adults. World Health Organization (2021)
258. Madhurantakam, S., Babu, K.J., Rayappan, J.B.B., Krishnan, U.M.: Nanotechnology-based electrochemical detection strategies for hypertension markers. Biosens. Bioelectron. **116**, 67–80 (2018)
259. Ismail, S.N.A., Nayan, N.A., Mohammad Haniff, M.A.S., Jaafar, R., May, Z.: Wearable two-dimensional nanomaterial-based flexible sensors for blood pressure monitoring: a review. Nanomaterials **13**, 852 (2023)
260. Ai, Y., Hsu, T.H., Wu, D.C., Lee, L., Chen, J.-H., Chen, Y.-Z., et al.: An ultrasensitive flexible pressure sensor for multimodal wearable electronic skins based on large-scale polystyrene ball@reduced graphene-oxide core–shell nanoparticles. J. Mater. Chem. C. **6**, 5514–5520 (2018)
261. Kim, J., Kim, W.S.: A paired stretchable printed sensor system for ambulatory blood pressure monitoring. Sens. Actuators A **238**, 329–336 (2016)
262. Kireev, D., Sel, K., Ibrahim, B., Kumar, N., Akbari, A., Jafari, R., Akinwande, D.: Continuous cuffless monitoring of arterial blood pressure via graphene bioimpedance tattoos. Nat. Nanotechnol. **17**, 864–870 (2022)
263. Luo, N., Dai, W., Li, C., Zhou, Z., Lu, L., Poon, C.C.Y., et al.: Flexible piezoresistive sensor patch enabling ultralow power cuffless blood pressure measurement. Adv. Func. Mater. **26**, 1178–1187 (2016)
264. Park, S.-J., Kim, J., Chu, M., Khine, M.: Flexible piezoresistive pressure sensor using wrinkled carbon nanotube thin films for human physiological signals. Adv. Mater. Technol. **3**, 1700158 (2018)
265. Bainbridge, W.S.: Social and ethical implications of nanotechnology. In: Bhushan, B. (ed.) Springer Handbook of Nanotechnology, pp. 1135–1151. Springer, Berlin, Heidelberg (2004)
266. Khalil, Y., Mahmoud, A.E.D.: Nanomaterial-based sensors for wearable health monitoring in bioelectronics nano engineering. J. Contemp. Healthc. Anal. **7**, 126–144 (2023)
267. Syedmoradi, L., Ahmadi, A., Norton, M.L., Omidfar, K.: A review on nanomaterial-based field effect transistor technology for biomarker detection. Microchim. Acta **186**, 739 (2019)
268. Viswambari Devi, R., Doble, M., Verma, R.S.: Nanomaterials for early detection of cancer biomarker with special emphasis on gold nanoparticles in immunoassays/sensors. Biosens. Bioelectron. **68**, 688–698 (2015)
269. Canali, S., Schiaffonati, V., Aliverti, A.: Challenges and recommendations for wearable devices in digital health: data quality, interoperability, health equity, fairness. PLOS Digit. Health **1**, e0000104 (2022)
270. Cai, L., Zhu, Y.: The challenges of data quality and data quality assessment in the big data era. Data Sci. J. **14**, 2 (2015)
271. Sim, I.: Mobile devices and health. N. Engl. J. Med. **381**, 956–968 (2019)
272. Ding, X., Clifton, D., Ji, N., Lovell, N.H., Bonato, P., Chen, W., et al.: Wearable sensing and telehealth technology with potential applications in the coronavirus pandemic. IEEE Rev. Biomed. Eng. **14**, 48–70 (2021)
273. Lomborg, S., Langstrup, H., Andersen, T.O.: Interpretation as luxury: heart patients living with data doubt, hope, and anxiety. Big Data Soc. **7**, 2053951720924436 (2020)
274. Prainsack, B.: Logged out: ownership, exclusion and public value in the digital data and information commons. Big Data Soc. **6**, 2053951719829773 (2019)
275. Hummel, P., Braun, M., Dabrock, P.: Own data? Ethical reflections on data ownership. Philos. Technol. **34**, 545–572 (2021)

276. Gupta, N., Fischer, A.R.H., Frewer, L.J.: Ethics, risk and benefits associated with different applications of nanotechnology: a comparison of expert and consumer perceptions of drivers of societal acceptance. NanoEthics **9**, 93–108 (2015)
277. Gupta, N., Fischer, A.R., van der Lans, I.A., Frewer, L.J.: Factors influencing societal response of nanotechnology: an expert stakeholder analysis. J. Nanopart. Res. **14**, 1–15 (2012)
278. Ochekpe, N.A., Olorunfemi, P.O., Ngwuluka, N.C.: Nanotechnology and drug delivery. Trop. J. Pharm. Res. (2009)
279. Wasti, S., Lee, I.H., Kim, S., Lee, J.H., Kim, H.: Ethical and legal challenges in nanomedical innovations: a scoping review. Front. Genet. **14** (2023)
280. Valverde, J.L.J.: Nanotechnology: risk assessment and risk management perspective **8**, 25–47 (2011)
281. Ganascia, J.-G.: The new ethical trilemma: security, privacy and transparency. C. R. Phys. **12**, 684–692 (2011)
282. Resnik, D.B., Tinkle, S.S.: Ethics in nanomedicine. Nanomedicine (Lond.) **2**, 345–350 (2007)
283. Khan, M.: How nanotechnology can boost drug discovery (2022)
284. Rani, A., Kumar, N., Kumar, M.: 2 Environmentally friendly green approaches and applications of nanoparticles. In: Metal Oxide–Based Carbon Nanocomposites for Environmental Remediation and Safety (2023)
285. El-Sayed, A., Kamel, M.: Advanced applications of nanotechnology in veterinary medicine. Environ. Sci. Pollut. Res. **27**, 19073–19086 (2020)
286. Thangavelu, L., Veeraragavan, G.R., Mallineni, S.K., Devaraj, E., Parameswari, R.P., Syed, N.H., et al.: Role of nanoparticles in environmental remediation: an insight into heavy metal pollution from dentistry. Bioinorg. Chem. Appl. 2022 (2022)

Raphael Kanyire Seidu is a Ph.D. candidate in Fashion and Textile Design in the School of Fashion and Textiles at The Hong Kong Polytechnic University, Hong Kong SAR. Prior to this, he gained both his MFA degree in Textile Design and BA. degree in Industrial Art (Textile option) from Kwame Nkrumah University of Science and Technology, Ghana. He has several publications in leading journals in textile, design, technology, and material science. His recent paper was adjudged the "Second-Best Paper" presented at the 12th Sustainable Education and Development Research Conference in Somanya, Ghana. His research interest focuses on retro-reflective materials, smart textiles, cloth culture, weaving, and print technologies. He has several papers and has a patent from his Ph.D. project which aims at developing smart interactive retro-reflective clothing for pedestrians' safety at night-time.

Anna Doersam received her B.Sc. degree in Clothing Technology from the University of Applied Science Berlin Germany in 2019 and obtained a M.Sc. Degree in Textile Technology from the University of Manchester, United Kingdom. Currently, Anna is pursuing her Ph.D. in Biomaterials at the University of Manchester, working within the Department of Materials at the Royce Institute. Her research interests include textile engineering, nanofiber technology, and biomaterials. Her current work focuses on the development of textile-based scaffolds for tubular tissue engineering applications. Anna is a published researcher in textile technologies for tissue engineering and a recipient of multiple scholarships and grants.

Emmanuel Abankwah Ofori is a postgraduate student studying material science and engineering at the Hubei University of Automotive Technology, Siyan, Hubei in China. Emmanuel's research interest focuses on sustainable environmental practices, textile science, polymers, and energy efficiency. Recently, he co-authored a paper that was adjudged the "Second-Best Paper" presented at the 12th Sustainable Education and Development Research Conference in Somanya, Ghana, and has actively participated in several research projects, working with renowned authors to explore various facets of environmental sustainability, adaptation to climate change, green dyeing, and sustainable renewable energy consumption, thanks to Dr. Benjamin Tawiah's guidance. Emmanuel is dedicated to fostering the next generation of environmental and scientific leaders as a passionate textile scientist and educator. Through his writings and research, Emmanuel continues to motivate his mentees, readers of his works, and colleagues in the field, fostering greater awareness of the planet's vulnerability and the need to protect its ecological vibrancy in order to achieve a sustainable future at large.

Shouxiang Jiang is currently a full Professor of fashion and textile design at The Hong Kong Polytechnic University, Hong Kong SAR. He teaches a broad range of subjects such as innovative textile design approach, advanced textile and fashion, and international fashion and textiles design. Prof. Jiang's research deals with innovative textile effects using physical and chemical treatment technologies. He is internationally renowned for his development of the metallized textiles. Leading a team that has developed a sputtering coating technology with research funding from the HKSAR government and sponsorships from the textile industries, Prof. Kinor Jiang has designed and launched an anhydrous system that uses metallising technologies to produce functional and decorative textiles for the production line. He holds 30 granted and filed patents based on the coating technology. Prof. Jiang has published more than 200 journal papers and more than 10 books. His designs and art projects have been exhibited worldwide and received more than 30 awards.

Chapter 4
The Role of Nanotechnology in Agriculture and Food Security

Leticia Donkor and Eric Banan-Mwine Daliri

Introduction

Agriculture has traditionally been considered a vital industry that provides food and contributes substantially to most societies and economies [1]. Agriculture is widely regarded as the backbone of the economies in developing countries [2]. It has been the primary supply of raw materials for food production for sustainable growth and development [3]. This sector has provided food for global populations throughout these decades, and its importance cannot be underestimated. The well-being of human health was highly dependent on the quality and availability of food. Access to food is a fundamental right to ensure food security, and it is when people always have access to food needed for healthy well-being [4]. However, there is rapid growth in the human population, and it is projected to exceed nine billion by 2050 [5, 6], which has led to the need for more food to be produced, resulting in a strain on natural resources [7]. Yadav et al. [8] stated that there are currently 815 million malnourished people, predicted to rise by 2050. According to Pingali [9], the agriculture sector in developing countries faces challenges in meeting the growing demand for dietary diversity due to rapid urbanization, increased incomes, and globalization of the food sector. Not only has the burgeoning population affected agriculture, but crop losses and pest infestations have also been significant barriers to global food security [10]. Changes in atmospheric conditions are closely linked to food security, especially for low-income countries such as Africa, which depend heavily on agriculture that relies

L. Donkor (✉)
Department of Food Process Engineering, School of Engineering Sciences, College of Basic and Applied Sciences, University of Ghana, Legon, Accra, Ghana
e-mail: ldonkor005@st.ug.edu.gh

E. B.-M. Daliri
Department of Biological Models, Institute of Biochemistry, Life Sciences Center, Vilnius University, Sauletekio Ave. 7, 10257 Vilnius, Lithuania

S. C. George and B. Tawiah (eds.), *Nanotechnology in Societal Development*, Advanced Technologies and Societal Change, https://doi.org/10.1007/978-981-97-6184-5_4

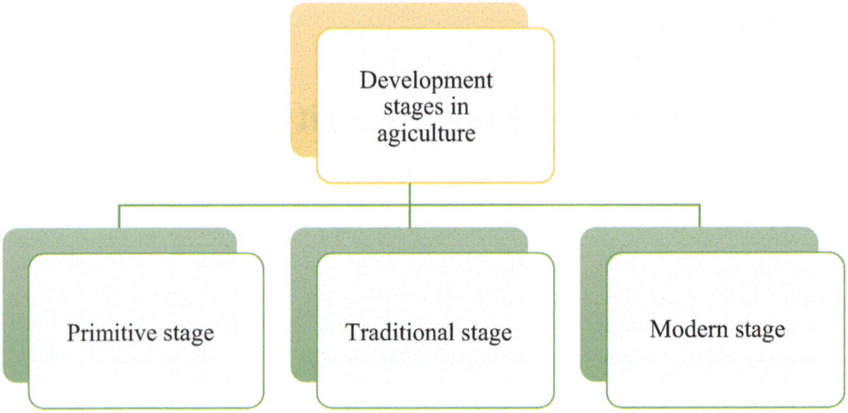

Fig. 4.1 Three main stages of agricultural development, advancing from left to right (*Source* Tang et al. [14])

on rainfall. As a result, it affects yield and increases harvest losses, thereby affecting availability and food security [11].

The global agricultural sector has undergone three major transformational stages that have contributed to its growth. The first stage dates to the 1950s and the 1960s [12] and is characterized by a transition from cultivation for home consumption to a cash economy [13]. The second stage was characterized by rapid commercialization, focusing on the cash economy, and the third stage saw advanced commercialization of both tree crop farming and the broader economy [13]. Tang et al. [14] also classified them broadly, as shown below (Fig. 4.1).

Although agriculture has experienced three stages of development, there is still a need for advanced technologies that make improvements possible. The production of food must be augmented by 70 to 100 percent to meet the growing population needs [15]. The United Nations (UN) indicated that modern agricultural practices have led to remarkable increases in productivity. Food security remains a basic problem, and there is a need to redouble modernized agriculture to double the food supply [16]. Therefore, advanced technologies are required to address these global problems. One such technology is nanotechnology, which is widely applied in agriculture and the promotion of food security. In agriculture and food security, nanotechnology offers excellent potential to enhance crop productivity, improve nutrient utilization, and develop novel techniques for food preservation, thereby leading to more sustainable and resilient food systems.

This chapter is structured to discuss the various nanotechnological applications in strengthening the agricultural and food production sectors to guarantee food availability, safety, and quality. This is preceded by some prospects and challenges faced with their application following the general schematic representation in Fig. 4.2.

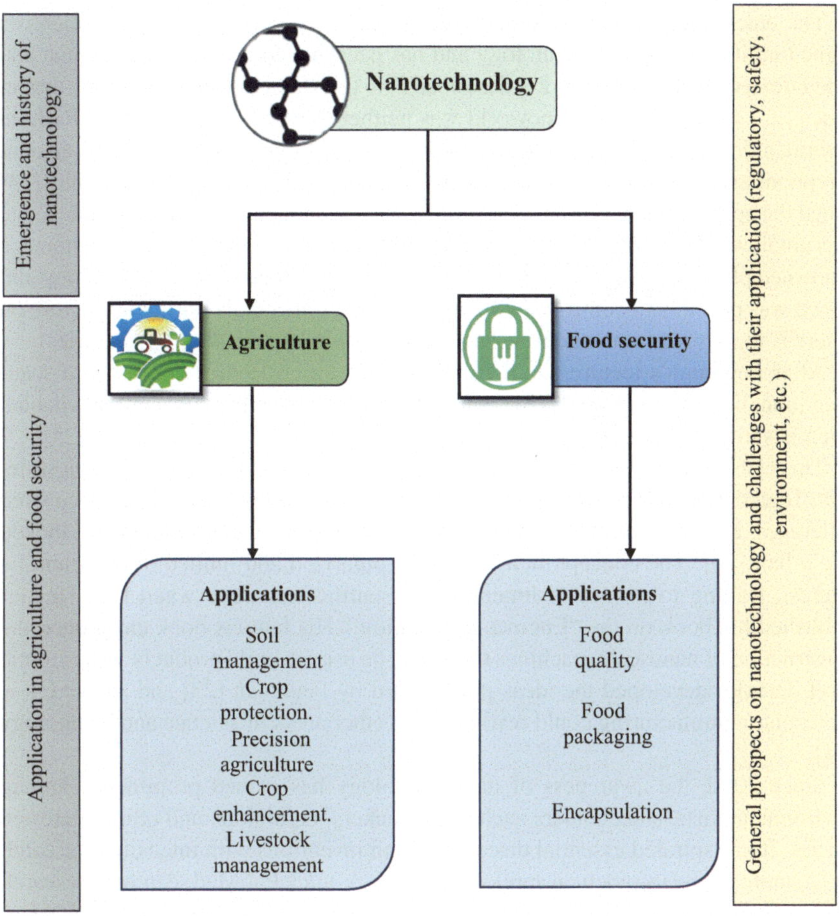

Fig. 4.2 Schematic overview of chapter.

Emergence and History of Nanotechnology

Nanotechnology is an interdisciplinary field [17] that has gained much recognition and considerable advancement in research over the past few decades [18, 19]. Nanotechnology uses nanoscience to create nanoparticles to improve industry and benefit humans [20]. This involves studying the fabrication and application of systems by manipulating these nanoparticles [18]. Nanoparticles are tiny particles varying from 1 to 100 nm in size [20]. The possibility of engineering and controlling matter at the smallest (atomic) and molecular level has made it possible to control them, paving the way for scientific and technological advancements with applications in diverse fields.

The emergence of nanotechnology dates back to the 1950s, as some scientists argue that it is not a new technology and has been linked to Richard Feynman and Eric Drexler. However, Selin [21] described Eric Drexler as a nanotechnology avatar. Jones [22] argued that the nanoworld was neither invented by him nor by Richard Feynman. Jones [22] further indicated the discovery of soap and Indian ink by ancient Egyptians and even the development of microscopes in the nanoworld. Tolochko [23] stated the application of this technology in the past centuries, although its knowledge was unknown. However, Melnik, and Shagalina [24] stated that Richard Feynman, a renowned physicist, was the first to openly express his ideas on nanotechnology and how it was possible to control and manipulate things on a scale during his lecture on "there is more room at the bottom," even though it existed in the past centuries.

After Feynman's lecture, Gerd Binnig and Heinrich Rohrer developed a scanning tunneling microscope (STM), which helped establish the concept that enabled scientists to visualize and manipulate individual atoms with high accuracy [25].

This caused some revolution, followed by the discovery of a unique carbon structure. To understand the mechanism of long-chain molecules, Kroto et al. discovered fullerenes, a carbon that bore a unique structure with 60 carbon atoms and looked like a ball [26]. The concept then became popularized and ignited much scientific interest, leading to its establishment as a scientific technique when Eric Drexler published his book on the "Engines of Creation". His famous book envisioned the construction of nanoscale machines that could be used to build products at the atomic level. Drexler developed the ideas put forward by Feynman [23] and showed how molecular manufacturing could revolutionize other areas of science and technology [24].

Since 2000, the awareness of nanotechnology has gained prominence among environmentalists, stakeholders such as lawmakers, regulators, and other interested parties. It has spiraled essential discoveries and inventions with intensified research on technology and its practical applications [23]. A quick PubMed search for research publications from 2000 to 2023 using specific terms like "nanotechnology", "nanotechnology AND drug AND delivery", "nanotechnology AND agriculture", "nanotechnology AND food AND security" yielded a total of 143,668 publications related to the field of nanotechnology (Fig. 4.3).

Figure 4.3 emphasizes the growing interest in research in this area and the need for more information on the technology. There is a need for research on nanotechnology in agriculture and food as it has been extensively applied in medicine (drug delivery) and other engineering fields.

Application of Nanotechnology

This innovative technology has had significant uses in areas such as medicine, transport, engineering, energy, and pharmaceuticals, and has proven its place in agricultural sciences and food agri-food industries [27].

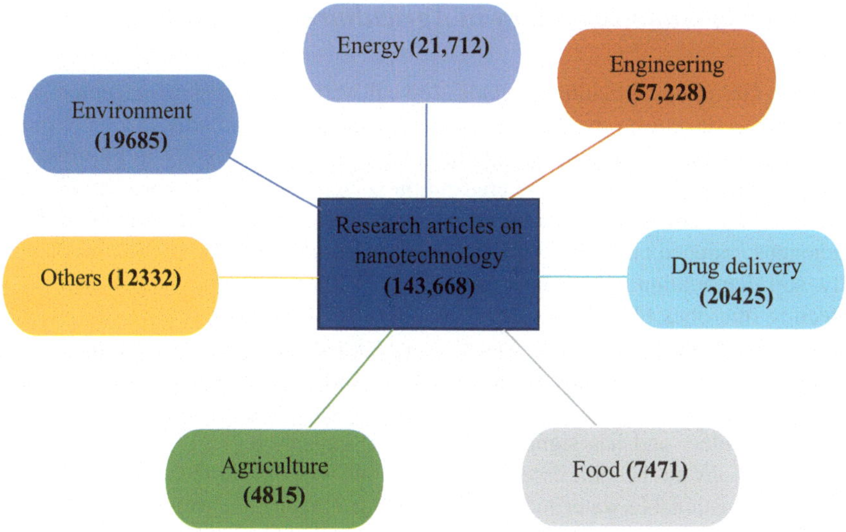

Fig. 4.3 Research articles on nanotechnology and their respective breakdowns in different disciplines from 2000 to 2023 in PubMed

In medicine, nanoparticles have contributed significantly to targeted drug delivery systems, making it possible for researchers to target specific cells or tissues with minimal side effects. It has also been used in regenerative medicine for tissue engineering and gene therapy applications [28]. Furthermore, magnetic resonance imaging (MRI) in diagnostic imaging is based on magnetic nanoparticles [29] and can visualize tissues for early disease detection [30].

It is widely used in engineering and has been applied in areas such as less-flammable quick-charging batteries and even for waste-converting materials in computers [31]. Nanotechnology is at the forefront of quantum computing research. The intersection of quantum computing and nanotechnology uses nanoscale materials to develop solid-state quantum computers that can solve complex problems at an exponentially faster rate. Hollenberg et al. [32] discussed the feasibility of manipulating and controlling atoms in a semiconductor crystal. Even in information technology, the possibility of making miniature forms of electronic components like semiconductor nanowires using nanotechnology has been instrumental in the evolution of that discipline [33].

However, their applications are widespread beyond medicine and engineering. Sections "The Role of Nanotechnology in Agriculture" and "Nanotechnology in Food" explore the diverse applications of nanotechnology in different fields, including agriculture, medicine, and engineering, with a specific focus on improving agriculture and ensuring food security. The intersection of nanotechnology and food security has the potential to revolutionize global agriculture. Using the unique properties of nanomaterials, researchers are developing novel solutions to address challenges in crop production, resource utilization, and food preservation.

The Role of Nanotechnology in Agriculture

The activities under agriculture include crop cultivation, livestock management, and utilization of resources, and they are essential in the global economy and the primary survival of humans. Sustainable agriculture has become necessary to improve the food systems and encourage the eradication of zero hunger. However, traditional agricultural practices are challenged with issues like agrarian waste, microbial plant pathogens, reduced cultivation due to urbanization, and many more [34]. Additionally, soil degradation, which is currently on the rise, results in oxygen depletion, affecting microbes [35]. Nanotechnology is a rapidly growing and emerging tool in modern agriculture that has gained considerable attention and is projected to be one of the most cost-effective tools in the future of sustainable agriculture [27, 36]. The concept of farm-to-fork is a complex system that involves various stages and stakeholders and has significantly used nanotechnology [17]. In agriculture, nanotechnology has become essential in enhancing the delivery of nutrients to plants, enhancing sustainable water management, and mitigating the impact of soil-borne diseases. These advances in agriculture have contributed immensely to increasing the yield of crops.

Another substantial application of nanotechnology in agriculture is precision farming, which maximizes crop yields by monitoring environmental factors and minimizing chemical input [37]. Furthermore, it has been instrumental in finding solutions to plant-related problems, and post-harvest losses.

The broad spectrum of its applications converges into distinct domains, each offering targeted solutions to specific challenges. The influence of nanotechnology broadly extends across crop improvement and protection, precision agriculture, soil management, pest control, precision agriculture, and livestock production and management, as shown in Fig. 4.4.

Nanoparticles in Soil Management

In agriculture, nanotechnology can revolutionize soil management using nanoparticles as delivery systems for agrochemicals and other substances to plants and their tissues to control the release of these chemicals [38]. Soil forms the fundamental basis for farmland ecosystems and is vital in sustaining the ecosystem by facilitating the process of crop production [39]. According to Fernández et al. [40], soils can be contaminated by pesticides, petroleum hydrocarbons, organic compounds, and heavy metals. Not only are soils contaminated by pesticides, but soil erosion can also result in nutrient loss. Soil erosion is a natural process that can carry away essential nutrients like potassium, phosphorus, nitrogen, and calcium, and this can hurt soil quality, crop growth, and the general health of the ecosystems [41].

In soil management, bioremediation techniques like using organisms, organic substances, or organic compounds can reduce the concentration of pollutants in the soil, groundwater, and air [42]. According to Deng et al. [43], phytoremediation

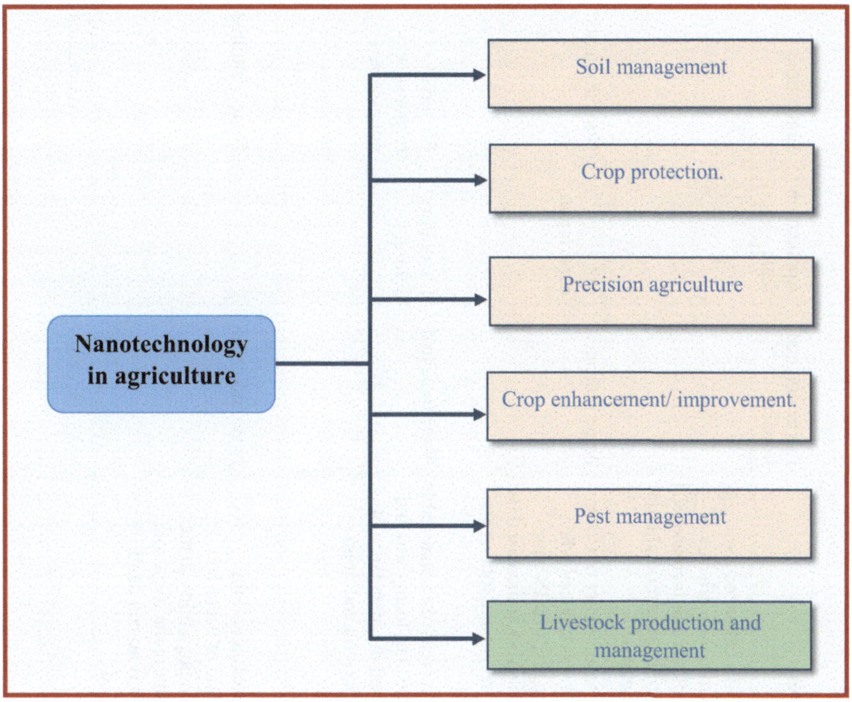

Fig. 4.4 Schematic representation of the possible role of nanotechnology in sustainable agriculture

happens when plants are used to remove impurities from contaminated environments. Soil pollutants are harmful substances in the soil that must be removed because they disrupt the ecosystems and can also put human health at risk throughout the food supply chain. The continuous practice of intensive cropping and pollution is causing a decline in the ability of the soil to support crop growth and output [44].

Nanomaterials are essential in soil remediation because they are essential for biotic and abiotic strategies and can detoxify and restore soil health [36]. Bhandari [45] suggested that nanomaterials such as nanoscale zero-valent iron (nZVI), carbon nanotubes, and nanofibers have been utilized for remediating various contaminants, including hydrocarbons, chlorinated compounds, heavy metals, and organic compounds. Also, nanomaterials can act as sources of nutrients, promoting the release of growth-required substances from the crop roots to enhance soil food web nutrient cycling and mineralization [44]. Additionally, these nanomaterials can precipitate soil pollutants and function as solid waste stabilizers to control soil erosion [46].

Cerium oxide nanoparticles and titanium oxide nanoparticles have been found to reduce cadmium from the soil [43, 47]. Table 4.1 summarizes some advanced applications of nanotechnology in soil management.

Table 4.1 Some summarized research studies that have been conducted on the implementation of nanotechnology in soil management, to improve agricultural practices.

Reference(s)	Focus	Nanomaterial/ nanoparticles	Findings	Environmental impact	Application scale	Limitations
Deng et al. [43]	Assess the potential use of titanium dioxide TiO_2 NPs to improve the removal of Cd from contaminated soils of ramie plant	TiO_2 NPs	TiO_2 NPs enhanced the phytoextraction efficacy of ramie plants and increased Cd absorption and accumulation in the leaves			
Ayub et al. [47]	Investigate the ability of cerium oxide nanoparticles to mitigate Cd stress in wheat grown in alkaline calcareous soil	CeO_2-NPs	The bioavailability of Cd in the soil was reduced when 1000 mg/kg of NPs were applied. Also, it increased the shoot weight and total grain yield	None indicated	Pot experiment	None discussed
Pide et al. [48]	Evaluate the impact of nano fertilizer and nano-plant hormone on the chemical properties and microbial population of two soil samples (one with a clay-loam texture and the other with a sandy-loam texture)	FertiGroe(r) N (FG-N), HormoGroe(r) auxin (HG-A)	FG-N increased exchangeable K in the clay-loamy soil and increased the abundance of some bacteria and fungi	None indicated	Laboratory setting	Bacterial response was not assessed
Calder et al. [49]	Assess how harmful AgNPs in sand and soil are toward a beneficial soil bacterium (Pseudomonas chlororaphis O6)	Ag-NPs	NP doses caused the reduction of bacterial reproducing ability in the sand, but in the loam soil, there was no cell death observed	None indicated		None Identified

(continued)

Table 4.1 (continued)

Reference(s)	Focus	Nanomaterial/ nanoparticles	Findings	Environmental impact	Application scale	Limitations
Borymski et al. [50]	Investigate the effect of ZnO and CuO NPs on soil microbial communities	ZnO-NPs, CuO-NPs	ZnO NPs were identified as less toxic than other NPs, and the minimum biocidal concentration was 10 μg/L. also, CuO NPs showed higher toxicity in yeast S. cerevisiae and C. albicans, than ZnO NPs	None indicated	Pot experiment	None Identified
Simonin et al. [51]	Assess the influence of TiO_2 NPs on soil bacterial abundance and carbon mineralization activity, and the correlation between the stability of TiO_2-NPs and their potential in the soil (sandy-loam, loam, and silty-clay)	TiO_2 NPs	The addition of NPs did not have any significant effect on the activity and abundance of the microbial populations in the soil; however, it significantly reduced the carbon mineralization rate			
Bellani et al. [52]	Evaluate the impact of TiO_2 on mineral nutrient availability in biosolids-amended farm soil, on the disturbance of associated soil bacterial population, and the mineral nutrition *Pisum sativum* L	TiO_2 NPs	When biosolids and TiO_2 were added to the soil, there were significant reductions in the potential availability of Mn, Fe, and P	None indicated	Pot experiment	None Identified

TiO_2—titanium dioxide, NPs—nanoparticles, K—potassium, CeO_2—cerium oxide, Cd—cadmium, Ag—silver, Mn—manganese, Fe—iron, ZnO—zinc oxide, CuO—copper oxide

Crop Improvement

Nanofertilizers

The use of fertilizers in agriculture is vital for promoting plant growth. However, traditional fertilizers can pose significant risks due to nutrient leaching and environmental pollution [53]. The emergence of nano-fertilizers is a promising solution. It has been effective due to its ability to slowly release active nutrients like copper, potassium, etc., at a slower rate as required by plants for growth, conserves the diversity of the beneficial microbiome, and is considered more efficient than conventional ones (Table 4.1).

Nanofertilizers are more efficient than conventional fertilizers. Another critical advantage of nano-fertilizers is their ability to reduce the loss of nutrients resulting from leaching due to their controlled slow release [55]. Recent studies have explored the use of chitosan nanoparticles (NPs) for enhanced delivery of nutrients to plants, with reduced environmental impacts [56, 57].

These NPs offer several advantages, including improved nutrient absorption and reduced environmental impacts. Furthermore, nano-fertilizers are promising because they increase nutrient efficiency and can reduce toxicity that may result from the over-application of any fertilizer [58].

Integrating nano-fertilizers and other innovative agricultural practices aligns with the broader goals of transitioning toward a more sustainable food system [59].

Plant Growth

Another aspect of nanotechnology on crop improvement is seed germination through priming, a process that stimulates biological mechanisms and makes seeds more tolerant to abiotic stresses [60]. The utilization of nanotechnology enables the resilience of crop and their yield. Additionally, it results in enhanced targeted delivery of bioactive compounds, making it possible to precisely mediate plant responses to stress. Zinc is a critical component for different enzymes necessary in plant germination and is instrumental in preserving the structural orientation of macromolecules and maintaining growth hormone production [61].

The effect of nanomaterials on the development of crops like coffee [57], potato [62], pearl millet [60], beans [63], and oil seed [64], has resulted in enhanced growth of these crops. Details of some applications of nano-fertilizers and other nanomaterial in agriculture have been summarized in Table 4.2.

Pest and Disease Control

Pest and disease infestations threaten agricultural productivity and jeopardize global food security. Harmful chemicals from pesticide applications often end up in water bodies, which may require investment to purify these waters to prevent adverse

Table 4.2 Summary of some applications of nano-fertilizers in the agricultural sector

Reference(s)	Focus	Nanoparticles	Findings	Plant	Application scale
Abdel-Aziz et al. [56]	Investigate the impact of applying chitosan nano-sized NPK fertilizer through foliar spray at different concentrations (10%, 25%, and 100%) on the chemical composition of wheat grains cultivated in different types of soil	Nano chitosan	Enhanced uptake of elements like phosphorus and potassium	Wheat	
Ha et al. [57]	Measuring the impact of chitosan NPs on the growth of coffee	Chitosan NPs	Improved nitrogen (17.04%), phosphorus (16.31%), and potassium (67.50%) in coffee leaves	Coffee seedlings	Greenhouse house condition
Kumar et al. [60]	Assess the impact of ZnO-NPs on the seed performance of pearl millet	ZnO-NPs	The growth of seeds exposed to ZnO-NPs increased by 20%, with increased seedling vigor and germination rate of 93%	Pearl millet	
Cyriac et al. [61]	Analyze the growth impact of ZnO-NPs on the development and root growth enhancement of pea and rice plants	ZnO-NPs	The germination index and root growth for rice and pea plants were improved significantly	Rice and pea plants	Laboratory scale
Abd El-Aziem et al. [62]	Efficacy of NPK nano fertilizer with no adverse retrogradation on yield and quality compared to nonnanofertilizer	Nano-N, nano-P, nano-K	The higher yield was recorded for plots that had foliage NPK nano fertilizers	Potato	Research farm
Rani et al. [63]	Assessed the effect of extracts from nanoparticles on seed germination and growth	ZnO and MgO NPs	Nanoparticles significantly increased the germination rate of the seeds	mung bean and cajan seed	Laboratory scale

(continued)

Table 4.2 (continued)

Reference(s)	Focus	Nanoparticles	Findings	Plant	Application scale
Sri Sindhura et al. [64]	Assess the potential changes in enzyme activity and physiological characteristics of plants after being treated with Zn-NPs	Zn-NPs	Enzyme activity increased, bacterial and fungi population significantly increased, and increments were recorded in leaf length, number of leaves (51.9–53%), and height (39.8–40.21%) when assessed after 30 and 60 days	Oil seed (*Arachis hypogea L*)	Pot culture experiment
Kottegoda et al. [65]	Study the slow release of nitrogen from nanohybrid urea-hydroxyapatite	Hydroxyapatite $[(Ca_{10}(PO_4)_6(OH)_2]$ NPs	Amine-carbonyl interactions resulted in the slow release of nitrogen		Laboratory scale
Kumar et al. [66]	Assessed the importance of zinc in onion seed growth enhancement	ZnO-MWCNT nanocomposite	Growth or seedlings improved with 15 mg/mL concentration of ZnO-MWCNT nanocomposite, with no adverse effect	Onion	Laboratory scale
Carrasco-Correa et al. [67]	Evaluate the effect of CaO NPs on fruits in the field	CaO NPs	The calcium content in the fruit peel and cell width both increased	Cherries, peaches, pears, apples, grapes, tomatoes, Paraguayan, and apricots	Field

(continued)

Table 4.2 (continued)

Reference(s)	Focus	Nanoparticles	Findings	Plant	Application scale
Burvey et al. [68]	Effectiveness of nano-fertilizer compared to conventional one, and its impact on plant growth	Fe_2O_3 loaded casein NPs	After a month, the iron-loaded casein nano-carriers made an evidential contribution to the growth of spinach and created matrices to regulate the release of bioactive substances	Spinach	Pot experiment
Karmous et al. [69]	Assessed the phytotoxic effects of Cu-NPs and ZnO-NPs on mung bean germinated under different concentrations, and their effect on the quality and safety	ZnO-NPs, Cu-NPs	10 mg/L concentrations of the NPs improved the nutritive properties (amino acids, essential minerals, antioxidant polyphenols) of the bean seeds. However, when concentrations increased between 100 and 2000 mg/L, the nutritional balance was altered with a decrease in macro (Ca, Na, Mg) and micronutrients (Fe, Cu, K, Zn, Mn)	Mung bean (Vigna radiata L)	

N—nitrogen, P—phosphorus, K—potassium, ZnO—zinc oxide, NPs—nanoparticles, MWCNT—multiwalled carbon nanotubes, MgO—magnesium oxide, Ag—silver, CaO—calcium oxide, Fe_2O_3—iron III oxide, Cu—copper, Fe—iron, Zn—zinc, Mn—manganese, Na—sodium impacts on human health. Also, these biological pressures can result in harvest losses, lower crop quality, limited yield factors, and economic difficulties for farmers. Effective management strategies, including advanced technologies like nanotechnology, have become necessary in mitigating these challenges sustainably for a more resilient food system.

Nanotechnology has revolutionized disease control in the search for more sustainable and efficient agricultural practices. This innovative approach, using unique properties of nanomaterials, reduces the effect of pests and diseases on crops and offers targeted and environmentally friendly solutions. Furthermore, NPs can easily absorb and spread through the leaves of plants, activate their defense mechanisms, and reduce the severity of diseases [70].

Nano pesticides are an essential application of nanotechnology in pest control, including the formulation of nanoscale pesticides. These nano-formulations exhibit controlled release mechanisms that ensure the gradual and lasting delivery of active substances. This not only improves the effectiveness of pest control but also minimizes the environmental risks resulting from using conventional pesticides [71]. One of the primary goals of nanopesticide formulation is to ensure the increased solubility of poorly active soluble ingredients [72].

Magnesium oxide nanoparticles have been highlighted as effective nanomaterials that can reduce bacterial populations in tomato plants [73]. It has been reported that silver (Ag) NPs exhibit activities against plant pathogens [74]. A summary of some nanomaterials explored in pest and disease control in agriculture has been presented in Table 4.3.

Nanosensors for Precision Agriculture

In the fast-paced world, technology has become inevitable, and its applications are numerous across many disciplines. Particularly in agriculture, the development of sensors is being employed to improve performance and productivity and reduce costs. Precision agriculture aims to optimize resource use, minimize environmental impact, and enhance crop yield by utilizing wireless networks and sensors. These sensors are embedded in the soil or attached to the crops for real-time data on parameters like pH, nutrient content, moisture, etc., to ease and improve monitoring [75]. Some studies have developed sensors that have made it less destructive and easy to collect moisture content data on soil [76, 77].

Additionally, agricultural activities have been enhanced with artificial intelligence (AI) to efficiently conduct activities like harvesting, weeding, and drone application [78]. These advanced technologies have been explored in estimating soil moisture content [76, 77] and even to explore weeding [79]. Magnesium oxide nanoparticles were assessed as antibacterial agents in plants [73]. Some of the studies conducted on applying nanosensors for improved precision agricultural activities have been summarized below (Table 4.4).

Table 4.3 Summary of some applications of nanotechnology in pest and disease control

Reference(s)	Focus	Nanomaterial	Findings	Plant	Application scale	Limitations
Pittarate et al. [72]	Assess the effect on nanoparticles on S. frugiperda larvae	Cu-NPs, KI-NPs, Ag-NPs	KI-NPs showed high insect larvicidal activity against the larvae, with 63–98% mortality at different concentrations. The NPs also did not have any adverse effects of the corn morphology or soil nutrient	Corn plants	Laboratory setting	None Identified
Imada et al. [73]	Assessed the potency of MgO NPs as antibacterial against *Ralstonia solanacearum* isolated from plants	MgO NPs	Concentrations of 0.05 or 0.1% exhibited strong antibacterial activity against R. *solanacearum in liquid medium. Also, tomato plants pretreated with MgO NPs shpwed significant reduction in bacterial population*	Tomato	Pot experiment	None discussed
Ayisigi et al. [74]	Assess the potency of Ag-NPs to treat soft rot disease in pepper and stimulate plant growth	Ag-NPs	Ag-NPs exhibited antimicrobial activity against Pectobacterium carotovorum, limited soft rot diseases in seedlings by 15%, and had a phyto-stimulative effect on seedling growth	Pepper	In vitro and pot culture	In vivo experiment was not done

(continued)

Table 4.3 (continued)

Reference(s)	Focus	Nanomaterial	Findings	Plant	Application scale	Limitations
Sathiyabama et al. [80]	Evaluate the impact of Cur-CnNPs as an antifungal agent against Fusarium oxysporum f. sp. cicero. The pathogen in chickpea plants	Cur-CnNPs	The nanoparticles showed a well-dispersed particle size that ranged between 20 and 30 nm. The plant also exhibited enhanced seedling vigor index	Chickpea	Laboratory setting	
Sofy et al. [81]	Investigate the capability of ZnO-NPs to trigger the resistance of tomato plants to tomato mosaic tobamvirus (ToMV) infection	ZnO-NPs	The applications of the NPs improved enzymatic activity, and foliar application helped to mitigate against adverse effect of the TomV infestation	Tomato	Pot experiment	
Hamed Derbalah and Elsharkawy [82]	Assess the potency of NiO-NS to induce resistance against cucumber mosaic virus (CMV) in cucumber plants	NiO-NS	Plants treated with NiO-NS by foliar spraying exhibited great reduction in CMV after 7- and 14-days inoculation	Cucumber	Pot experiment	

Ag—silver, NPs—nanoparticles, Cu—copper, KI—potassium iodide, Ag—silver, Cur-CuNPs—curcumin copper nanoparticles, MgO—magnesium oxide, ZnO—zinc oxide, NiO-NS—nickel oxide nanostructures

Table 4.4 Summary of some applications of nanotechnology in precision agriculture

Reference(s)	Focus	Nanomaterial	Findings	Environmental impact	Application scale
Nagahage et al. [76]	Automated soil moisture monitoring	Four (4) capacitive soil moisture sensor	Sensor to sensor variability	No adverse effect	Laboratory soil mixture
Stevanato et al. [77]	To estimate soil moisture content	Cosmic ray neutron sensing (SRNS)	Integration time for moisture measurement was 24 hh with reduced statistical uncertainty	No adverse effect	Campus site
Reiser et al. [79]	Sensor-robot for weeding	Sono sensor	Energy efficient, with performance of tilled area as 82%	No adverse effect	Vineyard

Nanotechnology is a catalyst for change in agriculture and provides many solutions to the sector's challenges. In soil remediation, nanomaterials demonstrate unparalleled abilities in soil detoxification, water purification, and sustainable agriculture.

The accuracy of nanoscale interventions in controlling pests and diseases revolutionizes agricultural practices, minimizes the impact on the environment, and maximizes the utilization of resources. Nanofertilizers increase crop growth and yield with a minimal ecological footprint.

Also, incorporating nanosensors in agriculture increases precision by providing real-time data on the conditions of the soil and crop health, making is to make informed decisions. As stakeholders work toward overcoming the complexity of global food security, nanotechnology promises a more sustainable, efficient, and resilient future for agriculture.

Nanotechnology in Livestock Production and Management

Livestock production plays a pivotal role in global agriculture and food security, and it contributes to livelihoods as well. Not only are these animals reared as protein sources, but they also support economies, serving as a source of income for many, thereby fostering socio-economic development [83, 84]. With growing concerns about the environment resulting from the contribution of the food systems,

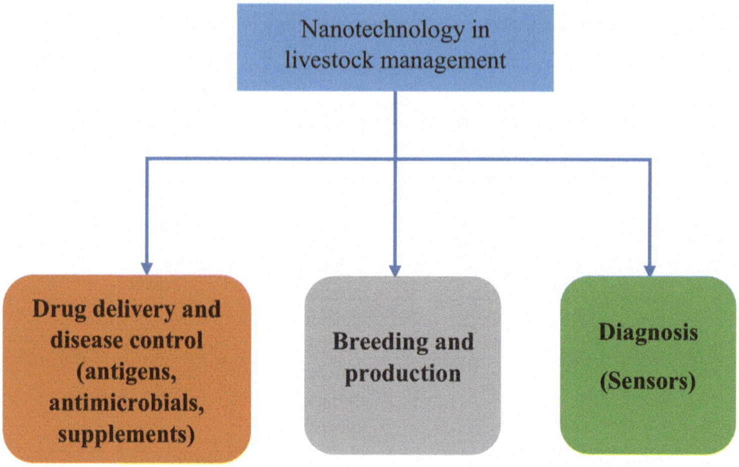

Fig. 4.5 Representation of some of the applications of nanotechnology in animal production and management

there is a need to ensure diversity and resilience in the face of environmental challenges. Nanotechnology holds significant promise for improving animal production worldwide by enhancing productivity, ensuring animal welfare, and improving sustainability practices [85].

Animal reproductive efficiency is crucial for productivity and profitability. The application of nanotechnological strategies in reproductive management includes biological, hormonal, nutritional, and disease control [86]. Also, some more applications of nanoparticles in livestock production include using nano biosensors, nanomedicines, nanofluids, and nano antibiotics [87]. A pictorial representation of the application of nanotechnology in livestock management, further discussed in the section is shown below in Fig. 4.5.

Diagnosis and Nanosensors

Nanochips that serve as diagnosing sensing devices are being used to observe and diagnose animal farming, such as monitoring the health of poultry and checking the physiological state of animals and their reproductive health. This sensing and monitoring assistance using nanotechnology-based sensors helps check the quality of protein produced. Moreover, they can be used to track contaminations from chemicals, additives, etc. [85].

Nanosensors can detect pathogens and diseases early, allowing timely intervention and preventing disease outbreaks. Rajwade [84] indicated the effectiveness of wearable sensors used in animal production to efficiently measure and monitor physical activity and parameters among livestock, detect any medical condition, for early remediation activities and to improve their health care.

Drug Delivery and Disease Control

A major issue plaguing livestock production is disease. This is because animal diseases can affect their performance, result in loss of productivity, and in the worst case, affect humans when consumed. These diseases could result from climate change, the immediate environment, and the atmosphere [88].

Regular vaccinations and drug application are necessary for animal production to control diseases and ensure thorough output. As such, ensuring that these drugs are well delivered for maximum effectiveness is crucial. Due to their unique properties, nanoparticles offer novel approaches for disease diagnosis, prevention, and treatment in livestock. To achieve this, antibacterials and antibiotics, for example, can be used as additives for their feed to protect animals from diseases, thereby increasing growth and reducing their production cycle [85]. Furthermore, in developing vaccines for disease control, nanoparticles can co-deliver antigens in veterinary medicine. Some NP transmitters for vaccine delivery include liposomes, micellar, and polymeric NPs [87, 89]. Nanoparticles can also deliver vaccines and antimicrobial agents more effectively, thereby bolstering the immune system of livestock and reducing the prevalence of infectious diseases.

Additionally, nanoparticles can be an effective means of treating ailments and diseases in livestock. Nanoparticles like silver can serve as carriers in drug delivery to ensure that drugs reach specific sites like the cell membranes for more efficacy [90]. Similarly, dendrimers, a nano platform, can deliver drugs that treat foot-and-mouth diseases and swine fever virus in piggery [89].

Breeding and Reproduction

Reproduction is paramount in livestock production because it ensures the sustainability and viability of the sector. The emergence of nanotechnology has enhanced productivity in livestock management.

Nanotechnology provides innovative solutions to improve animal breeding and production practices. Nanotechnology can contribute to developing advanced reproductive technologies that can enhance breeding efficiency and genetic selection processes in animal breeding. The technology can also support breeding for desired traits in livestock. According to Hashem [86], nanotechnology has enabled the delivery of reproductive hormones like pheromones to improve efficiency and effectiveness.

Additionally, nanoparticles can release hormones and other therapeutic agents to control reproduction and improve animal fertility rates [91]. Moreso, in breeding and reproduction, applying nanotechnology to ensure optimized nutrient delivery results in growth and improved well-being, resulting in increased productivity.

Studies on the use of nanotechnology in animal farming are to identify infections and diseases [92–95] and serve as antibiotics [100, 107, 110]. A summary of some applications of nanotechnology to improve livestock management is presented (Table 4.5).

Table 4.5 Application of nanotechnology in livestock management

Aim	Nanomaterial	Used case	References
Quick detection of infectious bursal disease virus	Ag NPs	Chicken (poultry)	Nurulfiza et al. [92]
Develop BCA technique to detect bluetongue virus	Au NPs	Ruminants	Yin et al. [93]
To detect foot-mouth disease virus using the rapid dot plot assay	Au NPs	Ruminants	Jain et al. [94]
Develop a sensor to quickly detection of avian influenza virus H5N1	Au NPs	Chicken	Karash et al. [95]
Biosensor for the detection of foot and mouth disease virus	Au NPs	Cattle, buffalo, calves	Hamdy et al. [96]
Identification of *Mycobacterium bovis*	Nano-artificial NOSE	Cattle	Peled et al. [97]
Effective detection of porcine reproductive and respiratory syndrome virus	Nanobodies (Nb 12, Nb 35)	Swine	Sun et al. [98]
Enhanced antibacterial activity	Ag NPs++ vancomycin		Kaur et al. [99]
Improving antibiotic application to treat *Staphylococcus xylosus* which causes mastisis	Chitosan NPs grafted CD-g-CS	Cattle	Zheng et al. [100]
Influence of NPs in treating *E. coli* infection (colisepticemia)	Ag NPs	Chicken	Awaad et al. [101]
Assess the immune capacity and protection efficacy of an inactive NDV vaccine encapsulated with NPs	Poly-(lactic-co-glycolic) acid (PLGA) NPs encapsulated NDV vaccine	Chicken	Kumar et al. [102]
Assess the immune response and efficacy of an inactive influenza virus vaccine	Poly- (D, L-lactic-co-glyolic) acid encapsulated CpG ODN	Chicken	Singh et al. [103]
Study the effect of different dietary CP levels and ZnO on the growth and performance of Japanese quails	Zn++ crude protein (CP)	Chicks (quail)	Elsayed et al. [104]
Assess the effect of ZnO NP supplements on the growth of layer chicks	ZnO NPs	Chicks	Mishra et al. [105]
To evaluate the long-term effects of NPs on rabbit reproductive activity and fertility	Ag NPs	Rabbit	Castellini et al. [106]
Assess the NPs as an alternative to antibiotics to preserve the fertilizing ability of bull semen	AgC NPs	Bull	Yousef et al. [107]

(continued)

Table 4.5 (continued)

Aim	Nanomaterial	Used case	References
Assess the impact of Ag NPs the quality of quail egg when used as alternative antibiotic	Ag NPs	Japanese quail	Farzinpour and Karashi [108]
To assess the effect of NPs on the reproductive tissue, the ovarian granule cells of cattle and their effect on the production of steroid hormones	Ag NPs	Cattle	Tabandeh et al. [109]
Assess the antimicrobial activity and toxicity effect of NPs on the *Staphylococcus aureus,* and subsequently, the sperm produced by swine	Ag NPs	Swine	Pérez-Duran et al. [110]
Assess the efficacy of Fe and Ag NPs to control the microbial proliferation during the processing of semen in ram, and consequently on the semen quality and functionality	Fe and Ag NPs	Ram	Tsakmakidis et al. [111]

BCA—bio-barcode amplification assay, Au—gold, Ag—silver, CD-g-CS—β-cyclodextrins, NDV—Newcastle disease virus, ODN—oligodeoxynucleotides, ZnO—zinc oxide, AgC—silver carbon, Fe—iron

Nanotechnology in Food

The food industry is very dynamic because these industries always strive to meet the changing trends in consumer dietary patterns. Although one of the utmost reasons for these industries is to make a profit, there is enormous competitiveness among them as they all attempt to explore newer technologies that will ensure the delivery of consistent quality and safe products to their consumers. Ensuring the security and quality of foods produced by industries remains a significant mission for manufacturers [112]. Food security remains one of the crucial goals global policymakers and stakeholders aim to achieve and ensure that everyone can always access nutritious and safe foods for their healthy lives. Recently, the environment has even been considered in the scope of the food supply chain. This is because the impact of food production on the environment in terms of its high contribution to global greenhouse emissions has raised concerns for more advanced technologies that consider the environment.

In food security, nanotechnology transforms food preservation and packaging, extending shelf life, and minimizing post-harvest losses. One of the significant causes of post-harvest loss in food is ethylene-induced ripening and microbial contamination, and this results in food loss, which affects food security [113]. Biocompatible bionanomaterials can be incorporated into bio-based packaging to offer a more sustainable alternative to conventional packaging [114]. Nanoscale innovations provide the prospect of creating a resilient and sustainable food supply chain, ensuring the availability of nutritious food for a growing global population. Currently, nanotechnology is used to develop nano-sized food ingredients and additives, enhance food quality, improve food packages and product development, and act as carriers for bioactive compounds, transporting them to their desired destination [112, 115, 116]. Interestingly, Bouwmeester et al. [115] further indicated that about 183 patents between 2009 and 2011 are related to nanotechnology in food. Broadly, the uses of nanotechnology in the food industry can be categorized as presented in Fig. 4.6, ensuring food security.

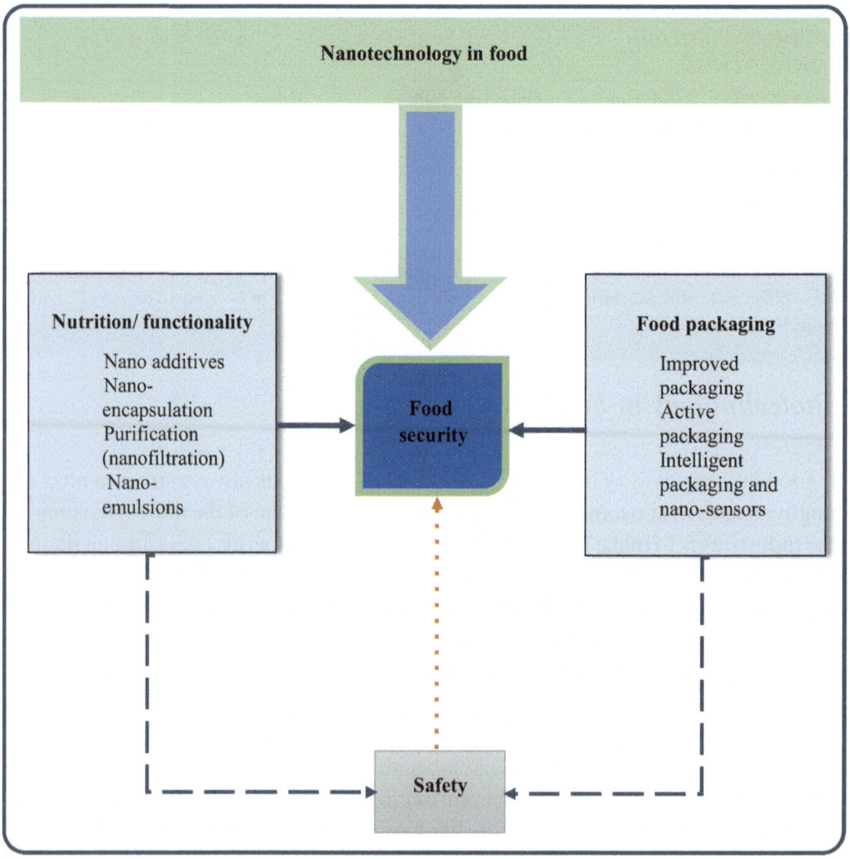

Fig. 4.6 An overview of some applications of nanotechnology in the food industry to ensure food security

Nanotechnology for Food Packaging

Food packaging is an integral aspect of food production because it keeps from contamination and ensures their safety in transportation and storage. Conventional food packaging materials from petroleum sources are unsustainable due to the difficulty of disposing of them [117]. Furthermore, waste disposal remains a global problem for the food supply chain [118]. Therefore, it is crucial to develop food packaging materials that are more sustainable.

Improved Packaging

These materials have significantly improved the overall quality, and improved packaging materials leveraging nanotechnological applications have reformed the food industry, enhancing overall safety and quality standards [119]. The integration of nanoparticles in food packaging materials contributes significantly to improving their performance and physical properties. These improvements include but are not limited to flexibility, durability, mechanical strength, gas barrier properties, and temperature resistance [120, 121].

Nanotechnology has paved the way for developing superior packaging materials with enhanced barrier properties, ensuring prolonged shelf life and preservation of food products. Incorporating nanoparticles allows for precise control over material characteristics, leading to packaging solutions that effectively protect against moisture, oxygen, and microbial contamination [122, 123].

In addition to traditional packaging approaches, edible coatings represent a promising frontier in enhanced packaging technology. These coatings are derived from polysaccharides, proteins, and lipids that are edible and serve as protective film layers applied directly on the food surfaces via coating methods [124]. Edible coatings offer several advantages, including improved food safety, extended shelf life, and reduced environmental impacts compared to conventional packaging materials [125, 126].

Active Packaging

Consumers' concerns for quality and health consciousness of diets are continuously increasing, and this has challenged all food industries to strive to produce products that satisfy their nutrient needs [119]. This packaging material often includes components to enhance sensory attributes, extend the shelf life of food, and maintain its quality [114]. Nanoparticles have the potential to be combined with specific antimicrobial agents and other metal nanoparticles to serve as active agents. This makes it possible to extend shelf life by inhibiting microbial growth [127]. Also, incorporating plant-based extracts into composite coatings can extend the shelf life of fruits and prevent bacterial growth [128]. Researchers have extensively used NPs to improve food packaging materials or develop innovative packages to preserve foods, ensure availability, reduce post-harvest losses, etc.

Using active packaging in the food industry can significantly reduce food waste, extending the shelf life of food materials [127]. Mei and Wang [129] noted the growing popularity of active packaging that utilizes nanotechnology, although they are yet to be commercialized by food industries. Active packaging can serve numerous functions that are not present in conventional packaging. The specific purpose of active packaging in the food industry varies based on the material used in the active packaging film. Depending on the use, active packaging can have antimicrobial properties, absorb oxygen, regulate moisture, or release ethylene [130]. These smart packages can be applied by dipping the food into the film-forming or spraying it onto the surface of the food. Or forming the film by casting to obtain a flat sheet that can be used to package the food [131]. When an antimicrobial agent is incorporated into food packages, they are slowly released onto the food surface and effectively help to prevent microbial growth during shelf life [132]. Also, microbial growth in food can be inhibited by using edible coatings made from biopolymers and nanoparticles, which will, in the end, extend its shelf life [121].

Intelligent Packaging

Chavashizadeh et al. [133] indicated that intelligent packaging, in another way, controls the efficiency of active packaging, even though both are different forms of packaging. Nanomaterials can function as intelligent packaging that interacts with the consumer on the quality stability of the food [120]. Nanosensors can be embedded in food packages to provide real-time monitoring of freshness by detecting gaseous emissions during spoilage, making it possible to prevent food waste and even offer a more accurate expiration date. These packaging materials utilize biological materials like proteins, fats, and carbohydrates [114]. For example, the ability of a nanomaterial to indicate a color change to signify a change in the quality of fresh produce like meat or vegetables. The advanced nanotechnology adds to conventional food packages is its ability to respond to external stimuli like temperature or moisture changes, acting as a form of active packaging.

Nanoparticles like silver are suitable semiconductor materials that can improve electrical conductivity, improve antibacterial activities, and improve the interaction with polypyrrole sensors for intelligent packaging [134]. Nanoparticles like TiO_2 are being consumed in developed countries, and they can used in food supplements and for edible packaging [135] and are being extensively studied.

Table 4.6 summarizes some studies that can use NPs for improved packaging and contributing to food security.

Table 4.6 Application of nanotechnology for improved smart and intelligent packaging.

Reference(s)	Matrix	Application method	Properties measured/ improved	Food used
Siripatrawan and Kaewklin [113]	Chitosan++ TiO_2	Casting	water vapor permeability and elongation to break decreased, and ethylene photodegradation was significantly enhanced. Also, adding UV enhanced the antimicrobial activity of the film against test microorganisms	
Friedrich et al. [128]	Cassava starch	Coating	Reduced respiration rate, higher vitamin C compared to control, delayed microbial growth	Strawberries
Kam et al. [132]	Gelatin++ durian leaf extract	Solvent-casting	The addition of 0.5 leaf extract increased the DPPH scavenging activity 17.6 times, composite film was able to retard oil oxidation. There was no improvement in water vapor permeability	Durian fruit pulp
Chavoshizadeh et al. [133]	Wheat gluten++ chlorophyll	Casting	Film composite chlorophyll has high tensile strength and elasticity ($p < < 0.05$), a reduced oxidative rate in oil, and exhibits antioxidant properties, preventing a color change in the oil against oxidant agents (HNO_3)	Sesame oil
Ghasemi et al. [134]	Conducting bacterial cellulose-polypyrrole-TiO_2-Ag nanocomposite	Solvent-casting	The sensitivity of sensors increased when TiO_2-Ag NPs were increased. High amounts of TiO_2-Ag NPs increased sensitivity to *S. aureus* and *E. coli.* Nanocomposite was able to rapidly detect bacterial growth	
Taechutrakul et al. [136]	Polylactic acid (PLA)		Antioxidant and oxygen (O_2) scavenging capacities, delayed fruit ripening	Banana
Nourmohammadi et al. [137]	Resveratrol combined with whey protein nanoclay composite	Coating	Significant decrease in microbial for cheese coated with nanocomposite	Cheese

(continued)

Table 4.6 (continued)

Reference(s)	Matrix	Application method	Properties measured/ improved	Food used
Rodriguez et al. [138]	Starch	Coating	Improved antioxidant and antimicrobial properties, increased elongation, and significant change in opacity	Mangoes
Correa-Pacheco et al. [139]	Bionanocomposite (edible propolis-chitosan film)	Coating	Decreased roughness, decreased water vapor permeability, strong inhibitory effect on *Listeria monocytogenes and E. coli,*	Strawberries
Stefanowska et al. [140]	Chitosan++ propolis and caffeine	Casting	99.13% antiradical activity showed activity against *E. coli, P. aeruginosa, and S. enterica.* The film also exhibited antifungal activity against *C. albicans*	
Pirsa and Shamusi [141]	Bacterial cellulose++ polypyrrole-ZnO nanocomposite (BC-PPy-ZnO)	Casting	Slowed down the reduced rate of odor and sensory, reduced speed of increasing pH. The film also exhibited antimicrobial and antioxidant activity	Chicken
Pirsa et al. [142]	Chitosan++ PPE++ MOE	Solvent-casting	Addition of MOE increased the water vapour permeability and increased antimicrobial effect on *B. cereus* and *E. coli*. Tensile strength also increased, and the film was sensitive to pH change, acting as a smart package	Cream cheese
Rezaei, Pirsa and Chavoshizadeh [143]	Wheat gluten++ ZnO		Nanocomposite film exhibited significant inhibitory effect against bacteria and fungi at wavelengths of 365 and 254 nm. Also, antioxidant activity increased when ZnO NPs increased	

(continued)

Table 4.6 (continued)

Reference(s)	Matrix	Application method	Properties measured/ improved	Food used
Zheng et al. [144]	PBAT/PLA++ natamycin++ nano-TiO_2	Extrusion	Including PBAT enhanced the tensile strength of the film (24.1 to 43.5 MPa), and elongation at break was also enhanced (85.8 to 258.2%). The composite film also exhibited a strong control effect on the decay of postharvest grapes, inhibiting pathogenic fungi (high inhibition rate of 95.72% against Penicillium expansum)	Grape
Cao et al. [145]	PBAT++ TiO_2Ag composite	Solvent-casting	Mechanical UV blocking (visible light of 400–800 nm), and barrier properties of the composite film were improved (5% ww inclusion of TiO_2Ag). Elongation at break increased to 790.69%, tensile strength increased to 14.25 MPa. Water barrier property was reduced by 37%. Film also exhibited good antibacterial performance against *E. coli* and *Staphylococcus aureus*)	Cherry tomato
Du et al. [146]	TiO_2++ THYb-CD composite film	Casting	Enhanced hydrophobicity, efficient in inhibiting microbial growth, tensile strength increased to 9.43 MPa but elongation to break reduced. Water solubility improved (97.17%). Film also exhibited excellent microbial inhibition rate (99%) against *E. coli* and S. aureus, and inhibited spoilage in tomatoes and strawberries	Strawberry and tomato

(continued)

Table 4.6 (continued)

Reference(s)	Matrix	Application method	Properties measured/ improved	Food used
Nguyen and Lee [147]	PVA++ CNCs++ TiO_2++ APE		Water vapor barrier property improved with the addition of CNCs to PVA, and high UV barrier property. The composite film tensile strength of (63.1 MPa), elongation of (631%), and elongation properties, but Young's modulus increased. Also, antioxidant and antimicrobial properties improved (due to the APE and TiO_2, respectively), reduced browning in potatoes, and less spoilage in tomatoes	Tomato and potato
Indumathi et al. [148]	Chitosan++ cellulose acetate phthalate++ ZnO NPs	Casting	Thermal stability and barrier properties were improved. Optimal tensile strength and stiffness were observed when the film was loaded with 5% (ww) ZnO. Surface wettability was lowered, biodegradability of the film ranged from 30 to 50 in 28 days and prolonged the shelf-life of black grapefruits up to 9 days	Black grapes
Ashrafi et al. [149]	Chitosan++ kombucha tea	Casting	DPPH scavenging activity increased, and total phenolic content also increased. UV light barrier property was improved. Also, water vapor permeability was enhanced, antioxidant activity increased and exhibited an inhibitory effect on *E. coli* and *S. aureus*. The film also delayed odor formation in minced meat	Minced beef

(continued)

Table 4.6 (continued)

Reference(s)	Matrix	Application method	Properties measured/ improved	Food used
Wang et al. [150]	Chitosan++ tea polyphenols (TP)	Casting	Water solubility increased, water vapor permeability reduced, total phenolic content increased, and DPPH-free scavenging activity was enhanced (with the addition of TP),	
Moreira et al. [151]	Chitosan++ sodium caseinate edible film	Casting	Significant antimicrobial inhibitory effects were observed in that of cheese and salami	Carrot, cheese, and salami

HNO_3—nitric acid, ZnO—zinc oxide, PPE—pomegranate peel extract, MOE—Melissa officinalis essence, TiO_2—titanium dioxide, Ag—silver, DPPH—2,2-diphenyl-1-picrylhydrazyl, PBAT—poly (butylene adipate-co-terephthalate), E. coli—Escherichia coli, THYB-CD—thymol-loaded b-cyclodextrin, PVA—polyvinyl alcohol, CNCs—cellulose nanocrystals, APE—apple peel extract, UV—ultraviolet. The ethylene photocatalytic degradation of the chitosan-TiO_2 nanocomposite films significantly increased ($P \leq 0.05$) with increasing TiO_2 concentration

Fig. 4.7 Schematic representation of an encapsulated bioactive compound

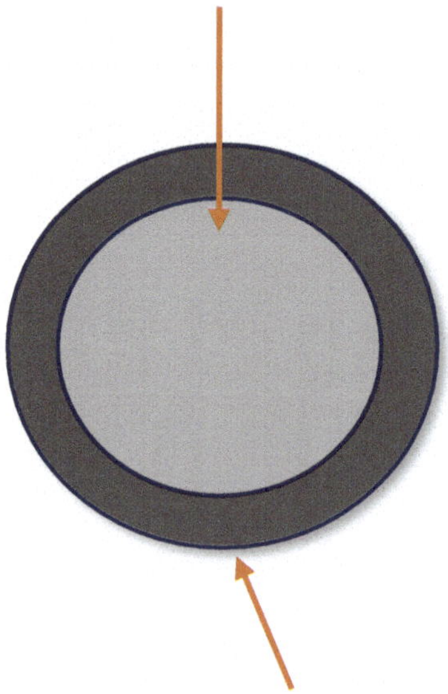

Nanoencapsulation

Nanotechnology is an evidential revolutionary for improving food security, explicitly improving nutrition and functionality. The development of microencapsulation dates back to the 1950s. Encapsulation is an important application of nanotechnology, which includes encapsulating bioactive compounds like carotenoids, essential oils, and antioxidants in nanoscale carriers [152]. The need to control the release of bioactive compounds in the gastrointestinal tract (GIT) is essential, and nanoencapsulation is vital in ensuring that. Nanoencapsulation is a technique used to release bioactive ingredients in a nanocarrier [153]. Nanoemulsion is one of the significant ways of encapsulation. Figure 4.7 shows a representation of nanoencapsulation.

Nano-emulsions

Nano-emulsion-based food delivery systems are also effective in improving the dispersion of bioactive compounds in food products, protecting them from any form of degradation and improving their bioavailability [154]. In nanoemulsion, the nanoscale droplets are dispersed in one liquid in another immiscible one and are an excellent means to deliver water-soluble food ingredients like lipophilic vitamins and fish oil [155]. This will, in the end, improve nutritional outcomes when consumed.

Nanoscale delivery systems can be made from different food components like lipids, proteins, and carbohydrates and can be used alone or together [152]. Nanoencapsulation has been extensively studied. Some of these applications include the encapsulation of β-carotene by nanodispersion [156], oil-in-water emulsification of β-carotene [157], fish oil encapsulation [158], encapsulation of omega 3–6–9 fatty acids-rich oil [159], and encapsulated fish oil powder [160].

Challenges and Prospects

Environmental and Health Concerns of Nanotechnology in Agriculture

The application of nanotechnology in managing animal reproduction is met with both practical and ethical challenges [86]. As with any transformative technology, careful consideration of ethical, environmental, and regulatory aspects is imperative to fully realize the prospective benefits of nanotechnology in agriculture and food security. Whereas nanotechnology is continuously proven to be important in agriculture and food security, there is a need for consideration and promotion of a secure and safe environment [35]. Even though the use of nanotechnology in agriculture and other fields is gaining momentum, there are concerns about its probable effect on human health and the environment, making it necessary for regulation to control these engineered materials [161, 162].

One primary concern revolves around the extensive surface area of engineered nanomaterials, which heightens the potential for interaction with the food matrix [163]. Nanoparticles applied in agriculture could end up in the gut of animals and humans through the food supply chain, and this raises more concerns from consumers [164–166]. For example, studies conducted by Karmous et al. [69] showed that higher concentrations (100–2000 mg/L) of ZnO and Cu NPs decreased phenolic compounds and nutrients like sodium, magnesium, calcium, copper, iron, manganese, and potassium content in mung beans. Also, though nanomaterials in agriculture are beneficial, their application can affect the microbial habitats in the soil and inhibit enzyme activities [55].

Moreover, another critical question is regarding the degradability of these nanomaterials, a topic that warrants urgent attention and clarification [163]. Chemical processes used to synthesize nanoparticles have some negative consequences with by-products that could be a source of environmental pollution, hence requiring more risk assessment studies [166].

Additionally, the lack of knowledge and public awareness of nanotechnology in food can limit its applicability since consumers may offer restrictions to their applications [120]. As much as consumers may have issues with their application in agriculture and food, commercializing the use of nanoemulsions could hold some challenges for industry players because of a suitable food-grade material, and an even more significant challenge would be transferring the laboratory scale set-ups into full-scale set-ups [163].

These issues underscore the complexity and importance of assessing the implications of nanotechnology in animal farming practices.

Prospects

There are high prospects for nanotechnology to significantly influence the development of sustainable agriculture and precision farming practices, which is promising. Also, nanotechnology is becoming instrumental in the post-harvest processes and, by extension extending the shelf life of perishable products using nanomaterials that can inhibit microbial growth. There are gaps in their application on a large scale and their toxicological impact on the ecosystem.

Additionally, the FAO/WHO [163] indicated the opportunities for further research to focus on identifying novel risk assessment strategies through proven testing methods to address gaps in the application of nanotechnology in agriculture and the food industry. Therefore, nanotechnology requires extensive research on the toxicity of these materials at the nanolevels.

Furthermore, there are great prospects for consumer awareness of the application of nanotechnology and the possibility of using it to transform the agricultural and food sectors. Also, regulatory bodies need to implement standards for measuring the safety of nanomaterials that are primarily included in food production.

References

1. Loizou, E., Karelakis, C., Galanopoulos, K., Mattas, K.: The role of agriculture as a development tool for a regional economy. Agric. Syst. **173**, 482–490 (2019). https://doi.org/10.1016/j.agsy.2019.04.002
2. Ghouri, M.Z., Khan, Z., Khan, S.H., Ismail, M., Aftab, S.O., Sultan, Q., Ahmad, A.: Nanotechnology: transformation of agriculture and food security. Bioscience **3**, 19 (2020)
3. Khan, F., Pandey, P., Upadhyay, T.K.: Applications of nanotechnology-based agrochemicals in food security and sustainable agriculture: an overview. Agriculture **12**(10), 1672 (2022)
4. Hulse, J.H.: Science, Agriculture, and Food Security. NRC Research Press, Ottawa (1995)
5. Kasturi, P.: Technology and food security. Humanomics **25**(2), 163–168 (2009)
6. Zain, M., Ma, H., Ur Rahman, S., Nuruzzaman, M., Chaudhary, S., Azeem, I., Mehmood, F., Duan, A., Sun, C.: Nanotechnology in precision agriculture: advancing towards sustainable crop production. Plant Physiol. Biochem. **206**, 108244 (2024). https://doi.org/10.1016/j.plaphy.2023.108244
7. Squires, V.: The Role of Food, Agriculture, Forestry, and Fisheries in Human Nutrition. Eolss Publishers Co., Ltd., Oxford (2011)
8. Yadav, N., Garg, V.K., Chhillar, A.K., Rana, J.S.: Recent advances in nanotechnology for the improvement of conventional agricultural systems: a review. Plant Nano Biol. **4**, 100032 (2023). https://doi.org/10.1016/j.plana.2023.100032

9. Pingali, P.: Chapter 74 agriculture renaissance: making "agriculture for development" work in the 21st century. In: Handbook of Agricultural Economics, vol. 4, pp. 3867–3894 (2010). https://doi.org/10.1016/S1574-0072(09)04074-2
10. Munir, N., Gulzar, W., Abideen, Z., Hancock, J.T., El-Keblawy, A., Radicetti, E.: Nanotechnology improves disease resistance in plants for food security: applications and challenges. Biocatal. Agric. Biotechnol. **51**, 102781 (2023). https://doi.org/10.1016/j.bcab.2023.102781
11. Arndt, C., Farmer, W., Strzepek, K., Thurlow, J.: Climate change, agriculture, and food security in Tanzania. Rev. Dev. Econ. **16**(3), 378–393 (2012)
12. Timmer, C.: Agriculture and economic development revisited. Agric. Syst. **40**(1–3), 21–58 (1992). https://doi.org/10.1016/0308-521X(92)90015-G
13. Barlow, C., Jayasuriya, S.K.: Stages of Development in Smallholder Tree Crop Agriculture. Australian National University, Research School of Pacific Studies, Department of Economics and National Centre for Development Studies (1986)
14. Tang, S., Zhu, Q., Zhou, X., Liu, S., Wu, M.: A conception of digital agriculture. In: IEEE International Geoscience and Remote Sensing Symposium, Toronto, Ontario, Canada, vol. 5, pp. 3026–3028 (2002). https://doi.org/10.1109/IGARSS.2002.1026858
15. Yadav, S.S., Hedge, V.S., Habibi, A.B., Dia, M., Verma, S.: Climate change, agriculture and food security. In: Yadav, S.S., Redden, R.J., Hatfield, J.L., Ebert, A.W., Hunter, D. (eds.) Food Security and Climate Change, pp. 1–3. John Wiley & Sons Ltd., Hoboken (2019)
16. UNEP: Organic Agriculture and Food Security in Africa. United Nations, New York, NY (2008)
17. Mousavi, S.R., Rezaei, M.: Nanotechnology in agriculture and food production. J. Appl. Environ. Biol. Sci. **1**(10), 414–419 (2011)
18. Norde, W.: Intermolecular interactions. In: Frewer, L.J., Norde, W., Fischer, A.R., Kampers, F.W.: Nanotechnology in the Agri-Food Sector: Implications for the Future. Wiley-CVH Verlag GmbH & Co. KGaA, Weinheim (2011)
19. Maine, E., Thomas, V.J., Bliemel, M., Murira, A., Utterback, J.: The emergence of the nanobiotechnology industry. Nat. Nanotechnol. **9**(1), 2–5 (2014). https://doi.org/10.1038/nnano.2013.288
20. Parkash, V., Gaur, A., Agnihotri, R.: Nanoparticles from fungal resources: importance and applications. In: Thangudurai, D., Sangeetha, J., Prasad, R. (eds.) Nanotechnology for Food, Agriculture, and Environment, pp. 1–3. Springer, Cham (2020)
21. Selin, C.: Expectations and the emergence of nanotechnology. Sci. Technol. Human Values **32**(2), 196–220 (2007)
22. Jones, R.A.: Soft Machines: Nanotechnology and Life. Oxford University Press, Oxford (2004)
23. Tolochko, N.K.: History of nanotechnology In: Encyclopedia of Life Support Systems (EOLSS) (2009)
24. Melnik, A.V., Shagalina, O.V.: History of Nanotechnology (2011)
25. Binnig, G., Rohrer, H.: Scanning tunneling microscopy. Surf. Sci. **126**(1–3), 236–244 (1983). https://doi.org/10.1016/0039-6028(83)90716-1
26. Kroto, H.W., Heath, J.R., O'Brien, S.C., Curl, R.F., Smalley, R.E.: C60: buckminsterfullerene. Nature **318**(6042), 162–163 (1985)
27. Sangeetha, J., Thangadurai, D., Hospet, R., Purushotham, P., Karekalammanavar, G., Mundaragi, A.C., Harish, E.R., et al.: Agricultural nanotechnology: concepts, benefits and risks. In: Prasad, R., Kumar, V. (eds.) Nanotechnology: An Agricultural Paradigm, p. 1. Springer, Singapore (2017)
28. Dwaine, E.F., Thanos, C.G.: Nanotechnology and medicine. Expert Opin. Biol. Ther. **3**(4), 655–663 (2003). https://doi.org/10.1517/14712598.3.4.655
29. Duncan, R.: Polymer conjugates as anticancer nanomedicines. Nat. Rev. Cancer **6**(9), 688–701 (2006)
30. Medintz, I.L., Uyeda, H.T., Goldman, E.R., Mattoussi, H.: Quantum dot bioconjugates for imaging, labelling and sensing. Nat. Mater. **4**(6), 435–446 (2005). https://doi.org/10.1038/nmat1390

31. Mobasser, S., Firoozi, A.A.: Review of nanotechnology applications in science and engineering. J. Civil Eng. Urban **6**(4), 84–93 (2016)
32. Hollenberg, L.C.L., Dzurak, A.S., Wellard, C., Hamilton, A.R., Reilly, D.J., Milburn, G.J., Clark, R.G.: Charge-based quantum computing using single donors in semiconductors. Phys. Rev. B **69**(11), 113301 (2004)
33. Cui, Y., Lieber, C.M.: Functional nanoscale electronic devices assembled using silicon nanowire building blocks. Science (2001). https://doi.org/10.1126/science.291.5505.851
34. Ingle, A.P., Mohanil, M.P., Saxena, S., Patil, V., Borse, S.: Nanotechnology: the new perspective in agroecosystem and sustainable agriculture. In: Ingle, A.P. (ed.) Nanotechnology in Agriculture and Agroecosystems, pp. 2–8. Elsevier, Amsterdam (2023)
35. Behl, T., Kaur, I., Sehgal, A., Singh, S., Sharma, N., Bhatia, S., Al-Harrasi, A., Bungau, S.: The dichotomy of nanotechnology as the cutting edge of agriculture: nano-farming as an asset versus nanotoxicity. Chemosphere **288**, 132533 (2022). https://doi.org/10.1016/j.chemosphere.2021.132533
36. Usman, M., Farooq, M., Wakeel, A., Nawaz, A., Cheema, S.A., Rehman, H.U., Ashraf, I., Sanaullah, M.: Nanotechnology in agriculture: current status, challenges and future opportunities. Sci. Total. Environ. **721**, 137778 (2020). https://doi.org/10.1016/j.scitotenv.2020.137778
37. Saranya, S., Aswani, R., Remakanthan, A., Radhakrishan, E.K.: Nanotechnology in agriculture. In: Panpette, D.G., Jhala, Y.K. (eds.) Nanotechnology for Agriculture: Advances for Sustainable Agriculture, pp. 5–10. Springer Nature Singapore Pte Ltd. (2019)
38. Thul, S.T., Sarangi, B.K., Pandey, R.A.: Nanotechnology in agroecosystem: implications on plant productivity and its soil environment. Expert Opin. Environ. Biol. **2**, 2–7 (2013)
39. Hu, Y., Cao, Y., Ma, C., Yan, W.: Nano-biochar as a potential amendment for metal(loid) remediation: implications for soil quality improvement and stress alleviation. J. Environ. Manage. **351**, 119658 (2024). https://doi.org/10.1016/j.jenvman.2023.119658
40. Fernández, M.D., Sánchez-Arguello, P.S., García-Gómez, C.: Soil pollution remediation. In: Encyclopedia of Toxicology, 4th edn., pp. 631–645 (2024). https://doi.org/10.1016/B978-0-12-824315-2.00002-6
41. Chaturvedi, S., Govindaraju, K., Vijayakumar, P., Elango, K., Kannan, M.: Nanomaterials as a tool for soil remediation in sustainable agriculture. Compr. Anal. Chem. **99**, 189–212 (2022). https://doi.org/10.1016/bs.coac.2021.11.006
42. Edgar, V., Hermes, P., Denisse, V.J., Andrea, P., Roberto, S.C., Ileana, V., Ingle, A.P., Fabián, F.: Nanotechnology for environmental remediation: A sustainable approach. In: Nanotechnology in Agriculture and Agroecosystems, pp. 297–346 (2023). https://doi.org/10.1016/B978-0-323-99446-0.00008-8
43. Deng, H., Lei, H., Luo, Y., Huan, C., Li, J., Li, H., He, F., Zhang, B., Yi, K., Sun, A.: The effects of titanium dioxide nanoparticles on cadmium bioaccumulation in ramie and its application in remediation of cadmium-contaminated soil. Alex. Eng. J. **86**, 663–668 (2024). https://doi.org/10.1016/j.aej.2023.12.019
44. Hasan, M.K., Shopan, J., Ahammed, G.J.: Nanomaterials and soil health for agricultural crop production: Current status and future prospects. In: Nanomaterials for Agriculture and Forestry Applications, pp. 289–312 (2020). https://doi.org/10.1016/B978-0-12-817852-2.00012-3
45. Bhandari, G.: Environmental Nanotechnology: Applications of Nanoparticles for Bioremediation. In: Prasad, R., Aranda, E. (eds.) Approaches in Bioremediation. Nanotechnology in the Life Sciences. Springer, Cham (2018). https://doi.org/10.1007/978-3-030-02369-0_13
46. Salem, T.A., Fetian, N.A., Elsheery, N.I.: Nanotechnology for polluted soil remediation. In: Nanotechnology for Agriculture: Advances for Sustainable Agriculture, pp. 285–305 (2019)
47. Ayub, M.A., Ahmad, H.R., ur Rehman, M.Z., Waraich, E.A.: Cerium oxide nanoparticles alleviates stress in wheat grown on Cd contaminated alkaline soil. Chemosphere **338**, 139561 (2023). https://doi.org/10.1016/j.chemosphere.2023.139561

48. Pide, J.L.V., Organo, N.D., Cruz, A.F., Fernando, L.M., Villegas, L.C., Delfin, E.F., Calubaquib, M.A.M., Madayag, R.E., Paterno, E.S.: Effects of nanofertilizer and nano-plant hormone on soil chemical properties and microbial community in two different soil types. Pedosphere **33**(5), 765–775 (2023)
49. Calder, A.J., Dimkpa, C.O., McLean, J.E., Britt, D.W., Johnson, W., Anderson, A.J.: Soil components mitigate the antimicrobial effects of silver nanoparticles towards a beneficial soil bacterium, Pseudomonas chlororaphis O6. Sci. Total. Environ. **429**, 215–222 (2012). https://doi.org/10.1016/j.scitotenv.2012.04.049
50. Borymski, S., Markowicz, A., Nowak, A., Matus, K., Dulski, M., Sułowicz, S.: Copper-oxide nanoparticles exert persistent changes in the structural and functional microbial diversity: a 60-day mesocosm study of zinc-oxide and copper-oxide nanoparticles in the soil-microorganism-nanoparticle system. Microbiol. Res. **274**, 127395 (2023). https://doi.org/10.1016/j.micres.2023.127395
51. Simonin, M., Guyonnet, J.P., Martins, J.M., Ginot, M., Richaume, A.: Influence of soil properties on the toxicity of TiO_2 nanoparticles on carbon mineralization and bacterial abundance. J. Hazard. Mater. **283**, 529–535 (2015). https://doi.org/10.1016/j.jhazmat.2014.10.004
52. Bellani, L., Siracusa, G., Giorgetti, L., Di Gregorio, S., Ruffini Castiglione, M., Spanò, C., Muccifora, S., Bottega, S., Pini, R., Tassi, E.: TiO_2 nanoparticles in a biosolid-amended soil and their implication in soil nutrients, microorganisms and Pisum sativum nutrition. Ecotoxicol. Environ. Saf. **190**, 110095 (2020). https://doi.org/10.1016/j.ecoenv.2019.110095
53. Singh, H., Sharma, A., Bhardwaj, S.K., Arya, S.K., Bhardwaj, N., Khatri, M.: Recent advances in the applications of nano-agrochemicals for sustainable agricultural development. Environ Sci Process Impacts **23**(2), 213–239 (2021)
54. Khan, M.N., Mobin, M., Abbas, Z.K., AlMutairi, K.A., Siddiqui, Z.H.: Role of nanomaterials in plants under challenging environments. Plant Physiol. Biochem. **110**, 194–209 (2017). https://doi.org/10.1016/j.plaphy.2016.05.038
55. Kalwani, M., Chakdar, H., Srivastava, A., Pabbi, S., Shukla, P.: Effects of nanofertilizers on soil and plant-associated microbial communities: emerging trends and perspectives. Chemosphere **287**, 132107 (2022). https://doi.org/10.1016/j.chemosphere.2021.132107
56. Abdel-Aziz, H., Hasaneen, M.N., Omar, A.: Effect of foliar application of nano chitosan NPK fertilizer on the chemical composition of wheat grains. Egypt. J. Bot. **58**(1), 87–95 (2018)
57. Ha, N.M.C., Nguyen, T.H., Wang, S.L., et al.: Preparation of NPK nanofertilizer based on chitosan nanoparticles and its effect on biophysical characteristics and growth of coffee in green house. Res. Chem. Intermed. **45**, 51–63 (2019). https://doi.org/10.1007/s11164-018-3630-7
58. Kumar, Y., Tiwari, K.N., Singh, T., Raliya, R.: Nanofertilizers and their role in sustainable agriculture. Ann. Plant Soil Res. **23**(3), 238–255 (2021)
59. Herrero, M., Thornton, P.K., Mason-D'Croz, D., Palmer, J., Benton, T.G., Bodirsky, B.L., Bogard, J.R., Hall, A., Lee, B., Nyborg, K., Pradhan, P.: Innovation can accelerate the transition towards a sustainable food system. Nat. Food **1**(5), 266–272 (2020) and many more papers under sustainability in agricultural production
60. Kumar, R., Dadhich, A., Dhiman, M., Sharma, L., Sharma, M.M.: Stimulatory effect of ZnO nanoparticles as a nanofertilizer in seed priming of pearl millet (Pennisetum glaucum) and their bioactivity studies. S. Afr. J. Bot. **165**, 30–38 (2024). https://doi.org/10.1016/j.sajb.2023.12.001
61. Cyriac, J., Melethil, K., Thomas, B., Sreejit, M., Varghese, T.: Synthesis of biogenic ZnO nanoparticles and its impact on seed germination and root growth of Oryza sativa L. and Vigna unguiculata L. Mater. Today Proc. **25**, 224–229 (2020). https://doi.org/10.1016/j.matpr.2020.01.107
62. Abd El-Azeim, M., Sherif, M., Hussien, M., Tantawy, I., Bashandy, S.: Impacts of nano- and non-nanofertilizers on potato quality and productivity. Acta Ecol. Sin. **40**(5), 388–397 (2020). https://doi.org/10.1016/j.chnaes.2019.12.007
63. Rani, P., Kaur, G., Rao, K.V., Singh, J., Rawat, M.: Impact of green synthesized metal oxide nanoparticles on seed germination and seedling growth of Vigna radiata (Mung Bean) and Cajanus cajan (Red Gram). J. Inorg. Organomet. Polym. Mater. **30**, 4053–4062 (2020)

64. Sri Sindhura, K., Prasad, T.N.V.K.V., Panner Selvam, P., Hussain, O.M.: Synthesis, characterization and evaluation of effect of phytogenic zinc nanoparticles on soil exo-enzymes. Appl. Nanosci. **4**, 819–827 (2014)
65. Kottegoda, N., Sandaruwan, C., Priyadarshana, G., Siriwardhana, A., Rathnayake, U.A., Arachchige, D.M.B., Kumarasinghe, A.R., Dahanayake, D., Karunaratne, V., Amaratunga, G.A.: Urea-hydroxyapatite nanohybrids for slow release of nitrogen. ACS Nano **11**(2), 1214–1221 (2017)
66. Kumar, V., Sachdev, D., Pasricha, R., Maheshwari, P.H., Taneja, N.K.: Zinc-supported multiwalled carbon nanotube nanocomposite: a synergism to micronutrient release and a smart distributor to promote the growth of onion seeds in arid conditions. ACS Appl. Mater. Interfaces **10**(43), 36733–36745 (2018)
67. Carrasco-Correa, E.J., Mompó-Roselló, Ò., Simó-Alfonso, E.F.: Calcium oxide nanofertilizer as alternative to common calcium products for the improvement of the amount of peel fruit calcium. Environ. Technol. Innov. **31**, 103180 (2023). https://doi.org/10.1016/j.eti.2023.103180
68. Burvey, P., Jain, P., Singh, A.: Strategic role of casein coated iron oxide nano-particles for delivering nano nutrients for enhance soil quality and their potential. Mater. Today Proc. **89**, 96–105 (2023). https://doi.org/10.1016/j.matpr.2023.05.739
69. Karmous, I., Tlahig, S., Loumerem, M., Lachiheb, B., Bouhamda, T., Mabrouk, M., Debouba, M., Chaoui, A.: Assessment of the risks of copper-and zinc oxide-based nanoparticles used in Vigna radiata L. culture on food quality, human nutrition and health. Environ. Geochem. Health, 1–17 (2022)
70. Mishra, P., Kumari, S., Saini, P., Meena, R.K.: Role of nanotechnology in management of plant viral diseases. Mater. Today Proc. **69**, 1–10 (2022). https://doi.org/10.1016/j.matpr.2022.06.355
71. Kah, M.: Nanopesticides and nanofertilizers: emerging contaminants or opportunities for risk mitigation? Front. Chem. **3**, 64 (2015)
72. Pittarate, S., Perumal, V., Kannan, S., Mekchay, S., Thungrabeab, M., Suttiprapan, P., Sengottayan, S., Krutmuang, P.: Insecticidal efficacy of nanoparticles against Spodoptera frugiperda (J.E. Smith) larvae and their impact in the soil. Heliyon **9**(5), e16133 (2023). https://doi.org/10.1016/j.heliyon.2023.e16133
73. Imada, K., Sakai, S., Kajihara, H., Tanaka, S., Ito, S.: Magnesium oxide nanoparticles induce systemic resistance in tomato against bacterial wilt disease. Plant Pathol. **65**(4), 551–560 (2016). https://doi.org/10.1111/ppa.12443
74. Ayisigi, M., Cokislerel, A., Kucukcobanoglu, Y., Yalcin, T., Aktas, L.Y.: Green synthesized silver nanoparticles for an effective control on soft rot disease pathogen Pectobacterium carotovorum and growth stimulation in pepper. Bul. J. Agric. Sci. **26**(3) (2020)
75. Ndlovu, N., Mayaya, T., Muitire, C., Munyengwa, N.: Nanotechnology applications in crop production and food systems. Int. J. Plant Breed **7**(1), 624–634 (2020)
76. Nagahage, E.A., Nagahage, I.S., Fujino, T.: Calibration and validation of a low-cost capacitive moisture sensor to integrate the automated soil moisture monitoring system. Agriculture **9**(7), 141 (2019). https://doi.org/10.3390/agriculture9070141
77. Stevanato, L., Baroni, G., Cohen, Y., Fontana, C.L., Gatto, S., Lunardon, M., Marinello, F., Moretto, S., Morselli, L.: A novel cosmic-ray neutron sensor for soil moisture estimation over large areas. Agriculture **9**(9), 202 (2019). https://doi.org/10.3390/agriculture9090202
78. Ullo, S.L., Sinha, G.R.: Advances in IoT and smart sensors for remote sensing and agriculture applications. Remote Sens. **13**(13), 2585 (2021). https://doi.org/10.3390/rs13132585
79. Reiser, D., Sehsah, E., Bumann, O., Morhard, J., Griepentrog, H.W.: Development of an autonomous electric robot implement for intra-row weeding in vineyards. Agriculture **9**(1), 18 (2019). https://doi.org/10.3390/agriculture9010018
80. Sathiyabama, M., Indhumathi, M., Amutha, T.: Preparation and characterization of curcumin functionalized copper nanoparticles and their application enhances disease resistance in chickpea against wilt pathogen. Biocatal. Agric. Biotechnol. **29**, 101823 (2020). https://doi.org/10.1016/j.bcab.2020.101823

81. Sofy, A.R., Sofy, M.R., Hmed, A.A., Dawoud, R.A., Alnaggar, A.E.A.M., Soliman, A.M., El-Dougdoug, N.K.: Ameliorating the adverse effects of tomato mosaic tobamovirus infecting tomato plants in Egypt by boosting immunity in tomato plants using zinc oxide nanoparticles. Molecules **26**(5), 1337 (2021). https://doi.org/10.3390/molecules26051337
82. Hamed Derbalah, A.S., Elsharkawy, M.M.: A new strategy to control cucumber mosaic virus using fabricated NiO-nanostructures. J. Biotechnol. **306**, 134–141 (2019). https://doi.org/10.1016/j.jbiotec.2019.10.003
83. Cerbu, C., White, J.C., Sabliov, C.M.: Nanotechnology in livestock: improving animal production and health. In: Nano-Enabled Sustainable and Precision Agriculture, pp. 181–213 (2023). https://doi.org/10.1016/B978-0-323-91233-4.00007-7
84. Rajwade, J.: An overview of myconanoparticles applications in veterinary medicine. In: Fungal Cell Factories for Sustainable Nanomaterials Productions and Agricultural Applications, pp. 657–691 (2023). https://doi.org/10.1016/B978-0-323-99922-9.00019-2
85. Poddar, K., Kishore, A.V.: Nanotechnology in animal production. In: Emerging Issues in Climate Smart Livestock Production, pp. 149–170 (2022). https://doi.org/10.1016/B978-0-12-822265-2.00009-0
86. Hashem, N.M.: Nanotechnology and reproductive management of farm animals: challenges and advances. Animals **11**(7), 1932 (2021). https://doi.org/10.3390/ani11071932
87. Singh, N., Awasthi, A., Kumar, P.: Applications and advances of nanotechnology in animal production. In: Bhateshwar, V., Rai, D.C., Datt, M., Kantwa, S.C. (eds.) Recent Advances and Applications in Livestock Production Management, pp. 125–128. The Agriculture Publication, Raipur (2022)
88. Nagpure, G., Singh, K.R., Singh, J., Singh, R.P.: Role of nanobiotechnology in maintaining a sterile environment for the livestock care management people. In: Nanobiotechnology for the Livestock Industry, pp. 83–97 (2023). https://doi.org/10.1016/B978-0-323-98387-7.00027-6
89. Mignani, S., Shi, X., Rodrigues, J., Tomás, H., Majoral, J.: Dendrimer nanoplatforms for veterinary medicine applications: a concise overview. Drug Discov. Today **27**(5), 1251–1260 (2022). https://doi.org/10.1016/j.drudis.2022.01.001
90. Adegbeye, M.J., Elghandour, M.M., Reddy, P.R.K., Alqaisi, O., Oloketuyi, S., Salem, A.Z., Asaniyan, E.K.: Potential of silver nanoparticles for veterinary applications in livestock performance and health. In: Silver nanomaterials for agri-food applications, pp. 657–683 (2021). https://doi.org/10.1016/B978-0-12-823528-7.00022-6
91. Hashem, N.M., Gonzalez-Bulnes, A.: State-of-the-art and prospective of nanotechnologies for smart reproductive management of farm animals. Animals **10**(5), 840 (2020). https://doi.org/10.3390/ani10050840
92. Nurulfiza, I., Hair-Bejo, M., Omar, A.R., Aini, I.: Immunochromatographic gold-based test strip for rapid detection of infectious bursal disease virus antibodies. J. Vet. Diagn. Invest. (2011). https://doi.org/10.1177/104063871102300220
93. Yin, H., Jia, M., Shi, L., Yang, S., Zhang, L., Zhang, Q., Wang, S., Li, G., Zhang, J.: Nanoparticle-based bio-barcode assay for the detection of bluetongue virus. J. Virol. Methods **178**(1–2), 225–228 (2011). https://doi.org/10.1016/j.jviromet.2011.05.014
94. Jain, B., Lambe, U., Tewari, A., Kadian, S.K., Prasad, M.: Development of a rapid test for detection of foot-and-mouth disease virus specific antibodies using gold nanoparticles. Virusdisease **29**, 192–198 (2018)
95. Karash, S., Wang, R., Kelso, L., Lu, H., Huang, T.J., Li, Y.: Rapid detection of avian influenza virus H5N1 in chicken tracheal samples using an impedance aptasensor with gold nanoparticles for signal amplification. J. Virol. Methods **236**, 147–156 (2016). https://doi.org/10.1016/j.jviromet.2016.07.018
96. Hamdy, M.E., Del Carlo, M., Hussein, H.A., Salah, T.A., El-Deeb, A.H., Emara, M.M., Compagnone, D., et al.: Development of gold nanoparticles biosensor for ultrasensitive diagnosis of foot and mouth disease virus. J. Nanobiotechnol. **16**, 1–12 (2018)
97. Peled, N., Ionescu, R., Nol, P., Barash, O., McCollum, M., VerCauteren, K., Koslow, M., Stahl, R., Rhyan, J., Haick, H.: Detection of volatile organic compounds in cattle naturally infected with Mycobacterium bovis. Sens. Actuators B Chem. **171–172**, 588–594 (2012). https://doi.org/10.1016/j.snb.2012.05.038

98. Sun, M., Sun, Y., Yang, Y., Zhao, M., Cao, D., Zhang, M., Xia, D., Wang, T., Gao, Y., Wang, S., Wang, H., Cai, X., An, T.: Multivalent nanobody-based sandwich enzyme-linked immunosorbent assay for sensitive detection of porcine reproductive and respiratory syndrome virus. Int. J. Biol. Macromol. **258**, 128896 (2024). https://doi.org/10.1016/j.ijbiomac.2023.128896
99. Kaur, A., Preet, S., Kumar, V., Kumar, R., Kumar, R.: Synergetic effect of vancomycin loaded silver nanoparticles for enhanced antibacterial activity. Colloids Surf. B **176**, 62–69 (2019). https://doi.org/10.1016/j.colsurfb.2018.12.043
100. Zheng, S.D., Zhang, Z.Y., Ma, J.X., Qu, Q.W., Bello-Onagshie, G., Qin, Y., Chen, X.-Y., Zhou, D.-F., Ding, W.-Y., Li, Y.H.: CD-g-CS nanoparticles for enhanced antibiotic treatment of Staphylococcus xylosus infection. Microb. Biotechnol. **15**(2), 535–547 (2022)
101. Awaad, M., Moustafa, K.M., Zoulfakar, S., Elhalawany, M., Mohammed, F., El-Refay, R., Morsy, E.: The role of silver nanoparticles in the reluctance of colisepticemia in broiler chickens. J. Appl. Poultry Res. **30**(2), 100155 (2021). https://doi.org/10.1016/j.japr.2021.100155
102. Kumar, B.A., Panickan, S., Bindu, S., Kumar, V., Ramakrishnan, S., Saxena, S., Shrivastava, S., Dandapat, S.: Immunogenicity and protective efficacy of an inactivated Newcastle disease virus vaccine encapsulated in poly-(lactic-co-glycolic acid) nanoparticles. Poult. Sci. **102**(6), 102679 (2023). https://doi.org/10.1016/j.psj.2023.102679
103. Singh, S.M., Alkie, T.N., Nagy, É., Kulkarni, R.R., Hodgins, D.C., Sharif, S.: Delivery of an inactivated avian influenza virus vaccine adjuvanted with poly(D, L-lactic-co-glycolic acid) encapsulated CpG ODN induces protective immune responses in chickens. Vaccine **34**(40), 4807–4813 (2016). https://doi.org/10.1016/j.vaccine.2016.08.009
104. Elsayed, A.A., Abol-Ela, S.S., Askar, A.A., Mohamed, L.A., El-Sayed, S.A., Ahmed, S.Y., Moustafa, A.A., Alagawany, M.: Supplementation of different zinc sources to low-CP diets and its effect on performance, carcass traits, liver and kidney functions, immunological, and antioxidant parameters of quail chicks. Poult. Sci. **100**(11), 101463 (2021). https://doi.org/10.1016/j.psj.2021.101463
105. Mishra, A., Swain, R.K., Mishra, S.K., Panda, N., Sethy, K.: Growth performance and serum biochemical parameters as affected by nano zinc supplementation in layer chicks. Indian J. Animal Nutr. **31**(4), 384–388 (2014)
106. Castellini, C., Ruggeri, S., Mattioli, S., Bernardini, G., Macchioni, L., Moretti, E., Collodel, G.: Long-term effects of silver nanoparticles on reproductive activity of rabbit buck. Syst. Biol. Reprod. Med. **60**(3), 143–150 (2014)
107. Yousef, M., Abdelhamid, H.N., Hidalgo, M., Fathy, R., Gómez-Gascón, L., Dorado, J.: Antimicrobial activity of silver-carbon nanoparticles on the bacterial flora of bull semen. Theriogenology **161**, 219–227 (2021). https://doi.org/10.1016/j.theriogenology.2020.12.006
108. Farzinpour, A., Karashi, N.: The effects of nanosilver on egg quality traits in laying Japanese quail. Appl. Nanosci. **3**, 95–99 (2013)
109. Tabandeh, M.R., Samie, K.A., Mobarakeh, E.S., Khadem, M.D., Jozaie, S.: Silver nanoparticles induce oxidative stress, apoptosis and impaired steroidogenesis in ovarian granulosa cells of cattle. Anim. Reprod. Sci. **236**, 106908 (2022). https://doi.org/10.1016/j.anireprosci.2021.106908
110. Pérez-Duran, F., Acosta-Torres, L.S., Serrano-Díaz, P.N., Toscano-Torres, I.A., Olivo-Zepeda, I.B., García-Caxin, E., Nuñez-Anita, R.E.: Toxicity and antimicrobial effect of silver nanoparticles in swine sperms. Syst. Biol. Reprod. Med. **66**(4), 281–289 (2020)
111. Tsakmakidis, I.A., Samaras, T., Anastasiadou, S., Basioura, A., Ntemka, A., Michos, I., Simeonidis, K., Karagiannis, I., Tsousis, G., Angelakeris, M., Boscos, C.M.: Toxic and Microbiological effects of iron oxide and silver nanoparticles as additives on extended ram semen. Animals **11**(4), 1011 (2021). https://doi.org/10.3390/ani11041011
112. Fen, L.B., Rashid, A.H.A., Nordin, N.I., Johan, M.R.: Applications and impacts of nanomaterials in food safety and quality. In: Preparation and Processing of Religious and Cultural Foods, pp. 131–161 (2018). https://doi.org/10.1016/B978-0-08-101892-7.00007-9

113. Siripatrawan, U., Kaewklin, P.: Fabrication and characterization of chitosan-titanium dioxide nanocomposite film as ethylene scavenging and antimicrobial active food packaging. Food Hydrocolloids **84**, 125–134 (2018). https://doi.org/10.1016/j.foodhyd.2018.04.049
114. Chausali, N., Saxena, J., Prasad, R.: Recent trends in nanotechnology applications of bio-based packaging. J. Agric. Food Res. **7**, 100257 (2022). https://doi.org/10.1016/j.jafr.2021.100257
115. Bouwmeester, H., Dekkers, S., Noordam, M.Y., Hagens, W.I., Bulder, A.S., De Heer, C., Ten Voorde, S.E., Wijnhoven, S.W., Marvin, H.J., Sips, A.J.: Review of health safety aspects of nanotechnologies in food production. Regul. Toxicol. Pharmacol. **53**(1), 52–62 (2009). https://doi.org/10.1016/j.yrtph.2008.10.008
116. Chellaram, C., Murugaboopathi, G., John, A., Sivakumar, R., Ganesan, S., Krithika, S., Priya, G.: Significance of nanotechnology in food industry. APCBEE Proc. **8**, 109–113 (2014). https://doi.org/10.1016/j.apcbee.2014.03.010
117. Donkor, L., Kontoh, G., Yaya, A., Bediako, J.K., Apalangya, V.: Bio-based and sustainable food packaging systems: relevance, challenges, and prospects. Appl. Food Res. **3**(2), 100356 (2023)
118. Adesegun, K.B., Majid, I., Hussain, S., Nanda, V.: Innovations and future trends in product development and packaging technologies. In: Functional and Preservative Properties of Phytochemicals, pp. 377–409 (2020). https://doi.org/10.1016/B978-0-12-818593-3.00013-0
119. Singh, D., Nanda, V.: Application of nanotechnology in food packaging and food quality. In: Dar, A.H., Nayik, G.A. (eds.) Nanotechnology Interventions in Food Packaging and shelf Life. CRC Press, Boca Raton (2023)
120. Zhao, S., Yue, C., Kuzma, J.: Consumer expectations and attitudes toward nanomaterials in foods. In: Handbook of Food Nanotechnology, pp. 705–733 (2020). https://doi.org/10.1016/B978-0-12-815866-1.00017-0
121. Wani, N.R., Dar, A.H., Dash, K.K., Pandey, V.K., Srivastava, S., Jan, S.Y., Deka, P., Sabahi, N.: Recent advances in the production of bionanomaterials for development of sustainable food packaging: a comprehensive review. Environ. Res. **237**, 116948 (2023). https://doi.org/10.1016/j.envres.2023.116948
122. Omerović, N., Djisalov, M., Živojević, K., Mladenović, M., Vunduk, J., Milenković, I., Knežević, N.Ž, Gadjanski, I., Vidić, J.: Antimicrobial nanoparticles and biodegradable polymer composites for active food packaging applications. Compr. Rev. Food Sci. Food Saf. **20**(3), 2428–2454 (2021). https://doi.org/10.1111/1541-4337.12727
123. Cheng, H., Xu, H., Julian McClements, D., Chen, L., Jiao, A., Tian, Y., Miao, M., Jin, Z.: Recent advances in intelligent food packaging materials: principles, preparation and applications. Food Chem. **375**, 131738 (2022). https://doi.org/10.1016/j.foodchem.2021.131738
124. Katiyar, V., Ghosh, T.: Nanotechnology in edible food packaging. In:Food Preservation Practices for a Sustainable Future. Springer Nature Singapore Pte Lrd (2021)
125. Chettri, S., Sharma, N., Mohite, A.M.: Edible coatings and films for shelf-life extension of fruit and vegetables. Biomater. Adv. **154**, 213632 (2023). https://doi.org/10.1016/j.bioadv.2023.213632
126. Nunes, C., Silva, M., Farinha, D., Sales, H., Pontes, R., Nunes, J.: Edible coatings and future trends in active food packaging–fruits' and traditional sausages' shelf life increasing. Foods **12**(17), 3308 (2023). https://doi.org/10.3390/foods12173308
127. Primožič, M., Knez, Ž, Leitgeb, M.: (Bio)Nanotechnology in food science—food packaging. Nanomaterials **11**(2), 292 (2021). https://doi.org/10.3390/nano11020292
128. Friedrich, J.C., Silva, O.A., Faria, M.G., Colauto, N.B., Gazzin, Z.C., Colauto, G.A., Caetano, J., Dragunski, D.C.: Improved antioxidant activity of a starch and gelatin-based biodegradable coating containing Tetradenia riparia extract. Int. J. Biol. Macromol. **165**, 1038–1046 (2020). https://doi.org/10.1016/j.ijbiomac.2020.09.143
129. Mei, L., Wang, Q.: Advances in using nanotechnology structuring approaches for improving food packaging (2020). https://doi.org/10.1146/annurev-food-032519-051804

130. Hajinezhad, S., Razavizadeh, B.M., Niazmand, R., Ghasemi, I.: Antimicrobial, mechanical, and physicochemical properties of ethylene vinyl alcohol (EVOH) extruded films blended with propolis. Int. J. Food Prop. **23**(1), 2020–2032 (2020)
131. Zhang, W., Rhim, J.: Titanium dioxide (TiO_2) for the manufacture of multifunctional active food packaging films. Food Packag. Shelf Life **31**, 100806 (2022). https://doi.org/10.1016/j.fpsl.2021.100806
132. Kam, W.Y.J., Mirhosseini, H., Abas, F., Hussain, N., Hedayatnia, S., Chong, H.L.F.: Antioxidant activity enhancement of biodegradable film as active packaging utilizing crude extract from durian leaf waste. Food Control **90**, 66–72 (2018)
133. Chavoshizadeh, S., Pirsa, S., Mohtarami, F.: Sesame oil oxidation control by active and smart packaging system using wheat gluten/chlorophyll film to increase shelf life and detecting expiration date. Eur. J. Lipid Sci. Technol. **122**(3), 1900385 (2020). https://doi.org/10.1002/ejlt.201900385
134. Ghasemi, S., Bari, M.R., Pirsa, S., Amiri, S.: Use of bacterial cellulose film modified by polypyrrole/TiO_2-Ag nanocomposite for detecting and measuring the growth of pathogenic bacteria. Carbohyd. Polym. **232**, 115801 (2020). https://doi.org/10.1016/j.carbpol.2019.115801
135. Rossi, M., Passeri, D., Sinibaldi, A., Angjellari, M., Tamburri, E., Sorbo, A., Carata, E., Dini, L.: Nanotechnology for food packaging and food quality assessment. Adv. Food Nutr. Res. **82**, 149–204 (2017). https://doi.org/10.1016/bs.afnr.2017.01.002
136. Taechutrakul, S., Piroonpan, T., Pasanphan, W.: Active film strips to extend the shelf life of fruits: multibranched PLA-gallic acid as an antioxidant/oxygen scavenger in a case study of bananas (Musa AAA group). J. Food Eng. **364**, 111794 (2024). https://doi.org/10.1016/j.jfoodeng.2023.111794
137. Nourmohammadi, A., Hassanzadazar, H., Aminzare, M., Hashemi, M.: The effects of whey protein/nanoclay biocomposite containing Thymus fedtschenkoi Ronniger essential oil and resveratrol on the shelf life of liqvan cheese during refrigerated storage. LWT **187**, 115175 (2023). https://doi.org/10.1016/j.lwt.2023.115175
138. Rodrigues, F.A.M., Dos Santos, S.B.F., de Almeida Lopes, M.M., Guimarães, D.J.S., de Oliveira Silva, E., de Souza, M.D.S.M., Ricardo, N.M.P.S., et al.: Antioxidant films and coatings based on starch and phenolics from Spondias purpurea L. Int. J. Biol. Macromol. **182**, 354–365 (2021)
139. Correa-Pacheco, Z.N., Bautista-Baños, S., Ramos-García, M.D.L., Martínez-González, M.D.C., Hernández-Romano, J.: Physicochemical characterization and antimicrobial activity of edible propolis-chitosan nanoparticle films. Prog. Org. Coat. **137**, 105326 (2019). https://doi.org/10.1016/j.porgcoat.2019.105326
140. Stefanowska, K., Woźniak, M., Sip, A., Mrówczyńska, L., Majka, J., Kozak, W., Ratajczak, I., et al.: Characteristics of chitosan films with the bioactive substances—caffeine and propolis. J. Funct. Biomater. **14**(7), 358 (2023)
141. Pirsa, S., Shamusi, T.: Intelligent and active packaging of chicken thigh meat by conducting nano structure cellulose-polypyrrole-ZnO film. Mater. Sci. Eng. C **102**, 798–809 (2019). https://doi.org/10.1016/j.msec.2019.02.021
142. Pirsa, S., Karimi Sani, I., Pirouzifard, M.K., Erfani, A.: Smart film based on chitosan/Melissa officinalis essences/pomegranate peel extract to detect cream cheeses spoilage. Food Addit. Contam. Part A **37**(4), 634–648 (2020)
143. Rezaei, M., Pirsa, S., Chavoshizadeh, S.: Photocatalytic/antimicrobial active film based on wheat gluten/ZnO nanoparticles. J. Inorg. Organomet. Polym. Mater. **30**(7), 2654–2665 (2020)
144. Zheng, Y., Jia, X., Zhao, Z., Ran, Y., Du, M., Ji, H., Pan, Y., Li, Z., Ma, X., Liu, Y., Duan, L., Li, X.: Innovative natural antimicrobial natamycin incorporated titanium dioxide (nano-TiO_2)/poly (butylene adipate-co-terephthalate) (PBAT)/poly (lactic acid) (PLA) biodegradable active film (NTP@PLA) and application in grape preservation. Food Chem. **400**, 134100 (2023). https://doi.org/10.1016/j.foodchem.2022.134100
145. Cao, C., Wang, Y., Zheng, S., Zhang, J., Li, W., Li, B., Guo, R., Yu, J.: Poly (butylene adipate-co-terephthalate)/titanium dioxide/silver composite biofilms for food packaging application. LWT **132**, 109874 (2020). https://doi.org/10.1016/j.lwt.2020.109874

146. Du, H., Min, T., Sun, X., Bian, X., Zhu, Z., Wen, Y.: Multifunctional film based on gelatin with titanium dioxide and thymol@β-cyclodextrins for fresh-keeping packaging. Food Biosci. **50**, 102168 (2022). https://doi.org/10.1016/j.fbio.2022.102168
147. Nguyen, S.V., Lee, B.: Multifunctional nanocomposite based on polyvinyl alcohol, cellulose nanocrystals, titanium dioxide, and apple peel extract for food packaging. Int. J. Biol. Macromol. **227**, 551–563 (2023). https://doi.org/10.1016/j.ijbiomac.2022.12.073
148. Indumathi, M., Saral Sarojini, K., Rajarajeswari, G.: Antimicrobial and biodegradable chitosan/cellulose acetate phthalate/ZnO nano composite films with optimal oxygen permeability and hydrophobicity for extending the shelf life of black grapefruits. Int. J. Biol. Macromol. **132**, 1112–1120 (2019). https://doi.org/10.1016/j.ijbiomac.2019.03.171
149. Ashrafi, A., Jokar, M., Mohammadi Nafchi, A.: Preparation and characterization of biocomposite film based on chitosan and kombucha tea as active food packaging. Int. J. Biol. Macromol. **108**, 444–454 (2018). https://doi.org/10.1016/j.ijbiomac.2017.12.028
150. Wang, L., Dong, Y., Men, H., Tong, J., Zhou, J.: Preparation and characterization of active films based on chitosan incorporated tea polyphenols. Food Hydrocolloids **32**(1), 35–41 (2013). https://doi.org/10.1016/j.foodhyd.2012.11.034
151. Moreira, R., Pereda, M., Marcovich, N.E., Roura, S.I.: Antimicrobial effectiveness of bioactive packaging materials from edible chitosan and casein polymers: assessment on carrot, cheese, and salami. J. Food Sci. **76**(1), M54–M63 (2011). https://doi.org/10.1111/j.1750-3841.2010.01910.x
152. Mahfoudhi, N., Ksouri, R., Hamdi, S.: Nanoemulsions as potential delivery systems for bioactive compounds in food systems: preparation, characterization, and applications in food industry. Emulsions, 365–403 (2016). https://doi.org/10.1016/B978-0-12-804306-6.00011-8
153. Assadpour, E., Jafari, S.M.: Importance of release and bioavailability studies for nanoencapsulated food ingredients. In: Release and Bioavailability of Nanoencapsulated Food Ingredients, pp. 1–24 (2020). https://doi.org/10.1016/B978-0-12-815665-0.00001-1
154. Donsì, F., Sessa, M., Mediouni, H., Mgaidi, A., Ferrari, G.: Encapsulation of bioactive compounds in nanoemulsion-based delivery systems. Procedia Food Sci. **1**, 1666–1671 (2011). https://doi.org/10.1016/j.profoo.2011.09.246
155. Fathi, M., Mozafari, M., Mohebbi, M.: Nanoencapsulation of food ingredients using lipid based delivery systems. Trends Food Sci. Technol. **23**(1), 13–27 (2012). https://doi.org/10.1016/j.tifs.2011.08.003
156. Tan, C., Nakajima, M.: β-Carotene nanodispersions: preparation, characterization and stability evaluation. Food Chem. **92**(4), 661–671 (2005). https://doi.org/10.1016/j.foodchem.2004.08.044
157. Yuan, Y., Gao, Y., Mao, L., Zhao, J.: Optimisation of conditions for the preparation of β-carotene nanoemulsions using response surface methodology. Food Chem. **107**(3), 1300–1306 (2008). https://doi.org/10.1016/j.foodchem.2007.09.015
158. Jafari, S.M., Assadpoor, E., Bhandari, B., He, Y.: Nano-particle encapsulation of fish oil by spray drying. Food Res. Int. **41**(2), 172–183 (2008). https://doi.org/10.1016/j.foodres.2007.11.002
159. Chang, C., Nickerson, M.T.: Encapsulation of omega 3-6-9 fatty acids-rich oils using protein-based emulsions with spray drying. J. Food Sci. Technol. **55**, 2850–2861 (2018)
160. Jafari, S.M., He, Y., Bhandari, B.: Role of powder particle size on the encapsulation efficiency of oils during spray drying. Dry. Technol. **25**(6), 1081–1089 (2007)
161. Takeuchi, M.T., Kojima, M., Luetzow, M.: State of the art on the initiatives and activities relevant to risk assessment and risk management of nanotechnologies in the food and agriculture sectors. Food Res. Int. **64**, 976–981 (2014). https://doi.org/10.1016/j.foodres.2014.03.022
162. Shallal, A.F., Owaid, M.N.: Challenges and hazards associated with nanotechnology in agriculture. In: Advances in Nanotechnology-Based Drug Delivery Systems, pp. 399–414 (2022). https://doi.org/10.1016/B978-0-323-88450-1.00009-0
163. FAO/WHO [Food and Agriculture Organization of the United Nations/World Health Organization]: FAO/WHO paper: state of the art on the initiatives and activities relevant to risk assessment and risk management of nanotechnologies in the food and agriculture sectors. Geneva, 56 pp (2013)

164. Zulfiqar, F., Navarro, M., Ashraf, M., Akram, N.A., Munné-Bosch, S.: Nanofertilizer use for sustainable agriculture: advantages and limitations. Plant Sci. **289**, 110270 (2019). https://doi.org/10.1016/j.plantsci.2019.110270
165. Meena, K., Meena, N., Meena, R. N., Choudhary, M., Meena, S., Kumar, S. (2023). Role of nanotechnology in organic agriculture. In: Organic Farming, 2nd edn, pp. 343–364. https://doi.org/10.1016/B978-0-323-99145-2.00015-X
166. Biswas, R., Alam, M., Sarkar, A., Haque, M.I., Hasan, M.M., Hoque, M.: Application of nanotechnology in food: processing, preservation, packaging and safety assessment. Heliyon **8**(11), e11795 (2022). https://doi.org/10.1016/j.heliyon.2022.e11795

Leticia Donkor is a doctoral candidate at the Department of Food Process Engineering at the University of Ghana, Legon, where her current research focuses on sustainable diets. has a Bachelor's degree in Nutrition and Food Science and a Master of Philosophy in Food Process Engineering, all from the University of Ghana. She has good experience in food safety and quality management systems through her engagements with various food industries. Leticia has several publications in peer review journals.

Eric Banan-Mwine Daliri is a research professor at the Department of Biological Models, Institute of Biochemistry, Vilnius University, Lithiuania. He obtained his PhD in Food Science and Biotechnology from Kangwon National University, Korea. Eric did his Postdoctoral fellowship in Korea and also in Vilnius University. He has more than 88 publications in peer review journals and 12 book chapters. He has good experience in food science and food safety and has several years of experience in the food industry and has won several awards in this field.

Chapter 5
Nanotechnology and the Environment: Opportunities and Challenges

Bismark Sarkodie, Benjamin Tawiah, and Bin Fei

Introduction

A healthy environment plays an essential part in sustaining a natural home for humans, plants, and animals, all of which are interdependent. It provides mankind with the resources they need to exist and contributes to the growth of technical items that improve our level of living. Industry modernization has enabled the development of novel solutions while employing widely available raw resources, assuring a sustainable future for everybody. Unfortunately, our environment is severely polluted with a wide range of toxins. These pollutants are caused by human activities and a variety of sectors, including paints, manufacturing, coating, mining, metallurgical, and nuclear. As a result, the air, water, and soil have been contaminated [1]. The ecosystem and human health can suffer greatly from pollution brought on by dangerous compounds including toluene, dyes, carbon monoxide, and antibiotics. For example, breathing in carbon monoxide can lead to lung issues and water supplies can get contaminated with dyes and toluene, which can kill aquatic life. In addition, a significant portion of environmental pollution is caused by the use of fossil fuels as a source of energy. Not only are toxic substances like sulfur dioxide and nitrogen oxide dangerous to human health, but they also seriously impair the ecosystem, contributing

B. Sarkodie
School of Textile and Clothing, Anhui Polytechnic University, Anhui Province 241000, Wuhu, China

B. Tawiah
Department of Industrial Art (Textiles), Kwame Nkrumah University of Science and Technology, Kumasi, Ghana
e-mail: benjamin.tawiah@knust.edu.gh

B. Tawiah · B. Fei (✉)
School of Fashion and Textiles (SFT), Hong Kong Polytechnic University, Hong Kong SAR, China
e-mail: bin.fei@polyu.edu.hk

S. C. George and B. Tawiah (eds.), *Nanotechnology in Societal Development*, Advanced Technologies and Societal Change, https://doi.org/10.1007/978-981-97-6184-5_5

to acid rain and climate change. By taking proactive steps to lessen pollution, we can create a safer and healthier environment for both current and future generations [2]. The development of cutting-edge technology for environmental remediation is essential to ensuring a healthy and sustainable environment. The problems facing our natural surroundings now present a significant chance to apply a wide range of materials to improve the ecosystem. As a result, removing toxins from our environment has been elevated to the global agenda. A chart detailing the industries that pollute the environment, the media they contaminate, and the illnesses that follow for people may be found in Fig. 5.1. The foundation for protecting the environment from the ever-increasing damage caused by the discharge of different contaminants is nanotechnology. To produce clean air, water, or soil, it entails creating nanoparticles with special properties that allow contact and reaction with contaminants.

In this chapter, we have examined challenges that create pollutants in the environment, opportunities for the application of nanotechnologies in solving using nanomaterials such as polymer-based, carbon-based, and metal/metal oxide nanomaterials in the elimination of pollutants from the environment. In addition, this chapter has analyzed and explained the processes used to create these nanomaterials. We have also discussed the physical, chemical, and the biological methods such as enzymatic degradation, photocatalysis, and adsorption methods used by nanomaterials to remove contaminants. Finally, the advantages of using nanomaterials to protect the environment along with the challenges and potential applications in the future are discussed.

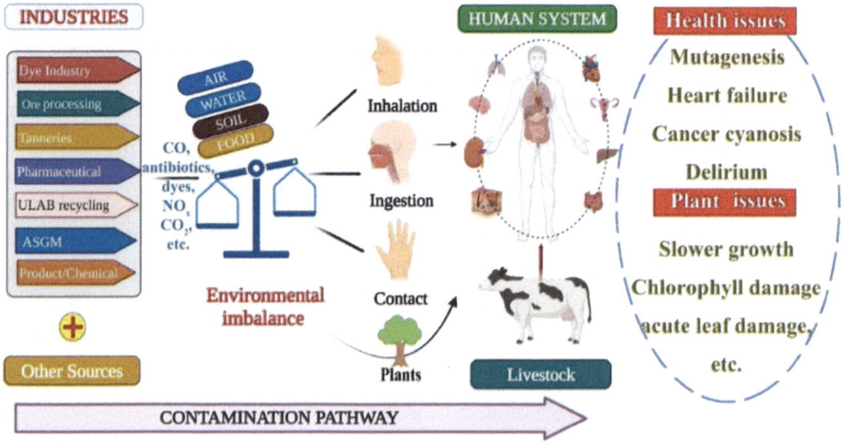

Fig. 5.1 The contamination pathway's journey and how it affects human health [3]. Reprinted with permission from Elsevier B.V. (On the left side are the industries that release pollutants into the environment (air water, soil, etc.), and the middle to the right side illustrate the pollutants' entry points in living organisms and their effects)

Nanotechnology

A vast array of knowledge and technologies, including those from physics, chemistry, biology, and engineering, are combined into the quickly developing discipline of nanotechnology. It entails the extremely tiny manipulation of matter at the nanoscale level to produce novel materials, apparatuses, and systems with special qualities and capacities. Because they make it possible to create more effective, efficient, and sustainable solutions, these advancements have the potential to completely transform several sectors, including electronics, healthcare, and energy. The creation of materials with extraordinary strength, flexibility, and conductivity like graphene and carbon nanotubes—is an example of how nanotechnology is being applied. The 1980s saw the public become more aware of nanotechnology as new technologies were used for commercial purposes [4]. By enabling the manipulation of materials as tiny as a billionth of a meter and opening new characteristics, nanotechnology transformed the field of material science. Nanoparticles, nanotubes/nanowires, and nanoplates/nanoribbons are the three categories into which nanomaterials are divided according to their size. Furthermore, they are classified as either organic or inorganic nanoparticles depending on their chemical makeup. Because their molecular makeup, size, shape, and surface chemistry may all be altered, nanomaterials have a wide range of highly adaptable features [5].

With the world facing serious difficulties that present possibilities for the creation and use of novel materials, new technologies are being looked to for direct or indirect solutions in the areas of energy, environment, health, and climate change. The United Nations held a historic summit in 2017 and demanded that all countries implement the Sustainable Development Goals immediately. Everyone would have a better future if everyone worked together to attain this aim [4]. Environmental challenges have been addressed through systems and applications developed as a result of the use of nanotechnology, particularly in the manufacture of nanomaterials, nanotubes, nanocomposites, nanofilters, and nanoparticles [6, 7].

Nanomaterials

Three types of nanomaterials are distinguished: those based on polymers, those based on carbon, and those based on metals or metal oxides (Fig. 5.2). Metal/oxide nanoparticles and carbon-based materials are frequently utilized in environmental remediation for a range of contaminants. A considerable decrease in carbon footprint and the creation of a greener environment are the results of the remarkable breakthroughs in the production and functionalization of nanomaterials [8].

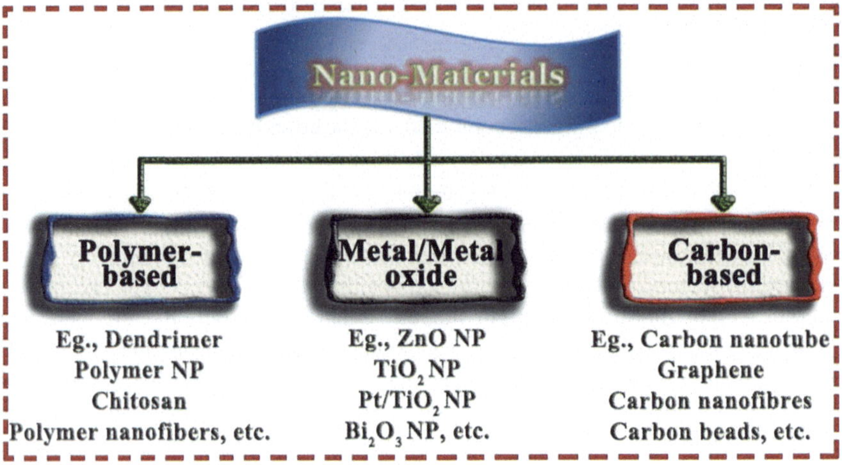

Fig. 5.2 Categories of nanomaterials: polymer-based, metal/metal oxide, and carbon-based nanomaterials (NP = nanoparticles)

Polymer-Based Nanomaterials

One class of extremely biodegradable and physiologically safe nanoparticles are polymer-based nanomaterials. These nanomaterials consist of poly (lactic-glycolic acid), polylactic acid, polyglutamic acid, and chitosan. Conversely, natural polymers are sustainable materials derived from various sources and may be converted into tiny inorganic molecules, water, and carbon dioxide. Chitosan, starch, alginate, cellulose, hyaluronic acid, and chondroitin sulfate are a few examples of natural polymer compounds [9].

Carbon-Based Nanomaterials

New carbon nanomaterials have been created in the last several decades. These materials, which include 0D fullerenes, 1D carbon nanotubes, and 2D graphene, are composed of sp2 hybridized carbon atoms. The characteristics of carbon nanomaterials are influenced by their atomic structures and interactions with other phases of materials. Therefore, to attain desired characteristics for particular applications, scientists concentrate on developing and functionalizing carbon nanomaterials at the nanoscale level. Due to their exceptional chemical and physical characteristics, including their high electroconductivity, huge surface area, chemical stability, excellent mechanical and thermal stability, and enhanced optical qualities, carbon nanomaterials have garnered a lot of interest. High-performance adsorbents like activated carbon are employed in a variety of applications, including separation, adsorption,

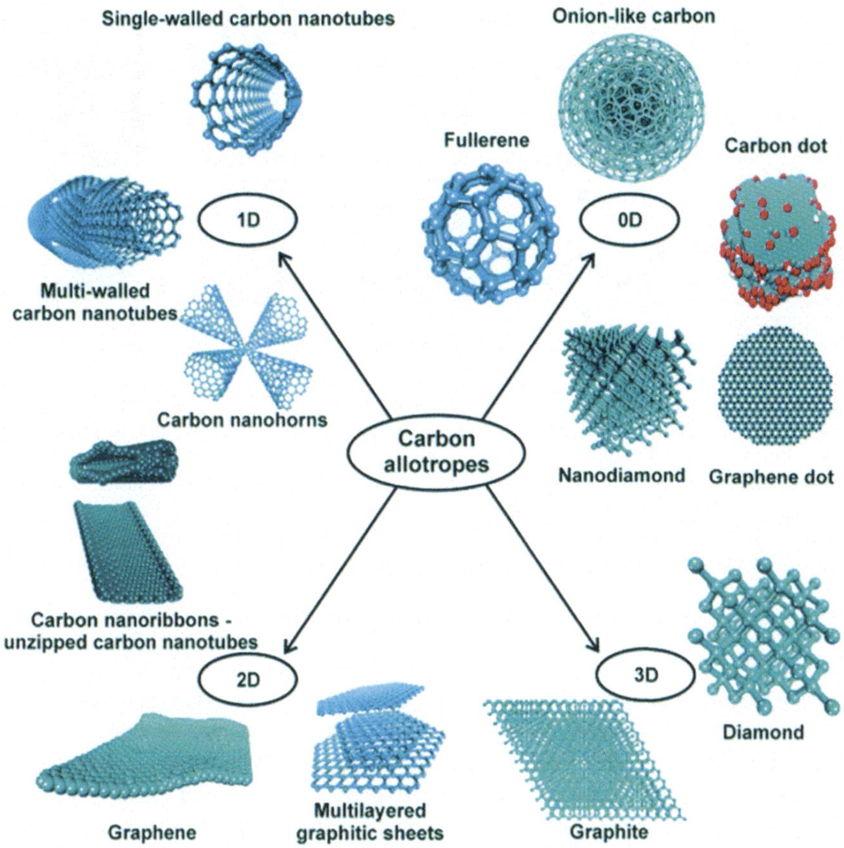

Fig. 5.3 Different forms of the carbon allotropes [11]. Reprinted from an open-access article by MDPI

and catalytic processes [5] for dichlorobenzene, ethylbenzene, dyes, Pb^{2+}, Zn^{2+}, Cd^{2+}, and Cu^{2+} [10]. The several kinds of carbon-based nanomaterials that find exciting applications in environmental remediation and other sectors are depicted in Fig. 5.3.

Metal–Organic Frameworks

A novel class of carbon-based nanomaterials with a huge specific surface area is called metal–organic frameworks, or MOFs. The distinct functional groups present in MOF oxides are essential for efficient adsorption processes. Numerous researchers have been investigating the use of MOFs to remove different kinds of contaminants from soil, air, and water in recent years. Because of the coordination interactions

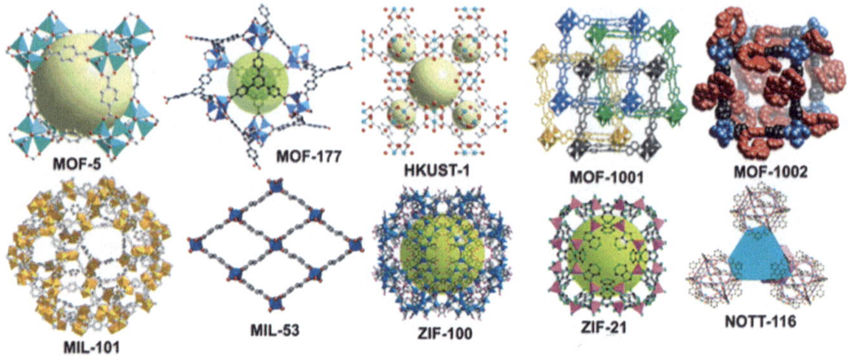

Fig. 5.4 Metal–organic framework (MOF) structures (reproduced from [2]. Reprinted from an open-access article by MDPI. (**HKUST-1**—Hong Kong University of Science and Technology or MOF-199 is made up of copper nodes with 1,3,5-benzenetricarboxylic acid struts between them; **MIL**—Matérial Institut Lavoisier, is comprised of trimeric chromium: octahedral clusters interconnected by 1,4-benzenedicarboxylates, resulting in a highly porous 3-dimensional structure; **ZIF**—Zeolitic imidazolate frameworks are a class of metal–organic frameworks that are topologically isomorphic with zeolites; **NOTT-116** is a metal–organic polyhedral framework that consists of two types of cages: a truncated octahedron (cage A) with six square windows and eight hexagonal faces and a truncated tetrahedron (cage B) with four triangular windows and four hexagonal faces)

between the constituents, metal–organic frameworks—porous hybrid materials made up of inorganic metal atoms and organic ligands have distinctive structural variety. Their potential as new adsorbents has piqued the curiosity of researchers worldwide. The inorganic parts contain the isolated polyhedral and/or small clusters such as in coordination chemistry. Whereas, the organic parts are di-, tri-, or tetra-dentate ligands [1]. Typical examples of MOF structures and SEM images are presented in Fig. 5.4. In the past few years, metal–organic frameworks (MOFs), due to their large surface area, tunable pore size, morphology, and functional pore surface, have been widely used in hydrogen storage, gas separation, and catalysis. It is also a promising adsorbent for sulfur removal [12].

Metal and Metal-Oxide Nanomaterials

The distinct electronic structures of metal and metal-oxide nanostructures dictate their characteristics as insulators, semiconductors, and conductors. These materials include transition metal ions with empty d-shells, which provide reactive electronic transitions, large band gaps, enhanced electrical properties, and elevated dielectric constants. Metal and metal-oxide nanostructures are very adaptable and helpful for a wide range of applications because of their unique form and size. They also display particular physicochemical qualities (e.g., large surface area, high porosity, excellent reactivity, etc.). Their high density of surface locations and nanostructures also

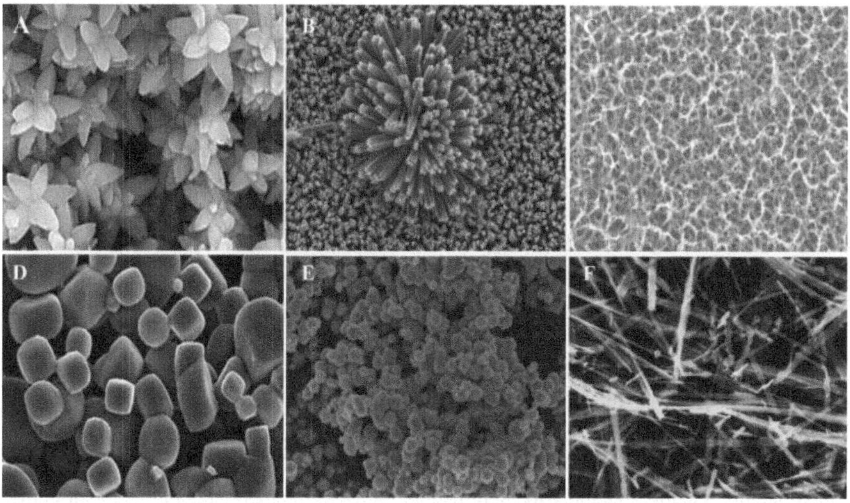

Fig. 5.5 TiO_2 nanoparticles in various shapes **a** nanoflowers, **b** nanorods, **c** nanosheets, **d** nanocubes, **e** nanospheres, and **f** nanobelt [13]. Reprinted with permission from Elsevier B.V

contributes to their broad applicability potential [14]. Scientists have recently made great progress in creating metal and metal-oxide nanoparticles to remove contaminants from the environment. Among the most prominent ones are Au/ZnO, CuO, ZnO, TiO_2, Pt/TiO_2, and Fe_2O_3 [15]. Among these, titanium dioxide is a favored semiconductor because of its air-purifying properties, affordability, accessibility, stability, and non-toxicity [15]. Also, zinc oxide's appealing qualities make it a popular semiconductor for use in photoelectron devices, catalysts, sensors, and other applications [16]. Metal- oxide nanostructures have a high surface area and strong activity, making them perfect for catalytic processes. Figure 5.5 shows the SEM images of metal oxides with different morphology.

Nanomaterials Fabrication Methods

The process used to fabricate nanomaterials varies depending on the intended use and characteristics. In general, a variety of factors, particularly the conditions and technique of synthesis, have a considerable impact on the chemical and physical properties of nanomaterials [8]. A few of the manufacturing techniques are co-precipitation, chemical vapor deposition, flame spray pyrolysis, electrospinning, hydrothermal method, sol–gel method, etc.

Hydro (solvo) thermal method: The hydrothermal method is a technique used to produce inorganic compounds in aqueous solutions at pressures and temperatures greater than ambient temperature. Solvothermal processes are typically conducted in organic solvents. For most purposes, autoclaves are closed reaction vessels (Fig. 5.6a) that are used to conduct reactions. The reaction mixture heats up to temperatures over the solvent's boiling point, which causes a significant increase in pressure inside the reactor. The combined impact of high temperature and pressure renders post-annealing techniques unnecessary for producing very crystalline materials. Controlling the nanomaterial's size and shape and guaranteeing a uniform composition are made possible by the use of organic additives [18].

Electrospinning: Nanofibrous membranes may be effectively created in an electric field using electrospinning. The basic electrospinning setup consists of a collector, spinneret, and high-voltage supply device (Fig. 5.6b). Many modifications have been made to the electrospinning setup and procedure to control the architecture of the electrospun nanofibers for particular applications [19].

Flame spray pyrolysis: This technique entails spraying an aqueous metal salt solution in a thin mist into a flame through a capillary. The droplet-form precursor solution burns and disintegrates in the flame, converting it to metal oxide by the agglomeration of metal atoms into nanoparticles [20].

Chemical Vapor deposition: Chemical vapor deposition (CVD) is a dependable chemical process for creating thin layers and 2D nanomaterials on a range of substrates. Among the well-known forms of CVD include pressure, plasma-enhanced, photo- and laser-assisted, and so on. The CVD procedure has been used to create a wide range of carbon-based nanomaterials, including graphene, carbon nanotubes, and carbon nanofibers [21]. In chemical vapor deposition, a high-temperature reaction between the substrate and vaporized precursor occurs in an evacuated chamber. The two primary methods for CVD are solid-sublimation precursor (such as molybdenum trioxide for MoS_2) and gas-phase precursor. Reactive gases are delivered to a receiving substrate in a heated furnace to generate functionalized nanomaterials in the gas-phase precursor technique. Methane/H2 is used to form graphene, while evaporated borazine/N_2 is used to form hexagonal boron nitride. In the second tactic, the desired material's surface is functionalized via solid compound sublimation.

Co-Precipitation Method: The precipitation process involves a reaction between a precursor and a precipitating reagent. Typically, the precursor of inorganic salts—nitrate and chloride salts, for example, is added dropwise to the precipitating agent solution (NaOH, NH4OH, etc.) or vice versa, depending on whether heat is present [23]. In the event of employing this technique, the particles often exhibit poor crystallinity. The application of heat facilitates the increase in crystallinity of the nanomaterial. Aside from the above-discussed method, the sol–gel approach is one of the other techniques (Fig. 5.7).

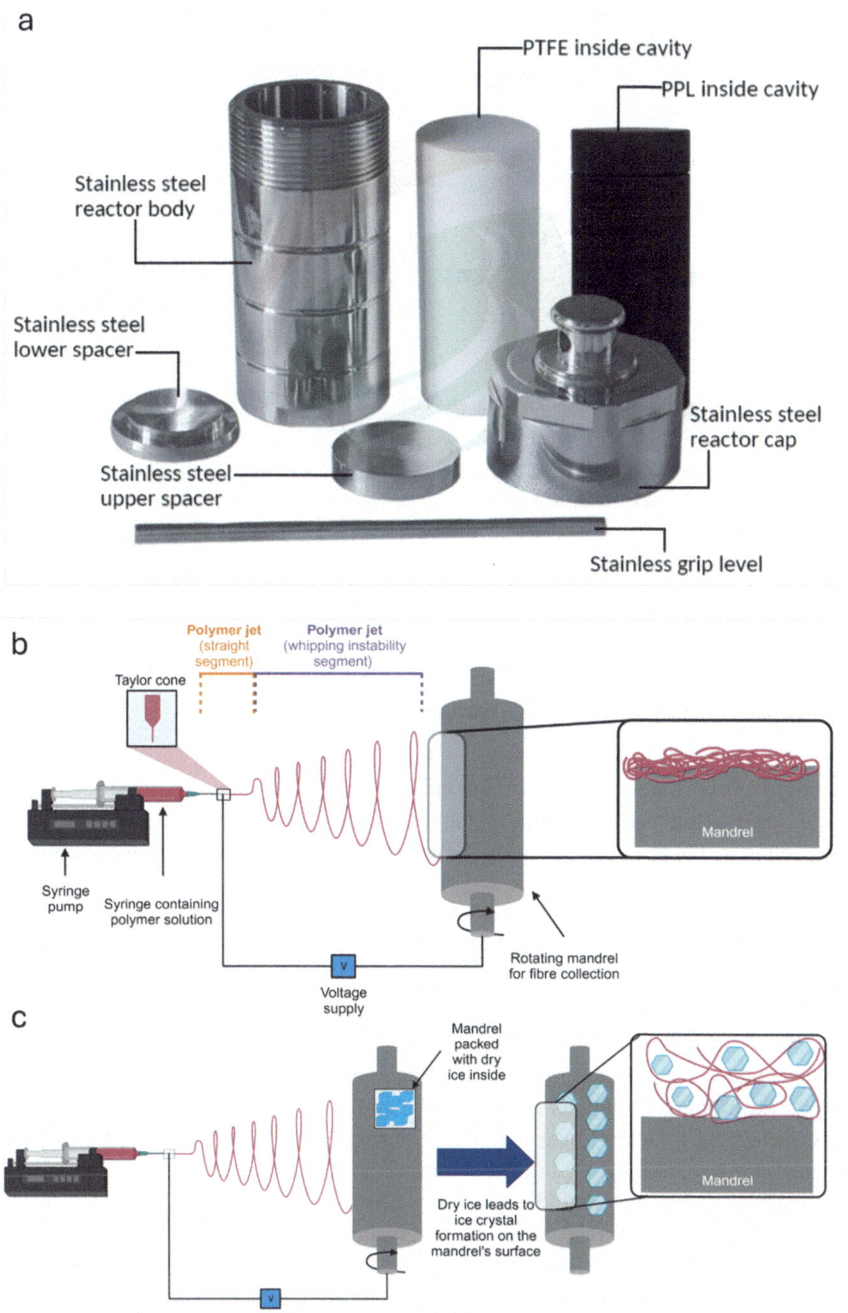

Fig. 5.6 **a** Image of autoclave vessel used for hydrothermal method, **b** conventional electrospinning, and **c** cryogenic electrospinning set-ups [17]. Open access article kind courtesy MDPI

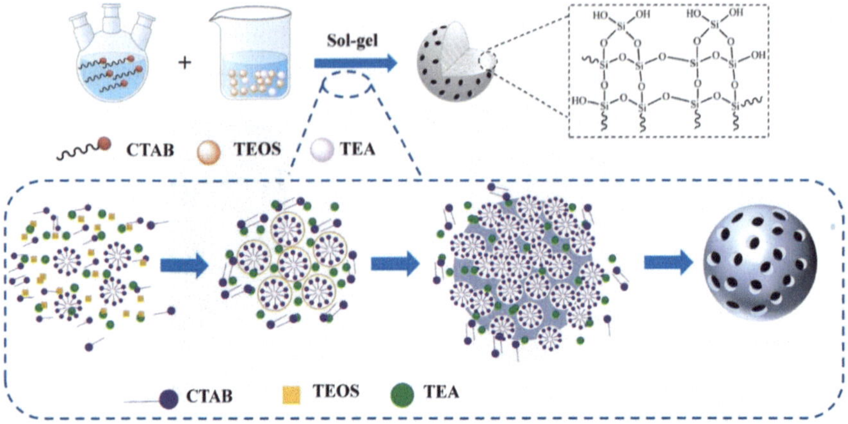

Fig. 5.7 The schematic illustration of worm-like porous nanosilica prepared by the sol–gel method [22]. Reprinted with permission from Elsevier B.V. (A solution of cetyltrimethylammonium bromide (CTAB) and triethylamine (TEA) was heated and stirred under high-speed stirring conditions to undergo a reaction. Tetraethyl orthosilicate (TEOS) was added dropwise to the above solution to undergo a reaction with CTAB and TEA to form a gel over some time. The precipitate is collected via centrifugation and washing using the ethanol–water liquid and then calcined to obtain the nanomaterial)

Strategies for Pollutants Removal

The application of nanomaterials concomitantly with the strategies used for pollutants removal from the environment. Generally, biological, physical, and chemical approaches are the three types of remediation technologies discussed in the following sections.

Physical Methods

Physical technology, sometimes referred to as membrane technology, is used to remove developing pollutants from aquatic systems. To eliminate residual pollutants, membranes made of materials with filtering qualities such as specific surface charge, pore size, and hydrophobic properties, are used. The classification of physical treatment technology includes reverse osmosis, forward osmosis, microfiltration, nano-filtration, and ultra-filtration. Although they may eliminate emergent pollutants by up to 99%, most of the techniques are yet to be used on a significant scale. Considering adsorption technology is one of the physical methods that are most frequently utilized, an in-depth investigation of many different adsorbents has been conducted by many researchers [7, 10]. In adsorption, gas or liquid solute builds up into an atomic or molecular layer on the surface of the adsorbent. Adsorption may be classified into two categories based on the nature of the interaction, usually termed chemisorption and physisorption, that exists between the adsorbent and the adsorbate. Adsorption of molecules or ions of pollutants has been demonstrated to be an effective way to treat water [24]. The benefits of nano adsorbents are numerous and include durability, recycling, low cost, environmentally friendly, and high adsorption capacity. The fundamental mechanism of adsorption entails migration to the surface and pores of adsorbent, and monolayer or multilayer accumulation of pollutants. Ion exchange, electrochemical potential, and surface characteristics play a crucial role in the adsorption mechanism. Positively charged adsorbents are more likely to absorb negatively charged pollutants through a combination of hydrogen bonds and electrostatic forces (Fig. 5.8). High efficiency of adsorption is demonstrated by nano adsorbents with mesoporous architectures and higher surface area [25].

Chemical Methods

Adept oxidation procedures are an unquestionable and exceptional chemical strategy for treating pollutants since they principally employ free radicals to mineralize the contaminants. The degradation process comprises the generation of radicals such as superoxide and hydroxyl species via reduction and oxidation, the interaction of free radicals with the molecules of pollutants, and the transfer of electrons to mineralized contaminants. Among them are Fenton and photocatalytic systems, which, in specific circumstances, aid in the breakdown of non-biodegradable contaminants from the environment [27]. Photocatalysis is one widely used green method. Theoretically, the initiation of a photocatalytic reaction on semiconductor nanomaterials involves the excitation and subsequent separation of carriers due to the migration of electrons from the valence band to the conduction band of the photocatalyst under the influence of light. Then, in the valence band and conduction band, respectively, positive holes and negative electrons are created. When the molecules interact with the electrons and holes, radicals are created that break down contaminants (Fig. 5.9). However, there is a possibility that electrons will return to the valence band and combine with holes as a result. This has a detrimental effect on photocatalytic degradation and charge separation efficiency. Because of this, scientists have spent a great deal of money developing innovative photocatalysts that are well-suited to slow down the rapid recombination of carriers. The efficiency of the nanomaterial in degrading contaminants is generally determined by the effectiveness of the carrier charges at each stage. To achieve high photocatalytic efficiency, semiconductors are hybridized into metal-doped transition metal oxides, mixed transition metal oxides, or other new nanocomposites with distinct electronic interactions and strategies to limit rapid carrier recombination [28].

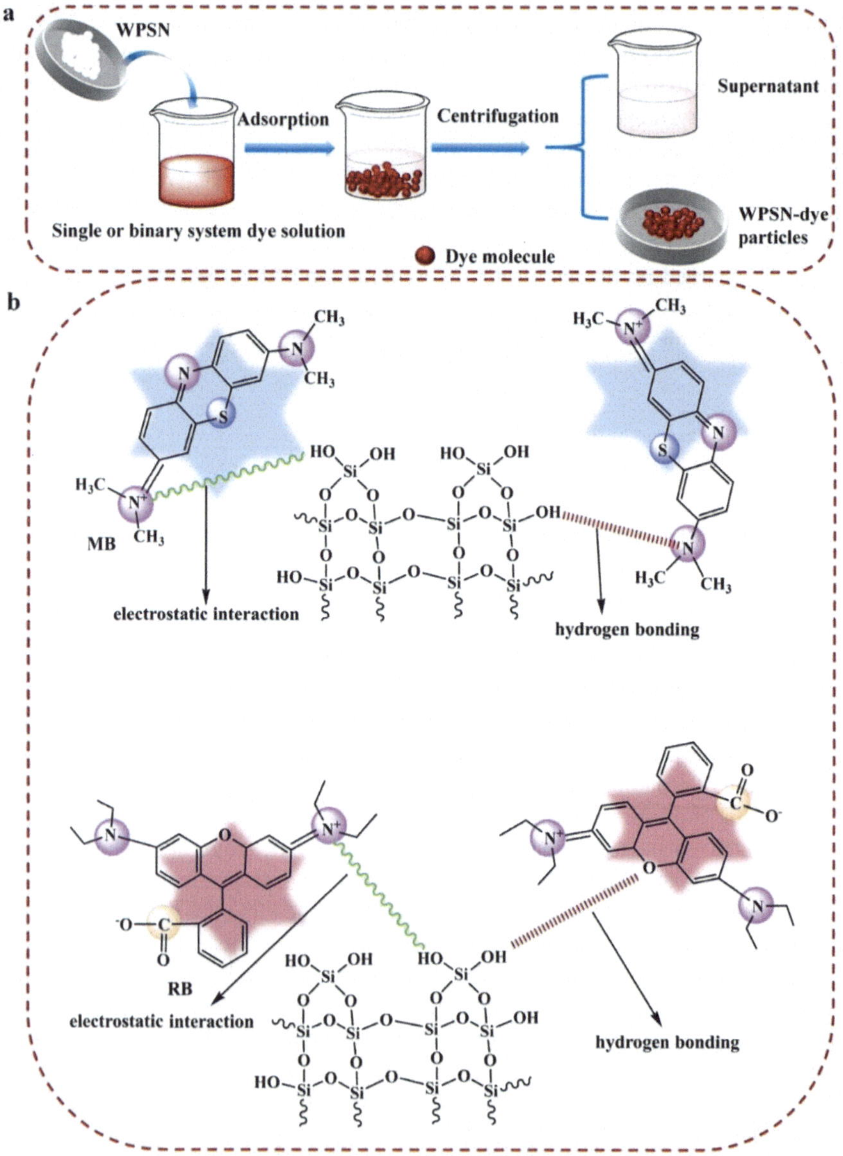

Fig. 5.8 **a** The flowchart of removing dyes from solution using worm-like porous nanosilica, and **b** the adsorption mechanism of worm-like porous nanosilica: illustration of methylene blue (MB) and Rhodamine B (RB) adsorption on nanomaterial via electrostatic attraction and hydrogen bonding [22]. Reprinted with permission from Elsevier B.V

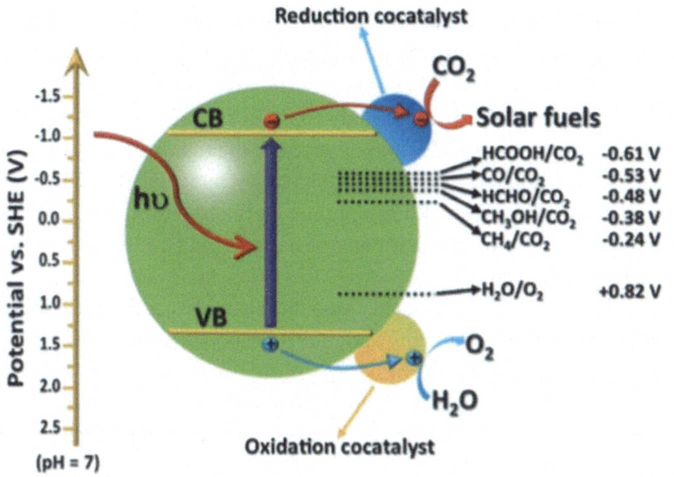

Fig. 5.9 Mechanism of photocatalytic reduction of CO_2 on semiconductor [26]. Reprinted with permission from WILEY-VCH Verlag GmbH & Co. KGaA, Weinheim 2018. (The figure shows band edge potentials at which molecules can be formed from CO_2 molecules through reduction by electrons and oxidized by holes in the conduction band and valence band, respectively)

Biological Methods

Biological processes are used in most of the conventional wastewater treatment procedures. Their base consists of the exploitation of microorganisms, such as fungi, bacteria, yeasts, and algae, which can convert some pollutants into phytonutrients. When oxygen is absent, anaerobic therapy occurs; when oxygen is present, aerobic treatment occurs. Anaerobic and aerobic conditions are commonly mixed for cleaning wastewater that contains a substantial number of organic contaminants. This results in a more comprehensive removal of pollutants than when employing a single-treatment approach [29]. The biological degradation approach eliminates or destroys contaminants more slowly than chemical and physical procedures. Because of this, research has focused a lot of attention on the application of physical or chemical therapy procedures [30].

Opportunities for Nanomaterials in the Environment

NanoSensors

Sensors are electronic devices that take in environmental input and react to it. In daily life, a multitude of sensors are used and classed based on the characteristics and amount of data they capture. Nanomaterial-enabled sensors, a novel technology, provide excellent pollution monitoring even at extremely tiny magnitudes of moles [31]. Nanosensors are essentially chemical, physical, or biological instruments with exceptionally high sensitivity and accuracy for quantifying or qualitatively assessing changes at the nanoscale. The exceptional sensitivity, strong monitoring power, and ability to test several ions and molecules at once are the main characteristics that have led to the high dependence on the data collected by nanosensors [6]. To overcome any casualty and impediment in experimental works and other activities, there is the need to determine if a specific contaminant exists at a field site and if its concentration is above the regulatory limit [31].

Pollutant detection equipment has become increasingly accessible and less expensive, which leads to proper monitoring and timely response to remediation of the environment from harmful pollutants [6]. A gas sensor is made up of transducers and receptors (receptor refers to the substance that engages in interactions with gas molecules) (Fig. 5.10). The sensing mechanism works by the change in parameters (namely, refractive index, mass, biological characteristics, heat, light, and other factors) when the sensor interacts with pollutant molecules. After the change in parameters, the transducer transforms the behavior into an electrical signal [32]. Owing to the nano-size and large surface area, nanomaterials offer substantial benefits for detection over traditional bulk materials. Nanostructured materials used in the production of nanosensors include carbon nanotubes, polymers, metal, and metal oxide nanostructures. On par with graphene and nano/mesoporous carbon, carbon nanotubes are used for a variety of electroanalytical purposes. They offer an effective and reactive surface for the interaction of analyte-transduction materials, leading to highly effective monitoring of pollutants [33].

Nanosensors play a critical role in mitigating factory and urban haze pollution by detecting minute amounts of air pollutants including carbon monoxide, nitrogen oxides, and volatile organic compounds (such as propane, acetone, and toluene).

Nanosensors have been developed to detect hydrogen at a location to prevent potential explosions. Other gases, such as flammable hydrogen gas, are extremely harmful in any setting. Because they can detect contaminants in water such as nitrates,

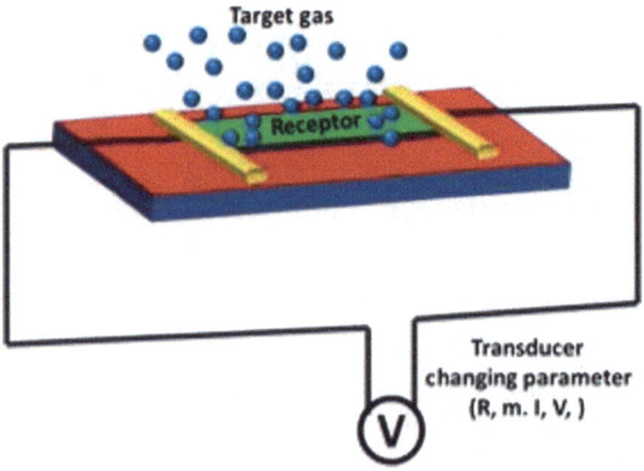

Fig. 5.10 Illustration of gas sensor composed or receptor and transducer [32] (The target gas or water pollutant reacts with the receptor, leading to the change in the parameters of the nanosensor. The change in the parameters is converted into an electrical signal to alert personnel in the environment)

heavy metals, phenolic compounds (including nitrophenol and bisphenol A), and many others, nanosensors also provide major benefits for controlling the quality of water. Because nanosensors allow for real-time examination, potential dangers may be quickly countered. Another area where nanosensors are used is in the evaluation of soil quality. They check the soil for dangerous substances and monitor the nutrients needed for farms to flourish properly [32].

Hao and coworkers fabricated MoS_2 flakes via magnetron sputtering. Exceptional detection of hydrogen was displayed by the manufactured MoS_2/Si sensor owing to the MoS_2 film's distinctive vertically oriented layer architecture, which provided a fast route for the electron transport chain. It demonstrated a good sensitivity of 108.7/101.9 s and a recovery time of 685.7% [34]. Agrawal and his colleagues also used chemical vapor deposition to develop a MoS_2 nanosensor. Alongside the quick response (14.3 s) even at ambient temperature, the sensitivity of the obtained MoS2 for hydrogen detection was discovered to be highly temperature-dependent, rising from around 1% to 11% as the temperature rises [35]. A heterostructure nanosensor founded on positive–negative heterojunction was developed by Zou and collaborators [36] to detect nitrophenol. The heterojunction structures can demonstrate adequate electrochemical efficiency within the context of pollution detection. The detection threshold was lowered to 12 nM due to the enhanced sensitivity and enhanced transportation of charges on the surface induced by the intrinsic electrical field phenomenon at the heterointerface of the graphitic carbon nitride and copper oxide in the nanocomposite. To detect 4-Nitrophenol electrochemically, Ag_2O-ZnO composite nanocones/gold electrode was developed by Chakraborty and coworkers [37]. It showed a low detection limit of 23 nM, a wide detection range of 0.4–26 μM and 28–326 μM, and a sensing range of 1.6 μA μM − 1 cm^{-2}. The composite exhibited extraordinarily rich electro-activities due to the formation of a positive–negative heterointerface between positive-type Ag2O and negative-type ZnO. Table 5.1 lists a few manufactured nanosensors together with the contaminants they can identify, their fabrication process, and their operational parameters. In general, all nanosensors exhibit high sensitivity even at low concentrations, rendering them practically useful for the identification and tracking of air and water contaminants. Remarkably, one important aspect of the detecting process is the passage of electrons from the target pollutant molecules to the different components inside the sensor. Since the main function of nanosensors is to identify harmful gases in any given environment, additional technologies are used to clean up pollution by breaking down harmful substances. The sections that follow have covered these nanomaterials.

Air Remediation

One of the main factors causing serious health problems and harm to plants is air pollution. Air pollution is the reduction in air quality caused by the buildup and release of both pollutants into the atmosphere. It has been shown that a number of heart and lung conditions are significantly influenced by ongoing exposure to contaminated air

Table 5.1 Nanosensors, target pollutants, fabrication method, and performances

Nanosensor	Pollutant	Fabrication method	Performances
Cu nanoparticles modified with a gold electrode	4-nitrophenol	Precipitation and dipping methods	Cu nanoparticles significantly enhanced electron transfer to the electrode, granting about tenfold signal than that of Cu-free gold electrode Low detection limit of 0.34 nM with 247.1 μA μM^{-1} cm^{-2}. [38]
CuO/SnO_2 mixed nanofibers	Hydrogen sulfide	Electrospinning method	Highly sensitivity was recorded owing to the remarkable effect of CuO A response 522 and a response time 1 s at 300 °C were reported[39]
Nanostructured copper-curcumin complex	4-Nitrophenol	In-situ electrochemical procedure	Charge transfer was very facile in the composite Detection limit of 68.2 nM and current sensitivity of 1.57 μA μM^{-1} cm^{-2} were reported[40]
Ag-doped ZnO nanoneedles	Acetone	Facile hydrothermal method	Maximum sensitivity of 30.233 was recorded at 370 °C [41]
Nickel dimethylglyoxime complex—gold nanoparticle modified glassy carbon electrode	o-nitrophenol and p-nitrophenol	Drop coating and electrodeposition methods	Detection limit of 0.58 μM and 0.103 μM were obtained for o-nitrophenol and p-nitrophenol, respectively [42]
Glassy carbon electrode modified with a reduced graphene oxide/ Fe_30_4 composite	4-nitrophenol	Modified hummers' method	Detection limit was 0.26 and 0.86 μM for differential pulse voltammetry and square wave voltammetry, respectively [43]
Pd nanoparticle decorated hollow SnO_2 nanofibers	Formaldehyde	Electrospinning and precipitation	Compared with the pure SnO_2 NFs gas sensor, the optimum operating temperature of Pd decorated hollow SnO_2 nanofibers gas sensor dropped to 160 °C from 180 °C [44] The response to 100 ppm formaldehyde was raised to 18.8 from 5.4. Furthermore, the response-recovery characteristics were enhanced significantly [44]

[45]. Pollutant emissions from indoor and outdoor sources are the two main categories of air pollution. According to estimates from the World Health Organization, seven million people worldwide lose their lives because of air pollution each year. Based on data from the World Health Organization, over nine out of 10 people breathe in air that contains more contaminants than usual [15]. Dust clouds in ventilation systems and other solid combustible materials are examples of particulate matter, which is a primary cause of indoor air pollution. However, the sources of outdoor air pollution include human activity, vehicle and industrial pollutants, and the combustion of fuels. Given the hazardous effects of air pollution, maintaining clean, high-quality air has become essential to maintaining both public health and the viability of civilization [46]. Extensive research has been conducted on the removal of common pollutants including sulfur, carbon monoxide, and carbon dioxide using nanomaterials such as metal–organic frameworks, metal oxide catalysts, etc.

Sulfur Removal: A novel MOF-199 nanomaterial was prepared by Sabine Achmann et al. [48] for the removal of sulfur. The MOF-199 is remarkably effective in removing sulfur from low-sulfur model gasoline and diesel due to its quick adsorption rates. It has been determined that MOF-199's high porosity increases the stability of the Cu-adsorbed sulfur complex through Van der Waals processes, which in turn helps to explain the sulfur's extraordinary removal. Chavan and associates [49] evaluated the properties of the hydrogen sulfide adsorption on the Ni domains of the coordination polymer based on nickel with open metal sites using spectroscopic, microcalorimetric, and computational investigations. The hydrogen sulfide evolves into conjugates on the nickel sites leading to the creation of species that can only be reversed by outgassing at 200 °C. The sample became dark green instead of yellow, showing that the electrical configuration of Ni sites is greatly impacted by hydrogen sulfide adsorption, which has been measured to have an adsorption value of 56–58 kJ mol^{-1}. Based on the collected data, the researchers concluded that, in comparison to previous MOFs tested for hydrogen sulfide adsorption, their adsorbent shows strong structure preservation and a higher absorption capacity rate. Wei et al. [50] suggested using a Ni-doped MoS2 monolayer as a possible gas adsorbent to remove sulfur dioxide and hydrogen sulfide molecules. The adsorption potential was assessed using an approach based on density functional theory. On the Ni-MoS2 monolayer, a detailed analysis of the adsorption energy and electron transfer of sulfur dioxide and hydrogen sulfide molecules was conducted. They found that by chemisorption, the molecules of sulfur dioxide and hydrogen sulfide had a large adsorption energy interaction with the Ni-MoS2 surface (up to -1.319 eV and -1.382 eV), respectively.

The adsorption behavior of hydrogen sulfide on three distinct nanomaterials and their composites on graphene, carbon nanotube, carbon nanocone, and Ti-decorated graphene were evaluated by Ashori and colleagues [51]. On unmodified graphene or carbon nanocone, hydrogen sulfide exhibited a low tendency to collect; on the surface of carbon nanotubes, however, it bonds dissociationally. In contrast to Ti-decorated carbon nanotubes and carbon nanocones, the adsorption of hydrogen sulfide on Ti-decorated graphene was substantially different. The main basis for this is the variations in the Ti-decorated graphene's capacity to adsorb hydrogen sulfide. While the sulfur atom quickly attaches itself to the Ti atom, the hydrogen sulfide molecule breaks down into hydrogen atoms, which then form hydrogen molecules. Generally, physisorption and chemisorption are two of the processes involved in the elimination of sulfur from the environment. Physisorption is the word used to describe the surface-level interaction that takes place following the adsorption of sulfur pollutants on the nanomaterial. Chemical absorption causes structural changes in the nano-adsorbent when a sulfur atom binds to the active metal component to create metal ions, such as copper sulfide, nickel sulfide, and water (Fig. 5.11). Sulfur pollutants are removed by specific routes determined by the design of the nano-adsorbent and the chemistry of certain sulfur compounds, such as reduced or oxidized sulfur compounds [52].

CO Removal: Carbon monoxide, a harmful gas, is mostly produced by automobiles released into the environment. About 90% of CO is released by vehicles in large cities, heavily originating from many private-owned automobiles. CO removal comprises the oxidation of CO into CO_2 or the methanation of CO into useful compounds. Given the unique architecture of metal oxide catalysts that enables the adsorption and activation of pollutant molecules on a high surface area, they are frequently employed in CO removal processes. Metal-support interaction, vacancy defect, high surface area, architecture, size of particles, metal distribution on support, heterojunction, electronic configurations of metal and support, etc., are the principal variables influencing the performance of metal-oxide catalysts. Fundamentally, the catalytic performance for the CO oxidation over the surfaces of transition metal oxides is dependent on the tendency of the metal surface to dissociate oxygen molecules and is re-occupied by the bond strength of the active oxygen species on the metal surface. In CO oxidation, the adsorption and dissociation of molecular oxygen into oxygen atoms is mostly considered as the rate-limiting step. Furthermore, the nature of the active site for surface reaction has been explored and debated for a long time. It is postulated that the active oxygen species could be atomic or molecular oxygen adsorbed on metal sites, oxygen adsorbed on the metal oxide support, most likely at the metal-support interface, or lattice oxygen of reducible transition metal oxide support. The types of catalytic mechanisms revealed in most studies related to CO oxidation include Mars–van Krevelen (M–vK) and Langmuir–Hinshelwood (L–H) mechanisms. In some instances, the M-vK mechanism may precede or concurrently occur with the L–H mechanism, since several vacancy defects, that are the oxygen adsorption sites, are formed after the M-vK reaction (Fig. 5.12).

Chen and coworkers [54] studied different contents of Cu loaded onto $CoFe_2O_4$ support to explore the best Cu loading content and the mechanism of CO oxidation. The CO oxidation performance was greatly improved upon Cu doping on the surface of $CoFe_2O_4$ catalysts, especially on 5 wt% Cu/$CoFe_2O_4$. Khder et al. [55] produced palladium-supported Mn_3O_4 nanomaterial for the removal of CO. It was determined that the addition of palladium nanoparticles substantially enhanced the low-temperature reduction of the Mn_3O_4 support. The full conversion of 4wt% CO at 20 °C was made possible by the great distribution of metallic Pd metal and oxidized Pd species on the support, which made it easier for the support to provide active lattice oxygen to significantly participatet in the reaction. The optimal load

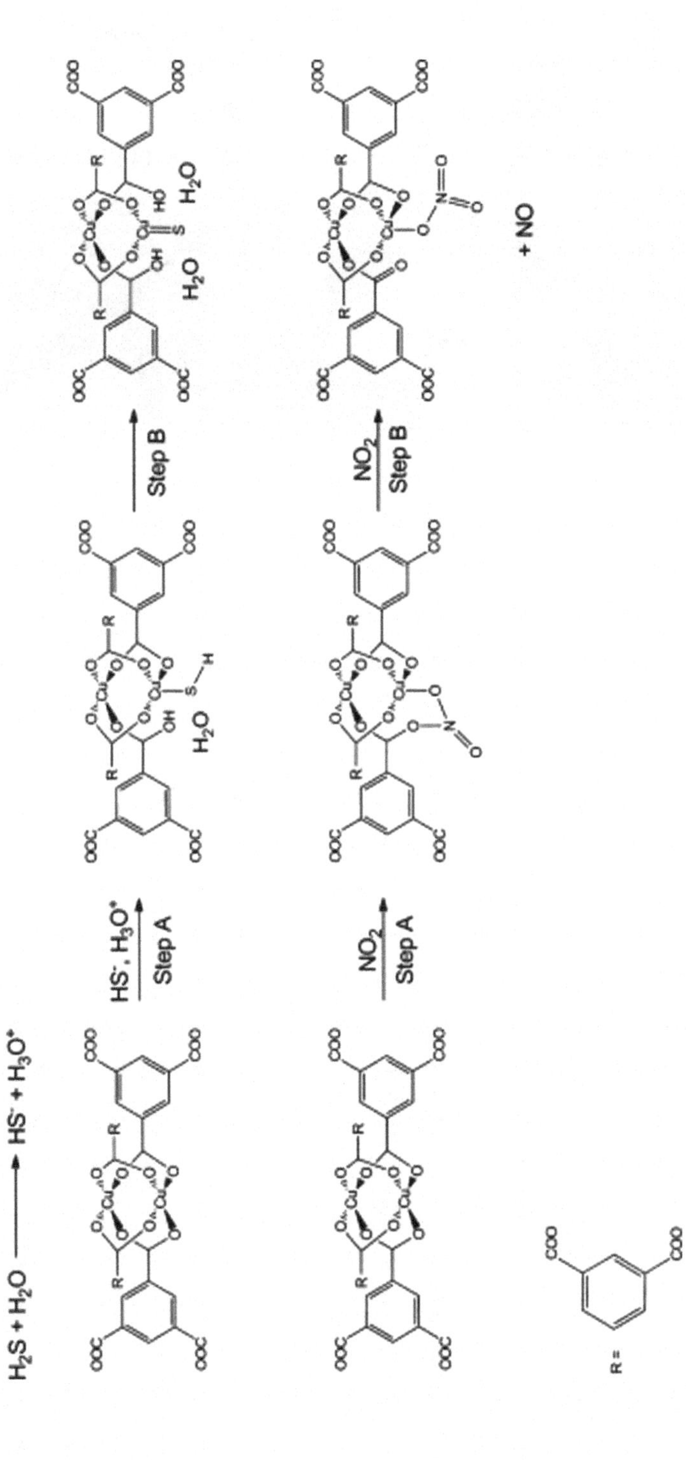

Fig. 5.11 A proposed mechanism for the adsorption of H_2S on HKUST-1 MOF [47] involving adsorption of hydrogen sulfide on the nanomaterial which dissociates into hydrogen atoms and copper sulfide

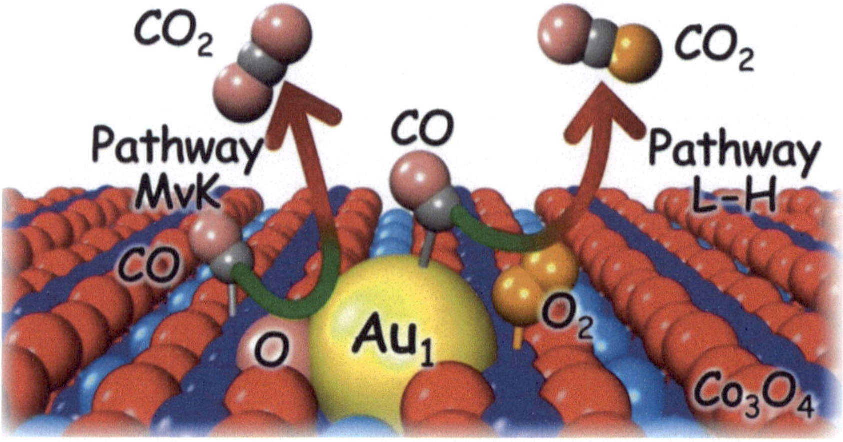

Fig. 5.12 Graphical illustration of the two common possible mechanisms that take place during the removal of CO from the environment [53]. Reprinted from American Chemical Society, Copyright 2020. (*M-vK pathway*: CO molecule adsorbed on the metal sites (Au) migrates to the metal-support interface (Au-Co_3O_4 interface) to react with lattice oxygen of the support to produce CO_2, generating an oxygen vacancy on the surface; molecular oxygen in the gas stream is adsorbed and dissociated into oxygen atoms at the oxygen vacancy sites; a second adsorbed CO molecule reacts with the activated oxygen atom, producing CO_2; *L–H pathway:* CO is adsorbed on the metal site and oxygen molecules in the gas stream are adsorbed on the support sites. The oxygen molecules dissociate into oxygen atoms, which react with adsorbed CO to produce CO_2)

of palladium that granted the best performance in eliminating CO was 4%. Alongside the good catalytic performance, the potential application of the nanomaterial in remediating the environment from CO pollutants was its good stability over 30 h. Composite catalyst nanomaterials such as Pd/CeO_2/γ-Al_2O_3, Pd/Fe_2O_3/γ-Al_2O_3, and Pd/MnO_2/γ-Al_2O_3[56], and Pd/γ-Fe_2O_3[57] has been investigated extensively. Other nanomaterials used to facilitate the oxidation of CO, their preparation and efficiencies are listed in Table 5.2.

Figure 5.13 depicts the pathways that have been examined for the CO hydrogenation to methanol and methane. Methane can be produced using the dissociative CO pathway, whereas both methane and methanol may be produced with the associative CO pathways. The dissociative route involves the initial dissociation of CO into surface species, O (*O) and C (*C), and then hydrogenation to produce CH_4 and H_2O. However, breaking C-O bonds is challenging due to its high energy barrier, which is mostly represented by the rate-determining step (RDS). In the associative route, intermediate *COH or *CHO are produced by CO hydrogenation, which happens before any dissociation, depending on where the first hydrogenation takes place. Following hydrogenation of the C-side, the formaldehyde isomer *CHOH, hydroxymethyl *CH_2OH, and ultimately methanol is produced, starting with the *COH intermediate. Any of the intermediates, *COH, *CHOH, and *CH_2OH, have the potential to break the C–O bond. This is followed by the hydrogenation of CHx and OHx to generate water and methane. Formaldehyde is formed by one further

Table 5.2 Nanomaterials, synthesis method, performance, and space velocity

Catalysts	Method	T_{100}	CO conc	GHSV (ml g^{-1} h^{-1})	Ref
Pd/CeO_2	Hydrothermal	125	1	120,000	[58]
CeO_2/ZnO	Deposition–precipitation	300	0.5	30,000	[59]
3DOM CeO_2-ZnO	Colloid crystal templated method	260	1	30,000	[60]
Ag-ZnS	Solvothermal approach	–	1	240,000	[61]
CuCe-nanosphere	Deposition–precipitation	122	1	80,000	[62]
Au/twin-brush-like ZnO	Deposition–precipitation + Hydrothermal	40	1	490,000	[63]
Cu^{2+} doped CeO_2	Deposition–precipitation	223	1	96,000	[64]
Cu0.29Ce0.71O_2-$_y$	Modified base precipitation method	203	2	4500	[65]
$Ce_{0.9}Pr_{0.05}Cu_{0.05}O_{2-\delta}$	Sol – gel method	150	1	9600	[66]
Cu/ZnO	Deposition–precipitation	330	0.5	8800	[67]
Cu_xO-Cu/ZnO	Ultrasonic spray pyrolysis method. + Deposition–precipitation	140	2	30,000	[68]
Porous Metal – Organic Polymer [Cu(5-methylisophthalate)]	Hydrothermal method	250	1	2000	[69]

hydrogenation on the C end of the *CHO species, followed by the production of methoxy and methanol by the subsequent hydrogenation processes. If the C–O link breaks inside the *CHO, *CH_2O, and *CH_3O intermediates, methane production is also conceivable. It is obvious that this reaction is quite complicated, and creating a catalyst that is highly selective to a single product is a significant task.

CO_2 Reduction: The concentration of carbon dioxide (CO_2) in the atmosphere has increased dramatically during the last 20 years, from 280 parts per million (ppm) before the preindustrial era to 410 parts per million (ppm) in recent years. One of the largest worldwide challenges facing humanity in the twenty-first century is anthropogenic carbon dioxide (CO_2) emissions. Global catastrophic disasters might occur if immediate action is not taken to reduce the atmospheric CO_2 level. Due to the gravity of this issue, some countries have chosen to implement carbon capture, storage, use, and sequestration technologies at both academic and industrial levels to reduce the amount of CO2 in the atmosphere [71]. Many approaches, including the development of carbon capture and storage technology and clean energy sources, have been put out to lower carbon footprint. Among these, the process of turning CO_2 into chemicals has been shown to be a desirable means of reducing CO_2 emissions while creating goods with additional value. The thermal method uses hydrogen as a reductant and high temperatures and pressures to transform CO_2 into useful compounds such as methane and methanol (see Eqs. 5.1–5.4).

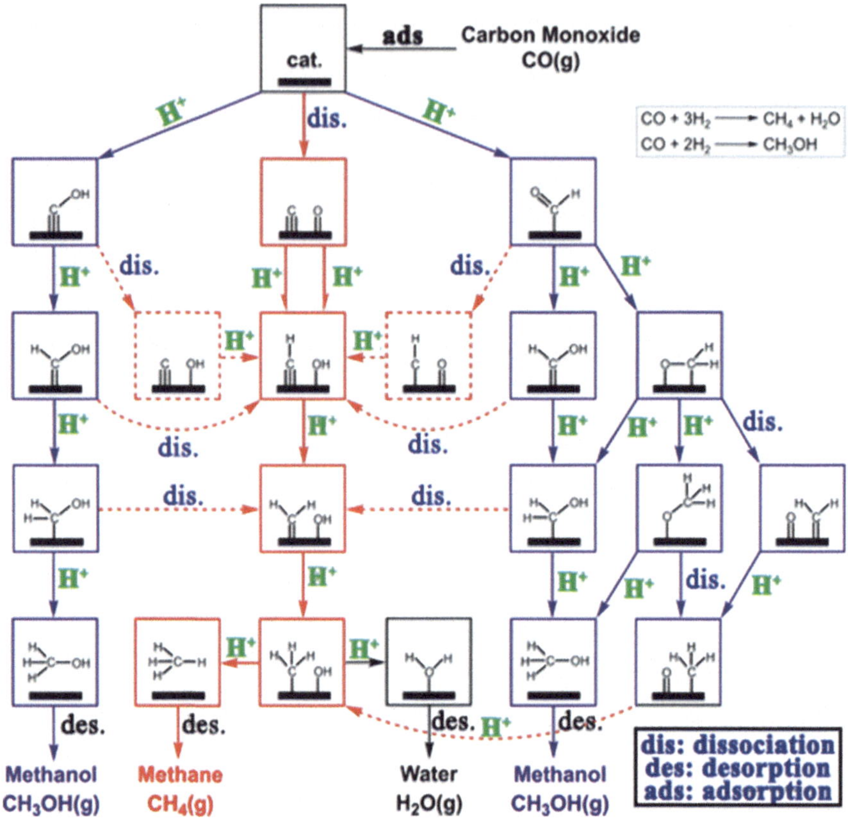

Fig. 5.13 Reaction pathways of CO hydrogenation into methane, methanol, and water [70]. Reprinted with permission from the American Chemical Society, copyright 2023

A viable method for converting CO_2 gas into fuel is via a photocatalytic reduction process. Water is mostly the source of electrons and protons for the reduction of CO_2 or CO_2 hydrogenation to constitute a closed cycle since H_2O is used in generating hydrogen reductant in the reaction and forms a major component in the products after hydrogenation of carbon pollutants (Eqs. 5.2–5.6). Consequently, the general formula presented in Eqs. 5.1–5.6 governs CO_2 reduction, methanation, etc., in eliminating CO_2 from the environment. Some CO_2 elimination process such as syngas is a combination of H_2O and CO that is produced when too much hydrogen is utilized in reducing CO_2 (as seen in Eq. 5.3). On a unique nanomaterial catalyst, the products of syngas, i.e., CO and H_2O, can be utilized to produce useful compounds such as methanol and methane, among others. Equations 5.2, 5.3, and 5.4 refer to the reverse water–gas shift reaction, CO_2 methanation, and the Fischer–Tropsch reaction, respectively. It is noteworthy that the set of mechanisms involved in the reduction or removal of CO_2 is influenced by factors such as the active sites and conduction and valence band potentials of nanocatalysts. Table 5.3 presents a series of nanomaterials in the form of catalysts used for transforming CO_2 into useful products or CO, the fabrication methods used for synthesizing the nanomaterials, the technology used in reducing/transforming/adsorbing CO_2, and their performance.

$$xCO_2 + yH_2 \rightarrow \text{Product} + ZO_2 \cdot CO_2 \quad \text{recycling reaction} \tag{5.1}$$

$$2H_2O \rightarrow 2H_2 + O_2 \quad \text{Water} - \text{Splitting reaction} \tag{5.2}$$

$$CO_2 + H_2 \rightarrow CO + H_2O \quad \text{Reverse water} - \text{gas shift} \tag{5.3}$$

$$CO_2 + 3H_2 \rightarrow CH_4 + H_2O \quad \text{Methanation} \tag{5.4}$$

$$CO_2 + 2H_2O \rightarrow CH_4 + 2O_2 \quad \text{Methanation} \tag{5.5}$$

$$nCO + (2n + 1)H_2 \rightarrow C_nH_{2n+2} + nH_2O \quad \text{Fischer} - \text{Tropsch} \tag{5.6}$$

Water Remediation

The textile industry uses a wide range of artificial colorants and produces large amounts of highly colored waste liquid, which interferes with plants' and aquatic life's ability to photosynthesize. The release of brightly colored waste liquid into water bodies caused by synthetic dyes damages plants' ability to photosynthesize and has an adverse effect on aquatic life because of the sluggish process of re-oxygenation. Moreover, marine life is impacted by chlorine in industrial effluent [82]. While global considerations primarily focused on water shortages, water usage effectiveness, and allocation problems, ineffective treatment (such as wastewater) and management resulted in extensive contamination of the drinking water system [1]. Many inorganic metallic elements, including mercury, cadmium, lead, chromium, copper, etc., are included in the group of inorganic hazardous elements. Pharmaceuticals, personal hygiene items, endocrine disruptors, pesticides, organic dyes, detergents, and common industrial organic wastes including phenolics, halogens, and aromatics are among the common organic pollutants found in water. Because organic pollutants are persistent and persistent in the environment, they have emerged as one of the most serious environmental problems, second only to heavy metals in water. Water pollution removal is controlled by several mechanisms, such as hydrogen bonding and electrostatic attraction (Fig. 5.14), etc. The pollutant is adsorbed on the active sites, which later undergoes degradation via electron transfer from the nanomaterial to the pollutants. Oxidant species also play a major role in the degradation of pollutants using chemical processes such as photocatalysis. The reduction

Table 5.3 Catalysts, fabrication methods, removal technology, and product yield/adsorption in CO_2 removal from the environment

Catalyst	Fabrication method	Removal technology	Product (CH_4, CO) yield
Ru/Al_2O_3	Precipitation	Photothermal catalysis	CH_4 yield (3.33 $mmol{\cdot}gcat^{-1}{\cdot}h^{-1}$) [72]
$NiRu/ZrO_2$	Precipitation	Photothermal catalysis	CH_4 yield (549.7 $mmol{\cdot}gcat^{-1}{\cdot}h^{-1}$) [73]
$Ni\text{-}La_2O_3/SiO_2$	Precipitation	Photothermal catalysis	CH_4 yield (42.9 $mmol{\cdot}gcat^{-1}{\cdot}h^{-1}$) [74]
Ru-Loaded Ultrathin Layered Double Hydroxides	Impregnationprocedure	Photothermal catalysis	CH_4 yield (277 $mmol{\cdot}gcat^{-1}{\cdot}h^{-1}$) [75]
$Ni\text{-}TiO_2$	Solvothermal method	Photothermal catalysis	CH_4 yield (95.7 $mmol{\cdot}gcat^{-1}{\cdot}h^{-1}$) [76]
$Fe_2Ce1\text{-}300$	Simple precipitation reaction	Photothermal catalysis	CO yield (19.61 $mmol{\cdot}gcat^{-1}{\cdot}h^{-1}$) [77]
Archetypal nickel phosphide/ SiO_2	Incipient wetness impregnation method	Photocatalysis	CO yield (960 $mmol{\cdot}gcat^{-1}{\cdot}h^{-1}$) [78]
Manganese-based metal–organic framework	Conventional solvothermal method	Adsorption	CO_2 uptake of 9.3 $mmol{\cdot}g^{-1}$ [79]
Cu–N, N′-dibenzoic acid-2,3-diaminoquinoxaline	Hydrothermal method	Adsorption	CO_2 uptake of 8.0 mmol. g^{-1} [80]

of oxygen molecules by electrons and the oxidation of water by holes into superoxide and hydroxyl radicals, respectively, dominate the reaction process [83]. Other factors such as acidity and alkalinity have an influence on the degradation process of nanomaterials. The respective conditions enable the production of more protons (H^+) and hydroxyl (OH) groups in the dye solution, respectively. These two species could promote or hinder the reaction process, considering the composition of the nanomaterial used in the remediation process. The presence of high hydroxyl groups mostly enhances the degradation of cationic pollutants on nanomaterial via electrostatic attraction since it promotes negative charges on the surface of the nanomaterial. On the other hand, numerous protons facilitate the good attraction of anionic pollutants

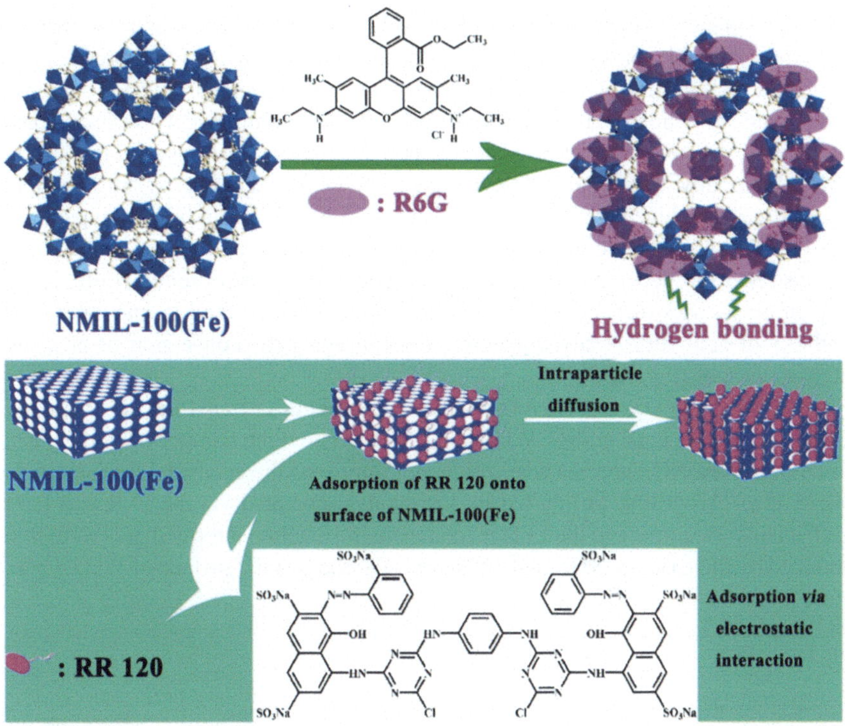

Fig. 5.14 Diagrammatic representation of the three dyes' adsorption method on Fe-NMIL-100. (The upper part shows the hydrogen bonding of the dye to Fe-NMIL-100 MOF nanomaterial. The lower part illustrates the adsorption of RR 120 dye on the surface of the nanomaterial which facilitates the intraparticle diffusion of pollutants into the pores of the nanomaterial [81]. Reprinted with permission from the American Chemical Society, copyright 2024

on positively-charged nanomaterials. It is worth noting that the interaction between these species and the nanomaterial is paramount to the reaction process.

Nitrogen Oxide Removal: The main methods for controlling nitrogen oxide (NOx) emissions include selective non-catalytic reduction (SCR), ionophore activation, adsorption, and selective catalytic reduction (SCR). Among them, the most efficient and environmentally benign is the elimination of NOx using catalytic materials and ammonia (NH_3) as a reducing agent (NH_3-SCR), which has been widely used in industry. Recent studies have shown that most catalysts for the NH_3-SCR reaction conformed to the Langmuir–Hinshelwood and Eley–Rideal reaction mechanisms. These mechanisms address the reactions of the adsorbed state with NOx and the gas phase with NH_3. Transitional metal oxides (CeO_2, MnOx, CoOx, VOx, FeOx, and CuOx) are commonly used as redox centers for the synergistic NO removal operation because of the different valence of the metal ions. The activation energy of the process can be considerably lowered by this electrical interaction, which also permits high mobility of electrons and lattice oxygen. The catalyst's metal ions can also serve as Lewis acid sites, which are essential for NH_3 adsorption and aid in the reaction's facilitation [84]. The presence of NOx has been addressed by the widespread application of a wide variety of designed heterogeneous catalysts. Iron-tungsten catalysts were found by Wang et al. to be more effective in eliminating NOx emissions, with over 90% conversion achieved across a broad temperature range of 225–450 °C [85]. Additionally, Wang and colleagues created WO_3-FeOx catalysts using a simple solvent-free technique for NO removal using NH3-SCR. Over a broad temperature range of 225–500 °C, the ideal WO_3–FeOx catalyst with the biggest surface area demonstrated the best catalytic activity and about 100% NOx removal efficiency [86]. In most cases, the creation of hybrid composites that tend to encourage the synthesis of acid sites is advantageous in encouraging the conversion of NOx on catalyst surfaces (Fig. 5.15).

According to in situ diffuse reflectance infrared Fourier transform spectroscopy (DRIFTS) experiments, adding WO_3 to Fe_2O_3 increased the number of acidic sites, which improved NOx species adsorption and improved the catalysts' performance. (Fig. 5.15a). According to Fig. 5.15b, the catalyst operated via the E–R and L–H mechanisms: According to the L–H mechanism, the reaction intermediates were first produced by the gaseous NH_3, and NO being adsorbed on the surface of WO3-FeOx

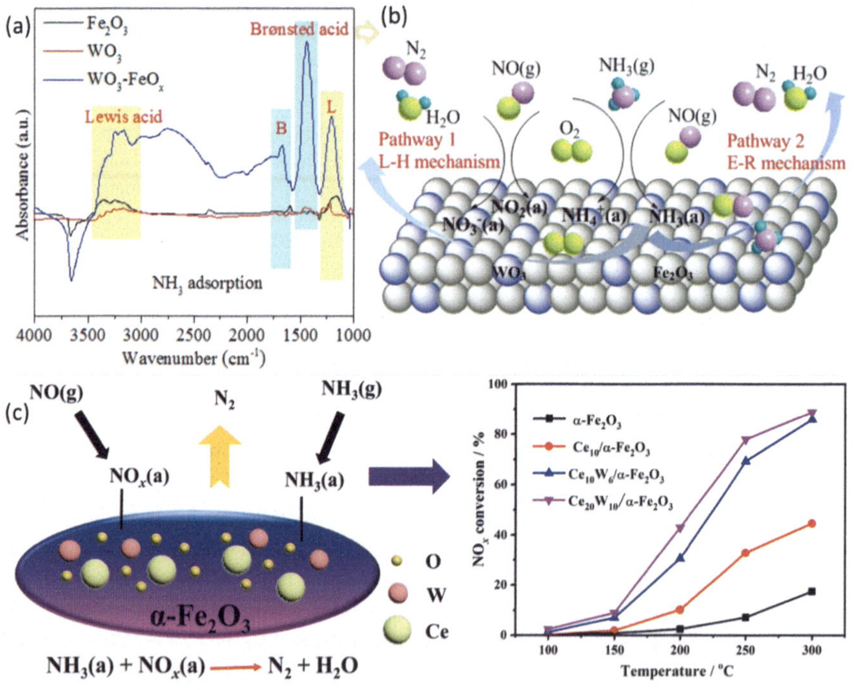

Fig. 5.15 **a**, **b** The reaction process and in situ DRIFTS result of WO_3-FeOx catalysts. Reprinted with permission from Ref. [86]. Catalytic surface acidity and alkalinity of Ce and W at outstandingly low-temperature catalytic reaction activity (Fig. 5.15c). Reprinted with permission from Ref. [87]

as NH3(a)/NH_4 + and NO_2/NO_3. Afterward, N_2 and H_2O are formed by the reaction between the two adsorbed species. By direct reaction, the adsorbed NH_3 species may produce N_2 and H_2O in the E–R mechanism when they encounter gaseous NO over WO_3-FeOx (Pathway 2). The presence of Ce elevated the concentration of active oxygen atoms and the location of mild alkalinity Tungsten (W) supplementation, on the other hand, dramatically raised the number of Brønsted acid sites and decreased the number of alkaline sites. The catalyst's surface acidity and alkalinity were ideally balanced by the addition of Ce and W, which produced outstanding low-temperature catalytic reaction activity (Fig. 5.15c) [87].

Dye Removal/Degradation: Dyes like rhodamine B (RhB), methylene blue (MB), methylene orange (MO), and crystal violet (CV), to mention a few, are often hard to get rid of from effluent. Most of them are eliminated from water using a range of physical and chemical techniques, such as the widely used green photocatalysis technology. For the treatment of wastewater, nanomaterials like carbon nanotubes (CNTs) are employed as adsorbents. It has been shown that carbonaceous materials have superior adsorption capabilities. CNTs are thought to be more beneficial for the adsorption of anionic azo dyes than cationic dyes because, according to zeta potential tests, they are positively-charged materials between 10 and 30 mV throughout a wide pH range [88]. Athari et al. [89] used a solvothermal approach to produce UiO-66 metal–organic framework as well as composite nanoparticles with carbonaceous materials including MWCNTs–COOH and graphene oxide (GO) with varied molar ratios (2.9 and 5.8 wt.%). Additionally, the experimental adsorption results showed that MWCNTs-UiO-66 nanocomposites could adsorb 98% of methyl red (MR) and MO, along with 72% of MB and 46% of malachite green (MG). Adsorbent removal of dye is largely influenced by the π–π stacking reaction that occurs between the aromatic rings of the dye molecules with π structures on the adsorbent surface, electrostatic attraction, and hydrogen bonds between the oxygen-containing functional groups of the adsorbents, the carbonyl functional groups of MR, and the sulfone functional groups of MO, respectively. Anionic dyes exhibited a higher rate of adsorption in the MWCNTs-UiO-66 nanocomposites than cationic dyes. Table 5.4 presents nanomaterials, their fabrication methods, and their application in removing specific water pollutants.

Antibiotics removal: Antibiotics encompass a wide range of compounds, such as vancomycin, clindamycin, amoxicillin, penicillin, ciprofloxacin, levofloxacin, cephalexin, cefuroxime, and cefixime. Its prevalence in natural ecosystems and wastewater treatment is partly caused by the indiscriminate use of antibiotics. Several nations, including Kenya, South Korea, Romania, France, Spain, China, and others in Latin America, have previously had antibiotics found in their home, hospitals, and industrial and agricultural wastewater. Antibiotic-resistant genes can spread and alter the microbiota's makeup and metabolism through interactions with antibiotics and their metabolites [98]. By using chemical cross-linking and in situ growth during freeze-drying to adsorb tetracycline hydrochloride, Xu et al. created recyclable bacterial cellulose/sodium alginate/zeolitic imidazolate framework-8 (BC/SA/ZIF-8) aerogel composites to mitigate the harm that antibiotics in wastewater cause to the environment. With a maximum adsorption capacity of 746.27 mg/g, the results show

Table 5.4 Nanomaterials, synthesis method, pollutant, removal techniques, and performance

Nanomaterials	Synthesis	Pollutant	Removal techniques	Findings
TiO_2 Degussa P25		RhB	Photocatalysis	55% degradation (0.2 g/L dye solution) at 120 min [90]
Bismuth-doped TiO_2 nanotubes	Sol–gel and hydrothermal methods	RhB	Photocatalysis	The degradation studies revealed that initial rate of photocatalytic degradation of RhB dye using Bi-TNT was 5.56, 4.16, 1.30 and 2.38 times higher compared to TNP, Bi-TNP, TNT and Degussa P-25 TiO_2 (P-25), respectively, under direct sunlight irradiation [91]
CeO_2 doped TiO_2 nanocomposites	Hydrothermal method	RhB	Photocatalysis	The highest absorption peak at 579 nm disappeared after 8 h due to the largest photocatalytic activity, and 99.89% of the RhB was broken down by solar light [92]
Chitosan (CS) supported gC_3N_4-MoS_2 hybrid nanocomposite	Hydrothermal method	RhB	Photocatalysis	Because of its small band gap (1.80 eV), gC_3N_4@CS-MoS_2 can effectively degrade RhB in the visible spectrum. It reached its maximum degradation efficiency of 99% after 40 min at pH 8 and a minimum dose of 60 mg [93]
Ni-Cu@MWCNTs	Polyol method	RhB	Photocatalysis	Ninety-eight percent degradation efficiency was achieved using heterogenous photocatalyst NiCu@MWCNTs under optimal circumstances (starting concentration of RhB dye ~0.04 mM, catalyst quantity ~0.01 g/ 50 mL, pH ~ 3 and temperature ~298 K, reaction duration ~50 min) [94]

(continued)

Table 5.4 (continued)

Nanomaterials	Synthesis	Pollutant	Removal techniques	Findings
$BiMnO_3$ nanoparticles	Hydrothermal method	RhB	Photocatalysis	Under UV irradiation, the degradation of RhB efficiency was 68% in 75 min. In H_2O_2 assisted visible light irradiation, the sample shows better performance (90% in 55 min)[95]
Cu-doped In_2S_3 hierarchical structures	Hydrothermal method	RhB	Photocatalysis	More than 90% dye degradation in 60 min. Superoxide radicals were crucial in its performance[96]
Worm-like porous nanosilica	Sol–gel method	RhB and MB	Adsorption	With ideal adsorption capacities of 552.00 mg/g and 537.00 mg/g for MB and RhB, respectively, it demonstrated superior selectivity performance and a high adsorption capacity to cationic dyes [22]
Recyclable porous nano-carbons (PNCs)	Carbonization	MB	Adsorption	For methylene blue, PNCs had a maximum removal effectiveness of 99%. PNCs have good degrading efficiency even after seven cycles, indicating their recyclability. The highest adsorption capacity using the Langmuir isotherm was around 1216.54 mg g^{-1}[97]

that the BC/SA/ZIF-8 aerogel composite exhibited superior tetracycline hydrochloride adsorption performance due to its much greater surface area of 165.24 m^2/g compared to that of BC/SA at 7.39 m^2/g[99].

Heavy Metal Ions Removal: Water contamination by heavy metals also plays a significant role in worldwide health and environmental issues. Heavy metals are any metallic elements that are more poisonous and have a larger density. The heavy metals that are most harmful and carcinogenic include mercury (Hg), lead (Pb), arsenic (As), chromium (Cr), cadmium (Cd), and thallium (Th) [100]. Because CNTs have excellent removal efficiencies, they are a potential adsorbent for the treatment of significant heavy metal pollutants such as Cd(II), Zn(II), and Pb(II) [88]. The adsorption capability of PAN@UiO-66-NH_2 nanofiber membrane for Cr (VI) ions was investigated and 312.25 mg/g of optimum adsorption capacity at 298 K was observed. The pseudo-second-order model and the Langmuir isotherm were both compatible with the adsorption process, according to kinetic tests. In five cycles of Cr (VI) elimination, the composite likewise demonstrated high reusability. PAN@UiO-66-NH_2 nanofiber membrane efficiently adsorbs Cr(VI) by electrostatic contact, according to Zeta potential, pH impact, and XPS examination [101]. Similarly, a straightforward chemical precipitation technique was used to produce calcium ferrite nanoparticles with super-paramagnetic behavior for the efficient removal of Cr (VI) from aqueous systems. The adsorption process was shown to entail intraparticle diffusion and followed second-order kinetics. The Langmuir isotherm was used to estimate the adsorption capacity, which was found to be as high as 124.11 mg/g. The adsorption process was shown to be exothermic, spontaneous, and viable by thermodynamic analysis. Experiments have demonstrated the effectiveness of nanoparticles in protecting the environment against a wide range of contaminants found in the air, water, and other media. In environmental remediation, the primary features of nanoparticles facilitate favorable interactions with contaminants and the stimulation of surface reactions [102].

Challenges and Prospects

This section presents the challenges regarding the characteristics, production, and application of nanomaterials in nanotechnology and the harmful effects posed to the environment, especially after the usage of nanomaterials. One of the major disadvantages of semiconductors is the fast electron/hole recombination, which is faster than the surface redox reactions and limits [63]. The challenge found in semiconductors has been a major hindrance to guaranteeing that there are enough radicals on the catalyst's surface to effectively remove contaminants from the surrounding air, as the radicals generated there have a brief lifespan. Because of this, there has been a persistent interest in the scientific community to engineer or modify semiconductors' electrical characteristics to increase the range of applications for them in environmental mitigation (Fig. 5.16).

Furthermore, the form, size, and diameter of nanomaterials are minuscule. When it comes to using nanomaterials, the one unanswered concern is whether they may unintentionally leak into the environment both during and after use. As with air treatment and many other applications, for example, some of the nanoparticles utilized in water treatment could be detected in the treated water. The health risks associated with nanomaterial residuals in water or air stem from the harmful impacts of nanoparticles. The risk to the human system is increased by the ease with which particles smaller than 2.5 mm can reach circulation. Nanoparticles can enter the human body through several routes, including the skin, food, and nasal passages. They can spread to several regions of the body, including the spleen, bone marrow, neurological system, and others, as seen in Fig. 5.15. The response that occurs

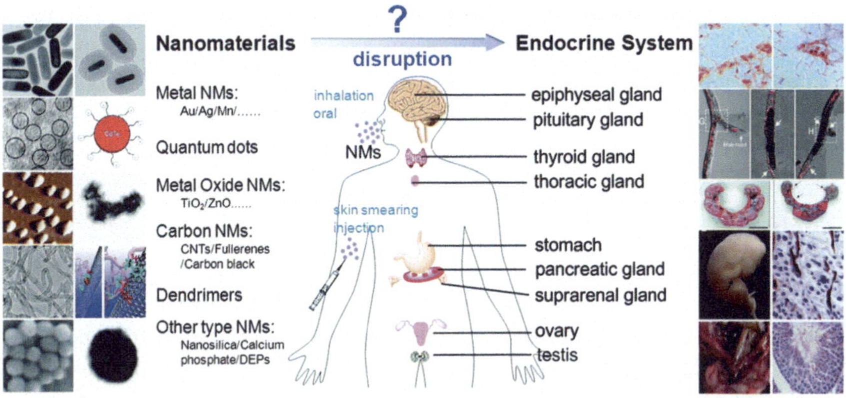

Fig. 5.16 Human endocrine functioning risks from nanoparticles that penetrate the body. [103]: The left image depicts nanomaterials; the center and right images represent endocrine organs and potential negative effects. Reproduced with permission from WILEY-VCH Verlag GmbH & Co. KGaA, Weinheim, 2013

when particles reach the human system is not clear. This is particularly devastating because the effects of reactive nanomaterials on the body are yet mostly understood. Human health risks associated with particulate matter have raised awareness of the potentially harmful effects of exposure to nanomaterials, including nanoparticles [7, 46]. As such, there may be restrictions on how widely nanotechnology may be used to clean up the environment after pollution. The effects of nanomaterial detritus on live organisms must be critically evaluated, i.e., their interactions with cells, tissues, organs, etc., to inform the development of potentially remediation-focused but non-reactive nanomaterials for the healthcare system. Additionally, to counteract the potentially fatal effects of nanoparticles on health, well-made and cozy devices must be developed and used. Thirdly, several uses for nanotechnology result in nanowaste. Getting rid of nano trash is one of the world's most difficult tasks. Industrial-scale production of nanomaterials has led to a rise in the generation and disposal of nanowaste in streams and landfills. Most nanomaterials are inorganic, making them minuscule in size, which makes handling nanowaste challenging. The usual techniques for treating nanowaste include recycling, burning, and filtering. Unfortunately, these techniques discharge the decomposed material into the surrounding environment, necessitating the development and use of ecologically acceptable alternative treatment technologies [104].

The development of advanced technologies in the future depends on nanotechnology advancements and the highest standards of environmental safety. Due in significant part to their research and development of novel materials and products, many of which contain nanomaterials, industries are the main source of pollution. This shows that while enterprises develop nanomaterial-related commodities for general usage in society, carbon emission happens throughout the production process (such as carbonization, flame spray pyrolysis, and other procedures). This implies that the procedure used to produce functional nanomaterials has been one of the main causes of pollution. Therefore, environmentally friendly manufacturing processes are required to manufacture functional nanomaterials. This offers a foundation for the broad research and use of sustainable manufacturing techniques.

Conclusion

The field of nanotechnology has yielded new materials with distinct features that fall inside the nanoscale, greatly impacting their applications. The problems and detrimental effects that pollutants from cars, industry, and human activity discharge into the environment pose to humans and plants present prospects for the widespread usage of nanomaterials. Water and air pollution remediation is the primary use for nanoparticles, including metal-oxide, carbon-based, and polymer-based nanomaterials with unique properties dependent on their composition. Large surface area, high porosity, excellent reactivity, and other features of nanoparticles vary greatly, and production techniques are important in customizing functional nanomaterials. The size, shape, and structure of a given nanomaterial have been discovered to be significantly impacted by a variety of fabrication techniques, including electrospinning, deposition–precipitation, sol–gel, flame spray pyrolysis, and many more. Nanosensors made from various polymers and metal/metal oxides have been employed in detecting dangerous and flammable gases or contaminants in air and water. Apart from nanosensors, other nanomaterials such as nano adsorbents, nanocatalysts, and photocatalytic nanomaterials have been created and utilized for the adsorption and/or degradation of pollutants found in air and wastewater. Additionally, through various reaction processes, these nanomaterials allow the conversion of harmful pollutants like CO, CO_2, and many others into beneficial chemicals like methane, methanol, hydrogen gas, etc. Despite the research being done on the creation and application of functional nanomaterials, their smaller size and lack of a clear disposal method make them a severe concern both during and after use. Because there is currently a dearth of comprehensive research regarding the extent of harm that nanomaterials may do to human and plant systems, it is possible for the particles to be breathed or to enter the system and cause death. In summary, protecting the environment both during and after the use of nanomaterials is a corollary of the development of nanotechnology. Therefore, investigation of highly functional nanomaterials and substantially incorporating the design of more environmentally friendly nanomaterials, production processes and nanowaste disposal would be interesting cornerstones under the umbrella of nanotechnology and the environment.

References

1. Ibrahim, A.O., Adegoke, K.A., Adegoke, R.O., AbdulWahab, Y.A., Oyelami, V.B., Adesina, M.O.: Adsorptive removal of different pollutants using metal-organic framework adsorbents. J. Mol. Liq. **333**, 115593 (2021)
2. Kampouraki, Z.-C., Giannakoudakis, D.A., Nair, V., Hosseini-Bandegharaei, A., Colmenares, J.C., Deliyanni, E.A.: Metal organic frameworks as desulfurization adsorbents of DBT and 4,6-DMDBT from fuels. Molecules **24**, 4525 (2019)
3. Saravanakumar, K., De Silva, S., Santosh, S.S., Sathiyaseelan, A., Ganeshalingam, A., Jamla, M., et al.: Impact of industrial effluents on the environment and human health and their remediation using MOFs-based hybrid membrane filtration techniques. Chemosphere **307**, 135593 (2022)
4. Pokrajac, L., Abbas, A., Chrzanowski, W., Dias, G.M., Eggleton, B.J., Maguire, S., et al.: Nanotechnology for a sustainable future: addressing global challenges with the international network4sustainable nanotechnology. ACS Nano **15**, 18608–18623 (2021)
5. Li, Z., Wang, L., Li, Y., Feng, Y., Feng, W.: Carbon-based functional nanomaterials: preparation, properties and applications. Compos. Sci. Technol. **179**, 10–40 (2019)
6. Aghababai Beni, A., Jabbari, H.: Nanomaterials for environmental applications. Results Eng. **15**, 100467 (2022)
7. Sun, H.: Grand challenges in environmental nanotechnology. Front. Nanotechnol., **1** (2019)
8. Li, K., An, X., Park, K.H., Khraisheh, M., Tang, J.: A critical review of CO2 photoconversion: catalysts and reactors. Catal. Today **224**, 3–12 (2014)
9. Han, J., Zhao, D., Li, D., Wang, X., Jin, Z., Zhao, K.: Polymer-based nanomaterials and applications for vaccines and drugs. Polymers **10**, 31 (2018)
10. Tahira, M., Saima, M., Rahmat, A., Abdul, N., Afsar, K.: Technologies for removal of emerging contaminants from wastewater. In: Muharrem, I., Olcay Kaplan, I. (eds.) Wastewater Treatment, p. Ch. 9. IntechOpen, Rijeka (2022)
11. Liu, W., Speranza, G.: Functionalization of carbon nanomaterials for biomedical applications. J. Carbon Res. **5**, 72 (2019)
12. Chen, Z., Ling, L., Wang, B., Fan, H., Shangguan, J., Mi, J.: Adsorptive desulfurization with metal-organic frameworks: a density functional theory investigation. Appl. Surf. Sci. **387**, 483–490 (2016)
13. Dey, S., Mehta, N.S.: Synthesis and applications of titanium oxide catalysts for lower temperature CO oxidation. Curr. Res. Green Sustain. Chem. **3**, 100022 (2020)
14. Yoon, Y., Truong, P.L., Lee, D., Ko, S.H.: Metal-oxide nanomaterials synthesis and applications in flexible and wearable sensors. ACS Nanosci. Au. **2**, 64–92 (2022)
15. Saleem, H., Zaidi, S.J., Ismail, A.F., Goh, P.S.: Advances of nanomaterials for air pollution remediation and their impacts on the environment. Chemosphere **287**, 132083 (2022)
16. Yang, C., Yang, S., Fan, H.-L., Wang, J., Wang, H., Shangguan, J., et al.: A sustainable design of ZnO-based adsorbent for robust H_2S uptake and secondary utilization as hydrogenation catalyst. Chem. Eng. J. **382**, 122892 (2020)
17. Crouch, D.J., Sheridan, C.M., Behnsen, J.G., D'Sa, R.A., Bosworth, L.A.: Cryo-electrospinning generates highly porous fiber scaffolds which improves trabecular meshwork cell infiltration. J. Funct. Biomater. **14**, 490 (2023)
18. Hagemann, H.H.R., Afshani, J.: Chapter 321—synthesis, luminescence and persistent luminescence of europium-doped strontium aluminates. In: Bünzli, J.C.G., Pecharsky, V.K. (eds.) Handbook on the Physics and Chemistry of Rare Earths, pp. 163–225. Elsevier (2021)
19. Long, Y.Z., Yan, X., Wang, X.X., Zhang, J., Yu, M.: Chapter 2—electrospinning: the setup and procedure. In: Ding, B., Wang, X., Yu, J. (eds.) Electrospinning: Nanofabrication and Applications, pp. 21–52. William Andrew Publishing (2019)
20. Nunes, D., Pimentel, A., Santos, L., Barquinha, P., Pereira, L., Fortunato, E., et al.: 2—synthesis, design, and morphology of metal oxide nanostructures. In: Nunes, D., Pimentel, A., Santos, L., Barquinha, P., Pereira, L., Fortunato, E., et al. (eds.) Metal Oxide Nanostructures, pp. 21–57. Elsevier (2019)

21. Choudhury, S., Paul, S., Goswami, S., Deb, K.: Chapter 2—methods for nanoparticle synthesis and drug delivery. In: Das Talukdar, A., Dey Sarker, S., Patra, J.K. (eds.) Advances in Nanotechnology-Based Drug Delivery Systems, pp. 21–44. Elsevier (2022)
22. Du, P., Zhang, J., Cai, Z., Ge, F.: High adsorption of cationic dyes from aqueous solution using worm-like porous nanosilica: isotherm, kinetics and thermodynamics. Mater. Today Commun. **35**, 105697 (2023)
23. Mishra, P.K., Mishra, H., Ekielski, A., Talegaonkar, S., Vaidya, B.: Zinc oxide nanoparticles: a promising nanomaterial for biomedical applications. Drug Discov. Today **22**, 1825–1834 (2017)
24. Singh, N.B., Nagpal, G., Agrawal, S., Rachna: Water purification by using adsorbents: a review. Environ. Technol. Innov. **11**, 187–240 (2018)
25. Ahmad, N.A., Goh, P.S., Zulhairun, A.K., Wong, T.W., Ismail, A.F.: The role of functional nanomaterials for wastewater remediation. In: Ismail, A.F., Goh, P.S. (eds.) Functional Hybrid Nanomaterials for Environmental Remediation. The Royal Society of Chemistry (2021)
26. Ran, J., Jaroniec, M., Qiao, S.Z.: Cocatalysts in semiconductor-based photocatalytic CO_2 reduction: achievements, challenges, and opportunities. Adv. Mater. (Deerfield Beach, Fla) **30** (2018)
27. Karthigadevi, G., Manikandan, S., Karmegam, N., Subbaiya, R., Chozhavendhan, S., Ravindran, B., et al.: Chemico-nanotreatment methods for the removal of persistent organic pollutants and xenobiotics in water—a review. Biores. Technol. **324**, 124678 (2021)
28. Guo, R.T., Wang, J., Bi, Z.X., Chen, X., Hu, X., Pan, W.G.: Recent advances and perspectives of core-shell nanostructured materials for photocatalytic CO(2) reduction. Small **19**, e2206314 (2023)
29. Coccia, M., Bontempi, E.: New trajectories of technologies for the removal of pollutants and emerging contaminants in the environment. Environ. Res. **229**, 115938 (2023)
30. Sarkodie, B., Amesimeku, J., Frimpong, C., Howard, E.K., Feng, Q., Xu, Z.: Photocatalytic degradation of dyes by novel electrospun nanofibers: a review. Chemosphere **313**, 137654 (2023)
31. Willner, M.R., Vikesland, P.J.: Nanomaterial enabled sensors for environmental contaminants. J. Nanobiotechnol. **16**, 95 (2018)
32. Kumar, M., Agrawal, A.V., Moradi, M., Yousefi, R.: Chapter 6—nanosensors for gas sensing applications. In: Abdeltif, A., Assadi, A.A., Nguyen-Tri, P., Nguyen, T.A., Rtimi, S. (eds.) Nanomaterials for Air Remediation, pp. 107–130. Elsevier (2020)
33. Javaid, M., Haleem, A., Singh, R.P., Rab, S., Suman, R.: Exploring the potential of nanosensors: a brief overview. Sensors Int. **2**, 100130 (2021)
34. Hao, L., Liu, Y., Gao, W., Liu, Y., Han, Z., Yu, L., et al.: High hydrogen sensitivity of vertically standing layered MoS_2/Si heterojunctions. J. Alloy. Compd. **682**, 29–34 (2016)
35. Agrawal, A.V., Kumar, R., Venkatesan, S., Zakhidov, A., Zhu, Z., Bao, J., et al.: Fast detection and low power hydrogen sensor using edge-oriented vertically aligned 3-D network of MoS2 flakes at room temperature. Appl. Phys. Lett. **111** (2017)
36. Zou, J., Deng, W., Jiang, J., He, X., Li, N., Fang, J., et al.: Built-in electric field-assisted step-scheme heterojunction of carbon nitride-copper oxide for highly selective electrochemical detection of p-nonylphenol. Electrochimica Acta **354** (2020)
37. Chakraborty, U., Bhanjana, G., Kaur, N., Sharma, R., Kaur, G., Kaushik, A., et al.: Microwave-assisted assembly of Ag_2O-ZnO composite nanocones for electrochemical detection of 4-Nitrophenol and assessment of their photocatalytic activity towards degradation of 4-Nitrophenol and Methylene blue dye. J. Hazard. Mater. **416** (2021)
38. Barman, K., Changmai, B., Jasimuddin, S.: Electrochemical detection of para-nitrophenol using copper metal nanoparticles modified gold electrode. Electroanalysis **29**, 2780–2787 (2017)
39. Katoch, A., Kim, J.-H., Kim, S.S.: CuO/SnO_2 mixed nanofibers for H_2S detection. J. Nanosci. Nanotechnol. **15**, 8637–8641 (2015)
40. Dinesh, B., Saraswathi, R.: Electrochemical synthesis of nanostructured copper-curcumin complex and its electrocatalytic application towards reduction of 4-nitrophenol. Sens. Actuators B Chem. **253**, 502–512 (2017)

41. Al-Hadeethi, Y., Umar, A., Ibrahim, A.A., Al-Heniti, S.H., Kumar, R., Baskoutas, S., et al.: Synthesis, characterization and acetone gas sensing applications of Ag-doped ZnO nanoneedles. Ceram. Int. **43**, 6765–6770 (2017)
42. Olorundare, F.O.G., Nkosi, D., Arotiba, O.A.: Voltammetric determination of nitrophenols at a nickel dimethylglyoxime complex—gold nanoparticle modified glassy carbon electrode. Int. J. Electrochem. Sci. **11**, 7318–7332 (2016)
43. Cheng, Y., Li, Y., Li, D., Zhang, B., Hao, R., Sang, S.: A Sensor for Detection of 4-nitrophenol based on a glassy carbon electrode modified with a reduced graphene oxide/Fe_3O_4 nanoparticle composite. Int. J. Electrochem. Sci. **12**, 7754–7764 (2017)
44. Lin, Y., Wei, W., Li, Y., Li, F., Zhou, J., Sun, D., et al.: Preparation of Pd nanoparticle-decorated hollow SnO_2 nanofibers and their enhanced formaldehyde sensing properties. J. Alloy. Compd. **651**, 690–698 (2015)
45. Mohamed, E.F., Awad, G.: Chapter six—development of nano-sensor and biosensor as an air pollution detection technique for the foreseeable future. In: Turan, N.B., Engin, G.O., Bilgili, M.S. (eds.) Comprehensive Analytical Chemistry, pp. 163–188. Elsevier (2022)
46. Han, S., Kim, J., Ko, S.H.: Advances in air filtration technologies: structure-based and interaction-based approaches. Mater. Today Adv. **9**, 100134 (2021)
47. Petit, C., Levasseur, B., Mendoza, B., Bandosz, T.J.: Reactive adsorption of acidic gases on MOF/graphite oxide composites. Microporous Mesoporous Mater. **154**, 107–112 (2012)
48. Achmann, S., Hagen, G., Hämmerle, M., Malkowsky, I., Kiener, C., Moos, R.: Sulfur removal from low-sulfur gasoline and diesel fuel by metal-organic frameworks. Chem. Eng. Technol. **33**, 275–280 (2010)
49. Chavan, S., Bonino, F., Valenzano, L., Civalleri, B., Lamberti, C., Acerbi, N., et al.: Fundamental aspects of H_2S adsorption on CPO-27-Ni. J. Phys. Chem. C **117**, 15615–15622 (2013)
50. Wei, H., Gui, Y., Kang, J., Wang, W., Tang, C.: A DFT study on the adsorption of H_2S and SO_2 on Ni doped MoS_2 monolayer. Nanomaterials **8**, 646 (2018)
51. Ashori, E., Nazari, F., Illas, F.: Adsorption of H2S on carbonaceous materials of different dimensionality. Int. J. Hydrogen Energy **39**, 6610–6619 (2014)
52. Soni Castro, P., Martinez Zuniga, G., Holmes, W., Buchireddy, P.R., Gang, D.D., Revellame, E., et al.: Review of the adsorbents/catalysts for the removal of sulfur compounds from natural gas. Gas Sci. Eng. **115**, 205004 (2023)
53. Lou, Y., Cai, Y., Hu, W., Wang, L., Dai, Q., Zhan, W., et al.: Identification of active area as active center for CO oxidation over single Au atom catalyst. ACS Catal. **10**, 6094–6101 (2020)
54. Chen, X., Wu, C., Guo, Z.: Synthesis of efficient $Cu/CoFe_2O_4$ catalysts for low temperature CO oxidation. Catal. Lett. **149**, 399–409 (2019)
55. Khder, A.E.R.S., Altass, H.M., Orif, M.I., Ashour, S.S., Almazroai, L.S.: Preparation and characterization of highly active Pd nanoparticles supported Mn_3O_4 catalyst for low-temperature CO oxidation. Mater. Res. Bull. **113**, 215–222 (2019)
56. Mao, X., Foucher, A., Stach, E.A., Gorte, R.J.: A study of support effects for CH_4 and CO oxidation over Pd catalysts on ALD-modified Al_2O_3. Catal. Lett. **149**, 905–915 (2019)
57. Wang, C., Wen, C., Lauterbach, J., Sasmaz, E.: Superior oxygen transfer ability of Pd/MnOx-CeO_2 for enhanced low temperature CO oxidation activity. Appl. Catal. B **206**, 1–8 (2017)
58. Choi, M.S., Jeong, H., Lee, H.: Re-dispersion of Pd-based bimetallic catalysts by hydrothermal treatment for CO oxidation. RSC Adv. **11**, 3104–3109 (2021)
59. Xie, Q., Zhao, Y., Guo, H., Lu, A., Zhang, X., Wang, L., et al.: Facile preparation of well-dispersed CeO_2–ZnO composite hollow microspheres with enhanced catalytic activity for CO oxidation. ACS Appl. Mater. Interfaces **6**, 421–428 (2014)
60. Mu, G., Liu, C., Wei, Q., Huang, Y.: Three dimensionally ordered macroporous CeO_2-ZnO catalysts for enhanced CO oxidation. Mater. Lett. **181**, 161–164 (2016)
61. In Choi, Y., Lee, S., Kim, S., Kim, Y.I., Cho, D.W., Khan, M.M., et al.: Fabrication of ZnO, ZnS, Ag-ZnS, and Au-ZnS microspheres for photocatalytic activities, CO oxidation and 2-hydroxyterephthalic acid synthesis (2016)

62. Wang, W.-W., Yu, W.-Z., Du, P.-P., Xu, H., Jin, Z., Si, R., et al.: Crystal plane effect of ceria on supported copper oxide cluster catalyst for CO oxidation: importance of metal-support interaction. ACS Catal. **7**, 1313–1329 (2017)
63. Liu, M.-H., Chen, Y.-W., Lin, T.-S., Mou, C.-Y.: Defective mesocrystal ZnO-supported gold catalysts: facilitating CO oxidation via vacancy defects in ZnO. ACS Catal. **8**, 6862–6869 (2018)
64. Liu, W., Liu, X., Feng, L., Guo, J., Xie, A., Wang, S., et al.: The synthesis of CeO_2 nanospheres with different hollowness and size induced by copper doping. Nanoscale **6**, 10693–10700 (2014)
65. Curran, C.D., Lu, L., Kiely, C.J., McIntosh, S.: Ambient temperature aqueous synthesis of ultrasmall copper doped ceria nanocrystals for the water gas shift and carbon monoxide oxidation reactions. J. Mater. Chem. A. **6**, 244–255 (2018)
66. Pu, Z.-Y., Liu, X.-S., Jia, A.-P., Xie, Y.-L., Lu, J.-Q., Luo, M.-F.: Enhanced activity for CO oxidation over Pr- and Cu-doped CeO_2 catalysts: effect of oxygen vacancies. The Journal of Physical Chemistry C. **112**, 15045–15051 (2008)
67. Milenova, K.I., Avramova, I.A., Nikolov, P.M., Kasabova, N.A., Ivanov, G.M.: Doped with Cu and Mn zinc oxide catalysts for oxidation of CO in noxious gas emissions. Bulg. Chem. Commun. **49**, 193–198 (2017)
68. Sarkodie, B., Hu, Y., Bi, W., Jiang, J., Li, C.: Promotional effects of CuxO on the activity of Cu/ZnO catalyst toward efficient CO oxidation. Appl. Surf. Sci. **548**, 149241 (2021)
69. Zou, R.-Q., Sakurai, H., Han, S., Zhong, R.-Q., Xu, Q.: Probing the lewis acid sites and CO catalytic oxidation activity of the porous metal−organic polymer [Cu(5-methylisophthalate)]. J. Am. Chem. Soc. **129**, 8402–8403 (2007)
70. Zhao, W., Xu, G., He, Z., Cai, C., Abild-Pedersen, F., Wang, T.: Toward carbon monoxide methanation at mild conditions on dual-site catalysts. J. Am. Chem. Soc. **145**, 8726–8733 (2023)
71. Wickramasinghe, S., Wang, J., Morsi, B., Li, B.: Carbon dioxide conversion to nanomaterials: methods, applications, and challenges. Energy Fuels **35**, 11820–11834 (2021)
72. Meng, X., Wang, T., Liu, L., Ouyang, S., Li, P., Hu, H., et al.: Photothermal conversion of CO_2 into CH_4 with H_2 over group VIII nanocatalysts: an alternative approach for solar fuel production. Angew. Chem. Int. Ed. **53**, 11478–11482 (2014)
73. Ding, X., Liu, X., Cheng, J., Li, D., Li, T., Jiang, Z., et al.: Boosted photothermal synergistic CO_2 methanation over Ru doped Ni/ZrO_2 catalyst: from experimental to DFT studies. Fuel **357**, 129779 (2024)
74. Zhang, G., Wu, S., Li, Y., Zhang, Q.: Significant improvement in activity, durability, and light-to-fuel efficiency of Ni nanoparticles by La_2O_3 cluster modification for photothermocatalytic CO_2 reduction. Appl. Catal. B **264**, 118544 (2020)
75. Ren, J., Ouyang, S., Xu, H., Meng, X., Wang, T., Wang, D., et al.: Targeting activation of CO_2 and H_2 over Ru-loaded ultrathin layered double hydroxides to achieve efficient photothermal CO_2 methanation in flow-type system. Adv. Energy Mater. **7**, 1601657 (2017)
76. Li, P., Zhang, S., Xiao, Z., Zhang, H., Ye, F., Gu, J., et al.: Ni-TiO2 catalysts derived from metal-organic framework for efficient photo-thermal CO_2 methanation. Fuel **357**, 129817 (2024)
77. Zhao, J., Yang, Q., Shi, R., Waterhouse, G.I.N., Zhang, X., Wu, L.-Z., et al.: FeO–CeO_2 nanocomposites: an efficient and highly selective catalyst system for photothermal CO_2 reduction to CO. NPG Asia Materials. **12**, 5 (2020)
78. Xu, Y.-F., Duchesne, P.N., Wang, L., Tavasoli, A., Ali, F.M., Xia, M., et al.: High-performance light-driven heterogeneous CO_2 catalysis with near-unity selectivity on metal phosphides. Nat. Commun. **11**, 5149 (2020)
79. Asghar, A., Iqbal, N., Noor, T., Kariuki, B.M., Kidwell, L., Easun, T.L.: Efficient electrochemical synthesis of a manganese-based metal–organic framework for H_2 and CO_2 uptake. Green Chem. **23**, 1220–1227 (2021)
80. Zhu, Y., Wang, Y.-M., Zhao, S.-Y., Liu, P., Wei, C., Wu, Y.-L., et al.: Three N-H functionalized metal-organic frameworks with selective CO_2 uptake, dye capture, and catalysis. Inorg. Chem. **53**, 7692–7699 (2014)

81. Duan, S., Li, J., Liu, X., Wang, Y., Zeng, S., Shao, D., et al.: HF-Free synthesis of nanoscale metal-organic framework NMIL-100(Fe) as an efficient dye adsorbent. ACS Sustain. Chem. Eng. **4**, 3368–3378 (2016)
82. Khan, W.U., Ahmed, S., Dhoble, Y., Madhav, S.: A critical review of hazardous waste generation from textile industries and associated ecological impacts. J. Indian Chem. Soc. **100**, 100829 (2023)
83. Lu, F., Astruc, D.: Nanocatalysts and other nanomaterials for water remediation from organic pollutants. Coord. Chem. Rev. **408**, 213180 (2020)
84. Wang, L., Zhou, S., You, M., Yu, D., Zhang, C., Gao, S., et al.: Research progress on metal oxides for the selective catalytic reduction of NOx with ammonia. Catalysts **13**, 1086 (2023)
85. Wang, H., Qu, Z., Dong, S., Xie, H., Tang, C.: Superior performance of Fe1−xWxOδ for the selective catalytic reduction of NOx with NH3: interaction between Fe and W. Environ. Sci. Technol. **50**, 13511–13519 (2016)
86. Wang, H., Ning, P., Zhang, Y., Ma, Y., Wang, J., Wang, L., et al.: Highly efficient WO3-FeOx catalysts synthesized using a novel solvent-free method for NH3-SCR. J. Hazard. Mater. **388**, 121812 (2020)
87. Ni, J., Peng, X., Yang, L., Zhang, K., Zhang, Y., Zhou, Y., et al.: Effects of cerium and tungsten addition on acid–base properties of spindle-like α-Fe_2O_3 in low-temperature SCR of NOx with NH3. J. Rare Earths **40**, 1734–1742 (2022)
88. Tan, K.B., Vakili, M., Horri, B.A., Poh, P.E., Abdullah, A.Z., Salamatinia, B.: Adsorption of dyes by nanomaterials: recent developments and adsorption mechanisms. Sep. Purif. Technol. **150**, 229–242 (2015)
89. Athari, M., Fattahi, M., Khosravi-Nikou, M., Hajhariri, A.: Adsorption of different anionic and cationic dyes by hybrid nanocomposites of carbon nanotube and graphene materials over UiO-66. Sci. Rep. **12**, 20415 (2022)
90. Bagbi, Y., Sarswat, A., Mohan, D., Pandey, A., Solanki, P.R.: Lead and chromium adsorption from water using L-Cysteine functionalized magnetite (Fe_3O_4) nanoparticles. Sci. Rep. **7**, 7672 (2017)
91. Natarajan, T.S., Natarajan, K., Bajaj, H.C., Tayade, R.J.: Enhanced photocatalytic activity of bismuth-doped TiO_2 nanotubes under direct sunlight irradiation for degradation of Rhodamine B dye. J. Nanopart. Res. **15**, 1669 (2013)
92. Kasinathan, K., Kennedy, J., Elayaperumal, M., Henini, M., Malik, M.: Photodegradation of organic pollutants RhB dye using UV simulated sunlight on ceria based TiO_2 nanomaterials for antibacterial applications. Sci. Rep. **6**, 38064 (2016)
93. Nikitha, M., Elanchezhiyan, S.S., Meenakshi, S.: Photodegradation of rhodamine-B in aqueous environment using visible-active gC3N4@CS-MoS2 nanocomposite. Environ. Res. **238**, 117032 (2023)
94. Tariq, M., Muhammad, M., Khan, J., Raziq, A., Uddin, M.K., Niaz, A., et al.: Removal of Rhodamine B dye from aqueous solutions using photo-fenton processes and novel Ni-Cu@MWCNTs photocatalyst. J. Mol. Liq. **312**, 113399 (2020)
95. Revathi, B., Balakrishnan, L., Pichaimuthu, S., Nirmala Grace, A., Krishna, C.N.: Photocatalytic degradation of rhodamine B using BiMnO3 nanoparticles under UV and visible light irradiation. J. Mater. Sci. Mater. Electron. **31**, 22487–22497 (2020)
96. Wang, Y., Zhu, Y., Guan, H., Hu, Y., Zhao, Y., Zhang, W., et al.: Cu-doped In2S3 hierarchical structures for photocatalytic organic dye degradation under visible light irradiation. Mater. Lett. **349**, 134843 (2023)
97. Krishnappa, B., Saravu, S., Shivanna, J.M., Naik, M., Hegde, G.: Fast and effective removal of textile dyes from the wastewater using reusable porous nano-carbons: a study on adsorptive parameters and isotherms. Environ. Sci. Pollut. Res. **29**, 79067–79081 (2022)
98. Langbehn, R.K., Michels, C., Soares, H.M.: Antibiotics in wastewater: from its occurrence to the biological removal by environmentally conscious technologies. Environ. Pollut. **275**, 116603 (2021)
99. Yang, X., Bismark, S., Wang, D., Liu, Z., Tao, Y., Sun, Z., et al.: Zeolitic imidazolate framework-8 modified bacterial cellulose/sodium alginate composite aerogel for efficient removal of tetracycline hydrochloride. J. Ind. Text. **53**, 15280837231204374 (2023)

100. Irshad, M.A., Sattar, S., Nawaz, R., Al-Hussain, S.A., Rizwan, M., Bukhari, A., et al.: Enhancing chromium removal and recovery from industrial wastewater using sustainable and efficient nanomaterial: a review. Ecotoxicol. Environ. Saf. **263**, 115231 (2023)
101. Wang, B.D., Sarkodie, B., Yang, X., Tao,Y., Mao, Z., Zhao, L.L., et al.: Reusable UiO-66-NH2 functionalized polyacrylonitrile nanofiber membrane for effective removal of Cr(VI) from water. Fibers Polym. (2024)
102. Debnath, A., Majumder, M., Pal, M., Das, N.S., Chattopadhyay, K.K., Saha, B.: Enhanced adsorption of hexavalent chromium onto magnetic calcium ferrite nanoparticles: kinetic, isotherm, and neural network modeling. J. Dispersion Sci. Technol. **37**, 1806–1818 (2016)
103. Lu, X., Liu, Y., Kong, X., Lobie, P.E., Chen, C., Zhu, T.: Nanotoxicity: a growing need for study in the endocrine system. Small **9**, 1654–1671 (2013)
104. Sharma, P., Guleria, P., Kumar, V.: Chapter 1—nanomaterial recycling: an overview. In: Rai, M., Nguyen, T.A. (eds.) Nanomaterials Recycling, pp. p. 3–19. Elsevier (2022)

Bismark Sarkodie is a Postdoctoral Fellow in Advanced Fiber Materials Engineering Research Center of Anhui Province, School of Textile and Clothing, Anhui Polytechnic University, China. He obtained his doctorate in School of Materials Science and Engineering, East China University of Science and Technology in 2022, and masters in School of Textiles Science and Engineering, Jiangnan University in the year 2018. His scientific interests include polymer nanomaterials, metal and metal-oxide nanomaterials and their application in the field of textiles and catalysis. His current research focus is on the synthesis of photocatalytic nanofibrous composites using spray synthesis and electrospinning strategies, and their application in degradation of pollutants.

Benjamin Tawiah is a lecturer at the Kwame Nkrumah University of Science and Technology. He obtained his Ph.D. from the Hong Kong Polytechnic University where he worked as postdoctoral Fellow and also served as research associate in the City University of Hong Kong. Dr. Tawiah has been engaged in basic and applied research in many areas including filament nanocomposites, flame retardant polymer composites and second-generation batteries for energy storage in grid systems.

Bin Fei is a full Professor of Textile Chemistry in the School of Fashion and Textiles (SFT), the Hong Kong Polytechnic University. He holds a B.S. degree from the Department of Applied Chemistry, Chinese University of Science & Technology and a Ph.D. from the Changchun Institute of Applied Chemistry, Chinese Academy of Sciences in 2003. With more than two decades' experience in polymer and catalysis research, Prof. Fei has been engaged in fundamental and applied research on health care and energy devices

Chapter 6
Nanotechnology, Global Development: Bridging the Digital Divide

Sujata Deb

Introduction and Background

Over the span of years, nanotechnology has emerged globally as a field with a wide range of disciplines and promises to benefit society with next-generation functional materials for far-ranging applications [1]. Nanomaterials are cores of nanoscience and nanotechnology. Nanomaterials have, by definition, at least one spatial dimension in the size range of 1 to 100 nm, and subsequently show superior properties compared to their bulk counterparts [2, 3], such as increased strength, flexibility, durability, outstanding magnetic, electrical, and optical properties. These novel properties open up possibilities for the development of enhanced materials for use in various industries, including aerospace, construction, and electronics. These innovative materials are replacing traditional manufacturing products with enhanced energy efficiency and performance, and reduced environmental impacts, thus contributing toward progress on virtually every aspect of life and bringing worldwide breakthroughs in many fronts. Global R&D collaboration in nanotechnology has also been reported, which identifies nanoelectronics and nanomaterials as significant collaborative fields [4]. The growing digitalization of the economy demands environmentally sustainable nanomaterials. Bio-based nanoparticles are emerging as an ecological alternative to synthetic chemical reagents and toxic metals in various industries [5]. Nanostructures are also significantly participating in achieving sustainable development goals in areas such as energy storage, environmental protection, healthcare, and digital connectivity and communication, through the development of functional materials and smart devices [6, 7]. In achieving the 17 Sustainable Development Goals (SDGs) outlined in the 2030 Agenda for Sustainable Development, nanomaterials play a critical role by enabling advancement in various industries [8]. Several studies report on the positive impact of the nanotechnology index in the energy sector which

S. Deb (✉)
Department of Physics, The Assam Royal Global University, Guwahati 781035, Assam, India
e-mail: sujata.deb@rgi.edu.in

S. C. George and B. Tawiah (eds.), *Nanotechnology in Societal Development*, Advanced Technologies and Societal Change, https://doi.org/10.1007/978-981-97-6184-5_6

largely affects economic growth and sustainable development globally. However, this impact is more pronounced in developed nations compared to developing ones [9]. This inequality creates a divide with respect to progress or accessibility between these nations. Thus, despite the significant contribution of nanotechnology in various fronts, developing ones face distinct challenges in reaping its benefits on equal footing with developed nations. A few of the factors involved in this major divide between developing and developed economics are the limited access to advanced capabilities and robust scientific and technical infrastructure by the former [10]. Reducing this gap between developing and developed nations remains a critical issue while bridging the digital divide.

It is therefore very much pivotal to ensure that the global development of nanotechnology is made equally accessible to the developing and developed nations. This will support the progression of bridging the prevailing digital divide in a greater way.

The Digital Divide

The term "*Digital Divide*" originated during the late twentieth century. In the present era of the fast-growing digital phase, access to emerging technologies has become indispensable for people and nations worldwide. However, inequality exists in access to these digital resources by billions of people, leading to disparity which is referred to as the digital divide. Further, there are many ways to measure this divide, and in fact, the digital divide is multifaceted in nature. This broad term, *digital divide*, includes digital variations between individuals, geographic areas, socioeconomic status, nanotechnological progresses, social resources, etc., [11] and comprises many factors such as availability of connectivity, digital literacy, access to equipment, quality of service and relevance of access and the additional divide includes other areas such as security, interconnectivity, etc. [12]. Reducing the digital divide is crucial for a sustainable digitalized society. However, this ongoing effort of digitalization poses many challenges, which need to be overcome to bridge the digital divide [13].

Although, the nations are inflated with overall connectivity rates, yet there are still evident inequalities in the availability of affordable and quality connectivity access in rural, and even urban areas, income and gender-based as well as shown in Figs. 6.1 and 6.2. The most recent data on digital divide with different metrics collected from the source: International Telecommunication Union (ITU) are displayed in Table 6.1, with certain potential limitations [14].

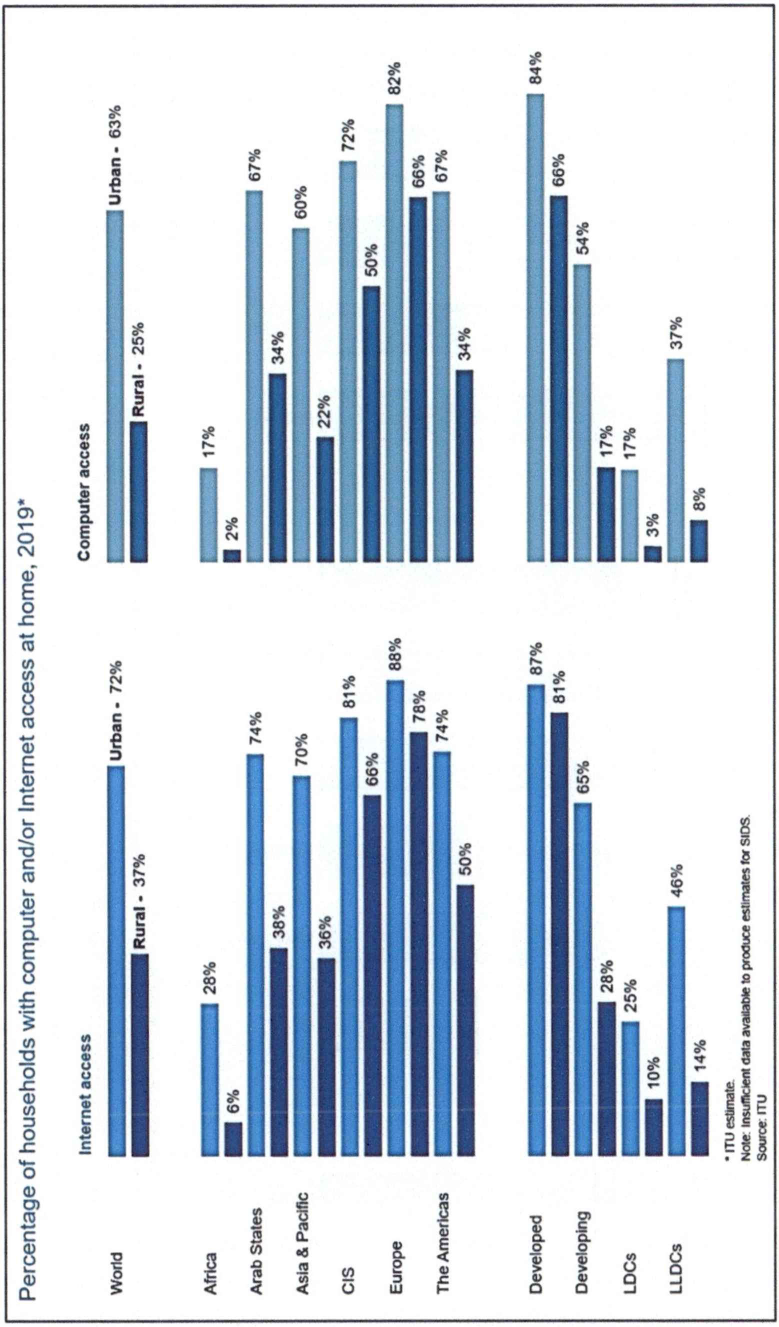

Fig. 6.1 Percentage of households with computer and/or internet access at home, 2019 [14]

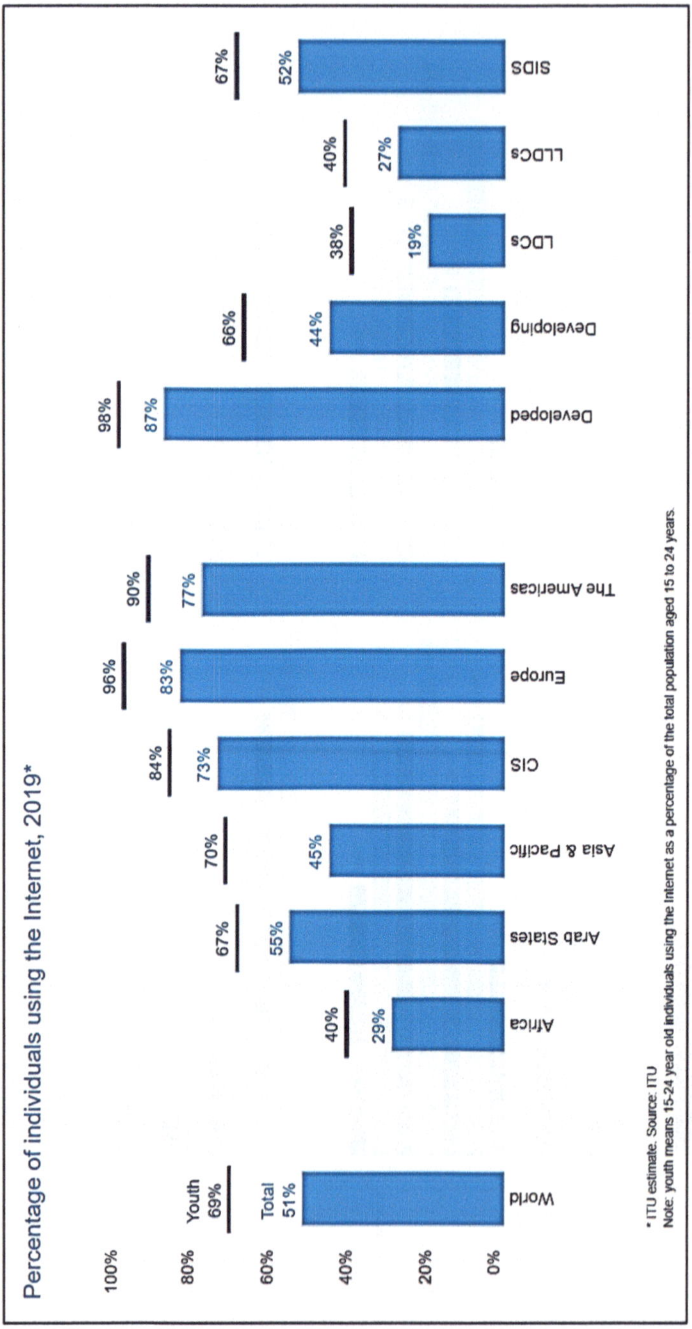

Fig. 6.2 Percentage of individuals using the internet, 2019 [14]

Table 6.1 The latest data* on infrastructure, access, and internet usage of some nations [13]

Nations	Infrastructure and access			Internet usage	
	Network coverage (%)	Mobile phone ownership (%)	Households with Internet access at home (%)	Individuals using the Internet (%)	Average monthly mobile broadband traffic per mobile broadband subscription
Algeria	98 (2022)	88 (2018)	74 (2018)	71 (2022)	6537 (2022)
Australia	100 (2022)	NA	86 (2017)	95 (2022)	13,474 (2022)
Brazil	92 (2022)	88 (2022)	80 (2022)	81 (2022)	4013 (2022)
China	100 (2022)	NA	NA	76 (2022)	14,563 (2022)
Bangladesh	100 (2022)	64 (2023)	44 (2023)	45 (2023)	6555 (2022)
South Africa	100 (2022)	78 (2019)	75 (2022)	75 (2022)	2682 (2022)
Georgia	100 (2021)	92 (2022)	88 (2022)	79 (2022)	8966 (2020)
Afghanistan	90 (2021)	47 (2022)	NA	18 (2020)	2077 (2020)
United States	100 (2021)	NA	87 (2019)	92 (2021)	NA
India	99 (2021)	NA	24 (2018)	46 (2021)	16972
United kingdom	100 (2022)	NA	95 (2020)	97 (2021)	NA

**Source* https://public.tableau.com/profile/ituint#!/vizhome/ITUFactsandFigures2020/ITU2020

Nanotechnology and Global Progress-Developing and Developed Countries

Nanotechnology has shown an impact on economic growth and sustainable progress of nations around the world. However, the nanotechnology index in the energy sector has been found to be stronger and greater in developed nations [15]. The three nations—United States, India, and China, are the top-tier in nanotechnology research and development, while Brazil occupies the fifteenth position. The prominent areas of research are energy, health science, wastewater treatment, and electronics [16]. A developing nation like Colombia has created network policies and innovation ecosystems to stimulate nanotechnology development, resulting in the formation of a National Nanotechnology Centre [17]. Malaysia, as a developing nation, is still in the initial stages of nanotechnology development compared to the European Union and the United States, but efforts are being made for strengthening its nanotechnology development policies and actively involving in global nanotechnology R&D [18, 19]. Although developing nations lag significantly behind in terms of nanotechnological advancement in developed countries, emerging research into nanotechnology holds significant promises for these nations. It offers innovative scientific and technical solutions to address the critical challenges they come across.

Nanotechnology facilitates the generation of advanced materials and systems with exclusive properties. For instance, materials that have at least one dimension on the nanometer scale offer opportunities for enhanced energy storage. The various means that can empower the deprived communities to access digital resources are the use of nanoelectronics, nanosatellites, and renewable energy options for enabling connectivity.

Nanomaterials present greatly enhanced properties compared to conventional battery and supercapacitor materials. They, in addition, ensure the occupation of entire intercalation sites available in the particle volume, which leads to high specific capacities and rapid ion diffusion [20]. These aspects make nanomaterial-based electrodes able to tolerate high currents, for creating high-energy and high-power energy storage solutions, which are not otherwise achievable with existing conventional materials. This could greatly benefit the underprivileged populations with unlimited accessibility to reliable sources of electricity, empowering them with superior energy storage options and enabling uninterrupted connectivity. Further, nanotechnology has the flair to narrow the digital divide by reducing the market price and upgrading the performance of nanoelectronic devices, making the technology more reachable to deprived populations.

Nanotechnology in Energy and Electronics Industry

The superior properties of nanomaterials contribute to more affordable and better-quality energy storage systems and production, such as the creation of efficient solar cells, durable batteries, and fuel cells, contributing to more sustainable energy solutions for developing nations [6]. Further, applications of nanotechnology in the energy industry include zinc oxide (ZnO) nanocones besides others which boost the efficiency of solar panels and nanoparticle materials for use in next-generation lithium-ion batteries.

Numerous products make use of nanoparticles, nanowalls, nanowires, nanocones, nanodots, and nanotubes. Today, nanotechnology exists in electrical devices, satellites, military hardware, cosmetics, microchips, industrial chemical catalysts, and many other products.

One of the driving factors in the area of nanoelectronics is the well-known Moore's law [21]. In the year 1965, Gordon E. Moore foretold that the number of transistors in an integrated circuit would approximately double every 18 to 24 months. Essentially, Moore's Law is a computing term; from focused observation of an emerging trend, Gordon Moore inferred that computing would vividly increase in power, and decline in relative cost, at an exponential pace. Moore's Law turned into a golden rule for the electronics industry, and a catalyst for innovation. As an Intel co-founder, Gordon paved the path for Intel to initiate our modern equipment. The evidence of the past couple of years supports the conclusion that Moore reached in 1965. The field of nanoelectronics attempts to enable the continued realization of this Law by using innovative methods and materials to form electronic devices with component sizes on the Nanoscale [22]. However, the continued size decrease of transistors is reaching the limitations of the materials' fundamentals. As per the reports by the financial information service in 2015, IBM was working on a computer chip that used 7 nm transistors; in 2017, it worked on a 5 nm chip [23]. The new chip with the "size of a fingernail" packs 30 billion 5 nm switches onto it. Adding to that, Samsung Electronics Co., Ltd. is making its most advanced chips smaller and more power-efficient than former generations [24].

The transfer of technology from academia to the industry and finally to the markets is being played by the financial sector. Sizeable investments are requisite for the development of new products and processes and also for the diffusion of new markets, especially during the seed stage. Both private and public sectors are spending constantly towards nanotechnology, but for any investor, it is vital that the science has a real-world practical use. Nanotechnology is playing an active role in the electronic industry and it is expected to have a substantial impact on the economy and society of the nations.

The U.S. government is making significant investments in nanotechnology to lead worldwide technological advancement. Due to the wide range of applications of nanomaterials, the trade is predicted to expand at a phenomenal rate. The market for nanoelectronics is principally controlled by molecular electronics and scaling up notably as a result of the development of Spintronics with Nanoelectronics, aiding toward the evolution of spin-based computing in the next generations.

According to the report by market researcher, the global nanotechnology in the electronics market was valued at $3.92 billion (US alone) in 2022 and is estimated to grow to $ 32 billion (US alone) by 2030. The market will grow by CAGR (Compound annual growth rate) of 31.98% annually during the period of analysis (2023–2030). A brief outline of the report is shown in Table 6.2. This research report on the global nanotechnology in the electronics market [25] has been divided into different segments based on types (1D—nanotubes, nanowires, 2D—graphene, and 3D—nanoflowers), applications, and region.

Nanomaterials offer exciting possibilities for enhancing electronic devices. If properly regulated and managed, the widespread use of nanomaterials in electronics manufacturing could result in reducing inequalities. There are a number of large companies as discussed in preceding paragraphs, taking long steps to utilize nanotechnology to help reduce the divide. Thus, nanotechnology has the potential to reduce the digital divide if the technology becomes widely accessible and affordable to the underprivileged population.

Affordable Satellite Through Nanotechnology

In the recent technology-based era, many developing nations cannot afford to have equal access to reliable connectivity. To address this issue, nanosatellite technology

Table 6.2 A brief outline of the market research coverage [25]

Metrics		Particulars
Base year		2022
Period of analysis		2023–2030
Market value	2022	USD 3.92 billion
	2030	Estimated to grow to USD 32 billion
	During the period of analysis (2023–2030)	CAGR of 31.98% per annum
Segments covered		By various categories, applications, geography, and regions like North America, Europe, Asia-Pacific, Latin America, the Middle East, and Africa

is being developed. Traditional tower network systems have their own limitations for remote and rural places. Nanosatellites have the potential to provide an affordable substitute to millions of devices and individuals living in developing nations. This is a mission being supported by the EU-funded Pearls Constellation project. These efforts emphasize expanding satellite-based solutions for global connection by increasing access to the digital world and bridging the digital divide. To make communication affordable, 200 constellation nanosatellites are provided by the Sky and Space Global (SAS) communication service. Each of these 200 satellites is located in a low Earth orbit. Using intersatellite link connectivity, the individual satellites have to communicate with each other. These satellites serve the function of both base station and router, creating a unique setup mesh in space. These nanosatellites are called Pearls. The Pearls Constellation aims to make communication accessible and affordable for everyone and everywhere. It operates in the S-band frequency range, enabling the nanosatellites to use a very small patch antenna or monopole with low power consumption. This effort of SAS is being provided by the EU funds [26, 27].

Reduced Price and More Accessibility

By reducing the cost of the nanomaterials or nanoproducts, it increases accessibility to nanotechnological innovations or benefits by the underprivileged populations, bridging the digital divide [28].

Enhanced Healthcare Treatments

Low-cost biosensors based on nanotechnology can detect heavy metals and pesticides present in water sources and remove water pollutants. Both developing and developed nations have crucial concerns with water contamination from antibiotics and other drugs; nanotech purifiers can filter these pharmaceuticals from water supplies and address these issues [29]. Nanotechnology enables targeted drug delivery systems, transforming medical treatments with more effective and precise treatments of diseases like cancer. It also facilitates the development of advanced medical imaging techniques for early disease detection and diagnosis, leading to more effective disease management for global health challenges. This could greatly revolutionize the medical field, most notably in developing nations [28, 30, 31].

Size Reduction of Electronic Component

Nanotechnology enables the development of smaller, faster, and more efficient electronic devices such as transistors, sensors, microprocessors, and memory chips. Many big companies are taking major steps in exploring nanoscale materials to enhance the efficiency of indispensable electronic devices and to assist reduce the digital divide [23].

Nanotechnology and the Case of South Africa

South Africa is one of the pioneering countries from the global South that have actively embraced nanotechnology with the goal of enhancing global competitiveness and foster sustainable economic growth.

Geography and Economy of South Africa-Overview

South Africa, also known as a middle-income country occupies the southernmost part of the African continent (Fig. 6.3) and is bordered by six countries: Botswana, Namibia, Zimbabwe, Mozambique, Lesotho, and Eswatini [32]. The land area of South Africa is big and has an area of 1,221,037 sq. km. The country is populated by 60.0 million people (2023), having 3 million inhabitants in its capital city, Pretoria; South Africa has two more capital cities—Cape Town and Bloemfontein. Life expectancy rate is 60 years for men and 67 years for women and its people speak several languages, with 12 official languages, which include isiZulu, isiXhosa, Afrikaans, English, Sepedi, siSwazi, Sesotho, Setswana, Xitsonga, Tshivenda, isiNdebele and sign language.

South Africa, despite being the strongest scientific and technologically developed country in Africa, also has a significant wealth inequality gap. It has a dual economy, plagued by the dynamics of imbalanced growth or progress, across different sectors.

Fig. 6.3 Map of South Africa

One category of the economy is described by the industrial, agricultural, manufacturing, mining, financial, and technologically sophisticated services sector, which drives the country's economic growth, and the other category includes poverty, inequality, and unemployment [33, 34].

Its economy was revolutionized in the late nineteenth century; however, presently the nation experiences several economic challenges: Inequality and rising poverty rates, inflation and rising fuel and food prices, low structural growth and unemployment, exacerbating socioeconomic challenges, electricity supply shortages, and limited or expensive connectivity. According to government statistics, as of 2023, 18.2 million people live in extreme poverty and the unemployment situation of the country remains challenging (32.1% in the fourth quarter of 2023). Its Gross domestic product (GDP) as reported by the World Bank is valued at 399 billion and 405.87 billion US dollars in 2023 and 2022, respectively [35]. GDP growth is 2.0% (2022) and 0.1% (2023), and based on the official data from International Monetary Fund-World Economic Outlook Database [36], it has been forecasted that the growth will be 1.8% in 2024. It has been predicted that its GDP will hit a new peak in 2028, worth 459.02 billion US dollars. Its GDP represents 0.17% (2023) of the global economy. However, a recent investigation shows that South Africa tops the list for development (R&D) associated with nanotechnology compared to other nations—Egypt and Nigeria on the continent, indicating the advancement of the nation [33]. Nanotechnology research and development in South Africa have surged over the past few decades. Most of the strategic nanotechnology research areas span material science, medicine, water, energy, and so on. Further collaborative work within disciplines or across disciplines and fusion of nanotechnology, artificial intelligence, biotechnology, cognitive sciences, and information technology is evident. Further South Africa is reporting publications on commercialization of nano-enabled products. The country is among the G20 and is the only African country to be in the list

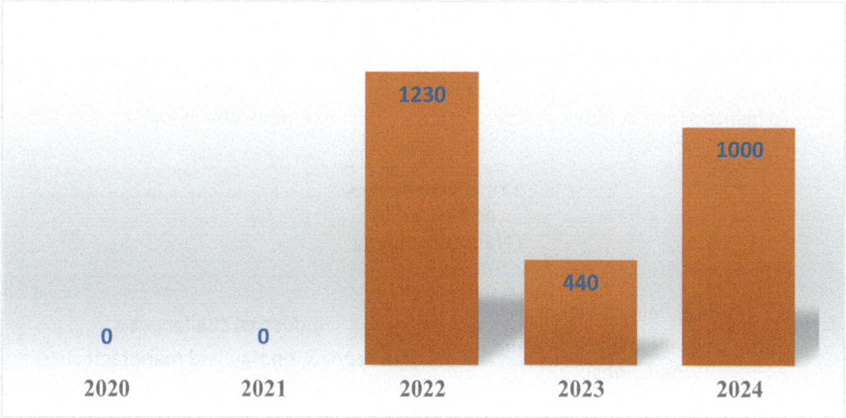

Fig. 6.4 South Africa: Commitments by Fiscal Year (in millions of dollars)**. ***Amounts include IBRD and IDA commitments* (South Africa Overview: Development news, research, data I World Bank)

of permanent members of the group. The commitments by Fiscal Year are shown in Fig. 6.4.

Critical Challenges and Nanotechnology Initiatives

Nanotechnology has the potential to contribute to sustainable development, economic growth, employment creation, and reduction of poverty in South Africa by addressing the critical challenges (Fig. 6.5), through the emerging innovative nano-based products and services, as mentioned under this section. Consequently, South Africa has proactively implemented various policies to disseminate its benefits [37].

South Africa is one of many middle-income countries according to World Bank figures [35], that has launched nanotechnology initiatives since around 2000. These initiatives include – the foundation of South African Nanotechnology Initiative (SANi) in 2003, the South Africa's Advanced Manufacturing Technology Strategy (AMTS) in 2004, launched by the Department of Science and Technology (DST) and a year later, South Africa became globally one of the first countries to publish an official nanotechnology strategy – the National Strategy on Nanotechnology (NSN), which was approved by Cabinet in 2005 and implemented in 2006. That identifies two developmental clusters as shown in Fig. 6.6.

The following organizations in South Africa are devoted to promoting and exploring nanoscience:

NanoLand South Africa
National Centre for Nano-structured Materials (NCNSM)
DST/Mintek Nanotechnology Innovation Centre (NIC).

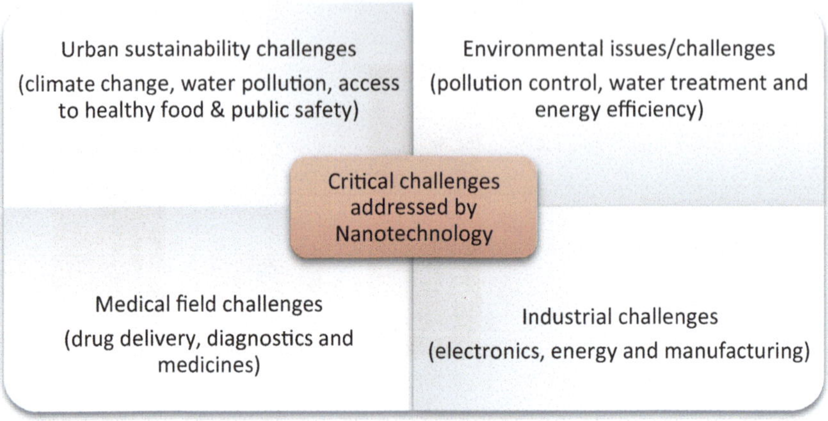

Fig. 6.5 Challenges faced by the people of South Africa

Fig. 6.6 Two developmental clusters

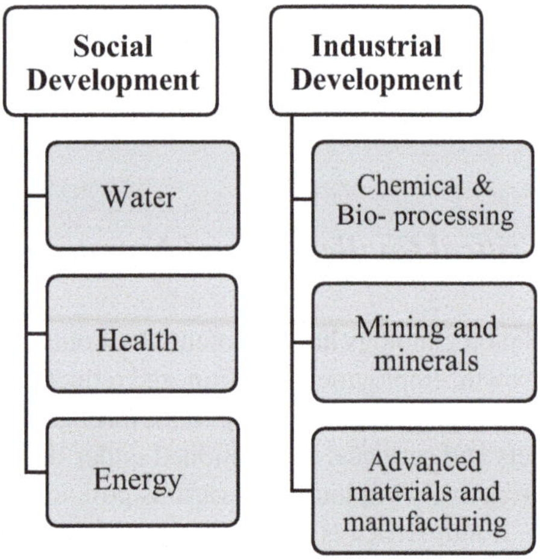

As a result of all these initiatives, the nanotechnology R&D capabilities of South Africa have amplified. Further, the launch of its first nanotechnology innovation center in 2007 boosted its technological and economic development, and significant investment and research were provided for nanotechnology in 2008. In 2011, South Africa became the first country in the African continent to own a new US$15 million electron microscopy center for nanoscience and nanotechnology research.

Over the past decades, the South African government has laid a strong foundation for research and product development in nanotechnology, establishing a policy framework and making significant investments in equipment and infrastructure,

Table 6.3 List of the key national facilities for nanotechnology in South Africa

Facility/Location	Equipment
• Centre for Nanotechnology Innovation • Rhodes University, Grahamstown	TOF–SIMS (time of flight-secondary-ion mass-spectrometer)
• Microscopy Centre for High Resolution Transmission Electron • Nelson Mandela Metropolitan University, Port Elizabeth	ARM (atomic resolution microscope)-TEM (transmission electron microscopy), STEM (scanning transmission electron microscope), FIB-SEM (focused ion beam-scanning electron microscope), AFM (atomic force microscopy), and nanoindenter
• National Centre for Nano-Structured Materials • Nelson Mandela Metropolitan University, Port Elizabeth	FIB-SEM, small- and wide-angle X-ray scattering, AFM, TEM, SEM
• National Cleanroom Facility DST-Mintek • Nanotechnology Innovation Centre (NIC), Pretoria	Clean room, TEM, SEM, AFM

research funding, and human capital development at universities and science councils throughout the country. Table 6.3 summarizes the key national facilities for nanotechnology research in South Africa [38].

The nanotechnology initiatives in South Africa serve as a vital case study for comprehending the impact of nanotechnology on economically poorer countries [39, 40]. This case study thus provides insights and shows how South Africa has good potential for growth and can expand its economy by embracing nanotechnology with innovation and research.

Potential Toxic Effects of Engineered Nanomaterials

Nanotechnology has the potential to offer numerous benefits to various sectors including industries and health care units. However, while offering numerous benefits, it also poses potential risks, such as the toxic effects of engineered nanomaterials on contamination of the environment and human health. Thus, it is of significance to cautiously consider the potential hazards associated with engineered nanomaterials and take necessary precautions to safeguard their usage in various applications, ensuring that their benefits outweigh any potential harm [41]. Therefore, bridging the digital divide through nanotechnology involves a balanced approach that considers both the opportunities and the challenges of this potential technology.

Security and Cultural Nuances in Bridging the Digital Divide Through Nanotechnology

Bridging the digital divide via nanotechnology development involves security considerations together with cultural nuances. As nanotechnology enables advanced electronic devices, ensuring their security becomes critical. Vulnerabilities in nanoscale components could lead to cyberattacks or unauthorized access. Developing robust security protocols and encryption mechanisms is essential. Nanotech-enabled devices collect and transmit data. Balancing data collection for connectivity with individual privacy rights is vital. Striking the right balance ensures that marginalized communities benefit without compromising their privacy [42].

Cultural norms, practices, and beliefs vary across different sections of the nations. Nanotechnology implementation must line up with regional values. For instance, introducing nanotech-based healthcare solutions should consider cultural perceptions of health and wellness. Bridging the digital divide necessitates education and awareness drives. These efforts should be culturally sensitive, addressing language barriers and tailoring messages to resonate with diverse populations. Ethical discussions around nanotechnology should involve diverse voices. Cultural perspectives on risk, benefit, and responsibility shape ethical frameworks. Informed consent and transparency are essential [43].

In summary, bridging the digital divide through nanotechnology necessitates a holistic approach that considers security and cultural context.

Conclusion and Prospects

Nanotechnology has emerged globally in both developed and developing nations. While the progress of developed nations is more pronounced than developing nations, efforts are being initiated by developing nations to strengthen their nanotechnology development policies and participate actively in nanotechnology R&D to stimulate their economic growth, bridging the digital divide. Reducing the digital gaps is crucial for a sustainable digitalized society. The exclusive properties of nanomaterials contribute to a more affordable and better-quality energy storage system, efficient solar cells, and durable batteries, contributing to more sustainable energy solutions for developing nations. Moreover, the deprived communities are now empowered to access digital resources. Thus, nanotechnology is playing an active role in the electronic industry and it is expected to have a substantial impact on the economy and culture of the nations. The U.S. government is making significant investments in nanotechnology for worldwide scientific and technological advancement. Owing to the wide range of applications of nanomaterials, the trade is foreseen to expand at a phenomenal rate. According to the report presented by market researcher, the global nanotechnology in the electronics market was valued at $3.92 billion (US alone in 2022) and is estimated to grow to $ 32 billion (US alone) by 2030. Many big companies are taking large steps in exploring nanoscale materials to enhance the efficiency of essential electronic devices, to help reduce the divide. Energy storage and connectivity are the issues that are also considered while addressing the digital divide. There is a possibility for an affordable Satellite with the use of nanotechnology. This technology can benefit the underprivileged population to improve their lives by providing "real access" to existing and emerging technologies. By reducing the cost of the nano products, accessibility to the nanotechnological benefits can be more by the underprivileged populations, bridging the digital divide. Enhanced water and medical treatments could greatly revolutionize the healthcare field, most notably in developing nations. The nanotechnology initiatives in South Africa serve as a vital case study for realizing the impact of nanotechnology on economically poorer nations. Thus, nanotechnology has the potential to reduce the digital divide if the technology becomes widely accessible and affordable to the deprived or underprivileged population. Bridging the digital gap is not only about accessibility but it is greatly about empowering nations, fostering their economic growth, and promoting social equity. While offering numerous benefits, nanotechnology also poses potential risks, Thus, it is of significance to cautiously consider the potential hazards associated with the engineered nanomaterials and take obligatory precautions to safeguard their usage in various applications. Further, bridging the digital divide through nanotechnology necessitates a holistic approach that considers security and cultural context.

References

1. Roco, M.C.: National nanotechnology initiative at 20 years: enabling new horizons. J. Nanopart. Res. **25**, 197 (2023). https://doi.org/10.1007/s11051-023-05829-9
2. Alivisatos, A.P.: Science **271**, 933–937 (1996)
3. Baig, N., Kammakakam, I., Falath, W.: Nanomaterials: a review of synthesis methods, properties, recent progress, and challenges. Mater. Adv. **2**, 1821–1871 (2021). https://doi.org/10.1039/D0MA00807A
4. Lo, C., Cho, H.C., Wang, P.W.: Global R&D collaboration in the development of nanotechnology: the impact of R&D collaboration patterns on patent quality. Sustainability (2020). https://doi.org/10.3390/SU12156055
5. Application of nanotechnology for eco-friendly and sustainable economic development. Advances in Chemical and Materials Engineering Book Series, pp. 1–24 (2023). https://doi.org/10.4018/978-1-6684-7358-0.ch001
6. Lin, A.: What is responsible development of nanotechnology **1220**, 111–119 (2016)
7. Pokrajac, L., et al.: ACS Nano **15**(12), 18608–18623 (2021). https://doi.org/10.1021/acsnano.1c10919
8. Yıldırım, M., Candan, Z.: Advanced renewable nanomaterials for sustainable development (2022). https://doi.org/10.3390/asec2022-13856
9. The effect of nanotechnology in the energy sector on economic growth and sustainable development. Int. J. Global Warm. (2023). https://doi.org/10.1504/ijgw.2023.128841
10. Anjorin, A.A.A.: Bridging the digital gap: addressing the 'AI divide' between developing and developed economies (2023). https://doi.org/10.31219/osf.io/rjasg
11. Pick, J., Sarkar, A.: Theories of the digital divide: critical comparison, pp. 3888–3897 (2016). https://doi.org/10.1109/HICSS.2016.484

12. van Dijk, J.A.G.M.: Digital divide research, achievements and shortcomings. Poetics **34**, 221–235 (2006). https://doi.org/10.1016/j.poetic.2006.05.004
13. Colom, A.: The digital divide. Inf. Commun. Soc. **23**(11), 1706–1708 (2020). https://doi.org/10.1080/1369118X.2020.1781916
14. https://public.tableau.com/profile/ituint#!/vizhome/ITUFactsandFigures2020/ITU2020
15. Rafiq,M.I.: J. Econ. Impact**5**(1), 69–75 (2023)
16. Santos, J.F., del Rocío Silva-Calpa, L., de Silva-Calpa, F.G., Pal, K.: Central countries' and Brazil's contributions to nanotechnology. Curr. Nanomater. 8 (2023). https://doi.org/10.2174/2405461508666230525124138
17. Mojica, D., Posada, E.: An innovative partnership for the development of nanotechnology in a developing country (2018)
18. Hasmin, N.O.R.A., Matmin, J., Basir, S.A.: National nanotechnology policies in malaysia, european union and united states: a comparative analysis. Quantum J. Soc. Sci. Humani. **3**(3), 72–92 (2022). https://doi.org/10.55197/qjssh.v3i3.148
19. Masron, T.A., Subramaniam, Y., Loganathan, N.: Does nanotechnology contribute to energy poverty reduction? Manag. Environ. Qual. Int. J. **34**(5), 1391–1405 (2023). https://doi.org/10.1108/meq-08-2022-0221
20. Pomerantseva, K., et al.: Energy storage: the future enabled by nanomaterials. Science 366, 8285 (2019). https://doi.org/10.1126/science.aan8285
21. Cao, G.: Nanostructures and Nanomaterials: Synthesis, Properties and Applications. Imperial College Press (2004)
22. Deb, S.: Ph. D. Thesis. Gauhati University (2017)
23. Stiffler, S.R., Ramachandran, R., Henson, W.K., Zamdmer, N.D., McStay, K., La Rosa, G., Parries, P.C.: Process technology for IBM 14-nm processor designs featuring silicon-on-insulator FinFETs. IBM J. Res. Dev. **62**(2/3), 11–21 (2018)
24. Nagapurkar, P., Nandy, P., Nimbalkar, S.: Cleaner chips: decarbonization in semiconductor manufacturing. Sustainability **16**(1), 218 (2023)
25. https://virtuemarketresearch.com/report/nanotechnology-in-electronics-market
26. Akhtar, M.S., Mathur, G., Kravchenko, O., Rakhra, M.: Connecting the unconnected: bridging the digital divide with affordable satellite through enabled smartphones. In: E3S Web of Conferences, vol. 453, p. 01054. EDP Sciences (2023)
27. Pearls constellation, Grant agreement ID: 867493. https://doi.org/10.3030/867493
28. Nanotechnology and the digital divide: a double-edged sword? Nano Mag.
29. Ali, Z., Ahmad, R.: Nanotechnology for water treatment. In: Dasgupta, N., Ranjan, S., Lichtfouse, E. (eds.) Environmental Nanotechnology, vol. 3. Environmental Chemistry for a Sustainable World, vol. 27. Springer, Cham (2020). https://doi.org/10.1007/978-3-030-26672-1_5
30. Recent advancements in nanotechnology: a human health perspectives, 1–17 (2022). https://doi.org/10.2174/9789815049961122020005
31. Jose, G.-M., Kimberly, H.-S.: Local development of nanotechnology-based diagnostics. Nat. Nanotechnol. (2021). https://doi.org/10.1038/S41565-021-00907-2
32. World Bank: World Bank: Open Data. World Bankruptcy, Washington, DC (2015)
33. Masara, B., van der Poll, J.A., Maaza, M.: A nanotechnology-foresight perspective of South Africa. J. Nanopart. Res. **23**, 92 (2021). https://doi.org/10.1007/s11051-021-05193-6
34. Maasdorp, G.: Globalisation, economic policy and the new dualism in the South African economy. SAJEMS NS **4**(3) (2001)
35. https://www.worldbank.org/en/country/southafrica/overview#2
36. https://www.imf.org/en/Publications/WEO/weo-database/2021/April
37. Youssef, A.M.A.R.: A nanotechnology foresight perspective for South Africa. Asian J. Manag. Sci. **11**(1), 29–40 (2022). https://doi.org/10.51983/ajms-2022.11.1.3108
38. Dube, A., Ebrahim, N.: The nanomedicine landscape of South Africa. Nanotechnol. Rev. **6**, 339–344 (2017). https://doi.org/10.1515/ntrev-2016-0108
39. Douglas, T.S.: Nanotechnology in South Africa—challenges in evaluating the impact on development Trust Saidi. S. Afr. J. Sci. **113**, 7–8 Pretoria (2017). https://doi.org/10.17159/sajs.2017/a0217

40. Sci. Public Policy **45**(5), 597–607 (2018). https://doi.org/10.1093/scipol/scx079; Advance Access Publication 8 Dec 2017
41. Salamanca-Buentello, F., Daar, A.S.: Nanotechnology, equity and global health. Nat. Nanotechnol. **16**(4), 358–361 (2021). https://doi.org/10.1038/S41565-021-00899-Z
42. Vassilakopoulou, P., Hustad, E.: Bridging digital divides: a literature review and research agenda for information systems research. Inf. Syst. Front. **25**(3), 955–969 (2023)
43. Wagner, A., de Clippele, M.-S.: Safeguarding cultural heritage in the digital era—a critical challenge. Int. J. Semiot. Law **36**, 1915–1923 (2023)

Dr. Sujata Deb is an Associate Professor and Associate Dean (Admission) in the Department of Physics at The Assam Royal Global University, Guwahati, Assam, India. She holds a Ph.D. and an M.Sc. in Physics, along with B.Ed. from Gauhati University. With over two decades of academic experience, Dr. Deb has made significant contributions to the field. Dr. Sujata Deb has several peer-reviewed research publications in international journals of repute to her credit. Dr. Deb is a life member of several professional bodies. As a journal reviewer and published author, she possesses a deep understanding of nanostructured materials and devices. Additionally, Dr. Deb holds several administrative positions at the University and actively collaborates on research projects. In addition to her research, Dr. Deb is a dedicated educator to undergraduate, graduate students and Ph.D. candidates, nurturing the next generation of researchers.

Chapter 7
Nanotechnology and Education: Preparing the Next Generation of Future Leaders

Benjamin Tawiah, Raphael Kanyire Seidu, George Kwame Fobiri, and Cynthia A. Chichi

Introduction

The field of nanotechnology has emerged as one of the most promising and rapidly advancing areas of science and technology. Its impact is visible in various fields, including health care, energy, electronics, and materials science [1, 2]. As we continue to witness the transformative power of nanotechnology, it becomes crucial to prepare the next generation of future leaders who can harness its potential and drive innovation in this field [3]. Nanotechnology education plays a vital role in equipping students with the knowledge and skills required to excel in the nanotechnology-driven world [4, 5]. It not only provides students with a deeper understanding of nanoscience and nanotechnology concepts but also fosters critical thinking, problem-solving, and innovation abilities. By integrating nanotechnology education into the curriculum, educational institutions can prepare students for the challenges and opportunities presented by this rapidly evolving field [6, 7].

Nanotechnology education offers a multitude of benefits for students and educators alike. By engaging students in hands-on activities and experiments, it enhances their understanding of fundamental scientific principles [8]. Moreover, it encourages

B. Tawiah (✉) · R. K. Seidu
School of Fashion and Textiles, The Hong Kong Polytechnic University, Hung Hom, Kowloon, Hong Kong SAR, China
e-mail: benjamin.tawiah@knust.edu.gh

B. Tawiah · C. A. Chichi
Department of Industrial Art (Fashion and Textiles), Kwame Nkrumah University of Science and Technology, Kumasi, Ghana

G. K. Fobiri
Department of Fashion Design and Textiles Studies, Kumasi Technical University, Kumasi, Ghana

S. C. George and B. Tawiah (eds.), *Nanotechnology in Societal Development*, Advanced Technologies and Societal Change, https://doi.org/10.1007/978-981-97-6184-5_7

interdisciplinary collaboration, as nanotechnology spans various scientific disciplines such as physics, chemistry, biology, and engineering [9, 10]. This interdisciplinary approach promotes creativity and fosters the development of holistic problem-solving skills. Furthermore, nanotechnology education nurtures a culture of innovation and entrepreneurship. By exposing students to cutting-edge research and real-world applications of nanotechnology, it inspires them to think critically, identify societal needs, and develop innovative solutions [11]. This entrepreneurial mindset prepares the next generation of leaders to address global challenges and drive economic growth through nanotechnology-driven innovation.

To ensure effective nanotechnology education, it is essential to adopt pedagogical strategies that promote active learning and engagement. Hands-on experiments, simulations, and project-based learning activities enable students to apply theoretical concepts and develop practical skills [7, 12]. Additionally, collaboration with industry partners, research institutions, and governmental organizations offers students invaluable opportunities to gain real-world insights and engage in cutting-edge research. Integration of nanotechnology education into existing STEM (Science, Technology, Engineering, and Mathematics) curricula is another crucial strategy [13]. By infusing nanotechnology concepts into subjects like physics, chemistry, biology, and engineering, students can develop a comprehensive understanding of the interdisciplinary nature of nanotechnology. This integration promotes a holistic scientific approach and enables students to leverage nanotechnology principles in various fields [4, 13].

While nanotechnology education offers immense potential, it also faces several challenges. Limited resources, lack of standardized curricula, and the rapid pace of technological advancements pose significant obstacles to delivering effective nanotechnology education [14, 15]. To address these challenges, collaboration between educational institutions, industry stakeholders, and policymakers is essential. This collaboration can facilitate the development of comprehensive and adaptable curricula, as well as the provision of necessary resources and infrastructure. More importantly, the future of nanotechnology education lies in empowering students to become knowledgeable, ethical, and responsible leaders in the field [15]. Emphasizing the ethical implications, safety considerations, and responsible use of nanotechnology is crucial in ensuring its sustainable and positive impact on society. Furthermore, by fostering diversity and inclusivity in nanotechnology education [16–18], we can unlock the full potential of talent from various backgrounds and perspectives.

This chapter, therefore, provides a comprehensive understanding of nanotechnology concepts, fostering innovation and entrepreneurship, and emphasizes the need for collaboration and interdisciplinary learning in empowering future leaders to shape the industry and drive global progress. This chapter, nanotechnology in education, will be very instrumental in preparing the next generation of leaders to tackle the challenges and opportunities of our rapidly advancing world.

Importance of Nanotechnology Education

The fabrication of functional structures at the nanoscale in the scientific field using chemical and physical techniques has widened the application of nanotechnology [19]. This technology has advanced and improved various aspects of human life in sports, textiles, and medicine communication, among others [20]. Such novel characteristics and properties exhibited by these materials have exposed potential possibilities from drug delivery to advanced smart electronics. The unique advantage of these materials to improve and enhance the quality of life has influenced educational researchers and school management to incorporate this novel field of nanotechnology into the curriculum of schools for students [21, 22] in various levels of education, up to the postgraduate level from the K-12 level [23, 24]. This integration and incorporation of nanotechnology into educational programmes have a lot of importance.

Promoting Scientific Literacy

As a growing science-based field with the use of advanced technological instruments, the education of students on nanotechnology is important to promote scientific or science literacy. The use of these words (scientific/science literacy) and their relationship to nanotechnology have been explained to highlight both technical skills and content knowledge [25]. With the advancement of nanotechnology to improve the quality of life, training on cutting-edge scientific concepts, interdisciplinary approaches, and real-world applications are important to help provide basic to in-depth knowledge in the field. Training of such a workforce will enhance their participation in the field of nanotechnology for continuous growth [26]. Promoting the scientific literacy of students entails the need for curriculum development towards effective training on technical know-how and skills set on relevant concepts. This education will help close the gap between the cutting-edge nanotechnology field to future workforce needs [27]. Such scientific literacy on nanotechnology will enhance student knowledge and skills in the field for informed opinions in a responsible way when issues of nanotechnology emerge [28]. This formal education for students provides the necessary knowledge and understanding of the social and environmental impacts of nanotechnology.

There have been studies to investigate how nanotechnology education fosters scientific literacy through interdisciplinary approaches. A recent study investigated using an evidence-based lesson approach for elementary high-ability students in [14] nanotechnology education in Taiwan. Students from the study showed great interest and developed effective knowledge of manometer concepts through a well-designed "age-appropriate curriculum". Subsequently, students are motivated to understand complex scientific concepts and basic science knowledge through nanotechnology [29]. Furthermore, with technological advancements, the issue of misinformation is prominent, hence, to promote an effective scientific literacy for students, critical teaching techniques are needed. These pedagogical practices such as a multistep framework for improved learning of scientific evidence by undergraduate students [30] and multi-week activities [31] were outlined in a recent study [32] to promote scientific literacy for effective decision-making.

Developing Critical Thinking Skills

The advancement of nanotechnology to fabricate new products for different applications with potential benefits and risks has raised serious concerns for critical thinking skills. Spyrtou et al. [33] have expressed that there is a potential development of the critical thinking skills of students with the inclusion of nanotechnology in their curriculum. Arguments on critical issues like interacting in an interdisciplinary space, risks and benefits of nanotechnology, and identifying original nanoscale images from "fake images", in nanotechnology education can develop the critical thinking skills of students. The interdisciplinary nature presented by nanotechnology education promotes teamwork from people of different fields to contribute and develop innovative solutions. It is a phenomenon that leads to the development of critical skills concerning the integration of knowledge and evaluation across varying fields for problem-driven solutions within the environment [34]. Alternatively, the issue of risks and benefits of nanotechnology to human beings and the environment is of great importance. This is because nanoscale materials are used for products for varying applications hence promoting the need for safety and health concerns in the environment or society [6]. The quest to develop innovative nanomaterials for societal use that have any threats fosters the development of critical skills for students. Herein, they are well encouraged and motivated to propose solutions for any problem or risk with serious effects on the quality of life. In this direction, these critical thinking skills give them multiple perspectives on challenges which leads to effective decisions with ethical norms or values.

Lastly, the advancement of technology in this modern era has influenced the production and spreading of "fake" nanoscale images. These are made possible due to the development of advanced microscopes that have promoted rapid visualization of nanomaterials (for nanoscale images) within the field. Such possibilities by modern advanced microscopes can be used to create misleading images. This phenomenon raises a serious question, "how do we know what is real when we see a picture of a nanoscale object?" [35]. Nanotechnology education helps develop the critical thinking skills of students so that they can easily identify real nanoscale images from misleading images. Students can make informed and accurate decisions based on understanding computer simulation, 3D models, nano images, and electron microscope photographs which comprise varying nanoscale representations [33]. Studies conducted by Rahmawati, Amalia and Budi [36], and Sebastian and Gimenez [37] have confirmed that nanotechnology education promotes the critical thinking skills of students by improving their understanding of chemical concepts, risk management of nanotechnology, and discovering new things based on identified problems for critical solutions. It is worth noting that the various arguments aforementioned indicate that nanotechnology education can significantly promote the development of critical thinking skills.

Innovation and Entrepreneurship

The continuous education of nanotechnology has influenced the discovery of new materials for products driving significant innovation in the field. Advanced knowledge gained from nanotechnology education provides a strong foundation for individuals or students to revolutionize various sectors of the economy be it health care, sports, energy, and electronics, among others. The discovery of these new technologies should be followed by patenting the idea and effective planning for a business [38]. Subsequently, the ability to understand and fabricate these nanomaterials is critical for breakthrough technologies that promote the establishment of small to medium-scale firms, hence fostering entrepreneurship. These discoveries allow entrepreneurs to start up new businesses [39]. The latter are normally established based on significant and novel results obtained from research findings at the university's laboratory; hence, these start-ups have a close link to the laboratory of the university [40]. Such a phenomenon promotes interdisciplinary collaboration from business with the science field where issues of funding and strategies to boost start-ups are addressed without any challenges. Education within this field further covers significant areas of understanding intellectual property laws and the dynamics of the market, which are important to commercializing innovations. This is because the potential technology impact is unknown and markets are uncertain [41].

Integrating Nanotechnology into Education

The field of nanotechnology has witnessed progressive growth where innovative solutions are presented for varying applications in the environment. The unique properties of these nanomaterials have been explored by industries and the research community to solve challenges in the medical, sports, and communication sectors, among others. This field provides practical applications "from enabling the next generation of most consumer goods to powering high-tech solutions for the most urgent global issues"' [42]. This rapid growth has influenced educational institutions to integrate nanotechnology courses into the school's curriculum: An approach that seeks to improve the critical thinking skills and understanding of scientific concepts by students for effective solutions to challenges in the environment. Integrating related courses in nanotechnology into the curriculum improves the scientific decisions of responsible generations and prepares the workforce for the future [43]. Integrating nanotechnology into education has much significance as explained in the next subheadings. A current search was conducted in the Scopus database to understand the research growth trajectory for nanotechnology in education. The relevant keywords were used with the search string TITLE-ABS-KEY ("nanotechnology" AND "education"), to obtain appropriate publications in nanotechnology education on 29 April 2024. Results show that the annual publication trend (Fig. 7.1a) of nanotechnology in education is fluctuating with higher document numbers in 2013, 2014, and 2006. These significantly high publication numbers are contributed by the top five (5) countries like the United States, India, France, the United Kingdom, and Germany (Fig. 7.1b). Such contributions show the significant commitment of these countries to integrate nanotechnology in education. These significant numbers are influenced by varying factors most especially funding support. Surprisingly, the National Science Foundation of the United States is the top contributing funding sponsor for nanotechnology in education (Fig. 7.1c). Additionally, the majority of these funding sponsors are from the United States. The United States recognizes the importance of science, technology, engineering, and mathematics (STEM) education in preparing students for future careers in high-tech fields. It is therefore not surprising that nanotechnology has been seen as a key area within STEM. As a result, efforts have been made to incorporate it into the curriculum to enhance students' understanding of fundamental scientific principles and technological advancements. Also, the U.S. has invested more than any other country in training and professional development programmes for educators to enhance their knowledge and skills in nanotechnology. This ensures that teachers have the necessary expertise to effectively integrate nanotechnology into their teaching, creating a strong foundation for student learning.

Curriculum Integration

The nature of the current Science, Technology, Engineering, and Mathematics (STEM) curricula across all levels, especially from Kindergarten through 12th grade, makes nanotechnology concept adaptation possible, although it requires strategic approaches. The curricula make room for interdisciplinary learning where students are allowed to solve problems by drawing concepts from STEM [44, 45]. The integration of technology-related topics is prevalent in the curricula. Presently, technology is adopted in various ways for teaching and learning at all levels of education [46–48]. The COVID-19 pandemic is also another catalyst that paved the way for technology adoption in the educational sector worldwide [49–53]. This subsequently will make easy adaptation of nanotechnology in education as the pandemic directed most educational settings around the globe to go digital and appreciate the use of technology in daily activities. It is observed that the nanotechnology concept is indirectly applied in diverse forms in many areas including the educational field, although not realized by users themselves [4], making its integration into existing STEM curricula a possible and commendable decision in the quest to raise well-equipped future generations.

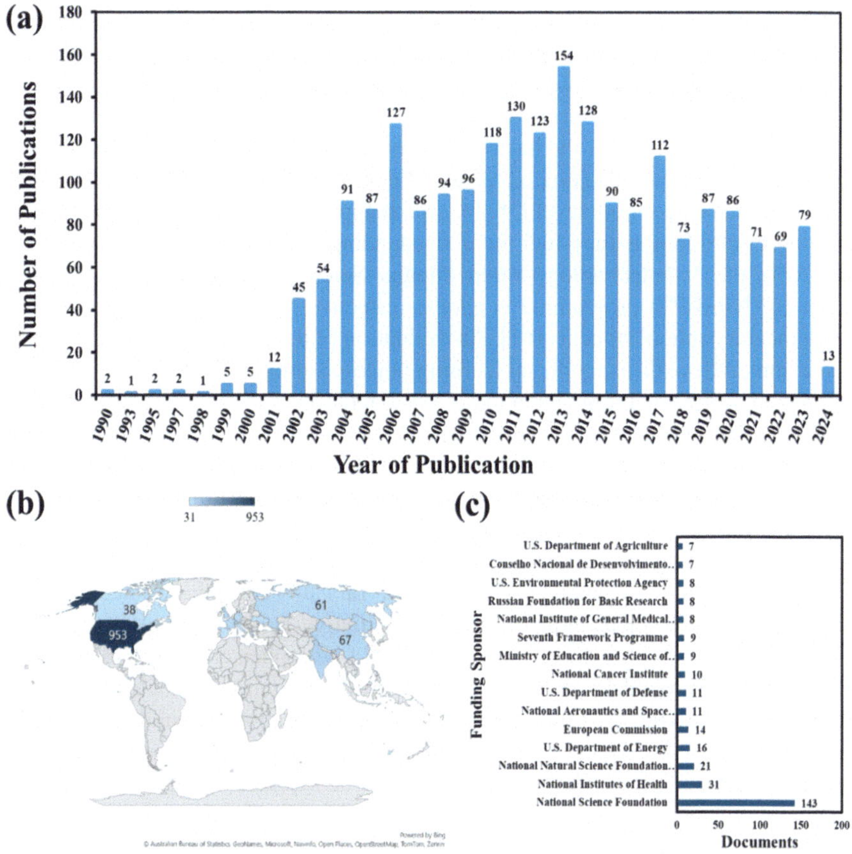

Fig. 7.1 **a** Research growth trajectory for nanotechnology in education. **b** Top contributing countries in the field. **c** Top contributing funding sponsors for nanotechnology in education

A study conducted by Spyrtou, Manou, and Peikos [20] to understand the readiness of teachers and students at the primary school level to accept the integration of nanotechnology at the basic school level revealed a very significant level of acceptance by the participants, which calls for strategic approaches to its integration in the STEM curricula.

To start with, nanotechnology could be integrated strategically into the existing curricula of basic education (K-6th grade for instance) through the design and implementation of easily adaptable teaching methods such as Teaching Sequence Learning (TSL) [7], which targets grade 6 students. The TSL approach focuses on exposing students to concepts such as *size-dependent properties* and *size and scale*. It is argued that students have a challenge of visibility and accessing nanoscale objects to easily comprehend the phenomena of nanotechnology in experimentations [54–56], hence the introduction of TSL to strategically inculcate *size and scale* concepts in students. The design of strategic teaching methods for the integration of nanotechnology may vary depending on the level of the students. Kindergarten curricula for instance could be lined with the exploration of sensory-based activities such as allowing students to touch various textures and sizes to help them be acquainted with scale which is regarded as the basics in nanotechnology [57]. Game-based educational applications are also crucial means of developing children's nanotechnology literacy at the basic school level [58–60]. Such applications could be adopted in the teaching of children as they fuel the understanding of nanotechnology fundamentals. Also, the Nanotechnology-based Popular Science Education Promotion and Teaching (NPSEPT) programme is another effective nanotechnology programme for elementary school students [61]. An evaluation of the programme as established by the authors reveals a significant learning outcome among the students.

Another worth-mentioning method is the adaptation of the 5E (Engage, Explore, Explain, Elaborate, and Evaluate) instructional model as a means of integrating nanotechnology in elementary (K-12) curricula [14]. The model is used for teaching in all subject areas and is grounded on the organization of the mind to help students apply scientific concepts for knowledge advancement [62]. It is frequently applied in science teaching and learning [63, 64]. The 5E nanotechnology model has been integrated into the science curriculum of Taiwan's education system and the initial impact assessment indicates that the initiative helped students understand nanometer concepts [14]. Similar concepts could be replicated in nanotechnology STEM curricula around the globe to help prepare the next generation of leaders. In the *engagement* of students' aspect in nanotechnology education, the students' interest could be captured in the area by highlighting how various fields are gaining popularity through nanotechnology. The subject could also be *explored* by allowing students to engage in nanoscale phenomena through experiments. To adapt the *explanation* concept in nanotechnology education, teachers could use multimedia resources such as videos and animations for a proper understanding of the concept being taught. Educators could also invite nanotechnology scientists to share their experiences with students under the *elaboration* aspect of the 5E model. Lastly, an *evaluation* could be done through self-assessment among students and by the teachers. Figure 7.2 shows the 5E model worth adopting in the teaching of nanotechnology in STEM programmes.

At the higher education level, the concept of nanotechnology could be integrated into the curricula still by adopting the 5E model presented in Fig. 7.1 in conjunction with other High School Research Experience Program (HSREP) models such as Extended Duration, Mentorship Focused, Inquiry-Driven Real-World, and Virtual SREPs [65]. This will help learners gain a deeper understanding of nanotechnology and subsequently use it in problem-solving from multidisciplinary perspectives. The design and implementation of such teaching models/methods, however, requires strong collaboration with nanotechnology scientists, the government, and policymakers for funds and effective delivery [66]. An approach to delivering the nanotechnology concept in higher education curricula is proposed where students are put in pairs to choose a topic on nanotechnology, conduct a review, present and defend orally with their peers as audience, and finally write an article following a template [67]. The

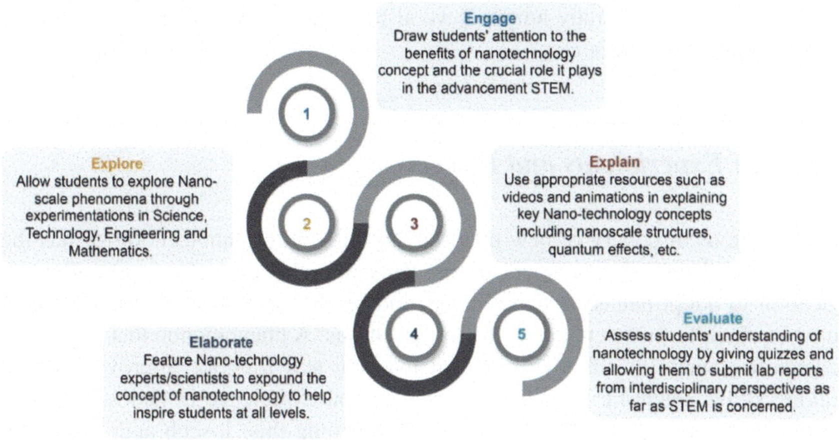

Fig. 7.2 The 5E teaching and learning model for integration of nanotechnology into education. *Source* Developed by authors

approach according to the findings of the study helped students grasp the concept of nanotechnology in addition to helping teachers evaluate the students' understanding level of nanotechnology and research competence. Again, the concept of Omni Nano Model (ONM) has been suggested as an effective approach to integrating nanotechnology into higher education to help science teachers deliver effectively [42]. The educational resources developed by Omni Nano are designed to provide students with a comprehensive understanding of the fundamental concepts, applications, and implications of nanotechnology, preparing them for further studies or careers in this rapidly evolving field. According to the information available on the Omni Nano website (https://omninano.org/) [68], the educational programmes cover the following key topics related to nanotechnology: "Introduction to Nanotechnology, specifically on the scale of the nanoscale, history and development of nanotechnology, applications of nanotechnology, nanomaterials and nanostructures, atoms, molecules, and crystal structures at the nanoscale, properties of nanomaterials (physical, chemical, optical, etc.), synthesis and characterization of nanomaterials and several aspects of nanoscale phenomena", etc.

Currently, there are professional development courses for teachers and instructors to help integrate nanotechnology into higher education curricula. Such courses help teachers in higher education gain more insight into nanotechnology through professional development courses, resulting in connecting the knowledge gained into the chemistry curriculum for a significant improvement in teaching and learning [69]. The adaptation of professional development would increase the competence level of educators/teachers to integrate nanotechnology in delivering STEM courses. It is worth mentioning that integrating nanotechnology into the existing curriculum is feasible at all levels from K-12 to higher education following appropriate strategies outlined in the present study among several teaching and learning methodologies employed in STEM education.

Hands-on Experiments and Projects

The continuous discovery of new materials in the field of nanotechnology has the potential to advance future applications. This provides critical opportunities for education to adopt hands-on experiments and projects for students to understand the core principles of science for effective solutions: A phenomenon that will allow students to engage in effective practical learning activities for improved critical thinking skills, scientific knowledge, and practical experience in nanoscale fabrication, characterization, and applications. Confirming this, Joseph and Teker [70] stipulated that there is a deeper learning process when students are involved in hands-on experimental projects hence contributing effectively to the comprehension of scientific concepts. Many perspectives from studies have been reported on the importance of having hands-on projects and experiments for nanotechnology education. For example, the nanotechnology in undergraduate education (NUE) programme was established to improve nanotechnology education for undergraduate students, by funding projects and research at the National Science Foundation (NSF) in the US. A ten-year review revealed that projects by NUE promoted hands-on activities with students benefiting and improving their knowledge by using instruments like microscopy during experiments [71]. Through this programme, relevant courses are developed and run by institutions to advance the teaching and learning of undergraduate students. A top-down nanoscale fabrication research-based course developed by Teri Odom and Vinayak Dravid at Northwestern University exposed students to understanding techniques in the experimental process (like soft lithography) and the use of instruments like the atomic force microscope in group projects [72, 73]. Students can further learn how to prepare nanocrystals with microwells using simple materials available for individual supervised projects in the bottom-up synthesis technique course [71]. Nanotechnology education through the NUE programme further motivated STEM students to learn more about nanotechnology and nanoscale and seek avenues to apply their knowledge after taking specialized courses at the University of North Dakota [74].

Subsequently, studies have reported on how to engage students in hands-on experiments with nanomaterials through laboratory-designed projects. Beginning college students were introduced to the field of nanoscience [75], where they worked with nanomaterials and scanning tunnelling microscopy. This strategy encourages an active physical engagement with scientific tools and materials. The attitude, knowledge, and experience of students are improved through these hands-on science projects. Other benefits such as communicating results, recording and interpreting data, and appropriate methods to handle tools and chemicals are experienced by students through hands-on projects [76, 77]. Subsequently, a hands-on activity called the nanopore technique developed for undergraduate students exposed them to data analysis, fabrication processes for biological nanopore artificial membranes, and how to set up instruments [78]. This hands-on learning approach improved students' understanding of single molecules and how to use their related tools for varying applications. To further advance students' understanding of fabricating nanoparticles, Bagaria, Dean [79] introduced K-12 students to nanoparticle-assembled capsules where materials are encapsulated (using safe and simple ingredients) to release food colouring dye. Students engaged in these hands-on experiments can understand non-covalent forces, charge screening and dissociation constants, and how to synthesize nanoparticle-assembled capsules. This further deepens their knowledge to conduct simple and safe self-assembly of nano-sized particles for useful functions.

In conclusion, adopting a more hands-on teaching and learning approach in nanotechnology education deepens students' knowledge and understanding of nanoscale materials and provides practical experiences to advance the field of nanotechnology.

Interdisciplinary Collaboration

Nanotechnology is increasingly encouraging and drawing different backgrounds into the field (Fig. 7.3). These interdisciplinary backgrounds prepare students to combine their field with nanotechnology to help address critical issues through a holistic approach. Interdisciplinary collaborations in the field of nanotechnology education are imperative, especially for students from different fields. For example, integrative learning and critical thinking skills were improved in non-science major students through a specialized designed nanotechnology course [80]. This course introduces students to core chemical principles of nanotechnology structured in modules. Some students' initial thoughts were that nanotechnology was used for futuristic technology; however after taking up the course, they understand that knowledge from nanotechnology could be applied to solve current challenges in society. Herein, these students easily connected real-world applications of nanotechnology with what they learned from the course. Alternatively, the development of minor subjects in nanotechnology education will expose students with interdisciplinary backgrounds to take up these courses to enhance their knowledge [28]. Students are provided with the opportunity to collaborate with other students from the fields of engineering and science to advance nanotechnology [81].

This approach will foster interdisciplinary collaboration and research from different fields to drive effective solutions to tackle complex problems in real-world situations [28]. As such, engineers and scientists forming an interdisciplinary team are required [28] since knowledge of the electrical, optical, mechanical, and magnetic properties of these nanoscale materials advances the technology [83]. Such collaboration welcomes skills from different fields to promote innovation and address complex scientific problems. The convergence of knowledge and expertise from different disciplines presents an opportunity to adopt effective solutions to various nanotechnology challenges. Aside from these, designing nanotechnology courses

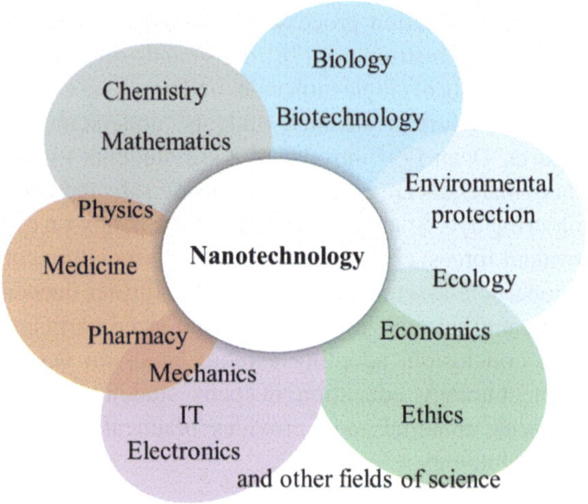

Fig. 7.3 Interdisciplinary nature of nanotechnology. Modified from Szczęsna-Antczak et al. [82], an open-access article by Fibres and textiles in Eastern Europe

with instructors from different fields to teach students of different backgrounds further promotes interdisciplinarity in nanotechnology education. For example, Winkelmann [84] reported on an interdisciplinary laboratory course designed for undergraduate freshmen with instructors from different fields at the Florida Institute of Technology. This provided multiple perspectives on the nanotechnology course from different fields for students to enhance their hands-on experience and knowledge.

Subsequently, conferences, workshops, and seminars also play a crucial role in advancing the knowledge of students from different backgrounds on nanotechnology. These platforms bring experts from different fields to present their current developments using nanotechnology for varying application areas. This approach exposes students to different methodologies which leads to interdisciplinary learning and skills development. Furthermore, these breakthroughs presented by leading experts inspire students to further advance their knowledge and critical thinking skills for innovative solutions in the field of nanotechnology. This highlights the key role seminars, workshops, and conferences play in integrating nanotechnology into education.

In conclusion, encouraging interaction and engagement of students from different fields produces the next generation of students and scientists equipped with interdisciplinary knowledge which are critical to tackling challenges in nanotechnology. This calls on various educational institutions to design programmes that promote interdisciplinarity in nanotechnology education.

Partnerships with Industry

The integration of nanotechnology into education promotes partnerships between educational institutions and industry stakeholders to provide students with access to state-of-the-art facilities, expert mentorship, and real-world industry experiences in nanotechnology. Additionally, courses in nanotechnology are designed to align with the standards and requirements of the industry. This approach produces students with the requisite knowledge and practical problem-solving skills to undertake hands-on experiments with nanoscale materials to advance the growth of the industry. Industry stakeholders can easily establish collaborative partnerships with laboratories in these universities for sponsored projects. There is however a seamless transition for these trained students from the university to the nanotechnology industry. Students are further provided with easy access to state-of-the-art facilities in these firms, equipping them with great insights for real-world applications of nano-based products on the market. Other opportunities such as mentorship programmes and internships are provided for students to learn the latest innovations in the field of nanotechnology through industry partnerships.

Educational Resources and Professional Development

The next generation of students is well equipped with a deep understanding and knowledge of nanotechnology due to the rich educational resources and professional development. Through advanced teaching strategies and facilities, instructors of these nanotechnology courses can transfer appropriate hands-on skills to students.

Nanotechnology Teaching Resources

With continuous research and innovations by scientists, several resources have been made available to support educators in teaching nanotechnology at all levels effectively. Winkelmann, Bernas, and Saleh [85] list useful sources such as books, journals, websites, professional organizations, and conference presentations by nanotechnology scientists to be at the teacher's disposal when searching for information to teach. Drawing inspiration from these resources makes teachers competent enough to deliver nanotechnology concepts in school. Nanotechnology outreach programmes also stand as effective methodologies to inspire both learners and facilitators in nanotechnology studies. For instance, Worsley and Heredia [86] posit that teachers learning outside their teaching environment gain the opportunities to learn novel things that are not accessible in school settings. Allowing students and teachers to learn outside their schools from nanotechnology experts and a well-equipped environment would go a long way to inspire both learners and teachers in the nanotechnology field. Again, with the advent of technology, modern methodologies such as the use of educative films and television series according to Baños, Blanco-Reina [87] are all useful resources to adopt in the teaching of STEM programmes effectively.

Moreover, simulation tools have come to enhance the visualization of processes in scientific explorations. Visualization tools have several benefits including enabling the user to change system settings and see the results without any negative side effects, getting rid of component flaws that have an impact on results, assisting individuals in exploring and comprehending ideas and problems at their own pace, etc. [66]. These desirable features of simulation tools make them effective resources for teaching and learning nanotechnology. Additionally, Kamali, Anderson, and Meyers [88] posit that Visual Reality (VR) is another technology that could help educators deliver nanotechnology effectively, even in institutions where expensive equipment is difficult to afford. This led to the creation of three virtual models including scanning electron microscope, plasma etching, and photolithography, which could be adopted by educational institutions in delivering nanotechnology programmes. Another useful learning tool to help students gain interest in nanotechnology is educational games such as the one developed by Fonseca, González [58] to catalyze nanotechnology studies. The video game, according to the developers, helps students, especially those at the middle and high school levels, gain insight into pursuing STEM programmes for personal and professional development. It could be said that teaching nanotechnology at various levels of education requires exceptional skills from teachers to effectively teach and inspire learners with the available resources.

Teacher Training and Professional Development

In an era of technological advancement, it is expedient for teachers to understand and apply new technologies such as nanotechnology in the educational sector for personal and professional advancement [46]. Ley, Tammets [89] noted that the integration of new technologies into education highly depends on the effective training of educators to help teachers innovate appropriate methodologies for effective teaching. As constant training and professional development courses are identified as key means of featuring nanotechnology in the existing curricula [69], employing the same measures by management is paramount to help equip teachers for a significant improvement in the education sector. Research shows the willingness of educators to accept the integration of nanotechnology in the present curricula [20], hence the need to develop and implement teacher training and professional development programmes to help teachers impart knowledge to students. Following this, Worsley and Heredia [86] explore a professional development programme in their study for teachers known as the "STEM after-school programme (ASP)" partnering with a local university to help teachers create science content that is more pertinent to students and easy to teach. Such programmes add value to the teacher's facilitation capabilities in addition to enriching the course contents.

More so, frequent organization of conferences on nanotechnology for teachers is encouraged to help equip educators with the knowledge and pedagogical skills required to teach nanotechnology concepts accurately. An example of such a conference is the NanoFlorida International Conference established to foster networking among nanoscientists, support cooperative research initiatives among institutions, establish robust industry-academia collaborations in the field of nanoscience, and highlight the accomplishments of trainees and students in the field of nanotechnology [90]. Such conferences inspire both teachers and students in the field of nanotechnology for massive innovations. Another instance is where both teachers and students develop a high interest in nanotechnology concepts after going through a training programme on nanotechnology [20]. Teachers gain in-depth knowledge from such professional development programmes to apply the same in their teaching careers. "Discover Nanotechnology" is another relevant professional development programme designed and implemented by Omni Nano to equip both teachers and students in the field of nanotechnology [42]. In this case, nanotechnology workshops are organized in numerous higher institutions. Embracing such initiatives by government, NGOs, and educational management authorities would help increase knowledge and interest in nanotechnology education among teachers and students.

Ethical, Social, and Environmental Considerations

Nanotechnology focuses on the fabrication of structures at the nanoscale level with functional properties using physical and chemical techniques [19]. The possibility of manipulating materials at a nanoscale provides the ability to create functional products with significant properties for various applications from medicine to the automotive industry [91–93]. Aside from the benefits nanotechnology brings to the different fields, its rapid development has raised serious ethical, social, and environmental concerns and considerations that have to be identified and addressed. It is important to raise serious questions about what happens to the environment or the human body when nanomaterials are disposed of into natural resources like lakes or rivers and when in contact with the human system [94]. This is important to limit unforeseen potential threats during the discovery and production of materials.

Ethical Implications

The application of nanomaterials to develop different products for use by humans raises some ethical implications. How these materials will interact with the biological system of the human body which may lead to unforeseen short to long-term health impacts is of great concern. Such challenges draw on a serious responsibility to ethically use these materials. The ethical considerations associated with nanotechnology include privacy concerns, potential health and safety risks, and responsible research and innovation. Nanotechnology is applied to produce various wearable devices and products used in other fields such as medicine, communication, and sports, among others. These products are widely used by humans for various purposes, for example, wearable devices are used to monitor the health of the wearer. Data from these devices are sensitive and widely used by healthcare providers for various treatments. This exposes the users to potential illegal surveillance monitoring, and misuse of personal information hence raising critical privacy concerns. This requires confidentiality and protection of data by the healthcare provider and the manufacturer; a breach of this could raise serious legal issues and mistrust in the use of the technology [95]. Another area of concern is the exposure of workers in production and research facilities where engineered nanoparticles are developed. This remains a first point of exposure to nanoparticles which could lead to societal exposure [96]; hence, maintaining strict safety guidelines at these facilities is of great importance. The absence of the aforementioned scientific knowledge on the negative effects of nanomaterials on human health when exposed could be of great concern. It is therefore important for periodic hazard assessment of the workplace to ensure the right safety protocols are maintained by workers to limit any risks or hazards.

Societal Impact

There are societal benefits concerning energy generation and storage, and medical treatment when nanotechnology is integrated into society. However, these benefits are met with certain societal concerns that should be addressed responsibly. For example, the use of nanotechnology in the area of health care raises some societal impact. Targeted nano-encapsulated medicine is used for drug delivery to target organs in the human body. The eventual effects of this approach on other organs in the human body raise some serious questions [97] hence raising uncertainties on the safety of the technique for drug delivery. Additionally, the immediate to long-term effects on the human body are unknown. Aside from this, nanotechnology has the potential to drive a significant economic impact and growth leading to the creation of new jobs and industries, and innovation. The manufacturing of large-scale nano-based products has pushed for millions of jobs [98]. The continuous advancement of nanotechnology has led to a significant increase in registered patents by 2020 signifying an economic contribution to research and development in the field [99]. These material discoveries have enhanced the development of technological products that have revolutionized various industries to drive growth in the economy.

Environmental Responsibility

The environmental impact of nanotechnology has also remained a significant area of concern due to their possible non-biodegradability, size, and chemistry during their distribution process in the environment [100]. The development and production stages lead to exposure of these nanomaterials to the environment [101]. This calls for responsible adherence to safety protocols to ensure that these nanomaterials are not released into the environment to cause any harm. Different products are made from nanomaterials which could be harmful to the environment during use and when disposed of. For example, some nanomaterials like quantum dots and nano silver can be potentially toxic due to the possibility of leaching into the environment and their bioaccumulative effect [102]. The potential toxicity of fullerenes to aquatic species [103], bacteria [104], flora and fauna have been the subject of many studies [105, 106]. These materials are widely used for wearable products which may be detrimental to the ecosystem, especially the aquatic. Fish, bacteria, algae, and other organisms are exposed to nanoparticles when disposed of, hence exposing the larger human population [107]. Aside from this, some commercial products like sunscreens and cosmetics containing these nanomaterials are on the market with no information on exposure data [101]. There is also the potential release of large compounds of nanomaterials when nano-engineered textiles are subjected to washing and drying [101]. Having exposure assessment information for consumers during usage will help to limit the rate of release to the environment.

Effective policies, regulations, and frameworks play a crucial role in promoting the environmental responsibility of nanotechnology. The U.S. Environmental Protection Agency, the National Science Foundation, and researchers in the field are providing critical information on the potential environmental risks regarding nanomaterials for consumers [108]. This raises serious concerns about sustainable approaches to adopt to manage and minimize the potential pollution and risks associated with nanomaterials. It is important to safeguard the environment through strict standards and policies. There have been critical initiatives developed by countries and institutions in the world to address the immediate to long-term issues of nanotechnology. One such example, Roco [109] has outlined key measures in the area of regulatory standards and coordinated measures to monitor the risks of exposure and provide the necessary information and training. Other key social responsibility approaches or frameworks are proposed to reflect on and raise awareness of the environmental and social impact of the research and production activities of nanotechnology [110]. The 2021 National Nanotechnology Initiative (NNI) strategic plan is one such key initiative to promote responsible research conduct and development of nanotechnology [111]. Such initiatives and frameworks help to foster the development of new nanomaterials and products while safeguarding the environment.

Future Perspective and Conclusion

Nanotechnology is inherently interdisciplinary, encompassing elements of physics, chemistry, biology, and engineering, among others; hence, educators will need to develop innovative teaching strategies and materials to effectively introduce these topics to students at different educational levels, from primary to tertiary. By seamlessly integrating nanoscale science and technology into existing subjects, students will gain a deeper understanding of the fundamental principles that underlie the world around them and be better equipped to tackle the challenges of the future.

It is anticipated that advancing nanotechnology education and training will require more laboratory hands-on experiences, because hands-on activities and experimentation are crucial for understanding the unique properties and behaviours of nanomaterials and nanostructures. With the continued development of nanoscale instrumentation and visualization tools, such as scanning probe microscopes and electron microscopes, students will have access to more interactive and engaging laboratory activities.

Additionally, the integration of virtual and augmented reality simulations can provide interactive learning experiences, even in the absence of specialized equipment, further enhancing the educational opportunities in this field. Alongside curriculum integration and laboratory experiences, the successful implementation of nanotechnology in education and training will require a focus on teacher training and professional development. Educators will need specialized training and ongoing support to stay informed about the latest developments in nanotechnology and to confidently incorporate these concepts into their teaching. Collaborative efforts between educational institutions, research centres, and industry partners can facilitate the development of comprehensive teacher training programmes, ensuring that educators are equipped to guide students through the complexities of this rapidly evolving field.

Furthermore, the multidisciplinary nature of nanotechnology requires educators from different fields, such as science, engineering, and technology, to work together to create a more cohesive and comprehensive understanding of this field. By fostering interdisciplinary collaboration, educational systems can better prepare students to tackle the complex problems that will arise in the years to come.

Finally, as nanotechnology finds applications in various industries, there will be a need for individuals with specialized knowledge and expertise. Educational and training programmes must adapt to this demand by providing relevant coursework, hands-on experiences, and opportunities for students to develop the necessary skills for careers in nanotechnology, ensuring that the workforce of the future is well-equipped to meet the challenges and opportunities presented by this dynamic and rapidly advancing field.

In conclusion, it must be emphasized that nanotechnology education plays a vital role in preparing the next generation of leaders to navigate the opportunities and challenges presented by this rapidly advancing field. By integrating nanotechnology into education, students can develop the necessary knowledge, skills, and mindset required to leverage nanotechnology's potential for innovation and address complex global issues. As educators, policymakers, and industry stakeholders collaborate to integrate nanotechnology into curricula, it is essential to emphasize hands-on experiences, interdisciplinary collaboration, and ethical considerations to shape future leaders who will drive the responsible and sustainable advancement of nanotechnology. By integrating nanoscale concepts into the curriculum, enhancing laboratory experiences, investing in teacher training, fostering interdisciplinary collaboration, and preparing students for the workforce, educational systems can harness the transformative potential of nanotechnology to shape the next generation of innovators and problem-solvers.

References

1. Nasrollahzadeh, M., et al.: An introduction to nanotechnology. In: Interface Science and Technology, pp. 1–27. Elsevier (2019)
2. Porter, A.L., Youtie, J.: How interdisciplinary is nanotechnology? J. Nanopart. Res. **11**, 1023–1041 (2009)
3. Roco, M.C., Williams, R.S., Alivisatos, P.: Nanotechnology Research Directions: IWGN Workshop Report: Vision for Nanotechnology in the Next Decade. Springer Science & Business Media (2000)
4. Ghattas, N.I., Carver, J.S.: Integrating nanotechnology into school education: a review of the literature. Res. Sci. Technol. Educ. **30**(3), 271–284 (2012)
5. Murcia, K.: Secondary school students' attitudes to nanotechnology: what are the implications for science curriculum development? Teach. Sci. **59**(3), 15–21 (2013)
6. Jones, M.G., et al.: Nanotechnology and nanoscale science: educational challenges. Int. J. Sci. Educ. **35**(9), 1490–1512 (2013)
7. Mandrikas, A., Michailidi, E., Stavrou, D.: Teaching nanotechnology in primary education. Res. Sci. Technol. Educ. **38**(4), 377–395 (2020)
8. Srinivas, K.: Need of nanotechnology in education. Sci. J. Educ. **2**(2), 58–64 (2014)
9. Agarwal, M., et al.: Impact of multidisciplinary nanotechnology curricula on engineering and science programs. In: 2014 ASEE Annual Conference & Exposition (2014)
10. Martirosyan, K.S., et al.: An integrated multidisciplinary nanoscience concentration certificate program for STEM education. J. Nano Educ. **5**(2), 154–163 (2013)
11. Martirosyan, K.S., Litvinov, D., Lyshevski, S.E.: NanoScience concentration program for science, engineering and technology curricula. In: 2012 12th IEEE International Conference on Nanotechnology (IEEE-NANO). IEEE (2012)
12. Blonder, R., Dinur, M.: Teaching nanotechnology using student-centered pedagogy for increasing students' continuing motivation. J. Nano Educ. **3**(1–2), 51–61 (2011)
13. Newberry, D.: Nanotechnology: exemplifing the essence of STEM education. In: 14th IEEE International Conference on Nanotechnology. IEEE (2014)
14. Yu, H.-P., Jen, E.: Integrating nanotechnology in the science curriculum for elementary high-ability students in Taiwan: evidenced-based lessons. Roeper Rev. **42**(1), 38–48 (2020)
15. Foley, E.T., Hersam, M.C.: Assessing the need for nanotechnology education reform in the United States. Nanotech. L. Bus. **3**, 467 (2006)
16. Harsh, M., et al.: The role of emerging technologies in inclusive innovation: the case of nanotechnology in South Africa. Sci. Publ. Policy **45**(5), 597–607 (2018)
17. Woodson, T.S., Hoffmann, E., Boutilier, S.: Evaluating the NSF broader impacts with the inclusion-immediacy criterion: a retrospective analysis of nanotechnology grants. Technovation **101**, 102210 (2021)
18. Foley, R., Wiek, A., Kay, B.: Nanotechnology development as if people and places matter. NanoEthics **11**, 243–257 (2017)
19. Hobson, D.W.: The commercialization of medical nanotechnology for medical applications. In: Prokop, A., Weissig, V. (eds.) Intracellular Delivery III: Market Entry Barriers of Nanomedicines, pp. 405–449. Springer International Publishing, Cham (2016)
20. Spyrtou, A., Manou, L., Peikos, G.: Educational significance of nanoscience–nanotechnology: primary school teachers' and students' voices after a training program. Educ. Sci. **11**(11), 724 (2021)

21. Mandrikas, A., Michailidi, E., Stavrou, D.: In-service teachers' needs and mentor's practices in applying a teaching-learning sequence on nanotechnology and plastics in primary education. J. Sci. Educ. Technol. **30**(5), 630–641 (2021)
22. Blonder, R., Yonai, E.: Exposing school students to nanoscience: a review of published programs. In: 21st Century Nanoscience–A Handbook: Public Policy, Education, and Global Trends, vol. **10** (2020)
23. Weiss, P.S.: A conversation with Prof. Flemming Besenbacher: innovator in nanoscience and nanoscience education. ACS Nano **2**(10), 1979–1983 (2008)
24. Greenberg, A.: Integrating nanoscience into the classroom: perspectives on nanoscience education projects. ACS Nano **3**(4), 762–769 (2009)
25. LeBlanc, S.: Nanotechnology as a tool for science and scientific literacy. In: Norris, P.M., Friedersdorf, L.E. (eds.) Women in Nanotechnology: Contributions from the Atomic Level and Up, pp. 15–27. Springer International Publishing, Cham (2020)
26. Roco, M.: Nanotechnology a frontier for engineering education. Int. J. Eng. Educ. **18**(5), 488–497 (2002)
27. Hart, D.: Closing the nanotechnology workforce gap [nano safety and education]. IEEE Nanatechnol. Mag. **6**(2), 27–28 (2012)
28. Kähkönen, A.-L., et al.: Interdisciplinary nature of nanoscience: implications for education. In: Winkelmann, K., Bhushan, B. (eds.) Global Perspectives of Nanoscience and Engineering Education, pp. 35–81. Springer International Publishing, Cham (2016)
29. Kumar, D.D., et al.: Science literacy strategies anchored in nanotechnology. Sch. Sci. Rev. **89**(329), 63 (2008)
30. Sarvary, M.A., Ruesch, J.M.: A multistep science literacy training framework in an introductory biology classroom: teaching how to find, evaluate, comprehend, and cite scientific evidence. J. Microbiol. Biol. Educ. **24**(1), e00197-e222 (2023)
31. Rholl, D.A., et al.: Making it matter: increasing student-perceived value of microbiology through reflective and critical news story analysis. J. Microbiol. Biol. Educ. **24**(1), e00163-e222 (2023)
32. Kelp, N.C., et al.: Developing science literacy in students and society: theory, research, and practice. J. Microbiol. Biol. Educ. **24**(2), e00058-e123 (2023)
33. Spyrtou, A., et al.: Facilitating Primary Student Teachers' Development of Critical Thinking Through a Nanotechnology Module. Springer International Publishing, Cham (2019)
34. Rowan, N., et al.: Critical thinking and interdisciplinary development fostering critical thinking in an interdisciplinary wellness coaching academic program. Eur. Sci. J. **11**(8) (2015)
35. Toumey, C., Cobb, M.: Nano in sight: epistemology, aesthetics, comparisons and public perceptions of images of nanoscale objects. Leonardo **45**(5), 461–465 (2012)
36. Rahmawati, Y., Amalia, R., Budi, S.: Challenging students' critical thinking skills: integrating socio-critical and problem-oriented approach in nanoscience and nanotechnology learning. Univ. J. Educ. Res. **8**(1), 98–104 (2020)
37. Sebastian, V., Gimenez, M.: Teaching nanoscience and thinking nano at the macroscale: nanocapsules of wisdom. Procedia Soc. Behav. Sci. **228**, 489–495 (2016)
38. Garcia-Martinez, J.: What can nanotech do for entrepreneurs and job creation? Jobs and the Future of work 2015. https://www.weforum.org/agenda/2015/07/what-can-nanotech-do-for-entrepreneurs-and-job-creation/. Last accessed 20 Apr 2024
39. Militaru, G., Niclescu, C.: The role of technology entrepreneurship in nanotechnology adoption: an exploratory study. U.P.B. Sci. Bull. **76**(4), 267–276 (2014)
40. Zucker, L.G., et al.: Minerva unbound: knowledge stocks, knowledge flows and new knowledge production. Res. Policy **36**(6), 850–863 (2007)
41. Yadav, S.K., Khan, Z.A., Mishra, B.: Impact of nanotechnology on socio-economic aspects: an overview. Rev. Nanosci. Nanotechnol. **2**(2), 127–142 (2013)
42. Curreli, M., Rakich, S.S.: An integrated approach to teaching and learning nanotechnology: the Omni Nano model. J. Res. Innov. Teach. Learn. **13**(1), 141–145 (2020)
43. Blonder, R., Sakhnini, S.: The making of nanotechnology: exposing high-school students to behind-the-scenes of nanotechnology by inviting them to a nanotechnology conference. Nanotechnol. Rev. **4**(1), 103–116 (2015)

44. Dominguez, A., et al.: Integration of physics and mathematics in STEM education: use of modeling. Educ. Sci. **14**(1), 20 (2024)
45. De Loof, H., et al.: Toward a framework of integrating ability: conceptualization and design of an integrated physics and mathematics test. Educ. Sci. **13**(3), 249 (2023)
46. Al-Mughairi, H., Bhaskar, P.: Exploring the factors affecting the adoption AI techniques in higher education: insights from teachers' perspectives on ChatGPT. J. Res. Innov. Teach. Learn. (2024) (ahead-of-print)
47. Ingram, E., et al.: Improving elementary pre-service teachers' science teaching self-efficacy through garden-based technology integration. Educ. Sci. **14**(1), 65 (2024)
48. Ravichandran, B., Shanmugam, K.: Adoption of EdTech products among college students: a conceptual study. Manag. Matters (2023) (ahead-of-print)
49. Crick, T.: Covid-19 and digital education: a catalyst for change? ITNOW **63**(1), 16–17 (2021)
50. Dash, G., et al.: COVID-19 and e-learning adoption in higher education: a multi-group analysis and recommendation. Sustainability **14**(14), 8799 (2022)
51. Ibrahem, U.M., et al.: The drivers of change for future learning: how teachers were taught in the COVID-19 crisis and what will come next? Sustainability **15**(20), 14766 (2023)
52. Sato, S.N., et al.: Navigating the new normal: adapting online and distance learning in the post-pandemic era. Educ. Sci. **14**(1), 19 (2024)
53. Zhao, Y.: COVID-19 as a catalyst for educational change. Prospects **49**(1), 29–33 (2020)
54. Peikos, G., et al.: Nanoscale science and technology education: primary school students' preconceptions of the lotus effect and the concept of size. Res. Sci. Technol. Educ. **41**(1), 89–106 (2023)
55. Swarat, S., et al.: A typology of undergraduate students' conceptions of size and scale: Identifying and characterizing conceptual variation. J. Res. Sci. Teach. **48**(5), 512–533 (2011)
56. Xie, C., Pallant, A.: The molecular workbench software: an innovative dynamic modeling tool for nanoscience education. In: Khine, M.S., Saleh, I.M. (eds.) Models and Modeling: Cognitive Tools for Scientific Enquiry, pp. 121–139. Springer, Netherlands, Dordrecht (2011)
57. Novak, M., Schwan, S.: Does touching real objects affect learning? Educ. Psychol. Rev. **33**(2), 637–665 (2021)
58. Fonseca, S., et al.: Educational nanotechnology video game to inspire middle and high school students to pursue STEM related professional careers. In: IEEE Frontiers in Education Conference (FIE) (2018)
59. Dorouka, P., Papadakis, S., Kalogiannakis, M.: The contribution of the health crisis to young children's nano-literacy through STEAM education. Hellenic J. STEM Educ. **2**(1), 1–7 (2021)
60. Ali, K., Burgos, D., Affouneh, S.: Educational loss at times of crisis: the role of games in students' learning in Palestine and Iraq. Sustainability **15**(6), 4983 (2023)
61. Lin, S.-Y., et al.: The effectiveness of a popular science promotion program on nanotechnology for elementary school students in I-Lan City. Res. Sci. Technol. Educ. **33**(1), 22–37 (2015)
62. Tegegne, T.A., Kelkay, A.D.: Comparative study of using 5E learning cycle and the traditional teaching method in chemistry to improve student understanding of water concept: the case of primary school. Cogent Educ. **10**(1), 2199634 (2023)
63. Ruiz-Martín, H., Bybee, R.W.: The cognitive principles of learning underlying the 5E Model of Instruction. Int. J. STEM Educ. **9**(1), 21 (2022)
64. Ha, V.T., et al.: Teaching science using argumentation-supported 5E-STEM, 5E-STEM, and conventional didactic methods: differences in the learning outcomes of middle school students. Education Sciences **13**(3), 247 (2023)
65. Ahmad, Z., Ammar, M., Al-Thani, N.J.: Pedagogical models to implement effective STEM research experience programs in high school students. Educ. Sci. **11**(11), 743 (2021)
66. Khan, A.S.: Application of simulation for emulating and visualizing nanotechnology. In: Vucinic, D., Rodrigues Leta, F., Janardhanan, S. (eds.) Advances in Visualization and Optimization Techniques for Multidisciplinary Research: Trends in Modelling and Simulations for Engineering Applications, pp. 307–316. Springer, Singapore (2020)
67. Bauer, J.: Teaching nanotechnology through research proposals. J. Chem. Educ. **98**(7), 2347–2355 (2021)

68. Curreli, M.: Omni Nano Curriculumn (2024). https://omninano.org/nanotechnology-stem-curriculum/. Last accessed 29 Apr 2024
69. Feldman-Maggor, Y., Tuvi-Arad, I., Blonder, R.: Development and evaluation of an online course on nanotechnology for the professional development of chemistry teachers. Int. J. Sci. Educ. **44**(16), 2465–2484 (2022)
70. Joseph, A.O., Teker, K.: The impact of hands-on experience in undergraduate nanotechnology education. MRS Online Proc. Libr. **1320**(1), 404 (2011)
71. Winkelmann, K.: A ten year review of the national science foundation nanotechnology in undergraduate education (NUE) program. J. Nano Educ. **6**(2), 109–116 (2014)
72. Meenakshi, V., Babayan, Y., Odom, T.W.: Benchtop nanoscale patterning using soft lithography. J. Chem. Educ. **84**(11), 1795 (2007)
73. Babayan, Y., Meenakshi, V., Odom, T.W.: Benchtop nanoscale patterning experiments. In: Nanotechnology in Undergraduate Education, pp. 177–188. American Chemical Society (2009)
74. Kaabouch, N., et al.: A multidisciplinary undergraduate nanoscience and nanotechnology program at the University of North Dakota (2018). arXiv preprint arXiv:1801.08205
75. Furlan, P.Y.: Engaging students in early exploration of nanoscience topics using hands-on activities and scanning tunneling microscopy. J. Chem. Educ. **86**(6), 705 (2009)
76. Faulconer, E., Kam, C.J.Y.: Service-learning in undergraduate general chemistry: a review. J. Exp. Educ. **46**(1), 32–51 (2023)
77. Laszcz, M., Dalvi, T.: Studying the affordances of a technology-based nanoscience module to promote student engagement in learning novel nanoscience and nanotechnology concepts at the middle school level. Res. Sci. Technol. Educ. **41**(2), 700–716 (2023)
78. Hu, Z.-L., et al.: A course of hands-on nanopore experiments for undergraduates: single-molecule detection with portable electrochemical instruments. J. Chem. Educ. **97**(12), 4345–4354 (2020)
79. Bagaria, H.G., et al.: Self-assembly and nanotechnology: real-time, hands-on, and safe experiments for K-12 students. J. Chem. Educ. **88**(5), 609–614 (2011)
80. Park, E.J.: Nanotechnology course designed for non-science majors to promote critical thinking and integrative learning skills. J. Chem. Educ. **96**(6), 1278–1282 (2019)
81. Schank, P., Krajcik, J., Yunker, M.: Can nanoscience be a catalyst for educational reform. In: Nanoethics: The Ethical and Social Implications of Nanotechnology, vol. 277, p. 290 (2007)
82. Szczęsna-Antczak, M., Kazimierczak, J., Antczak, T.: Nanotechnology-methods of manufacturing cellulose nanofibres. Fibres Text. East. Eur. **2**(91), 8–12 (2012)
83. Feather, J.L., Aznar, M.F.: Nanoscience Education, Workforce Training, and K-12 Resources. CRC Press (2018)
84. Winkelmann, K.: Practical aspects of creating an interdisciplinary nanotechnology laboratory course for freshmen. J. Nano Educ. **1**(1), 34–41 (2009)
85. Winkelmann, K., Bernas, L., Saleh, M.: A review of nanotechnology learning resources for K-12, college and informal educators. J. Nano Educ. **6**(1), 1–11 (2014)
86. Worsley, T.E., Heredia, S.: Exploring STEM teacher learning in out-of-school settings: sites for learning about relevance. Educ. Sci. **13**(3), 305 (2023)
87. Baños, J.E., et al.: Beyond lectures and practical courses: teaching pharmacology using imaginative pedagogical tools. Pharmacol. Res. **202**, 107130 (2024)
88. Kamali, R., Anderson, J.D., Meyers, M.: Comparison of virtual reality versus reality: effects on student learning using virtual technology on nanotechnology education. In: 2019 ASEE Annual Conference & Exposition (2019)
89. Ley, T., et al.: Adopting technology in schools: modelling, measuring and supporting knowledge appropriation. Eur. J. Teach. Educ. **45**(4), 548–571 (2022)
90. Mohapatra, S.S., et al.: Advances in translational nanotechnology: challenges and opportunities. Appl. Sci. **10**(14), 4881 (2020)
91. Anselmo, A.C., Mitragotri, S.: Nanoparticles in the clinic: an update. Bioeng. Transl. Med. **4**(3), e10143 (2019)

92. Rickerby, D.G., Morrison, M.: Nanotechnology and the environment: a European perspective. Sci. Technol. Adv. Mater. **8**(1–2), 19 (2007)
93. Malik, S., Muhammad, K., Waheed, Y.: Nanotechnology: a revolution in modern industry. Molecules **28**(2), 661 (2023)
94. Chen, R., Chen, C.: Environment, health and safety issues in nanotechnology. In: Bhushan, B. (ed.) Springer Handbook of Nanotechnology, pp. 1559–1586. Springer, Heidelberg, Berlin (2017)
95. Burak, A.: The promise and challenges of using nanobots in medicine. Tech Label 2023. https://relevant.software/blog/nanobots-in-medicine/. Last accessed 27 Feb 2024
96. Schulte, P.A., Salamanca-Buentello, F.: Ethical and scientific issues of nanotechnology in the workplace. Cien. Saude Colet. **12**, 1319–1332 (2007)
97. Bruce, D., Ethical implications of nanomedicine: implications of imagining new futures for medicine. In: Ge, Y., et al. (eds.) Nanomedicine: Principles and Perspectives, pp. 251–269. Springer, New York (2014)
98. Seear, K., Petersen, A., Bowman, D.: The social and economic impacts of nanotechnologies: a literature review. Final Report Prepared for the Department of Innovation, Industry, Science and Research (2009)
99. Bandyopadhyay, S., Dorbala, H.C., Mandal, S.: An overview of the impact of nanotechnology on economy and business. In: Sarkar, B., Sonawane, A. (eds.) Biological Applications of Nanoparticles, pp. 201–216. Springer Nature Singapore, Singapore (2023)
100. Stern, S.T., McNeil, S.E.: Nanotechnology safety concerns revisited. Toxicol. Sci. **101**(1), 4–21 (2007)
101. Thomas, T., et al.: Research strategies for safety evaluation of nanomaterials, part VII: evaluating consumer exposure to nanoscale materials. Toxicol. Sci. **91**(1), 14–19 (2006)
102. Benn, T.M., Westerhoff, P.: Nanoparticle silver released into water from commercially available sock fabrics. Environ. Sci. Technol. **42**(11), 4133–4139 (2008)
103. Oberdörster, E.: Manufactured nanomaterials (fullerenes, C_{60}) induce oxidative stress in the brain of Juvenile largemouth bass. Environ. Health Perspect. **112**(10), 1058–1062 (2004)
104. Lyon, D.Y., et al.: Bacterial cell association and antimicrobial activity of a C60 water suspension. Environ. Toxicol. Chem. **24**(11), 2757–2762 (2005)
105. Yoon, K.-Y., et al.: Susceptibility constants of Escherichia coli and Bacillus subtilis to silver and copper nanoparticles. Sci. Total. Environ. **373**(2), 572–575 (2007)
106. Morones, J.R., et al.: The bactericidal effect of silver nanoparticles. Nanotechnology **16**(10), 2346 (2005)
107. Glisovic, S., et al.: Emerging technologies and safety concerns: a condensed review of environmental life cycle risks in the nano-world. Int. J. Environ. Sci. Technol. **14**(10), 2301–2320 (2017)
108. Dreher, K.L.: Health and environmental impact of nanotechnology: toxicological assessment of manufactured nanoparticles. Toxicol. Sci. **77**(1), 3–5 (2004)
109. Roco, M.C.: Environmentally responsible development of nanotechnology. Environ. Sci. Technol. **39**(5), 106A-112A (2005)
110. Caballero-Díaz, E., Simonet, B.M., Valcárcel, M.: The social responsibility of nanoscience and nanotechnology: an integral approach. J. Nanopart. Res. **15**(4), 1534 (2013)
111. National Nanotechnology Initiative: National Nanotechnology Initiative Strategic Plan (2021). https://www.nano.gov/2021strategicplan. Last access 20 Apr 2024

Dr. Benjamin Tawiah received his B. A. Industrial Art (Textile Option) and M. Eng. in Textile Engineering from the Kwame Nkrumah University of Science and Technology, Ghana, and the Jiangnan University, China, respectively. Prior to his Master of Engineering degree, he earned a postgraduate Diploma in Management Information Systems from the Ghana Institute of Management and Public Administration (GIMPA) and obtained his Ph. D. at the Hong Kong Polytechnic

University where he also worked as a postdoctoral Fellow. He also worked as a research associate at the City University of Hong Kong. He is a lecturer at the Kwame Nkrumah University of Science and Technology. His research focuses on flame retardant polymer composites, advanced energy electrochemical energy storage systems, and flexible fibre-based current collectors for zinc ion batteries.

Raphael Kanyire Seidu is a Ph. D. candidate in Fashion and Textile Design at the School of Fashion and Textiles at The Hong Kong Polytechnic University, Hong Kong. Before this, he gained both his M. F. A. degree in Textile Design and B. A. degree in Industrial Art (Textile option) from Kwame Nkrumah University of Science and Technology, Ghana. He is also a Visiting Research Student at Nottingham Trent University (NTU) in the UK. He has several publications in leading journals in textile, design, technology, and material science. His recent paper was adjudged the "Second-Best Paper" presented at the 12th Sustainable Education and Development Research Conference in Somanya, Ghana. His research interest focuses on retro-reflective materials, smart textiles, cloth culture, weaving, and print technologies. He has several papers and has filed for a patent from his Ph. D. project which aims at developing smart interactive retro-reflective clothing for pedestrians' safety at night-time.

George Kwame Fobiri is an Assistant Lecturer at the Department of Fashion Design and Textiles Studies, Kumasi Technical University. He holds a bachelor's degree (B.A. Industrial Art-Textiles) and a master's degree (M. F. A. Textile Design), all from the Department of Industrial Art, Kwame Nkrumah University of Science and Technology. His research interests cover textile printing and dyeing, fashion, studio-based textile art, CAD, product design, etc.

Dr. Cynthia Akua Chichi is a lecturer in the Department of Industrial Art, KNUST, and has served on various committees at different levels of the University community. She has a Ph.D. in Fashion Design Technology but before that, she obtained a bachelor's degree in Industrial Art (Textiles) in 2006 and went on to pursue an M. A. in Design Innovation (Textiles) in 2008, having served as a National Service Person in the same Department in 2007. She worked as a Research Assistant from 2010 to 2011 at Coventry University, UK. Her area of specialization is in Pattern and Garment Technology, Sustainable Fashion, and Printed Textile Design Applications.

Chapter 8
The Intersection of Nanotechnology and Biotechnology Implications for Human Health

Akash A. Podutwar, Purva U. Chandorkar, Anuruddha R. Chabukswar, Satish A. Polshettiwar, and Swati C. Jagdale

Introduction

In the vast landscape of scientific exploration, the convergence of nanotechnology and biotechnology has emerged as a ground-breaking frontier with profound implications for human health. Nanotechnology, the manipulation of materials at the nanoscale, and biotechnology, the application of biological systems and organisms for technological advancements, are two dynamic fields that, when intertwined, offer unprecedented opportunities to revolutionize healthcare, diagnostics, and therapeutic interventions. This intersection has spurred a wave of research and innovation, bringing forth a new era in medicine where the amalgamation of nanoscale precision and biological specificity holds the promise of addressing some of the most challenging health issues of our time [1].

The nanoscale, typically defined as dimensions ranging from 1 to 100 nanometers, provides unique properties and behaviors distinct from those observed in macroscopic materials. At this scale, materials exhibit quantum effects, increased surface area-to-volume ratios, and enhanced reactivity, opening avenues for novel applications in medicine. Concurrently, biotechnology harnesses the power of living organisms and biological processes, enabling the manipulation of cellular and molecular functions with remarkable precision. The synergy between these fields is increasingly evident in the development of innovative tools, diagnostics, and therapeutic modalities, shaping the landscape of modern health care [2]. The utilization of nanotechnology in medicine has witnessed remarkable progress, particularly in the areas of diagnostics, drug delivery, and imaging. Nanoscale materials, such as

A. A. Podutwar · P. U. Chandorkar · A. R. Chabukswar · S. A. Polshettiwar · S. C. Jagdale (✉)
School of Health Sciences and Technology, Department of Pharmaceutical Sciences, Dr. Vishwanath Karad MIT World Peace University, MIT Campus, Kothrud, Pune, India
e-mail: swati.jagdale@mitwpu.edu.in

S. C. George and B. Tawiah (eds.), *Nanotechnology in Societal Development*, Advanced Technologies and Societal Change, https://doi.org/10.1007/978-981-97-6184-5_8

nanoparticles and nanocomposites, are engineered to interact with biological systems at the cellular and molecular levels. This capability is exemplified by advancements in targeted drug delivery systems, where nanocarriers can transport therapeutic agents directly to the site of action, minimizing systemic side effects. For instance, liposomal formulations like Doxil have demonstrated enhanced drug delivery and reduced toxicity in the treatment of certain cancers [3].

Furthermore, nanotechnology plays an important role in the advancement of imaging technologies that provide unprecedented resolution for disease detection. Quantum dots, for example, exhibit unique optical properties that enhance imaging contrast and enable real-time tracking of cellular processes [4]. Additionally, multifunctional nanoparticles can serve as contrast agents in various imaging modalities, such as Positron Emission Tomography (PET), Computed Tomography (CT), and Magnetic Resonance Imaging (MRI), allowing for more accurate and comprehensive diagnostics [5].

Biotechnology, on the other hand, is rooted in the manipulation of biological systems to develop products and technologies that benefit society. From the early days of genetic engineering to the current era of synthetic biology, biotechnology has evolved into a diverse and dynamic field. By harnessing the capabilities of living organisms, scientists can engineer proteins, modify genes, and manipulate cellular processes to achieve a myriad of objectives [6].

Biotechnology has already contributed significantly to health care, offering solutions ranging from recombinant insulin production to gene therapies. The ability to manipulate the fundamental building blocks of life has paved the way for innovative treatments and diagnostic tools. However, the complexity of biological systems presents challenges that nanotechnology is uniquely poised to address [6, 7].

History of Biotechnology

Karl Ereky, a Hungarian engineer, coined the term "biotechnology" in 1919 to describe the use of living organisms for the creation of products from raw materials. Biotechnology encompasses various disciplines that involve manipulating living cells or their molecular components, employing diverse tools and technologies for applications focused on human well-being. It embodies the fusion of biological science and engineering, wherein living organisms, cells, or their constituents are essential for the synthesis of products and the delivery of services. Key subdivisions within biotechnology encompass industrial biotechnology, medical biotechnology, agricultural biotechnology, environmental biotechnology, marine biotechnology, and food biotechnology [8].

Traditional biotechnology involves the application of living organisms for specific purposes, exemplified by processes like bread and cheese making. In contrast, modern biotechnology encompasses techniques utilizing cellular molecules such as biologics, monoclonal antibodies, and DNA. Before delving into the technical strides of recombinant DNA technology and DNA, it is imperative to establish a fundamental comprehension of DNA and its functional roles. DNA recombination has been instrumental in sequencing the human genome, serving as a cornerstone for emerging fields such as bioinformatics, nanomedicine, and individualized therapy. Genetic engineering has brought about a revolutionary transformation in medicine, facilitating the large-scale production of biochemically pure and highly effective versions of substances naturally produced by the human body [9].

The field of biotechnology showcases a multitude of applications and holds a crucial position in producing tissue-engineered products, recombinant pharmaceuticals, regenerative medicines including gene therapy, stem cells, and also various other biotechnological products designed to improve human health and well-being. Biotechnological methodologies are instrumental in the large-scale production of purified bio-therapeutic agents. Within the crucial domain of pharmaceutical biotechnology, animals or hybrid cells, including tumor cells, leukocytes, or prokaryotic and mammalian cells, are employed to develop safer, effective, and cost-efficient iterations of conventional biopharmaceuticals [10].

Numerous bio-therapeutic agents employed in clinical settings are derived by using the application of biotechnology in pharmaceutical development. Insulin stands out as one of the earliest recombinant drugs, marking a pivotal moment in the field. The inaugural bio-engineered medication, a recombinant version of human insulin, was approved by the United States Food and Drug Administration (US FDA) in 1982. Before this breakthrough, insulin was derived by providing a limited supply of pancreas tissue from beef or pork. Introducing the human insulin gene into bacteria facilitated the production of substantial quantities of life-saving insulin. A notable advancement is the potential for diabetic patients to inhale insulin, potentially disregarding the necessity for injections. Inhaled insulin products, exemplified by Exubera®, obtained approval from the US FDA, although Exubera was later withdrawn. Alternatives like Technosphere® insulin (Afrezza®) are currently under study and may offer relief from prandial insulin requirements. Afrezza consists of premeal insulin powder delivered through an oral inhalation cartridge [8, 11, 12].

In 1997, the FDA approved a noteworthy increase in biotech drugs, surpassing the combined approvals from preceding years. The FDA has sanctioned a multitude of recombinant drugs designed to address diverse human health conditions. These encompass ailments such as AIDS, anemia, various cancers (including kidney, colorectal, ovarian cancers, leukemia, and Kaposi's sarcoma), specific circulatory disorders, and various hereditary conditions (like Gaucher's disease, Turner's syndrome, familial hypercholesterolemia, severe combined immunodeficiency disease, hemophilia A, and cystic fibrosis). Additionally, approvals extend to conditions like multiple sclerosis, human growth hormone deficiency, hepatitis C, diphtheria, hepatitis B, genital warts, and diabetic foot ulcers. Presently, the catalog includes over 100 vaccines and recombinant drugs. Harnessing their effectiveness, safety, and cost-effectiveness, molecular diagnostic tests and recombinant vaccines possess considerable potential for tackling persistent diseases in developing countries, such as leishmaniasis (a tropical infection causing lesions and fever) and malaria [8, 11–13].

History of Nanotechnology

The term "nanotechnology" was coined and defined in 1974 by Norio Taniguchi, a Japanese scientist. He described nanotechnology as the manipulation of materials through processes involving separation, consolidation, and deformation, conducted at the scale of one atom or one molecule [14]. The golden era of nanotechnology commenced in the 1980s with the ground-breaking discovery of fullerenes by Kroto, Curl, and Smalley. The field of nanotechnology saw further advancements when another Japanese scientist, Iijima, made significant contributions by developing carbon nanotubes [15].

Nanotechnology is progressing steadily, paving the way for innovative advancements in research and applications. Its influence is poised to extend across various facets of life and sciences, with a particular emphasis on medicine. In this field, the significance lies in areas such as diagnosis, drug delivery systems within the body, and the treatment of diseases. Nanomaterials have proven successful in several medical domains, including precise medical diagnosis, the effective and efficient delivery of pharmaceuticals, and the creation of artificial cells [16].

Nanomedicine has arisen as a specialized application of nanotechnology within the healthcare system, offering novel solutions to unresolved challenges in medicine. This entails the utilization of nano-based materials across diverse medical contexts, encompassing the early diagnosis and treatment of various diseases, as well as applications in tissue engineering and regenerative medicine. Some nanoscale materials have already found clinical use, contributing to the establishment of the field of nanomedicine. Positioned as a pivotal component of the broader nanotechnology domain, nanomedicine operates at the intersection of biology, chemistry, engineering, and medicine, presenting more effective tools for preventing and treating a range of diseases. The integration of nanomedicine-based approaches has opened up new possibilities for biomedical engineers and clinicians involved in the prevention, diagnosis, and treatment of severe ailments such as cancer [17].

Presently, several nanomedicine products have received approval from the FDA and are available in the market. According to a report by Bobo et al., fifty-one nanomedicines have obtained FDA approval, and an additional seventy-seven products are undergoing clinical trials. Remarkably, approximately 40% of the trials documented on clinicaltrials.gov began in either 2014 or 2015. The report emphasizes that the predominant FDA-approved materials in nanomedicine include polymeric, liposomal, and nanocrystal formulations. However, recent emphasis has been placed on the development of more intricate materials, including protein-based nanoparticle micelles, and a diverse array of inorganic and metallic particles currently undergoing clinical trials [18, 19].

Indeed, nano-based drugs like Caelyx®, Doxil®, Transdrug®, and Abraxane® are currently available in the market and are employed in the treatment of cancer. These drugs leverage nanotechnology to enhance their effectiveness in delivering therapeutic agents to targeted areas, contributing to more efficient cancer treatment strategies [20].

Nanobiotechnology: Biotechnology's Integration with Nanoscale Systems

Biotechnology's integration with nanoscale systems extends beyond passive interactions to actively harness the capabilities of biological entities for diagnostic and therapeutic purposes. One remarkable example is the field of nano-biosensors, where nanomaterials are combined with biological receptors to manufacture highly selective and sensitive devices for the detection of biomolecules. This intersection enables the development of point-of-care diagnostics with rapid and accurate detection capabilities, paving the way for early disease diagnosis and personalized medicine [21].

Moreover, the marriage of biotechnology and nanotechnology has given rise to the concept of nanobiotechnology, a field that seeks to exploit the inherent properties of nanoscale materials to interface with biological systems. Nano-sized materials, such as gold nanoparticles and carbon nanotubes, can be functionalized with biological molecules, facilitating their interaction with specific cellular targets. This has significance for gene therapy, as nanocarriers can carry therapeutic genes to specific cells, providing a more precise and targeted approach to treating genetic problems [22].

As we navigate this confluence of the nanotechnology and the biotechnology, nanobiotechnology emerges as a dynamic and multidisciplinary field, poised to reshape the landscape of health care and beyond. The ability to engineer materials with precision at the nanoscale, coupled with an understanding of biological intricacies, positions nanobiotechnology as a catalyst for innovation with the potential to address some of the most pressing challenges facing humanity. However, the transformative potential of nanobiotechnology is not without challenges and ethical considerations. The toxicity of nanomaterials and the potential long-term effects within biological systems demand thorough investigation [23]. Additionally, the ethical implications of manipulating biological entities at the nanoscale, particularly in gene editing and synthetic biology, necessitate a thoughtful and responsible approach [24].

Applications of Nanotechnology in Human Health

The pharmaceutical and biotechnology industries are increasingly urged to embrace nanobiotechnology. Nanobiotechnology is anticipated to play a pervasive role in the pharmaceutical design process, encompassing formulations aimed at enhancing absorption efficiency to its applications in detecting drug efficacy during testing trials.

a. **Diagnosis**

A primary objective in nanobiotechnology involves developing innovative methodologies for early-stage disease diagnosis, utilizing cost-effective materials and more advanced equipment than currently available. Substantial research efforts are underway in this field, with numerous studies exploring the application of metal and semiconductor nanoparticles in biomedical contexts [25]. Aaron et al. have successfully demonstrated the efficacy of 25 nm gold nanoparticles, conjugated with anti-epidermal growth factor receptor monoclonal antibodies, as in vivo targeting agents for imaging cancer markers, particularly epidermal growth factor receptors. The utilization of these gold nanoparticles has resulted in a significant enhancement of signal contrast compared to other agents employing antibody-fluorescent dye combinations [26].

An alternative avenue for molecular biomedical applications is using aptamers alongside antibodies. Aptamers, being chemically stable, facilitate the production of single-stranded nucleic acid molecules. A noteworthy application of aptamers in diagnostics is evident in the work of Niedzwiecki and colleagues. Their study combines nano-pores and aptamers to achieve highly sensitive detection of a singular molecule, specifically, the nucleocapsid protein 7 (NCp7), a protein biomarker associated with the Human Immunodeficiency Virus (HIV-1) virus. This innovative approach showcases the potential of aptamers in advancing molecular diagnostics with heightened sensitivity [27].

A compelling advancement in the diagnosis of tumor is the emergence of RNA nanoparticles as a promising tool. Researchers, such as Shu and colleagues, have made noteworthy progress in developing a toolkit for creating stable RNA nanoparticles. Their work has resulted in the successful generation of 14 homogeneous RNA nanoparticles. Significantly, these nanoparticles demonstrate an exclusive in vivo targeting of cancer, exhibiting minimal accumulation in normal organs and tissues. This innovation holds immense potential for enhancing the precision and efficacy of tumor diagnosis, marking a notable stride in the field of medical diagnostics [28].

In a study conducted by Chen et al., dendrimer-trapped cubic iron oxide nanoparticles (IONPs) have been coupled with HSP90α mRNA linked to a quenched fluorophore. Once these nanoparticles are internalized by cells that overexpress HSP90α, the fluorescence is reactivated, enabling their precise localization with high specificity using both fluorescence imaging and MRI [29, 30].

The exploration of exosomal RNAs as diagnostic biomarkers has been instigated by the discovery that exosomes contain RNA molecules. Exosomal RNAs are notably shielded from degradation by RNases due to the protective lipid bilayer membrane, enabling their reliable detection in blood, and rendering them ideal diagnostic biomarkers. Among the various substances transported by exosomes, microRNA (miRNA) has garnered significant attention due to its intricate involvement in regulating the cancer microenvironment, which includes processes such as angiogenesis, cell proliferation, and metastasis. The role of miRNA in modulating cellular behaviors both locally and in distant recipient cells is currently undergoing thorough investigation [31].

In a recent development, Exosome Diagnostics has introduced a urine-based liquid biopsy test named ExoDx, which is commercially available. This test aims to provide risk assessments for aggressive prostate cancer in patients. Notably, the ExoDx test was granted FDA Breakthrough Device Designation in 2019. It utilizes

RNA copy numbers of ERG (ETS-related gene), PCA3 (prostate cancer antigen-3), and SPDEF (ETS transcription Factor) to calculate a predictive score, correlating with the likelihood of a patient developing prostate cancer [32, 33].

b. **Gene Therapy**

The prevalent strategy for rectifying defective genes involves the insertion of a normal gene into the genome at a non-specific location, replacing a malfunctioning gene. Another strategy involves the targeted substitution of an abnormal gene with a normal variant via selective reverse mutation, effectively restoring the gene to its typical functional state. Notably, mammalian cells, with diameters typically measured in microns, house organelles within the nano-meter range. This discrepancy in scale becomes pivotal when considering the use of nano-devices. Capitalizing on nano-devices provides a distinct advantage, as their smaller size facilitates easier penetration into cells compared to larger counterparts. This heightened capability not only enables improved interaction with cells but also opens up the possibility for a different and potentially more effective mode of interaction in the context of gene correction [34].

The application of nanotechnology in gene therapy has the potential to transform the field by substituting current viral vectors with nano-sized gene carriers that may be less immunogenic. This shift in delivery mechanisms opens up new avenues for the successful introduction of nanoscale objects in the context of delivering repaired genes or replacing incorrect ones. The use of nanoscale gene carriers represents a promising approach to enhance the precision and efficiency of gene therapy while addressing concerns related to immune response, marking a significant stride in the evolution of this therapeutic modality [35].

Hedtrich Sarah et al. conducted research focused on loading lipid nanoparticles (LNPs) with Cas9 ribonucleoprotein (RNP) and mRNA for gene editing in primary human keratinocytes. They explored various lipid-to-RNP and nitrogen-to-phosphate ratios to optimize gene editing efficiency. Applying LNPs topically in skin models, the team observed successful gene editing outcomes without inducing inflammation. This study introduces a novel LNP-based strategy for gene editing in skin disorders, demonstrating promising results. Future investigations are crucial to validate the therapeutic benefits and safety profile of this approach, highlighting the importance of further exploration and validation in the realm of gene editing for skin diseases [36].

c. **Drug Delivery**

Drug delivery scientists encounter substantial challenges, including tackling issues like low bioavailability, high molecular size, and poor solubility in clinical candidates. Additional hurdles involve administering drugs to pediatric and geriatric populations, as well as delivering proteins and peptides. The prevailing focus in drug delivery is on advancing ideal, safe, and effective non-invasive methods for drug administration. Overcoming these challenges is crucial for advancing drug delivery technologies and ensuring the successful and efficient administration of therapeutic agents in diverse clinical scenarios [37].

The nanobiotechnology sector plays a crucial role in addressing challenges associated with drug delivery, offering solutions as follows [38]:

(a) Particle size reduction to the nano-meter range enhances the surface area, improving the dissolution rate of drugs.
(b) Nano-meter-sized drugs improve solubility.
(c) Ongoing initiatives aim to create non-invasive methods for drug administration, diminishing the dependence on injectable drugs.
(d) Formulations utilizing nanoparticles offer alternatives for formulations with less stability and shorter shelf life.
(e) Nanotechnology improves the solubility of poorly soluble drugs, enhancing absorption capability, bioavailability, and release rates for large molecules. Additionally, it reduces optimal doses, thereby improving safety margins by minimizing side effects.
(f) Principles of nanobiotechnology assist in developing sustained and controlled release formulations, contributing to enhanced patient compliance.

Controlled delivery systems play a critical role in augmenting the therapeutic safety and effectiveness of drugs. They accomplish this by delivering medications to the precise location at a rate regulated by the demands of the physiological environment. This precision in drug delivery not only reduces toxicity but also minimizes side effects associated with systemic drug distribution [39].

One promising avenue for controlled drug delivery is the utilization of Electrospun Nanofibers (ESNF). These nanofibers present a compelling delivery vehicle due to their 3D nano-sized features, closely mirroring the structural characteristics of the extracellular matrix. This resemblance to the natural cellular environment enhances their compatibility with biological systems [40]. The technique involves the incorporation of biological molecules (DNA, RNA, antibodies, etc.) into the nanofibers, which can be achieved through methods such as emulsion or direct integration into a polymer solution using a syringe pump as illustrated in Fig. 8.1. This flexibility in incorporating therapeutic agents ensures a tailored and controlled release of drugs, allowing for a more precise modulation of drug concentration at the target site. In essence, electrospun nanofibers stand as a promising strategy in advancing controlled drug delivery, capitalizing on their nano-sized features to mimic and interact effectively with the intricate biological milieu [41, 42].

In a study conducted by Liang et al. in 2005, the research focused on non-viral gene delivery in vitro using nanofibrous scaffolds and investigated factors influencing successful transfection. Polylactide-co-glycolide (PLGA) ESNF scaffolds were employed for loading DNA particles, utilizing two distinct preparation methods:

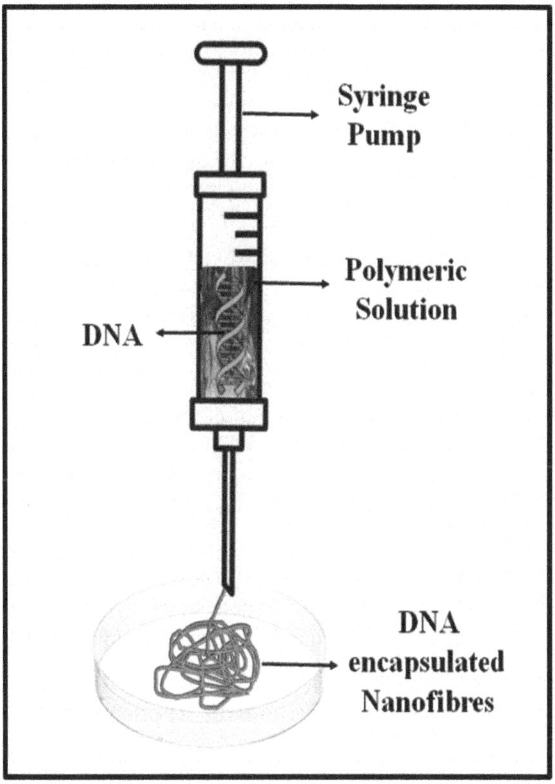

Fig. 8.1 Method of preparation of electrospun nanofiber using syringe pump

(a) DNA condensation occurred in a solvent mixture, followed by encapsulation within a triblock copolymer of poly(lactide)-b-poly (ethylene glycol)-b-poly(lactide) to form micelles. Subsequently, the micellar mixture was electrospun together with PLGA to produce a nonwoven nanocomposite and nanofibrous scaffold using coaxial electrospinning techniques, resulting in a core–shell structure.
(b) DNA was directly combined with PLGA and electrospun accordingly.

The study revealed that DNA not encapsulated within the copolymer experienced degradation during the electrospinning process, consequently exhibiting no transfection capability. Conversely, DNA encapsulated within the copolymer remained largely intact, attributed to the protective core–shell structure, facilitating successful transfection. This comparison underscores the challenge of preserving the bioactivity of incorporated additives within electrospun structures. Therefore, careful consideration of the intended biological function is essential when incorporating biological agents into such structures [43].

d. **Drug Delivery for Cardiac Therapy**

Addressing the challenges associated with cardiac repair strategies, particularly the limitations in poor cell engraftment, differentiation, and survival, necessitates innovative approaches. In the quest to improve cardiac function after injury, a novel approach involves designing self-assembling peptide nanofibers for the sustained delivery of Insulin-like Growth Factor-1 (IGF-1). IGF-1 is a vital parameter for the growth and differentiation of cardiomyocytes in the myocardium. The strategy employed utilizes a "biotin sandwich" approach to overcome the hurdles associated with conventional cell injection methods [44].

In this technique, a biotinylated IGF-1-streptavidin complex is formed and subsequently binds to biotinylated self-assembling peptides. This engineered biotin sandwich technique not only facilitates the IGF-1 binding but also allows the self-assembling of peptides into nanofibers in the myocardium, addressing the issues of poor cell engraftment and survival. In experimental studies on rats, the application of cell therapy coupled with the delivery of IGF-1 through biotinylated nanofibers has shown promising results in improving systolic function after induced myocardial infarction. This innovative approach rooted in nanobiotechnology holds promise as a significant advancement in cardiac repair, potentially elevating the effectiveness of cell therapy for individuals experiencing myocardial infarction. It's noteworthy that this innovative strategy has progressed to clinical trials, underscoring its translational potential and significance in advancing cardiac regenerative medicine [45, 46].

i. **Drug Delivery for Orthopedic Application**

Implant loosening, often triggered by immune responses to foreign bodies like titanium alloy implants, poses a significant challenge in clinical practice, hindering osteogenesis and leading to complications such as aseptic loosening. Presently, there is a lack of effective solutions to address this issue. A potential breakthrough lies in the use of exosomes, specifically those carrying microRNA (miRNA), as nanocarriers to mitigate the immune response and promote osteogenesis [47, 48]. In a study led by Liu et al., the focus was on investigating the impact of exosomes overexpressing miR-181b (Exo-181b) on M2 macrophage polarization, inflammation inhibition, and osteogenesis promotion, both in vitro and in vivo. The initial findings of the study demonstrated that Exo-181b significantly enhanced M2 macrophage polarization and suppressed inflammation by modulating the Protein Kinase-c-delta (PRKCD) gene and activating the p-AKT pathway. In the in vivo experiments, Exo-181b showcased its ability to enhance M2 polarization, reduce inflammatory responses, and promote osteointegration. Importantly, the study delved into the underlying mechanisms, revealing that enhanced M2 polarization indirectly facilitated migration and osteogenic differentiation by secreting crucial factors like Bone Morphogenetic Protein-2 (BMP-2) and VEGF. The study's key findings point to Exo-181b as a powerful regulator of the immune response and osteogenesis. By activating the PRKCD/AKT signaling pathway, Exo-181b effectively suppressed inflammatory responses by promoting M2 macrophage polarization. This dual action, both in vitro and in vivo, showcases the potential of Exo-181b as a therapeutic tool for enhancing osteointegration in the context of implantation [49].

ii. Drug Delivery in Cancer

The effective and precise administration of pharmaceutical agents plays a pivotal role in the therapeutic management of cancer, aiming to optimize drug efficacy while minimizing adverse effects. A plethora of nanobiotechnological strategies, predominantly leveraging nanoparticles, have been investigated to facilitate the targeted delivery of anticancer drugs. Among these innovative approaches are formulations that entail encapsulating drugs within nanoparticles. Colloidal gold, owing to its exceptional chemical properties, emerges as a promising avenue for targeted drug or gene delivery to specific cells. The chemical and physical attributes of colloidal gold facilitate the binding of numerous protein molecules onto a single gold nanoparticle. For instance, Tumor Necrosis Factor-α (TNF-α) can efficiently adhere to gold nanocrystals and be precisely directed toward tumors in experimental animals through intravenous administration. This targeted delivery results in the localized manifestation of TNF-α's cytotoxic effects within the tumor, thereby minimizing harm to normal tissues. The described technology is presently undergoing preclinical development for anticancer drugs, underscoring its potential to advance targeted therapeutic approaches [50].

Utilizing nanoparticle formulations for anticancer drugs holds the potential for increased effectiveness against cancer metastases. In a mammary cancer mouse model, the oral delivery of α-Tocopherol Ether-linked Acetic acid (α-TEA) encapsulated in liposomes or biodegradable poly-(D, L-lactide-co-glycolide) nanoparticles resulted in a noteworthy reduction in tumor burden. Both formulations successfully diminished micro-metastatic tumor foci in lymph nodes and lungs, with the nanoparticle formulation exhibiting superior efficacy in mitigating metastases [51].

Nanoparticle formulations facilitate the precise delivery of genes encoding immunomodulatory cytokines to distant metastatic sites in cancer. For example, intravenous administration of nanoparticles carrying the gene for granulocyte/macrophage colony-stimulating factor demonstrated targeted delivery and localized expression in patients having metastatic cancer. This strategy led to the recruitment of substantial tumor-infiltrating lymphocytes, underscoring the potential of nanoparticle-mediated targeted gene delivery in the context of cancer metastasis [52].

Nanobiotechnology plays an important role in advancing the identification of molecular biomarkers, participating in the prediction of prognosis assessment and disease development, and supervising the treatment outcomes at an individualized level. An illustrative example involves conjugating Quantum Dots (QDs) with a monoclonal anti-HER2 antibody (Trastuzumab). This innovative approach enables the single molecular in vivo imaging of breast cancer cells within a 3D microscopic system. By utilizing QDs coupled with anti-HER2 antibodies, the nanobiotechnological application allows for the visualization of cancer cells expressing the HER2 protein at subcellular resolution in vivo. This capability provides a powerful tool for personalized decision-making in chemotherapy, as it allows clinicians to discern the HER2-positive/HER2-negative tumor status. This information, obtained through single molecular imaging facilitated by nanobiotechnology, contributes to more precise and personalized treatment strategies for patients with breast cancer [53].

Wang et al. developed a gentle and self-sustaining drug delivery mechanism comprising DNAzyme nanosponges (NSs) and pH-responsive zinc oxide nanoparticles (ZnO NPs). These robust DNAzyme NSs were equipped with a repetitive tandem aptamer sequence, sgc-8c, facilitating efficient and tumor-specific drug delivery into cells. Upon cellular uptake, the acidic environment triggers the release of Zn2+ ions from the ZnO NPs, thereby inducing precise fragmentation of the DNAzyme NSs and controlled drug release. This innovative, nonviolent DNAzyme NS system serves as a versatile platform for drug delivery, enhancing therapeutic efficacy through the synergistic combination of ZnO NPs and doxorubicin (DOX) constructs. The integration of self-cleaving DNA nanostructures with therapeutic ZnO NPs and self-hydrolyzing DNAzymes provides valuable insights for developing DNA-based therapeutic platforms, with potential applications in creating multifunctional theranostic platforms by incorporating additional therapeutic sequences [54].

Tissue Engineering: Repair and Regeneration

The rising trend in life expectancy, coupled with the challenges associated with the use of autologous grafts, allografts, or xenografts, has prompted scientists globally to intensify efforts in exploring alternative approaches. Consequently, research in this domain is focused on applying the ideologies of cell engineering and transplantation to develop biological alternatives. Tissue engineering at the nanoscale aims to achieve this by creating biomaterials that guide interactions within cells and their microenvironment through the generation of nanoscale molecular signals with biological significance [55, 56].

Nanofibrous scaffolds emerge as highly promising substrates for providing physical guidance in the regeneration of nerves, especially in traversing larger nerve gaps within the peripheral nervous system. When considering the morphology of biomaterials for nerve regeneration, the orientation of fibers is a critical parameter, drawing significant attention in research endeavors [57]. In a study by Subramanian et al., a comparison was made between longitudinally aligned and random scaffolds constructed from poly (lactic-co-glycolic acid) for cultivating Schwann cells, essential players in nerve regeneration. The study's results, obtained through morphological and cell proliferation assays, underscore the advantages of aligned fibers. The longitudinally aligned nanofibers were found to guide Schwann cells more effectively, resulting in a superior proliferation rate compared to their randomly oriented counterparts. Furthermore, the evaluation of physicochemical parameters demonstrated that the aligned nanofibers exhibited more favorable outcomes for nerve regeneration compared to the random nanofibers [58].

Mesenchymal Stem/Stromal Cells (MSCs) present a promising avenue for treating degenerative conditions where normal tissue repair and replacement are lacking. In a biotechnological exploration, Kondo and colleagues propose another approach involving MSC-engineered cell sheets. These sheets can be engineered into 3D structures *in vitro* using thermos-responsive tissue culture ware, facilitating the harvesting of sheets for subsequent transplantation to affected tissues. The therapeutic aids of this approach are multifaceted. MSC-engineered cell sheets induce paracrine signaling, promoting neovascularization, and contributing to the regeneration of cartilage. Kondo and colleagues review both pre-clinical and clinical shreds of evidence promoting the safe utilization of MSC-containing cell sheets for tissue regeneration and cartilage repair [59].

With the advancement of biomedical technology, the integration of Graphene-Based Materials (GBMs) into treatments for Spinal Cord Injuries (SCI) shows promise for enhancing neuroprotective and neuroregenerative approaches aimed at improving neural recovery. For instance, employing sophisticated genetic engineering principles, Leibinger et al. achieved successful administration of AAV-hyper-interleukin-6 into the sensorimotor cortex of paralyzed adult mice with complete SCI. Their findings demonstrated effective transneuronal transduction of motoneurons to various brain regions spanning the spinal cord's length. This resulted in notable and sustained regeneration of neuronal pathways, including the corticospinal and raphe spinal tracts, ultimately leading to the restoration of locomotor function in both hind limbs [60].

Chen et al. conducted an investigation into the impact of DiGeorge syndrome critical region 8 (Dgcr8) deficiency on intestinal regeneration and explored the therapeutic potential of miR-200 in promoting epithelial repair. Their study elucidated that Dgcr8 deficiency impedes crypt regeneration and cell proliferation following injury, resulting in heightened immune cell infiltration and nuclear factor kappa-light-chain-enhancer of activated β cells signaling, indicative of compromised intestinal barrier function. The absence of miR-200 family members in Dgcr8-deficient cells activated the p53 pathway. Notably, oral administration of miR-200 via lipid nanoparticles effectively stimulated epithelial regeneration in Dgcr8-deficient mice. These findings underscore the pivotal role of epithelial miRNAs, particularly miR-200, in facilitating the regeneration of injured intestinal epithelium. The research proposes a potential therapeutic strategy for enhancing intestinal repair processes through targeted delivery of miR-200 [61].

Pathogen Detection

Detecting pathogens in a sample for analysis can be challenging when they are present in low numbers. Conventional methods, such as the estimation of colony count, can be labor-intensive and time-consuming, taking anywhere from 24 h to 7 days to complete. This poses substantial challenges for maintaining the quality of semi-perishable foods. Moreover, these approaches could result in an underestimation of pathogen numbers, given that microorganisms have the capacity to enter viable but not culture due to environmental stress [62].

Progress in nanomaterial manipulation has enabled the binding of various biomolecules, including proteins, bacteria, nucleic acids, and toxins. A significant advantage of employing nanomaterials in biosensing lies in their substantial surface area, which allows for the immobilization of a larger number of biomolecules. As a result, bioavailability at reaction site increases for interactions with a target molecule. The combination of this property with the outstanding optical and electronic characteristics of nanomaterials makes them particularly suitable for label-free detection. Furthermore, nanomaterials play a crucial role in the modification of biosensors characterized by improved sensitivities and enhanced response times [63, 64].

Single-walled carbon nanotubes (SWCNTs) offer a versatile platform for studying protein interactions, including both protein–protein and surface-protein interactions, and hold promise for developing highly accurate electronic biomolecule detectors. To address issues related to non-specific protein binding, polyethylene oxide chains are immobilized on the nanotubes. A systematic approach involves selectively recognizing and binding target proteins by coupling their unique receptors to polyethylene oxide-functionalized nanotubes. These configurations are appealing as they eliminate the need for labeling, allowing the entire experiment to be conducted in the solution phase. When combined with the sensitivity of nanotube electronic devices, this approach results in more precise electronic sensors capable of detecting clinically relevant biomolecules, such as antibodies associated with autoimmune disorders in humans [65].

For centuries, silver salts and elemental silver have been recognized for their antimicrobial properties in both preventive and curative health care. The effectiveness of silver salts and complexes, particularly ionic silver, in fighting against microbes largely depends on the interaction of metal ions with various bio-macromolecules. Cationic silver, in particular, selectively binds to the negatively charged elements present in nucleic acids and proteins. This interaction results in structural variations along with bacterial cell wall deformations, membranes as well as nucleic acids [66, 67].

Moreover, silver ions are widely acknowledged for their capacity to interact with diverse electron-donating functional groups, such as phosphates, thiols, imidazoles, hydroxyls, amines, and indoles. Consequently, there is a hypothesis that silver ions, when binding to DNA, obstruct transcription, whereas their binding to cell surface components disrupts the synthesis of Adenosine Triphosphate (ATP) and bacterial respiration. Furthermore, several studies imply that silver ions interfere with the respiratory chain of microorganisms, notably in the region of cytochrome oxidase and Nicotinamide Adenine Dinucleotide (NADH) succinate dehydrogenase, which is an enzyme complex within bacterial cells [68].

Gram Identification and Rapid Bacterial Detection

The prompt detection and identification of bacteria are crucial in cases of bacterially induced sepsis, where timely intervention can significantly impact patient outcomes.

Traditional culture-based methods for bacterial detection are time-consuming, taking anywhere from 10 h to several days. In this study, a novel microfluidic device is used, featuring a bacterially activated coating made from macrophage membranes on nanowired silicon (Si) surfaces. This system enables rapid detection and gram identification of bacteria present in contaminated blood samples. When bacterial suspensions containing gram-negative or gram-positive bacteria were perfused through the microfluidic device with membrane-coated adsorbent surfaces, even low levels of bacteria (<10 CFU/mL) were successfully detected as shown in Fig. 8.2. Subsequent fluorescence staining enabled gram identification, aiding in the selection of appropriate antibiotics. In mixed suspensions containing *Escherichia coli* and *Staphylococcus aureus*, the proportions of gram-negative and gram-positive bacteria detected mirrored those in the original suspensions through the utilization of Confocal Laser Scanning Microscopy (CLSM). Validation of this method was conducted using 15 human blood samples spiked with various bacterial strains or combinations, with evaluation by two blinded observers. The results showed a 100% accuracy rate, indicating the potential of this platform for the rapid diagnosis of bacterial sepsis within a total processing time of 1.5 h [69].

Nanorobotics

Nanorobotics, a subset of nanotechnology, delves into the design, programming, manufacturing, and control of robots at the nanoscale, typically ranging from 0.1 to 10 μm. A nanobot, an enormously small robot tailored for precise tasks at the nanoscale, is a key component of nanorobotics. Mostly they react to external stimuli

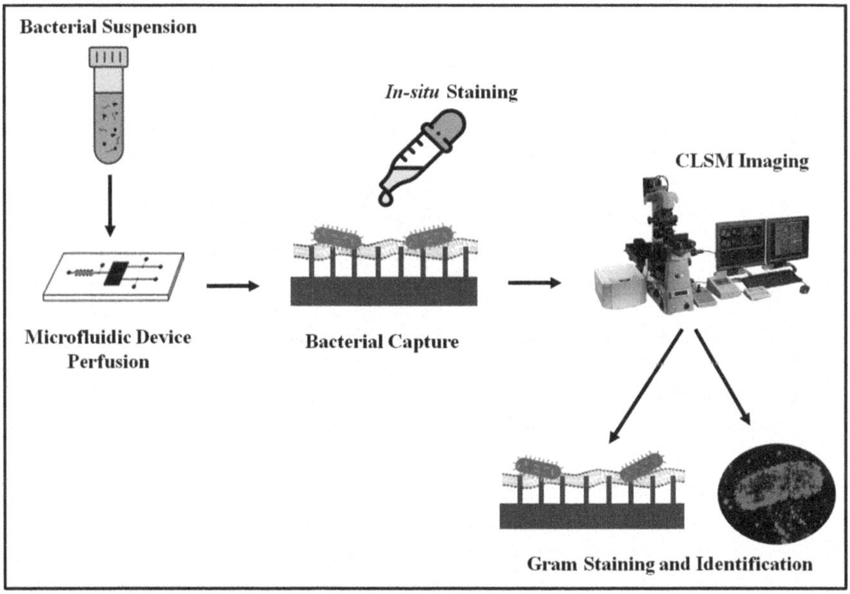

Fig. 8.2 Preparation of bacterial suspension and counting of bacteria using CLSM

and release drugs present in them, as shown in Fig. 8.3. Drawing inspiration from the bacteria biological model, nanobots are constructed primarily using carbon, often in the form of carbon or fullerene nanocomposites. While still in the research and development stage, nanobots hold immense potential for medical applications, introducing a new set of tools for treating diseases and augmenting the biological system in humans. These nanobots exhibit the capability to perform specific tasks at the molecular and cellular levels [70, 71]. Several examples of nanorobots with their unique characteristics are mentioned in Table 8.1 [72].

Anti-HIV Nanorobots

HIV is notorious for compromising the immune system, leaving the host susceptible to various illnesses. This immune system breakdown leads to the progression of Acquired Immunodeficiency Syndrome (AIDS), ultimately resulting in a fatal outcome. The HIV specifically targets white blood cells (WBCs), converting them into HIV-infected cells. As a result, the immune system becomes compromised, contributing to the patient's demise. Tiny robotic agents would intervene by restoring AIDS-affected WBCs to their original, uninfected state. The process would be represented as follows:

$$\text{Hosts WBC} + \text{HIV} = \text{Infected WBC}$$

$$\text{Infected WBC} + \text{Nanorobot} = \text{Restored WBC}$$

The nanobots play a crucial role in converting the HIV-affected white blood cells back to their original form, thereby maintaining a stable and functional immune system [73].

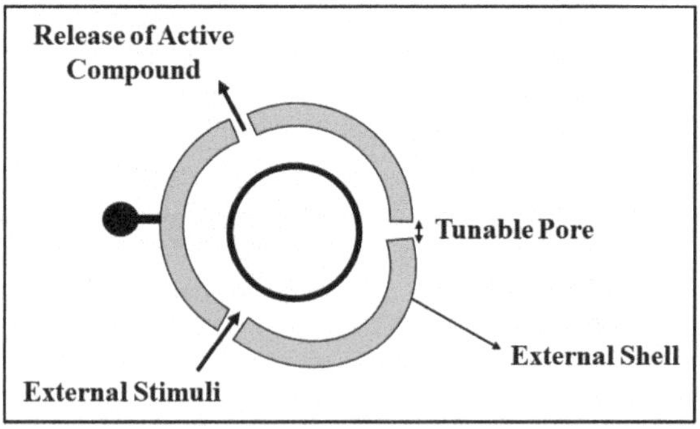

Fig. 8.3 Schematic representation of nanorobot

Table 8.1 Some classical examples of nanorobots with their characteristics

Sr. No	Type of nanorobots	Characteristics
1	Clottocytes	They exhibit immediate hemostatic biological activity, referred to as artificial mechanical platelets. Additionally, they facilitate the transportation of substances aiding the coagulation process
2	Respirocyte	This nanorobot functions similarly to an artificial red blood cell, transporting oxygen. Its energy is derived from naturally occurring serum glucose within the body
3	Microchips	Nanorobots are equipped with microchips capable of transmitting electrical signals upon detecting disease-specific molecules. The advantages include minimal power requirements and straightforward operation
4	Pharmacyte	A medical nanorobot with a diameter ranging from 1 to 2 μm employs molecular indicators or chemotactic sensors to ensure the accuracy of its targeting system. Upon completing tasks, these nanorobots can be eliminated or recovered via centrifugal nanapheresis
5	Microbiovores	The nanorobot designed for nanomedical applications adopts a fat, spheroidal shape, with dimensions measuring 3.4 μm along its primary axis and 2.0 μm along its minor axis. It exhibits phagocytic capabilities nearly 80 times more proficient than other macrophages
6	Chromallocyte	The renovation machine initially evaluates cellular conditions by scrutinizing substances, actions, and functions. Subsequently, these repair machines demonstrate the capability to comprehensively overhaul the entire cell

Nanorobots in Hematology

The utilization of nanobots, known as clottocytes, has demonstrated effectiveness in reducing clotting time and minimizing blood loss. In certain patients, irregular blood clotting may occur, necessitating treatment with corticosteroids. However, corticosteroids are linked to side effects including hormonal secretion, lung damage, and allergic reactions. Clottocytes present a promising alternative for addressing this abnormality without inducing these side effects [74].

The natural process of blood clotting typically spans a duration of 2–5 min. Nanotechnology has demonstrated the potential to abbreviate clotting time and minimize blood loss. Some patients exhibit irregular blood clot formation, a condition often addressed with corticosteroid medication. However, corticosteroid treatment can lead to adverse effects, including hormonal imbalances, lung damage due to blood platelet aggregation, and allergic reactions. The concept of a theoretically engineered clottocyte refers to an artificial mechanical platelet designed to achieve hemostasis within approximately 1 s. This clottocyte, a spherical nanorobot powered by serum-oxyglucose and measuring about 2 μm in diameter, houses a tightly folded fiber mesh onboard. Its response time is estimated to be 100–1000 times faster than the natural hemostatic process. The biodegradable fiber mesh, coated with a soluble

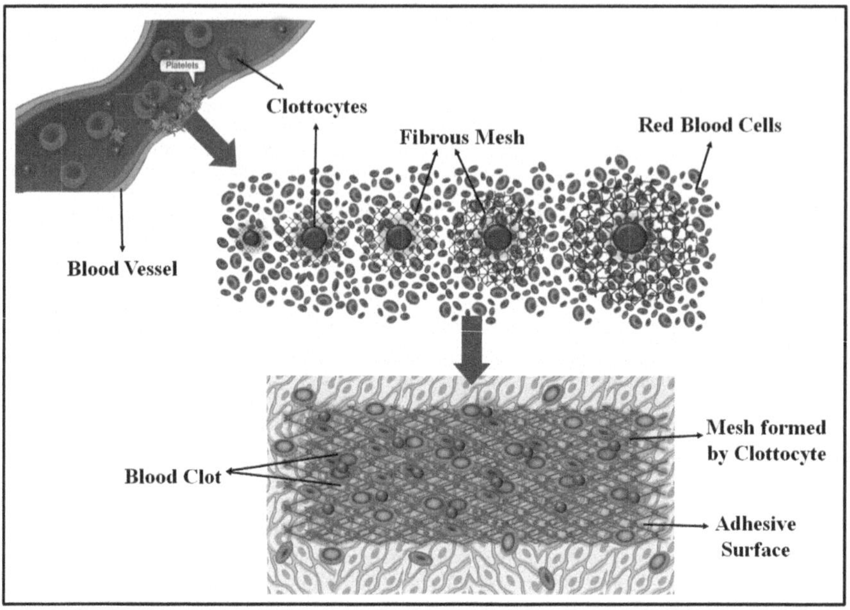

Fig. 8.4 Mechanism of formation of a blood clot by clottocyte

film, dissolves upon contact with plasma to reveal an adhesive surface (as depicted in Fig. 8.4). Upon entry into a damaged blood vessel, clottocyte-mounted sensors swiftly detect changes in partial pressure, indicating potential blood loss from the body. Positioned approximately 75 μm from the air-serum interface, the clottocyte receives oxygen molecules via diffusion through the serum at body temperature. This detection triggers the rapid transmission of signals to neighboring clottocytes via acoustic pulses, facilitating a meticulously orchestrated device-enabled cascade for prompt intervention [75].

Nanorobots in the Detection and Treatment of Cancer

The diagnosis and treatment of cancer can be significantly improved through the use of nanobots. Unlike conventional drugs, the nanobots exhibit high site-specificity, which means they are programmed to exclusively detect and act upon diseased cells while sparing healthy cells. This targeted approach results in minimal side effects. Nanorobots equipped with embedded biosensors hold particular promise for the early detection of tumor cells within a patient's body during the initial stages of development. This precision in targeting and sensing enhances the effectiveness of cancer diagnosis and treatment while mitigating adverse effects on healthy tissues [76]. Certain nanorobots are constructed using engineered DNA strands that are specifically designed to fold into a particular tertiary structure. When these nanobots bind to their intended target, the conformation of the DNA nanobot undergoes a transition from a closed state to a tertiary state. This transition triggers the release of the stored therapy, allowing for precise and controlled therapeutic delivery [77].

Nanorobots in Dentistry

Nanorobots play a crucial role in various dental applications, including desensitization, anesthesia, orthodontics, enhanced durability, repairs, and aesthetics. Administered through toothpaste or mouthwash, nanorobots cover all subgingival surfaces, metabolizing organic pollutants into safe and tasteless vapors. They actively locate and eliminate harmful germs, ensuring the safety of mechanical elements when ingested. Intelligent nanorobots in mouthwash recognize and eradicate harmful germs, promoting a balanced oral ecosystem. These invisible nanorobots work harmoniously on multiple dental tasks, utilizing tissue engineering, regeneration, and genetic engineering for healing severely damaged teeth. Overall, nanorobotic dental approaches offer advanced and comprehensive oral care [78].

Nanobots, when coated with precise proteins, can effectively target and bind to specific pathogens for infection treatment. In root canal procedures, employing a miniature camera allows accurate visualization of the root, minimizing guesswork. Hypersensitive teeth, characterized by the increased number and diameter of dentinal tubules, can be addressed by nanobots with selective ablation or occlusion capabilities. This targeted approach prevents stimuli within minutes, providing a promising solution to alleviate hypersensitivity [79].

Nano-biosensors

A nano-biosensor, also known as a nano-sensor, is distinguished by its nano-meter scale dimensions. It comprises a bio-sensitive layer that incorporates biological recognition elements, either composed of these elements or covalently attached to a transducer. The primary purpose is to induce a physicochemical change through the interaction between the bioreceptor and the target analyte, which can then be translated into a computable effect, such as an electrical signal [80]. Bioreceptors play a vital role in ensuring specificity in biosensor technologies, enabling the binding of the specific analyte with minimal interference by other elements in complex samples. These biological sensing elements may encompass molecular species, such as antibodies, proteins, nucleic acids, and enzymes, or utilize living biological elements like cells, tissues, or whole organisms, employing a biochemical mechanism for recognition. Several detection techniques, including electrochemical, optical, or mass-sensitive methods, can be employed in biosensors. Applications of nano-biosensors span from detecting microorganisms in diverse samples and monitoring metabolites in body fluids to identifying diseases like cancer and facilitating glucose monitoring [81].

In Wuhan, China, Lui Hu et al. conducted a study utilizing nano-biosensor technology for COVID-19 detection, demonstrating remarkable sensitivity. Their research exhibited a 100% sensitivity rate in both training and test groups, highlighting the potential of this innovative approach for early and accurate virus diagnosis. This advancement holds promise for bolstering diagnostic capabilities and potentially easing the burden on healthcare systems grappling with the pandemic. Through the utilization of nano-biosensor technology, they have introduced a valuable tool for swift and efficient COVID-19 case identification, offering a solution to the challenges posed by the virus's widespread transmission [82].

Polley Nacarun et al. have undertaken research aimed at enhancing sensitivity in nanoscale sensors through the implementation of directed analyte transport. Their study centered on the utilization of polyNIPAM monolayers within nanopump sensors, responsive to temperature fluctuations. The photo thermomechanical nanopump sensor they engineered integrates a thermoresponsive polymer for both pumping and plasmonic response. By introducing a flow-through NP concept, they achieved heightened sensitivity and improved detection limits, eliminating the necessity for a separate pumping apparatus. This innovative approach underscores the potential of advanced sensor technologies capable of efficiently detecting and analyzing analytes at the nanoscale level, with promising applications across biotechnology, environmental monitoring, and healthcare sectors [83].

Recent Advances in Nanobiotechnology

Some recent research works in the field of nanobiotechnology are mentioned in Table 8.2.

Global Market Scenario of Nanobiotechnology [84]

Nanobiotechnology is attracting considerable investments globally, particularly from businesses and governments. In the healthcare domain, it revolutionizes disease diagnosis and treatment methodologies. Notably, nanoparticles facilitate precise drug delivery to cancer cells, minimizing traditional chemotherapy side effects. This targeted therapeutic approach enhances patient outcomes and creates new pharmaceutical markets. Additionally, nanobiotechnology supports the creation of rapid diagnostic tools for early disease detection, enabling timely interventions and reducing healthcare expenses.

Pfizer, a prominent pharmaceutical company, has strategically invested in nanobiotechnology research and startups, focusing on advancing drug delivery

Table 8.2 Recent advances in nanobiotechnology

Sr. no	Technique	Description	References
1	Metallic nanostructures	Metallic nanostructures have been created for potential applications in Surface Enhanced Raman Spectroscopy (SERS). Additionally, these nanostructures can be designed to possess photo-thermal properties, contributing to their utility in laser ablation therapy. The primary objective of nanobiotechnology is to advance the development of multifunctional nano-devices with enhanced sensitivity and significant clinical applications	[60]
2	Nano-plasmonic sensors	Nano-plasmonic sensors play a crucial role in detecting biomolecules, contributing significantly to diverse scientific fields of medicine	[68]
3	Physical antibodies of black silicon	Trials are underway for the creation of physical antibodies composed mainly of black silicon. These innovative antibodies aim to combat bacteria, showcasing the potential of this unique material in advancing the field of antibacterial agents	[69]
4	Silver nanoparticles	Silver nanoparticles have demonstrated potent antibacterial properties, making them effective in self-sanitizing toothbrushes. Their potential extends to potential future applications in toothpaste, offering enhanced antimicrobial benefits for oral hygiene	[70]
5	Cerium oxide nanoparticles	The intravitreal injection of cerium oxide nanoparticles exhibited encouraging outcomes as it effectively suppressed neovascularization in VLDLr knockout mice. This positive impact was linked to the inhibition of the VEGF cascade and the activation of antioxidant gene expression pathways. These results imply that nanoceria has the potential to become a therapeutic option for addressing diabetic retinopathy	[71]
6	Nanobodies	In the ongoing battle against COVID-19, scientists are exploring nanobodies as a potent weapon. These are compact antibodies, smaller than those naturally occurring in the human body. The development of nanobodies represents a promising approach to enhance the fight against the virus	[72]

(continued)

Table 8.2 (continued)

Sr. no	Technique	Description	References
7	mRNA lipid nanoparticles	Lipid nanoparticles (LNPs) show significant promise in delivering nucleic acids, including mRNAs. Utilizing microfluidic mixers, mRNA-encapsulated LNPs are crafted and efficiently delivered to the cytoplasm. This process triggers an immune response, highlighting the potential of LNPs in facilitating mRNA-based therapeutic interventions	[73]
8	3D nano-printing	The biosensing system is created by utilizing 3D nano-printing to produce electrodes, which are then coated with nanoflakes made of reduced-graphene-oxide (rGO). Subsequently, specific viral antigens are immobilized onto the rGO nanoflakes, allowing for quick detection of COVID-19 antibodies within a mere 10 s	[74]
9	Transferrin-conjugated solid lipid nanoparticles	In a study addressing cerebral malaria treatment, solid lipid nanoparticles (SLNs) conjugated with transferrin were employed. The results demonstrated a significant enhancement in the delivery of quinine to the brain compared to both unconjugated SLNs and a conventional drug solution. This underscores the promising effectiveness of employing targeted drug delivery in this context	[75]

methods and improving therapeutic results. Notable initiatives involve collaborations with nanotechnology firms to develop targeted drug delivery systems for cancer treatment and vaccine innovation.

List of Patents in the Field of Nanobiotechnology

Some of the recent patents in the field of nanobiotechnology with their status are listed in Table 8.3.

Table 8.3 List of patents in the field of nanobiotechnology

Sr. no	Patent no	Title	Description	Status	References
1	CN108602849B	RNA ligand-displaying exosomes for specific delivery of therapeutic agents to cells by RNA nanotechnology	RNA nanoparticles possess the capability to direct extracellular vesicles toward a specific cell by means of a targeting component. These extracellular vesicles might additionally carry functional elements, which could be harnessed for therapeutic or diagnostic purposes	Granted (2022–10–21)	[77]
2	US10343903B2	Cationic polymer-coated mesoporous silica nanoparticles and uses thereof	The submicron entity can exhibit either a spherical or non-spherical shape and could incorporate a cationic polymer or co-polymer coating on the surface of its silica structure. Additionally, this submicron entity may incorporate an oligonucleotide, enabling the delivery of said oligonucleotide to a cell. Furthermore, the submicron structure might also contain a therapeutic agent, facilitating its delivery to a cell. In certain instances, both an oligonucleotide and a therapeutic agent could be utilized in tandem. For instance, in cases where the oligonucleotide functions as a siRNA, the composition could be employed to reduce cellular resistance to the therapeutic agent by inhibiting the translation of a resistance gene	Granted (2019–07–09)	[78]
3	CN108601823A	Composition for being modified dendrimer nanoparticles vaccine delivery and method	Dendrimer nanoparticles, abbreviated as "MDNP," are introduced into the cells of a subject. These MDNPs demonstrate a capacity to significantly promote the expansion of T cells possessing specificity for intracellular antigens, while also augmenting the response of antigen-specific antibodies	Pending	[79]

(continued)

Table 8.3 (continued)

Sr. no	Patent no	Title	Description	Status	References
4	JP2018516847A	Compositions and methods for delivering biopolymer drugs	The current innovation concerns nanoparticles combined with biopolymers designed to address, prevent, or alleviate a range of disorders, along with methods for their creation. Specifically, this invention fabricates compositions consisting of nanoparticles, such as synthetic high-density lipoprotein (sHDL), which transport biopolymers like nucleic acids, peptides, glycolipids, and so forth	Pending	[80]
5	CN107683132A	Targeted nanocarrier for the targeted delivery of drugs of gene therapy	The current invention pertains to a targeted nanocarrier, also known as nano medication, which is specifically designed or independently utilized for facilitating gene transfer or genome editing using instruments such as plasmids or restriction enzymes (e.g., Zinc finger nucleases, CRISPR/Cas systems, and TALENs), as well as for gene silencing or post-transcriptional control of gene expression using molecules like microRNA, siRNA, mRNA, ASON, and sense oligonucleotides. This invention encompasses a range of methods for delivering these components to various mammalian cell species	Pending	[81]

(continued)

Table 8.3 (continued)

Sr. no	Patent no	Title	Description	Status	References
6	EP2805713B1	Magnetic nanoparticle-samirna complex and method for preparing same	The current invention concerns a SAMiRNA-magnetic nanoparticle complex, comprising SAMiRNA, a nanoparticle composed of a double-stranded oligo RNA structure combined with both hydrophilic and hydrophobic materials to enhance the effective delivery of double-stranded oligo RNA, particularly useful in cancer therapy, infectious diseases treatment, and similar applications. Additionally, the complex incorporates magnetic nanoparticles for diagnostic purposes. The invention also encompasses a method for preparing this complex, delivering the double-stranded oligo RNA using it, as well as technologies for therapy and/or diagnosis of diseases, including cancer and infectious diseases	Granted (2018–10-10)	[82]

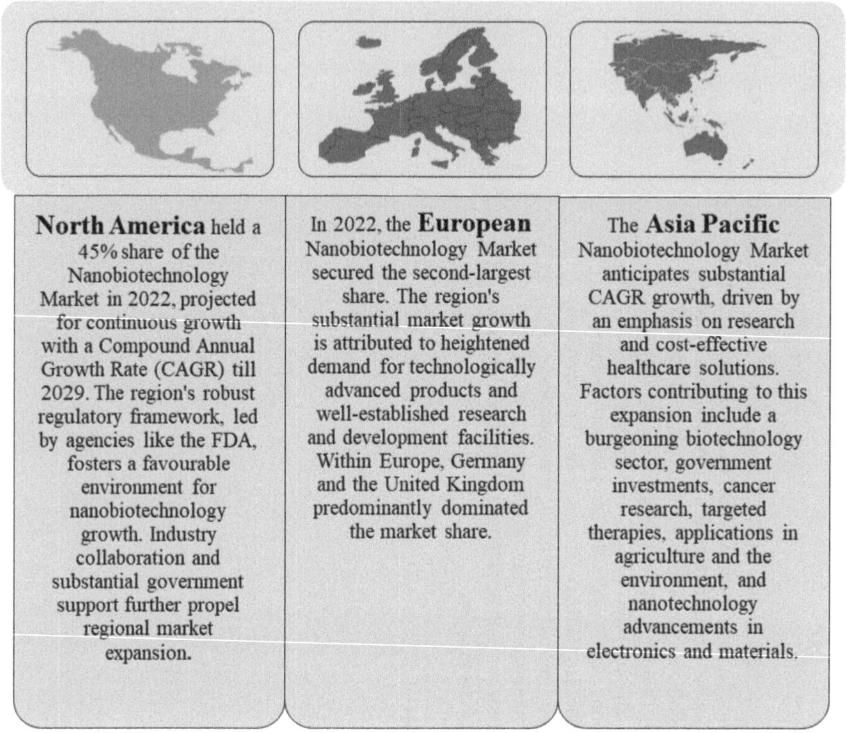

Fig. 8.5 Market regional insights of nanobiotechnology

Market Regional Insights of Nanobiotechnology [84]

See Fig. 8.5.

Leading Nanobiotechnology Product Manufacturers in the World

Table 8.4 mentions some of the leading manufacturers of nanobiotechnological products in regions like North America, Europe, and Asia Pacific.

Table 8.4 Leading nanobiotechnology product manufacturers in the world [84]

North America	Europe	Asia Pacific
Nami Therapeutics (US)	Ablynx (Belgium)	National University of Singapore (Singapore)
Celgene Corporation (US)	SkyePharma Pharmaceuticals (France)	Chinese Academy of Sciences (China)
Nanophase Technologies Corporation (US)	Nano Bridging Molecules SA (Switzerland)	National Institute of Advanced Industrial Science and Technology (Japan)
Sigma-Aldrich (US)	XanTecbioanalytics GmbH (Germany)	Indian Institute of Science (India)
KGaA (US)	Nanobiotix (France)	Korea Advanced Institute of Science and Technology (South Korea)
Johnson and Johnson (US)		Riken Institute (Japan)
Abbott Laboratories (US)		
Novartis AG (US)		

Challenges and Ethical Considerations

Despite the immense potential of the intersection between nanotechnology and biotechnology, several challenges and ethical considerations accompany the development and implementation of these technologies in the realm of human health. The potential toxicity of nanomaterials, their long-term effects on the environment, and the ethical implications of manipulating biological processes at the molecular level are subjects of ongoing scrutiny and debate. Maintaining a balance between innovation and safety is imperative to safeguard the responsible development and deployment of nanobiotechnological solutions. The emergence of potentially disruptive technologies, including nanotechnology, introduces greater challenges in establishing new regulatory requirements. Global authorities are faced with the complex task of thoroughly assessing the potential risks associated with the widespread adoption of this modern technology and determining appropriate governmental responses. The dynamic nature of such innovations requires careful consideration and adaptation of regulatory outlines to ensure the safe and accountable implementation of these advancements on a global scale. Moreover, issues related to regulatory frameworks and standardization pose significant hurdles in the translation of these technologies from the laboratory to clinical settings [23, 24, 85].

Biotechnology, while holding the potential for significant positive impacts, also carries the risk of causing harm. Potential harm may manifest physically through unforeseen consequences of the technology. Ethical considerations in biotechnology often align with a rights-based approach, which identifies areas where individuals possess rights. Each right is accompanied by a corresponding duty or responsibility. For instance, if individuals have a right to health care, it implies that someone bears the responsibility to provide the necessary healthcare resources. This ethical framework seeks to balance the benefits and potential risks associated with biotechnological advancements [86].

The dynamic nature of nanobiotechnology requires adaptable regulatory approaches that can keep pace with rapid advancements while ensuring patient safety and efficacy. Collaborative efforts among scientists, regulators, and ethicists are essential to establish guidelines that foster innovation while safeguarding against potential risks. Ultimately, the most significant obstacle lies in the commercialization of nanobiotechnology, encompassing challenges such as effective scalability, perseverance, funding, innovation, and scarce resources [87].

Future Perspectives

As the intersection of nanotechnology and biotechnology continues to evolve, the future holds exciting prospects for human health. The development of smart nano systems capable of real-time monitoring, diagnostics, and therapeutic interventions heralds a new era in personalized medicine. Tailoring treatments to individual genetic profiles, minimizing side effects, and improving overall patient outcomes are tangible goals within reach [88].

The ongoing convergence of these two dynamic fields not only presents opportunities for addressing current healthcare challenges but also opens new avenues for scientific discovery. Nanobiotechnology has the potential to reshape our understanding of disease mechanisms, facilitate early detection, and enable precise interventions at the molecular level. Collaborative efforts across disciplines, from chemistry and physics to biology and medicine, will be pivotal in unlocking the full potential of this interdisciplinary frontier [89].

Despite occasional disputes, this technology holds substantial promise for the future. It has the potential to drive innovations and perform a crucial role in several biomedical applications, spanning from targeted delivery of drug and gene therapy to molecular imaging and further to biomarkers and biosensors. A significant current research focus involves targeted drug therapy and the development of methods for early diagnosis and disease treatment [90].

Conclusion

As a conclusion, nanotechnology's intersection with biotechnology positions as a witness to the transformative power of interdisciplinary collaboration. The marriage of nanoscale precision with biological specificity has already yielded innovative solutions in medicine, and the journey has only just begun. With responsible research, ethical considerations, and continued dedication to overcoming challenges, the amalgamation of nanotechnology and biotechnology holds the key to a future where the healthcare industry is not only more effective but also more personalized and patient-centric. Nanobiotechnology is currently in its nascent stages, with the interdisciplinary field making strides in bringing the science of minuscule devices closer to practical application. Its applications in medicine are extensive, with developments like drug delivery systems marking just the initial phase of a transformative era. The integration of nanobiotechnology knowledge into areas such as gene delivery, drug delivery, stem cell-based therapy, and cancer diagnostics has ushered in a realm of groundbreaking innovations in the medical science domain. In a nutshell, nanobiotechnology is an evolving scientific field poised to captivate humanity with its pioneering advancements.

References

1. Contera, S., De La Serna, J.B., Tetley, T.D.: Biotechnology, nanotechnology and medicine. Emerg. Top Life Sci. **4**, 551–554 (2021). https://doi.org/10.1042/ETLS20200350
2. Singh, A.V., Laux, P., Luch, A., Sudrik, C., Wiehr, S., Wild, A.M., et al.: Review of emerging concepts in nanotoxicology: opportunities and challenges for safer nanomaterial design. Toxicol. Mech. Methods **29**, 378–387 (2019). https://doi.org/10.1080/15376516.2019.1566425
3. Barenholz, Y.: Doxil®—the first FDA-approved nano-drug: lessons learned. J. Control. Release **160**, 117–134 (2012). https://doi.org/10.1016/j.jconrel.2012.03.020
4. Michalet, X., Pinaud, F.F., Bentolila, L.A., Tsay, J.M., Doose, S., Li, J.J., et al.: Quantum dots for live cells, in vivo imaging, and diagnostics. Science **307**(80), 538–544 (2005). https://doi.org/10.1126/science.1104274
5. Jokerst, J.V., Lobovkina, T., Zare, R.N., Gambhir, S.S.: Nanoparticle PEGylation for imaging and therapy. Nanomedicine **6**, 715–728 (2011). https://doi.org/10.2217/nnm.11.19.Nanoparticle
6. Afzal, H., Zahid, K., Ali, Q., Sarwar, K., Shakoor, S., Nasir, U., et al.: Role of biotechnology in improving human health. J. Mol. Biomark. Diagn. **07**, 1–7 (2016). https://doi.org/10.4172/2155-9929.1000309
7. Awais, M., Alam, F., Sarwar, R.: Current status of biotechnology in health. Am. Eurasian J. Agric. Environ. Sci. **7**, 210–220 (2010)
8. Gupta, V., Sengupta, M., Prakash, J., Tripathy, B.C.: Basic and Applied Aspects of Biotechnology, pp. 1–520 (2016). https://doi.org/10.1007/978-981-10-0875-7
9. Khan, S., Ullah, M.W., Siddique, R., Nabi, G., Manan, S., Yousaf, M., et al.: Role of recombinant DNA technology to improve life. Int. J. Genomics (2016). https://doi.org/10.1155/2016/2405954
10. Guzman, C.A., Feuerstein, G.Z.: Pharmaceutical biotechnology. Curr. Opin. Biotechnol. **15**, 503–505 (2004). https://doi.org/10.1016/S0958-1669(04)00147-8
11. Bentahar, S., Abada, R., Ykhlef, N., Soumia, B., Rofia, A., Nadia, P.Y.: Biotechnology: definitions, types and main applications. Yamer Digit, 563–575 (2023). https://doi.org/10.37896/YMER22.04/49
12. Kepplinger, E.E.: FDA's expedited approval mechanisms for new drug products. Biotechnol. Law Rep. **34**, 15–37 (2015). https://doi.org/10.1089/blr.2015.9999
13. Nascimento, I.P., Leite, L.C.C.: Recombinant vaccines and the development of new vaccine strategies. Braz. J. Med. Biol. Res. **45**, 1102–1111 (2012). https://doi.org/10.1590/S0100-879X2012007500142
14. Bayda, S., Adeel, M., Tuccinardi, T., Cordani, M., Rizzolio, F.: The history of nanoscience and nanotechnology: From chemical-physical applications to nanomedicine. Molecules **25**, 1–15 (2020). https://doi.org/10.3390/molecules25010112

15. Hulla, J.E., Sahu, S.C., Hayes, A.W.: Nanotechnology: history and future. Hum. Exp. Toxicol. **34**, 1318–1321 (2015). https://doi.org/10.1177/0960327115603588
16. Filipe, J.A.: Nanotechnology and medicine improvement * 32 | Part A: Appl. Nat. Sci. Int. J. Acad. Res. **7**, 32–37 (2015). ISSN: 2075-4124. https://doi.org/10.7813/2075-4124.2015/7-2/A.5
17. Kargozar, S., Mozafari, M.: Nanotechnology and nanomedicine: start small, think big. Mater. Today Proc. **5**, 15492–15500 (2018). https://doi.org/10.1016/j.matpr.2018.04.155
18. Bobo, D., Robinson, K.J., Islam, J., Thurecht, K.J., Corrie, S.R.: Nanoparticle-based medicines: a review of FDA-approved materials and clinical trials to date. Pharm. Res. **33**, 2373–2387 (2016). https://doi.org/10.1007/s11095-016-1958-5
19. Dobrovolskaia, M.: Pre-clinical immunotoxicity studies of nanotechnology_formulated drugs: challenges, considerations and strategy. J. Control. Release **220**, 571–583 (2015). https://doi.org/10.1016/j.jconrel.2015.08.056.Pre-clinical
20. Rodríguez, F., Caruana, P., De la Fuente, N., Español, P., Gámez, M., Balart, J., et al.: Nano-based approved pharmaceuticals for cancer treatment: present and future challenges. Biomolecules **12**, 1–27 (2022). https://doi.org/10.3390/biom12060784
21. Kulkarni, M.B., Ayachit, N.H., Aminabhavi, T.M.: Recent advancements in nanobiosensors: current trends, challenges, applications, and future scope. Biosensors **12**, 1–30 (2022). https://doi.org/10.3390/bios12100892
22. Davis, M.E., Chen, Z., Shin, D.M.: Nanoparticle therapeutics: an emerging treatment modality for cancer. Nat. Rev. Drug Discov. **7**, 771–782 (2008). https://doi.org/10.1038/nrd2614
23. Nel, A., Xia, T., Madler, L., Li, N.: Toxic potential of materials at the nanolevel. Science **311**(80), 622–627 (2006). https://doi.org/10.1126/science.1114397
24. O'Mathúna, D.P.: Bioethics and biotechnology. Cytotechnology **53**, 113–119 (2007). https://doi.org/10.1007/s10616-007-9053-8
25. De Morais, M.G., Martins, V.G., Steffens, D., Pranke, P., Da Costa, J.A.V.: Biological applications of nanobiotechnology. J. Nanosci. Nanotechnol. **14**, 1007–1017 (2014). https://doi.org/10.1166/jnn.2014.8748
26. Aaron, J., Travis, K., Harrison, N., Sokolov, K.: Dynamic imaging of molecular assemblies in live cells based on nanoparticle plasmon resonance coupling. Nano Lett. **9**, 3612–3618 (2009). https://doi.org/10.1021/nl9018275
27. Aaron, J., Nitin, N., Travis, K., Kumar, S., Collier, T., Park, S.Y., et al.: Plasmon resonance coupling of metal nanoparticles for molecular imaging of carcinogenesis in vivo. J. Biomed. Opt. **12**, 034007 (2007). https://doi.org/10.1117/1.2737351
28. Shu, Y., Haque, F., Shu, D., Li, W., Zhu, Z., Kotb, M., et al.: Fabrication of 14 different RNA nanoparticles for specific tumor targeting without accumulation in normal organs. RNA **19**, 767–777 (2013). https://doi.org/10.1261/rna.037002.112
29. Mukherjee, P., Khademhosseini, A., Seaberg, J., Montazerian, H., Hossen, M.N., Bhattacharya, R.: Hybrid nanosystems for biomedical applications. ACS Nano **15**, 2099–2142 (2021). https://doi.org/10.1021/acsnano.0c09382
30. Chen, Z., Peng, Y., Xie, X., Feng, Y., Li, T., Li, S., et al.: Dendrimer-functionalized superparamagnetic nanobeacons for real-time detection and depletion of HSP90α mRNA and MR imaging. Theranostics **9**, 5784–5796 (2019). https://doi.org/10.7150/thno.36545
31. Tenchov, R., Sasso, J.M., Wang, X., Liaw, W.S., Chen, C.A., Zhou, Q.A.: Exosomes nature's lipid nanoparticles, a rising star in drug delivery and diagnostics. ACS Nano **16**, 17802–17846 (2022). https://doi.org/10.1021/acsnano.2c08774
32. McKiernan, J., Donovan, M.J., O'Neill, V., Bentink, S., Noerholm, M., Belzer, S., et al.: A novel urine exosome gene expression assay to predict high-grade prostate cancer at initial biopsy. JAMA Oncol. **2**, 882–889 (2016). https://doi.org/10.1001/jamaoncol.2016.0097
33. Clair, D.: FDA grants breakthrough device designation to bio-techne's ExoDx prostate (IntelliScore) (EPI) Test. Bio-Techne Corp (2019). https://www.exosomedx.com/news-events/fda-grants-breakthrough-device-designation-bio-technes-exodx-prostate-intelliscore-epi. Last access 26 Feb 2024

34. Kompella, U.B., Amrite, A.C., Pacha Ravi, R., Durazo, S.A.: Nanomedicines for back of the eye drug delivery, gene delivery, and imaging. Prog. Retin. Eye Res. **36**, 172–198 (2013). https://doi.org/10.1016/j.preteyeres.2013.04.001
35. Sahoo, S.K., Parveen, S., Panda, J.J.: The present and future of nanotechnology in human health care. Nanomed. Nanotechnol. Biol. Med. **3**, 20–31 (2007). https://doi.org/10.1016/j.nano.2006.11.008
36. Bolsoni, J., Liu, D., Mohabatpour, F., Ebner, R., Sadhnani, G., Tafech, B., et al.: Lipid nanoparticle-mediated hit-and-run approaches yield efficient and safe in situ gene editing in human skin. ACS Nano **17**, 22046–22059 (2023). https://doi.org/10.1021/acsnano.3c08644
37. Kumar Dash, D., Kant Panik, R., Kumar Sahu, A., Tripathi, V.: Role of nanobiotechnology in drug discovery, development and molecular diagnostic. Appl. Nanobiotechnol., 1–13 (2020). https://doi.org/10.5772/intechopen.92796
38. Kumari, A., Yadav, S.K., Yadav, S.C.: Biodegradable polymeric nanoparticles based drug delivery systems. Colloids Surf. B Biointerfaces **75**, 1–18 (2010). https://doi.org/10.1016/j.colsurfb.2009.09.001
39. Saiz, E., Zimmermann, E.A., Lee, J.S., Wegst, U.G.K., Tomsia, A.P.: Perspectives on the role of nanotechnology in bone tissue engineering. Dent. Mater. **29**, 103–115 (2013). https://doi.org/10.1016/j.dental.2012.08.001
40. Cunha, C., Panseri, S., Antonini, S.: Emerging nanotechnology approaches in tissue engineering for peripheral nerve regeneration. Nanomed. Nanotechnol. Biol. Med. **7**, 50–59 (2011). https://doi.org/10.1016/j.nano.2010.07.004
41. Amna, T., Shamshi Hassan, M., Nam, K.T., Bing, Y.Y., Barakat, N.A.M., Khil, M.S., et al.: Preparation, characterization, and cytotoxicity of CPT/Fe_2O_3-embedded PLGA ultrafine composite fibers: a synergistic approach to develop promising anticancer material. Int. J. Nanomed. **7**, 1659–1670 (2012). https://doi.org/10.2147/IJN.S24467
42. Lin, G.G., Scott, J.G.: Nanotechnology in drug delivery and tissue engineering: from discovery to applications Jinjun. Nano Lett. **100**, 130–134 (2012). https://doi.org/10.1021/nl102184c.Nanotechnology
43. Cleeton, C., Keirouz, A., Chen, X., Radacsi, N.: Electrospun nanofibers for drug delivery and biosensing. ACS Biomater. Sci. Eng. **5**, 4183–4205 (2019). https://doi.org/10.1021/acsbiomaterials.9b00853
44. Zhang, M., Ai, W.W., Mei, Z.L., Hu, Y.H., Zhang, Z.L.: Delivery of biotinylated IGF-1 with biotinylated self-assembling peptides combined with bone marrow stem cell transplantation promotes cell therapy for myocardial infarction. Exp. Ther. Med. **14**, 3441–3446 (2017). https://doi.org/10.3892/etm.2017.4982
45. Davis, M.E., Hsieh, P.C.H., Takahashi, T., Song, Q., Zhang, S., Kamm, R.D., et al.: Local myocardial insulin-like growth factor 1 (IGF-1) delivery with biotinylated peptide nanofibers improves cell therapy for myocardial infarction. Proc. Natl. Acad. Sci. U.S.A. **103**, 8155–8160 (2006). https://doi.org/10.1073/pnas.0602877103
46. Jain, K.K.: Applications of Biotechnology in Cardiovascular Therapeutics (2011). https://doi.org/10.1007/978-1-61779-240-3
47. Kohli, N., Ho, S., Brown, S.J., Sawadkar, P., Sharma, V., Snow, M., et al.: Bone remodelling in vitro: Where are we headed? A review on the current understanding of physiological bone remodelling and inflammation and the strategies for testing biomaterials in vitro. Bone **110**, 38–46 (2018). https://doi.org/10.1016/j.bone.2018.01.015
48. Tao, S.C., Guo, S.C., Zhang, C.Q.: Modularized extracellular vesicles: the dawn of prospective personalized and precision medicine. Adv. Sci. **5** (2018). https://doi.org/10.1002/advs.201700449
49. Liu, W., Yu, M., Chen, F., Wang, L., Ye, C., Chen, Q., et al.: A novel delivery nanobiotechnology: engineered miR-181b exosomes improved osteointegration by regulating macrophage polarization. J. Nanobiotechnol. **19**, 1–18 (2021). https://doi.org/10.1186/s12951-021-01015-y
50. Lu, Y., Mahato, R.I.: Pharmaceutical perspectives of cancer therapeutics. Pharm. Perspect. Cancer Ther., 1–693 (2009). https://doi.org/10.1007/978-1-4419-0131-6

51. Wang, P., Jia, L., Sanders, B.G., Kline, K.: Liposomal or nanoparticle α-TEA reduced 66cl-4 murine mammary cancer burden and metastasis. Drug Deliv. **14**, 497–505 (2007). https://doi.org/10.1080/10717540701606418
52. Gordon, E., Levy, J., Reed, R., Liu, L., Wendler, C., Hall, F.: Targeting metastatic cancer from the inside: a new generation of targeted gene delivery vectors enables personalized cancer vaccination in situ. Int. J. Oncol. **33**, 665–675 (2008). https://doi.org/10.3892/ijo_00000052
53. Takeda, M., Tada, H., Higuchi, H., Kobayashi, Y., Kobayashi, M., Sakurai, Y., et al.: In vivo single molecular imaging and sentinel node navigation by nanotechnology for molecular targeting drug-delivery systems and tailor-made medicine. Breast Cancer **15**, 145–152 (2008). https://doi.org/10.1007/s12282-008-0037-0
54. Wang, J., Wang, H., Wang, H., He, S., Li, R., Deng, Z., et al.: Nonviolent self-catabolic DNAzyme nanosponges for smart anticancer drug delivery. ACS Nano **13**, 5852–5863 (2019). https://doi.org/10.1021/acsnano.9b01589
55. Atala, A.: Technology insight: applications of tissue engineering and biological substitutes in urology. Nat. Clin. Pract. Urol. **2**, 143–149 (2005). https://doi.org/10.1038/ncpuro0121
56. Wheeldon, I., Farhadi, A., Bick, A.G., Jabbari, E., Khademhosseini, A.: Nanoscale tissue engineering: spatial control over cell-materials interactions. Nanotechnology **22** (2011). https://doi.org/10.1088/0957-4484/22/21/212001
57. Bellamkonda, R.V.: Peripheral nerve regeneration: an opinion on channels, scaffolds and anisotropy. Biomaterials **27**, 3515–3518 (2006). https://doi.org/10.1016/j.biomaterials.2006.02.030
58. Subramanian, A., Krishnan, U.M., Sethuraman, S.: Fabrication of uniaxially aligned 3D electrospun scaffolds for neural regeneration. Biomed. Mater. **6** (2011). https://doi.org/10.1088/1748-6041/6/2/025004
59. Kondo, M., Kameishi, S., Grainger, D.W., Okano, T.: Novel therapies using cell sheets engineered from allogeneic mesenchymal stem/stromal cells. Emerg. Top Life Sci. **4**, 677–689 (2021). https://doi.org/10.1042/ETLS20200151
60. Girão, A.F., Serrano, M.C., Completo, A., Marques, P.A.A.P.: Is graphene shortening the path toward spinal cord regeneration? ACS Nano **16**, 13430–13467 (2022). https://doi.org/10.1021/acsnano.2c04756
61. Wei, X., Yu, S., Zhang, T., Liu, L., Wang, X., Wang, X., et al.: MicroRNA-200 loaded lipid nanoparticles promote intestinal epithelium regeneration in canonical MicroRNA-deficient mice. ACS Nano **17**, 22901–22915 (2023). https://doi.org/10.1021/acsnano.3c08030
62. Oliver, J.D.: Recent findings on the viable but nonculturable state in pathogenic bacteria. FEMS Microbiol. Rev. **34**, 415–425 (2010). https://doi.org/10.1111/j.1574-6976.2009.00200.x
63. Gilmartin, N., O'Kennedy, R.: Nanobiotechnologies for the detection and reduction of pathogens. Enzyme Microb. Technol. **50**, 87–95 (2012). https://doi.org/10.1016/j.enzmictec.2011.11.005
64. Lahiff, E., Lynam, C., Gilmartin, N., Diamond, D.: The increasing importance of carbon nanotubes and nanostructured conducting polymers in biosensors. Anal. Bioanal. Chem. **398**, 1575–1589 (2010). https://doi.org/10.1007/s00216-010-4054-4.
65. Jain, K.K.: The role of nanobiotechnology in drug discovery. Adv. Exp. Med. Biol. **655**, 37–43 (2009). https://doi.org/10.1007/978-1-4419-1132-2_4
66. Abu-youssef, M.A.M., Soliman, S.M., Langer, V., Gohar, Y.M., Hasanen, A.A., Makhyoun, M.A., et al.: Synthesis, crystal structure, quantum chemical calculations, DNA interactions and antimicrobial activity of Ag (2-amino-3-methylpyridine) 2] NO3 and [Ag (pyridine-2-carboxaldoxime) NO3], 9788–9797 (2010)
67. Cavicchioli, M., Massabni, A.C., Heinrich, T.A., Costa-Neto, C.M., Abrão, E.P., Fonseca, B.A.L., et al.: Pt(II) and Ag(I) complexes with acesulfame: crystal structure and a study of their antitumoral, antimicrobial and antiviral activities. J. Inorg. Biochem. **104**, 533–540 (2010). https://doi.org/10.1016/j.jinorgbio.2010.01.004
68. Dallas, P., Sharma, V.K., Zboril, R.: Silver polymeric nanocomposites as advanced antimicrobial agents: classification, synthetic paths, applications, and perspectives. Adv. Colloid Interface Sci. **166**, 119–135 (2011). https://doi.org/10.1016/j.cis.2011.05.008

69. Liu, S., Wang, H., Yu, L., Ren, Y., Bouma, H.R., Liu, J., et al.: Rapid bacterial detection and gram-identification using bacterially activated, macrophage-membrane-coated nanowired-SI surfaces in a microfluidic device. Nano Lett. **23**, 8326–8330 (2023). https://doi.org/10.1021/acs.nanolett.3c02686
70. Rifat, T., Hossain, M.S., Alam, M.M., Rouf, A.S.S.: A review on applications of nanobots in combating complex diseases. Bangladesh Pharm. J. **22**, 99–108 (2019). https://doi.org/10.3329/bpj.v22i1.40081
71. Walait, M., Rehman Mir, H., Noor, T., Rani, K., Aslam, J., Khalid, K., et al.: Nanobiotechnology: current applications and future scope (2022)
72. Kong, X., Gao, P., Wang, J., Fang, Y., Hwang, K.C.: Advances of medical nanorobots for future cancer treatments. J. Hematol. Oncol. **16**, 1–45 (2023). https://doi.org/10.1186/s13045-023-01463-z
73. Manasjyoti Bhuyan, S.B.: Nanorobots—a panacea to HIV. Int. Res. J. Eng. Technol. **03**, 1309–1315 (2016)
74. Manjunath, A., Kishore, V.: The promising future in medicine: nanorobots. Biomed. Sci. Eng. **2**, 42–47 (2014). https://doi.org/10.12691/bse-2-2-3
75. Boonrong, P., Kaewkamnerdpong, B.: Canonical PSO based nanorobot control for blood vessel repair. Int. J. Biomed. Biol. Eng. **5**, 428–478 (2011)
76. Sivasankar, M.: Brief review on nano robots in bio medical applications. Adv. Robot. Autom. **01**, 1–4 (2012). https://doi.org/10.4172/2168-9695.1000101
77. Dietz, H., Douglas, S.M., Shih, W.M.: Folding DNA into twisted and curved nanoscale shapes. Science **325**(80), 725–730 (2009). https://doi.org/10.1126/science.1174251
78. Padovani, G.C., Feitosa, V.P., Sauro, S., Tay, F.R., Durán, G., Paula, A.J., et al.: Advances in dental materials through nanotechnology: facts perspectives and toxicological aspects. Trends Biotechnol. **33**, 621–636 (2015). https://doi.org/10.1016/j.tibtech.2015.09.005
79. Saadeh, Y., Vyas, D.: Nanorobotic applications in medicine: current proposals and designs. Am. J. Robot. Surg. **1**, 4–11 (2014). https://doi.org/10.1166/ajrs.2014.1010
80. Khazaei, M., Hosseini, M.S., Haghighi, A.M., Misaghi, M.: Nanosensors and their applications in early diagnosis of cancer. Sens. Bio-Sensing Res. **41**, 100569 (2023). https://doi.org/10.1016/j.sbsr.2023.100569
81. Chaudhari, Y.S., Mody, H.R., Kokardekar, R.R., Paranjpe, P.P., Bahuguna, Y.M.: Nanobiotechnology: the emerging field in pharmaceutical sciences. J. Pharm. Res. **3**, 1297–1299 (2010)
82. Shan, B., Broza, Y.Y., Li, W., Wang, Y., Wu, S., Liu, Z., et al.: Multiplexed nanomaterial-based sensor array for detection of covid-19 in exhaled breath. ACS Nano **14**, 12125–12132 (2020). https://doi.org/10.1021/acsnano.0c05657
83. Polley, N., Sardar, S., Werner, P., Gersonde, I., Kanehira, Y., Bald, I., et al.: Photothermomechanical nanopump: a flow-through plasmonic sensor at the fiber tip. ACS Nano (2022). https://doi.org/10.1021/acsnano.2c09938
84. Maximize Market Research: Nanobiotechnology market : global industry analysis and forecast 2023–2029 n.d. https://www.maximizemarketresearch.com/market-report/nanobiotechnology-market/207986/. Last accessed 10 Jan 2024
85. Luqueño Fabián, F.: Advantages and drawbacks of the nanotechnology and biotechnology toward shaping a global sustainable development. Biosci. Biotechnol. Res. Asia **16**, 693–695 (2019). https://doi.org/10.13005/bbra/2784
86. Bruce, D.: Ethical and social issues in nanobiotechnologies: Nano2Life provides a European ethical "think tank" for research in biology at the nanoscale. EMBO Rep. **7**, 754–758 (2006). https://doi.org/10.1038/sj.embor.7400762
87. Qamar, S.A., Asgher, M., Khalid, N., Sadaf, M.: Nanobiotechnology in health sciences: current applications and future perspectives. Biocatal. Agric. Biotechnol. **22**, 101388 (2019). https://doi.org/10.1016/j.bcab.2019.101388
88. Hamburg, M., Collins, F.: The path to personalized medicine. N. Engl. J. Med. **363**, 301–304 (2010). https://doi.org/10.1056/NEJMp1006304
89. Fakruddin, M., Hossain, Z., Afroz, H.: Prospects and applications of nanobiotechnology: a medical perspective. J. Nanobiotechnol. **10**, 1–8 (2012). https://doi.org/10.1186/1477-3155-10-31

90. Sahoo, S.K., Labhasetwar, V.: Nanotech approaches to drug delivery and imaging. Drug Discov. Today **8**, 1112–1120 (2003). https://doi.org/10.1016/S1359-6446(03)02903-9

Akash A. Podutwar, a distinguished scholar and current Ph.D. candidate in the School of Health Science and Technology at MIT-World Peace University, Pune, holds a Master of Pharmacy in the same discipline from the institution. Acknowledged for his academic prowess, he has published review articles in reputable journals such as the "Journal of Pharmaceutical Research International" and the "International Journal of Recent Scientific Research," including contributions to esteemed publication houses like "IGI Global." These publications underscore his commitment to advancing pharmaceutical research and providing valuable insights to the academic community.

In addition to his scholarly achievements, Mr. Podutwar has showcased innovation through a design patent titled "Novel Integrated Dissolution-Disintegration Apparatus," successfully filed in 2022. This patent reflects his dedication to pioneering solutions in pharmaceutics, bridging theoretical knowledge with practical applications.

Mr. Akash Podutwar's academic pursuits and patent achievements highlight his fervor for pushing the boundaries of pharmaceutical research. As a dedicated scholar and researcher, he continues to make significant contributions to the discipline, leaving an enduring mark on the landscape of pharmaceutics and academic excellence.

Purva U. Chandorkar, a diligent Ph.D. Scholar in the Department of Pharmaceutical Sciences at MIT-WPU, holds a Master's degree in Pharmaceutics from the same institution. Her academic prowess is evidenced by three publications in esteemed journals such as the "International Journal of Pharmacy, Biology and Applied Sciences," "International Journal of Recent Scientific Research", and the "Research Journal of Biotechnology". Additionally, she has authored a book chapter in IGI Global's publication titled "Cutting-Edge Applications of Nanomaterials in Biomedical Sciences".

Demonstrating her collaborative research efforts, Ms. Chandorkar has successfully filed a design patent alongside her colleagues. Her active engagement in the scholarly community extends to international e-poster presentation conferences hosted by institutions like "SNJB" and "BNCP" College of Pharmacy, where she presented two posters. Purva Chandorkar's academic pursuits reflect a commitment to excellence and a noteworthy contribution to the field of pharmaceutical sciences.

Anuruddha R. Chabukswar is working as Professor and Head of Pharmaceutical Chemistry at School of Health Sciences and Technology, Pharmaceutical Sciences Dr Vishwanath Karad MIT World Peace University, Pune, India. Dr. A. R. Chabukswar has rich academic and research experience of 24 years in the field of Pharmaceutical Education and research. He has published more than 80 papers in various national and international journals. He has filed 5 patents and 1 patent has been granted. He has 2 major research projects to his credit from DST and AICTE, New Delhi, India. He has guided 04 Ph.D. students and many UG and PG students in his career. His research areas include drug design, synthesis, analytical method development, and formulation studies.

Satish A. Polshettiwar is working as HoD and Asst. Professor at School of Health Sciences and Technology MIT-WPU Pune. He has published more than 50 research and review articles in peer-review journals. He has filed 7 Indian patents and 2 patents have been granted. He has published 3 textbooks on Pharm Microbiology and Pharm Engineering.

Swati C. Jagdale is a highly accomplished academic and researcher in the field of pharmaceutics. She earned her B.Pharm. (1998), M.Pharm. (2000) from Pune University, and Ph.D. (2009) from Shivaji University Kolhapur. Currently, she holds the prestigious position of Professor in the Department of Pharmaceutics at the School of Health Sciences and Technology in Pune. Dr. Jagdale's significant contributions to the field of pharmaceutics are widely recognized, with more than 80 research articles published in reputed national and international journals, along with five authored books and ten book chapters. She holds one Indian patent and two patents from the Republic of South Africa, with an additional five Indian patents currently in process. She has also received funding for her research projects from the Council of Scientific and Industrial Research (CSIR) (Amount 21.36 Lakh) and AICTE-MODROB grant (Rs.17.22 Lacs), in addition to completing two minor research projects at Savitribai Phule Pune University (4.9 Lakhs).

As an experienced guide, Dr. Jagdale has successfully mentored 75 M. Pharm. students and 06 Ph.D. students, with four M. Pharm. and three Ph.D. students currently working on their projects under her guidance. Dr. Jagdale has presented her research work at various national and international conferences and also serves as a member of the Board of Studies under Science and Technology for Faculty of Pharmaceutics at Savitribai Phule Pune University. Her research article on "Formulation and Evaluation of Gastro-retentive Drug Delivery System of Propranolol Hydrochloride" was one of the top five downloaded articles in AAPS PharmSciTech in 2012. She had also previously served as a member of the Board of Studies for Faculty of Pharmaceutics at the University of Pune.

Chapter 9
Nanotechnology, Artificial Intelligence, and the Future of Manufacturing

Promila Sharma, Varinder Singh, Anita Thakur, and Ashok Kumar Malik

Introduction

Over the years, a variety of technology sectors have appreciated ironic development in the twenty-fir century. With industrial applications spanning from farming to health, it wouldn't be erroneous to argue that nanotechnology convergence with artificial intelligence is the most important and extensively used technique [1].

Through the application of diverse approaches, a multitude of industries are working on advance usage of AI and nanotechnology to satisfy the needs and expectations of the people they serve. Manufacturing has always been a component of many different sectors, including gaming, electronics, textiles, and many others. With time, physical labor was replaced by digital and technological advancements in the manufacturing sectors. Employment firms that specialize in manufacturing state that although technology has made labor less intensive, human contribution is still necessary. Manufacturing will need to quickly adapt to new technologies in the future, and this will need hiring qualified personnel to ensure the efficient operation of every aspect of the manufacturing process. Let's examine the emerging trends in nanotechnology that have transformed machine learning-driven industries.

Artificial learning along with incorporated analytical systems is being used in the fast-expanding field of nanotechnology to adapt new technological forms. The field of artificial intelligence exhibits great promise for the synthesis, characterization, refining, testing, and prediction of nanoparticle activity in vitro and in vivo [2]. There are still many challenges and impediments on the path of discovery that was started in the experimental laboratories. However, ongoing advances and research in

P. Sharma (✉) · V. Singh · A. Thakur
Department of Chemistry, RIMT University, Mandi, Gobindgargh 147301, Punjab, India
e-mail: promilaattri07@gmail.com

P. Sharma · A. K. Malik
Department of Chemistry, Punjabi University, Patiala 147002, Punjab, India
e-mail: ashokmalik@pbi.ac.in

S. C. George and B. Tawiah (eds.), *Nanotechnology in Societal Development*, Advanced Technologies and Societal Change, https://doi.org/10.1007/978-981-97-6184-5_9

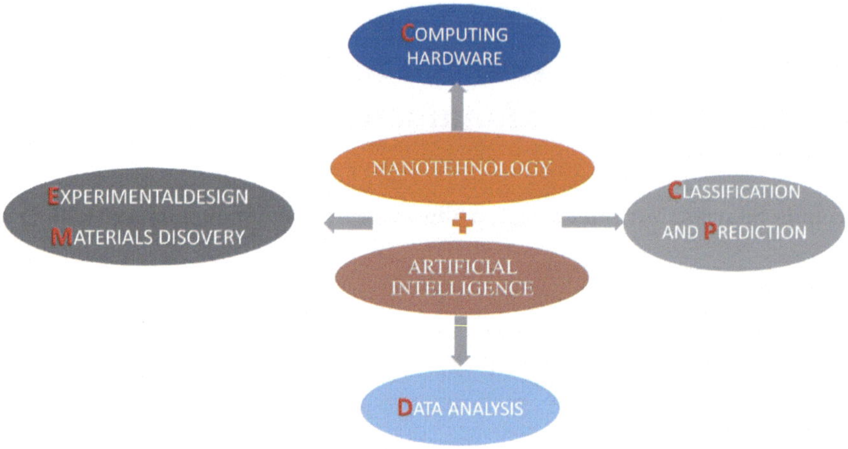

Fig. 9.1 Examples of developments based on integration of AI and Nanotechnology

automated processes, nanotechnology, and AI integration provide hope that we are headed in a favorable pathway as shown in Fig. 9.1. Examples of these developments include.

- Computers with intelligent algorithms: This will help create similar cognitive patterns in machines so they can act like people and execute suitable actions to prevail over difficult circumstances [3].
- Development of expert systems: Expert system development is the process of building intelligent, automated machines that can advise humans on what to do.
- Applications across numerous disciplines: Artificial Intelligence will facilitate the application of various areas, including health care, psychology, medical science, ethics, and the natural sciences. The influence of artificial intelligence (AI) on drug research and design, the application of machine learning and statistical programming techniques, and the setup and refinement of simulations built around nanotechnologies are all covered in this section of the book. It also illustrates the opportunities and difficulties related to adopting cutting-edge methods and modern technology.

An Overview of the Working of AI and Nanotechnology

AI and nanotechnology are rising simultaneously; thus, collaboration is vital. Artificial Intelligence utilizes nanotechnology to collect data that would be hard to get otherwise. It is essential to comprehend the domains of artificial intelligence and nanotechnology in order to get the most out of this collaboration. Innovative nanostructured materials and nanocomposites unquestionably offer superior performance over macro/bulk equivalents and a plethora of applications. Due to their numerous

Fig. 9.2 Understanding of computer science and programming with the help of AI

uses across several industries, these two technologies are generating a lot of interest from the public.

The term artificial intelligence refers to a broad category within computer science that includes technologies as represented in Fig. 9.2 [4, 5].

These technologies hold the potential to enable machines to emulate human intelligence in order to carry out a variety of complex operations. Large and variable datasets are used to train machine learning and deep learning models, which allow intelligent models to recognize, categorize, or forecast patterns for a given input. It is not, however, restricted to the tasks that are given because many solutions for a broad range of real-world issues have been made possible by further advancements in the amendment. Statistical patterns, computational intelligence, object identification, classification, and prediction are all recognized by AI systems.

The implementation of AI tools in the field of nanotechnology has significantly increased in the past few years, with applications including simulations, nanocomputing, and scanning probe microscopy. A particular method known as functional recognition has been employed to distinguish the local actions from spectroscopic reactions, despite the existence of numerous AI algorithms. An Artificial Neural Network (ANN) has been found to have a significant role in identifying the specific action of the imaged materials, streamlining the procedure, and permitting the consideration of just relevant variables. Functioning at the nanoscale level involves several fundamental simulation concerns. Since authentic optical data cannot be obtained at the nanoscale, numerical models are typically used to interpret nanoscale images [6]. On the other hand, several kinds of programs are used for these kinds of image representations, and, oftentimes, it can become a complicated process to incorporate several parameters for precise representations.

In this sense, AI efficiently creates simulations and easier-to-understand result interpretations. In non-linear cognitive systems, formulation, process factors, and controlled release are integrally related; hence, optimized controlled release information and interconnected networks are necessary.

AI methods are constantly being presented to address different nanotechnology challenges. With an emphasis on cutting down on computational time and optimizing parameter estimation, prediction, and system simulation, the three main domains in creating principles of nanoscale simulation are designing nanosystems, designing nanocomputing, and designing AI methods. With human-like intelligence systems, artificial intelligence (AI) facilitates the extraction and processing of quantitative data.

The increased development and availability of medical imaging data in recent years has encouraged research into clinical applications of AI. The utilization of human knowledge to solve complicated problems more quickly and efficiently has made artificial intelligence (AI) an increasingly potent weapon. Promising developments in artificial intelligence and machine learning offer a revolutionary potential for drug discovery, formulation, and medication testing. Through the application of AI algorithms that examine vast amounts of biological data, such as proteomics and genomes, scientists can pinpoint targets linked to disease and predict how these targets may interact with possible therapy candidates. This renders the ability to approach drug discovery in a more effective and targeted manner, which boosts the possibility of successful medication approvals. AI can also save development costs by streamlining the research and development process. Machine learning algorithms can predict the pharmacokinetics and toxicity of drug candidates and aid in designing experiments.

Need for Collaborative Working of AI and Nanotechnology

Integrating intelligence into nanotechnology poses tremendous challenges in terms of control, communication, power, and safety at such minute scales. It is essential to ensure the moral and responsible application of intelligent nanotechnology in order to mitigate potential dangers, such as unexpected outcomes, effects on the environment, privacy difficulties, and safety hazards. It is imperative to effectively address the observed shortcomings through suitable modifications, while simultaneously acknowledging that offering enough training can be problematic.

Two of the most cutting-edge and popular topics in technology today are artificial intelligence (AI) and nanotechnology. Furthermore, even though they appear to be two entirely unrelated topics, they have a lot in common. From inventing more inventive materials to generating novel production techniques, these two technologies have the potential to completely transform the world as we know it. The majority of scientific domains can benefit from the use of nanoscience and nanotechnology, two overlapping fields of human endeavor that examine the properties and uses of nanoscale materials. Given its prevalence in a variety of industries, including robotics and medical diagnostics, nanoscience and nanotechnology can be utilized in conjunction with artificial intelligence [7].

Modern methods like artificial intelligence (AI), which have shown promise in the fields of drug discovery and optimization, are the result of ongoing technological advancements. Nevertheless, despite its many benefits, nanomedicine faces a number of difficulties. The extracellular matrix (EPR) effect, the size and composition of drug delivery systems, problems with drug reservoir design, biocompatibility, drug concentrations, and toxicity are some of the most frequently brought up issues [8]. While considering nanorobots as potential future theranostic applications, the fabrication challenges are another important factor to consider. Noise and unknown parameters, which could result in an inability to give the right quantity of medication and the inability to distinguish between tumor and healthy cells, are the drawbacks of using nanorobots for drug delivery. Artificial Intelligence (AI) has promise in addressing various nanotechnology-related problems, including complicated data processing and analysis, as well as aiding in medication design and discovery. Furthermore, by integrating AI in this way, the difficulties associated with precise dose distribution can be addressed. Furthermore, AI is very good at genetic programming and can find a variety of intricate patterns to detect additional information on cancer genetics. Fuzzy logic, decision trees, and ANNs are essential elements of contemporary drug development procedures. While ANNs are very good at classification, pattern recognition, and prediction, other kinds of networks can be very useful as well, like supervised associating networks. These are particularly useful when using response surface methodology, which makes it possible to investigate relationships between explanatory and response variables more quickly. Reinforcement learning is another cutting-edge AI technique that has attracted interest in silicomedicine. The main explanation for this is that new AI systems are less reliant on datasets and are capable of learning from their surroundings. A supervised learning technique that can distinguish between normal and pathological patient profiles and locate missing characteristics from datasets was recently proposed in a study. At first, it was believed that artificial intelligence (AI) and associated fields would not benefit much from the complex nanotechnology systems. Nonetheless, this is no longer the case due to breakthroughs and the expanding application of AI. These days, research encompasses developments in the characterization, design, and production of nanomaterials enhanced by AI.

Some General Aspects of AI and Nanotechnology in Future Manufacturing

The Medical Advancement

Over time, parallel developments in AI and nanotechnology have revealed similarities in their mutual support and cooperation. The medical industry will surely be among the major beneficiaries of the fusion of AI with nanotech. The precision, command, enhanced signal, etc., provided by AI, can be very helpful to fields employing or researching the application of nanotechnologies, such as organ regeneration from stem cells and microscopy [9]. Furthermore, there is a significant variation among individuals concerning the dosages required to achieve therapeutic synergy and the extent of drug exposure required to achieve the best possible therapy results. An assortment of adaptable carriers is being utilized by the field of nanomedicine to enhance the localization and targeting of drug delivery to illness locations. Recently, these carriers have been used as entry points for combinatorial nanotherapy, which targets several disease pathways at once to significantly improve therapeutic efficacy. However, combination therapy employing nanomedicine will suffer the same obstacles that standard drug development roadmaps do. Among these are the maintenance of pharmaceutical synergy, enhanced targeted effectiveness, and logically planned medication exposure. Finding medication synergy is a good beginning toward better treatment results; however, the goals of optimization and synergy are fundamentally different. To achieve global treatment optimization, a significant amount of medication and dose parameter space is opened up by combination therapy design and needs to be balanced.

Within the same patient, these factors also change over time. A redesigned combination therapy landscape that will be significantly improved by therapeutics modified by nanotechnology will be made possible by the recent studies discussed in this work, and the emergence of AI will play a critical role in reconciling this space into an actionable treatment response.

Since AI can agnostically optimize treatment in a mechanism-independent manner, the convergence of AI and nanomedicine can be exploited to address these challenges. Through the rational coordination of all these factors—drug choosing, medication picking, stimuli-responsive material selection, frequency and amplitude stimulus, etc., with outputs such as efficacy and safety, which are dependent on targeting and localization effects—artificial intelligence (AI) has the potential to boost combination nanotherapy for patients as well as populations [10].

Microscopy

Despite the significant advancements in recent years, the atomic force microscopy (AFM) still has difficulty generating high-quality signals from scanning instruments [11].

The major problem happens to be the several complicated, hard-to-decipher, and varied tip-sample interactions, something microscopes depend on. But, as you might have guessed, the use of AI can help deal with the issue.

Functional recognition imaging (FR-SPM), an AI approach, is based on directly identifying local actions through measured spectroscopic reactions. The emphasis of this procedure will be on simplifying neural network modeling input data through a combination of ANNs [12] and principal component analysis (PCA) [13].

Research

Everything gets more challenging when the particles in question are in the range of nanometers, from the extensive scope of the investigation to the measurement of the variables, recording, and interpretation of the results. Artificial intelligence (AI) can be used in these situations to simplify and improve the process, including data optimization, technology-specific algorithm design, parameter estimation, and experiment interpretation. It follows that the incorporation of AI into research divisions at numerous nanotechnology facilities is not surprising given the desire for increased precision and efficiency.

Introducing Better and Sustainable Food Alternatives

According to the experts, the forthcoming fundamental advancement in human food evolution may result in sustainable solutions because of the integration of nanotechnology and AI. The efforts to build plant-based meat, which in terms of taste, smell, and look mimics the meat but without any animal product [14].

The community of vegetarians has been debating synthetic meat for some time now. However, as every fundamental flavor and aroma of food is determined by its micro-components, which are formed of a mix of particular molecules like proteins, lipids, etc., it becomes very challenging to replicate the same, especially when avoiding the use of animal products [15].

A few companies successfully implemented the method to replicate liquid food, including scrambled eggs, mayo, etc., given the complexity of replicating solid foods. However, it is not far from reality when the alternatives to livestock farming using AI and nanotech are made into accessible and scalable products [16, 17].

Challenges to global food security include climate change, population growth, competing needs for land for biofuel production, and deteriorating soil quality. It takes audacious new methods and the integration of information from other disciplines, including materials science and informatics, to find sustainable solutions. Exciting potential for sustainable food production arises from the merging of nanotechnology and artificial intelligence in precision agriculture, as farmers use these technologies to react in real time to changes in crop development. Designing nanoscale agrochemicals that combine the best safety and functionality profiles will be made possible by combining current models for nutrient cycling and crop productivity with nano informatics techniques to optimize targeting, uptake, delivery, nutrient capture, and long-term impacts on soil microbial communities [18].

Nanocomputing

Many nanoscale structures, including proteins and DNA, can be used to create nanocomputing. There are numerous approaches to building nanocomputers, which can result in a variety of nanocomputer types [19]. Electronic nanocomputing: Using nanolithography to build tiny circuits; nanocomputers can be electronic. These would function analogously to modern microcomputers. Intelligent sensors based on nanoelectronics offer a singular chance to interact with the physical environment at submicron and smaller sizes. Some believe that future computers will most likely be electrical nanocomputers. Mechanical nanocomputing: This uses just mechanical principles and focuses on the smallest moving components. The information would be encoded by small moving parts called nanogears in mechanical nanocomputers. Mechanical nanocomputers are capable of controlling or operating nanorobots [20]. However, there is debate surrounding mechanical nanocomputer technology, with some experts claiming it to be impracticable.

Chemical/Biochemical Nanocomputing: A chemical computer is a computer that uses chemical bonding to process data. Chemical structures and their interactions are the information they process and store. Chemical structure and interaction data would be processed and stored by both chemical and biochemical nanocomputers.

Quantum nanocomputing: As artificial general intelligence (AGI), nanotechnology, and quantum computing become more accessible, the field of medical aesthetics is set to undergo a technological revolution. Advances in quantum computing, nanoscale interventions, and AI-driven personalized therapy protocols are revolutionizing the field of medical aesthetics. Although these technologies promise exceptional accuracy and productivity, their integration presents significant ethical and legal issues. As we set out to reinvent beauty and health care through the convergence of cutting-edge technology, finding a balance between innovation and prudent deployment will be essential [21]. Quantum physics is the foundation of quantum nanocomputers. Every item of data will be stored in them as a quantum state of the computer. They integrate concepts from computer science, quantum physics, and classical information theory. Quantum computers use quantum bits, also known as qubits, to encode information, as opposed to traditional digital computers that store information in bits. Quantum computers, according to certain people, hold the key to real artificial intelligence. Unpredictability is the primary issue with quantum nanocomputers.

Prominent corporations including Motorola, Intel, IBM, and HP are making significant investments in nanocomputer research and development.

Applications of Intelligent Nanotechnology

Intelligent nanotechnology refers to the integration of nanotechnology with intelligent systems or artificial intelligence (AI) to create sophisticated nanoscale devices, materials, or systems capable of autonomous or semi-autonomous operation and decision-making at the nanometer scale. The advancement of intelligent nanotechnology holds promise for groundbreaking innovations across various industries, although its practical implementation is still in the early stages and requires extensive research and development. Artificial Intelligence (AI) and nanotechnology are two rapidly advancing fields that have the potential to revolutionize various industries and aspects of everyday life. Their convergence opens up new possibilities and applications across different domains.

Here are some potential aspects and applications of intelligent nanotechnology:

AI-Assisted Nanoscale Manufacturing and Precision Engineering

Intelligent nanotechnology can enable the precise manipulation and assembly of nanoscale materials and structures. This could involve nanobots or nanoscale machinery guided by AI algorithms to fabricate intricate structures. They are using AI to optimize and control nanomanufacturing processes, ensuring accuracy, efficiency, and scalability in the production of nanoscale materials and devices. Quality Control and AI algorithms can enhance quality control and streamline production in nanotechnology-based industries.

Predictive Maintenance

AI algorithms analyze data from sensors embedded in manufacturing equipment to predict and prevent potential breakdowns, reducing downtime and maintenance costs. Maintenance innovations can be further leveraged to anticipate product performance degradation and to autonomously manage and optimize product service requirements as more software and integrated intelligence are included in industrial goods and systems. Cyber-physical systems, cloud computing, big data, artificial intelligence, and the industrial Internet of Things (IIoT) are some of the key enabling technologies that are helping to improve system performance, achieve near-zero downtime, guarantee predictive maintenance, and more. Starting with system operations and availability, we have the essentials. By using cutting-edge sensing and data processing methods for diagnostics and prognosis, machine availability is maximized. Systems are dynamically modeled to optimize operations. To mitigate the detrimental effect on productivity, maintenance events are carefully scheduled. Numerous revolutionary technologies have emerged throughout the last 20 years. To create links between traditional maintenance analytics and digitalized smart manufacturing systems, these technologies are employed as enablers of Intelligent Maintenance Systems [22–24].

Nanofabrication Techniques

AI-guided nanotechnology facilitates precise manufacturing at the atomic or molecular level, enabling the production of nanoscale components and devices with high accuracy. The world over, a growing number of electrical, photonic, energy, and biological devices are utilizing nanofabrication to create useful micro- and nanofeatures. These unique device creations stem from the incorporation of high-quality, reasonably priced micro/nanofeatures into 3D designs [25–27].

Nanoscale 3D Printing

Integration of AI and nanotechnology can enhance 3D printing technologies to produce intricate structures with nanoscale precision. According to the analysis, the main factors propelling computer-aided design (CAD) forward are the need for a large range of goods based on a single item input, interactive simulation, and direct design-to-manufacturing integration. According to the authors, the goal of integrating AI into CAD is to produce various software with computational and analytical capabilities to provide an intelligent design environment in the design and manufacturing domains. AI offers a simple way to reuse components that already exist. In the manufacturing sector, 3D printing technology has started to show enormous potential uses and potential benefits in the aerospace, automotive, biomedical, energy, and other industries. This is made possible by the design freedom provided by CAD tools. Research and development in the areas of designs, unique process modeling, optimization, and control are needed to expand the applications of AI-integrated CAD systems [28–30].

Nanomaterials Design

AI can simulate and predict the properties of nanomaterials, enabling the creation of innovative materials with enhanced strength, conductivity, or other specific properties tailored to manufacturing needs. AI algorithms can optimize manufacturing processes at the nanoscale, improving efficiency, reducing waste, ensuring high-quality production, and designing and simulating the properties of new nanomaterials, enhancing their strength, conductivity, or other desired traits. Revolutionary next-generation paradigms made possible by AI have made it possible to greatly speed up every step of the material discovery process and make it easier to explore the vast design space. Also, a few combinatorial synthesis methods within the framework of these measures highlight areas that could use more work. These potentials range from early high-throughput experimentation deployment to groundbreaking AI-powered automated decision-making demonstrations [31–33].

Sensing and Detection

Nanoscale sensors, when combined with intelligent AI, can detect and respond to extremely small-scale changes, aiding in applications such as environmental monitoring, healthcare diagnostics, or even in detecting minute defects in manufacturing, integrating nanosensors capable of detecting minute changes in various environments or biological systems. AI algorithms process the data these nanosensors collect, allowing for real-time analysis, pattern recognition, and decision-making in various fields.

Nanoscale Sensors and AI Analytics

With the advent and use of precision medicine, which is based on AI-enabled medical nanosensors, a new era in diagnostics and therapeutic decision-making is now underway. The combination of state-of-the-art sensing technologies and data fusion algorithms driven by artificial intelligence and sensing technologies will offer a viable approach to within-the-hospital and distant patient surveillance. This equipment will be able to independently and consistently identify a large variety of psychological and pharmacological manifestations of illnesses and problems associated with aging [34]. Nanosensors combined with AI-driven analytics allow for real-time monitoring of manufacturing processes, detecting defects at the micro or nanoscale and ensuring quality control. Nanotechnology facilitates the development of highly sensitive sensors and imaging devices at the nanoscale. AI algorithms can process vast amounts of data generated by these sensors, enhancing accuracy and extracting meaningful information [35, 36].

Within the very short period that nanosensors and AI combined, their chemistry has significantly increased. Two general parameters can be used to classify optical and electronic nanosensors:

(a) **Structure and design**

The numerous nanomaterials and nanomanufacturing procedures with exclusive properties as shown in Table 9.1 are implemented for the manufacturing of nanosensors. At the nanoscale, substances display unique properties that are not noticeable in bulk form, and nanosensors capitalize on these characteristics. Nanosensors are slowly finding applications across a variety of domains, such as medical science and biotechnology, explosive detection, water treatment, the food sector, electronic goods and optics, and forensics. The most important advantages of nanosensors are their rapidity, minimal sample amount required, reliability, prevention competence, and economics.

(b) **Function and Uses**

Nanosensors are becoming more and more involved in practically every facet of daily life. Numerous nanosensors can identify potentially toxic substances or bacteria in

Table 9.1 Properties of material used in manufacturing of nanosensors

Category	Properties		
Materials	Nanotubes	Nanocomposites	Nanoparticles
Medicine	Biocompatible nanoparticles	Nanorobotics	–
Optics and engineering	Surface plasmon	Polarization	Photodetectors
Food industry	Antibacterial	Packaging	Preservatives
Energy and power	Solar cell	Thermal cell	–
Physics	Surface tension	Resonance	Molecular electronics

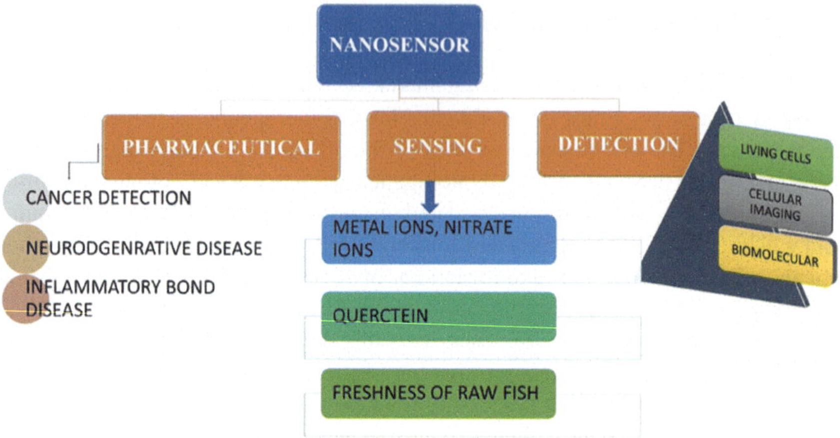

Fig. 9.3 Applications of these AI-based nanosensors

meals, beverages, and the surrounding environment. Nanosensors have a significant impact on medical science. For instance, a range of nanosensors are utilized in targeted medication delivery, protein and genome detection, and cancer detection. Homeland security is one area where deployable sensors are used. In order to prevent military deaths in combat, drones are now equipped with a variety of chemical nanosensors to identify the presence of noxious gases. Some applications of these AI-based nanosensors are represented in Figure 9.3.

Nanorobotics and Surgery

The fundamental building blocks of nanorobots are nanoelectromechanical systems (NEMS). The size of these nanorobotic devices is similar to that of biological cells and organelles. The design, manufacturing, and programming technology for this nanorobotics is referred to as nanorobotics. It is a multidisciplinary topic that calls for highly skilled contributions from a variety of scientific and technological disciplines, including engineering, biotechnology, physics, chemistry, biology, medicine, and pharmaceutical sciences [37]. Nanotechnology enables the creation of nano-sized sensors, actuators, or even nanobots that can interact with biological systems. AI assists in interpreting complex biological data, facilitating personalized medicine, disease diagnosis, and targeted treatments, developing nanoscale robots that can perform specific tasks within the body or in various industries, controlled and guided by AI algorithms. AI can manage and direct these nanobots to execute precise actions, such as targeted drug delivery, cell manipulation, and performing intricate procedures [38, 39].

Nanorobots carried out many types of surgery as shown in Fig. 9.4 with the use of artificial intelligence and nanobiotechnology. Because of their small size, these tiny

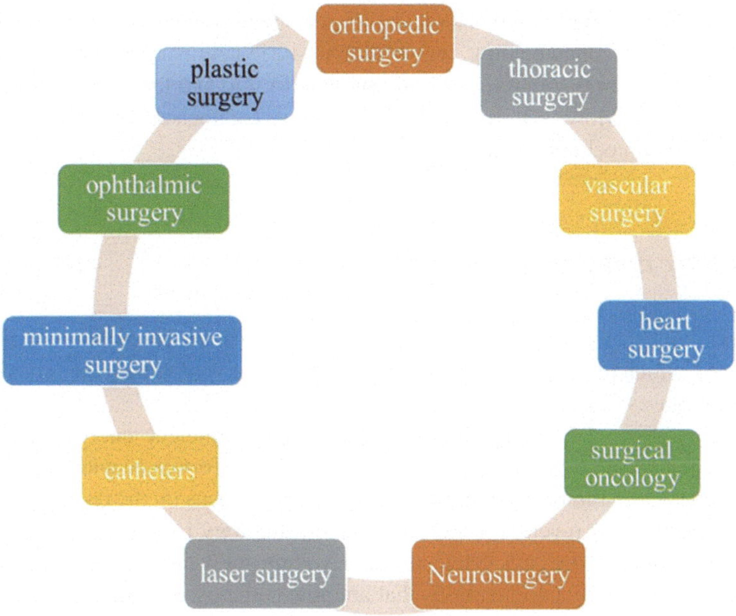

Fig. 9.4 Types of surgery carried out with the help of Nanorobotics

robots can be put into the body via catheters or the vascular system. They can carry out accurate intracellular surgery that is not feasible for a human hand to conduct with the help of the external guidance and supervision of the surgeon.

Using coded ultrasound signals, an onboard computer may perform several functions such as pathology identification, diagnosis, and lesion removal or repair by nanomanipulation, all while maintaining communication with the supervising surgeon. Malignancies are now apparent during surgery thanks to the use of nanoparticles, which enables doctors to completely eradicate or even just visually identify them. Cancers can already be completely removed or even visually checked for metastases throughout the body, thanks to the use of nanoparticles during surgery. By employing in-body sensors to detect and eliminate bacteria before a patient exhibits any symptoms, nanobot seeks to upend the core tenet of contemporary medicine. Nanorobotics can be used to remove malignant cells and unhealthy plaques, repair damaged tissues, monitor biological processes molecularly, and improve human health and functionality. Biochips, microbes, positional nanoassembly, and nucleic acid robots are examples of nanorobots. The structural shape of the diamond gives it strength. Robots made of tiny parts are designed to be put into the circulatory system of humans. A signal produced at the cellular level, force, or motion powers the robot. Precise intracellular surgery can be performed with outside guidance and oversight. Nanorobots are being used in clinical settings for patient monitoring, targeted medicine delivery, disease diagnosis (MRI, nuclear imaging, and ultrasound), and therapy.

As an outcome, the use of nanotechnology together with AI in surgery holds great potential for wound healing, regrowth of tissue in neurosurgery, blood clotting, bone rafting, vertebral spinal disks, and implants for hip and knee repair surgeries, among other uses. Moreover, the surgical instruments (scalpels and needles) are most impacted by nanotechnologies.

Agriculture and Energy Storage

The agriculture sector is one of the largest and biggest resource-intensive enterprises worldwide, which produces approximately more than 55% of the global emission of greenhouse gases [40].

Intelligent nanotechnology might lead to highly efficient energy-harvesting devices or nanoscale batteries. AI could optimize the performance and longevity of these systems. We summarize agriculture trends with a focus on the recent advances of AI applications in revelation involving nanotechnology in this section. Furthermore, by pointing out the gaps in the current development, we address the constraints and difficulties of the AI applications that are now being used in this field.

Precision Farming

Precision Agriculture is an administration policy that utilizes communication and information technology for farm management. It is a key to improving productivity by using the best agricultural practices and optimal usage of resources. Agriculture faces diverse challenges due to soil degradation, climate variation, and increasing costs. AI can analyze data from nanosensors embedded in the soil or on plants to optimize irrigation, fertilization, and pest control, leading to improved crop yields.

In recent years, agriculture has seen increasing use of intelligent machines at nanoscales. Agronomy is using AI-based nanomaterials to grow crops that are more nutritious, manage pests, evaluate ground as well as development under certain circumstances, interpret information to assist farmers and improve other aspects of food management tasks. It pointed out various steps as given below.

- Selecting the best seed
- Meterological conditions
- The best time to plant seeds
- Weather forecast data
- More production with fewer resources
- Time management for all the process-Irrigation, Harvesting, Consumer reach etc.

Farmers have access to cutting-edge data and analytics tools using nanotechnology-AI, which will minimize adverse environmental impacts while boosting superior agricultural practices, increasing efficiency, and reducing trash in the manufacturing of crops and energy.

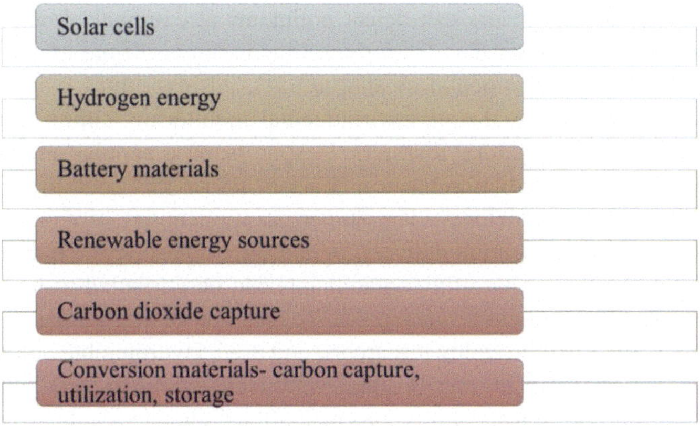

Fig. 9.5 Applications of Nano-AI-enabled nanomaterials for farming

AI-based nanotechnology has revolutionized many professions, and the agriculture sector is the latest one to experience the impact of this trend. Entrepreneurs are creating many technologies to facilitate farmers' cultivation and soil monitoring for health [18, 41, 42].

Energy Efficacy

Alternative applications in nano-AI-enabled agriculture include anticipating nanotechnology modifications over various weather and soil situations, as well as the clustering and projection of the efficacy of fertilizers and agrichemicals that are facilitated by AI and nanotechnology. The best methods for applying nano-agrichemicals might be integrated with crop cultivation methods. To attain carbon neutrality in the future, zero-carbon energy and negative emission technologies are essential, and nanomaterials have been vital in the advancement of these technologies as listed in Fig. 9.5.

The development of nanomaterials has been greatly impacted by the introduction and use of artificial intelligence (AI) as part of the materials research framework in recent times, owing to the rapid rise of data. Breakthrough future-oriented concepts made possible by AI have made it possible to greatly speed up every step of the material discovery process and make it easier to explore the vast domain of design.

Environmental Monitoring and Remediation

Intelligent nanotechnology could be utilized in environmental remediation, such as nanoscale devices capable of targeting and neutralizing pollutants in water or air.

Nanotechnology-based sensors can detect pollutants at extremely low concentrations. AI algorithms process data from these sensors for real-time monitoring and decision-making in environmental cleanup efforts. AI algorithms combined with nanosensors can provide real-time environmental monitoring, aiding in the early detection of pollution or natural disasters. Despite being a popular and affordable treatment option, bioremediation has a number of issues that need to be resolved for optimal results and efficiency. The numerous applications of nanotechnology in diverse fields like medication delivery, agriculture, optics, space industries, and environmental cleanup of harmful contaminants can be attributed to its birth as a research topic and its numerous advantages. To clean up polluted areas and restore the ecosystem, NPs are thoroughly researched and used. Hydrophobic pollutants can be properly sequestered from matrix solution by NPs, thanks to their high surface area and appealing sorption capability; this makes the pollutants available to bacteria for effective biodegradation [43].

Security and Defense

The merging of AI and nanotechnology is seen as the next great revolution, following the invention of machine learning. It exhibits similarities with the earlier century of industrialization, which saw the development of robots as well as textiles. As nanomaterial science is enabling engineering and multidisciplinary, have many applications in all fields of science to observe, measure, manipulate, assemble, control, and manufacture matter at the nanometer scales. The globe is currently facing complex security issues that need to be resolved in order to protect people's lives, property, and sovereignty [44]. The huge security risks posed by worldwide environmental degradation are a prime example.

Promoting novel theoretical models and organization to explain different strategies for the development of nanotechnology and emerging science programs, understanding their effect on privacy, permitting mechanisms for nations to oversee the implications of its inventiveness, and informing security, defense, and foreign policy are the ultimate goals of collaboration of nanotechnology and AI. As the fields of nanotechnology and artificial intelligence (AI) develop, public transparency in the defense and business sectors will become increasingly important for fostering international trust. In the commercial sector, transparency is essential to allay public concerns and mistrust regarding both nanomanufacturing as well as merchandise offered by nanotechnology. Honesty will help to distinguish facts from illusions when it comes to the social effects of contemporary technologies. While analyzing AI's possible use with nanotechnology, it is important to take into account the possibility of particular tactical, and operational uses across the security and defense domains. Robust scientific case studies that integrate this information with the purpose and intent that will propel military capability are required. These case studies will also assist in breaking down whether anything is meant to be used offensively, defensively, or altogether for multiple reasons [45].

Recent developments in nanoscale science and AI that help accomplish national security objectives are examined, and potential paths toward developing nanodefense through enhanced production, water- and armor-proof materials, nanomedicine, and nanosensors/nano-spies were noted. These scientific methodologies are currently being used by the majority of the military in developed nations for research, projects, and applications. When these technologies work together, the army may receive lighter and stronger equipment, support nanomedicines, stop bleeding, apply bandages to cure wounds, and use silver-packed foods for biological and antibiotic sensing [46]. Even with all of the advancements and improvements in nanotechnology, there are still some concerns associated with its creation, use, and disposal. It also emphasized how state-of-the-art advancements in nanotechnology could be used in defense, as illustrated in Fig. 9.6.

In this situation, self-renewing nanobots, or nanomachines, run rampant and conquer the entire globe. With the ultimate objective of facilitating international comparisons, this work devises a standard quantitative methodology for the establishment of strategically important nanotechnology and emerging science programs. It is known that some elements—such as institutional, ideational, and technical—are crucial to the function and importance of new technologies.

A lot of nations have launched significant R&D initiatives, mostly for civilian purposes, in anticipation of the profound advances in nanotechnology that are soon to come. The military's R&D initiatives, especially the US government's, are frequently ignored. For the time being, the possible military uses of AI and nanotechnology have not been presented systematically or thoroughly. These applications could include robotics, weapons, miniature computers, satellites, nanosensors, and more in the years to come [47].

Block Chain (BC) Technology

The Internet of Things and IoNT working together with AI and ML-based decision-making systems (DMSs) hold the potential to produce significantly worthy, innovative services that provide the immediate and accurate perspectives needed to run more ecologically friendly smart farming operations. Blockchain (BC) technology has emerged as a potentially viable approach for developing completely distributed DMSs with these kinds of functionalities. The utilization of BC technology facilitates the operation of an inviolable and shared ledger, thereby enabling the secure dissemination of information to all parties involved in the distribution of food, including clients, food vendors, and wholesalers. This allows direct communication between organizations without the need for brokers, which may jeopardize security. Supply chains for nanomaterials can be made more transparent, traceable, and secure by fusing blockchain technology with artificial intelligence and nanotechnology [48].

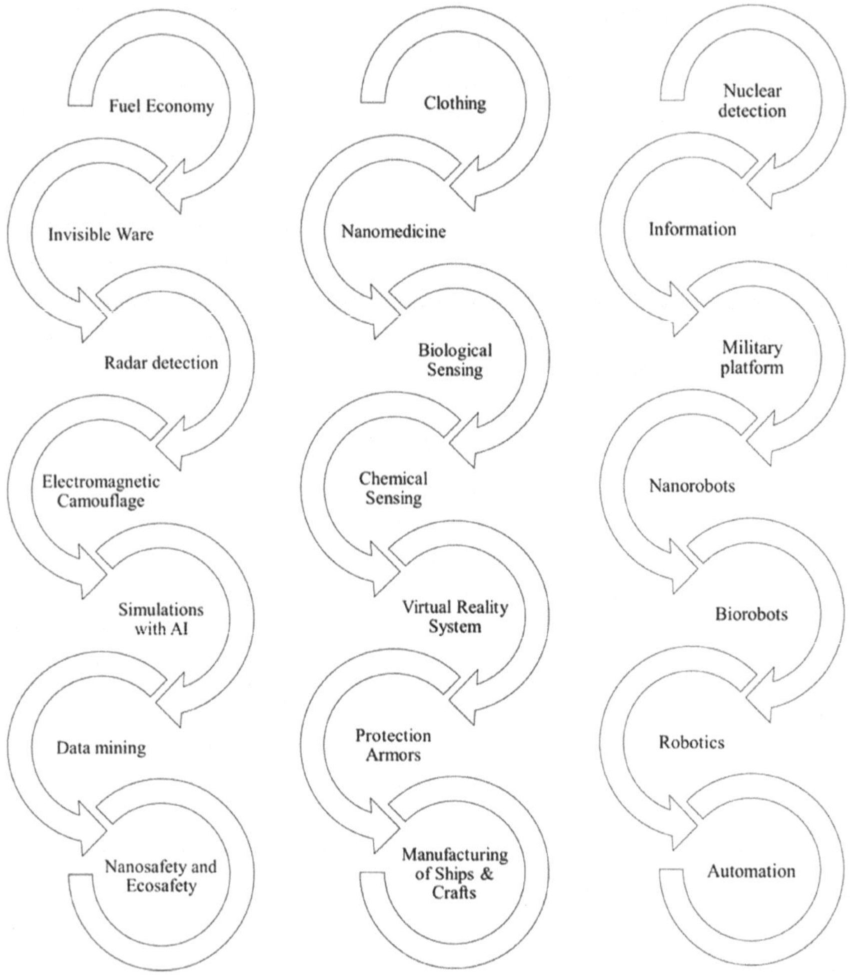

Fig. 9.6 Applications of AI and nanotechnology in security and defense

Medicine and Healthcare

To ensure a greater therapeutic effect, drug delivery incorporates a number of strategies, including drug formulation, manufacturing processes, bioactive ingredient storage, and transportation to the intended areas. In order to enhance the medical profession's physiochemical characteristics and combat the instability of bioactive chemicals, which can seriously impair the pharmacokinetic qualities, it is imperative to optimize the drug delivery procedures. In an effort to optimize drug delivery, artificial intelligence (AI) techniques have recently been used in the design, characterization, and manufacturing of drug delivery nanosystems. The bulk of these techniques are used to analyze and interpret biological and genetic data. Artificial intelligence (AI) in nanotechnology and pharmaceutical research has drastically cut down on false-positive rates, pharmacodynamic and pharmacokinetic profile evaluation, and drug discovery time and expense. Still, there are difficulties with high processing power, availability, upkeep, moral dilemmas, and the dependability of AI-enabled results.

Assuming, in the meantime, that AI is being explored for nanomedicine, then, to guarantee that strong models are accessible for the prediction of therapeutic synergies, the identification of suitable chemical combinations, and biomarker imaging assisting in correct drug distribution and treatment efficacy, issues like overfitting, validation, and bias must be addressed. The availability of sizable datasets containing various sets of clinical data to train AI models and optimize for precise drug administration is another prevalent issue. Although AI is not currently being used for medication development and delivery, successful AI integration has been witnessed in several emerging fields, including bioinformatics and the pharmaceutical industries, in recent years. Despite the difficulties, experts have observed that AI is accelerating the manufacturing of drugs, involving everyday problems.

Comparable to how the distribution of medicines has been constrained, artificial intelligence (AI) holds promise for improving drug delivery in the future.

The combination of nanotechnology with artificial intelligence (AI) offers novel approaches to problems in pharmaceutical sciences and wellness. Every aspect of the pharmaceutical industry that has chosen to use AI, from formulation designs to clinical trials to breakthroughs in cancer diagnosis, has reduced human workload and produced the desired results quickly [49]. The result has opened the door for earlier clinical translation by, for example, accurately and optimally predicting the properties of materials and how they interact with biological systems.

Future Aspects

The merging of artificial intelligence and nanotechnology exhibits enormous potential in several domains, portending breakthroughs that may profoundly influence diverse facets of human existence, including nanorobotics. The potential for creating extremely sophisticated, accurate, and effective solutions that can handle challenging problems in health care, electronics, materials science, environmental sustainability, and other fields is enormous when AI and nanotechnology come together. Artificial Intelligence (AI) and nanotechnology have a huge potential to revolutionize many parts of the industrial business in the future. When AI and nanotechnology combine, their potential is increased and extremely accurate and complex solutions can be developed for a wide range of industries.

As with any medical research or treatment, precision is crucial—it becomes even more so when applying nanotechnology. AI is helpful not only because it supports the complex programming of the nanobots but also because it makes the transformation of stem cells into bone cells through command conceivable, which can be used to cure a variety of illnesses. Additionally, the AI commands can direct the nanobots to create and manipulate stem cells in order to produce human organ replacements or even repair damaged ones.

To treat diseases like cancer, nanobots can use the massive amount of data provided by AI to distinguish between healthy and unhealthy cells. By using the vast amount of data, a model that distinguishes between the look of infected and normal cells may be trained, allowing nanobots to selectively target and eliminate infected cells. The advancement and use of precision medicine have initiated a transformative shift in the identification and treatment of cancer patients. Clinics all around the world offer cancer multigene panel scans that are tailored for certain cancer types in an effort to help with therapy decisions. To fully realize the promise of precision medicine, however, additional advancements and considerations need to be made.

This development process is still ongoing, and there doesn't seem to be an end in sight. Although the media and professional futurologists portray a different picture of the future, it is safe to say that artificial intelligence (AI) will play a bigger role in it than it does now, whether we like it or not. This time, there is no need to worry about a second AI winter because (a) the field of AI applications is much broader; (b) AI is already very successful in many domains in everyday life (speech processing like Alexa, Siri, etc., DeepL translator, etc., or ChatGPT, the newest system from OpenAI that impresses with amazing performance), and (c) entire AI ecosystems are already being adopted by biotechnology companies.

Lastly, we hope and believe that our study has discovered unique research gaps and research topics conducive to continued fruitful inquiry by building a rich research agenda. This will help push the boundaries of innovation research and prepare the next generation of innovation scholars to take on new scientific problems. It will also inform how (technology) innovation management research will develop over the next ten years. More specifically, by identifying and addressing important research questions, the proposed research agenda will assist innovation scholars who are interested in digital technologies—Artificial Intelligence in particular—and help advance the ongoing scientific discussion on digital technologies in the innovation field. It is possible that the suggested study agenda may serve as a helpful basis for cooperative multidisciplinary and interprofessional investigations into focal fields.

Future goals include open science, open data, and AI fairness—AI ecosystems that benefit everyone on the earth.

Some of the world's most brilliant thinkers working on nanotechnology ethics are brought together in this book. These volumes highlight the new health, environmental, ethical, and legal challenges brought forth by convergent and developing nanoscale technologies. We strongly advise all professionals in the sectors of research, education, and industry to carefully review these cutting-edge works on ethics in nanoscale science and technology.

This chapter discusses the unquestionably digital future of manufacturing, which will be molded by a confluence of technological developments. It examines the difficulties that manufacturers are currently facing in detail and looks ahead to what lies ahead for the industry. Reviewing cutting-edge manufacturing technologies comes first. It takes into account the factories of the future as well as the advancements in manufacturing technologies. It draws attention to the difficulties that the manufacturing sector will face going forward. It discusses how manufacturing will develop worldwide based on AI and nanotechnology.

Cultural and Privacy Concerns with Nanotechnology in AI and Manufacturing

Artificial Intelligence plays a practical role in the field of nanoscience by helping to create nanomaterials and devices and by interpreting experimental techniques. This interaction drives advances in science and technology and creates a cross-disciplinary environment full of opportunities and uses. The dynamic nexus of AI and nanotechnology offers countless opportunities for growth in industries and leads us into an era of limitless creativity. The synergistic cooperation of artificial intelligence (AI) and nanotechnology is at the forefront of scientific and technical developments, from the synthesis of innovative materials with previously unheard-of qualities to the development of advanced nanorobotics and sustainable solutions for global concerns. The concrete impact on our daily lives is highlighted by the revolutionary applications in nanomedicine as well as customized material solutions. Although the future seems bright, it is imperative to acknowledge and address the difficulties associated with combining AI and nanotechnology. The significance of ethical and socially acceptable development is emphasized by drawing lessons from the past, especially in the area of nanotechnology. The blend of nanotechnologies with intelligent machines (AI) offers revolutionary opportunities in several fields, including health care, industry, the environment, and ethics. This synthesis explores the complex interactions between various technologies and looks at their effects on society, ethics, laws, and the environment.

The integration of nanotechnology and AI spurs creativity, but it also raises issues of privacy as shown in Fig. 9.7, responsibility, and bias that must be carefully addressed. To shape a responsible future, equitable access, workforce changes, and transparent governance become essential. Ethics in health care include considerations of patient privacy, equity, and accountability in AI-powered decision-making. Collaboration is essential to navigating employment changes as sectors shift, and safeguarding the environment necessitates sustainable and energy-efficient procedures. This combination shows a way forward where innovation and moral principles work together to propel advancement.

Additionally, there may be ethical questions about the implementation of cutting-edge surveillance technology made readily accessible by machine learning and nanosensors. The concurrent impact brought about by AI and nanotechnology extends to nanocomputing, nanomaterials, and nanodevices, providing expanded computational capacity and architectures for sophisticated machine learning techniques. For instance, as the graphic illustrates, if AI systems are not properly constructed, they could jeopardize people's safety. Artificial intelligence (AI) has

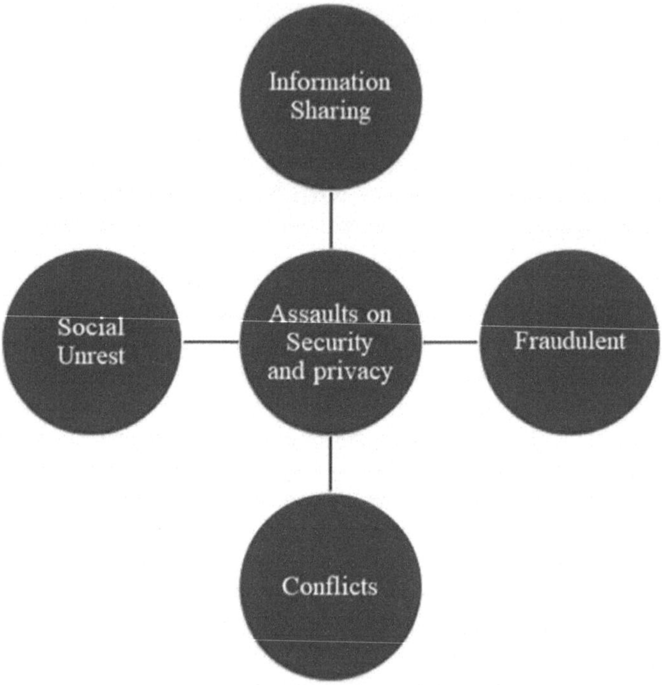

Fig. 9.7 Major Privacy concerns due to the integration of AI and nanotechnology

the potential to surpass human intelligence at some point, which would mean that AI systems might someday take over our lives in a variety of ways. It was discovered that AI might be used for a variety of nefarious purposes, including disinformation campaigns and cyberattacks.

According to the many National Academies of Sciences, Technologies, and Medicine, priority might need to be given to AI safety research. The ethics of AI in health care, according to Talwar and Osofsky, 2020, must be carefully studied since AI systems may be utilized for bias against particular groups of people or decide things that are not ethically acceptable. In order to successfully venture into the unknown at this scientific crossroads, we must approach these advances with a dedication to moral principles and the welfare of society.

References

1. Hrvat, F., et al.: Artificial intelligence in nanotechnology: recent trends, challenges and future perspectives. In: International Conference on Medical and Biological Engineering. Springer (2021)
2. Egorov, E., et al.: Robotics, microfluidics, nanotechnology and AI in the synthesis and evaluation of liposomes and polymeric drug delivery systems. Drug Deliv. Transl. Res. **11**(2), 345–352 (2021)
3. Dulta, K., et al.: Nanotechnology and applications. In: Applications of Computational Intelligence in Multi-Disciplinary Research, pp. 129–141. Elsevier (2022)
4. Hu, T., et al.: Advances in portable heavy metal ion sensors. Sensors **23**(8), 4125 (2023)
5. Soori, M., Arezoo, B., Dastres, R.: Artificial intelligence, machine learning and deep learning in advanced robotics, a review. Cogn. Robot. (2023)
6. Elahi, M., et al.: A comprehensive literature review of the applications of AI techniques through the lifecycle of industrial equipment. Discov. Artif. Intell. **3**(1), 43 (2023)
7. Sim, S., Wong, N.K.: Nanotechnology and its use in imaging and drug delivery. Biomed. Rep. **14**(5), 1–9 (2021)
8. Harun-Or-Rashid, M., et al.: Recent advances in micro-and nano-drug delivery systems based on natural and synthetic biomaterials. Polymers **15**(23), 4563 (2023)
9. Moore, J.A., Chow, J.C.: Recent progress and applications of gold nanotechnology in medical biophysics using artificial intelligence and mathematical modeling. Nano Express **2**(2), 022001 (2021)
10. Ho, D., Wang, P., Kee, T.: Artificial intelligence in nanomedicine. Nanoscale Horiz. **4**(2), 365–377 (2019)
11. Ge, M., et al.: Deep learning analysis on microscopic imaging in materials science. Mater. Today Nano **11**, 100087 (2020)
12. Xu, W., Wang, J., Yan, X.: Advances in memristor-based neural networks. Front. Nanotechnol. **3**, 645995 (2021)
13. Sacha, G.M., Varona, P.: Artificial intelligence in nanotechnology. Nanotechnology **24**(45), 452002 (2013)
14. Qin, C., et al.: Artificial olfactory biohybrid system: an evolving sense of smell. Adv. Sci. **10**(5), 2204726 (2023)
15. Belluco, S., et al.: State of art of nanotechnology applications in the meat chain: a qualitative synthesis. Crit. Rev. Food Sci. Nutr. **58**(7), 1084–1096 (2018)
16. Khan, R.: Artificial intelligence and machine learning in food industries: a study. J. Food Chem. Nanotechnol. **7**(3), 60–67 (2022)
17. Ansari, M.A.: Nanotechnology in food and plant science: challenges and future prospects. Plants **12**(13), 2565 (2023)
18. Zhang, P., et al.: Nanotechnology and artificial intelligence to enable sustainable and precision agriculture. Nat. Plants **7**(7), 864–876 (2021)
19. Ursyn, A.: Graphical Thinking for Science and Technology Through Knowledge Visualization. IGI Global (2019)
20. Sadiku, M.N., Kotteti, C.M., Musa, S.M.: Nano Computing: An Introduction (2019)
21. Rana, T.M., Zohuri, B.: The future beauty: artificial general intelligence driven nanotechnology and quantum computing in medical aesthetics. J. Clin. Bio Med. Adv. **2**(4), 01 4 (2023)
22. Lee, J., et al.: Intelligent maintenance systems and predictive manufacturing. J. Manufact. Sci. Eng. **142**(11) (2020)
23. Pech, M., Vrchota, J., Bednář, J.: Predictive maintenance and intelligent sensors in smart factory. Sensors **21**(4), 1470 (2021)
24. Winkler, D.A.: Role of artificial intelligence and machine learning in nanosafety. Small **16**(36), 2001883 (2020)
25. Betancourt, T., Brannon-Peppas, L.: Micro-and nanofabrication methods in nanotechnological medical and pharmaceutical devices. Int. J. Nanomed. **1**(4), 483–495 (2006)

26. Kumar, P.P.P., Kaushal, S., Lim, D.K.: Recent advances in nano/microfabricated substrate platforms and artificial intelligence for practical surface-enhanced Raman scattering-based bioanalysis. TrAC Trends Anal. Chem., 117341 (2023)
27. Tadokoro, Y., et al.: Artificial-intelligence-assisted mass fabrication of nanocantilevers from randomly positioned single carbon nanotubes. Microsyst. Nanoeng. **9**(1), 32 (2023)
28. Hunde, B.R., Woldeyohannes, A.D.: Future prospects of computer-aided design (CAD)–a review from the perspective of artificial intelligence (AI), extended reality, and 3D printing. Results Eng. **14**, 100478 (2022)
29. Zhu, Z., et al.: 3D-printed multifunctional materials enabled by artificial-intelligence-assisted fabrication technologies. Nat. Rev. Mater. **6**(1), 27–47 (2021)
30. Xu, W., et al.: 3D printing-enabled nanoparticle alignment: a review of mechanisms and applications. Small **17**(45), 2100817 (2021)
31. Tsai, H.-F., Podder, S., Chen, P.-Y.: Microsystem advances through integration with artificial intelligence. Micromachines **14**(4), 826 (2023)
32. Chen, H., et al.: AI for nanomaterials development in clean energy and carbon capture, utilization and storage (CCUS). ACS Nano (2023)
33. Gregoire, J.M., Zhou, L., Haber, J.A.: Combinatorial synthesis for AI-driven materials discovery. Nat. Synth., 1–12 (2023)
34. Haick, H., Tang, N.: Artificial intelligence in medical sensors for clinical decisions. ACS Nano **15**(3), 3557–3567 (2021)
35. Zhang, Z., et al.: Advances in machine-learning enhanced nanosensors: from cloud artificial intelligence toward future edge computing at chip level. Small Struct., 2300325 (2023)
36. Manickam, P., et al.: Artificial intelligence (AI) and internet of medical things (IoMT) assisted biomedical systems for intelligent healthcare. Biosensors **12**(8), 562 (2022)
37. Patel, S.S., Patel, P.N.: A brief review on nanorobotics applications in medicine and future prospects. Asian J. Res. Pharm. Sci. **13**(1) (2023)
38. Ojha, S., et al.: Micro and nano robotics-assisted targeted drug delivery, surgery and radiotherapy for cancer treatment. Curr. Cancer Ther. Rev. **20**(1), 18–25 (2024)
39. Hussain, S., et al.: Nanorobots: the future of healthcare. In: AIP Conference Proceedings. AIP Publishing (2023)
40. Dinesh, R.A., Shanmugam, J., Biswas, K.: Integration of technology and nanoscience in precision agriculture and farming. In: Contemporary Developments in Agricultural Cyber-Physical Systems, pp. 149–171. IGI Global (2023)
41. Alzoubi, L., Aljabali, A.A., Tambuwala, M.M.: Empowering precision medicine: the impact of 3D printing on personalized therapeutic. AAPS PharmSciTech **24**(8), 1–33 (2023)
42. Singh, R.K., Berkvens, R., Weyn, M.: AgriFusion: an architecture for IoT and emerging technologies based on a precision agriculture survey. IEEE Access **9**, 136253–136283 (2021)
43. Patowary, R., Devi, A., Mukherjee, A.K.: Advanced bioremediation by an amalgamation of nanotechnology and modern artificial intelligence for efficient restoration of crude petroleum oil-contaminated sites: a prospective study. Environ. Sci. Pollut. Res., 1–26 (2023)
44. Choudhary, B., et al.: Nanotechnology in defence and security. Emerg. Appl. Nanomater. **141**, 151–168 (2023)
45. Kosal, M.E.: Military applications of nanotechnology: implications for strategic security I. Project on Advanced Systems and Concepts for Countering WMD (PASCC) (2014)
46. Abed, M.S., Jawad, Z.A.: Nanotechnology for defence applications. In: Nanotechnology for Electronic Applications, pp. 187–205. Springer (2022)
47. Altmann, J.: Military Nanotechnology: Potential Applications and Preventive Arms Control. Routledge (2007)
48. Vimalajeewa, D., et al.: Block chain and internet of nano-things for optimizing chemical sensing in smart farming (2020). arXiv preprint arXiv:2010.01941
49. Das, K.P.: Nanoparticles and convergence of artificial intelligence for targeted drug delivery for cancer therapy: current progress and challenges. Front. Med. Technol. **4**, 1067144 (2023)

Dr. Promila Sharma is presently working as an Assistant Professor at the Department of Chemistry, under the School of Basic, Applied, and Biosciences at RIMT University, Mandi Gobindgarh, Punjab. She has a total academic experience of 4 years in teaching. She obtained her M.Sc. degree with specialization in inorganic chemistry from Punjabi University, Patiala. She qualified for NET in chemical sciences. She also has a research experience of 5 yearsyears with seven publications in peer-reviewed journals and two papers in international conferences and has one book chapter to her credit.

Her research interest is synthesizing and characterizing heterocyclic and bis-heterocyclic compounds. In medicinal chemistry, heterocyclic compounds play an essential role by demonstrating different biological activity. Further emphasis is on their application as sensors to develop new methodologies and remediation for detecting metals and explosive materials. Computational studies using DFT are also her expertise to shed light on the design and development of synthesized hybrids and to correlate the photophysical properties of synthesized probes with their experimental characteristics.

Dr. Varinder Singh, an Assistant Professor in the department of chemistry, at RIMT University, Mandi Gobindgarh, Punjab, earned his Ph.D. in synthetic organic and medicinal chemistry from Punjabi University, Patiala, following an M.Sc. in organic chemistry from the same institution. Specializing in organic synthesis, computational chemistry, DFT studies, and analytical chemistry, Dr. Singh is an asset to the Department of Chemistry. His wealth of knowledge enriches both academic and research endeavors at the university, where he actively contributes to the advancement of scientific understanding. With a passion for diverse aspects of chemistry, Dr. Varinder Singh plays a crucial role in shaping the intellectual landscape within the department, fostering a culture of excellence and innovation.

Dr. Anita Thakur is working as an Assistant Professor at the Department of Chemistry, under the School of Basic, Applied, and Biosciences at RIMT University, Punjab. She has a total academic experience of 4 years in teaching. She obtained her M.Sc. Degree with specialization in physical chemistry from Punjabi University, Patiala. After that, she joined M.Phil. in 2012 and Ph.D. (2015 2019) from the same institute. She also has a research experience of 5 years with seven publications in peer-reviewed journals and one paper at in an international conference and has one book to her credit.

Dr. Anita has made significant contributions to the field of surface chemistry, nanotechnology, and water and wastewater treatment through her research publications. Her studies have not only expanded scientific understanding but also hold promise for the development of potential therapeutic interventions.

Prof. Ashok Kumar Malik He is an expert in analytical chemistry, electrochemistry, and inorganic chemistry. He published 300 research articles with 7000 citations and an H-index of 40. He guided 25 students to an award-winning Ph.D. degree in chemical sciences, and 8 students are still working in his research group. He did more than 15 research projects (Humboldtian), Top 2% of World-Ranking-Scientists by Stanford-Elsevier 2023, Ex-Head, Department of Chemistry, Co-ordinator GIAN Program, Punjabi University, Patiala, Patiala–147 002, Punjab, India. Web page: Profile_P13X36C7E.pdf (Punjabiuniversity.ac.in).

Chapter 10
Nanotechnology and Water: Emerging Solutions for a Global Crisis

Benjamin Tawiah, Anna M. Abraham, Justin J. Kappen, Nikhi Maria Raju, C. Surej Rajan, and Soney C. George

Introduction

Nanotechnology, a fast expanding discipline, has enormous potential in tackling one of the most critical global concerns of our time: the water crisis [1]. Urbanization, climate change, and population growth are all contributing to the rising demand and falling supply of drinking water. Many of the water filtration techniques now in use are unfriendly and harmful to the environment. Nonetheless, nanotechnology has transformed methods for cleaning up contaminated water and ensures that clean water is available [2]. At the foundation of nanotechnology is the manipulation and designing of materials at the nanoscale, which generally ranges from 1 to 100 nm [3]. The materials exhibit unique and enhanced properties at this scale, leading to several applications. Nanotechnology has unmatched opportunities for improving the effectiveness, efficiency, and sustainability of water purification processes in the context of water treatment [4]. Filtration systems based on nanomaterials are the finest devices to purify water [5] whereas conventional filters are not as competitive as that of nanofilters [6]. The distinct nanofeatures of nanomaterials allow them to remove pollutants selectively. Pathogens are among the many contaminants that they can eliminate.

B. Tawiah
Department of Industrial Art, Kwame Nkrumah University of Science and Technology, KNUST, PMB, Kumasi, Ghana

A. M. Abraham · J. J. Kappen · N. M. Raju · C. S. Rajan · S. C. George (✉)
Center for Nanoscience and Technology, Amal Jyothi College of Engineering Kanjirappally, Koovappally, Kerala, India
e-mail: soneycgeorge@amaljyothi.ac.in

N. M. Raju · C. S. Rajan
Department of Chemical Engineering, Amal Jyothi College of Engineering Kanjirappally, Koovappally, Kerala, India

S. C. George and B. Tawiah (eds.), *Nanotechnology in Societal Development*, Advanced Technologies and Societal Change, https://doi.org/10.1007/978-981-97-6184-5_10

Moreover, nanotechnology makes it possible to create sophisticated water treatment methods that go beyond filtering. By functionalizing them with certain attributes, such as photocatalytic or antimicrobial capabilities, nanomaterials can actively break down or eliminate impurities found in water [7]. Light-activated nanoparticles can convert the contaminants through appropriate reactions to innocuous products [8]. Similarly, antimicrobial nanomaterials, such as AgNPs can arrest the maturity and multiplication of microbial things, ensuring the microbial safety of water. With their high surface area-to-volume ratio, AgNPs can come into increased contact with microbes, disrupting their cell membranes and leading to cell death [9, 10]. Also, titanium dioxide nanoparticles (TiO_2NPs) have photocatalytic properties that can be harnessed for microbial inactivation. Active oxygen species generated from UV-irradiated TiO_2NPs will damage the genetic material of microbes [11, 12]. Recently, carbon nanotubes (CNTs) have been found to physically puncture bacterial cell walls, leading to cell lysis [13]. Additionally, their strong adsorption capabilities allow them to trap and remove bacteria from water. Graphene oxide (GO) and other compounds based on graphene have shown promise as antibacterial agents due to their capacity to cause damage to bacterial cell membranes [14].

Nanotechnology provides opportunities for efficient and sustainable water desalination, a critical process for addressing freshwater scarcity in coastal regions [14]. Reverse osmosis is one of the energy-intensive traditional desalination techniques that frequently produces enormous amounts of brine waste, which is bad for the environment [16]. On the other hand, desalination methods based on nanomaterials, like capacitive deionization, forward osmosis, and nanoparticle membranes, may need less energy [17].

The notion of nanotechnology and its application to water treatment are introduced at the beginning of the chapter. It demonstrates how better water treatment procedures and higher-quality water can result from the engineering and manipulation of materials at the nanoscale level. The wide variety of nanomaterials used in water treatment applications is next covered, along with an explanation of each one's properties, potential, and uses as nanofilters to remove contaminants, such as heavy metals, organic pollutants, and microbes from water sources. Along with the challenges associated with the development and application of nanoscale materials, a brief introduction to the techniques of nanomaterials synthesis and manufacturing for water treatment is presented. Furthermore, the advantages and disadvantages of nanotechnology-based water treatment systems, as well as their inherent novel properties, such as higher selectivity, efficiency, and lower energy consumption, and the risks and challenges that need to be considered for safe implementation are discussed. The chapter concludes with a last review of the direction that nano water research is taking now and in the future. The declaration emphasizes the necessity of ongoing innovation and interdisciplinary cooperation to realize the full potential of nanotechnology in offering affordable and sustainable clean water solutions.

Fundamentals of Nanomaterials

The study of the synthesis, composition, and characteristics of materials with a nanometre-scale size is the focus of the multidisciplinary discipline of nanotechnology. This scale represents a significant departure from the macroscopic attributes and underlying principles associated with such materials. By 2020, a wide range of nanotechnology applications in computers, medicine, and industry as well as more knowledge and environmental conservation will be possible because of the convergence of nanoscale research, engineering, and nano systems (18). Nanoparticles exhibit distinct size-dependent features, owing to their small size and vast surface area. The periodic boundary conditions of a crystalline particle are disrupted when a particle reaches the nanoscale with a characteristic length that is near or smaller than the de Broglie wavelength or the wavelength of light. Nanoparticles have distinct physical properties that differentiate them apart from bulk materials, enabling their novel applications (19). It is possible to classify and organize nanomaterials according to their morphology, size, form, chemical makeup, condition, and dimensionality as shown in Fig. 10.1.

Properties of Nanoscale Materials

Nanoscale materials are structures with nanometre scale dimensions, typically ranging from 1 to 100 nm. At this scale, materials have distinct characteristics and behaviours that differ dramatically from their bulk counterparts. Higher surface area-to-volume ratios, enhanced surface reactivity, and quantum effects are the causes of these characteristics. Comparing nanomaterials to bulk materials, the ratio of surface area to volume is noticeably higher. More reactivity is produced by this expanded surface area, and this reactivity is particularly helpful for surface-enhanced spectroscopies, sensing, and catalysis. Because NPs have a higher surface area-to-volume ratio (SA/V) and more reactive sites than bigger NPs or bulk materials, they work better in catalytic and sensing applications. Sharma et al. suggested a method based on surface oxidation/reduction for determining Nano Sphere size using SA/V(24) using electrochemical dissolution of Au51(24). Certain nanomaterials, like graphene and carbon nanotubes, have remarkable mechanical qualities, like high stiffness and tensile strength. These qualities make them ideal for reinforcing composite materials and creating lightweight but strong structural components.

Quantum confinement effects are an important feature of nanomaterials. This property causes materials at the nanoscale to have size-dependent properties. Because of the quantum confinement phenomenon, recent research has concentrated on creating nanomaterials with unique physical and chemical properties. Arul et al. proposed the synthesis of very uniform and well distributed at the quantum level CeO_2 in hexane (22) with the ability to modify materials' electrical, optical, magnetic, and mechanical characteristics. Quantum dots, for example, have size-dependent

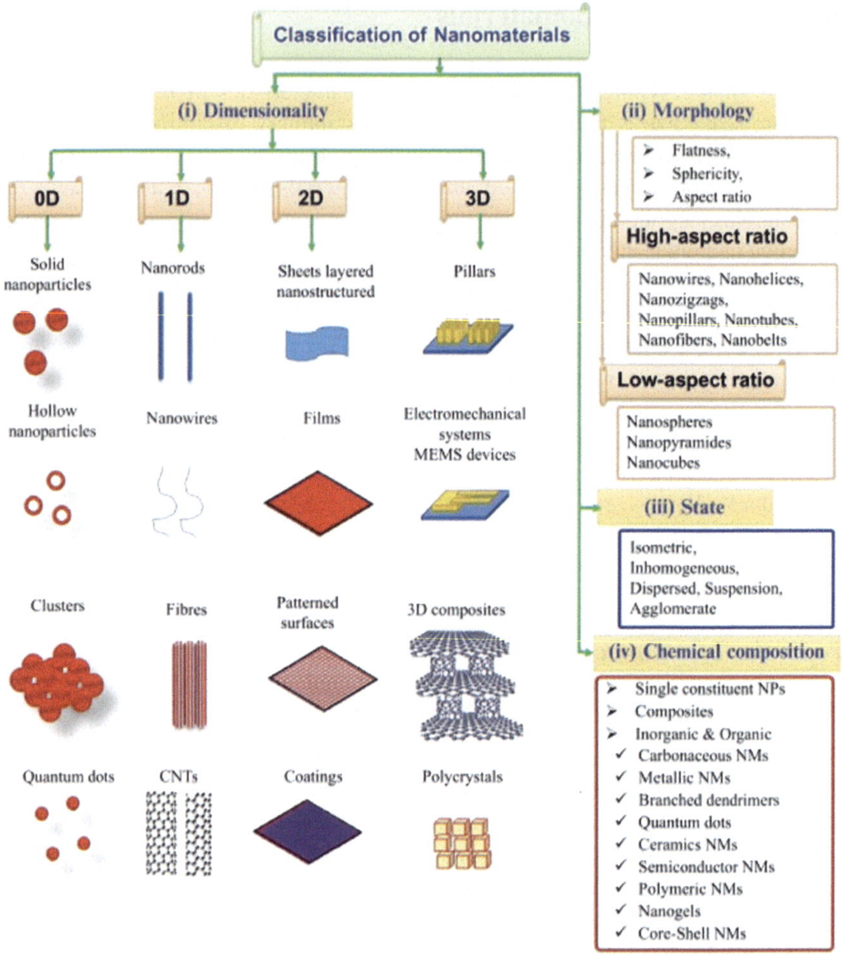

Fig. 10.1 Classification of nanomaterials [15]

optical properties, with a changing band gap resulting in tunable emission colours. Nowadays, semiconductor quantum dots have been investigated to possess strong and adjustable fluorescence emission properties, which render them suitable for biosensing and bioimaging, and to various other applications (23).

Nanomaterials frequently exhibit novel optical features, such as plasmonic resonances and quantum confinement effects. Applications in light-emitting diodes (LEDs), photovoltaics, and nanophotonic are made possible by these characteristics. Certain metal nanoparticles and carbon nanotubes are examples of nanomaterials that have a high electrical conductivity. This characteristic is useful for a variety of electronic applications, including transparent conductive films, flexible electronics, and high-performance transistors. They can have higher thermal conductivity than bulk materials. This feature is advantageous for thermal management applications, such as heat sinks, thermoelectric devices, and thermal barrier coatings. Nanoparticle-fluid blends may be best for heat transfer in various devices, including micro devices with small components and flow pathways, due to their small size. Nanoparticles' small size makes it possible to act as a lubricant when in contact with solid surfaces (25).

Nanomaterials' surface chemistry can be tuned to produce precise chemical reactivity and stability. Nanoparticle surface functionalization with diverse functional groups enables fine control over interactions with target molecules, making it suitable for drug delivery, sensors, and catalytic applications. Nanomaterials, such as magnetic nanoparticles and nanowires, can have distinct magnetic properties, such as superparamagnetic and increased magnetic anisotropy. These features are used in magnetic storage devices, MRI contrast agents, and magnetic hyperthermia for cancer treatment. The prospective uses of superparamagnetic nanoparticles in biosensing, targeted drug delivery, and MRI contrast agents have drawn a lot of interest. Superparamagnetic nanoparticles (SPIONs, < 20 nm) show thermal fluctuation at room temperature. SPIONs can align in the direction of an external magnetic field and reach magnetic saturation at a magnitude that is significantly greater than any known biological entity (26). Because of these characteristics, nanoscale materials are very versatile and useful in many different areas, such as electronics, healthcare, energy, environmental remediation, and aerospace. However, harnessing these features for water treatment necessitates a thorough understanding of synthesis, characterization, and manipulation methodologies. Furthermore, careful evaluation of potential environmental and health consequences is critical for the responsible creation and utilization especially in the commercial production of clean water for consumption.

Methods of Synthesizing Nanomaterials for Water Treatment

Several processes are used for synthesizing nanomaterials, each having unique benefits and generating materials with various features (27). This section will provide a brief overview of the methods of nanomaterial synthesis for water treatment.

(a) **Chemical Vapour Deposition (CVD):** A method in which precursor gases react to generate solid thin films or nanoparticles on a substrate surface. It is commonly employed to create thin films, nanowires, and graphene. The procedure is frequently carried out at high temperatures and/or pressures, allowing for the controlled development of nanomaterials with specific properties. CVD enables the production of materials with various physical, tribological, and chemical properties by adjusting experimental conditions, such as substrate material, temperature, composition of the reaction gas mixture, and gas flow. CVD can also achieve uniform thickness and properties of coatings on substrates with complex shapes, thanks to its excellent throwing power. Moreover, it allows for localized or selective deposition on patterned substrates. A simple schematic diagram is shown in Fig. 10.2.

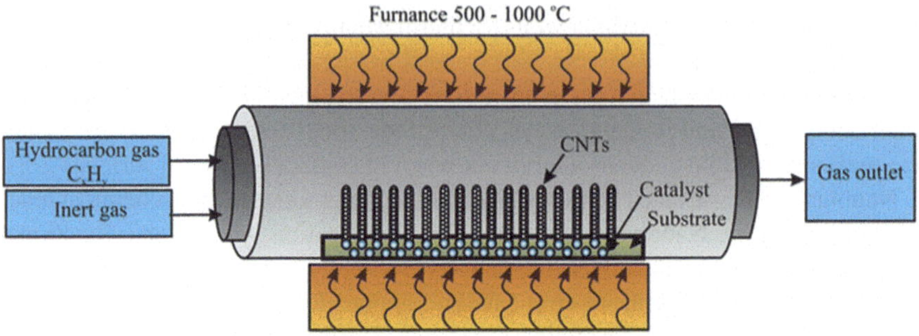

Fig. 10.2 Schematic illustration of CVD CNT processing [16]. An open access article by Springer CC BY 4.0 license

There are many different types of CVD processes, including atmospheric pressure CVD (APCVD), low-pressure CVD (LPCVD), ultrahigh vacuum CVD (UHVCVD), aerosol-assisted CVD (AACVD), direct liquid injection CVD (DLICVD), microwave plasma-assisted CVD (MPCVD), and more. Each type has its own pressure range, description, and specific applications. This method allows for the control of the size, composition, and structure of the synthesized nanomaterials. Lee et al. introduced an improved method for attaching TiO_2 nanoparticles to glass beads using the CVD technique [16]. It offers a flexible and scalable process for creating a large variety of nanomaterials with unique properties for variety of uses, including energy storage, electronics, catalysis, and medical devices.

(b) **Sol–Gel Method:** This is another widely used technique for creating nanomaterials, especially metal oxides, and hybrid organic–inorganic materials. It includes converting a precursor solution into a gel-like colloidal suspension, which is then dried and thermally treated to produce the required nanomaterials. It is widely used in the manufacture of metal oxides, especially silica nanoparticles. Particles produced by these chemical processes have precise dimensions, compositions, and structures, which makes them useful for a variety of applications, such as electrical devices, sensing, and catalysis. Sol–gel synthesis is more cost-effective than physical approaches due to its low processing temperature and energy requirements [17]. A typical process of nanoparticle synthesis using the sol–gel method is illustrated in Fig. 10.3.

Chemical precipitation: This method involves reacting iron salts with a base in the presence of a surfactant, resulting in the formation of iron oxide nanoparticles. Magnetite NPs are generally being synthesized by this technique (30). A typical synthesis is illustrated in Fig. 10.3. This process is easy, cheap, and can generate wide variety of nanoparticles. However, to achieve the desired nanoparticle characteristics—such as size, size distribution, and crystallinity—careful management of the reaction conditions is necessary. Recently, green nanoparticle synthesis using natural resources, such as plant extracts or microorganisms, to reduce iron salts and form iron oxide nanoparticles has gained considerable attention due to the quest for

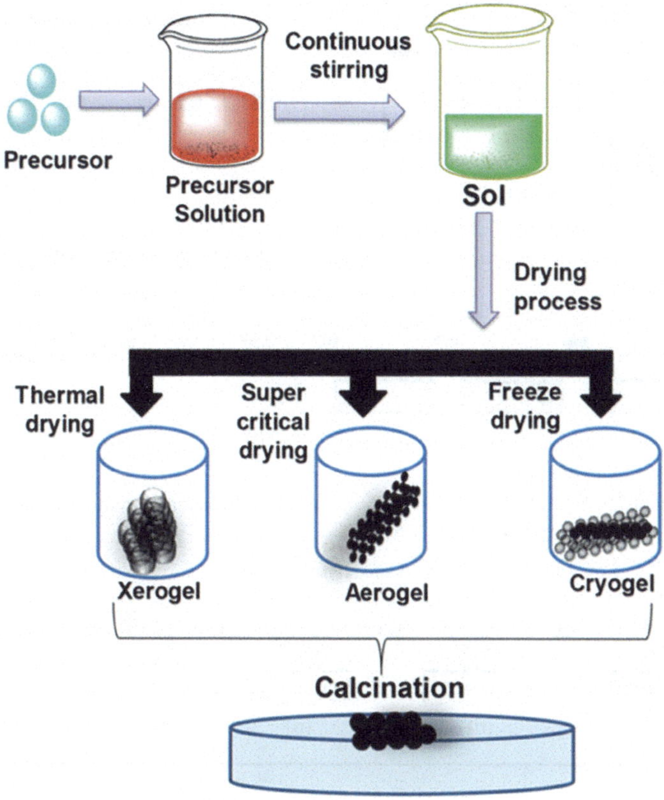

Fig. 10.3 Sol–gel synthesis method. Reproduced from [17], an open-access article by MDPI under CC BY license 4.0

sustainability in all spheres of life. This strategy is especially crucial for treating water. In this sense, a unique environmentally friendly method has been used to synthesize functionalized iron oxide MNPs that have the capacity to absorb various organic contaminants from plants [18]. The results showed that green nanoparticles could absorb above 80% effluent efficiency under a wide range of conditions in a very short time and as high as 94% under optimum conditions. This result illustrates the application of green particles for water treatment purposes (Fig. 10.4).

(c) **Hydrothermal/ Solvothermal Synthesis**: Hydrothermal synthesis is a useful method for creating a range of nanomaterials, including metal sulphides, oxides, and other chemicals. The above method uses high-temperature and high-pressure aqueous conditions that encourage the synthesis and growth of nanomaterials. When compared to other methods and processes, it permits the production of nanomaterials at temperatures ranging from ambient temperature to extremely high temperatures with comparatively little material loss. A schematic illustration of the hydrothermal process is shown in Fig. 10.5. It

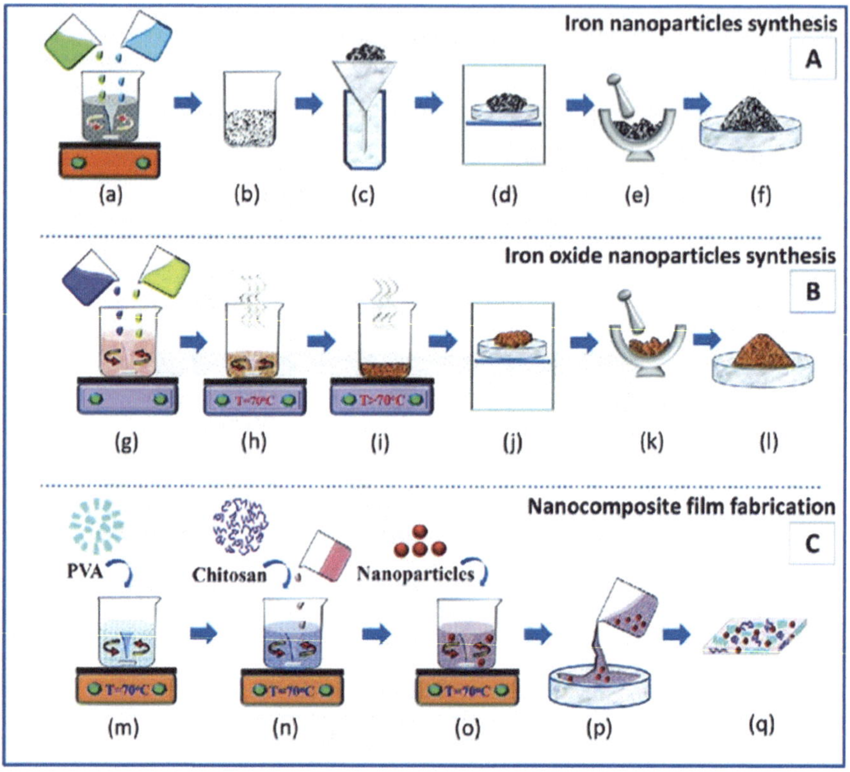

Fig. 10.4 Typical synthesis procedure of iron nanoparticles and their possible composite formation. Reproduced from [19] under CC BY-NC-ND 4.0 2018 by Elsevier

enables the manufacture of nanoparticles, nanowires, and nanostructures with precise size, shape, and crystallinity. It has an array of uses, spanning catalysis, energy storage, sensors, medicinal devices, and environmental cleanup.

(d) **Electrospinning:** This method produces nanofibers by delivering a high voltage to a polymer solution or melt, giving in a charged jet that solidifies into fibres when collected on a grounded substrate as shown in Fig. 10.6. NPs-electro spun fibres offer many potential applications due to their flexibility, free-standing nature, and ability to be tailored to specific polymers and NPs. Electrospinning NPs having anisotropic structures like nanorods, and nanowires align them within the threads, minimizing Gibbs free energy. Thus, electrospinning is able to be used for both fabrication and self-assembly, making it versatile.

(e) **Ball milling**: Ball milling is a versatile nanomaterial synthesis method that employs the top-down approach by using high-energy collisions between balls and a material within a chamber to break it down into nanometre-sized particles. This technique offers versatility in synthesizing various nanomaterials,

Fig. 10.5 A schematic diagram of the solvothermal nanoparticle synthesis. Reproduced [20] with permission from Springer

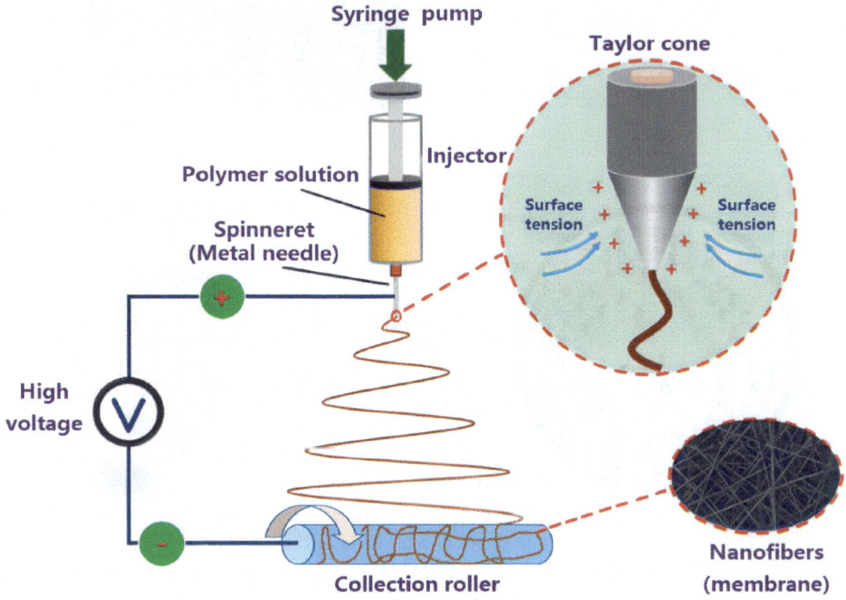

Fig. 10.6 Electrospinning method of nanofiber production. Reproduced from [21], an open access article by MDPI under CC BY license 4.0

including metallic, multi-metallic, alloyed, ceramic, and composite materials. Fundamentally, it involves the grinding of materials in a ball mill to achieve a reduction in particle size and potentially influence the characteristics. It is the mechanical grinding of bulk materials in the presence of a grinding medium to generate nanoparticles (Fig. 10.7).

This is a quick and easy way to make nanocrystalline powders with unique characteristics. It does not require high temperature treatment and can produce huge amounts (34). It is frequently used to create metallic nanoparticles and composites. It is widely employed in a variety of domains, including materials science, chemistry, nanotechnology, and biomedical engineering, to create nanomaterials for use in catalysis, energy storage, sensor technology, and drug delivery.

(f) **Template-Assisted Synthesis**: Template-assisted synthesis of nanomaterials is a versatile and effective approach for the controlled fabrication of nanostructures

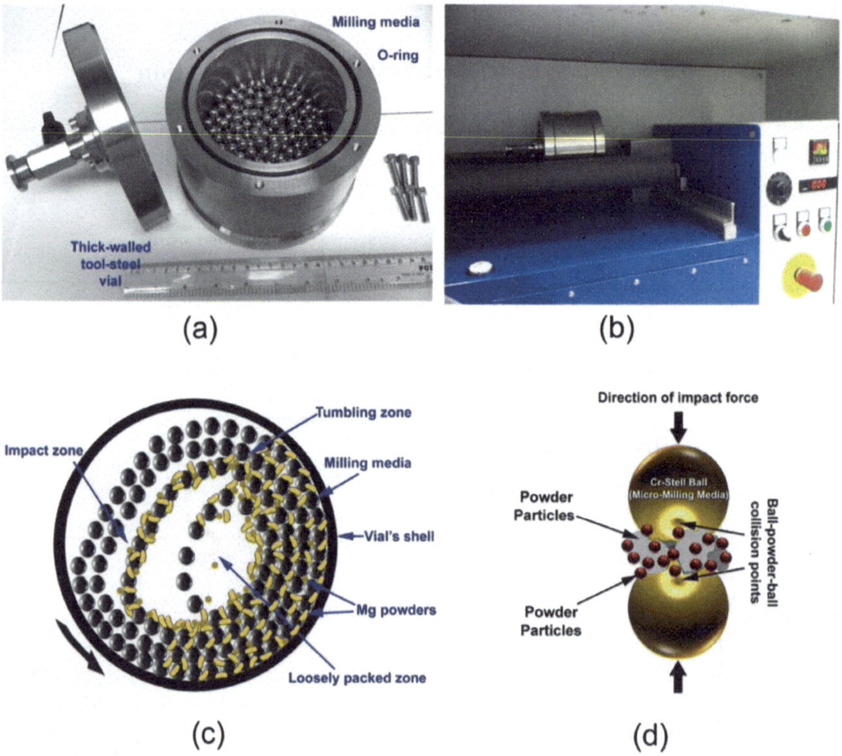

Fig. 10.7 Laboratory-scale 1 L roller mill setup, **a** The milling balls and vial, **b** The ball milling process, which utilizes a roller mill, **c** Schematic representations of ball positions and movement inside the vial of a tumbler mall mill operating in dynamic mode, and **d** An example of a normal ball-powder-ball collusion during a low-energy tumble. Reproduced from [22], an open access article by MDPI under CC BY license 4.0

with specific shapes, sizes, and properties. It involves the use of templates, which are materials or molecular systems that provide a framework or scaffold for the growth of desired nanostructures. These templates can be classified into two main categories: hard templates and soft templates.

Hard templates, such as porous membranes or inorganic materials, have well-defined structures and pore geometries. They serve as moulds or frameworks for the deposition or growth of nanomaterials. The hard template method often involves the filling of the template with precursor materials, followed by the removal or dissolution of the template to obtain the desired nanostructures [23]. Examples of hard templates include anodized aluminium oxide (AAO) membranes, silica templates, and track-etched polymer membranes. Hard templates offer precise control over the size, shape, and arrangement of nanostructures due to their well-defined templates, making them suitable for various applications, such as nanofabrication and nanoelectronics.

Soft templates, on the other hand, are composed of flexible organic molecules, surfactants, or block copolymers. These soft templates interact with the precursor materials through weak non-covalent bonds, such as hydrogen bonding or van der Waals forces. The soft-templating method offers advantages, such as easy synthesis, low cost, and straightforward removal of the template, as they are typically soluble in solvents. The use of soft templates enables the control of the nanomaterials' morphology and size at the nanoscale.

The template-assisted synthesis of nanomaterials offers several benefits. First, it allows the precise control of the size, shape, and composition of the nanostructures, which is crucial for tailoring their properties and functionalities. Second, templates provide a scaffold or framework for the growth of nanomaterials, leading to a high degree of structural uniformity and regularity. Third, template-assisted synthesis methods are versatile and applicable to wide range of materials, including metals, metal oxides, semiconductors, and polymers. Lastly, the use of templates can simplify the synthesis process and enhance the reproducibility of the desired nanostructures (Fig. 10.8).

Nanomaterials Applied in Water Purification.

The unique properties of nanomaterials make it extremely promising for use in a range of water treatment applications. Currently, different nanoparticles are progressively being used as adsorbents for water purification. Nanotechnology holds enormous potential as the most viable technique to tackle persistent or emerging pollutants [25]. Figure 10.9 shows some of the typical nanomaterials used for water purification with a brief description of some of each under this section.

(i) Nanoparticles: Nanoparticles of different materials are used to purify water. These materials include metal oxides (like iron oxide and titanium dioxide), carbon-based materials (like carbon nanotubes and graphene oxide), and other

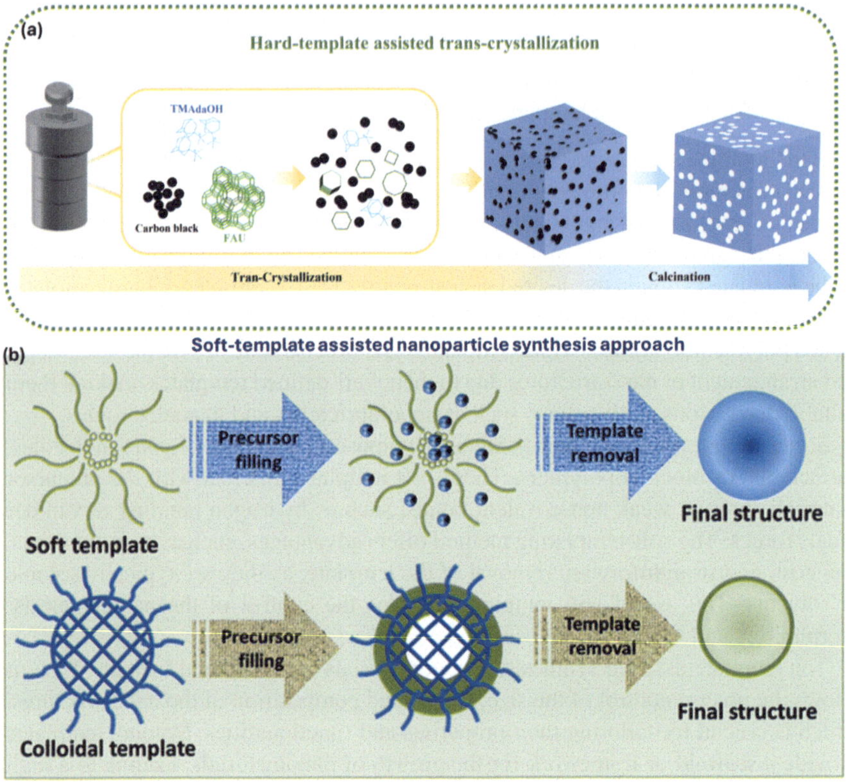

Fig. 10.8 **a** Schematic illustration of hard template-assisted synthesis. Reproduced from [23], an open access article by MDPI under CC BY licence. **b** Schematic illustration of soft and colloidal assisted template synthesis. Reproduced from [24], an open access article by the Royal Society of Chemistry under CC-BY-NC licensed under a Creative Commons Attribution-Noncommercial 3.0 Unported Licence

nanomaterials (like silver nanoparticles). They are suitable for adsorption, catalysis, disinfection, and membrane filtration applications [26].

(ii) Nanocomposites: They are hybrid materials incorporating nanoparticles scattered throughout a matrix material. Because of its greater versatility, smaller installation footprints, cheaper cost compared to inorganic membranes, and straightforward pore generation mechanism, polymeric membranes are the most widely employed for water treatment. Conventional nanocomposite, thin-film nanocomposite (TFN), thin-film composite (TFC) with nanocomposite substrate, and surface-located nanocomposites are among the several polymeric membranes used for water treatment [27]. These materials can combine the features of several nanoparticles to enhance performance in water treatment applications. For example, metal oxide and polymer nanocomposites can be utilized for adsorption or membrane filtration.

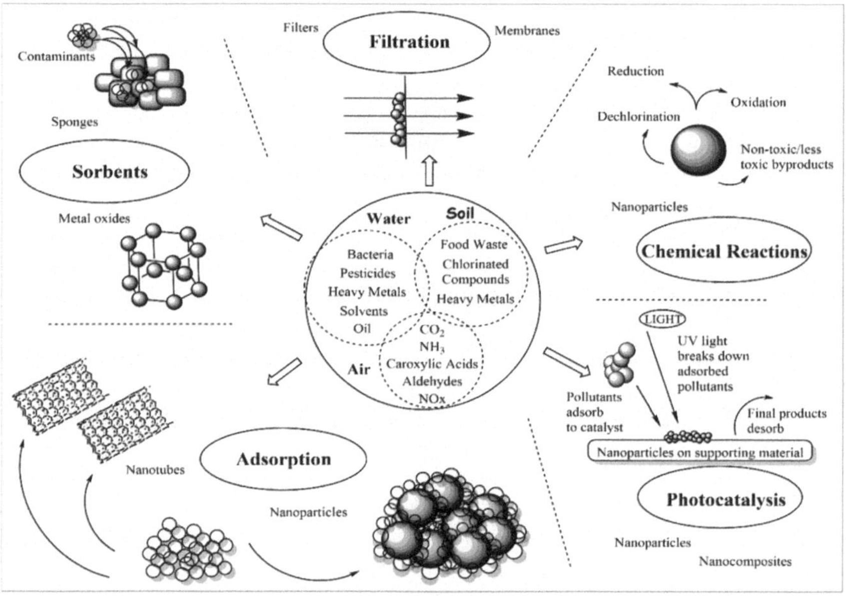

Fig. 10.9 Some typical nanomaterials are used in water purification. Reproduced from [25], an open access article by MDPI under CC BY licence

(iii) Carbon Nanotubes (CNTs): CNTs are perfect for a variety of water treatment applications due to their high surface area and mechanical strength. They are appropriate to remove pollutants, as membrane materials, or as catalyst supports in advanced oxidation processes. CNT is an effective adsorbent for various types of water pollutants, including biological (algae, viruses, bacteria, and fungi) and inorganic (heavy metals, ammonia, and salts) as well as chemical (detergents, insecticides, herbicides, fats, grease, volatile organic compounds, perchlorate, drug molecules, and colours). The splitting reactions of water pollutants and photocatalysis are accelerated by CNT-reinforced composite/hybrid materials [28].

(iv) Graphene and Graphene Oxide (GO): Advancements in graphene materials show potential for water purification. GO and graphene are perfect for treating water because of their superior surface area, electrical conductivity, and mechanical strength. They can be used to break down contaminants as catalysts, adsorbents, and membrane filters [29].

(v) Metal–Organic Frameworks (MOFs): MOFs are porous materials containing both organic ligands and metal ions or clusters. They are composed of varying porosity and surface chemistry, making them outstand for adhesion, catalysis, and membrane separation in water treatment. The various matrix MOFs implemented for water treatment and filtration are Thin Film Composite Membrane (TFN), Mixed Matrix Membrane (MMM), and Porous Matrix Membrane (PMM) [30].

(vi) Nanofibers: Polymer or carbon nanotube-based nanofibers can filter water and remove contaminants through membranes. With hole sizes ranging from tens of nanometres to several micrometres, an interconnected open pore structure, and great gas permeability, electro spun nanofibrous membranes are incredibly transparent. The greater porosity of nanofiber membranes allows for a larger surface area per unit volume, making them suited for liquid filtration applications like distillation [31].

The use of these nanomaterials for removing pathogens, heavy metals, organic pollutants, and new contaminants from water help reduce water scarcity and provide clean water for all. More importantly, these microscopic particles provide innovative solutions to conventional water treatment issues that enhance environmental sustainability and water quality. That notwithstanding, their potential environmental and health consequences ought to be studied extensively to ensure they possess no advert effect to living organisms in the future [32].

Nanomaterials-Based Filtration Systems

The use of nanoparticle membranes improves the mechanical strength, water permeability, separation effectiveness, and fouling reduction is common in water filtration systems. Nanomaterials-based water filtration systems enhance the ability to remove a wide range of contaminants [33]. The high surface area-to-volume ratio of nanomaterials, such as nanoparticles, nanotubes, and nanofibers, enhances their adsorption and catalytic capabilities. This allows them to effectively capture and remove a diverse array of pollutants, including heavy metals, organic compounds, and even pathogens, with remarkable efficiency. Moreover, nanomaterials-based membranes and filters have demonstrated superior performance in desalination and water-softening processes. Graphene-based membranes and carbon nanotube-based filters, for instance, have exhibited exceptional selectivity and permeability, leading to more efficient and energy-saving desalination techniques [34]. This is particularly important in regions facing water scarcity, where the ability to produce clean, potable water from saline or brackish sources is crucial.

Nanomaterials also play a crucial role in disinfection and antimicrobial applications within water filtration systems. Certain nanomaterials, such as silver nanoparticles and metal oxide nanoparticles, possess potent antimicrobial properties, making them effective in eliminating pathogenic microorganisms from water [35]. By incorporating these nanomaterials into filtration systems, an additional layer of protection against microbiological contaminants can be provided, ensuring the safety and quality of the purified water. Furthermore, the catalytic properties of nanomaterials, such as titanium dioxide (TiO_2) nanoparticles, can be leveraged for the degradation of organic pollutants. When exposed to UV or visible light, these photocatalytic nanomaterials generate reactive oxygen species that break down various organic contaminants, contributing to a more comprehensive water purification process [36].

Nanomaterials-based water filtration systems also offer self-cleaning and antifouling capabilities. Coatings of superhydrophobic or super hydrophilic nanomaterials can be applied to filtration membranes or surfaces, reducing the buildup of contaminants, and extending the lifetime of the filtration system [37]. This feature is particularly valuable in maintaining the efficiency and longevity of water treatment systems, especially in challenging environments. Solid materials called adsorbents are used to absorb gases or dissolved chemicals; the molecules that are adsorbed are known as adsorbates. Because of their huge specific surface area, nano adsorbents show a substantially higher rate of adsorption for organic molecules as compared to granular or powdered activated carbon. They hold out a lot of potential for the development of novel, faster, more efficient techniques for decontaminating environments that eliminate both organic and inorganic contaminants, including heavy metals and micropollutants. For instance, CNTs have a surface chemistry that can be adjusted and highly accessible adsorption sites. CNTs must be stabilized in an aqueous solution to prevent aggregation that would lower the active surface area because of their hydrophobic surface [38]. They can be applied to preconcentrate and detect pollutants, as well as for the adsorption of persistent pollutants. CNTs could absorb metal ions through chemical bonding and electrostatic attraction. Adsorption-based technologies are easy to use and quick to implement for point-of-use water filtration systems, but their capacity to remove salt is limited. Conventional desalination processes, on the other hand, are energy-consuming and technically difficult. These innovative CNTs are predicted to provide enhanced filtration, disinfection, and desalination capabilities in next-generation potable water purification systems [39]. It is possible to use polymeric nano adsorbents, including dendrimers (repeatedly branched molecules) to remove heavy metals and organics as demonstrated by Nishu and Sudesh [40] shown in Fig. 10.10. The internal hydrophobic shells can absorb organic substances, while the fitted outer branches can absorb heavy metals. The bio adsorbent is harmless, biocompatible, and biodegradable. They can remove up to 99% of certain dyes. Since the early 1980s, zeolites have been associated with silver atoms. Zeolite's high porosity makes it possible for it to hold nanoparticles, such as silver ions [41, 42].

Nanoscale metal oxides are effective adsorbents for the removal of heavy metals and radionuclides and hold significant potential as activated carbon alternatives. Due to their superparamagnetic properties, several of these tiny metal oxides, such as nano maghemite and nano magnetite; allow for easier recovery and separation using a low-gradient magnetic field. Slurry reactors and adsorptive media filters may both use them [43, 44]. Nano-titanium dioxide (TiO_2) has minimal human toxicity and excellent chemical stability at an affordable price. The primary benefit of nano-TiO_2 coatings over nano silver is their almost infinite lifespan, as TiO_2 acts as a catalyst that doesn't alter when organic chemicals and microorganisms break down. Using iron nanoparticles in nano remediation has many benefits. Iron (III) and iron (II) were reduced to iron at the nanoscale using borohydride as a reductant. The width of the zero-valent iron nanoparticles ranges from 10 to 100 nm. They are composed of the standard exterior and centre. Nanoscale zero-valent iron (nZVI) is the environmental

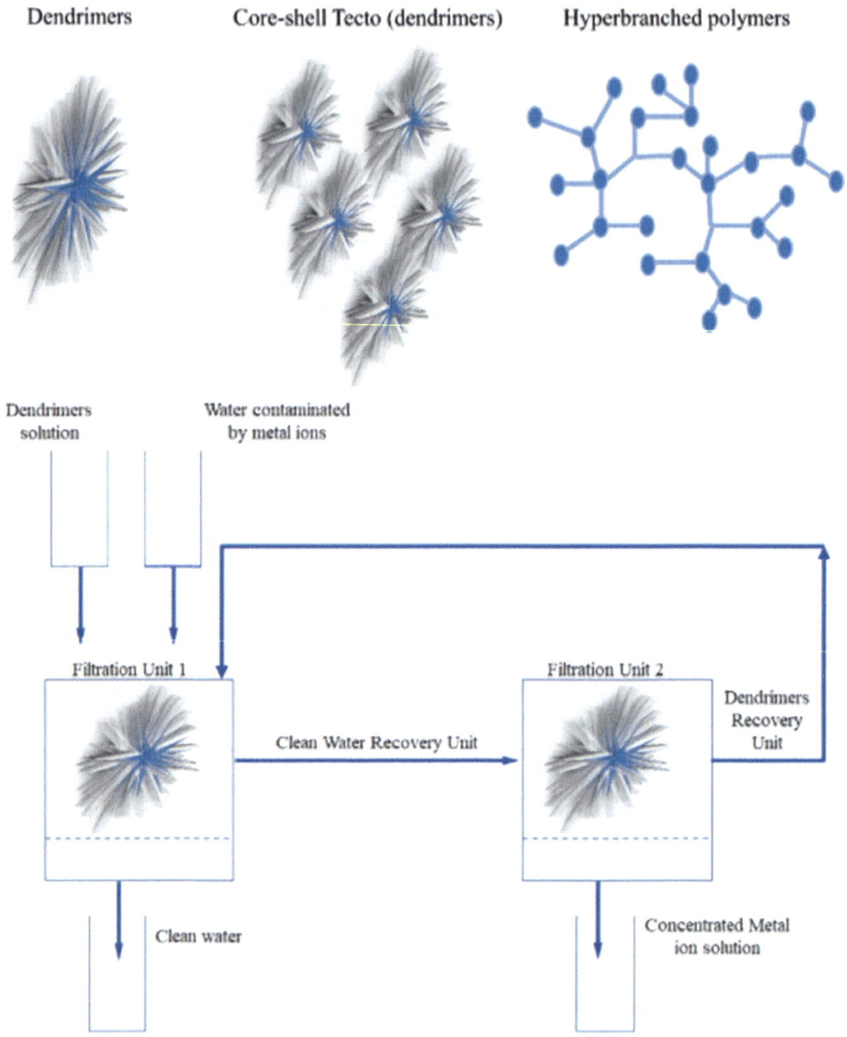

Fig. 10.10 Metal ion recovery from aqueous solutions using improved dendrimer filtration. Reproduced from [40], an open access article by Elsevier under an open article by Elsevier under CC BY-NC-ND 4.0 DEED

nanotechnological approach to groundwater cleansing that has been studied the most [45, 46].

It has been sufficiently demonstrated that metallic irons are especially effective at destroying a wide range of common pollutants at the nanoscale, including pesticides, dyes, trihalomethanes, brominated and chlorinated methanes, chlorinated benzenes, chlorinated ethenes, and other polychlorinated hydrocarbons. A crucial nanomaterial for water filtration is two-dimensional graphene, or 2D graphene, which comprises reduced graphene oxide (GO), and pure graphene. These nanoparticles are widely recognized for their exceptional qualities as adsorbents, antifouling agents, and photocatalytic materials. Graphene nanoparticle desalination has attracted a lot of attention. Due to their remarkable ability to take up sodium cations, 2D graphene nanomaterials are very effective in taking out salts from water. It has been said that several microbes are immune to the antibacterial properties of graphene and graphene oxide [2]. A summary of water treatments with nanomaterial by different mechanisms has been provided in Table 10.1.

The focus of the new era has been on advanced oxidation processes (AOPs) as trustworthy, alternative processes that have the potential to oxidize and mineralize a variety of organic molecules. Strong and extremely powerful oxidizing radicals are produced by AOPs [40]. Photocatalysis, a well-known AOP, is an efficient approach to increase the biodegradability of persistent organic pollutants and eradicate both current and newly emergent microbial illnesses.

Photocatalytic oxidation is a family of processes in which an activated catalyst produces highly reactive radical species. Several 2DM-based photocatalysts, such as metal oxide, carbon nitride, graphene, and chalcogenide nanosheets, have recently been investigated for various uses. These photocatalysts will act as decontaminants and disinfectants [40]. The key reactions and wavelengths of AOPs are shown in Fig. 10.11. Nanoparticle synthesis is the technique of generating free-standing, very thin membranes from nanoparticles. Nanoparticle membranes are constructed via filtering, blown-film extrusion, a drying-mediated self-assembly approach, and by segregation of nano films. Nanomembranes from gold nanoparticles as well as ultra-thin freestanding membranes are also used. High weight-to-weight ratios and very porous structures with exceptional pore interconnectivity are features of nanofibers. The exceptional mechanical and film-forming qualities of cellulose make it useful in a variety of industries. The cellulose surface is also highly safe to work with and simple to change, which makes surface modification easier.

Revolutionary methods adopted by nanotechnology ease the process of water purification. For use in filtration, adsorption, and photocatalysis in water treatment, a wide range of nanomaterials have been created, such as nanoparticles, nanofibers, and nano biocides [48]. There are several nanomaterials, such as nano iron, nano zeolites, photocatalysts besides dendrimers are effective materials in purifying water. To further enhance the performance of water filtration, nanomaterials can be integrated into hybrid and multi-stage systems. By combining nanomaterials with other technologies, such as adsorption, ion exchange, and membrane filtration, the unique properties of nanomaterials can be leveraged to improve the overall efficiency, versatility, and cost-effectiveness of the water treatment process [49].

As research and development in the field of nanomaterials-based water filtration systems continue to evolve, one can expect to see even more innovative and sustainable solutions to address the global water crisis. These advanced materials have the potential to revolutionize the way we purify and manage water resources, paving the way for a future where clean and accessible water is available to all.

Table 10.1 Summary of water treatments with nanomaterial by different mechanisms [47]

Nanoparticles (NPs)	Target analyte	Treatment mechanism	Limitation	Positive aspect
TiO_2	Organic pollutants	Photocatalysis	Higher operation cost Tough to recovery, sludge production	Toxic less, Water insolubility, photostability
Fe	Heavy metals, anions, organic pollutants	Reduction, adsorption	Tough to recovery, sludge Production, difficult sludge disposal, Health risk	In situ water remediation, less cost harmless to handle
Bimetallic NPs	Dichlorination	Reduction, adsorption	Tough to recovery, sludge Production	Higher reactivity
Nanofiltration and nanomembranes	Organic and inorganic substances	Nanofiltration	High cost, membrane Fouling	Low pressure
Magnetite NPs	Heavy metals, organic Compounds	Adsorption	Outside magnetically field, needed for Separation	Easy separation, no sludge production
Metal-sorbing vesicles	Heavy metals	Adsorption	Re-use option, higher selective uptake profile, better metal affinity	Difficult to maintain stability
Micelles	Organic pollutants	Adsorption	In situ treatment, excellent affinity for hydrophobic	Costly
Dendrimers	Heavy metals, organic Pollutants	Encapsulation	Easy separation, renewable, high binding no sludge production	Costly
Nanoclay	Heavy metals, anions, organic pollutants	Adsorption	Lower coat, Exclusive structures, long stability, recycle, Higher sorption capacity	Sludge production

Advanced Water Treatment Processes

Advanced water treatment processes are used to remove contaminants from water beyond the capabilities of conventional treatment methods. These processes can be used to treat both drinking water and wastewater, and they can remove a wide variety of contaminants, including bacteria, viruses, pharmaceuticals, and heavy metals. Types of advanced water treatment processes include reverse osmosis, membrane

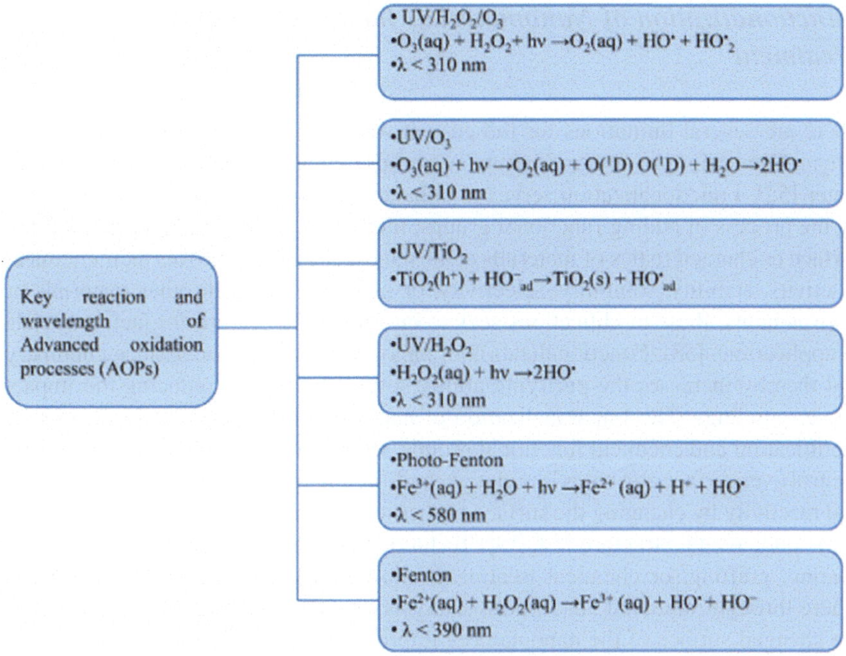

Fig. 10.11 Illustrates AOPs that produce hydroxyl radicals through radiation Reproduced from [40], an open article by Elsevier under CC BY-NC-ND 4.0 DEED

filtration, water oxidation, carbon adsorption, ozone, and ultraviolet (UV) disinfection [50]. The reverse osmosis process uses a semipermeable membrane to remove dissolved salts, organic molecules, and other contaminants from water. The membrane filtration process uses a variety of membranes to remove contaminants based on their size and charge, whereas the oxidation process uses oxidants to break down contaminants into smaller, less harmful molecules. Carbon adsorption process uses activated carbon to remove organic contaminants from water. In water disinfection, the ozone is used to remove organic contaminants whereas the UV light is used to kill bacteria and viruses in water [51].

Possible applications of advanced water treatment processes are often used to treat surface water sources, such as rivers and lakes, to make them safe for drinking. Advanced water treatment processes can also be used to treat wastewater to a high standard, allowing it to be reused for irrigation or other purposes. Moreso, advanced water treatment processes can be used to treat water for use in industrial processes, such as cooling water or boiler feedwater.

Functionalization of Nanomaterials for Advanced Water Treatment

There are several limitations for the current water remediation process due to the inherent limitations by some of the nanomaterials for precise impurities removal from water [52]. Functionalization serves to enhance the potency of impurities removal by the process of adding functional groups, molecules, or nanoparticles to alter the surface or characteristics of materials at the nanoscale. This improves their chemical reactivity, stability, solubility, selectivity, or compatibility with other materials or environments, these modifications seek to customize the features of nanomaterials to applications [53]. Functionalization of nanoparticles alters the surface chemistry and thereby increases the purifying ability several fold while reducing the impact on surroundings. The functionalization of nanomaterials relies heavily on surface modification and chemical functional groups, where surface modification gives fine control over surface attributes like charge, hydrophilicity/hydrophobicity, roughness, and reactivity by changing the surface properties of nanomaterials without affecting their fundamental structure [54, 55]. Techniques for surface modification, such as coating, grafting, or chemical treatment, allow molecules to modify the surface where through improved adsorption, catalysis, or membrane separation processes, the changed surface of the nanomaterial improves its interaction with target pollutants in water. Unique chemical qualities are conferred by the chemical moieties. Functional groups can create hydrogen bonds, align with metal ions, or participate in chemical processes to boost affinity and thereby speed up purification. A brief insight into some of the functionalization techniques are discussed briefly.

Chemical Modification: Bond formation with functional groups or molecules to the surface of nanomaterials via chemical reactions like amination, esterification, or salinization is known as chemical modification. This is one of the few common ways of functionalization [55]. Consequently, it offers exact control over the kind and number of functional groups attached to the nanomaterial, resulting in customized characteristics and improved functions.

Surface Coating: It is a thin layer of functional material that adds to the stability, improves dispersibility, compatibility with the environment, and allows for the introduction of functionalities into nanomaterials without changing their fundamental structure [56].

Doping: Doping involves introducing dopant atoms or ions into the lattice of nanomaterials to alter their electronic or optical properties. Doping can be achieved through ion implantation, chemical vapour deposition, or solid-state diffusion. It alters nanomaterials' electronic band structure, conductivity, and catalytic activity, resulting in increased performance in photocatalysis, sensing, and energy conversion [57].

Ligand Exchange: Ligand exchange involves replacing the native ligands or capping agents on the surface of nanomaterials having specific ligands to modify nanomaterial surfaces to enhance their solubility, dispersibility, and reactivity in different solvents or environments [58].

Photocatalysis for Pollutants Degradation in Water Treatment

Industries produce effluent that is contaminated with a range of organic substances, one of which being dyes [59]. For a lot of years, the dyes in the wastewater have been seriously harming the ecosystem. Dyes were used in the dying process, but a certain percentage of dye carried away with contaminants. The complex structure and synthetic origin of the colours make extensive and economical treatment of the dyeing wastewater an undeniably difficult problem to address. Diverse methods have been proposed to examine their capacity to eliminate dye. Among these methods is semiconductor photocatalysis, which is thought to be a useful approach to break down organic dyes using sunlight. The technology is ecologically friendly and cost-effective [60].

Basic principle of pollutant photocatalysis

During photocatalysis, the stimulated electron moves from valence band to conduction band and thereby conduction band receives an electron which is reducing, while the valence band receives a hole which is oxidizing as shown in Fig. 12a. The primary advantage of heterogeneous photocatalysis is its ability to occur in ambient environments with adequate air oxygen serving as an oxidant [61]. Photocatalysis has emerged as a highly effective method for degrading pollutants in wastewater, offering numerous applications across various sectors. Titanium dioxide (TiO_2) or zinc oxide (ZnO) is activated by light energy to generate reactive oxygen species (ROS) that facilitate pollutant degradation. In the context of dye degradation, photocatalysis demonstrates remarkable efficiency and versatility [62]. The reactive species oxidize the chromophore groups and thereby initiates the degradation process. This oxidative attack leads to the cleavage of dye molecules into smaller fragments, ultimately resulting in their mineralization to carbon dioxide, water, and inorganic ions [63]. What makes photocatalysis particularly attractive for dye degradation is its ability to address the diverse range of synthetic dyes with varying chemical structures and properties.

Whether it's azo, anthraquinone, or triphenylmethane dyes, photocatalysis can effectively degrade them without the need for additional chemical reagents or complex treatment processes. This versatility ensures that photocatalysis remains a viable solution for treating wastewater from different industrial sectors where dyes are commonly used. Moreover, photocatalysis offers several advantages specifically in the context of dye wastewater treatment. Firstly, the process operates under mild conditions, reducing the energy requirements and environmental footprint associated with treatment. Secondly, photocatalysis promotes the complete mineralization of dyes, eliminating the formation of harmful byproducts and ensuring thorough pollutant removal [65]. Moreover, the photocatalysts are reused several times without affecting its property. A visible-NIR-responsive decatungstate charge-transfer salt hybrid material was also demonstrated in Fig. 12b, where the effective separation of charge-carriers by local surface plasmon resonance was demonstrated through the assembly of g-$C_3N_4H_x$ + cation and decatungstate anion [64].

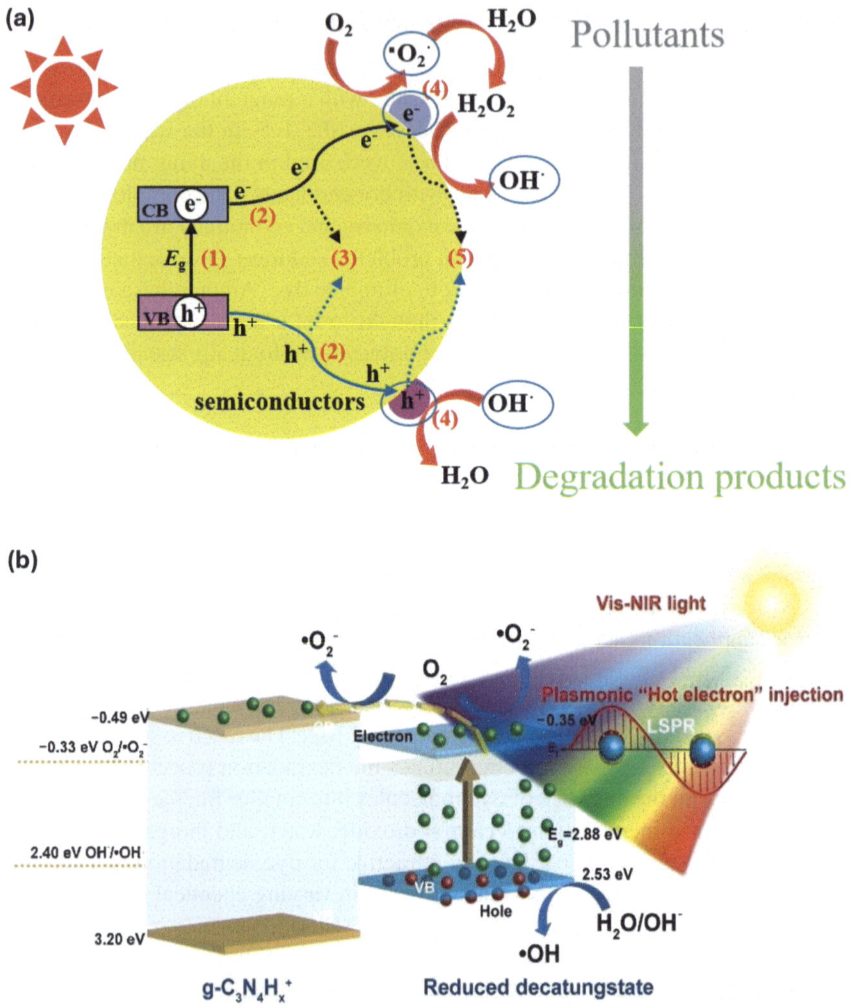

Fig. 10.12 **a** heterogeneous photocatalytic process of pollutant degradation, **b** Schematic illustration of g-$C_3N_4H_x$ + cation photocatalyst under Vis–NIR light excitation. Reproduced from [64], an open access article by MDPI under CC BY license

Therefore, photocatalysis emerges as a highly efficient and environmentally friendly approach for degrading synthetic dyes in wastewater. Its ability to address the complex nature of dye pollutants, coupled with its versatility, efficiency, and sustainability, positions photocatalysis as a key technology in the quest for cleaner water resources. Continued research and development efforts aimed at optimizing photocatalytic materials and reactor designs will further enhance its efficacy and broaden its applications in dye wastewater treatment.

Antimicrobial Nanomaterials for Pathogen Inactivation and Disinfection

One of the main causes of mortality and persistent diseases is bacterial infection. Because they are effective and affordable, antibiotics have long been the treatment of choice for bacterial illnesses. Prolonged use of antibiotics will create a situation of drug resistant bacterial strains and the appearance of super-bacteria [66]. These bacteria have been shown to have the NDM-1 super-resistance gene. Three bacterial targets are targeted by the main classes of antibiotics currently in use: the machinery involved in the creation of cell walls, translation, and DNA replication [67]. Regrettably, each of these ways of action can be overcome by bacteria developing resistance. Nanoparticles which are antimicrobial in nature have gained lot of interest in recent years. TiO_2 nanoparticles exhibit excellent biocidal effects and since they may contribute harmless nanoparticle to control microorganisms. Copper nanoparticles have antimicrobial properties and thereby several microorganisms are neutralized by it. Therefore, Cu nanoparticles find applications in the healthcare sector [68]. It has been found that carbon nanotubes have trouble spreading in water, which prevents enough CNT and microbe interaction for the disinfection process. High dispersion can boost the CNTs' ability to interact with bacterial cells, which will enhance their antibacterial properties [69]. The effectiveness of CNTs as a disinfectant can be enhanced by additional procedures, such as functionalization. CNTs' surface function can enhance their meagre antibacterial capabilities. Water treatment has made use of nanocomposites made of metal particles and carbon nanotubes, which show promise as powerful antibacterial agents with a synergistic impact [66]. To increase water decontamination and disinfection, silver nanoparticles were successfully grafted on MWCNT and eliminated germs without reducing the flow rate of membrane filtration. No trace of viruses was found to flow through the nanofilter with this composite [70].

Nanomaterials for Contaminant Removal

Nanomaterials have emerged as a revolutionary solution in the field of environmental remediation, particularly in the purification of wastewater and the removal of various pollutants. These materials, with their unique properties and remarkable reactivity, have demonstrated remarkable potential in addressing the pressing challenges of environmental contamination. Nanoparticles composed of materials, such as iron, gold, iron oxide, nickel oxide, copper oxide, zinc oxide, and titanium have all been successfully employed in the removal of contaminants from water, showcasing their versatility and efficiency [71].

Beyond their ability to remove traditional pollutants, nanomaterials have also shown promise in addressing emerging contaminants, such as persistent organic compounds and microplastics. The combination of nanomaterials with biochar has proven to be a particularly effective and sustainable approach, offering a multifaceted solution to the complex problem of environmental remediation. The application of nanomaterials extends beyond water purification, as they are also utilized for the elimination of toxic gases and heavy metals from the environment. Nano sensors made from materials like carbon nanotubes, tin oxide, platinum, copper, silver, and carbon quantum dots have demonstrated their effectiveness in this domain, underscoring the diverse capabilities of nanomaterials [72, 73].

One of the key factors driving the adoption of nanomaterials in contaminant removal is their cost-effectiveness and high efficiency. Compared to conventional methods, nanomaterials offer a more economical and sustainable solution, making them an attractive option for various industries and environmental organizations [73]. The remarkable capabilities of nanomaterials have positioned them as a game-changing tool in the field of environmental remediation. Their ability to effectively remove a wide range of contaminants, from heavy metals to emerging pollutants, coupled with their cost-effectiveness and sustainability, makes them an indispensable component in the ongoing efforts to address the pressing environmental challenges we face today. As research and development in this field continues to evolve, the potential of nanomaterials to revolutionize the way water contaminants cleanup and protect remains immense.

Contaminants Commonly Found in Water and Their Health Implications

Water is a basic need for humans and living organisms. Therefore, it should not be contaminated with pollutants that may bring negative effects to human beings as well as ecosystems. Water becomes nondrinkable due to the presence of contaminants and microorganisms. Table 10.2 shows some typical water pollutants, their sources, and their effect on living organisms.

Adsorption and Catalysis Mechanisms of Nanomaterials for Contaminant Removal

For the removal of toxic metals from the wastewater, the most suitable method is the adsorption mechanism due to its high efficiency. In the adsorption process, the contaminants are adsorbed on to the surface of a material known as the adsorbent [74]. The extensively used adsorbents are activated carbon, zeolites, etc., but the efficiency of these materials is relatively low. Recent developments in nanotechnology have helped overcome this hurdle due to surface modification. As a result, a variety of efficient, low-cost, and eco-friendly nanomaterials with unique functionalities have been proposed lately. The advantage of nanostructured materials over conventional

Table 10.2 Pollutant types and their effects on humans

Type of pollutant	Name of the pollutant	Source of pollutant	Pollution effect
Inorganic pollutants	Nitrates	Nitrogenous manure, wastewater, septic system seepage, urban drainage discard, industrial effluents	Methemoglobinemia in newborns, mortality by asphyxiation, and stomach cancer in adults
	Phosphate	Sewage water and industrial wastewaters	Eutrophication
	Heavy metals	Polluted industrial water	Affects brain, lungs, kidney, liver, blood composition
	Halogens: chlorides, bromides, and fluorides		Dental fluorosis, Skeletal fluorosis, Arthritis, bone damage, osteoporosis, muscular damage, fatigue, joint-related problems, nausea, vomiting, diarrhoea, and abdominal pain. It also causes negative effects on kidneys and nervous system and hearing loss
Organic pollutants	Endocrine Disruptive Chemicals (EDC):	Pesticides and dyes	Headache, vomiting, dizziness, muscle weakness, shortness of breath, skin rash
	Phenolic compounds	Industries, agricultural and domestic activities	Cause damage to the red blood cells and the liver, even at low concentrations
Pesticides		Human exposure to human causes cancers, and heart, respiratory and neurological diseases	Human exposure to human causes cancers, and heart, respiratory, and neurological diseases When these compounds enter the body, they undergo transformation into moieties which is more harmful as compared to the original compounds
Pharmaceutical		Drug manufacturing industries	Endocrine disturbance, antibiotic resistance development, and toxicity to aquatic creatures

(continued)

Table 10.2 (continued)

Type of pollutant	Name of the pollutant	Source of pollutant	Pollution effect
Pathogens	Virus and bacteria, such as Cryptosporidium, Campylobacter, Salmonella, Shigella, can lead to water-borne diseases	Storm water runoff, pollution from agriculture, and municipal sewage	Cholera, typhoid, dysentery, polio, and hepatitis are some of the common diseases which are seen in human beings caused by the water pollutants
Suspended solids and sediments			Harm the gills of fishes and originate asphyxiation
Radioactive compounds		Nuclear power plants and industrial effluents. Another radioactive element found in water is radium	Radiation exposure results in several physiological diseases in humans, such as chromosomal disruption, cancer, leukaemia, genetic mutations, osteonecrosis, and cataracts

materials is its high surface area which is a reason for higher efficiency and fast adsorption rate in water treatment [75]. An ideal adsorbent for wastewater treatment purposes:

- should be environment friendly.
- should have high sorption capacity and high selectivity.
- the adsorbed pollutants can be easily removed from its surface.
- should be recyclable.

The most common adsorbents used include carbon nanotubes (CNTs), graphene, ferric oxide (Fe_3O_4), manganese oxide (MnO_2), titanium oxide (TiO_2), magnesium oxide (MgO) and zinc oxide (ZnO) [75]. The adsorption process is a surface-related occurrence where the adsorbate accumulates on the surface of the adsorbent. It can be defined as a mass transfer operation by which a substance is transferred from a liquid phase to the surface of a solid. This process is prominent when a solution carrying an absorbable solute contacts a solid characterized by a highly porous surface structure. The reason for this deposition or concentration of solute molecules on to the surface of the solid adsorbent is the intermolecular forces of attraction between the liquid and solid phases [76]. The surface atoms of the adsorbent lack complete surrounded by other adsorbent atoms, this unique surface characteristic makes them capable of attracting and accumulating adsorbates from the solution. The specific bonding characteristics vary depending on the species involved, but the adsorption process is typically categorized into two main types: physisorption and chemisorption. Physisorption refers to the adsorbate being attached to the surface through weak van der Waals forces. On the other hand, chemisorption involves a stronger attachment mechanism, where the adsorbate is either tethered through covalent bonding or held in place due to electrostatic attraction.

The adsorption mechanism is affected by several variables, including pH, temperature, contact time, and the initial concentration of the pollutant. By charging the molecule or the functional groups attached, the pH influences the adsorbent surface and increases the attraction between the adsorbate and the adsorbent. The minimal amount of time needed for a specific dye concentration to interact with the adsorbent is determined by the contact time [76]. The point at which equilibrium is reached is known as the contact time. Adsorption becomes steady after a certain point, indicating that all the adsorbent sites have been occupied and that no more pollution can be absorbed. The initial organic pollutant concentration is directly related to adsorption sites and aids in determining the adsorbent's adsorption capacity. To determine how effective an adsorbent is and how much is required for the full adsorption of a pollutant, the amount of adsorbent is studied. The measurement of adsorption capacity is also necessary.

The adsorption isotherm models present the amount of solute adsorbed per unit weight of adsorbent as a function of the equilibrium concentration in the bulk solution at constant temperature. Different isotherm models include Langmuir and Freundlich, Temkin, Harkin–Jura, and Dubinin–Radushkevich, among which Langmuir and Freundlich's models are more commonly used [77].

Adsorption of pollutants by nanoparticles

Nanomaterials have the high reactivity, strong sorption, and increased surface-area-to-volume ratio that make them ideal for wastewater treatment. Environmental pollutants are extracted from aqueous media using several kinds of metal oxide nanoparticles (NPs).

Silver nanoparticles are most frequently utilized in the treatment of water and wastewater, which is typically not achievable by traditional technologies. This is due to silver's low toxicity and well-documented antibacterial action. Antimicrobial nanofibers and nanocomposites are made using Ag NPs integrated into various polymer types [78]. Due to their ease of synthesis, large surface area, and noteworthy removal efficiency, iron nanoparticles and various forms of iron oxide, including crystalline, amorphous, and goethite, have been employed as adsorbents by numerous researchers to remove metallic pollutants from aqueous solutions and industrial effluents. Zero-valent iron nanoparticles have also been used to remove environmental contaminants from industrial effluents. Zinc oxide NPs have recently received attention because they exhibit a large surface area and high catalytic activity. These NPs exhibit antibacterial, anti-corrosive, antifungal, and UV filtering properties, and have biomedical and therapeutic applications [79]. As it contains highly reactive carboxyl and amino groups, chitosan is thought to be the ideal material for treating industrial effluents. Since chitosan is extremely soluble in acidic media, its use as a sorbent is limited. However, the adsorption capability can be somewhat enhanced by the addition of metal oxide nanoparticles. Because of their huge specific surface area and hollow and layered structure, carbon nanotubes (CNTs) are the most widely employed nanomaterials and are an effective adsorbent for hazardous compounds.

Attaching various materials to their surface is required to increase their efficiency. There are four subclasses of dendritic polymers: dendrigrafts, dendrons, dendrimers, and random hyperbranched polymers. Among these, dendrimers are highly significant from a scientific and technological standpoint. A novel group of symmetric, hyperbranched, globular, nano-spherical, and three-dimensional macromolecules is represented by dendrimers [80]. Dendrimers are particularly appealing as functional materials for water treatment due to their desirable physicochemical characteristics, which include their strong recyclability, high biocompatibility, degradability, and adaptable hosting capability.

Nanotechnology in Water Desalination

At the heart of the nanotechnology revolution in water desalination lies the unique properties of nanomaterials. These materials, with their exceptionally high surface area-to-volume ratios and enhanced reactivity, have the potential to significantly improve the performance and efficiency of desalination processes. Nanomaterials, such as carbon nanotubes, graphene, and metal–organic frameworks (MOFs) have garnered considerable attention due to their remarkable ability to facilitate the transport and separation of water molecules from saline or brackish sources [81]. One of the key advantages of nanotechnology-based desalination is its potential to reduce the energy consumption associated with conventional desalination methods, such as reverse osmosis and thermal distillation. Nanomembranes, for instance, have been engineered to exhibit superior water permeability and selectivity, reducing the pressure and energy requirements needed for the separation of salts and other contaminants from water. This not only contributes to cost savings but also aligns with the growing global emphasis on sustainability and environmental preservation.

Moreover, nanomaterials have demonstrated their effectiveness in addressing the issue of membrane fouling, a common challenge in desalination systems. The incorporation of nanoparticles into membrane structures can enhance their antifouling properties, leading to increased operational efficiency, reduced maintenance, and extended membrane lifespans. This, in turn, translates to improved water quality, reduced environmental impact, and more sustainable desalination operations. Beyond their application in membranes, nanomaterials have also shown promise in the realm of desalination pre-treatment and post-treatment processes. Nanomaterials can be utilized for the efficient removal of heavy metals, organic contaminants, and microorganisms, ensuring the production of high-quality water that meets rigorous health and environmental standards [81–83]. The versatility of nanotechnology in water desalination extends further, as researchers have explored the integration of nanomaterials with renewable energy sources, such as solar and wind power. This convergence holds the potential to unlock new frontiers in sustainable desalination, reducing the carbon footprint and increasing the accessibility of clean water in remote or off-grid regions.

As the global demand for clean water continues to rise, the role of nanotechnology in water desalination becomes increasingly crucial. With its ability to enhance efficiency, reduce energy consumption, and address the challenges of membrane fouling and contaminant removal, nanotechnology holds the promise of revolutionizing the way we approach water desalination. As research and development in this field continue to advance, the future of water desalination looks increasingly bright, with nanotechnology poised to play a transformative role in addressing the world's water scarcity challenges.

Water Desalination Methods and Their Limitations

Water desalination is a process that removes salt and other contaminants from saltwater or brackish water to create fresh water suitable for many applications like drinking, agriculture, industrial use, etc. Figure 10.13 describes the desalination process. Thermal desalination and membrane processes are major desalination methods. Thermal desalination uses the distillation principle, which involves evaporating water, collecting steam, and then condensing it [83]. Multi-Stage Flash Distillation (MSF) and Multi-Effect Distillation (MED) are examples for thermal desalination. The membrane desalination is based on the phenomena of filtration. It is also known as the pressure desalination method. Here porous membranes were used. Water passes through a vessel having a membrane at high pressure. Here it filters the salt content from water. Electrodialysis and reverse osmosis are the best examples for membrane desalination. A few of the desalination methods are described briefly in this section.

Multi-Stage Flash Distillation: MSF produces freshwater from seawater using a thermal-based desalination process. Multiple flashing and condensation stages are

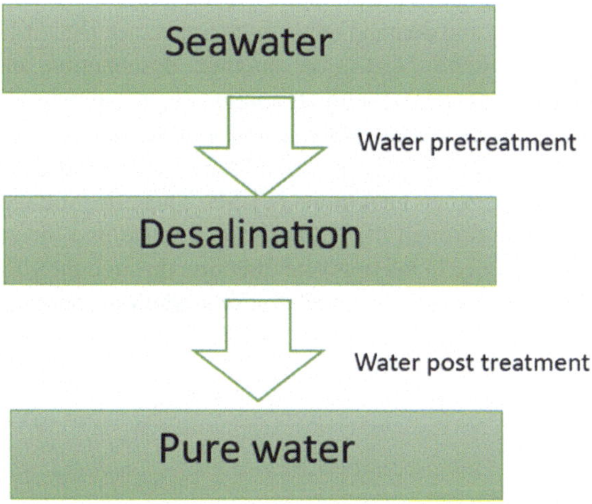

Fig. 10.13 Basic concept of water desalination

involved in this operation. As a first step the incoming seawater is heated which is then pumped with a high pressure and then heated to boiling. When it reaches the boiling temperature, the pressure is suddenly reduced which is known as flashing. This sudden pressure drop causes the water to vaporize or "flash" into steam. This produced steam is then converted into liquid by cooling in the condensation stage. MSF technology accounts for nearly 40% of global desalination capabilities. This method needs high energy as heat for evaporation. This makes it less energy-efficient and causes high environmental impact. MSF also faces scaling issues. MSF distillation plants are normally large and complex facilities, so this becomes difficult for small scale applications. Another limitation is its temperature dependence. When seawater temperature changes MSF distillation performance also can be varied.

Multiple effect distillation: MED works on the basic theory of evaporation and condensation at decreased pressure. The basic principle of MED is like MSF, but the plant design is different. There are different stages to remove salt from seawater. In every step, salt water is heated within the tubes using steam. At this stage, a major portion of water condensed and some water evaporated. After that steam passes to the next level of tubes. Here some more clean water evaporates from salt water. Like this, every stage finds the energy from the prior stage. When compared with MSF, MED is thermodynamically more advanced [83, 84]. Energy consumption, environmental impact, scaling issues, high capital cost, etc., are the major limitations.

Reverse osmosis: Reverse osmosis (RO) is the popular membrane-based desalination method. A semipermeable membrane is used here to eliminate salt and other unwanted compounds from salt water. In RO pressure should be applied to water on the side where the membrane contains salt water. By doing this it helps to overcome the natural osmotic pressure and prevent the flow of salt across the membrane. At the same time, it allows clean water to easily pass through the membrane. Major limitations of RO are high energy requirements and issues due to high membrane folding. RO requires a high amount of energy at high pressures. This can lead to high operational costs and environmental concerns. Membranes used in RO are subject to fouling, hence the pollutants develop on the membrane surface, which lowers the efficiency and demand constant maintenance [85, 86].

Electrodialysis: ED is another desalination method where ion exchange membranes (IEMs) with an applied electric potential difference is used. Here saltwater passes through both anionic and cationic membranes. Anionic membranes are known as anion exchange membrane (AEM) and cationic membranes are known as cation exchange membrane (CEM). Then, the voltage is delivered to electrodes by travelling through the IEMs. Anions move in a direction through the AEM, while cations move in the opposite direction through the CEM. This produces fresh water. Energy consumption, membrane degradation, continuous maintenance of the system, high initial capital cost, etc., are the main limitations of electrodialysis [87, 88].

Environmental and Ethical Considerations

The continuous research, manufacture, and release of nanomaterials into the environment during wastewater treatment processes raises concerns about their potential impact on aquatic ecosystems, soil, and groundwater. Some nanomaterials may accumulate in the food chain, potentially affecting the health of organisms. Moreover, ethical issues regarding the safety and fair labour conditions of researchers and workers in nanotechnology for water treatment ought to be given serious attention [89]. Some specific aspects of environmental and ethical dimensions are discussed in this subsection.

Potential Environmental Impacts of Nanotechnology in the Treatment of Water

Presently, the potential environmental implications on aquatic organisms, plants, and humans cannot be accurately quantified because there's not sufficiently long enough life cycle analysis of these materials on flora and fauna. Nanomaterials are inadvertently released into water sources like lakes, rivers, and oceans. These nanoparticles may also have an impact on potable water, but the effects are not fully understood. As a result, these nanomaterials may or may not be harmful to aquatic or human life. Aqueous habitats support a vast range of microorganisms. As a result, when nanomaterials reach watery settings, they frequently interact with microbes. Before commercializing nanotechnology for water and wastewater treatment, the potential environmental consequences of its application should be considered.

Nanomaterials in water do not directly affect humans, but they may influence humans indirectly and the ecology directly [90]. So, the influence of nanomaterials on water must be considered. Nanoparticles may be released into the environment during the manufacturing, usage, and disposal of nanomaterials used in water treatment. These nanoparticles may survive in the environment and amass in ecosystems, endangering aquatic creatures and potentially entering the food chain and drinking water sources. It is imperative to know about the long-term impact of these nanoparticles on humans and the environment. Furthermore, nanoparticles have the potential to bioaccumulate in aquatic creatures, which means they can accumulate in their tissues over time. This bioaccumulation has the potential to harm both species and ecosystems [90, 91]. Furthermore, while nanotechnology-based water treatment methods can be highly effective at removing contaminants, some nanoparticles may interact with water chemistry or other treatment chemicals in unexpected ways, potentially reducing treatment efficacy or producing harmful byproducts. The different ways by which nanomaterials can enter our system is shown in Fig. 10.14.

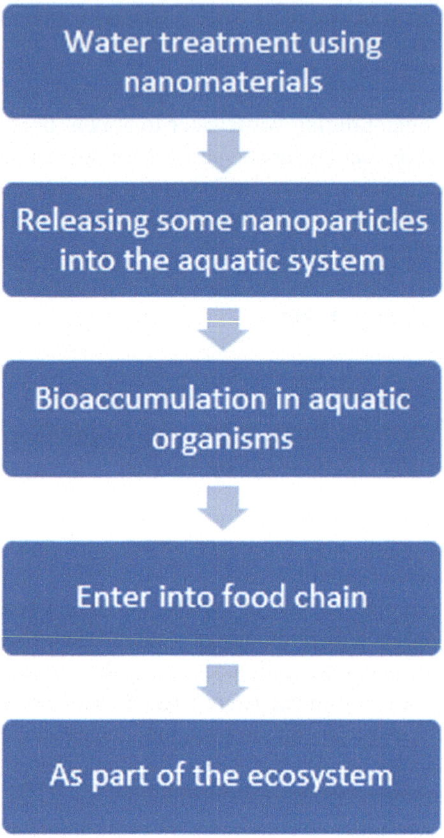

Fig. 10.14 Shows the flow chart of nanoparticles entering the ecosystem

Safety and Toxicity Concerns of Nanomaterials

With the increased use of nanoparticles, it is very important to address the concerns regarding the nanoparticle's toxicity. If the utilizing nanoparticles are not biodegradable, they will remain as such for a long period. Exposing these nanoparticles to aquatic organisms can lead to genotoxic and cytotoxic effects and accumulate on them, which have chances to enter the human body through the food chain. In general, when discussing the mechanism of toxicity, nanoparticles can produce highly reactive oxygen species (ROS) [91]. It can also produce free radicals directly or indirectly via oxidative enzymatic pathways. These will result in oxidative stress. The free radicals could be superoxide radical anions or hydroxyl radicals. The size of nanoparticles has a considerable impact on their toxicology. Nanoparticles with diameters ranging from 1–100 nm can behave differently from larger particles made of the same material. This size-dependent effect may cause different interactions with biological systems, potentially increasing toxicity. Some nanoparticles have been observed to collect in specific tissues or organs upon exposure, raising concerns about their bioaccumulation potential and long-term health implications. They can cross biological barriers including cell membranes, thereby impairing cellular function, and producing inflammation, oxidative stress, or genotoxicity [92].

Moreso, surface functionalization of nanoparticles can potentially lead to toxicity because functional groups or coatings may deteriorate over time, affecting the behaviour and toxicity of nanoparticles in biological and environmental systems. Certain types of nanoparticles, such as metallic and metal oxide nanoparticles (e.g., silver, titanium dioxide), have aroused special concerns due to their antibacterial capabilities and extensive use in consumer goods. These nanoparticles may contaminate the environment and harm non-target creatures. Furthermore, determining the safety of nanoparticles is difficult due to a lack of standardized testing procedures and regulatory rules relevant to nanotechnology. Traditional toxicity testing methods may fail to capture the unique features and interactions of nanoparticles, prompting the development of novel approaches to nanotoxicology. The harmful effects of nanoparticles on aquatic systems can be mitigated by using large nanoparticles in water treatment. Some typical effects of specific nanoparticles are described briefly in this section.

Carbon nanotubes: CNTs are not soluble in water, so they are deposited as sediments, which is hazardous to aquatic organisms in water. The oxidation performance of enzymes in water molecules can be altered by CNTs which could be harmful to bacteria. Carbon nanotubes can also alter the DNA repair machinery. Additionally, the structure, length, and diameter of carbon nanotubes have a considerable impact on their toxicity. The toxicity of carbon nanotubes increases with their length. According to studies, CNTs can enhance lead and copper toxicity in the environment. Because of their surface structure, some CNTs are very reactive and can transfer other molecules into the cell body. Carbon nanotubes cannot survive alone in the actual world, and because these nanomaterials have a restricted amount of adsorption to other pollutants in ecology, it is impossible to exactly assess the environmental harm caused by CNTs.

Graphene-based nanomaterial: Graphene oxide has been shown to have harmful effects on green algae, including the formation of reactive oxygen species and membrane degradation. These graphene-related nanomaterials will undoubtedly be released into the environment during the manufacturing, transportation, use, and disposal processes. When these particles are discharged into water, they have the chance to interact with various types of biological and physicochemical elements, resulting in significant detrimental effects on the environment, with implications at the ecological unit level. Adsorption abilities are an important factor to consider when assessing the environmental danger of graphene-based nanomaterials in the water system. These functionalized GO nanoparticles are extremely difficult to remove from wastewater after heavy metal ion adsorption due to their higher water solubility and smaller size.

Silver nanoparticles: Aqueous species may be harmed when the silver nanoparticles released from commodities into an aquatic system. Based on the water chemistry and sediment properties the toxicity of silver nanoparticles may vary. Trace metals may be associated with colloids and particles, which play an important role in heavy metal transport and are controlled by silver nanoparticles. Silver nanoparticles may act as a coagulant, depositing ions and colloids containing Cd, Pb, Zn, and Cu from the sediment's pore water, resulting in a drop in heavy metal concentrations. Conversely, a portion of the silver nanoparticles may become stabilized in pore water and form a silver-mercury amalgam, which could constitute a substantial sort of mercury transformation in the sediment. Mercury-silver amalgam NPs may significantly improve mercury transport over longer distances when discharged into surface water. Moreover, silver particles mainly accumulate in mammals' livers and brains. So, if they reach in the aquatic environment, causing nanotoxicity in aquatic creatures that may then enter the human body via the food chain.

To address these safety and toxicity concerns, ongoing research is aimed at understanding the mechanisms of nanoparticle toxicity, developing predictive models for assessing their safety, and establishing strong regulatory frameworks to ensure the responsible development and use of nanomaterials in a variety of applications. Collaboration among scientists, regulators, industry stakeholders, and policymakers is required to properly address these issues and maximize the benefits of nanotechnology while minimizing possible dangers to human health and the environment.

Ethical and Regulatory Considerations

There are a lot of ethical issues based on nanotechnology related to its manufacturing activities, use of the nanomaterials and the final products, diffusion, regulation and nanotechnology research and development. Nanotechnology is a potential area that can contribute benefits to the world. But like other new technologies nanotechnology also raises serious social and ethical issues at the time of its introduction and implementations. These issues have effects on both human and environmental sides. So, it is important to analyse and study the possible ethical issues before it is commercialized.

Currently our knowledge regarding the effect of nanoparticles in humans and the environment is not comprehensive enough. This is a serious issue which can expose humans and the environment to toxic nanoparticles. Hazardousness of nanoparticles and exposure of risk are the major concerns regarding nanotechnology. Biological and chemical effects on humans and environment are the first concern while, the issues related to leakage, spillage, circulation, and concentration of these particles can cause harmful effects on both human and environment [92, 93]. Surface area, size, shape, reactivity, etc., are properties that make nanoparticles unique. These helpful properties can also make them harmful in certain situations, for example, these can make them able to enter the food chain and drinking water supply.

Nanotechnology has many environmental, social, and economic advantages but at the same time, the possible risk associated with nanoparticles to humans and the environment needs serious consideration. It is important to identify the full risk of this technology and balance the potential benefits to address water scarcity.

Before employing nanoparticles on a big scale for water treatment, researchers must overcome the obstacles. Furthermore, additional information is needed to close the knowledge gap about the effects of nanotechnology on humans and the environment. Researchers need to answer the question, "How can we prepare nanotechnology safely?".

Ethical Implications and Societal Acceptance

The use of nanotechnology in water treatment presents several ethical questions and the need to gauge societal acceptance. Ensuring that everyone has access to clean water is one of the most significant ethical obligations. Although water treatment systems based on nanotechnology might offer superior purification processes, there is a worry that their high cost or unavailability may prevent them from being used in regions with inadequate resources. To resolve this ethical dilemma, it is imperative to guarantee the accessibility and affordability of water treatment systems based on nanotechnology [94]. Environmental justice should be considered while using nanotechnology in water treatment. A disproportionate amount of environmental degradation and a lack of access to clean water are experienced by historically excluded people. It is imperative to ascertain if the implementation of nanotechnology-driven water treatment solutions intensifies pre-existing environmental inequalities or advances environmental justice by tackling water quality concerns in underprivileged communities.

To identify and prevent any dangers associated with the use of nanomaterials in water treatment, thorough risk evaluations must be completed. This is another ethical concern. This includes examining nanoparticles' potential environmental and health implications throughout their lifecycle, from manufacture to disposal, and putting in place suitable safety measures to reduce hazards to human health and the environment [95].

The use of nanotechnology in water treatment by society depends on openness and significant public involvement. It is imperative to elucidate the advantages, risks, and uncertainties associated with nanotechnology-based water treatment to all relevant parties, such as legislators, advocacy groups, and community members. Involving the public in decision-making procedures promotes trust and guarantees that societal ideals and issues are appropriately handled. Evaluating the impact and long-term viability of water treatment systems based on nanotechnology is crucial. Determining the environmental impact of producing nanomaterials, assessing the potential for unforeseen or unintentional environmental effects, and assessing the long-term viability of water treatment facilities and technologies are some examples of ethical considerations [96].

In general, a multidisciplinary strategy involving research, stakeholder involvement, ethical analysis, and policy creation is required to address the moral implications and societal acceptability of nanotechnology in water treatment. If these considerations are taken into account beforehand, we can ensure that nanotechnology helps handle water challenges in a way that is both morally and socially acceptable.

Future Directions and Conclusion

Water treatment procedures could undergo a radical transformation thanks to the enormous potential of nanotechnology. Utilizing the special qualities of nanomaterials to solve the problems of water scarcity and pollution has seen tremendous progress over time. Solutions based on nanotechnology have demonstrated exceptional effectiveness in eliminating impurities, boosting treatment system performance, and replenishing clean water. Future developments in nanotechnology could bring water treatment to new heights in several fascinating areas. These areas will be the primary focus of future study and development.

Enhanced Contaminant Removal: With the use of nanotechnology, it may be possible to create highly effective and selective adsorbents that can remove a variety of pollutants from water. In order to improve the selectivity and adsorption capability of nanomaterials, future research can concentrate on enhancing their surface chemistry and porosity. Furthermore, there is a lot of potential for the creation of multifunctional nanomaterials that can target several pollutants at once.

Membrane Technology Advancements: Water treatment requires membrane-based filtering methods, and nanotechnology can significantly enhance these methods' effectiveness. Graphene, carbon nanotubes, and nanocomposites are examples of nanomaterials that can be included into membranes to improve their permeability, selectivity, and fouling resistance. In the future, efforts can focus on investigating novel fabrication methods and scalable production procedures to bring these membranes augmented by nanomaterials to the market.

Photocatalysis and Advanced Oxidation Processes: As previously noted, nanomaterials possessing photocatalytic characteristics, including nanoparticles of titanium dioxide, have demonstrated significant promise in the degradation of organic contaminants by sophisticated oxidation techniques. Subsequent investigations will concentrate on creating new nanomaterials and refining their characteristics to increase their stability, selectivity, and photocatalytic performance. Additionally, investigating the synergistic effects of various nanomaterials and combining them with renewable energy sources can open the door to the development of environmentally friendly and energy-efficient water treatment devices.

It is crucial to highlight that nanotechnology has already significantly improved the treatment of water, and there is a great deal of room for growth in this field going forward. Researchers can create water treatment systems that are both extremely effective and sustainable by utilizing the special features of nanoparticles. To guarantee the successful application of nanotechnology-based solutions, issues including scalability, cost-effectiveness, and possible environmental effects must be resolved. It is imperative that researchers, legislators, industry players, and regulatory agencies work together to expedite the transfer of nanotechnology advancements into practical uses. Nanotechnology has the potential to supply communities worldwide with clean, safe, and easily accessible water as we investigate these future avenues, thereby promoting a sustainable future for all.

Exploration of Emerging Trends and Future Applications

In the dynamic field of water treatment, exploring the latest developments and potential applications of nanotechnology reveals a landscape brimming with opportunity and optimism. With the incorporation of nanomaterials into various facets of water management, such as cutting-edge filtration systems and inventive purification methods, we find ourselves on the cusp of profound transformation. As scientists and visionaries continue to push the boundaries of what is achievable, a new era in water treatment emerges—one characterized by sustainability, effectiveness, and inclusivity. Through leveraging the capabilities of nanotechnology and promoting interdisciplinary cooperation, we pave the path toward a future where access to clean water is not merely a necessity but a basic human entitlement. Amidst the intricate challenges posed by water issues on a global scale, nanotechnology serves as a beacon of hope, illuminating the way forward toward a world where every droplet matters and every community flourishes.

Recommendation for Further Research and Development in the Field

It is evident that despite notable progress, there are abundant opportunities for further research and advancement. Looking ahead, it is crucial to prioritize specific areas to propel the field towards greater innovation and effectiveness. Firstly, deepening our understanding of the environmental and social impacts of emerging technologies is essential for responsible progression. This involves conducting thorough evaluations of potential risks and benefits to ensure outcomes that are sustainable and fair. Furthermore, encouraging interdisciplinary cooperation among researchers, industry players, and policymakers will facilitate the exchange of ideas and the translation of research discoveries into practical applications. Additionally, investing in educational and professional development initiatives will nurture the next generation of problem solvers equipped with the necessary skills and expertise to tackle complex issues. Lastly, harnessing emerging technologies such as artificial intelligence and machine learning holds significant promise for expediting research processes and optimizing resource management. By prioritizing these suggestions and maintaining a commitment to ongoing collaboration and innovation, we can unlock fresh opportunities within the field, paving the path towards a future marked by sustainable progress and societal prosperity.

Potential Impact of Nanotechnology in Solving the Global Water Crisis

Examining new developments in nanotechnology offers a hopeful story with hope and promise for solving the global water dilemma. Considering the enormous problems with pollution and water scarcity, the application of nanotechnology appears to be a promising development. Advanced filtration and creative purification are only two of the ways that nanomaterials have the power to completely transform the water treatment industry. The ability to manage water resources in an efficient, economical, and sustainable manner is made possible by the accuracy that nanoscale engineering provides. Moreover, partnerships across several disciplines are encouraged by the collaborative character of nanotechnology research, which amplifies the impact of innovation and guarantees comprehensive answers to complex water concerns. All people can have access to clean and safe water in the future if we embrace the potential of nanotechnology and cultivate a collaborative and innovative culture. The revolutionary potential of nanotechnology offers a glimmer of hope as we work through the complexity of the global water problem, highlighting our ability as a group to overcome even the most difficult obstacles for the benefit of both the environment and humanity.

Acknowledgements Author (SCG) is gratefully acknowledge the financial support from ANERT, KERLA (ANERT-RD/28/2022-TB) and Department of Science and Technology, New Delhi under FIST Scheme (SR/FST/COLLEGE-346/2018). Author (AMA) acknowledges fellowship from APJ Abdul Kalam Technological University, Kerala, India.

References

1. Santhosh, C., et al.: Role of nanomaterials in water treatment applications: a review. Chem. Eng. J. **306**, 1116–1137 (2016)
2. Cantarella, M., Impellizzeri, G., Privitera, V.: Functional nanomaterials for water purification. La Rivista del Nuovo Cimento **40**, 595–632 (2017)
3. Mansoori, G.A., Soelaiman, T.F.: Nanotechnology--An introduction for the standards community. ASTM International (2005)
4. Nasrollahzadeh, M., et al.: An introduction to nanotechnology. In: Interface science and technology, pp. 1–27. Elsevier (2019)
5. Iravani, S.: Nanomaterials and nanotechnology for water treatment: recent advances. Inorg. Nano-Metal Chem. **51**(12), 1615–1645 (2021)
6. Ahmad, A., Azam, T.: Water purification technologies. In: Bottled and packaged water, pp. 83–120. Elsevier (2019)

7. Chenab, K.K., et al.: Water treatment: functional nanomaterials and applications from adsorption to photodegradation. Mater. Today Chem. **16**, 100262 (2020)
8. Ogunsona, E.O., et al.: Engineered nanomaterials for antimicrobial applications: a review. Appl. Mater. Today **18**, 100473 (2020)
9. Yu, Y. et al.: Purifying water with silver nanoparticles (AgNPs)-incorporated membranes: Recent advancements and critical challenges. Water Res. 118901 (2022)
10. Khan, S.A., et al.: Leveraging the potential of silver nanoparticles-based materials towards sustainable water treatment. J. Environ. Manage. **319**, 115675 (2022)
11. Yang, X., et al.: Influence of titanium dioxide nanoparticles on functionalities of constructed wetlands for wastewater treatment. Chem. Eng. J. **352**, 655–663 (2018)
12. Mousa, S.A., et al.: Enhanced the photocatalytic degradation of titanium dioxide nanoparticles synthesized by different plant extracts for wastewater treatment. J. Mol. Struct. **1250**, 131912 (2022)
13. Morsi, R.E., et al.: Multifunctional nanocomposites of chitosan, silver nanoparticles, copper nanoparticles and carbon nanotubes for water treatment: antimicrobial characteristics. Int. J. Biol. Macromol. **97**, 264–269 (2017)
14. Tom, A.P.: Nanotechnology for sustainable water treatment—a review. Mater. Today Proc. **10**, 16–78 (2021)
15. Alkaç, I.M. et al.: 2 - Nanomaterials and their classification, in Nanomaterials for Direct Alcohol Fuel Cells, F. Şen, Editor. Elsevier. pp. 17–33 (2021)
16. Zaytseva, O., Neumann, G.: Carbon nanomaterials: production, impact on plant development, agricultural and environmental applications. Chem. Biolog. Technol. Agricult. **3**(1), 17 (2016)
17. Singh, C., et al.: Bimetallic Au–Ag Nanoparticles: advanced nanotechnology for tackling antimicrobial resistance. Molecules **27**(20), 7059 (2022)
18. Getahun, Y.W., et al.: Green synthesized superparamagnetic iron oxide nanoparticles for water treatment with alternative recyclability. J. Mol. Liq. **356**, 118983 (2022)
19. Hoque, M.A., et al.: Fabrication and comparative study of magnetic Fe and α-Fe2O3 nanoparticles dispersed hybrid polymer (PVA + Chitosan) novel nanocomposite film. Resul. Phys. **10**, 434–443 (2018)
20. Bugalia, A., Gupta, V.: Structural, morphological and spectroscopic studies of Bi–Ca co-doped SnTe. J. Mater. Sci. Mater. Electron. **34**(33), 2190 (2023)
21. Li, C., et al.: Fabrication of unidirectional water permeable PS/PET composite nanofibers modified with silver nanoparticles via electrospinning. Membranes **13**(3), 257 (2023)
22. El-Eskandarany, M.S., et al.: Mechanical milling: a superior nanotechnological tool for fabrication of nanocrystalline and nanocomposite materials. Nanomaterials **11**(10), 2484 (2021)
23. Yang, F., et al.: Hard template-assisted trans-crystallization synthesis of hierarchically porous Cu-SSZ-13 with enhanced NH3-SCR performance. Catalysts **13**(8), 1217 (2023)
24. Poolakkandy, R.R., Menamparambath, M.M.: Soft-template-assisted synthesis: a promising approach for the fabrication of transition metal oxides. Nanoscale Adv. **2**(11), 5015–5045 (2020)
25. Mathur, J., et al.: Nanomaterials for water remediation: an efficient strategy for prevention of metal(loid) Hazard. Water **14**(24), 3998 (2022)
26. Savage, N., Diallo, M.S.: Nanomaterials and water purification: opportunities and challenges. J. Nanopart. Res. **7**, 331–342 (2005)
27. Yin, J., Deng, B.: Polymer-matrix nanocomposite membranes for water treatment. J. Membr. Sci. **479**, 256–275 (2015)
28. Islam, M.H., et al.: Graphene and CNT-based smart fiber-reinforced composites: a review. Adv. Func. Mater. **32**(40), 2205723 (2022)
29. Kumari, P., et al.: Recent advances in application of the graphene-based membrane for water purification. Mater. Today Chem. **22**, 100597 (2021)
30. Elrasheedy, A., et al.: Metal organic framework based polymer mixed matrix membranes: review on applications in water purification. Membranes **9**(7), 88 (2019)

31. Botes, M., Eugene Cloete, T.: The potential of nanofibers and nanobiocides in water purification. Crit. Rev. Microb. **36**(1), 68–81 (2010)
32. Nayl, A.A., et al.: Review of the recent advances in electrospun nanofibers applications in water purification. Polymers **14**(8), 1594 (2022)
33. Wen, Y., et al.: Polymeric nanocomposite membranes for water treatment: a review. Environ. Chem. Lett. **17**(4), 1539–1551 (2019)
34. Sun, X.-F., et al.: Graphene oxide–silver nanoparticle membrane for biofouling control and water purification. Chem. Eng. J. **281**, 53–59 (2015)
35. Naseem, T., Durrani, T.: The role of some important metal oxide nanoparticles for wastewater and antibacterial applications: a review. Environm. Chem. Ecotoxicol. **3**, 59–75 (2021)
36. Anucha, C.B., et al.: Titanium dioxide (TiO_2)-based photocatalyst materials activity enhancement for contaminants of emerging concern (CECs) degradation: in the light of modification strategies. Chem. Eng. J. Adv. **10**, 100262 (2022)
37. Bayer, I.S.: Superhydrophobic coatings from ecofriendly materials and processes: a review. Adv. Mater. Interf. **7**(13), 2000095 (2020)
38. Zhang, S., et al.: The impacts of aggregation and surface chemistry of carbon nanotubes on the adsorption of synthetic organic compounds. Environ. Sci. Technol. **43**(15), 5719–5725 (2009)
39. Apul, O.G., Karanfil, T.: Adsorption of synthetic organic contaminants by carbon nanotubes: a critical review. Water Res. **68**, 34–55 (2015)
40. Nishu, Kumar, S.: Smart and innovative nanotechnology applications for water purification. Hybrid Adv. **3**, 100044 (2023)
41. Margeta, K., et al.: Natural zeolites in water treatment–how effective is their use. Water Treat. **5**, 81–112 (2013)
42. Kallo, D.: Applications of natural zeolites in water and wastewater treatment. Rev. Mineral. Geochem. **45**(1), 519–550 (2001)
43. Chakraborty, A., Pal, A., Saha, B.B.: A critical review of the removal of radionuclides from wastewater employing activated carbon as an adsorbent. Materials **15**(24), 8818 (2022)
44. Kumar, R., Rauwel, P., Rauwel, E.: Nanoadsorbants for the removal of heavy metals from contaminated water: current scenario and future directions. Processes **9**(8), 1379 (2021)
45. Lu, H.-J., et al.: Mechanism, synthesis and modification of nano zerovalent iron in water treatment. Nanoscale **8**(19), 9962–9975 (2016)
46. Crane, R.A., Scott, T.B.: Nanoscale zero-valent iron: future prospects for an emerging water treatment technology. J. Hazard. Mater. **211**, 112–125 (2012)
47. Rana, N., Bassi, A.: Role of Nanomaterials in the Treatment of Wastewater. In: Modern Nanotechnology: Volume 1: Environmental Sustainability and Remediation, pp. 125–144. Springer (2023)
48. Butt, B.Z.: Nanotechnology and waste water treatment. Nanoagronomy, pp. 153–177 (2020)
49. Amin, M.T., Alazba, A.A., Manzoor, U.: A review of removal of pollutants from water/wastewater using different types of nanomaterials. Adv. Mater. Sci. Eng. **2014**, 1–24 (2014)
50. Giwa, A., et al.: Recent advances in advanced oxidation processes for removal of contaminants from water: a comprehensive review. Process. Saf. Environ. Prot. **146**, 220–256 (2021)
51. Rizzo, L., et al.: Consolidated vs new advanced treatment methods for the removal of contaminants of emerging concern from urban wastewater. Sci. Total. Environ. **655**, 986–1008 (2019)
52. You, J., et al.: A review of amino-functionalized magnetic nanoparticles for water treatment: features and prospects. J. Clean. Prod. **281**, 124668 (2021)
53. Chan, C.C., Gallard, H., Majewski, P.: Fabrication of amine-functionalized magnetite nanoparticles for water treatment processes. Springer (2014)
54. Gao, F.: An overview of surface-functionalized magnetic nanoparticles: preparation and application for wastewater treatment. Chemistry Select **4**(22), 6805–6811 (2019)
55. Makvandi, P., et al.: Functionalization of polymers and nanomaterials for water treatment, food packaging, textile and biomedical applications: a review. Environ. Chem. Lett. **19**, 583–611 (2021)

56. Bouazizi, N., et al.: Advances in amine-surface functionalization of inorganic adsorbents for water treatment and antimicrobial activities: a review. Polymers **14**(3), 378 (2022)
57. Kanakaraju, D., et al.: Recent progress of Ag/TiO2 photocatalyst for wastewater treatment: doping, co-doping, and green materials functionalization. Appl. Mater. Today **27**, 101500 (2022)
58. Berbar, Y., et al.: Effect of functionalized silica nanoparticles on sulfonated polyethersulfone ion exchange membrane for removal of lead and cadmium ions from aqueous solutions. J. Water Proc. Eng. **32**, 100953 (2019)
59. Zhu, D., Zhou, Q.: Action and mechanism of semiconductor photocatalysis on degradation of organic pollutants in water treatment: a review. Environ. Nanotechn. Monit. Manag. **12**, 100255 (2019)
60. Belver, C., et al.: Semiconductor photocatalysis for water purification. In: Nanoscale materials in water purification, pp. 581–651. Elsevier (2019)
61. Ibhadon, A.O., Fitzpatrick, P.: Heterogeneous photocatalysis: recent advances and applications. Catalysts **3**(1), 189–218 (2013)
62. Zhang, H., Shan, Y., Dong, L.: A comparison of TiO2 and ZnO nanoparticles as photosensitizers in photodynamic therapy for cancer. J. Biomed. Nanotechnol. **10**(8), 1450–1457 (2014)
63. Khalid, A.D., et al.: Functional bioinspired nanocomposites for anticancer activity with generation of reactive oxygen species. Chemosphere **310**, 136885 (2023)
64. Ren, G., et al.: Recent advances of photocatalytic application in water treatment: a review. Nanomaterials **11**(7), 1804 (2021)
65. Singh, S., Srivastava, V.C., Mall, I.D.: Mechanism of dye degradation during electrochemical treatment. J. Phys. Chem. C **117**(29), 15229–15240 (2013)
66. Li, Q., et al.: Antimicrobial nanomaterials for water disinfection and microbial control: potential applications and implications. Water Res. **42**(18), 4591–4602 (2008)
67. Ojha, A.: Nanomaterials for removal of waterborne pathogens: opportunities and challenges. Waterborne pathogens, pp. 385–432 (2020)
68. Zhu, J., et al.: Polymeric antimicrobial membranes enabled by nanomaterials for water treatment. J. Membr. Sci. **550**, 173–197 (2018)
69. Liu, D., Mao, Y., Ding, L.: Carbon nanotubes as antimicrobial agents for water disinfection and pathogen control. J. Water Health **16**(2), 171–180 (2018)
70. Booshehri, A.Y., Wang, R., Xu, R.: The effect of re-generable silver nanoparticles/multi-walled carbon nanotubes coating on the antibacterial performance of hollow fiber membrane. Chem. Eng. J. **230**, 251–259 (2013)
71. Ansari, A.A., et al.: Nanomaterials as a cutting edge in the removal of toxic contaminants from water. Mater. Chem. Phys. **295**, 127092 (2023)
72. Saharan, P., et al.: Removal of water contaminants by iron oxide nanomaterials. J. Nanosci. Nanotechnol. **14**(1), 627–643 (2014)
73. Saxena, R., Saxena, M., Lochab, A.: Recent progress in nanomaterials for adsorptive removal of organic contaminants from wastewater. ChemistrySelect **5**(1), 335–353 (2020)
74. Ghodke, S.A., et al.: Nanomaterials for adsorption of pollutants and heavy metals: introduction, mechanism, and challenges. In: Handbook of nanomaterials for wastewater treatment, pp. 343–366. Elsevier (2021)
75. Tan, K.B., et al.: Adsorption of dyes by nanomaterials: recent developments and adsorption mechanisms. Sep. Purif. Technol. **150**, 229–242 (2015)
76. Awad, A.M., et al.: Adsorption of organic pollutants by nanomaterial-based adsorbents: an overview. J. Mol. Liq. **301**, 112335 (2020)
77. Majd, M.M., et al.: Adsorption isotherm models: a comprehensive and systematic review (2010–2020). Sci. Total. Environ. **812**, 151334 (2022)
78. Tee, G.T., Gok, X.Y., Yong, W.F.: Adsorption of pollutants in wastewater via biosorbents, nanoparticles and magnetic biosorbents: a review. Environ. Res. **212**, 113248 (2022)
79. Spoială, A., et al.: Zinc oxide nanoparticles for water purification. Materials **14**(16), 4747 (2021)

80. Nyabadza, A., et al.: Advanced nanomaterials and dendrimers in water treatment and the recycling of nanomaterials: a review. J. Environ. Chem. Eng. **12**(3), 112643 (2024)
81. Bhoj, Y., et al.: Recent advancements in practices related to desalination by means of nanotechnology. Chem. Phys. Impact **2**, 100025 (2021)
82. Nanotechnology in water treatment: an emerging trend: Intern. J. Nucl. Desalin. **4**(2), 184-188 (2010)
83. Feria-Díaz, J.J., et al.: Commercial thermal technologies for desalination of water from renewable energies: a state of the art review. Processes **9**(2), 262 (2021)
84. Qiblawey, H.M., Banat, F.: Solar thermal desalination technologies. Desalination **220**(1), 633–644 (2008)
85. Greenlee, L.F., et al.: Reverse osmosis desalination: Water sources, technology, and today's challenges. Water Res. **43**(9), 2317–2348 (2009)
86. Qasim, M., et al.: Reverse osmosis desalination: a state-of-the-art review. Desalination **459**, 59–104 (2019)
87. Campione, A., et al.: Electrodialysis for water desalination: a critical assessment of recent developments on process fundamentals, models and applications. Desalination **434**, 121–160 (2018)
88. Al-Amshawee, S., et al.: Electrodialysis desalination for water and wastewater: a review. Chem. Eng. J. **380**, 122231 (2020)
89. Meerganz von Medeazza, G.L.: "Direct" and socially-induced environmental impacts of desalination. Desalination **185**(1), 57–70 (2005)
90. Malakar, A., et al.: Nanomaterials in the environment, human exposure pathway, and health effects: a review. Sci. Total. Environ. **759**, 143470 (2021)
91. Anawar, H.M., Strezov, V.: Chapter 2 - Transport, fate, and toxicity of the emerging and nanomaterial contaminants in aquatic ecosystems: Removal by natural processes, in Emerging and Nanomaterial Contaminants in Wastewater, A.K. Mishra, H.M.D. Anawar, and N. Drouiche, Editors, Elsevier, pp. 43–62 (2019)
92. Sani-Kast, N., et al.: Addressing the complexity of water chemistry in environmental fate modeling for engineered nanoparticles. Sci. Total. Environ. **535**, 150–159 (2015)
93. Schaumann, G.E., et al.: Understanding the fate and biological effects of Ag- and TiO2-nanoparticles in the environment: The quest for advanced analytics and interdisciplinary concepts. Sci. Total. Environ. **535**, 3–19 (2015)
94. Inbathamizh, L., et al.: Nanotechnology: ethical impacts, health issues, and safety issues, in modern nanotechnology: Volume 2: green synthesis, sustainable energy and impacts. J.A. Malik and M.J. Sadiq Mohamed, Editors, Springer Nature Switzerland: Cham, pp. 455–477 (2023)
95. Street, A., Duncan, J.S., Savage, N.: Chapter 30 - nanotechnology in water: societal, ethical, and environmental considerations. In: Savage, N., et al. (eds.) Nanotechnology applications for clean water, pp. 453–462. William Andrew Publishing, Boston (2009)
96. Gardner, J., Dhai, A.: Nanotechnology and Water: Ethical and Regulatory Considerations, in Application of Nanotechnology in Water Research, pp. 1–20 (2014)

Chapter 11
Nanotechnology and Materials Science: Innovations for Industry and Infrastructure

Yuyang Zhou, Jingjing Wang, Chunguang Liang, and Feiyang Zheng

Introduction

In recent years, with the continuous development of micro-nano processing methods, its application range is becoming wider and wider. This technology extends the limits that humans have not broken for decades, opening a window into the microscopic world of tiny, special machines and particles with particular properties. The United States was the first country to initiate a formal program for government funding of research in nanotechnology. The National Nanotechnology Initiative (NNI) was launched in 2000 and prompted USD 22 billion on nanotechnology including approximately USD1.5 billion in the President's budget at 2016 [1]. Nanotechnology has been used in several industries to help improve the quality and comfort of people's homes. Due to its large surface area and high adjustability, it has been widely used in electronics, optics, biomedicine, and other fields; it has been broadly used in cosmetics and environmental remediation. Nanomaterials are widely used because of their versatility and multi-performance. For example, an ordinary 5-nm protein is already comparable in size to the smallest nanoparticles scientists have ever produced.

Y. Zhou (✉) · C. Liang · F. Zheng
National Engineering Laboratory for Modern Silk, College of Textile and Clothing Engineering, Soochow University, Suzhou 215123, China
e-mail: yuyangzhou@suda.edu.cn

C. Liang
e-mail: 494065585@qq.com

F. Zheng
e-mail: zhengfeiyang2022@163.com

J. Wang (✉)
Suzhou Mingwang New Material Technology Co. Ltd, Suzhou 215010, China
e-mail: wangjingjing_sz@hotmail.com

S. C. George and B. Tawiah (eds.), *Nanotechnology in Societal Development*, Advanced Technologies and Societal Change, https://doi.org/10.1007/978-981-97-6184-5_11

In this context, medical and pharmaceutical research has been enhanced by the development of research results. These include cosmetics, drug delivery, imaging, medical diagnostics, etc.; in many ways, such as tissue engineering, they have been widely used in biology. In medicine, functional magnetic materials can be used to extract high and small molecular substances, and carbon materials can also be used to improve the transport ability of various cells. In the cosmetic field, new drug carriers are expected to achieve controlled release of active ingredients. For food use, researchers have developed a new iron/zinc complex that effectively improves iron absorption without changing the color and taste of food. CNTs can be added to yarn to obtain bulletproof, lightweight, highly conductive and other multifunctional textiles, and it is also expected to use its super-hydrophobic surface to achieve the purpose of self-cleaning. Lithium-ion batteries used in hybrid vehicles can be 10 times more efficient in energy transfer when nano-devices are used. Future buildings will use active smart windows that generate electricity and materials characterization elate light transmission, or silicon nanofiber-coated wood as ultrahydrophobic materials. This chapter focuses application of nanotechnology in the following industries, namely, chemicals, pharmaceuticals, food, textile, electronics, and infrastructure related to transportation, energy, water, environment. At the appropriate place, the academic background and representative research results of the above technological innovation are described. Various characterization methods and performance indexes commonly used at present are summarized. Finally, the challenges, future trends/potentials, and impacts on the industry and infrastructure of nanotechnology are discussed.

Manufacturing Technologies and Applications

Nanotechnology, a scientific technology that involves the production of substances from individual atoms and molecules, is gradually changing our world. It studies the properties and applications of materials with structural dimensions ranging from 1 to 100 nm. By precisely manipulating atoms and molecules, we can obtain new materials with specific functions, thereby promoting innovation and development in the manufacturing industry.

The application of nanotechnology has become increasingly widespread in the manufacturing industry. Firstly, nanotechnology has shown enormous potential in the development of new materials. Through nanotechnology, we can manufacture new materials with excellent physical and chemical properties, such as high-strength and high toughness nanocomposites, as well as nanoelectronic materials with special optoelectronic properties. These new materials have broad application prospects in fields, such as aerospace, automotive, and electronics. Secondly, significant achievements have been made in the application of nanotechnology in the fields of microelectronics and power. Nanoscale electronic components can greatly improve the integration and computational speed of circuits, making electronic devices more efficient and energy efficient. At the same time, nanotechnology has also provided possibilities for the miniaturization and efficiency of power equipment, such as nanogenerators, nanobatteries, etc., bringing infinite possibilities for the future energy field. In addition, the application of nanotechnology in biology and medicine is also worth paying attention to. The emergence of nanoproducts, such as nanomedicines and nanobiosensors has provided new means for the diagnosis and treatment of diseases. Nanotechnology can also be used to manufacture artificial organs and tissues, bringing significant changes to the medical field. Meanwhile, nanotechnology also has extensive applications in fields, such as chemistry and environmental monitoring. Nanocatalysts can improve the efficiency of chemical reactions and reduce environmental pollution; Nanosensors can monitor harmful substances in the environment in real-time, ensuring people's health.

Overall, nanotechnology, with its unique advantages, is driving the rapid development of the manufacturing industry. However, the application of nanotechnology also faces some challenges, such as the safety issues of nanomaterials and the production costs of nanotechnology. Therefore, we need to focus on the safety and sustainability of nanotechnology applications while conducting in-depth research, in order to achieve long-term stable development of nanotechnology.

Medical and Pharmaceutical Industry

Nanomaterials are an important part of life system at nanometer scale. In many living substances, including proteins, viruses, etc., are carried out at the nanoscale, so it is an important means to understand life systems and life processes. Nanomedicine, or the healthcare landscape of nanotechnology in terms of market tendencies, includes products based on nanocarriers containing approved drugs, innovative drug candidates, or nanodiagnosis components. The nanomedicine market is expected to grow from USD 219,850 million in 2020 to USD 461,252 million by 2026, with a CAGR of 11.9% [2]. The nanomedicine market is usually segmented by application (diagnosis, drug delivery, biomaterials, and tissue regeneration, among others), diseases (cardiovascular diseases, infection diseases, neurodegenerative disorders, cancer, and others), and geography (Europe, North America, Asia–Pacific, and others) [2]. In the field of biotechnology, nanotechnology has two important meanings: one is to imitate the structure and function of organisms, and the other is to manipulate and modify biological systems, such as cells and subcellular organs, protein molecules, receptors, cytokines, etc. Because of electrostatic attraction, such as hydrogen bonding, hydrophobicity, Van der Waals force, it is easy to make biological macromolecules adsorbed to the particle surface. Based on this, a novel biosensor based on nanomaterials is proposed in this project, such as glucose sensor, DNA sensor, targeted drug delivery, etc.

With the development of life science, personalized, quantitative drug therapy has become possible. Nanobiology plays an important role in the medical field. Diagnosis is a very important step in health care. Whether from the effect or overall cost to consider, early diagnosis, the treatment effect for patients is better. Nanomedicine is a new subject formed by the cross fusion of nanotechnology and medicine, which utilizes the science of nanotechnology in the preclusion and cure of various diseases using nanoscale materials, such as biocompatible nanoparticles and nanorobots, for various applications including, diagnosis, delivery, sensory, or actuation purposes in a living organism. Nanomedicine applications are grouped below in three interrelated areas: analytical/diagnostic tools, drug delivery, and regenerative medicine [3]. The ultimate goal of nanomedicine is to develop new drugs that can be used to prevent and treat many diseases. Among them, there are drug delivery platform, imaging system, biochip and probe; needle, which can be used for painless infusion or intravenous infusion [4, 5].

Nanomaterials are a key link in biosensors [6, 7]. When it has specific photoelectric properties, it has excellent photoelectric properties, which provides a new way to construct new biosensor devices. In the aqueous phase, the biosensor is a biological macromolecule containing DNA, enzymes, antibodies, etc.; and an analytical device associated with or integrated into the physicochemical sensor [8]. At present, biosensor technology has been applied to a variety of major diseases (such as tumor diagnosis) [9], organophosphorus (pesticide), pathogenic bacteria [10–12], cholesterol in blood [13], blood sugar, etc.

In the field of medicine, it can also be used as a controlled-release carrier of effective ingredients [14]. Drugs with very low solubility possess various biopharmaceutical delivery issues, however, this could be overcome by the application of nanotechnology approaches in the drug delivery mechanism. There are two ways through which nanostructures deliver drugs: passive and self-delivery. In the former, drugs are incorporated in the inner cavity of the structure mainly via the hydrophobic effect. When the nanostructure materials are targeted to a particular site, the intended amount of the drug is released because of the low content of the drugs which is encapsulated in a hydrophobic environment. Conversely, in the latter, the drugs intended for release are directly conjugated to the carrier nanostructure material for easy delivery [15]. Drug delivery systems based on nano-carriers can achieve efficient loading, efficient cell penetration, and so on. To achieve efficient delivery, a method must be devised to fuse drugs with nanomaterials so that they can be used as an effective mode of drug delivery [16]. In recent years, with the deepening of research, its core contents mainly include targeted drug delivery to reduce adverse reactions; this project will provide optimized topological configurations for medical material implantation, surgical auxiliary devices, nano-robots, and high-throughput anti-tumor therapy [17]. Nanomaterials have also received extensive attention in medical repair and implant materials [18]. Developing biomedical endoplants or scaffolds that resemble natural organisms or organs has always been a challenging subject.

Cosmetic Industry

At present, the largest scale of nano-related research and development is aimed at the ingredients of beauty products. The cosmetic industry is attaining the benefits of nanotechnology by developing nanoparticles for the enhanced performance and bioavailability of active components in cosmetics, sunscreens, anti-aging creams, moisturizers, and perfumes [19]. At present, there are three aspects of nanotechnology in cosmetic manufacturing process: formulation, packaging, and manufacturing equipment [20]. Cosmetic nanocarriers, which do not possess any biological activity of their own, are employed to deliver vitamins, antioxidants, chemical UV filters, anti-acne or anti-aging substances in the target layer of skin where they are then dissolved or eroded, after which the effect of the active ingredients is no different from that of the molecular form.

The current or proposed nanotechnologies are: nano-silver, zinc oxide, magnesium oxide, etc., gas and moisture insulation (montmorillonite), and through a variety of Radio Frequency Identification (RFID) tags to achieve the monitoring of temperature, humidity, and other environmental factors in food storage. The application of nanotechnology in the manufacturing process may not be obvious to the users, but it can bring great benefits to the manufacturing process. Stickless machine materials, easy-to-clean surfaces, low-friction details, and higher energy efficiency are all benefits of beauty products made using nanotechnology. Current mainstream nanotechnology products, such as Ag, C, Ti/TiO_2, Si/SiO_2, Zn/ZnO, Au, etc., are widely used in physical UV protection, skin whitening, hair and teeth, and so on [20]. The well-known cosmetic companies Boots, Avon, The Body Shop, L'Oréal, Nivea, and Unilever are implementing nanoparticles in the formulations of cosmetics. Along with sunscreens TiO_2 is used in lip balms, foundations, day creams as UV filters so as to protect the skin from the carcinogenic effects of UV radiations. Besides TiO_2, ZnO, zirconiumoxide (ZrO_2), Ceriumoxide (CeO_2) nanoparticles are used in cosmetics as UV filters as the particles are able to scatter and reflect the UV radiations. The most important features for the use of these nanoparticles include their defined size and surface area. This ultimately determines the efficacy of UV filtration in sunscreens [21]. The synthesis of nanomaterials is possible with the help of nanotechnology. These nanomaterials can be used in cosmetics to impart new characteristics to the cosmetics. The varied types of nanomaterials with various characteristics may create different benefits. The types of nanomaterials that are used include nanocapsules, dendrimers, nanoemulsions, nanoliposomes, nano-hydroxyapatite materials, which have been used in cosmetics for various purposes, such as oral care, protective carriers, and delivering ingredients through the skin. The brief outline of their methods of synthesis has been summarized in Table 11.1.

Food

Nanotechnology has been widely used in the food industry and can greatly assist in the identification, processing, and processing of nanostructures. The material can improve its solubility in human body, improve its bioavailability, and realize the controlled release of targeted drugs. Nanomaterials can also be used as anti-caking agents, nano-additives, transportation system of nutritious food, and so on. At the same time, it has been widely concerned by people for its excellent

Table 11.1 A brief outline of different nanomaterials synthesis methods

Classification	Synthesis method	Size range	Refs
Liposomes	Sonication/ Micro-fluidization	100 to 200 nm	[22]
Noisome	Ether injection method Thin-film hydration technique Reverse-phase evaporation technique	50 to 1000 nm 200 to 1000 nm 3460 to 3610 nm	[23, 24]
Dendrimers	Divergent method Convergent method	1 nm up to 10 nm	[25, 26]
Nano-emulsions	Microfluidization High pressure homogenization	20–200 nm	[27, 28]
Solid lipid nanoparticles and Nanostructured lipid carriers	Solvent emulsification-evaporation technique Microemulsion technique	50 nm–1000 nm	[29]
Nano-Hydroxyapatite	Wet chemical precipitation Technique Ultrasound-assisted technique	20–80 nm long and 2–5 nm thick	[30, 31]
Nanocapsules	Miniemulsion polymerization Chemical vapour condensation method	10 nm–1000 nm	[32, 33]
Nanotubes	Chemical vapour deposition Electrochemical anodization	1.4 nm	[34, 35]
Nanospheres	Solvo-thermal routes	100–1000 nm	[36]

performance when it is applied to food packaging. The application of nanotechnology to the food industry can facilitate the delivery of food to all parts of the world by increasing the shelf life of most food. In addition, nanotechnology also focuses on fresh food, through the nano-delivery system in the process to ensure the health of the food. The delivery system is used to transport food supplies. Nanotechnology has been widely used in various industries, especially in agricultural products and food industry, which is widely used in primary manufacturing, food processing and packaging, food additives, and so on [37]. As far as original products are concerned, nanotechnology has been widely used in various aspects, such as nanosensors that monitor the health of livestock. It is used for the prevention, diagnosis, and treatment of livestock diseases. Nanometer fertilizer. And insecticide nanocapsules. In food processing, preparations of nanocapsules and nanoemulsions, including adding, absorbing or dispersing compounds. They are used as additives, preservatives, spices, and food additives so that the ingredients in the food can last

Table 11.2 An illustration of various types of nanostructures used in the food sector so far

Nanostructures	Characteristic features	Refs
Biopolymeric nanoparticles	It solubilizes water-insoluble compounds in the hydrophobic interior, high solubility, low toxicity	[39–41]
Liposomes	Since liposome surrounds an aqueous solution inside a hydrophobic membrane, it can be used as delivery vehicle for hydrophobic molecules or hydrophilic molecules	[42, 43]
Nanoemulsions	(i) Greater stability to droplet aggregation and gravitational separation; b-Carotene-based nanoemulsion (ii) Higher optical clarity; and, (iii) increased oral bioavailability	[44]
Nanocomposites	Great alternative to conventional food packaging, biodegradable and cost effective. To preserve the quality of fresh foods during extended storage	[45, 46]
Inorganic nanoparticles	They display good encapsulation capability and their rigid surfaces allow controlled functionalization	[47, 48]
Hydrogels	Can be easily placed into capsules, protects drugs from extreme environments, and to deliver them in response to environmental stimuli, such as pH and temperature	[49]
Solid lipid nanoparticles	To preserve the quality of fresh foods during extended storage	[50]

longer in the digestive tract. The research shows that it is of great significance to apply nanotechnology to food packaging [37].

Clay and silicate nanoplatelets, starch nanocrystals, carbon nanotubes, metal nanoparticles are the nanomaterials often applied to food [38]. Their functions in food industry include Controlled and/or burst release of substances in active and functional food packaging technologies and intelligent food packaging; Formulation of active packaging technologies based on bioplastics, such as more efficient antioxidant, oxygen scavenging or antimicrobial biopackaging, which has more direct implications in increasing packaged foods quality and safety; Important issues associated with the use of bioplastics, such as the non-intended migration of plastic components to foods, can also be potentially reduced by the use of nanoparticles. Table 11.2 summarizes various nanostructures that have been employed in food science.

Textile

Nanotechnology is a new developing discipline, which is expected to be applied to material science, material processing technology, machinery, electronics, optics, medicine, and in various scientific and technological fields, such as energy, aerospace, plastics, textile, and so on. Although this method is still in its infancy, it has been proven to be very effective in improving the performance of the fabric and attracting worldwide attention. In the textile field, the new use of nanotechnology makes it have better characteristics, and it is possible to develop new functions. New or enhanced features can be generated using nanotechnology to improve or improve features. Nanomoieties can induce stain repellence, wrinkle-freeness, static elimination, and electrical conductivity in fibers without compromising their comfort and flexibility. Nanomaterials also offer a wider application potential to create connected garments that can sense and respond to external stimuli via electrical, color, or physiological signals [51]. The information provided in this sector is not comprehensive, but it does reflect the current domestic and international research, including the business behavior of nanofabrics. In all aspects of science and technology, nanotechnology and nanomaterials-based process applications are growing rapidly. In a variety of uses, the textile industry also feels the benefits of nanotechnology. All kinds of products combined with nanotechnology and textile fabrics, from nanocomposites, nanofibers to polymer intelligent coatings, not only achieve high performance, but also give new properties to a variety of traditional textile fabrics. In the textile field, the biggest advantage of the development of nanotechnology is its good repeatability, high reliability, and good toughness. In the textile field, such as dyeing, finishing, coating, etc., through surface modification to make it have better characteristics.

Nano-Tex, a subsidiary of Burlington Industrial Group in the United States, is an important research work. Since then, other textile companies around the world have developed nanotechnology on a global scale. For example, paint is a common process for applying nanoparticles to fabrics. Fabric surface modification coatings are generally composed of nanoparticles, surfactants, additives, and carriers. There are a variety of ways to paint the fabric, including spraying, transfer, washing, flushing, flushing, and filling. Among them, the most common one is filling. The nanoparticles can be adhered to the textiles by adjusting the nanoparticles to appropriate air pressure and speed, drying, and hardening them with a buffer pad. Nanotechnology is applied to the fabric to make it waterproof, anti-pollution, anti-wrinkle, anti-microbial, anti-static, ultraviolet, flame retardant, dyeing properties, and so on. In view of its wide range of uses, this article focuses on only a few of its famous features [52].

Nanotechnology has a wide range of applications in the field of textiles because of its excellent durability, comfort, and sanitary properties. Because of its large specific surface area and strong binding ability with textiles, the durability of the required fabric is enhanced, and the finished fabric has high durability. Due to the use of nanotechnology, textile products are more widely used, its characteristics are: antibacterial, anti-ultraviolet, easy to clean, water-resistant, anti-pollution, anti-odor. Because silver nanoparticles can kill bacteria and suppress bad taste, they can be used in socks and sweatshirts [53].

Infrastructure

At present, nanotechnology is widely used in the construction industry. Traditional building materials (such as paint / paint, glass, etc.) are modified to give them self-cleaning, antibacterial, pollution reduction, and other characteristics, while the use of "new" materials (such as silica (SiO_2) aerogel suitable for efficient or translucent thermal insulation, or the use of nano-capsule rust inhibitor, etc.). Nanotechnology has the potential to reduce building energy saving, environmental protection, energy saving, and so on. Many innovations based on nanotechnology will have huge and profound consequences and have been widely concerned by countries all over the world, such as "Green Star", and so on.

At present, nanotechnology and materials used in the construction industry are constantly developing from concept to industrialization. However, people in the industry know very little about how to apply nanotechnology to buildings. It is impossible to distinguish the existing from the theoretically feasible, which leads to a lot of misunderstandings about nanotechnology. A study by Rilem, based on the results of the study by builders, points out that customers who want to complete the project in the cheapest way in the shortest possible time, think nanotechnology is too expensive and too complex. On the contrary, people's understanding and attention to nanotechnology are also increasing rapidly, such as chemical industry, automobile, energy, and so on. In the construction industry, the negative impression is the development trend at present and in the future, which is of great significance to the success or failure of the construction enterprise and the social progress it brings.

Construction Industry

The purpose of the National Nanotechnology Program (NNI) is to transfer new research results from the laboratory to commercial or public welfare. National and local institutions in some regions are already all over the United States, and they are conducting collaborative research on nanotechnology. Many large companies like IBM, Intel, Motorola, Lucent, Boeing, Hitachi, Mitsubishi, Pfizer, Corning and Ford, etc., have made important contributions to the growth and growth of nanotechnology. A study conducted by Delphi in the 1990s shows that only the construction industry regards nanotechnology as the most promising technology. Although more than 1 million articles have been published on nanotechnology between 1990 and 2010, the construction industry accounts for only a small portion. There is little research on the potential areas of applying nanotechnology to the construction industry and identifying the benefits and obstacles of research and development in this area. The construction industry lags behind in the automotive, computer, electronics, and other industries; the research and development in the fields of national defense, medicine, biotechnology, and other fields are favored and invested by many industries and venture capital institutions. It's just that not long ago, the construction industry did not know the importance of nanotechnology to industry. After a period of time, all aspects of the construction process will be rapidly revolutionized by nanotechnology and will create new opportunities to strengthen and support the industry. The application of nanotechnology to the construction industry can be divided into five aspects: (1) current nanoconstruction methods; (2) nanoscale materials with excellent properties; (3) nano-films and coatings; (4) intelligent structures with embedded nano-sensors; (5) renewable energy and nano-household appliances. Some of these industries involve the construction industry, and nanotechnology can enhance the properties of materials and components or add other properties to some conventional structural components. Such interactions could push the construction industry toward a wider use of nanostructures.

In the past, the attitude of the construction industry toward innovation and change has been conservative. However, in recent decades, many new technologies have been introduced and effectively integrated, thus making the building more green, efficient, and intelligent. The use of this new technology is likely to involve concrete materials, gypsum or building paint, windows or new insulation materials. The mechanical properties of cement-based materials can be improved by adding nanoparticles. A nanoparticle can also be used to repair mortar shells or to heal itself in the future. Others can also be used to clean the air or the surface of objects. A new type of waterproof, anti-mildew agent and anti-mildew agent has been developed; materials, such as paint can be selected by the construction industry according to their properties.

Energy

Nanotechnology has expanded a variety of areas of the oil and gas industry, including exploration, drilling, refining, and sales. Because of its light weight, corrosion resistance, and high strength, it plays a very important role in oil and gas drilling equipment, especially in oil and gas drilling equipment. Nanotechnology has become a breakthrough factor thanks to innovative surveillance and more sensitive nanosensors. In the extreme environment, conventional detection methods, such as sensors have great limitations. Sensors for imaging, measuring, and controlling oil and gas reservoirs and oil fields will improve the work of oil and gas exploitation, drilling, transportation, and storage. In addition, nanotechnology may also be used in the oil and gas industry to develop new polymer materials that are economical and environmentally friendly. The viscosity reduction of crude oil by nano-technology is of great significance to improve the development and transportation efficiency of crude oil. The application of nanomaterials to catalysts can bring several benefits and improve the economic benefits of the process. The improvement features of the catalyst include: (a) it has a high surface area-volume ratio, thus improving the performance of the catalyst in treatment; (b) because of the high fluidity of the reactant in the reactor, there is more chance of contact; (c) the conversion operation lasts longer because the nano-catalyst is loaded into the medium without the need to replace the catalyst; (d) High-intensity, lasting and continuous production.

Worldwide energy needs to be about 30 trillion watts by 2050. In order to protect our environment, we must improve energy efficiency and develop clean energy. If we want to find this problem, there must be a breakthrough in science and technology, and there must be a real revolution. Therefore, it is very meaningful to use nanomaterials to solve the above problems. Some aspects of the use of nanotechnology to promote energy efficiency believe that it has good environmental effects. In addition, another use of nanomaterials in energy is production, distribution, and storage. For example, carbon nanotube electrodes have been used to make high-energy lithium batteries, solar cells, fuel cells, and other electrochemical applications. The lithium-ion battery constructed with micro-nanocomposite structure has excellent energy storage performance and stability. Because of its high energy density, it has a wide range of applications in mobile devices, electric vehicles, and other fields. The use of nanomaterials can also produce smaller volume and lighter mass, which will attract more attention. Another example of the positive impact of nanostructures is the use of nanomaterials in solar cells. For example, solar cells based on nanoscale semiconductors have been developed using natural photosynthesis. Solar energy is one of the most potential new generations of new energy in the world. The solar cells prepared with nanomaterials can not only reduce the use of materials, but also greatly reduce the cost of manufacturing and installation of devices, but also greatly improve the photoelectric conversion efficiency of devices.

Fuel cell is a kind of sustainable and sustainable conversion electrochemical device composed of electrolyte, anode, and cathode. In terms of vehicle drive, fuel cells can replace mineral energy and reduce human dependence on gasoline. However, its effectiveness needs to be further improved. In this system, the speed of transfer through the electrode is the key factor affecting its efficiency. On this basis, the effects of electrode materials, electrode materials, and electrode materials under different process conditions on the performance of the electrode were studied. Compared with other conventional electrodes, the fuel cell prepared with nanomaterials is superior in both conversion efficiency and cycle performance. Nanotechnology can improve the efficiency of fuel cells by specially designing different catalysts.

Hydrogen is the highest content in nature, and its molecules contain huge chemical energy, which is used in electric power production. It lays a solid foundation for the development of hydrogen fuel cells. Hydrogen can be used instead of mineral energy to drive vehicles and power stations. The most important point is that the use of hydrogen as fuel will not cause environmental pollution. Therefore, the first problem to be solved in the research and development of fuel cell is how to improve its hydrogen storage capacity. By chemical or physical connection between hydrogen and the storage material, hydrogen can be compressed and then reversibly stored in a reservoir. Nanotechnology makes memory more potential. Macroporous materials can effectively absorb hydrogen through its pores or composite complexes and can recombine with hydrogen in the lattice. Carbon materials (such as activated carbon, carbon nanofibers, carbon nanotubes) are the most potential hydrogen adsorption materials. The excellent properties of carbon materials are mainly reflected in their morphology and structure, such as granular, porous structure, special interaction between carbon and gas molecules, and so on.

Water

Traditional chemical disinfectants are difficult to meet the urgent need of no residual disinfection byproducts, distributed or single point water supply to develop efficient sterilization and microbiological regulation methods. Some natural and synthetic nanomaterials have unique antibacterial effects, including: (1) using their own characteristics (such as titanium oxide, zinc oxide, fullerenes, etc.);(2) degrading the outer membrane of microorganisms (such as polypeptides, chitosan, carboxyl fullerenes, etc.). Carbon nanotubes, zinc oxide, and silver nanoparticles (NAG) are used to block energy transfer (e.g. NAG and aqueous fullerene nanoparticles) and affect biological enzyme activity and DNA (e.g. chitosan).

Environmental

In recent years, nanotechnology has received more and more attention, and its environmental protection effects have been increasingly applied to many fields, such as: as a new type of environmental protection catalyst, as a non-refrigeration thermoelectric material; nanocomposites with high photoelectric conversion efficiency and light weight, which can be used in vehicles; tiny devices that can reduce material loss; nanosensors that do not need wet chemical detection, etc.

In environmental protection, nanotechnology innovations also bring great applications. Nano-based new nanomaterials can be used to monitor and purify pollutants in soil, groundwater, and other water bodies, thus improving the quality of ecosystems. Nanomaterials have important application value in the fields of environmental monitoring, chemical industry, pesticide, etc. In recent years, efficient removal of heavy metals, synthetic dyes, organisms, and other organic pollutants has been widely used in water, air, and other fields. It is a new material with high efficiency and high selectivity. Dendrimer, molecular sieve, and carbon nanotube are the main materials in common use. In addition, other nanomaterials represented by ZnO are also used to treat environmental pollution. The use of nano-thin films, nano-adsorbent, and nanomaterials, and other materials to achieve effective control of the environment, to achieve effective control of motor vehicle exhaust, industrial waste gas.

In addition, there will be better use of water treatment. Due to the consumption of fresh water and environmental degradation, resulting in a shortage of fresh water, people have turned their attention to another source of drinking water. However, because the salt content is too high for people to eat, it must be desalinated to remove the salt; providing a new source of drinking water for people. Desalination costs are high, and nano-membrane technology has great potential. In addition, nanomaterial-based materials, due to their unique characteristics, can also improve their ecological requirements. Although positive environmental effects are rarely associated with the use of nanoproducts, they have been identified as negative. They promote environmental protection and climate change by saving raw materials and energy. For example, the use of nanocomposites can improve the durability of products to mechanics, heat, climate, etc., thus increasing the life of products. Studies have shown that adding nanomaterials, such as nano-silicon, nano-carbon wires, and nano-carbon wires to tires can reduce rolling resistance and thus save fuel. In addition, the material can also reduce vehicle quality, reduce fuel consumption, and achieve energy saving and emission reduction. Nanocoatings are resistant to contamination or easy to clean, do not require frequent cleaning, and save energy and water during equipment cleaning.

The use of nanostructured new insulation materials for buildings can not only improve the energy efficiency of buildings, but also reduce the heating and cooling energy consumption of buildings. In the chemical production process, the most remarkable feature of nanotechnology is that it can effectively improve energy utilization. In recent years, with the rapid development of nanotechnology, the development and application of catalysis theory has attracted more and more attention. Nanomaterials can be used as a new catalytic material to improve reaction efficiency and reduce environmental pollution. To a certain extent, catalysts can effectively regulate the formation and cleavage of chemical bonds. In the fields of industrial production and transportation, its energy conversion and environmental protection are of great significance. In the field of chemical engineering, energy utilization, and resource utilization can be effectively improved through improved nano-catalytic materials. Expectations are high for products and processes that utilize nanotechnology to create and store energy. Luminescent materials composed of nanoplastics and organic pigments can effectively improve the efficiency of light energy conversion.

Nanotechnology in Materials Characterization

Size of Nanoparticles

Particle size is one of the most basic and important test methods for nanoparticles. This factor can determine the size and distribution of particles and can also determine whether particles are microns or nanometers. Electron microscopy is a method of determining particle size, and diffraction is used to determine solid samples. The particles in fluid were detected by photoelectron spectroscopy and centrifugation. It is complicated to analyze gaseous particles by image analysis method, and there are some limitations. Because of its advantages of speed and accuracy, particle testing method based on particle scanning is adopted.

Surface Area of Nanoparticles

Surface area is another important characteristic to study the properties of nanoparticles. The ratio of surface area to volume is the key factor to determine the properties of materials. Beta analysis is the most popular method to estimate surface area at present. Although it is sufficient to determine the surface area of liquid particles by direct titration, this method takes time and effort. Therefore, the MRI spectrum (also known as MRI) is used. The nanoparticles in the gas phase were characterized by differential movement rate analyzer (DMA) and modified SMPS. This factor can determine the size and distribution of particles, as well as whether the particles are microns or nanometers. Electron microscope is a method used to determine the size and distribution of particles. The solid samples were characterized by laser diffraction using the morphology information of particles and aggregates obtained by SEM and TEM.

The particles in the fluid were detected by photoelectron spectroscopy (XPS) and centrifugation. The use of image analysis method to analyze gaseous particles is more complex, and there are some limitations. Because of the advantages of speed and accuracy, the particle testing method based on particle scanning is adopted.

The Composition of nanoparticles

The purity and properties of particles are closely related to the composition or composition of particles. Due to the excess of secondary or harmful elements in nanoparticles, it will not only cause the decline of material properties, but also cause secondary reaction and environmental pollution. XPS (XPS) method is generally used to determine the components of materials. In some studies, the particles are hydrolyzed and then detected by mass spectrometry, atomic emission spectrometry, and ion chromatography. Gaseous particles can be collected by electrostatic or filtration methods and analyzed by spectral or wet methods.

Surface morphology of nanoparticles

Nanoparticles have a variety of forms, and their surfaces can be arranged in a variety of forms, these methods are the key to give full play to their properties. It has many shapes, some are spherical, some are flat, some are cylindrical. Having a tubular, conical, or irregular shape. Its surface may be crystalline or amorphous, or uniform or irregular. At present, the commonly used image processing techniques are mainly used, such as scanning electron microscope (SEM) and transmission electron microscope (TEM). The liquid particles are scanned by electron microscopy, the liquid particles are captured by electrostatic action or filtration, and the gaseous particles are attached to the surface of the particles for further detection.

Surface charge number of nanoparticles

On this basis, the interaction between nanoparticles and targets was studied. One of the most used methods of Zeta potentiometer is to determine its charge in aqueous phase and its dispersibility in water. The charge of nanoparticles in gas phase was measured by differential moving rate analyzer (DMA).

Crystallography of Nanoparticles

Crystallography is the study of the arrangement of atoms and molecules in crystals and other substances. The crystal structure was characterized by powder X-ray, electron or neutron diffraction.

Future Perspective and Challenges

With the development and application of polymer, textile, and other new functional materials, people have carried out extensive and in-depth research on them. In a word, the introduction of nanotechnology into the textile industry will certainly lead to a revolution in scientific and technological textiles. However, considering that the commercialization of nanotechnology-based industries may become a commercial fact, we still have to be careful.

Green Chemistry Opportunities and potentials in Nanotechnology

To improve the production efficiency of the chemical industry chain and reduce its damage to the human body and ecology, it is a very important research topic. This study focuses on the influence of the principles and methods of green chemistry on the whole process of chemicals from design to treatment. In recent years, with the rapid development of China's economy, more and more countries have begun to pay attention to its research and apply it to real life. Both inventions are driven by a wide range of creativity and are committed to ushering in a new era of scientific and technological development. This paper discusses the application of contemporary environmental protection science and technology and its impact on the environment, and looks forward to the possibility of multi-disciplinary integration. The 12 principles of environmental chemistry published in 1998 are recognized by people and provide a framework for designers and decision makers. The above guidelines are implemented because of the accidental damage to the environment caused by people over the past few decades, as well as the consequences for human health caused by the manufacture and application of chemicals. However, the preparation principle of artificially prepared nanomaterials has been used up to now. With the widespread concern around the world, the research of nanomaterials is still in its infancy, and its potential economic, social, and environmental effects may be applied quickly and safely. Under the guidance of the concept of "green economy", people can better understand all kinds of social and economic relations between ecosystem and ecosystem. "Green" means to provide a new method of environmental protection and energy saving for the social development of our country through research achievements in the field of materials and engineering.

Environmental Risks of Nanotechnology

At present, there is major research on how nanomaterials have an impact on our environment around the world. Some are natural, some are man-made, and some are industrialized. These are indisputable facts. The purpose of this project is to study the migration and transformation of organic pollutants in water, the exposure mechanism of organic pollutants in water, sediment, and soil and their harm to organisms. Some nanoparticles can be dissolved in water and redeposited with less toxic minerals; this is because they have lost their original nature. Because of its extremely low content in organisms, its role in organisms, especially in the food chain, is not clear. This project intends to study the migration and migration of complex nanomaterials in the environment and their effects on organic compounds in the food chain, and establish a corresponding risk model to achieve its predictability to the environment. The research in this area should be interdisciplinary.

Atmosphere, ocean, river, and other environmental media have an important influence on the toxicological effects of atmosphere and ultra-fine particles. On this basis, the interaction mechanism between NPs and water body was discussed, and its potential effect on human health was discussed. On the basis of this understanding, it is possible to develop an efficient domestic drinking water treatment strategy. At the same time, the project will also study the interaction mechanism between NPs and water body, and explore its ecological harm to water body.

At present, there are still many problems in the efficient use of nano-bacteriostatic materials, including its dispersion and retention in water, the persistence of bacteriostatic properties, and so on. However, this advantage is weakened when it is reunited in aqueous solution. Nano-particles, such as titanium dioxide will reunite violently after being added to the water. Although nC60 has good stability in pure water, it is also affected by salt. On the other hand, the uniformly distributed nanoparticles contain very small particles. When it is used as a slurry, it can be captured and recycled by an efficient separation process downstream (for example, membrane filtration). By immobilizing nanoparticles on the reactor or filter membrane, there is no need for separation. However, compared with the slurry reactor, because of its limited surface area, it is difficult to obtain light, resulting in poor overall germicidal efficacy. In addition, the nanoparticles will escape from the processing system to the finished water. In addition, because all the nanoparticles are stored in the processing system, no fungicide residues are produced. Therefore, in the distribution system, they need to cooperate with secondary disinfectants to supply residues [54].

Human Safety Issues

The extensive use of nanomaterials in cosmetic products has been attributed to its multi-faceted nature. The use of lipid and polymeric nanoparticles as delivery systems has been preferred because of its biocompatible nature as well as their ease of biodegradation. However, there do exist toxicity factors upon interaction with the human body which should be taken into consideration while using nanomaterials so abundantly in the cosmetic industry [55, 56]. For example, implications of the use of metal-based nanoparticles in cosmetic products on human health have been reported. Traces of different metal nanoparticles, such as copper, iron, cobalt, nickel, chromium, cadmium, lead, arsenic, mercury, etc., have been detected in cosmetics which pose a severe threat to human health. It is due to their small size, these nanoparticles can easily penetrate through the dermal tissue thereby entering blood via which it can reach almost every vital organ leading to a much severe response after its accumulation [57]. In another study, the effect of polyvinylpyrrolidone-functionalized silver nanoparticles of size 15 nm on human keratinocytes has been evaluated. It is observed that exposure to silver nanoparticles remarkably decreases cell viability, metabolic rate as well as severely impairs its migration and proliferation. Furthermore, prolonged exposure of silver nanoparticles results in caspase 3/7 activation and DNA damage thereby triggering genotoxicity as well as cytotoxicity [58]. Moreover, further research focusing on newer ways to reduce nanoparticle toxicity should be undertaken which in turn would increase its applicability in the cosmetic industry.

References

1. Chang, E.H., et al.: Nanomedicine: past, present and future - a global perspective. Biochem. Biophys. Res. Commun. **468**(3), 511–517 (2015)
2. Domingues, C., et al.: Where is nano today and where is it headed? A review of nanomedicine and the dilemma of nanotoxicology. ACS Nano **16**(7), 9994–10041 (2022)
3. Boulaiz, H., et al.: Nanomedicine: application areas and development prospects. Int. J. Mol. Sci. **12**(5), 3303–3321 (2011)

4. Papazoglou, E.S., Parthasarathy, A.: Synthesis lectures on biomedical engineering. Bionanotechnology **2**(1), 1–139 (2007)
5. Lee, Y.-C., Moon, J.-Y.: chapter 9 bionanotechnology in pharmaceuticals. Introduction to Bionanotechnology, pp. 149–169 (2020)
6. Huang, X., Zhu, Y., Kianfar, E.: Nano biosensors: properties, applications and electrochemical techniques. J. Market. Res. **12**, 1649–1672 (2021)
7. Neethirajan, S., et al.: Nano-biosensor platforms for detecting food allergens – new trends. Sens. Bio-Sens. Res. **18**, 13–30 (2018)
8. Christopher, F.C., et al.: Recent advancements in rapid analysis of pesticides using nano biosensors: a present and future perspective. J. Clean. Prod. **269**, 122356 (2020)
9. Nejad, H.E., Mir, A.: Supersensitive and tunable nano-biosensor for cancer detection. IEEE Sens. J. **19**(13), 4874–4881 (2019)
10. Pavlov, V., Xiao, Y., Willner, I.: Inhibition of the Acetycholine esterase-stimulated growth of au nanoparticles: nanotechnology-based sensing of nerve gases. Nano Lett. **5**(4), 649–653 (2005)
11. Bao, J., et al.: 3D graphene/copper oxide nano-flowers based acetylcholinesterase biosensor for sensitive detection of organophosphate pesticides. Sens. Actuators, B Chem. **279**, 95–101 (2019)
12. Alahi, M., Mukhopadhyay, S.: Detection methodologies for pathogen and toxins: a review. Sensors **17**(8), 1885 (2017)
13. Arya, S.K., Datta, M., Malhotra, B.D.: Recent advances in cholesterol biosensor. Biosens. Bioelectron. **23**(7), 1083–1100 (2008)
14. Halwani, A.A.: Development of pharmaceutical nanomedicines: from the bench to the market. Pharmaceutics **14**(1), 106 (2022)
15. Patra, J.K., et al.: Nano based drug delivery systems: recent developments and future prospects. J. Nanobiotechnol. **16**(1), 71 (2018)
16. Riehemann, K., et al.: Nanomedicine—challenge and perspectives. Angew. Chem. Int. Ed. **48**(5), 872–897 (2009)
17. Pelaz, B., et al.: Diverse applications of nanomedicine. ACS Nano **11**(3), 2313–2381 (2017)
18. Arsiwala, A., Desai, P., Patravale, V.: Recent advances in micro/nanoscale biomedical implants. J. Control. Release **189**, 25–45 (2014)
19. Kaul, S., et al.: Role of nanotechnology in cosmeceuticals: a review of recent advances. J. Pharm. (Cairo) **2018**, 3420204 (2018)
20. Mihranyan, A., Ferraz, N., Strømme, M.: Current status and future prospects of nanotechnology in cosmetics. Prog. Mater. Sci. **57**(5), 875–910 (2012)
21. Yadwade, R., Gharpure, S., Ankamwar, B.: Nanotechnology in cosmetics pros and cons. Nano Express **2**(2), 022003 (2021)
22. Panahi, Y., et al.: Recent advances on liposomal nanoparticles: synthesis, characterization and biomedical applications. Artif. Cells Nanomed. Biotechnol. **45**, 788–799 (2017)
23. Yeo, P., et al.: Niosomes: a review of their structure, properties, methods of preparation, and medical applications. Asian Biomed. **11**, 301–314 (2018)
24. Ge, X., et al.: Advances of non-ionic surfactant vesicles (niosomes) and their application in drug delivery. Pharmaceutics **11**, 55 (2019)
25. Abbasi, E., et al.: Dendrimers: synthesis, applications, and properties. NRL **9**, 1–10 (2014)
26. Gopidas, K., Whitesell, J., Fox, M.: Nanoparticle-cored dendrimers: synthesis and characterization. J. Am. Chem. Soc. **125**, 6491–6502 (2003)
27. Qian, C., McClements, D.: Formation of nanoemulsions stabilized by model food-grade emulsifiers using high-pressure homogenization: factors affecting particle size. Food Hydrocoll. **25**, 1000–1008 (2011)
28. Jafari, S., He, Y., Bhandari, B.: Optimization of nano-emulsions production by microfluidization. Eur. Food Res. Technol. **225**, 733–741 (2007)
29. Jaiswal, P., Gidwani, B., Vyas, A.: Nanostructured lipid carriers and their current application in targeted drug delivery. Artif. Cells Nanomed. Biotechnol. **44**, 27–40 (2016)
30. Abidi, S.S.A., Murtaza, Q.: Synthesis and characterization of nano-hydroxyapatite powder using wet chemical precipitation reaction. J. Mater. Sci. Technol. **30**, 307–310 (2014)

31. Poinern, G.E., et al.: Synthesis and characterisation of nanohydroxyapatite using an ultrasound assisted method. Ultrason. Sonochem. **16**, 469–474 (2009)
32. Tiarks, F., Landfester, K., Antonietti, M.: Preparation of polymeric nanocapsules by miniemulsion polymerization. Langmuir **17**, 908–918 (2001)
33. Wang, Z., et al.: Characterization and magnetic properties of carbon-coated cobalt nanocapsules synthesized by the chemical vapor-condensation process. Carbon **41**(1751–1758) (2003)
34. Andrews, R., et al.: Multiwall carbon nanotubes: synthesis and application Acc. Chem. Res. **35**(1008–1017) (2002)
35. Macák, J., Tsuchiya, H., Schmuki, P.: High-aspect-ratio TiO2 nanotubes by anodization of titanium. Angew. Chem. Int. **44**, 2100–2102 (2005)
36. Sawant, S.Y., et al.: Synthesis of submicron size hollow carbon spheres by a chemical reduction—solvothermal method using carbon tetrachloride as carbon source Mater. Mater. Lett. **63**, 2339–2342 (2009)
37. Villena de Francisco, E., García-Estepa, R.M.: Nanotechnology in the agrofood industry. J. Food Eng. **238**, 1–11 (2018)
38. Sahani, S., Sharma, Y.C.: Advancements in applications of nanotechnology in global food industry. Food Chem. **342**, 128318 (2021)
39. Pathakoti, K., Manubolu, M., Hwang, H.M.: Nanostructures: current uses and future applications in food science. J. Food Drug Anal. **25**(2), 245–253 (2017)
40. Zambrano-Zaragoza, M., et al.: Nanosystems in edible coatings: a novel strategy for food preservation. Int. J. Mol. Sci. **19**(3), 705–729 (2018)
41. Tazhbayev, Y., et al.: Synthesis and comparative study of nanoparticles derived from bovine and human serum albumins. Polymers **12**(6), 1301 (2020)
42. Zarrabi, A., et al.: Nanoliposomes and tocosomes as multifunctional nanocarriers for the encapsulation of nutraceutical and dietary molecules. Molecules **25**(3), 638 (2020)
43. Vélez, M.A., et al.: Effect of lyophilization on food grade liposomes loaded with conjugated linoleic acid. J. Food Eng. **240**, 199–206 (2019)
44. Silva, H.D., Cerqueira, M.Â., Vicente, A.A.: Nanoemulsions for food applications: development and characterization. Food Bioproc. Technol. **5**(3), 854–867 (2012)
45. Yola, M.L., Atar, N.: Simultaneous determination of β-agonists on hexagonal boron nitride nanosheets/multi-walled carbon nanotubes nanocomposite modified glassy carbon electrode. Mater. Sci. Eng. C **96**, 669–676 (2019)
46. Tibolla, H., et al.: Banana starch nanocomposite with cellulose nanofibers isolated from banana peel by enzymatic treatment: In vitro cytotoxicity assessment. Carbohyd. Polym. **207**, 169–179 (2019)
47. Tang, F., Li, L., Chen, D.: Mesoporous silica nanoparticles: synthesis, biocompatibility and drug delivery. Adv. Mater. **24**(12), 1504–1534 (2012)
48. dos Santos, C.A., Ingle, A.P., Rai, M.: The emerging role of metallic nanoparticles in food. Appl. Microbiol. Biotechnol. **104**(6), 2373–2383 (2020)
49. Ali, A., Ahmed, S.: Recent advances in edible polymer based hydrogels as a sustainable alternative to conventional polymers. J. Agric. Food Chem. **66**(27), 6940–6967 (2018)
50. Katouzian, I., et al.: Formulation and application of a new generation of lipid nano-carriers for the food bioactive ingredients. Trends Food Sci. Technol. **68**, 14–25 (2017)
51. Yetisen, A.K., et al.: Nanotechnology in textiles. ACS Nano **10**(3), 3042–3068 (2016)
52. Hosne Asif, A.K.M.A., Hasan, M.Z.: Application of nanotechnology in modern textiles: a review. Intern. J. Curr. Eng. Technol. **8**(02) (2018)
53. Singh, N.A.: Nanotechnology innovations, industrial applications and patents. Environ. Chem. Lett. **15**(2), 185–191 (2017)
54. Li, Q., et al.: Antimicrobial nanomaterials for water disinfection and microbial control: potential applications and implications. Water Res. **42**(18), 4591–4602 (2008)
55. Katz, L., Dewan, K., Bronaugh, R.: Nanotechnology in cosmetics. Food Chem. Toxicol. **85**, 127–137 (2015)
56. Muller, H., Shegokar, R., Keck, M.: 20 years of lipid nanoparticles (SLN&NLC): present state of development & industrial applications. Curr. Drug Discov. Technol. **8**, 207–227 (2011)

57. Borowska, S., Brzóska, M.: Metals in cosmetics: implications for human health. J. Appl. Toxicol. **35**, 551–572 (2015)
58. Szmyd, R.E.A.: Effect of silver nanoparticles on human primary keratinocytes. Biol. Chem. **394**, 113–123 (2013)

Yuyang Zhou obtained PhD degree from Soochow University in 2018. He worked as a visiting scholar at the University of Boras (Sweden) during 2017–2018, and as postdoctoral research fellow at The Hong Kong Polytechnic University (HKSAR) and University College Dublin (Ireland) during 2019–2021. He joined Soochow University in Oct. 2021. He published 29 first/corresponding authored research articles (Total citation > 1000, Accumulative IF 228), and has an H-index of 19. He received National Natural Science Foundation of China, Opening Project of Key Laboratory, etc. His research focuses on sustainable functional composites based on bio-based materials and cleaner technology.

Ms Wang is a Senior Engineer in environmental engineering. She has over ten years' experience in developing new materials and techniques for environmental remediation, and has saved over 20 million RMB cost from manufacturing companies under her reformation suggestions.

Mr Liang is now a master's student in the research group of Dr Zhou. His research focuses on the synthesis of micro-nano materials for antimicrobial finishing of textiles.

Mr Liang is now a master student in the research group of Dr Zhou. His research focuses on the development of bio-based micro-nano spheres for functional materials.

Chapter 12
The Role of Nanotechnology in Space Exploration and Colonization

Peeyush Phogat, Shreya, Ranjana Jha, and Sukhvir Singh

Abbreviations

CNT	Carbon Nanotubes
MWCNT	Multi-Walled Carbon Nanotubes
SWCNT	Single-Walled Carbon Nanotubes
GPa	GigaPascals
ISRU	In-Situ Resource Utilization
TE	Transverse Electric
RTGs	Radioisotope Thermoelectric Generators
NTP	Nuclear Thermal Propulsion
TENGs	Triboelectric Nanogenerators
EVAs	Extravehicular activities

Introduction

Space exploration and colonization have long captured the human imagination, with ambitions of establishing a permanent presence beyond Earth's confines. Achieving this vision, however, demands innovative solutions to a myriad of challenges, from spacecraft construction to life support systems. This chapter delves into the pivotal role of nanotechnology in addressing these challenges, offering insights into how nanoscale science and engineering are transforming the landscape of space endeavours.

P. Phogat (✉) · Shreya · R. Jha · S. Singh
Department of Physics, Netaji Subhas University of Technology, Dwarka, Sector-3, New Delhi 110078, India
e-mail: peeyush.phogat@gmail.com

S. C. George and B. Tawiah (eds.), *Nanotechnology in Societal Development*, Advanced Technologies and Societal Change, https://doi.org/10.1007/978-981-97-6184-5_12

In the vast expanse of the cosmos, the realms of space exploration and colonization beckon humanity with unparalleled opportunities and challenges. As we stand on the precipice of a new era, the significance of venturing beyond our celestial home cannot be overstated [1, 2]. Space exploration represents the pinnacle of scientific achievement, pushing the boundaries of our understanding of the universe and our place within it. The desire to unravel the mysteries of distant galaxies, explore uncharted planets, and comprehend the origins of the cosmos has fuelled the human spirit for centuries. However, the quest for space exploration is no longer confined to the realm of scientific curiosity alone; it has evolved into a strategic imperative for the future of our species [3]. The vision of establishing human colonies beyond Earth is no longer relegated to the realm of science fiction; it is a tangible and pressing reality. With an ever-growing population, dwindling resources, and a planet vulnerable to the impacts of climate change, the imperative to extend our reach into the cosmos has become more urgent than ever before [4]. In this unfolding narrative of cosmic exploration and habitation, the significance of space exploration and colonization becomes a catalyst for technological innovation, international collaboration, and the pursuit of sustainable solutions. It represents a bold leap into the unknown, where the challenges are as vast as the opportunities [5]. This journey goes beyond the scientific method; it is a testament to human resilience, adaptability, and the indomitable spirit that has propelled us from the caves to the stars. As we embark on this odyssey into the cosmos, the echoes of pioneers and explorers resonate with each thrust of a rocket into the inky depths of space. This is not merely a quest for knowledge; it is a pilgrimage towards the future—a future where humanity is no longer confined to a single pale blue dot, but one that extends its roots into the celestial tapestry, embracing the cosmos as our collective home [6]. The stage is set, the stars await, and the significance of space exploration and colonization is etched into the very fabric of our aspirations and the destiny of generations yet unborn.

In the expansive realm of space exploration, the integration of nanotechnology stands as a paradigm-shifting frontier, poised to redefine the limits of our cosmic ambitions. As humanity extends its reach beyond Earth's confines, the challenges of space travel become increasingly intricate, demanding innovative solutions that go beyond traditional approaches. It is within this context that the growing importance of nanotechnology emerges, offering a transformative toolkit to navigate the complexities of space-related challenges. At the heart of this scientific revolution lies the manipulation of matter at the nanoscale—a scale so minuscule that it ventures into the realm of individual atoms and molecules. This manipulation bequeaths materials with extraordinary properties, enabling the creation of spacecraft that are not only stronger and lighter but also equipped with capabilities that were once the domain of science fiction. From fortifying spacecraft structures to crafting propulsion systems, sensors, and instruments with unprecedented precision, nanotechnology unveils a new frontier in space exploration [7, 8]. This introduction sets the stage for an exploration into the multifaceted role of nanotechnology in surmounting the hurdles that lie between humanity and the cosmos. As we delve into the intricacies of nanomaterials, self-replicating nanobots, and their diverse applications, we embark on a journey that not only promises to enhance the efficiency and safety of space missions but also paves the way for the realization of grand visions—colonizing distant celestial bodies and unlocking the mysteries of the universe. The narrative that follows navigates through the synergy of nanotechnology and space exploration, illuminating the path toward a future where the once-daunting challenges of the cosmos become surmountable through the power of the infinitesimally small.

Nanomaterials: Building Blocks for Exploring the Cosmos

Nanomaterials have revolutionized all the industries, spanning from electronics, energy, healthcare, and aerospace. The field of nanomaterials is dynamic and constantly evolving, with ongoing research aiming to unlock new possibilities and harness the exceptional properties of these nano materials for practical applications, such as suitable materials for spacecraft and space shuttles. Nanomaterials, with their unique properties, have played a transformative role in spacecraft construction, addressing the challenges of space exploration and colonization [9–11]. Among various nanomaterials, carbon-based nanomaterials such as carbon nanotubes (CNTs), graphene, multi-walled carbon nanotubes (MWCNTs) and their nanocomposites are the most researched due to their unique properties [12–17]. CNTs and graphene, renowned for their extraordinary tensile strength, stand out as the two strongest materials known, surpassing even steel. Single-walled carbon nanotubes (SWCNTs) exhibit a tensile strength of about 100 GigaPascals (GPa) to 300 GPa, whereas MWCNTs can have tensile strengths of approximately 20 GPa to 60 GPa [18–22]. Graphene, being a single layer of carbon atoms arranged in a hexagonal lattice, is extraordinarily strong and has a tensile strength of around 130 GPa [23–25]. Both CNTs and graphene exhibit tensile strengths notably higher than most materials, including steel. However, in comparison, CNTs are deemed best suited for space exploration. Their exceptional strength-to-weight ratio, lightweight nature, and resilience make carbon nanotubes particularly advantageous for spacecraft construction and other applications in the demanding space environment [26, 27]. Not only do CNTs possess higher strength, but they also known for their lightweight nature, making them ideal for aerospace applications. Reduced mass is critical for spacecraft, as it affects launch costs and fuel consumption. CNTs have a lightweight nature, but they also have enhanced thermal and electrical conductivity [28, 29]. CNTs have high thermal conductivity, aiding in the dissipation of heat generated during space travel. This property is essential for maintaining stable temperatures within the spacecraft and protecting sensitive instruments. Not only can CNT be used as spacecraft materials, but they also have potential for radiation shielding [30, 31]. CNTs are also recognized for their radiation resistance, which is crucial factor in protecting both the spacecraft's structure and sensitive electronics from the effects of cosmic rays and solar radiation [32, 33].

Spacecraft structures face numerous challenges during their journeys through space, including extreme temperatures, micro-meteoroid impacts, and the need for lightweight yet durable materials. Nanomaterials, such as CNTs, graphene, and their nanocomposites have emerged as transformative materials in addressing these challenges, playing a pivotal role in strengthening spacecraft structures. Instead of bulk materials, nanomaterials have an exceptional strength-to-weight ratio, while reinforced composite materials in polymers, metal matrices, ceramic matrices, and carbon matrices, can result in the composites gain enhanced mechanical properties that can withstand extreme temperatures. This reinforcement improves the overall strength and durability of the spacecraft structure. Spacecraft endure prolonged missions, subjected to cyclic loading and potential wear. Nanomaterials contribute to the fatigue resistance of spacecraft structures, ensuring they maintain structural integrity over extended periods [34, 35]. Their durability is essential for the success of long-duration missions. Micro-meteoroid impacts pose a significant threat to spacecraft. Nanomaterials and nanocomposites with their exceptional toughness, can absorb and dissipate impact energy effectively [36–38]. Incorporating these materials into spacecraft components provides enhanced protection against collisions with micrometeoroids and space debris. The temperature variations in space, ranging from extreme cold in the shadow of celestial bodies to intense heat in direct sunlight, can stress spacecraft materials. Nanomaterials and their composites in the form of ceramic metrix composites, carbon /carbon matrix (C/C) composites contribute to improved thermal management by efficiently conducting and dissipating heat, preventing structural damage due to temperature extremes [39–42]. Moreover, their creep and fatigue strength is better. C/C is developed specifically for the parts that must operate in extreme temperature ranges. C/C Composites of carbon matrix reinforced with carbon yarn fabric, 3-D woven fabric, 3-D braiding, etc., are very commonly used for these temperatures. C/C Composites meet applications from rocket to aerospace because of their ability to maintain and even increase their structural properties at extreme temperatures. Aluminum and titanium have been traditionally employed for their lightweight, heat resistance, and potential wear properties [43–45]. While CNT is a potential alternative, there is an ongoing search for materials that offer improved properties for space exploration missions. The utilization of nanomaterials in spacecraft construction reduces overall mass, leading to lower fuel requirements for launch [46, 47]. This reduction in launch mass is directly associated with reduced launch costs, a critical factor for space missions operating within budget constraints. Lightweight nanomaterials are particularly advantageous in the construction of satellites, where payload capacity is limited.

Spacecraft and astronauts are exposed to various forms of radiation, including cosmic rays and solar radiation. Prolonged exposure poses health risks, making effective radiation shielding critical [48, 49]. Traditional materials like aluminum and polyethylene have been used for their radiation-blocking properties [50–52]. However, improvements are sought for extended missions and deeper space exploration. Carbon nanotubes and graphene exhibit excellent radiation-absorption properties. These nanomaterials can be integrated into spacecraft structures to enhance radiation shielding capabilities. Nanomaterials are particularly effective in blocking ionizing radiation, preventing damage to biological tissues and sensitive electronic components [53, 54]. Nanomaterials and their nanocomposites can be seamlessly integrated into the structural components of spacecraft, providing continuous and

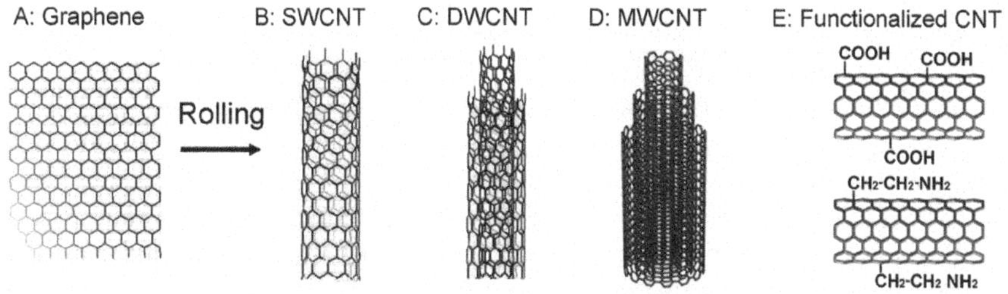

Fig. 12.1 **a** Single layer of graphene when rolled, **b** the names are derived from the number of cylinders, known as single-walled carbon nanotubes (SWCNT), **c** the doubling of wall is known as DWCNT and **d** the presence of multiple wall will create multi-walled carbon nanotubes (MWCNT), **e** Functionalized CNTs reproduced with permission from [55] copyright © 2023 BioMed Central Ltd unless otherwise stated. Part of Springer Nature

efficient radiation shielding without significant increases in mass. Nanomaterials might be produced from resources available on other celestial bodies, reducing the need for transporting shielding materials from Earth (Fig. 12.1).

Propulsion Systems: Nanoparticles and Solar Sails

Space exploration requires propulsion systems that are not only efficient but also capable of addressing the challenges posed by vast distances and the need for extended mission durations. Nanoparticles, at the forefront of nanotechnology, have emerged as a promising avenue for enhancing propulsion systems in space exploration [56, 57]. There are various ways in which nanoparticles contribute to the advancement of propulsion technology and the potential implications for future space missions. One of the primary applications of nanoparticles in space exploration is their integration into rocket propellants. The addition of nanoparticles, such as aluminum or boron, to traditional propellants brings about several advantages such as; increased combustion efficiency [58–61]. Nanoparticles serve as combustion catalysts, promoting more efficient burning of propellants, these result in higher thrust and improved fuel efficiency, critical for long-distance space missions [62]. Another significant advantage of nanoparticle is thermal management. Nanoparticles can enhance the thermal properties of rocket propellants, enabling better control over temperature gradients during combustion [62]. This is particularly significant in preventing overheating and ensuring the stability of propulsion systems. Additionally, nanoparticles can facilitate thrust vectoring, allowing for more precise control of the spacecraft's direction. This contributes to improved navigation and manoeuvrability, crucial for complex space missions. The integration of nanoparticles into rocket propellants represents a significant advancement in space exploration propulsion technology. The enhanced combustion efficiency, improved thermal management, and thrust vectoring capabilities offered by nanoparticles open new possibilities for the success of future space missions, especially those involving extended durations and intricate maneuvers [62].

The exploration of deep space has perpetually faced constrained imposed by the limitations of traditional propulsion systems. In recent years, there has been a paradigm shift in space exploration with the introduction of innovative solar sails constructed from nanomaterials [63, 64]. These sails exploit the extraordinary properties of nanotechnology to harness the pressure of solar radiation, offering a sustainable and efficient means of propulsion for space missions. Traditional solar sails rely on large, lightweight sheets of reflective material to capture and utilize the momentum of photons emitted by the Sun [65, 66]. However, the incorporation of nanomaterials introduces a new dimension to this concept, revolutionizing the design and efficiency of solar sails. Nanomaterials, such as graphene and carbon nanotubes, offer an unparalleled combination of strength and lightness [67, 68]. These properties make them ideal candidates for constructing solar sails that are not only robust and durable but also significantly lighter than traditional materials. The unique optical properties of nanomaterials enhance the efficiency of photon capture [69–73]. Their ability to interact with a broad spectrum of light enables more effective utilization of solar radiation, providing a higher thrust-to-mass ratio compared to conventional sails. Nanomaterials provide a level of versatility in design that was previously unattainable. Solar sails can be engineered at the nanoscale, allowing for intricate patterns and structures that optimize photon absorption and propulsion efficiency [74–76]. Traditional sails require thickness to maintain structural integrity, adding unnecessary mass. Nanomaterial-based sails, on the other hand, can be engineered at the nanoscale, reducing thickness while maintaining strength. This thinness contributes to the overall reduction in the mass of the spacecraft. Nanomaterials exhibit remarkable radiation resistance, crucial for enduring the harsh conditions of space [10, 32]. This resilience ensures the prolonged functionality of solar sails over extended missions, mitigating the impact of space radiation.

Nanomaterial-based solar sails provide a sustainable propulsion solution for deep space missions. By continuously harnessing solar radiation, these sails eliminate the need for traditional propellants, offering an energy-efficient, stable, and long-term means of travel. The lightweight and responsive nature of nanomaterial-based sails enhances the spacecraft's manoeuvrability. Precise control over the orientation of the sail allows for efficient course corrections and trajectory adjustments during the mission. With their potential for high thrust and efficiency, nanomaterial-based solar sails open up possibilities for interstellar exploration. The extended lifespan of these sails ensures sustained propulsion over the vast distances involved in interstellar journeys. When exposed to intense laser beams, nanoparticles can generate thrust through a phenomenon known as the photonic laser thrust. This concept, still in experimental stages, holds the potential for high-speed propulsion without the need for traditional propellants. The advancements facilitated by nanoparticles in propulsion systems and solar sails open the door to more ambitious and extended missions. This includes interplanetary exploration and potentially interstellar missions, where the continuous and sustainable propulsion provided by solar sails becomes increasingly advantageous for covering vast distances (Fig. 12.2).

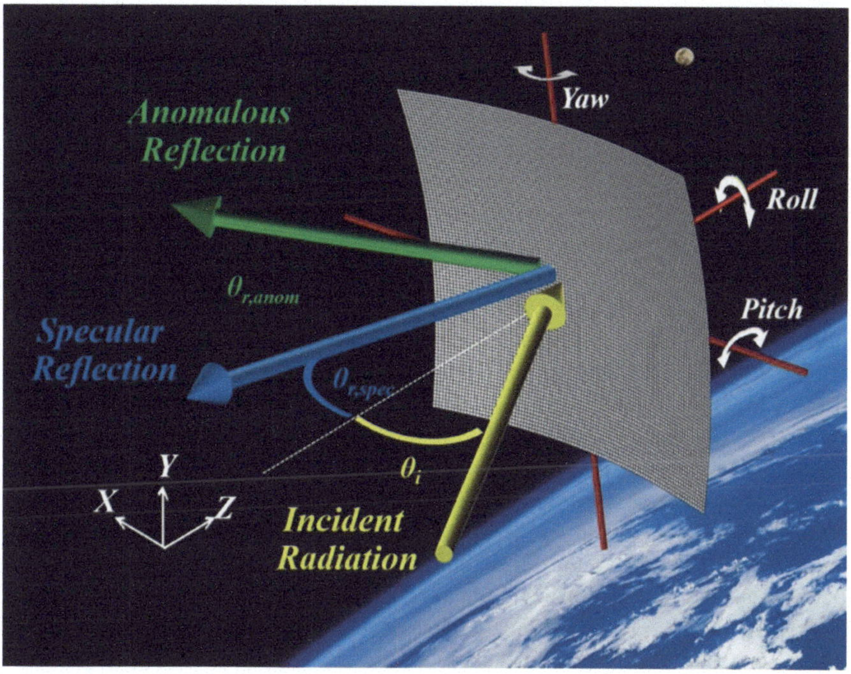

Fig. 12.2 Schematic of solar sail system interacting with oncoming transverse electric polarized electromagnetic field. Background image was provided by NASA reproduced with permission from [66] copyright © 2023 Springer Nature Limited

Sensors and Instruments: Miniaturization and Precision

Nanotechnology facilitates the development of miniature sensors and nanoelectronics, capable of withstanding the harsh conditions of space. These sensors empower spacecraft with the capacity for precise data collection and imaging, opening new frontiers in astronomical observations and scientific discoveries. Space exploration demands cutting-edge technologies to collect precise data in the harsh and vast environment beyond Earth. Miniature sensors and nanoelectronics play a pivotal role in this endeavour, offering solutions that are compact, efficient, and capable of withstanding the challenges of space [77, 78]. Miniature Sensors and nanoelectronics are characterized by their small size, highly efficient and lightweight design. These attributes are crucial for space missions where every gram counts. The reduced size also allows for easy integration into spacecraft, satellites, and rovers without compromising overall mission objectives [79, 80]. Miniature sensors encompass a wide range of functionalities, including imaging, spectroscopy, temperature sensing, and atmospheric analysis [81, 82]. Their versatility enables scientists and engineers to tailor sensor arrays to specific mission goals, providing a wealth of data on celestial bodies, atmospheres, and cosmic phenomena. Miniature sensors excel in remote sensing applications, capturing data from distant locations without direct physical contact [58, 83, 84]. This capability is vital for exploring planets, moons, asteroids, and other celestial bodies, where direct human interaction is not feasible. These sensors contribute to autonomous exploration by providing real-time data and feedback to spacecraft systems. This autonomy allows for adaptive decision-making, enhancing the efficiency of space missions.

Nanoelectronics, operating at the nanoscale, typically have lower power requirements compared to traditional electronics. This characteristic is critical for space missions where energy conservation is paramount. Low-power nanoelectronics extend the lifespan of spacecraft and maximize the duration of scientific data collection. The space environment exposes electronic components to radiation, which can be detrimental to conventional electronics. Nanoelectronics, with their inherent radiation resistance, ensure the reliability of systems in the presence of cosmic rays and solar radiation [85, 86]. Nanoelectronics enable high-speed data processing, facilitating quick analysis of collected data. This is particularly important for time-sensitive observations and decision-making during space exploration missions. Advancements in nanotechnology have revolutionized the field of space exploration, particularly in the development of imaging instruments for astronomical observations and remote sensing applications [87]. These innovations enable scientists and researchers to gather precise data, unravel mysteries of the cosmos, and enhance our understanding of distant celestial bodies. Nanotechnology contributes to improved imaging instruments in space exploration. Nanomaterials play a pivotal role in improving the optics of imaging instruments [88, 89]. Nanostructured materials, such as metamaterials and plasmonic devices, can manipulate light at the nanoscale, leading to improved resolution and sensitivity in telescopic systems. Adaptive optics, utilizing nanomaterials like deformable mirrors with nanoscale actuators, correct distortions caused by atmospheric turbulence [90]. This technology is crucial for ground-based telescopes and space observatories, ensuring clearer and sharper images of distant celestial objects. Quantum technologies, a subset of nanotechnology, are revolutionizing imaging instruments through quantum sensing and quantum imaging techniques [91]. Quantum entanglement and superposition principles allow for unprecedented levels of precision in measuring faint signals, contributing to high-fidelity astronomical observations.

Nanotechnology facilitates the development of nanoscale spectrometers capable of analyzing the composition of distant objects with exceptional precision. This is invaluable for remote sensing applications, enabling scientists to identify specific molecules or elements on planets, moons, and other celestial bodies [92]. Nanomaterials contribute to the reduction of noise and enhancement of signal strength in imaging instruments. This results in improved signal-to-noise ratios, allowing for the detection of faint signals from distant objects, which is essential for observing celestial bodies in deep space. In the era of miniaturized satellites, nanotechnology enables the development of nanoscale imaging instruments for small satellites (nanosatellites or CubeSats) [93–95]. These compact devices contribute to cost-effective space missions and provide valuable data for scientific research. Nanotechnology aids in the development of multi-spectral and hyperspectral imaging systems. These advanced

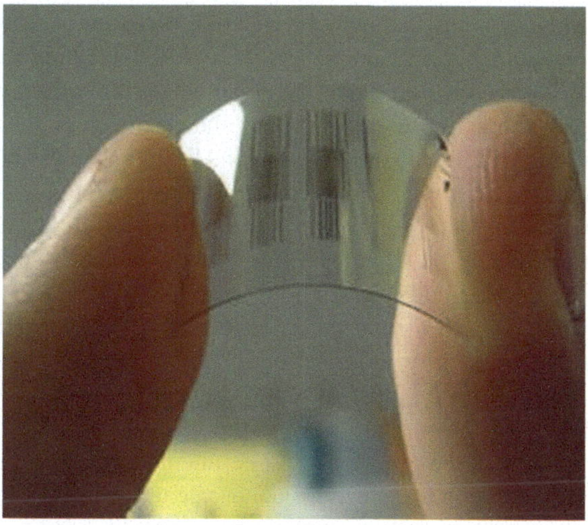

Fig. 12.3 Photograph of a nanosensor made on a transparent plastic piece with flexibility for multiple applications reproduced with permission from [96] copyright © 2017 Elsevier Inc

instruments capture data across multiple wavelengths, enabling scientists to analyze the composition, temperature, and atmospheric conditions of distant planets and moons. Continued advancements in nanotechnology hold the promise of even more sophisticated imaging instruments. From the utilization of advanced nanocomposites for telescope construction to the integration of artificial intelligence algorithms for real-time data analysis, the future of space-based imaging is self-assured for ground-breaking developments (Fig. 12.3).

Resource Utilization: Mining the Cosmos

Space colonization and exploration pose the challenge of sourcing essential resources from celestial bodies beyond Earth. Nanoparticles, operating at the smallest scale, introduce a revolutionary approach to resource extraction. Resource utilization delves into the applications, advantages, and challenges of using nanoparticles for resource extraction in the context of space colonization and exploration. To colonize space, humans as a species will require many resources, which can be obtained through Lunar or asteroid mining [97, 98]. The utilization of nanoparticles for mining lunar regolith represents a pioneering frontier in space exploration and colonization. Nanomaterials, particularly engineered nanoparticles, offer unique advantages in extracting precious metals and volatiles from the Moon's regolith [99, 100]. These nanoparticles can be tailored to interact selectively with specific minerals, facilitating efficient extraction processes. For instance, gold and platinum group metals, vital for advanced technologies, can be targeted with precision, maximizing resource utilization. Additionally, nanomaterials enhance the separation of volatiles such as water ice, crucial for sustaining future lunar colonies [101, 102]. The introduction of nanotechnology in lunar mining holds promise for resource efficiency, reduced environmental impact, and the feasibility of establishing sustainable bases on the Moon. By harnessing the versatility of nanomaterials, space exploration ventures can optimize resource utilization, laying the groundwork for a new era of extra-terrestrial mining and human presence beyond Earth. In the vastness of space, asteroids present valuable reservoirs of minerals and metals essential for sustaining future colonies and supporting space missions. Nanoparticles, with their unique properties, can play a pivotal role in developing advanced sensors for prospecting, enabling precise identification of mineral compositions from a distance [103, 104]. Concurrently, nanobots, equipped with drilling and sampling capabilities at the nanoscale, offer unprecedented versatility in microgravity environments. These nanobots can navigate the intricate topography of asteroids, extracting samples with precision. Furthermore, nanotechnology facilitates the processing of these raw materials directly on-site, minimizing the need for resource-intensive transportation back to Earth [105, 106]. By harnessing the power of nanoscale technologies, the prospecting and extraction processes become more efficient, sustainable, and responsive to the challenges of space exploration and colonization, propelling humanity into a new era of extra-terrestrial resource utilization [107]. In the pursuit of sustainable space colonization and exploration, the concept of In-Situ Resource Utilization (ISRU) emerges as a crucial strategy [108, 109]. ISRU involves utilizing resources available on celestial bodies, such as the Moon or Mars, rather than relying solely on Earth-bound supplies [110–112]. This approach not only reduces the dependency on resupply missions but also contributes to long-term sustainability. Nanotechnology plays a significant role in enabling and enhancing various aspects of ISRU [113, 114].

Nanoparticles play a crucial role in thermal and mechanical processes designed for extracting water ice from celestial bodies. Additionally, nanomaterial-based filtration systems contribute to purifying the extracted water. This combination of technologies presents an unparalleled opportunity to revolutionize resource extraction on celestial bodies, laying the foundation for sustainable space colonization and exploration. As technology advances and challenges are addressed, the integration of nanoparticles into space missions promises to redefine humanity's relationship with the cosmos, offering a pathway to self-sufficiency beyond the confines of Earth. Space colonization and exploration demand innovative solutions for life support systems, and advanced filtration systems play a pivotal role in ensuring a sustainable supply of clean water and breathable air for astronauts. Nanotechnology, with its precision and efficiency, has paved the way for the development of cutting-edge filtration systems that address the unique challenges of the space environment. Nanomaterials, such as carbon nanotubes, graphene, and nanocomposites, are at the forefront of revolutionizing filtration membranes [115–117]. Their nanoscale structures provide a high surface area and unique properties that enhance filtration efficiency. These materials can be tailored to filter out specific contaminants, ensuring the production of purified water and clean air in space habitats [118, 119].

For suitable water filtration technologies, there is a need of nanoparticle-based filtration technique, as nanoparticles can be incorporated into water filtration systems to capture and remove contaminants at the molecular level. This technology enables the removal of pollutants, microorganisms, and even traces chemicals from water sources. Not limited to filtration alone, nanotechnology has improved the performance of traditional water purification methods like Nano filtration and reverse osmosis [120–122]. Nanoscale membranes enhance the efficiency of separating impurities, allowing for the production of high-quality, potable water. Nanoparticles not only provide enhanced water purification but also has self-cleaning technique, such as Nano coatings on filter surfaces can create self-cleaning properties, reducing the frequency of maintenance and replacement [123, 124]. This is crucial for long-term space missions where resource efficiency is paramount. Alongside water purification, air filtration is also of utmost priority need of the hour for extended space missions. Using nanomaterials based filtration for air can be done through nanofiber filters and coatings made from nanomaterials are employed in air purification systems [125, 126]. These filters effectively capture airborne particles, including dust, allergens, and microorganisms, ensuring a clean and breathable atmosphere within space habitats [127, 128]. Photoactive nanomaterials, when exposed to light, can generate reactive oxygen species that help break down airborne pollutants. This technology offers an additional layer of air purification, promoting a healthier living environment for astronauts. Advanced filtration systems, empowered by nanotechnology, are indispensable for achieving self-sufficiency in water and air supply during space colonization and exploration. As these technologies continue to evolve, the prospect of establishing sustainable habitats on other celestial bodies becomes increasingly attainable, ensuring the well-being of astronauts and the success of future space missions (Fig. 12.4 and Table 12.1).

Energy Harvesting: Power from the Cosmos

For the sustainable colonization of humankind in space, there is a regular requirement for energy. To harness this energy, there is a need of energy harvesting systems, including nanogenerators that convert vibrations, heat, and light into electrical energy, offer a continuous power supply for space colonies. This promises autonomy and sustainability in energy generation, reducing reliance on Earth's power sources. Space missions, whether for exploration or colonization, rely heavily on efficient power generation and storage systems to ensure the success and sustainability of endeavors beyond Earth. In the vastness of space, where sunlight, distance, and resource constraints pose unique challenges, innovative technologies are crucial for meeting energy needs. For the power generation and storage solutions at the forefront of space exploration and colonization we will require different kind of energy source.

Energy generation can be accomplished through solar arrays and solar sails, as discussed above. However, there are many promising alternatives for ensuring energy generation, including nuclear power. *Radioisotope Thermoelectric Generators (RTGs)* are vital power sources in space missions, providing reliable and long-lasting electrical energy in environments where other conventional power sources may be impractical or insufficient [144, 145]. RTGs operate based on the principle

Fig. 12.4 An AI-generated image of what future asteroid mining would look like by using nanotechnology (image generated using site www.canva.com)

of thermoelectricity. They use the heat produced by the natural radioactive decay of isotopes, such as plutonium-238, to generate electrical power through thermocouples. RTGs are known for their longevity and reliability. The decay of isotopes occurs at a predictable rate, allowing for a steady and continuous power output over an extended period. This makes RTGs suitable for deep space missions where sunlight is limited, such as missions to outer planets and their moons [146, 147]. RTGs have been employed in numerous space missions, including the Apollo missions to the Moon, the Voyager and Pioneer probes exploring the outer solar system, and the Mars rovers (e.g., Curiosity and Perseverance). They serve as a dependable and independent power source, especially when solar power may not be feasible or sufficient. RTGs are particularly valuable in environments with harsh conditions, such as those encountered in outer space. They operate independently of sunlight and are not affected by factors like dust storms on Mars or the extended darkness in certain

Table 12.1 Potential resources from Moon, Mars, and Near-Earth Asteroids and their potential applications. Data taken from [143] Copyright © 2022 The Author(s)

Potentially exploitable resources	Mining and processing operations	Techno-optimistic application	Refs
The Moon			
Regolith (Lunar Surface)	Bulk surface excavation; material handling; dry beneficiation (magnetic, electrostatic or size separation); extraction; ilmenite reduction; molten salt electrolysis; fluorination; acid digestion; reduction of Al_2O_3; magma electrolysis; carbothermal reduction; reduction of magnesium oxide with ferrosilicon; vapour pyrolysis; vacuum distillation) product storage; waste management	Radiation shielding (human habitats) & construction by additive manufacturing; waterless concrete & aggregate; production of oxygen (lifesupport, rocket oxidiser), Fe/Al (infrastructure), Ti (aerospace), Si (photovoltaic cells, silane as a possible rocket fuel), Mg (external structures, replacement parts)	[129–133]
Water (Polar ice; implanted by solar wind in lunar regolith; hydrated minerals)	Diverse methods include strip mining; in-situ sublimation; regolith devolatization; ilmenite reduction; condensation; electrolysis	Life (drinking, personal hygiene, agriculture); processing; propellant; production of H_2 and O_2	
Hydrogen and Helium 3 He, 4 He (Lunar surface; implanted by solar wind in regolith)	Bulk surface mining operation; material handling; beneficiation; regolith degassing by heating (> 700 ◦C); product storage; waste management	Propellant (H); reactant (H); fuel for future nuclear fusion reactors (3 He)	
Near-Earth Asteroids (NEAs)			
Platinum-group elements (M-type NEAs, dominantly NickelIron alloys)	Bulk surface mining operation; underground and/or in-situ extraction; fragmented pieces collected in a bag surrounding the asteroid	Sale in Earth-based market	[134–137]
Base metals, e.g., iron and nickel (M-type NEAs)	Bulk surface mining operation; underground and/or in-situ extraction	Construction for satellite industry; off-Earth habitation	
Water and other volatiles (C-type asteroids; near-Earth comets)	Multiple proposed methods (e.g., heating by solar thermal processes)	Life processing, propellant, etc	
Mars			
Regolith (Martian Surface)	Bulk surface mining operation	Construction by 3D printing; production of waterless concrete	[138–142]

(continued)

Table 12.1 (continued)

Potentially exploitable resources	Mining and processing operations	Techno-optimistic application	Refs
Water (Water ice at poles; hydrated minerals; liquid brines)	Diverse methods include ice mining; drilling-based water extraction; regolith devolatization; electrolysis of brine	Life processing, propellant, etc	
Oxygen and other atmospheric gases (Martian atmosphere)	Mars Oxygen In-Situ Resource Utilization Experiment (MOXIE)	Life processing, propellant, etc	

regions of space. RTGs provide reliability and steady power output, however RTGs have low power density and to overcome that difficulty *Nuclear Thermal Propulsion (NTP)* is used. NTP is an advanced propulsion technology that utilizes nuclear reactions to heat a propellant, such as hydrogen, to high temperatures [148]. The NTP system works by transferring heat from the reactor directly to a gaseous hydrogen propellant. Heated hydrogen expands through a nozzle to provide thrust to propel a spacecraft.

NTP offers several advantages over traditional chemical rocket propulsion systems and holds promise for future space missions [149, 150]. Nuclear thermal propulsion is about twice as efficient as chemical rockets and can shorten travel times while delivering greater payloads. For example, a trip to Mars can be reduced by a quarter. NTP systems provide a significantly higher specific impulse compared to chemical rockets. Specific impulse is a measure of propulsion efficiency, and higher values mean more efficient use of propellant. The increased specific impulse of NTP allows for faster travel and more efficient maneuvering in space. The higher efficiency of NTP systems can result in reduced travel times for space missions [151, 152]. This is particularly advantageous for crewed missions to destinations like Mars, where minimizing the time spent in space is essential for astronaut safety and well-being. NTP systems can be designed with reusable components, reducing the overall cost of space exploration. Reusability is a critical factor in making space travel more economically sustainable.

While energy generation is essential for powering space colonies, energy storage is equally critical for maintaining a reliable and continuous power supply, managing intermittent energy sources, addressing peak energy demands, and ensuring resilience in the face of unforeseen challenges or emergencies. The combination of effective energy generation and storage systems is vital for the success and sustainability of space colonies. For ensuring continuity in energy storage lithium-ion and other advanced battery technologies can be used for immediate energy needs i.e., powering critical systems during periods of low or no sunlight [153, 154]. However, for a long time period, hydrogen fuel cells and advanced capacitors play an important role. Nanomaterials play an advance role in both of these energy storage solutions.

Hydrogen fuel cells offer a sustainable source of power for extended missions. Nanomaterials, such as platinum nanoparticles, are commonly used as catalysts in the electrodes of hydrogen fuel cells [155, 156]. Their high surface area-to-volume ratio and unique electronic properties enhance the kinetics of the hydrogen oxidation reaction at the anode and the oxygen reduction reaction at the cathode, improving overall cell efficiency. Nanomaterials, especially nanostructured catalysts can enhance the mass transport properties within the fuel cell [157]. This is critical for facilitating the transport of reactants (hydrogen and oxygen) to the catalyst sites, improving reaction kinetics, and reducing mass transport limitations [158, 159]. Nanomaterials enable the engineering of advanced catalyst structures, such as nanoalloys and core–shell nanoparticles. These structures exhibit enhanced catalytic activity and stability, addressing challenges related to catalyst degradation and reducing the reliance on expensive noble metals. For advanced capacitors rapid energy release during peak demands complementing batteries in hybrid energy storage systems. Nanomaterials, such as nanoparticles and nanotubes, provide an incredibly high surface area per unit volume. This characteristic is beneficial for increasing the capacitance of capacitors [160]. As the surface area expands, so does the ability of the capacitor to store and release electrical energy. The unique properties of nanomaterials contribute to the longevity and durability of capacitors [161]. Nanocomposite materials can withstand repeated charge and discharge cycles, leading to longer life cycle and increased reliability in various applications [162]. Nanomaterials are instrumental in the development of supercapacitors, a type of capacitor with higher energy density compared to traditional capacitors. Supercapacitors find applications in energy storage systems, transportation, and renewable energy technologies [163] (Fig. 12.5).

Advancements in materials, such as nanotechnology, for improved efficiency, is a novel concept for exploring novel concepts like space-based solar power for enhanced energy generation. *Interplanetary power grids* are such an example—The potential development of interplanetary power grids for sustainable colonization—Creating a networked approach to power distribution in the cosmos. Nanogenerators *(a small device that converts mechanical or thermal energy into electricity)* play a significant role in space missions and colonization efforts due to their ability to harvest and convert various forms of energy into electrical power at the nanoscale. In space, there are abundant energy sources that can be harnessed, such as vibrations, thermal gradients, and solar radiation. Nanogenerators provide a means to capture and convert these ambient energy sources into electrical power, contributing to the sustainability of space missions. Nanogenerators offer a way to reduce reliance on traditional battery systems. In space missions, where the weight and size of batteries are critical factors, nanogenerators can supplement or replace batteries for various applications, ensuring a more sustainable and long-lasting power supply. There are three types of nanogenerators: Piezoelectric, Triboelectric, and Thermoelectric (Table 12.2).

Fig. 12.5 An AI-generated image of how solar panels can be attached to satellite and space station for energy generation I space (image generated using site www.canva.com)

Table 12.2 Comparison of energy harvesting methods

Method	Pros	Cons
Solar power	Abundant sunlight	Ineffective in deep space or shaded areas
Radioisotope thermoelectric generators (RTGs)	Reliable, long-lasting power source	Limited availability of isotopes
Nuclear Fusion	Abundant fuel, sustainable	Technological challenges, experimental
Piezoelectric nanogenerators	Harness mechanical vibrations	Limited power output, scalability
Thermoelectric generators	Utilize temperature differentials	Efficiency dependent on temperature gradient
Kinetic energy harvesting	Capture energy from spacecraft motion	Limited energy storage capacity

Piezoelectric Nanogenerators

Piezoelectric nanogenerators represent a ground breaking technology that harnesses mechanical vibrations to generate electrical energy at the nanoscale [164]. This innovation holds significant promise, especially in the realm of space exploration. Here, we delve into the applications of piezoelectric nanogenerators, focusing on their integration into spacecraft structural components and their potential role in powering sensors and instruments. Piezoelectric nanogenerators offer a novel approach to address the energy needs of spacecraft by tapping into the mechanical vibrations experienced during various operational phases [165, 166]. This technology can be seamlessly integrated into structural components, capitalizing on the inherent vibrations and deformations that spacecraft encounter during launch, maneuvering, and orbital activities. The versatility of piezoelectric nanogenerators extends to power critical sensors and instruments on board spacecraft, offering a sustainable and self-sufficient energy solution.

Remote Sensing Instruments

Sensors responsible for remote sensing, imaging, and data collection can benefit immensely from the continuous and sustainable power supply derived from piezoelectric nanogenerators. These nanogenerators, harnessing the piezoelectric effect at the nanoscale, convert mechanical vibrations or strain into electrical energy, providing a reliable source for powering these critical instruments. The uninterrupted operation facilitated by piezoelectric nanogenerators is particularly crucial for space missions where traditional power sources might be limited or impractical. In the vastness of space, where resources are finite and the ability to replenish or recharge is challenging, the endurance of remote sensing instruments becomes a critical factor. The incorporation of piezoelectric nanogenerators ensures that these instruments can function continuously, allowing for extended and enhanced scientific observations. This is particularly relevant for missions that involve long-duration space exploration or the study of distant celestial bodies, where the availability of sunlight or other conventional power sources may be inconsistent. Moreover, the compact nature of piezoelectric nanogenerators aligns seamlessly with the stringent size and weight constraints of space missions. Their small form factor allows for easy integration into the design of spaceborne instruments without compromising the overall efficiency or functionality. This adaptability is crucial for spacecraft, probes, or rovers where every gram of weight and every cubic centimeter of space is meticulously planned. The benefits extend beyond prolonged operational capabilities; the continuous power supply ensures a more stable and consistent stream of data collection. Uninterrupted observations empower scientists and researchers to capture a more comprehensive dataset, unveiling nuances and patterns that might be missed in shorter observation windows. This, in turn, leads to more robust scientific analyses and a deeper understanding of the extraterrestrial environments under scrutiny. In essence, the marriage of remote sensing instruments with piezoelectric nanogenerators heralds a new era in space exploration, where the challenges of energy scarcity are mitigated, and the quest for knowledge is propelled further into the cosmic unknown. The application of nanotechnology in this context not only ensures the reliability of scientific instruments but also contributes to the efficiency and success of space missions, unlocking the full potential of remote sensing in the pursuit of unraveling the mysteries of our universe [167].

Communications Systems

Nanogenerator-powered systems mark a pioneering leap in ensuring the seamless operation of communication instruments within the vast expanse of space [168, 169]. The unique capability of nanogenerators to harvest energy from ambient vibrations becomes particularly invaluable in sustaining the reliable functioning of communication devices across extended space missions. In the dynamic and often unpredictable environment of space, where traditional power sources may be limited, nanogenerators emerge as a resilient solution. By efficiently converting vibrational energy generated during routine communication activities, these nanoscale power generators offer a continuous and sustainable power supply. The intricate dance of nanogenerators with communication systems transcends the conventional boundaries of power constraints. In the absence of readily available power sources, such as sunlight or conventional batteries, nanogenerators step in to fill the void. Their ability to capture and convert ambient vibrations into usable electrical energy ensures that communication instruments remain operational even in the face of extended mission durations, where the preservation of power reserves is paramount. Moreover, the incorporation of nanogenerator-powered systems introduces a layer of self-sufficiency to communication networks in space. By harnessing energy from the very activities that necessitate communication—be it the hum of machinery or the subtle vibrations inherent to spacecraft movement—these systems reduce dependency on external power sources. This newfound autonomy contributes not only to the sustainability of space missions but also to the resilience and adaptability of communication systems in the unforgiving conditions of the cosmos.

As humanity ventures further into deep space, where conventional power sources may become impractical, the marriage of nanogenerators and communication systems emerges as a strategic alliance. This synergy ensures that the lines of communication remain open, facilitating not only the exchange of critical information between spacecraft and Earth but also enabling collaborative efforts among interstellar probes. In this way, nanogenerator-powered communication systems become a testament to the versatility and efficiency of nanotechnology in addressing the multifaceted challenges of space exploration.

Environmental Sensors

Environmental Sensors: Spacecraft equipped with environmental sensors for monitoring radiation, temperature, or atmospheric conditions can leverage piezoelectric nanogenerators to sustainably power these instruments [170, 171]. This is particularly advantageous in scenarios where continuous data collection is imperative. The utilization of piezoelectric nanogenerators in conjunction with environmental sensors exemplifies a harmonious synergy of nanotechnology and space exploration. These nanogenerators, typically composed of piezoelectric materials such as zinc oxide or lead zirconate titanate, harness ambient vibrations or mechanical stresses encountered during space travel to generate electrical power. In the vacuum of space or on celestial bodies with minimal atmosphere, conventional power sources may face limitations. However, piezoelectric nanogenerators offer a decentralized and reliable energy solution. As the spacecraft maneuvers through space or encounters gravitational forces, the inherent mechanical deformations create electrical charges within the piezoelectric material, thus providing a continuous and sustainable source of power for environmental sensors. This innovative approach not only addresses the challenges of powering instruments in resource-constrained environments but also extends the operational lifespan of the spacecraft. Continuous monitoring of environmental conditions is pivotal for ensuring the well-being of astronauts, assessing the health of spacecraft components, and gathering critical data for scientific research. Whether navigating through deep space or stationed on distant planets, the integration of piezoelectric nanogenerators with environmental sensors demonstrates the adaptability and versatility of nanotechnology in meeting the unique demands of space exploration. In the quest for comprehensive knowledge about the cosmos, these advancements underscore the importance of sustainable energy solutions. As we push the boundaries of human exploration further into the cosmos, the marriage of piezoelectric nanogenerators and environmental sensors becomes not just a technological innovation but a testament to humanity's ingenuity in overcoming the challenges of the final frontier. This confluence of nanotechnology and space exploration continues to pave the way for unprecedented discoveries and sustainable, long-term missions beyond Earth's orbit (Fig. 12.6).

Triboelectric Nanogenerators

Triboelectric nanogenerators (TENGs) are innovative devices that harness mechanical energy from friction to generate electricity. They operate on the basic principle of triboelectric effect, which involves the generation of electric charge when two materials come into contact and then separate. The triboelectric effect occurs when two materials with different electronegativities come into contact and electrons are transferred from one material to the other. This transfer of electrons creates a potential difference between the materials, leading to the generation of static electricity. The mechanism can be summarized as follows:

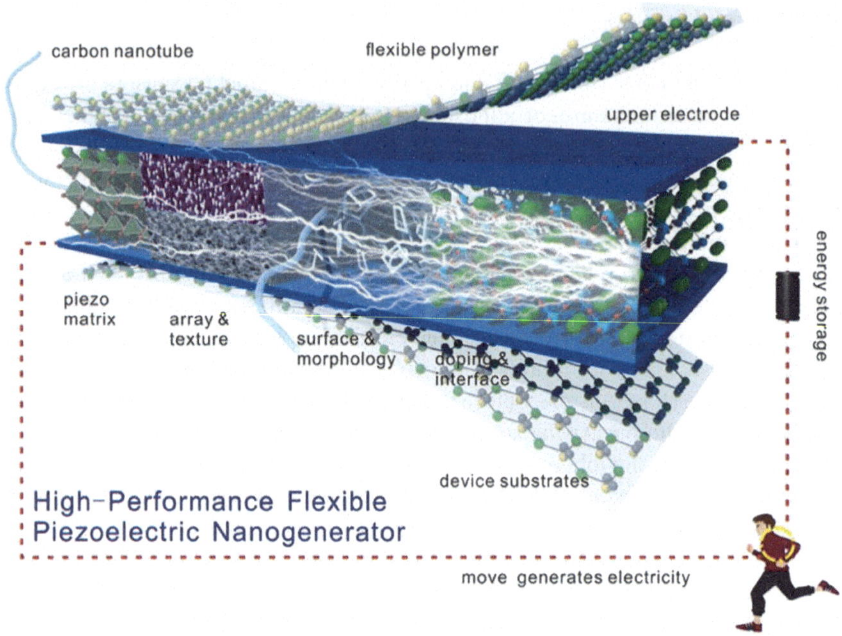

Fig. 12.6 Flexible Piezoelectric Nanogenerator reproduced with permission from [172] copyright © 2018 Elsevier Ltd

When two materials with different surface properties come into contact and then separate, one material tends to gain electrons (becoming negatively charged) while the other material loses electrons (becoming positively charged). This phenomenon is known as contact electrification and is a fundamental aspect of the triboelectric effect. The separation of charges between the two materials creates an electric field, resulting in the accumulation of charges on their surfaces. This accumulation of charges forms an electric potential, which can be harnessed to generate electrical power. When an external circuit is connected between the positively and negatively charged materials, electrons flow through the circuit, generating an electric current. This current can be used to power electronic devices or stored in batteries for later use. Triboelectric nanogenerators convert mechanical energy from various sources, such as human motion, vibration, or wind, into electrical energy through the triboelectric effect. The mechanical motion causes the materials in the TENG to come into contact and separate, inducing the generation of electric charges and producing electricity. TENGs can be designed in various configurations, including vertical contact-separation, lateral sliding, and single-electrode modes. Each configuration utilizes specific materials and structures to optimize the triboelectric effect and maximize energy generation efficiency (shown in Fig. 12.7).

Triboelectric nanogenerators have diverse applications, including self-powered sensors, wearable electronics, environmental monitoring devices, and energy harvesting systems for portable electronics. Their flexibility, scalability, and ability

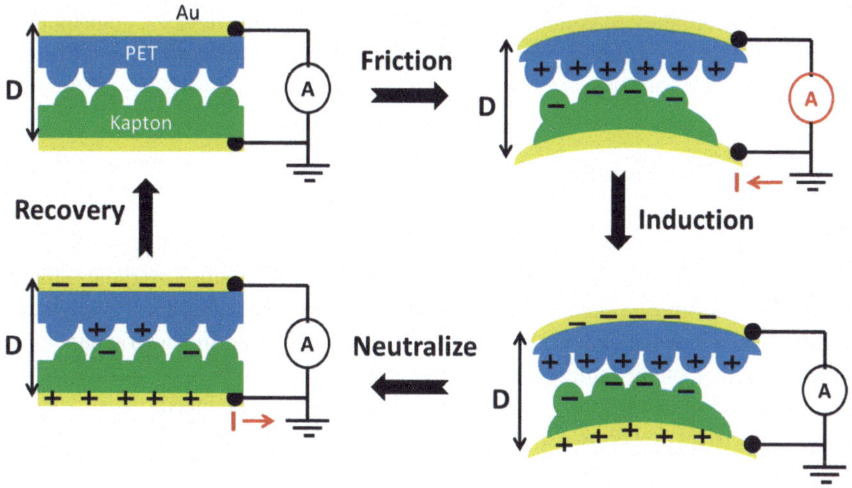

Fig. 12.7 Showing working mechanism of flexible Triboelectric Nanogenerators (TENGs)

to generate electricity from mechanical motion make them promising candidates for powering various devices and systems in both indoor and outdoor environments. By harnessing this fundamental principle, TENGs offer a sustainable, eco-friendly and versatile approach to energy harvesting, with broad applications across different industries and domains.

Triboelectric Nanogenerators (TENGs) represent a cutting-edge technology in the realm of energy harvesting, particularly in the context of space exploration. These devices capitalize on the principles of friction and contact electrification to convert mechanical energy into electrical power [173]. Several applications within the domain of space missions and colonization showcase the versatility and potential of Triboelectric Nanogenerators: TENGs exploit the phenomenon of triboelectrification, where two dissimilar materials come into contact and generate an electric charge through friction. In space environments, where mechanical interactions are prevalent, TENGs offer a means of harnessing energy from routine movements, vibrations, or contact between materials [173]. This provides a sustainable and self-sufficient power source for various devices and sensors. The flexibility and lightweight nature of TENGs make them ideal candidates for integration into space suit fabrics [174]. By incorporating these nanogenerators into the very fabric of space suits, energy can be harvested from the astronaut's movements and interactions with the suit. This approach transforms the space suit into a self-powering system, reducing the reliance on external power sources and increasing the autonomy of astronauts during extravehicular activities (EVAs) or surface explorations on other celestial bodies. Within the confines of spacecraft, where relative motion is inherent, TENGs can be strategically placed to capture energy from mechanical vibrations, structural deformations, or movements within the spacecraft. This harvested energy can then be utilized to power onboard systems, sensors, or communication devices, contributing to the overall energy efficiency and sustainability of space missions.

Thermoelectric Nanogenerators

Thermoelectric Nanogenerators also represent a cutting-edge technology with transformative potential in space exploration. Harnessing temperature gradients, these nanogenerators offer a range of applications, from spacecraft thermal management to converting waste heat into usable electrical power [175]. Thermoelectric nanogenerators capitalize on the temperature differentials inherent in space environments. In the vacuum of space, where temperature variations are pronounced, these devices generate electrical power by exploiting the Seebeck effect [175, 176]. As heat flows from one side of the nanogenerator to the other, a voltage is generated, providing a novel means of energy harvesting in the vastness of the cosmos. In the vacuum and extreme temperatures of space, efficient thermal management is critical for spacecraft. Thermoelectric nanogenerators play a dual role by not only converting temperature gradients into electrical power but also by actively participating in thermal regulation. By mitigating temperature extremes on spacecraft surfaces, these nanogenerators contribute to the overall thermal balance, preventing overheating (using heat pumps based on the peltier effect) or chilling of critical components [175]. One of the ground breaking applications of thermoelectric nanogenerators lies in their ability to convert waste heat into valuable electrical power. Spacecraft operations generate considerable heat as a by product, which is often wasted. Thermoelectric nanogenerators can efficiently capture and convert this waste heat into electricity, offering a sustainable and efficient solution for powering on-board systems (Fig. 12.8).

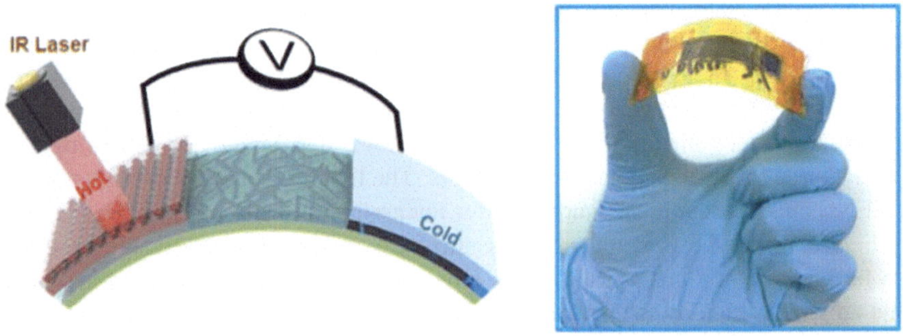

Fig. 12.8 Flexible photo-thermoelectric nanogenerator based on MoS_2/PU photothermal layer. Image reproduced with permission from [178] copyright © 2018 Elsevier Ltd

Nanomaterials Revolutionizing Space Suit Design for Colonization and Exploration

Space suits are the frontline defence for astronauts, shielding them from the harsh conditions of space, including extreme temperatures, micro-meteoroids, and radiation. The integration of nanomaterials into space suit design represents a transformative leap forward, enhancing both performance and safety for future space exploration and colonization endeavours.

Space Suit with Lightweight, High Strength, Durable Construction and Can Withstand Extreme Temperatures

Nanomaterials are at the forefront of revolutionizing space suit design for colonization and exploration, offering a host of benefits that address key challenges faced by astronauts in the harsh environment of space. One of the most significant advantages of nanomaterials is their lightweight nature, which allows for the creation of space suits that are considerably lighter than their traditional counterparts. By incorporating nanofibers, carbon nanotubes, and graphene into the fabric layers of the suit, engineers can achieve remarkable reductions in weight without compromising on strength or durability. Furthermore, nanomaterials boast exceptional strength-to-weight ratios, lower specific gravity, improved stiffness, and toughness and stable at extreme temperatures making them ideal for reinforcing space suit construction. Carbon nanotubes, in particular, exhibit tensile strengths hundreds of times greater than steel, ensuring that the suit can withstand the rigors of space travel and extravehicular activities (EVAs) without succumbing to tears or punctures. This enhanced strength not only improves the overall durability of the suit but also enhances astronaut safety by minimizing the risk of structural failures in critical moments [179]. In addition to their lightweight and high strength properties, nanomaterials offer unparalleled durability, capable of withstanding the harsh conditions encountered in space. Whether facing micrometeoroid impacts, radiation exposure, or temperature extremes, nanocomposite materials reinforced with nanoparticles/nanofibers/nanotubes/nanosheets provide robust protection against external hazards, ensuring the longevity of the space suit throughout extended missions. Nanocoatings applied to the suit's surface further enhance its resistance to radiation damage and thermal degradation, safeguarding both the integrity of the garment and the well-being of the wearer. Moreover, nanomaterials exhibit exceptional resilience to extreme temperatures, making them well-suited for space exploration missions where temperature differentials can be extreme. Graphene, known for its excellent thermal conductivity and stability, enables space suits to maintain optimal temperature regulation, even in the face of fluctuating environmental conditions. By harnessing the unique thermal properties of nanomaterials, engineers can design space suits that provide reliable protection against both heat and cold, ensuring astronaut comfort and safety throughout their missions [12, 180].

Radiation Shielding

Radiation poses a significant threat to astronauts during space missions, particularly during long-duration missions beyond low Earth orbit where exposure to galactic cosmic rays and solar radiation is heightened. Traditional space suit materials offer limited protection against these harmful radiation sources. However, nanomaterials are revolutionizing space suit design by enhancing radiation shielding capabilities. Nanomaterials, such as carbon nanotubes (CNTs) and nanocomposites, possess unique properties that make them effective radiation shields. Their dense molecular structure and high atomic number enable them to absorb and deflect incoming radiation particles, reducing the dose absorbed by astronauts. By incorporating nanomaterials into the fabric layers of space suits, engineers can enhance their radiation shielding effectiveness without significantly increasing the bulk or weight of the garment. Furthermore, nanomaterial-based coatings applied to the outer surface of space suits can provide an additional layer of protection against radiation. These coatings can reflect or scatter incoming radiation particles, further reducing the astronaut's exposure. Additionally, nanomaterials can be tailored to target specific types of radiation, offering customized protection against different sources of space radiation. The use of nanomaterials for radiation shielding not only enhances astronaut safety but also enables longer-duration space missions beyond Earth's protective magnetosphere. By mitigating the risks associated with radiation exposure, nanotechnology plays a crucial role in advancing human exploration of the cosmos and paving the way for future space colonization efforts [50, 53].

Thermal Regulation

Nanomaterials are revolutionizing space suit design by enhancing thermal regulation capabilities, crucial for astronauts' comfort and safety during space exploration. Traditional space suits struggle to maintain optimal body temperature in the extreme thermal environments of space, leading to discomfort and potential health risks. However, nanomaterials offer innovative solutions to this challenge. One approach involves incorporating phase-change materials (PCMs) into the suit's fabric. These nanocomposite materials can absorb and release thermal energy as the astronaut's surroundings change, helping to regulate body temperature within a comfortable range. Additionally, nanocoatings with tailored thermal properties can be applied to the suit's outer layers to reflect or absorb heat, depending on the environmental conditions. Furthermore, nanofibers such as carbon nanotubes (CNTs) and graphene possess high thermal conductivity, allowing for efficient heat dissipation away from the body. By integrating these nanofibers into the suit's layers, engineers can enhance heat transfer properties, preventing overheating during strenuous activities and maintaining thermal equilibrium in microgravity environments. Moreover, nanomaterial-based insulating layers can provide protection against temperature extremes, insulating the astronaut from the harsh cold of space while preventing overheating in direct sunlight. This improved thermal insulation reduces energy consumption by the suit's environmental control systems, extending battery life and enhancing overall mission sustainability [74, 75].

Advanced Sensors and Monitoring

In addition to enhancing the structural integrity and durability of space suits, nanomaterials are facilitating the integration of advanced sensors and monitoring systems into the garment itself. These sensors enable real-time monitoring of vital signs, environmental conditions, and suit integrity, providing astronauts with critical data to ensure their safety and well-being during space missions. Nanotechnology enables the development of miniature sensors capable of precise data collection in the confined space of a space suit. Nanoelectronics, such as nanowires and quantum dots, offer high sensitivity and accuracy for detecting changes in temperature, pressure, humidity, and radiation levels. These sensors can be embedded within the fabric layers of the space suit, allowing for continuous monitoring without interfering with astronauts' mobility. Furthermore, nanomaterials provide the structural framework for sensor integration, offering a lightweight and flexible platform for sensor deployment. Carbon nanotubes and graphene, with their exceptional strength and conductivity, serve as ideal substrates for sensor arrays and wiring systems within the suit. These nanomaterial-based sensors can detect physiological parameters such as heart rate, respiration rate, and body temperature, providing valuable insights into astronauts' health status in real-time. Moreover, nanomaterials enable the development of smart textiles capable of adaptive responses to environmental changes. Nanoparticles embedded in the fabric can adjust their properties in response to temperature fluctuations, providing thermal regulation and comfort to the wearer. Additionally, nanocoatings applied to the suit's surface can repel moisture and contaminants, maintaining hygiene and prolonging the suit's lifespan in space. Overall, the integration of advanced sensors and monitoring systems into space suits revolutionizes astronauts' ability to monitor their physiological status and environmental conditions during space exploration missions. Nanomaterials play a pivotal role in enabling this advancement, offering lightweight, durable, and multifunctional solutions for enhanced safety and performance in the harsh environment of space [90].

Self-Healing Materials

Innovations in nanomaterials have paved the way for the development of self-healing materials, which hold tremendous potential for revolutionizing space suit design for colonization and exploration. These materials possess the remarkable ability to autonomously repair damage caused by wear and tear, micrometeoroid impacts, or other environmental hazards encountered in space. By incorporating self-healing nanomaterials into the fabric layers of space suits, engineers can enhance the durability and longevity of the garments, reducing the need for frequent repairs or replacements during extended missions. This self-healing capability not only improves the reliability of space suits but also ensures the safety and well-being of astronauts by maintaining the integrity of their protective gear in the challenging conditions of space [180, 181]. As a result, self-healing materials represent a significant advancement in space suit technology, offering enhanced resilience and performance for future exploration endeavors beyond Earth's orbit (Fig. 12.9).

Enhanced Mobility and Flexibility

In addition to their lightweight and durable properties, nanomaterials are also revolutionizing space suit design by enhancing mobility and flexibility for astronauts during extravehicular activities (EVAs) in space. Traditional space suits, with their bulky and rigid construction, often impose limitations on astronauts' movements, hindering their ability to perform tasks efficiently in microgravity environments. However, nanomaterials such as carbon nanotubes (CNTs) and graphene offer unique advantages in terms of flexibility and mechanical properties. By incorporating these nanomaterials into the fabric layers and structural components of space suits, engineers can create garments that provide optimal mobility while maintaining the necessary level of protection against space hazards. Nanofibers, for example, can be woven into flexible textiles that allow for greater range of motion without compromising on strength or durability. This flexibility enables astronauts to bend, twist, and manoeuvre more easily during EVAs, improving their overall comfort and efficiency in space. Furthermore, nanocomposite materials reinforced with nanoparticles can enhance the elasticity and resilience of space suit components, providing astronauts with greater freedom of movement without sacrificing safety or reliability. Graphene-based materials, known for their exceptional strength and flexibility, are particularly well-suited for reinforcing joints and articulation points in space suits, ensuring optimal mobility in the challenging conditions of space. By leveraging the unique mechanical properties of nanomaterials, space suit designers can create garments that strike the perfect balance between protection and mobility, empowering astronauts to perform complex tasks with precision and agility during space exploration missions. As humanity looks towards the colonization of other celestial bodies, the

Fig. 12.9 An AI-generated image of what future space suits would look like with durable CNT-based space suits and sensors and self-repairing material (image generated using site www.canva.com)

role of nanotechnology in enhancing space suit mobility will be crucial in enabling the success of future extra-terrestrial endeavours.

Anti-Bacterial and Anti-Contamination Properties

Nanomaterials offer space suits an additional layer of protection through their inherent anti-bacterial and anti-contamination properties. These materials possess surfaces that can resist the adhesion of bacteria, viruses, and other contaminants, thereby reducing the risk of microbial growth and contamination within the suit. By incorporating nanostructured coatings or additives into space suit fabrics, engineers can create surfaces that repel microorganisms and prevent biofilm formation. This feature is crucial for maintaining astronaut health during long-duration missions, where exposure to pathogens in confined spaces can pose significant risks. Furthermore, the anti-contamination properties of nanomaterials help preserve the integrity of scientific experiments conducted in space by minimizing the introduction of foreign particles or microbes into controlled environments. Overall, the integration of nanotechnology enhances the cleanliness and hygiene of space suits, ensuring the well-being of astronauts and the success of space missions [74].

Smart Fabrics for Communication

The integration of smart fabrics enhances situational awareness and safety for astronauts operating in remote and hazardous environments. For instance, Nano sensors woven into the fabric can monitor vital signs like heart rate, body temperature, and oxygen levels, providing immediate feedback to mission control on the astronaut's health status. Additionally, conductive fibres integrated into the fabric enable wireless communication systems, allowing astronauts to communicate with each other and receive instructions from mission control without the need for bulky external devices. Furthermore, smart fabrics offer the flexibility to incorporate other communication technologies, such as antennas or RFID tags, directly into the suit's material. This seamless integration minimizes the need for additional hardware attachments, reducing weight and enhancing mobility for astronauts during EVAs. By leveraging nanotechnology-enabled smart fabrics, space suits can evolve into intelligent, interactive systems that optimize communication and safety for astronauts during space exploration and colonization endeavours (Fig. 12.10).

Self-Replicating Nanobots and Future Prospects

The concept of self-replicating nanobots, often depicted in speculative discussions about advanced nanotechnology, involves the creation of tiny robotic entities capable of reproducing themselves autonomously. While this idea remains largely theoretical and faces numerous scientific, ethical, and practical challenges, it sparks intriguing discussions about the potential future of nanotechnology. Self-replicating nanobots would theoretically possess the ability to manipulate molecules at the atomic or molecular level, allowing them to assemble copies of themselves using raw materials from their environment [179, 184, 185]. This process mimics the way biological organisms replicate, albeit at a much smaller scale. The nanobots would need to operate autonomously, making decisions about replication, resource acquisition, and functionality without direct human intervention. This autonomy could involve sophisticated programming and decision-making algorithms. To sustain their activities, self-replicating nanobots would require a means of harvesting energy from their environment [186]. This could involve extracting energy from ambient sources or

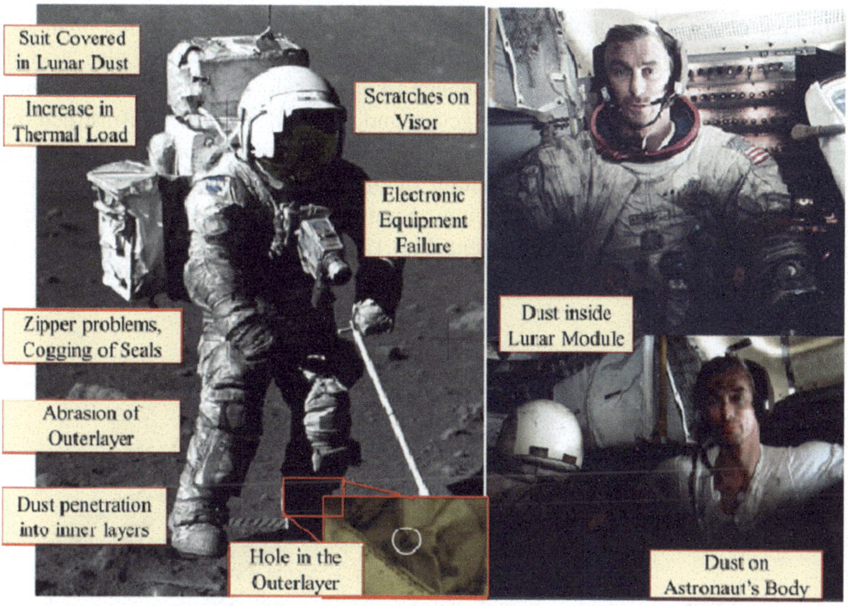

Fig. 12.10 Self-Cleaning spacesuits for lunar dust protection (reproduced with permission from [183] copyright 2018 Published by Elsevier Ltd)

utilizing specific energy-harvesting mechanisms integrated into their design. Self-replicating nanobots would need to include error correction mechanisms to ensure accurate replication [187]. Additionally, the concept suggests the possibility of introducing variations or improvements during the replication process, leading to an evolutionary aspect in their design. Sophisticated programming would be essential to control the behavior of self-replicating nanobots. This programming could include predefined tasks, communication protocols, and safety mechanisms to prevent unintended consequences [188]. The envisioned applications of self-replicating nanobots span a wide range, from medical applications (such as targeted drug delivery or tissue repair) to environmental remediation and space exploration. The theoretical concepts explore how these nanobots could fulfill specific purposes based on their programming and design.

The concept of self-replicating nanobots holds intriguing potential for transformative applications in space mining and infrastructure development. Self-replicating nanobots could be deployed to autonomously extract valuable resources from asteroids, moons, or other celestial bodies [189]. These nanobots would replicate themselves, creating a network that systematically mines and processes materials, providing a sustainable source of resources for space colonies. Self-replicating nanobots could be programmed to construct infrastructure in space. From assembling habitats to building landing pads or solar arrays, these nanobots could work collaboratively to create structures without the need for extensive human intervention. This approach could significantly reduce the costs and time associated with traditional construction methods. In space environments where repairing or maintaining infrastructure is challenging, self-replicating nanobots could play a vital role. They could detect and repair damage to spacecraft, habitats, or other structures autonomously, ensuring the longevity and reliability of space infrastructure. Self-replicating nanobots could be deployed for exploration missions, autonomously mapping and analyzing the composition of distant celestial bodies. This would provide valuable data for future mining operations and help in selecting suitable locations for infrastructure development (Fig. 12.11).

Fig. 12.11 An AI-generated image self-replicating nanobot (image generated using site www.canva.com)

Conclusion

See Table 12.3.

Nanotechnology stands as a transformative force, propelling space exploration into new frontiers and shaping the future of human endeavours beyond Earth. Throughout our discourse, we have unveiled the multifaceted role of nanotechnology in revolutionizing space exploration and colonization. Nanotechnology profoundly impacts space exploration, from spacecraft development with advanced nanomaterials to enhanced propulsion systems and miniature sensors for precise data collection. The integration of nanomaterials and their nanocomposites in spacecraft addresses structural challenges, utility in extreme environments, and weight constraints. Moreover, it enhances radiation shielding, contributing to astronaut safety on extended missions. The promise of self-replicating nanobots further extends nanotechnology's influence on space mining and infrastructure development. Nanoparticles in propulsion systems and innovative solar sails present sustainable

Table 12.3 Table summarizing various studies on the role of nanotechnology in space exploration and colonization

Study title	Key findings
Nanomaterials for lightweight spacecraft	Explored the use of nanomaterials in spacecraft construction for weight reduction and enhanced performance
Nanoengineered propulsion systems	Investigated the application of nanotechnology in developing advanced propulsion systems for deep space exploration
Nanosensors for space environmental monitoring	Developed miniature sensors based on nanotechnology for monitoring radiation levels, temperature, and atmospheric conditions in space
Nanomaterials for radiation shielding	Analyzed the effectiveness of nanomaterials in providing radiation shielding for spacecraft and habitats during long-duration space missions
Self-assembling nanobots for space exploration	Explored the potential of self-assembling nanobots in autonomously conducting tasks such as exploration, repair, and resource extraction in space environments
Nanotechnology in life support systems	Investigated the integration of nanotechnology in life support systems for recycling water, generating oxygen, and removing carbon dioxide in space habitats
Nanomedicine for astronaut health	Explored the use of nanomedicine in developing targeted drug delivery systems and medical treatments for astronauts during space missions
Nanomaterials for sustainable resource utilization	Examined the application of nanomaterials in enhancing resource utilization and recycling processes in space colonies

solutions, reshaping the dynamics of space travel. Miniaturized sensors, enabled by nanotechnology, empower spacecraft with unprecedented capabilities for data collection and imaging in the harsh conditions of space. This is important to mention here that nanotechnology extends its influence to resource utilization, life support systems, and biomedical applications, ensuring the well-being of astronauts during space exploration and colonization. Nanomaterials promise breakthroughs in energy harvesting, contribute to advanced space suit technologies, and offer solutions for the challenges posed by microgravity environments. As we peer into the future, the concept of self-replicating nanobots emerges as a revolutionary force in space mining and infrastructure development. These nanobots hold the potential to autonomously mine resources, construct structures, and adapt to changing environments, redefining the boundaries of what is achievable in the cosmos. Finally, it is concluded that the marriage of nanotechnology and space exploration is not merely a technological collaboration but a testament to human ingenuity and adaptability. It opens up possibilities, enables pioneering solutions, and empowers humanity to extend its reach into the vast unknown. The synergy between nanotechnology and space exploration propels us toward a future where the dreams of space colonization become not just aspirations but tangible realities. As we continue to unlock the potential of the materials at nanoscale, we are ushering in an era where the cosmos becomes a canvas for human innovation and the stars beckon us to explore, adapt, and thrive.

Conflicts of Interest or Competing Interests:

No conflicts of interest exist among the authors.

Acknowledgements The authors gratefully acknowledge essential resources provided by Prof. Anant Srivastava Hon.Vice Chancellor of Netaji Subhas University of Technology.

During the preparation of this work, the authors used ChatGPT in order to polish language and avoid grammatical errors. After using this tool, the authors reviewed and edited the content as needed and take full responsibility for the content of the publication. Some images were also generated using AI.

Authors Contributions: Peeyush Phogat: Conceptualization, Writing - original draft and Formal analysis. Shreya: Writing - original draft. Ranjana Jha: Supervision. Sukhvir Singh: review & editing.

References

1. Levchenko, I., Xu, S., Mazouffre, S., Keidar, M., Bazaka, K.: Mars colonization: beyond getting there. in *Terraforming Mars*, pp. 73–98 (2021)
2. Landry, K. S., Morey, J. M., Bharat, B., Haney, N. M., Panesar, S. S.: Biofilms—impacts on human health and its relevance to space travel. Microorganisms **8**(7) (2020). https://doi.org/10.3390/microorganisms8070998

3. Levchenko, I., Bazaka, K., Belmonte, T., Keidar, M., Xu, S.: Advanced materials for next-generation spacecraft. Adv. Mater. **30**(50), 1802201 (2018). https://doi.org/10.1002/adma.201802201
4. Umbrello, S., Baum, S.D.: Evaluating future nanotechnology: The net societal impacts of atomically precise manufacturing. Futures **100**, 63–73 (2018). https://doi.org/10.1016/j.futures.2018.04.007
5. Bastani, B., Fernandez, D.: Intellectual property rights in nanotechnology. Thin Solid Films **420–421**, 472–477 (2002). https://doi.org/10.1016/S0040-6090(02)00843-X
6. Aithal, P. S., Aithal, S.: Opportunities & challenges for green technologies in 21 st century, **I**(I), 818–828 (2016)
7. Bainbridge, W.S.: Motivations for space exploration. Futures **41**(8), 514–522 (2009). https://doi.org/10.1016/j.futures.2009.04.021
8. Aithal, P. S., Aithal, S.: Ideal technology concept & its realization opportunity using nanotechnology. **4**(2), 153–164 (2015)
9. Novikov, L. S., Voronina, E. N.: Potential space applications of nanomaterials BT - protection of materials and structures from the space environment, pp. 139–147 (2017)
10. Thibeault, S.A., Kang, J.H., Sauti, G., Park, C., Fay, C.C., King, G.C.: Nanomaterials for radiation shielding. MRS Bull. **40**(10), 836–841 (2015). https://doi.org/10.1557/mrs.2015.225
11. Arepalli, S., Moloney, P.: Engineered nanomaterials in aerospace. MRS Bull. **40**(10), 804–811 (2015). https://doi.org/10.1557/mrs.2015.231
12. Bellucci, S., Balasubramanian, C., Micciulla, F., Rinaldi, G.: CNT composites for aerospace applications. J. Exp. Nanosci. **2**(3), 193–206 (2007). https://doi.org/10.1080/17458080701376348
13. Samareh, J.A., Siochi, E.J.: Systems analysis of carbon nanotubes: opportunities and challenges for space applications. Nanotechnology **28**(37), 372001 (2017). https://doi.org/10.1088/1361-6528/aa7c5a
14. Watson, K. A., Connell, J. W.: Chapter 19 - polymer and carbon nanotube composites for space applications. L. B. T.-C. N. Dai, Ed. Amsterdam: Elsevier, pp. 677–698 (2006)
15. Atar, N., et al.: Atomic-oxygen-durable and electrically-conductive CNT-POSS-polyimide flexible films for space applications. ACS Appl. Mater. Interf. **7**(22), 12047–12056 (2015). https://doi.org/10.1021/acsami.5b02200
16. Siochi, E.J.: Graphene in the sky and beyond. Nat. Nanotechnol. **9**(10), 745–747 (2014). https://doi.org/10.1038/nnano.2014.231
17. Liu, X., Pu, J., Wang, L., Xue, Q.: Novel DLC/ionic liquid/graphene nanocomposite coatings towards high-vacuum related space applications. J. Mater. Chem. A **1**(11), 3797–3809 (2013). https://doi.org/10.1039/C3TA00764B
18. Kim, H., et al.: Tensile properties of millimeter-long multi-walled carbon nanotubes. Sci. Rep. **7**(1), 9512 (2017). https://doi.org/10.1038/s41598-017-10279-0
19. Montazeri, A., Javadpour, J., Khavandi, A., Tcharkhtchi, A., Mohajeri, A.: Mechanical properties of multi-walled carbon nanotube/epoxy composites. Mater. Des. **31**(9), 4202–4208 (2010). https://doi.org/10.1016/j.matdes.2010.04.018
20. Li, F., Cheng, H.M., Bai, S., Su, G., Dresselhaus, M.S.: Tensile strength of single-walled carbon nanotubes directly measured from their macroscopic ropes. Appl. Phys. Lett. **77**(20), 3161–3163 (2000). https://doi.org/10.1063/1.1324984
21. Ogata, S., Shibutani, Y.: Ideal tensile strength and band gap of single-walled carbon nanotubes. Phys. Rev. B **68**(16), 165409 (2003). https://doi.org/10.1103/PhysRevB.68.165409
22. Yu, M.-F., Files, B.S., Arepalli, S., Ruoff, R.S.: Tensile loading of ropes of single wall carbon nanotubes and their mechanical properties. Phys. Rev. Lett. **84**(24), 5552–5555 (2000). https://doi.org/10.1103/PhysRevLett.84.5552
23. Kashani, H., Ito, Y., Han, J., Liu, P., Chen, M.: Extraordinary tensile strength and ductility of scalable nanoporous graphene. Sci. Adv. **5**(2), eaat6951 (2023). https://doi.org/10.1126/sciadv.aat6951
24. Mazilova, T.I., Sadanov, E.V., Mikhailovskij, I.M.: Tensile strength of graphene nanoribbons: an experimental approach. Mater. Lett. **242**, 17–19 (2019). https://doi.org/10.1016/j.matlet.2019.01.051

25. Lee, C., Wei, X., Kysar, J. W., Hone, J.: Measurement of the elastic properties and intrinsic strength of monolayer graphene. Science (80-.). **321**(5887), 385–388 (2008). https://doi.org/10.1126/science.1157996
26. Manohara, H., Toda, R., Lin, R. H., Liao, A., Mojarradi, M.: Carbon nanotube-based digital vacuum electronics and miniature instrumentation for space exploration. in *Proc.SPIE*, vol. 7594, p. 75940Q (2010). https://doi.org/10.1117/12.844076
27. Huo, C., Liang, F., Sun, A.: Review on development of carbon nanotube field emission cathode for space propulsion systems. High Volt. **5**(4), 409–415 (2020). https://doi.org/10.1049/hve.2019.0257
28. Jang, J.G., Kim, T.-H., Kim, S.H., Hong, J.-I.: Enhancing thermoelectric performance of single-walled carbon nanotube/reduced graphene oxides composites with small organic molecules as a novel additive. New J. Chem. (2023). https://doi.org/10.1039/D3NJ03568A
29. Ma, P.-C., et al.: Enhanced electrical conductivity of nanocomposites containing hybrid fillers of carbon nanotubes and carbon black. ACS Appl. Mater. Interf. **1**(5), 1090–1096 (2009). https://doi.org/10.1021/am9000503
30. Li, Z., et al.: PMMA/MWCNT nanocomposite for proton radiation shielding applications. Nanotechnology **27**(23), 234001 (2016). https://doi.org/10.1088/0957-4484/27/23/234001
31. Acedillo, S. M. M.: Evaluation of the application of carbon nanotubes for radiation shielding. in *AIAA SPACE 2015 Conference and Exposition*, American Institute of Aeronautics and Astronautics (2015)
32. Wang, H., et al.: A material with high electromagnetic radiation shielding effectiveness fabricated using multi-walled carbon nanotubes wrapped with poly(ether sulfone) in a poly(ether ether ketone) matrix. J. Mater. Chem. **22**(39), 21232–21237 (2012). https://doi.org/10.1039/C2JM35129C
33. Cha, J.-H., Jang, W.-H., Sarath Kumar, S. K., Noh, J.-E., Choi, J.-S., Kim, C.-G.: Functionalized multi-walled carbon nanotubes/hydrogen-rich benzoxazine nanocomposites for cosmic radiation shielding with enhanced mechanical properties and space environment resistance. Compos. Sci. Technol. **228**, 109634 (2022). https://doi.org/10.1016/j.compscitech.2022.109634
34. Miller, K.J.: Materials science perspective of metal fatigue resistance. Mater. Sci. Technol. **9**(6), 453–462 (1993). https://doi.org/10.1179/mst.1993.9.6.453
35. Lawrence, F.V., Ho, N.J., Mazumdar, P.K.: Predicting the fatigue resistance of welds. Annu. Rev. Mater. Sci. **11**(1), 401–425 (1981). https://doi.org/10.1146/annurev.ms.11.080181.002153
36. Sun, L., Gibson, R.F., Gordaninejad, F., Suhr, J.: Energy absorption capability of nanocomposites: a review. Compos. Sci. Technol. **69**(14), 2392–2409 (2009). https://doi.org/10.1016/j.compscitech.2009.06.020
37. Ingrole, A., Aguirre, T.G., Fuller, L., Donahue, S.W.: Bioinspired energy absorbing material designs using additive manufacturing. J. Mech. Behav. Biomed. Mater. **119**, 104518 (2021). https://doi.org/10.1016/j.jmbbm.2021.104518
38. McKittrick, J., et al.: Energy absorbent natural materials and bioinspired design strategies: a review. Mater. Sci. Eng. C **30**(3), 331–342 (2010). https://doi.org/10.1016/j.msec.2010.01.011
39. Meziani, M.J., et al.: Boron nitride nanomaterials for thermal management applications. ChemPhysChem **16**(7), 1339–1346 (2015). https://doi.org/10.1002/cphc.201402814
40. Mebarek-Oudina, F., Chabani, I.: Review on nano enhanced PCMs: insight on nePCM application in thermal management/storage systems. Energies **16**(3) (2023). https://doi.org/10.3390/en16031066
41. He, M., Zhao, B., Yue, X., Chen, Y., Qiu, F., Zhang, T.: Infrared radiative modulating textiles for personal thermal management: principle, design and application. Nano Energy **116**, 108821 (2023). https://doi.org/10.1016/j.nanoen.2023.108821
42. Al-kayiem, H.H., Lin, S.C., Lukmon, A.: Review on Nanomaterials for thermal energy storage technologies, pp. 60–71 (2013)
43. Peters, M., Kumpfert, J., Ward, C.H., Leyens, C.: Titanium alloys for aerospace applications. Adv. Eng. Mater. **5**(6), 419–427 (2003). https://doi.org/10.1002/adem.200310095

44. Anil Kumar, V., Gupta, R.K., Prasad, M.J.N.V, Narayana Murty, S.V.S.: Recent advances in processing of titanium alloys and titanium aluminides for space applications: a review. J. Mater. Res. **36**(3), 689–716 (2021). https://doi.org/10.1557/s43578-021-00104-w
45. Siva Kumar, C., Mayanna, S.M., Mahendra, K.N., Sharma, A.K., Uma Rani, R.: Studies on white anodizing on aluminum alloy for space applications. Appl. Surf. Sci. **151**(3), 280–286 (1999). https://doi.org/10.1016/S0169-4332(99)00290-1
46. Sharma, A.K.: Anodizing titanium for space applications. Thin Solid Films **208**(1), 48–54 (1992). https://doi.org/10.1016/0040-6090(92)90946-9
47. Hilpert, E., Hartung, J., von Lukowicz, H., Herffurth, T., Heidler, N.: Design, additive manufacturing, processing, and characterization of metal mirror made of aluminum silicon alloy for space applications. Opt. Eng. **58**(9), 92613 (2019). https://doi.org/10.1117/1.OE.58.9.092613
48. Prasad, S.G., Parthasaradhi, K., Bloomer, W.D., Al-Najjar, W.H., McMahon, J., Thomson, O.: Aluminum, copper, tin and lead as shielding materials in the treatment of cancer with high-energy electrons. Radiat. Phys. Chem. **53**(4), 361–366 (1998). https://doi.org/10.1016/S0969-806X(98)00130-3
49. Dean, P.J., Lightowlers, E.C., Wight, D.R.: Intrinsic and extrinsic recombination radiation from natural and synthetic aluminum-doped diamond. Phys. Rev. **140**(1A), A352–A368 (1965). https://doi.org/10.1103/PhysRev.140.A352
50. Shin, J.W., et al.: Polyethylene/boron-containing composites for radiation shielding. Thermochim. Acta **585**, 5–9 (2014). https://doi.org/10.1016/j.tca.2014.03.039
51. El-Sayed Abdo, A., Ali, M.A.M., Ismail, M.R.: Natural fibre high-density polyethylene and lead oxide composites for radiation shielding. Radiat. Phys. Chem. **66**(3), 185–195 (2003). https://doi.org/10.1016/S0969-806X(02)00470-X
52. Aldhuhaibat, M.J.R., Amana, M.S., Aboud, H., Salim, A.A.: Radiation attenuation capacity improvement of various oxides via high density polyethylene composite reinforcement. Ceram. Int. **48**(17), 25011–25019 (2022). https://doi.org/10.1016/j.ceramint.2022.05.154
53. Yi, X., Shen, M., Liu, X., Gu, J.: Emerging strategies based on nanomaterials for ionizing radiation-optimized drug treatment of cancer. Nanoscale **13**(33), 13943–13961 (2021). https://doi.org/10.1039/D1NR03034E
54. Güven, O.: Radiation-assisted synthesis of polymer-based nanomaterials. Appl. Sci. **11**(17) (2021). https://doi.org/10.3390/app11177913
55. Jackson, P. et al.: Bioaccumulation and ecotoxicity of carbon nanotubes, pp. 1–21 (2013)
56. Phogat, P., Shreya, R. Jha, Singh, S.: Electrochemical analysis of thermally treated two dimensional zinc sulphide hexagonal nano-sheets with reduced band gap. Phys. Scr. **98**(12), 125962 (2023). https://doi.org/10.1088/1402-4896/ad0d93
57. Kim, H.S., Kim, S.H.: Additive manufacturing and combustion characteristics of polyethylene oxide/aluminum/copper oxide-based energetic nanocomposites for enhancing the propulsion of small projectiles. Nanomaterials **13**(6) (2023). https://doi.org/10.3390/nano13061052
58. David, A.O., Knoll, A.K.: Experimental demonstration of an aluminum-fueled propulsion system for cubesat applications. J. Propuls. Power **33**(5), 1320–1324 (2017). https://doi.org/10.2514/1.B36330
59. Maggi, F., Dossi, S., Paravan, C., DeLuca, L.T., Liljedahl, M.: Activated aluminum powders for space propulsion. Powder Technol. **270**, 46–52 (2015). https://doi.org/10.1016/j.powtec.2014.09.048
60. Ingenito, A., Bruno, C.: Using aluminum for space propulsion. J. Propuls. Power **20**(6), 1056–1063 (2004). https://doi.org/10.2514/1.5132
61. Gany, A., Timnat, Y.M.: Advantages and drawbacks of boron-fueled propulsion. Acta Astronaut. **29**(3), 181–187 (1993). https://doi.org/10.1016/0094-5765(93)90047-Z
62. Li, T., et al.: Denary oxide nanoparticles as highly stable catalysts for methane combustion. Nat. Catal. **4**(1), 62–70 (2021). https://doi.org/10.1038/s41929-020-00554-1
63. Singh, A.V., et al.: Emerging application of nanorobotics and artificial intelligence to cross the BBB: advances in design, controlled maneuvering, and targeting of the barriers. ACS Chem. Neurosci. **12**(11), 1835–1853 (2021). https://doi.org/10.1021/acschemneuro.1c00087

64. Colombo, C., Miguel, N., Gkolias, I.: Modulating solar sail control for end-of-life disposal with solar sails. Int. Symp. Sol. Sail. (2019)
65. Fleury-Frenette, K.: 61st International Astronautical Congress 2010, IAC 2010. Université de Liège - ULiège > CSL (Centre Spatial de Liège) (2010)
66. Ullery, D.C., et al.: Strong solar radiation forces from anomalously reflecting metasurfaces for solar sail attitude control. Sci. Rep. **8**(1), 10026 (2018). https://doi.org/10.1038/s41598-018-28133-2
67. Santoli, S.: Carbon nanotube membrane solar sails a challenge for extremely fast space flight. J. M. Marulanda, Ed. Rijeka: IntechOpen, p. Ch. 30 (2010)
68. Scalia, T., Bonventre, L., Terranova, M.L.: From protosolar space to space exploration: the role of graphene in space technology and economy. Nanomaterials **13**(4) (2023). https://doi.org/10.3390/nano13040680
69. Phogat, P., Shreya, R. Jha, Singh, S.: Impedance study of zinc sulphide quantum dots via one step green synthesis. Mater. Sci. Forum. **1099**, 119–125 (2023). https://doi.org/10.4028/p-G1CCxq
70. Shreya, P. Phogat, R. Jha, S. Singh.: Elevated refractive index of MoS2 amorphous nanoparticles with a reduced band gap applicable for optoelectronics BT - recent advances in mechanical engineering. in *Recent Advances in Mechanical Engineering*, B. Sethuraman, P. Jain, and M. Gupta, Eds. Singapore: Springer Nature Singapore, pp. 431–439 (2023)
71. P. Phogat, Shreya, R. Jha, S. Singh.: Optical and microstructural study of wide band Gap ZnO@ZnS core--shell nanorods to be used as solar cell applications. in *Recent Advances in Mechanical Engineering*, pp. 419–429 (2023)
72. Shreya, P. Phogat, Jha, R., Singh, S.: Microwave-synthesized γ-WO3 nanorods exhibiting high current density and diffusion characteristics. Transit. Met. Chem. **48**(3), 167–183 (2023). https://doi.org/10.1007/s11243-023-00533-y
73. Phogat, P., Shreya, Jha, R., Singh, S.: Diffusion controlled features of microwave assisted ZnS/ZnO nanocomposite with reduced band gap. ECS J. Solid State Sci. Technol. **12**(3), 034004 (2023). https://doi.org/10.1149/2162-8777/acc426
74. Sharma, S., Phogat, P., Jha, R., Singh, S.: Electrochemical and optical properties of microwave assisted MoS 2 nanospheres for solar cell application **12**(3), 66–72 (2023). https://doi.org/10.12720/sgce.12.3.66-72
75. Kumar, T., Shreya, P. Phogat, Sahgal, V., Jha, R.: Surfactant-mediated modulation of morphology and charge transfer dynamics in tungsten oxide nanoparticles. Phys. Scr. **98**(8), 085936 (2023). https://doi.org/10.1088/1402-4896/ace566
76. Dipti, P. Phogat, Shreya, D. Kumari, Singh, S.: Fabrication of tunable band gap carbon based zinc nanocomposites for enhanced capacitive behaviour. Phys. Scr. **98**(9), 95030 (2023). https://doi.org/10.1088/1402-4896/acf07b
77. Cerofolini, G.F., et al.: Strategies for nanoelectronics. Microelectron. Eng. **81**(2), 405–419 (2005). https://doi.org/10.1016/j.mee.2005.03.041
78. Robertson, C., Shipton, R.D., Gray, D.R.: Miniature sensors using high density screen printing. Sens. Rev. **19**(1), 33–36 (1999). https://doi.org/10.1108/02602289910255559
79. Urban, M., Nentvich, O., Stehlikova, V., Baca, T., Daniel, V., Hudec, R.: VZLUSAT-1: nanosatellite with miniature lobster eye X-ray telescope and qualification of the radiation shielding composite for space application. Acta Astronaut. **140**, 96–104 (2017). https://doi.org/10.1016/j.actaastro.2017.08.004
80. Guo, P., et al.: Porphyrin nanoassemblies via surfactant-assisted assembly and single nanofiber nanoelectronic sensors for high-performance H2O2 vapor sensing. ACS Nano **8**(4), 3402–3411 (2014). https://doi.org/10.1021/nn406071f
81. Shashurin, V.D., Vetrova, N.A., Pchelintsev, K.P., Kuimov, E.V., Meshkov, S.A.: Designing radio electronic systems for space purposes optimal by the criterion of reliability based on ultra-high-speed heterostructure nanoelectronics devices. AIP Conf. Proc. **2171**(1), 150004 (2019). https://doi.org/10.1063/1.5133302
82. Santoli, S.: Hyper-interspersed nano/MEMS-architecture design for new concepts in miniature robotics for space exploration. Acta Astronaut. **44**(2), 117–122 (1999). https://doi.org/10.1016/S0094-5765(99)00037-5

83. Ziegler, M.M., Stan, M.R.: Design and analysis of crossbar circuits for molecular nanoelectronics. in Proceedings of the 2nd IEEE Conference on Nanotechnology, pp. 323–327. https://doi.org/10.1109/NANO.2002.1032256 (2002)
84. Sansone, F., Francesconi, A., Olivieri, L., Branz, F.: Low-cost relative navigation sensors for miniature spacecraft and drones. IEEE Metrology for Aerospace (MetroAeroSpace) **2015**, 389–394 (2015). https://doi.org/10.1109/MetroAeroSpace.2015.7180688
85. Oyubu, O.A., Kazeem, O.U.: An overview of nanoelectronics and nanodevices. J. Eng. Stud. Res. **26**(3), 165–172 (2020). https://doi.org/10.29081/jesr.v26i3.220
86. Haselman, M., Hauck, S.: The future of integrated circuits: a survey of nanoelectronics. Proc. IEEE **98**(1), 11–38 (2010). https://doi.org/10.1109/JPROC.2009.2032356
87. Chou, S.: Integration and innovation in the nanoelectronics era. in ISSCC. 2005 IEEE International Digest of Technical Papers. Solid-State Circuits Conference, pp. 36–41, vol. 1 (2005). https://doi.org/10.1109/ISSCC.2005.1493858
88. Wang, X., Yang, L., (Georgia) Chen, Z., Shin, D.M.: Application of nanotechnology in cancer therapy and imaging. CA. Cancer J. Clin. **58**(2), 97–110 (2008). https://doi.org/10.3322/CA.2007.0003
89. Brongersma, M.L., Zia, R., Schuller, J.A.: Plasmonics – the missing link between nanoelectronics and microphotonics. Appl. Phys. A **89**(2), 221–223 (2007). https://doi.org/10.1007/s00339-007-4151-1
90. Fountain, T.J., Duff, M.J.B., Crawley, D.G., Tomlinson, C.D., Moffat, C.D.: The use of nanoelectronic devices in highly parallel computing systems. IEEE Trans. Very Large Scale Integr. Syst. **6**(1), 31–38 (1998). https://doi.org/10.1109/92.661242
91. Hartshorn, C.M., et al.: Nanotechnology strategies to advance outcomes in clinical cancer care. ACS Nano **12**(1), 24–43 (2018). https://doi.org/10.1021/acsnano.7b05108
92. Yu, B., Meyyappan, M.: Nanotechnology: role in emerging nanoelectronics. Solid State Electron. **50**(4), 536–544 (2006). https://doi.org/10.1016/j.sse.2006.03.028
93. Lemmer, K.: Propulsion for CubeSats. Acta Astronaut. **134**, 231–243 (2017). https://doi.org/10.1016/j.actaastro.2017.01.048
94. Selva, D., Krejci, D.: A survey and assessment of the capabilities of cubesats for earth observation. Acta Astronaut. **74**, 50–68 (2012). https://doi.org/10.1016/j.actaastro.2011.12.014
95. Shiroma, W.A., et al.: CubeSats: a bright future for nanosatellites. Cent. Eur. J. Eng. **1**(1), 9–15 (2011). https://doi.org/10.2478/s13531-011-0007-8
96. Dahman, Y.: Chapter 6 - Nanopolymers**By Yaser Dahman, Kevin Deonanan, Timothy Dontsos, and Andrew Iammatteo. in *Micro and Nano Technologies*, Y. B. T.-N. and F. M. for E. Dahman, Ed. Elsevier, pp. 121–144 (2017)
97. L.P.J.A., Fei-Yue, W.: Intelligent excavator control system for lunar mining system. J. Aerosp. Eng. **8**(10), 16–24 (1995). https://doi.org/10.1061/(ASCE)0893-1321(1995)8:1(16)
98. Thangavelautham, J., Chandra, A., Jensen, E.: Autonomous robot teams for lunar mining base construction and operation. IEEE Aeros. Conf. **2020**, 1–16 (2020). https://doi.org/10.1109/AERO47225.2020.9172811
99. Papike, J.J., Simon, S.B., Laul, J.C.: The lunar regolith: chemistry, mineralogy, and petrology. Rev. Geophys. **20**(4), 761–826 (1982). https://doi.org/10.1029/RG020i004p00761
100. Langevin, Y., Arnold, J.R.: The evolution of the lunar regolith. Annu. Rev. Earth Planet. Sci. **5**(1), 449–489 (1977). https://doi.org/10.1146/annurev.ea.05.050177.002313
101. Klenner, F., et al.: Analogue spectra for impact ionization mass spectra of water ice grains obtained at different impact speeds in space. Rapid Commun. Mass Spectrom. **33**(22), 1751–1760 (2019). https://doi.org/10.1002/rcm.8518
102. Poch, O., Pommerol, A., Jost, B., Carrasco, N., Szopa, C., Thomas, N.: Sublimation of water ice mixed with silicates and tholins: Evolution of surface texture and reflectance spectra, with implications for comets. Icarus **267**, 154–173 (2016). https://doi.org/10.1016/j.icarus.2015.12.017
103. Deep Space Industries: Asteroid Mining, no. 384 (2016). https://deepspaceindustries.com/mining/

104. Hein, A.M., Saidani, M., Tollu, H.: Exploring potential environmental benefits of asteroid mining. Proc. Int. Astronaut. Congr. IAC, vol. 2018-Octob, pp. 1–7 (2018)
105. Andrews, D.G., et al.: Defining a successful commercial asteroid mining program. Acta Astronaut. **108**, 106–118 (2015). https://doi.org/10.1016/j.actaastro.2014.10.034
106. Zacny, K., Cohen, M.M., James, W.W., Hilscher, B.: Asteroid mining. in *AIAA SPACE 2013 Conference and Exposition*, American Institute of Aeronautics and Astronautics (2013)
107. Hein, A.M., Matheson, R., Fries, D.: A techno-economic analysis of asteroid mining. Acta Astronaut. **168**, 104–115 (2020). https://doi.org/10.1016/j.actaastro.2019.05.009
108. Muscatello, A., Santiago-Maldonado, E.: Mars in situ resource utilization technology evaluation. in *50th AIAA Aerospace Sciences Meeting including the New Horizons Forum and Aerospace Exposition*, American Institute of Aeronautics and Astronautics (2012)
109. Chen, H., Sarton du Jonchay, T., Hou, L., Ho, K.: Integrated in-situ resource utilization system design and logistics for Mars exploration. Acta Astronaut. **170**, 80–92 (2020). https://doi.org/10.1016/j.actaastro.2020.01.031
110. Starr, S.O., Muscatello, A.C.: Mars in situ resource utilization: a review. Planet. Space Sci. **182**, 104824 (2020). https://doi.org/10.1016/j.pss.2019.104824
111. Sridhar, K.R., Finn, J.E., Kliss, M.H.: In-situ resource utilization technologies for Mars life support systems. Adv. Sp. Res. **25**(2), 249–255 (2000). https://doi.org/10.1016/S0273-1177(99)00955-2
112. S.G.B., L.W.E.: Progress made in lunar in situ resource utilization under NASA's exploration technology and development program. J. Aerosp. Eng. **26**(1), 5–17. https://doi.org/10.1061/(ASCE)AS.1943-5525.0000208 (2013)
113. Anand, M., et al.: A brief review of chemical and mineralogical resources on the Moon and likely initial in situ resource utilization (ISRU) applications. Planet. Space Sci. **74**(1), 42–48 (2012). https://doi.org/10.1016/j.pss.2012.08.012
114. Ellery, A.: Sustainable in-situ resource utilization on the moon. Planet. Space Sci. **184**, 104870 (2020). https://doi.org/10.1016/j.pss.2020.104870
115. Chadha, U. et al.: A review of the function of using carbon nanomaterials in membrane filtration for contaminant removal from wastewater. Mater. Res. Express **9**(1) (2022). https://doi.org/10.1088/2053-1591/ac48b8
116. Liu, G., Jin, W., Xu, N.: Graphene-based membranes. Chem. Soc. Rev. **44**(15), 5016–5030 (2015). https://doi.org/10.1039/C4CS00423J
117. Rashid, M.H.-O., Ralph, S.F.: Carbon nanotube membranes: synthesis, properties, and future filtration applications. Nanomaterials **7**(5) (2017). https://doi.org/10.3390/nano7050099
118. Aghigh, A., Alizadeh, V., Wong, H.Y., Islam, M.S., Amin, N., Zaman, M.: Recent advances in utilization of graphene for filtration and desalination of water: a review. Desalination **365**, 389–397 (2015). https://doi.org/10.1016/j.desal.2015.03.024
119. Faccini, M. et al.: Electrospun carbon nanofiber membranes for filtration of nanoparticles from water. J. Nanomater. **2015** (2015). https://doi.org/10.1155/2015/247471
120. Bowen, W.R., Mukhtar, H.: Characterisation and prediction of separation performance of nanofiltration membranes. J. Memb. Sci. **112**(2), 263–274 (1996). https://doi.org/10.1016/0376-7388(95)00302-9
121. Bandehali, S., Parvizian, F., Moghadassi, A., Hosseini, S.M.: Chapter 5 - Nanomaterials for the efficient abatement of wastewater contaminants by means of reverse osmosis and nanofiltration. in *Micro and Nano Technologies*, B. Bonelli, F. S. Freyria, I. Rossetti, and R. B. T.-N. for the D. and R. of W. P. Sethi, Eds. Elsevier, pp. 111–144 (2020)
122. Khan, A.A., et al.: Metal oxide and carbon nanomaterial based membranes for reverse osmosis and membrane distillation: a comparative review. Environ. Res. **202**, 111716 (2021). https://doi.org/10.1016/j.envres.2021.111716
123. Saleem, H., Zaidi, S.J.: Nanoparticles in reverse osmosis membranes for desalination: a state of the art review. Desalination **475**, 114171 (2020). https://doi.org/10.1016/j.desal.2019.114171
124. Van der Bruggen, B., Mänttäri, M., Nyström, M.: Drawbacks of applying nanofiltration and how to avoid them: a review. Sep. Purif. Technol. **63**(2), 251–263 (2008). https://doi.org/10.1016/j.seppur.2008.05.010

125. Wang, C., et al.: Silk nanofibers as high efficient and lightweight air filter. Nano Res. **9**(9), 2590–2597 (2016). https://doi.org/10.1007/s12274-016-1145-3
126. Wang, J., Kim, S.C., Pui, D.Y.H.: Investigation of the figure of merit for filters with a single nanofiber layer on a substrate. J. Aerosol Sci. **39**(4), 323–334 (2008). https://doi.org/10.1016/j.jaerosci.2007.12.003
127. Tang, X., Wang, X., Yang, L., Zhang, Z., Luo, J., Wang, J.: Multifunctional nickel nanofiber for effective air purification: PM removal and NO reduction from automobile exhaust. J. Mater. Sci. **55**(14), 6161–6171 (2020). https://doi.org/10.1007/s10853-020-04383-2
128. Zhang, Q., Welch, J., Park, H., Wu, C.-Y., Sigmund, W., Marijnissen, J.C.M.: Improvement in nanofiber filtration by multiple thin layers of nanofiber mats. J. Aerosol Sci. **41**(2), 230–236 (2010). https://doi.org/10.1016/j.jaerosci.2009.10.001
129. Anand, M.: Lunar water: a brief review. Earth Moon Planets **107**(1), 65–73 (2010). https://doi.org/10.1007/s11038-010-9377-9
130. Barker, D.C.: Lunar and off earth resource drivers, estimations and the development conundrum. Adv. Sp. Res. **66**(2), 359–377 (2020). https://doi.org/10.1016/j.asr.2020.04.001
131. Haym, B., Sohrob, M., Zachary, P.: Magnesium as an ISRU-derived resource for lunar structures. Earth and Space, pp. 175–182 (2012). https://doi.org/10.1061/9780784412190.020
132. Crawford, I.A.: Lunar resources: a review. Prog. Phys. Geogr. Earth Environ. **39**(2), 137–167 (2015). https://doi.org/10.1177/0309133314567585
133. Cannon, K.M., Britt, D.T.: A geologic model for lunar ice deposits at mining scales. Icarus **347**, 113778 (2020). https://doi.org/10.1016/j.icarus.2020.113778
134. Hasnain, Z., Lamb, C.A., Ross, S.D.: Capturing near-earth asteroids around earth. Acta Astronaut. **81**(2), 523–531 (2012). https://doi.org/10.1016/j.actaastro.2012.07.029
135. Simakova, E., Coenen, C.: Visions, hype, and expectations: a place for responsibility. in *Responsible Innovation*, pp. 241–267 (2013)
136. Gertsch, R., Gertsch, L.S., Remo, J.L.: Mining near-earth resources. Ann. N. Y. Acad. Sci. **822**(1), 511–537 (1997). https://doi.org/10.1111/j.1749-6632.1997.tb48362.x
137. Dahl, C., Gilbert, B., Lange, I.: Mineral scarcity on earth: are asteroids the answer. Miner. Econ. **33**(1), 29–41 (2020). https://doi.org/10.1007/s13563-020-00231-6
138. Gayen, P., Sankarasubramanian, S., Ramani, V.K.: Fuel and oxygen harvesting from Martian regolithic brine. Proc. Natl. Acad. Sci. **117**(50), 31685–31689 (2020). https://doi.org/10.1073/pnas.2008613117
139. Fischer, E., Martínez, G.M., Rennó, N.O.: Formation and persistence of brine on mars: experimental simulations throughout the diurnal cycle at the phoenix landing site. Astrobiology **16**(12), 937–948 (2016). https://doi.org/10.1089/ast.2016.1525
140. Arvidson, R.E. et al.: Spirit mars rover mission: overview and selected results from the northern Home Plate Winter Haven to the side of Scamander crater. J. Geophys. Res. Planets **115**(E7) (2010). https://doi.org/10.1029/2010JE003633
141. Feldman W.C. et al.: Global distribution of near-surface hydrogen on Mars. J. Geophys. Res. Planets **109**(E9) (2004). https://doi.org/10.1029/2003JE002160
142. Kornuta, D., et al.: Commercial lunar propellant architecture: a collaborative study of lunar propellant production. REACH **13**, 100026 (2019). https://doi.org/10.1016/j.reach.2019.100026
143. Moore, K.R., et al.: The out-of-this-world hype cycle: progression towards sustainable terrestrial resource production. Resour. Conserv. Recycl. **186**, 106519 (2022). https://doi.org/10.1016/j.resconrec.2022.106519
144. Ambrosi, R.M., et al.: European radioisotope thermoelectric generators (RTGs) and radioisotope heater units (RHUs) for space science and exploration. Sp. Sci. Rev. (2019). https://doi.org/10.1007/s11214-019-0623-9
145. Ritz, F., Peterson, C.E.: Multi-mission radioisotope thermoelectric generator (MMRTG) program overview. in *2004 IEEE Aerospace Conference Proceedings (IEEE Cat. No.04TH8720)*, vol. 5, pp. 2950–2957 (2004). https://doi.org/10.1109/AERO.2004.1368101
146. Holgate, T.C., Bennett, R., Hammel, T., Caillat, T., Keyser, S., Sievers, B.: Increasing the efficiency of the multi-mission radioisotope thermoelectric generator. J. Electron. Mater. **44**(6), 1814–1821 (2015). https://doi.org/10.1007/s11664-014-3564-9

147. O'Brien, R.C., Ambrosi, R.M., Bannister, N.P., Howe, S.D., Atkinson, H.V.: Safe radioisotope thermoelectric generators and heat sources for space applications. J. Nucl. Mater. **377**(3), 506–521 (2008). https://doi.org/10.1016/j.jnucmat.2008.04.009
148. B.S.K.: Nuclear thermal propulsion: past accomplishments, present efforts, and a look ahead. J. Aerosp. Eng., **26**(2), 334–342 (2013). https://doi.org/10.1061/(ASCE)AS.1943-5525.0000313
149. Kim, T., Houts, M.G.: NASA's nuclear thermal propulsion project. in *AIAA SPACE 2015 Conference and Exposition*, American Institute of Aeronautics and Astronautics (2015)
150. Gabrielli, R.A., Herdrich, G.: Review of nuclear thermal propulsion systems. Prog. Aerosp. Sci. **79**, 92–113 (2015). https://doi.org/10.1016/j.paerosci.2015.09.001
151. Borowski, S.K., McCurdy, D.R., Packard, T.W.: Nuclear thermal propulsion (NTP): a proven growth technology for human NEO/Mars exploration missions. IEEE Aerospace Conf. **2012**, 1–20 (2012). https://doi.org/10.1109/AERO.2012.6187301
152. Ludewig, H., Powell, J.R., Todosow, M., Maise, G., Barletta, R., Schweitzer, D.G.: Design of particle bed reactors for the space nuclear thermal propulsion program. Prog. Nucl. Energy **30**(1), 1–65 (1996). https://doi.org/10.1016/0149-1970(95)00080-4
153. Wakihara, M.: Recent developments in lithium ion batteries. Mater. Sci. Eng. R Reports **33**(4), 109–134 (2001). https://doi.org/10.1016/S0927-796X(01)00030-4
154. Kim, T., Song, W., Son, D.-Y., Ono, L.K., Qi, Y.: Lithium-ion batteries: outlook on present, future, and hybridized technologies. J. Mater. Chem. A **7**(7), 2942–2964 (2019). https://doi.org/10.1039/C8TA10513H
155. Yang, J., et al.: In situ deposition of platinum nanoparticles on bacterial cellulose membranes and evaluation of PEM fuel cell performance. Electrochim. Acta **54**(26), 6300–6305 (2009). https://doi.org/10.1016/j.electacta.2009.05.073
156. Rao, C.V., Viswanathan, B.: Monodispersed platinum nanoparticle supported carbon electrodes for hydrogen oxidation and oxygen reduction in proton exchange membrane fuel cells. J. Phys. Chem. C **114**(18), 8661–8667 (2010). https://doi.org/10.1021/jp101481g
157. Guo, K.W.: Green nanotechnology of trends in future energy: a review. Int. J. Energy Res. **36**(1), 1–17 (2012). https://doi.org/10.1002/er.1928
158. Karthik Pandiyan, G., Prabaharan, T.: Implementation of nanotechnology in fuel cells. Mater. Today Proc. **33**, 2681–2685 (2020). https://doi.org/10.1016/j.matpr.2020.01.368
159. Dahiya, J., Phogat, P., Hooda, A., Khasa, S.: Investigations of Praseodymium doped LiF-ZnO-Bi2O3-B2O3 glass matrix for photonic applications. AIP Conf. Proc. **2995**(1), 20065 (2024). https://doi.org/10.1063/5.0178197
160. Shah, R., Zhang, X., Talapatra, S.: Electrochemical double layer capacitor electrodes using aligned carbon nanotubes grown directly on metals. Nanotechnology **20**(39), 395202 (2009). https://doi.org/10.1088/0957-4484/20/39/395202
161. Sarno, M.: Chapter 22 - Nanotechnology in energy storage: the supercapacitors. in *Catalysis, Green Chemistry and Sustainable Energy*, vol. 179, A. Basile, G. Centi, M. De Falco, and G. B. T.-S. in S. S. and C. Iaquaniello, Eds. Elsevier, pp. 431–458 (2020)
162. Goikolea, E., Mysyk, R.: Chapter four - nanotechnology in electrochemical capacitors. in *Micro and Nano Technologies*, L. M. Rodriguez-Martinez and N. B. T.-E. N. in R. E. S. S. Omar, Eds. Boston: Elsevier, pp. 131–169 (2017)
163. Wei, L., Tian, K., Zhang, X., Jin, Y., Shi, T., Guo, X.: 3D porous hierarchical microspheres of activated carbon from nature through nanotechnology for electrochemical double-layer capacitors. ACS Sustain. Chem. Eng. **4**(12), 6463–6472 (2016). https://doi.org/10.1021/acssuschemeng.6b01227
164. Deng, W., Zhou, Y., Libanori, A., Chen, G., Yang, W., Chen, J.: Piezoelectric nanogenerators for personalized healthcare. Chem. Soc. Rev. **51**(9), 3380–3435 (2022). https://doi.org/10.1039/D1CS00858G
165. Xu, Q., Wen, J., Qin, Y.: Development and outlook of high output piezoelectric nanogenerators. Nano Energy **86**, 106080 (2021). https://doi.org/10.1016/j.nanoen.2021.106080
166. Briscoe, J., et al.: Measurement techniques for piezoelectric nanogenerators. Energy Environ. Sci. **6**(10), 3035–3045 (2013). https://doi.org/10.1039/C3EE41889H

167. Wang, X.: Piezoelectric nanogenerators—Harvesting ambient mechanical energy at the nanometer scale. Nano Energy **1**(1), 13–24 (2012). https://doi.org/10.1016/j.nanoen.2011.09.001
168. Wang, Z.L., Song, J.: Piezoelectric nanogenerators based on zinc oxide nanowire arrays. Science (80-.) **312**(5771), 242–246 (2006). https://doi.org/10.1126/science.1124005
169. Lu, L., Ding, W., Liu, J., Yang, B.: Flexible PVDF based piezoelectric nanogenerators. Nano Energy **78**, 105251 (2020). https://doi.org/10.1016/j.nanoen.2020.105251
170. Hu, D., Yao, M., Fan, Y., Ma, C., Fan, M., Liu, M.: Strategies to achieve high performance piezoelectric nanogenerators. Nano Energy **55**, 288–304 (2019). https://doi.org/10.1016/j.nanoen.2018.10.053
171. Briscoe, J., Dunn, S.: Piezoelectric nanogenerators – a review of nanostructured piezoelectric energy harvesters. Nano Energy **14**, 15–29 (2015). https://doi.org/10.1016/j.nanoen.2014.11.059
172. Dahman, Y.: Strategies to achieve high performance piezoelectric nanogenerators. in *Micro and Nano Technologies*, Y. B. T.-N. and F. M. for E. Dahman, Ed. Elsevier, pp. 121–144 (2017)
173. Wang, Z.L.: Triboelectric nanogenerator (TENG)—sparking an energy and sensor revolution. Adv. Energy Mater. **10**(17), 2000137 (2020). https://doi.org/10.1002/aenm.202000137
174. Wu, C., Wang, A.C., Ding, W., Guo, H., Wang, Z.L.: Triboelectric nanogenerator: a foundation of the energy for the new Era. Adv. Energy Mater. **9**(1), 1802906 (2019). https://doi.org/10.1002/aenm.201802906
175. Yang, Y., et al.: Pyroelectric nanogenerators for harvesting thermoelectric energy. Nano Lett. **12**(6), 2833–2838 (2012). https://doi.org/10.1021/nl3003039
176. Yang, Y., et al.: Thermoelectric nanogenerators based on single Sb-doped ZnO Micro/Nanobelts. ACS Nano **6**(8), 6984–6989 (2012). https://doi.org/10.1021/nn302481p
177. Yang, Y., Lin, Z.-H., Hou, T., Zhang, F., Wang, Z.L.: Nanowire-composite based flexible thermoelectric nanogenerators and self-powered temperature sensors. Nano Res. **5**(12), 888–895 (2012). https://doi.org/10.1007/s12274-012-0272-8
178. He, M., et al.: A flexible photo-thermoelectric nanogenerator based on MoS2/PU photothermal layer for infrared light harvesting. Nano Energy **49**, 588–595 (2018). https://doi.org/10.1016/j.nanoen.2018.04.072
179. Kumari, D., Shreya, P. Phogat, Dipti, S. Singh, Jha, R.: Enhanced electrochemical behavior of C@CdS Core-Shell heterostructures. Mater. Sci. Eng. B. **301**, 117212. https://doi.org/10.1016/j.mseb.2024.117212 (2024)
180. Pour, G.B., Ashourifar, H., Aval, L.F., Solaymani, S.: CNTs-supercapacitors: a review of electrode nanocomposites based on CNTs, graphene, metals, and polymers. Symmetry (Basel) **15**(6), 1–16 (2023). https://doi.org/10.3390/sym15061179
181. Shreya, A. Yadav, Khatri, R., Jain, N., Bhandari, A., Puri, N.K.: Double zone thermal CVD and plasma enhanced CVD Systems for deposition of films/coatings with eminent conformal coverage BT - advances in manufacturing technology and management. in *Advances in Manufacturing Technology and Management*, R. M. Singari, P. K. Jain, and H. Kumar, Eds. Singapore: Springer Nature Singapore, pp. 273–283 (2023)
182. A. Yadav, Shreya, Puri, N.K.: Preliminary observations of synthesized WS2 and various synthesis techniques for preparation of nanomaterials. in *Advances in Manufacturing Technology and Management*, Springer Science and Business Media Deutschland GmbH, pp. 546–556 (2023)
183. Manyapu, K.K., Peltz, L., De Leon, P.: Self-cleaning spacesuits for future planetary missions using carbon nanotube technology. Acta Astronaut. **157**, 134–144 (2019). https://doi.org/10.1016/j.actaastro.2018.12.019
184. Hortelão, A.C., Patiño, T., Perez-Jiménez, A., Blanco, À., Sánchez, S.: Enzyme-powered nanobots enhance anticancer drug delivery. Adv. Funct. Mater. **28**(25), 1705086 (2018). https://doi.org/10.1002/adfm.201705086
185. Smith, A., Turney, P., Ewaschuk, R.: Self-replicating machines in continuous space with virtual physics. Artif. Life **9**(1), 21–40 (2003). https://doi.org/10.1162/106454603321489509

186. Andhari, S.S., et al.: Self-propelling targeted magneto-nanobots for deep tumor penetration and pH-responsive intracellular drug delivery. Sci. Rep. **10**(1), 4703 (2020). https://doi.org/10.1038/s41598-020-61586-y
187. Muthukumaran, G., Ramachandraiah, U., Samuel, D.G.H.: Role of nanorobots and their medical applications. Adv. Mater. Res. **1086**, 61–67 (2015). https://doi.org/10.4028/www.scientific.net/AMR.1086.61
188. Drexler, K.E.: Nanotechnology: from feynman to funding. Bull. Sci. Technol. Soc. **24**(1), 21–27 (2004). https://doi.org/10.1177/0270467604263113
189. Moses, M.S., Chirikjian, G.S.: Robotic self-replication. Annu. Rev. Control. Robot. Auton. Syst. **3**(1), 1–24 (2020). https://doi.org/10.1146/annurev-control-071819-010010

Mr. Peeyush Phogat is currently pursuing his Ph.D. in the Department of Physics at Netaji Subhas University of Technology. His current area of research is synthesis and characterization of materials, focusing particularly on applications in solar energy and capacitors. Holding a master's degree with specialization in condensed matter physics, He is driven by the pursuit of uncovering novel materials and their distinct properties, meticulously tailored for effective solar energy capture and the design of efficient capacitors. His scholarly endeavors have resulted in significant contributions towards the development of new and advanced materials for solar applications, which are evidenced by his 7 publications in esteemed SCI/SCIE journals and authorship of 2 book chapters within the renowned Springer book series. His expertise extends comprehensively into the domain of capacitors, where his research showcases a remarkable ability to engineer materials that optimize energy storage and discharge capabilities.

Ms. Shreya is currently pursuing her Ph.D. in Physics at Netaji Subhas University of Technology, where she is making significant contributions to the field of photovoltaic devices based on nanomaterials. Her profound interest revolves around exploring the exceptional properties of nanomaterials and propelling renewable energy technologies through their unparalleled potential. She is particularly focused on the properties and applications of two-dimensional materials like transition metal dichalcogenides. She has authored six SCI/SCIE/ Scopus journal publications and four book chapters in highly esteemed Springer book series of conference proceedings.

Prof. Ranjana Jha Department of Physics, Netaji Subhas University of Technology, has been actively involved in research and teaching in the field of Solar Energy Materials and Applied Physics. Received Lifetime Achievement Award in Science and several International and National awards for academics and research in Solar Energy Utilization, Materials and Device fabrication. She is the founder and in-charge of Research Lab for Energy Systems at Department of Physics. Her research interests are Solar Energy Materials and Solar Energy Utilization, Single Crystal Formation, Development of Nano-structured thin film Solar cells, and Characterization of Energy Materials for Device Applications. Several postgraduate students have been awarded Ph.D. degree and at present many students are pursuing their respective Ph.D. under her supervision in Solar Energy materials and utilization and published their research work in SCI Indexed journals under the umbrella of Research Lab for Energy Systems. Published 67 manuscripts in reputed International / National Journals. Attended/ presented papers as an author, co-author, and corresponding author in 123 International and National conferences. There are 05 published eBook Chapter of Springer Nature and 05 book chapters. She has also been Conference Chair in 'International Conference on Advanced Materials for Emerging Technologies' (ICAMET-2023) during May 04-06 organized by Department of Physics, Netaji Subhas University of Technology, Delhi, India.

Presently Professor Emeritus at Department of Physics, Netaji Subhash University of Technology, Dwarka, New Delhi, India. Ex. Emeritus Scientist at CSIR-National Physical Laboratory, New Delhi India, Ex Sr Principal Scientist, and Head Electron and Ion Microscopy Group, CSIR-National Physical Laboratory, New Delhi India. He was also a Faculty Member (Professor) in Academy of Scientific and Innovative Research (ACSIR) at National Physical Laboratory, New Delhi. Dr. Sukhvir Singh is Ph.D in physics from Jamia Millia Islamia, New Delhi, India. He served as a Sr. Principal Scientist & Head of Electron and Ion Microscopy group at CSIR-National Physical Laboratory, New Delhi.

Synthesis and characterization of bulk and nanostructured thin films of compound semiconductors, metal oxides, nanocomposites and investigation of their structural, optical, electronics, and electrical properties in order to ascertain their application for various sensors and device applications are the main areas of his research. The development of standard reference materials (SRM) for the calibration of SEM and TEM is another area of his specialized research under Emeritus Scientist Project of CSIR. Worked on various CSIR network projects and US-funded projects in different areas of leading research. In addition, testing and calibration of all types temperature measuring equipments and maintenance of National Standards in the country at par with the International Standards for Temperature Measurements. He has also established a Centre for Nano Scale Science at CSIR-National Physical Laboratory, New Delhi where highly sophisticated and state-of-the-art advance equipment, such as XRD/HRXRD, TEM/HRTEM, SEM/EDS, AFM, TOF-SIMS, PL spectrometer and highly sophisticated specimen preparation tools for TEM and SEM are installed and serving the nation in various areas of the current research. He has published more than 150 research papers in SCI journals of high repute. He has three no US, South Africa and Asia patents granted in his name.

Chapter 13
Nanotechnology in Defense and Social Justice: Opportunities and Risks

S. K. Suja and S. Mathiya

Introduction

The annual global amount spent towards the military has increased at a higher rate. Nearly US$ 1917 billion was utilized towards military purposes during the year 2019, 3.6% higher than in the year 2018 [1]. The expenditure towards the world military rises to $1.8 trillion for the year 2018 [2]. Military personnel perform their duty in threatening and dangerous environments [1]. Despite the stability of the security situation in the world military, several unstable factors affecting the military security of the world are increasing [3]. Currently, the severity of the situation in military war is high, where few countries are at war, leading to human causality and financial loss. Death of nearly 3,00,000 people in the Syrian conflict (2011–2021) was reported [4]. At the start of 2022, the military of Ukraine and Russia had a conflict. Soldiers and people facing the war directly suffer a lot of casualties. Donbas region in Ukraine faced 100 to 200 deaths of soldiers every day. The soldier's safety is important in sustaining the combat power of the war situation. Hence to enhance combat effectively and to protect the soldiers, personal protective equipment is essential [3]. Breakthrough materials and their combinations are required to enhance the performance of military applications. Nanotechnology became the agenda of 'hype' amidst the scientific community owing to its contribution to modern life. The new generation of defence has been heralded by engineered nanomaterials. Materials designed for military purposes should assure various requirements enclosing speed, surviving ability, precision, and cost-effectiveness. This is illustrated by the cases where, by the use of nanomaterials, ships might travel 75 knots faster, and lightweight military tanks travel up to 75mph [5].

S. K. Suja (✉) · S. Mathiya
Department of Chemistry, Lady Doak College, Tamil Nadu Madurai 625 002, India
e-mail: suja@ldc.edu.in

S. Mathiya
e-mail: mathiya@ldc.edu.in

S. C. George and B. Tawiah (eds.), *Nanotechnology in Societal Development*, Advanced Technologies and Societal Change, https://doi.org/10.1007/978-981-97-6184-5_13

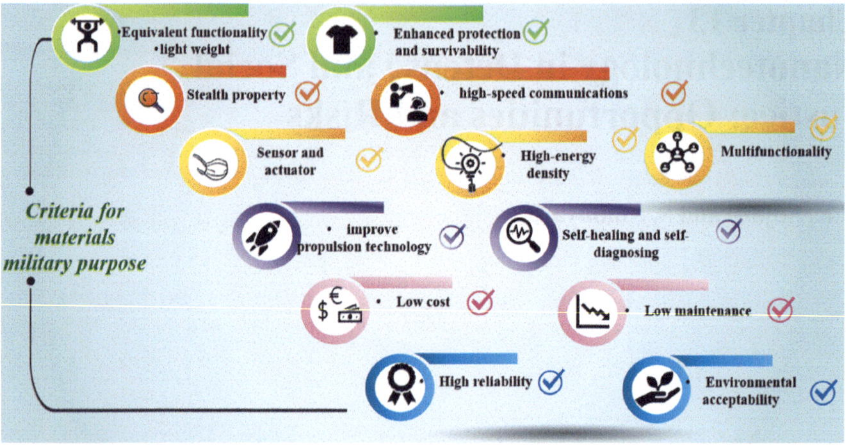

Fig. 13.1 Criteria for materials used for military purposes [5]

The criteria involving specific properties of materials that are required for defense usage are given in Fig. 13.1

Nanotechnology in Defense

Ballistics

Ballistics is the study related to projectile propulsion, the flight, and the impact in a medium. Different materials possess varying effects for penetration of the pellets into the human body. Armour has been used during various periods. Dhal, the first armour used in India for sword fighting to protect against arrows and spears. Ballistic impact (BI), a dynamic and complex event occurs within 50–200 μs. In body armour, mobility and performance are the most important [6]. The Ballistic impact can result in one among the following [7]:

(a) Less energy is transferred to the target than its absorption capacity. Hence projectile drop-off takes place with the bounce back, resulting in partial perforation.
(b) Perforation of the target by the projectile with zero residual velocity, which is the ballistic limit, indicates that the target has absorbed the total kinetic energy.
(c) Even after the complete perforation of the target, there exists residual velocity. Therefore, it states the higher initial kinetic energy compared to the energy that the target will be able to absorb

Steel and ceramic are the basic components in traditional body armour, which are well known for their rigidity and heaviness. Additionally, they fail to protect the joints, neck, and shoulders. These limitations were addressed by shifting to textile-based armours [6]. The parameters significantly influencing them are (i) material (ii) yarn (iii) fabric (iv) testing conditions (v) projectile [6]. Ballistic impact (BI) is a complex experience in composites that is influenced by (i) fabric (ii) material property (iii) target conditions and (iv) projectile parameters [6, 7]. Soft-natured armour based on silk was employed by the Japanese. These body armours work well against fast-moving projectiles (120 m/s) of low impact but not well against high impact velocity of 180 m/s. During the second world war, nylon flak for soft body armor was introduced and protected well from armament fragments, but ceased to resist bullets from pistols and rifles [8]. Purposeful motive towards the reduction of weight of soft body armour and improve the fitting level. The body armour of good standard costs quite a few thousand dollars. It is essential to reduce the cost from a commercial perspective [9]. Sustainable, lightweight, smart, excellently performing body armor to be developed using nanohybrid-based advanced armor system to meet the challenging needs [6, 9]. This could be met through the structural hybridization implemented in the body armor, via (i) inter-ply level and (ii) intra-ply level [10].

Inter-ply techniques comprise various fabric layers, several orientations, and differences in stacking sequences and intra-ply techniques include hybrid weaving, nano-coatings, and fiber or fabric surface modification to meet challenging threat situations. These modifications were achieved by coating various nano-reinforcements. Nano coatings are multifaceted in their application by enhancing the ballistic resistance and aid in producing smart body armor. Carbon nanotubes (CNTs), Nano clays (NCs), graphene, and nanosilica (NS) are employed as reinforcing nanomaterials for improving ballistic impact [6] (Fig. 13.2).

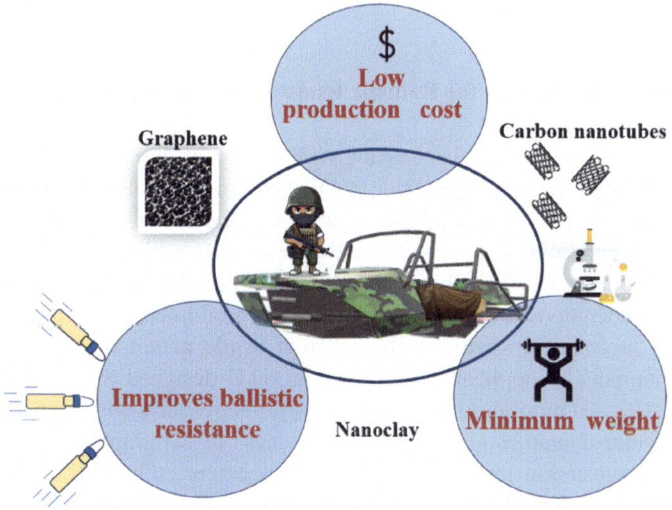

Fig. 13.2 Requirements of armoured vehicles

CNT for Ballistic Resistance

Aramid fibres possess good properties as materials with low density, high strength, and moduli showing high stability, making them as suitable candidates for applications requiring impact resistance [11]. The global market for armoured vehicles shows strong growth and is expected to grow faster up to 2025. At present, the armoured vehicles are made of steel. Even then upgradation of the armoured vehicles is much required because (i) the projectile caliber has increased and (ii) difficulty in handling armour with increased weight. Hence steel is considered to be a prohibited material for armoured military vehicles and a huge demand for lightweight armour. To provide maximum ballistic protection with less weight and minimum production cost, novel composite materials are fabricated [12]. Metal matrix composites (MMC) along with CNT are studied. MMC reinforced with CNTs are found to possess enhanced mechanical properties, with lightweight armour protection. Tsirogiannis et al. studied composite reinforced nanomaterials for protecting light and medium armoured vehicles in the military and their interaction with ballistic projectiles. Materials, as such or in combination with adequate thickness can lead to projectile defeat. Investigation on composite armour materials reinforced by nanomaterials for protecting light and medium armoured vehicles in the military was carried out by Tsirogiannis et al. Four armoured materials possessing front hybrid fibre-based reinforced polymer, strike-faced based on ceramic, intermediate layer with fibre reinforced polymer, and MMC-reinforced backplate. The backplate of the composite material is a hybrid of MMC-reinforced CNTs. All the plates exhibited good mechanical and ballistic properties compared to traditional armour plates. It was found that aluminium alloy 2024-CNT3 displayed the best protection. The projectile did not perforate the plate [12].

Graphene-Based Material for Ballistic Resistance

The ballistic property of graphene oxide was studied by Silva et al. [13] by reinforcing the solution of Graphene Oxide (GO) over aromatic polyimide aramid fabric (Twaron®brand). The samples exhibited ballistic performance against 9 mm ammunition. The aramid fabric coated with two filtrations, GO exhibited resistive force up to 50% more than the originally received unmodified fabric [13]. Graphene-based reinforced nanocomposites are materials with potential impact protection owing to outstanding mechanical strength. The inherent brittle nature of 2D geometry of graphene hampers its application. Zhang et al. [14] designed 3D graphene-based nanocomposite and monitored the wave propagation property with high velocity BI using molecular dynamics. The inherent brittleness of 2D graphene was subsided by effective conversion into 3D structural arrangement embedded within copper moiety. This also effectively dissipates the impact energy by quickly dislocating the propagation in the copper moiety during the early phase of the impact.

Nanoclay for Ballistic Resistance

Nanoclays are layered structures consisting of silicates with a high aspect ratio. Ballistic properties of composites have been studied by optimizing the concentration of nanoparticles [7]. Nanoclay incorporated composite material possesses properties like (i) good mechanical strength (ii) energy absorbing capacity mainly at impact velocities near to the ballistic limit. This is illustrated by nanoclay closite 30B on E-glass. A 5% increase in energy absorption capacity is noticed upon the addition of 5 wt% nanoclay. As the nanoclay content is enhanced to 7 wt%, the energy absorbed also increases [15].

Stealth Technology in Defense

Stealth technology integrates several methods and techniques to make it 'invisible' to the opponents during battle. Military vehicles, equipment, and weapons are mainly constructed with metals, so they can be located easily by RADAR systems. Radar assimilation materials are employed in modern military tactics for the assimilation of electromagnetic (EM) waves. These materials with good conductivity, along with suitably designed materials incorporated can exhibit reflection and show scattering towards EMI shielding [16].

Recently, the development and usage in electronic instruments and devices have made life much easier for humans. The electronic equipment and gadget generate unnecessary electromagnetic energy [17]. Electromagnetic radiation pollution has affected human health, aviation, and industries of the medical sector. Malfunctioning and deterioration of electronic gadgets and the EM pulse can potentially be used as military weapons for neutralizing the enemy's network [18]. The electrical field and magnetic field produced by the electrical and electronic devices disturb the working of adjacent electronic devices called electromagnetic interference [17]. Therefore, EMI shielding products are required to solve the issue of EM wave attenuation for military purposes, especially in military radar stealth [17, 19]. The EMI as a result of electronic communication, leads to channel instability, affects the quality, and also may leak essential information, thereby endangering national security [19]. EMI-SE for commercial and military applications should be at least 20 dB and 30 dB respectively [20, 21]. EMI shields involve exclusively absorption along with or without reflection of EM waves as shown in Fig. 13.4 (Fig. 13.3).

The reflection of the electromagnetic waves leads to the bouncing of the EM waves off the walls of the shield. SE materials with absorption of the EM radiation are preferred over reflection, as the reflected radiation may [22] (i) exert a detrimental effect on the environment, especially on the living system (ii) reach the Radar making the system detected in case of low visibility (stealth). Materials with outstanding electroconductibility provide the nanocomposite film with EMI shielding efficiencies [21].

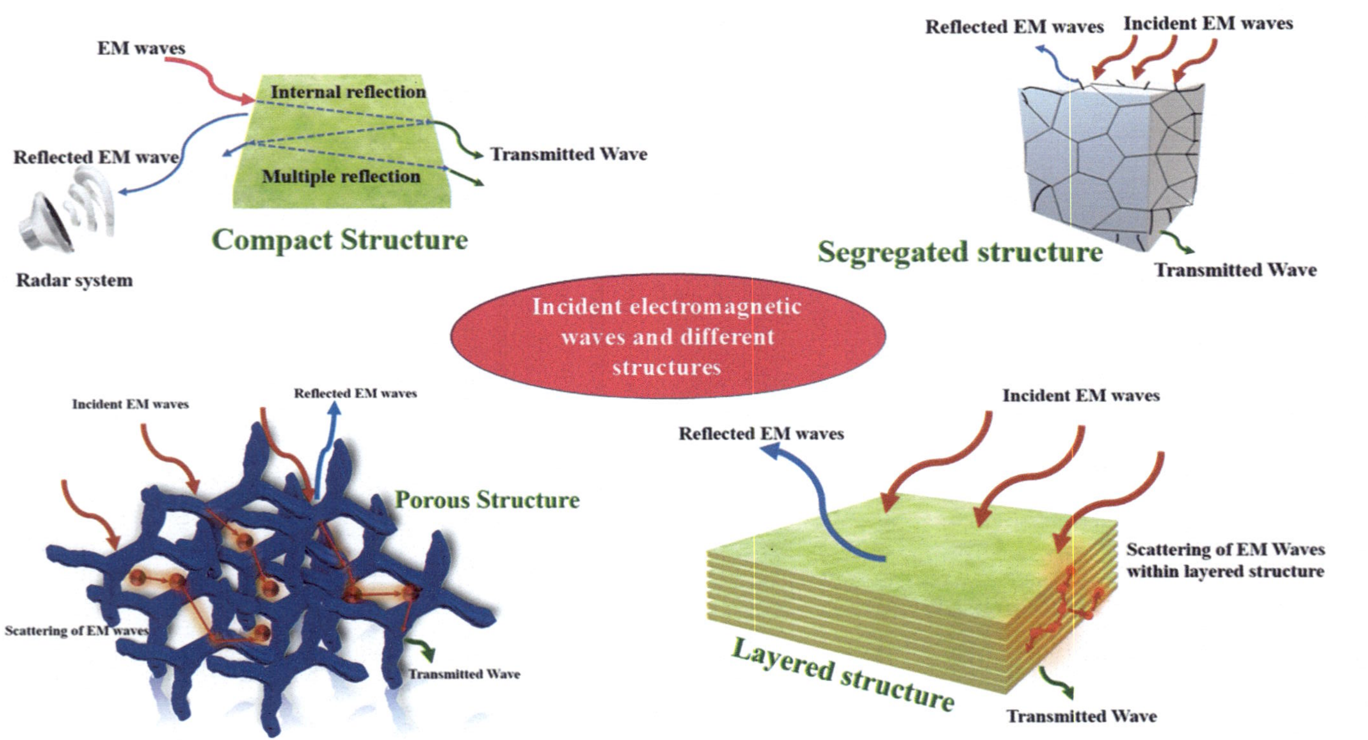

Fig. 13.3 Mechanism of EMI shielding in materials with different structures

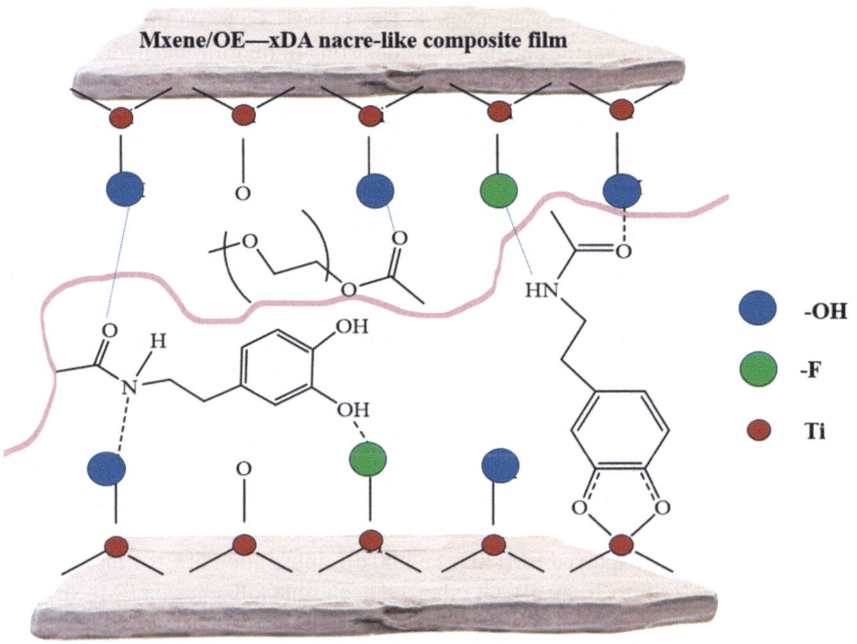

Fig. 13.4 $Ti_3C_2T_x$ MXene/catechol functionalized polymer nanocomposite film (modified from reference [21])

MXene-Based Shielding Material

$Ti_3C_2T_x$ owing to its unique properties including electrical, optical, electrochemical, and mechanical properties and especially their functional property with flexible free-standing films of $Ti_3C_2T_x$ exhibits great scope for application in areas of electromagnetic wave shielding and absorption [21, 23]. Nanosheets of $Ti_3C_2T_x$ MXene were first reported in the year 2011 [24]. Liu et al. have prepared $Ti_3C_2T_x$ nanocomposite films using polyacrylate polymers containing various catechol moieties constructed via the nacre-mimetic strategy as shown in Fig. 13.4.

$Ti_3C_2T_x$ MXene of 5.0 μm thickness film exhibited good SE at 8.2 GHz. The EMI SE of the nanocomposite films decreases upon increasing the polymer content provided they are higher than 40 dB [21]. The reasons are the nanosheets of $Ti_3C_2T_x$ are uniformly and completely coated with the insulating polymer. The high conductive nature of the nanosheets of $Ti_3C_2T_x$ in small polymer content exhibits a shielding effect towards the EM waves efficiently. The mechanism involved in the shielding process is that the dissipation of EM waves via direct reflection happens when they encounter the nanocomposite film owing to the mismatch in the impedance. The EM waves not reflected pass via $Ti_3C_2T_x$ the lattice of possessing high electron density creating ohmic loss. The EM waves survived so far enter the interior of the laminar film (nacre-like) generating several scatterings within the interlayer, thereby the EM waves are absorbed and further attenuated. The Ti in $Ti_3C_2T_x$ on the flip side can create localized dipoles via the end moieties consisting of fluoride and hydroxyl, etc., leading to polarisation loss of EM waves. Additionally, EMI SE and the electrical conductivity are proportional to each other [21].

In the present scenario, the portable electronic gadget is widely used for information transmission and military purposes [25]. To cater to the practical requirements of flexible devices with thermoregulating properties and EMI shielding performance is an issue. Thermoregulating substances bring about electrical to thermal transformation, which is essential in low-temperature surroundings. Conventional methods of heating consume a lot of energy. Accommodating the insulation layer requires a large area and adds volume and weight to the electronic device. Thermoregulation by Joule heating is highly advantageous. Hence fabrication of multifunctional material with good flexibility is required [26]. MXene/cellulose/GO composite-based wearable film was fabricated by Li et al. [27] which demonstrates photo-/electrothermal and EMI shielding performance. Chang et al. [26] prepared pristine MXene film and composite films where MXene is incorporated with CNT and CNT/GO. They compared their EM interference shielding performance. It was found that pristine MXene film displayed EMI SE of 34.00 dB. Integration of CNT into MXene increased the EMI SE to 36.29 dB. GO showed negligible influence on EMI SE. The composite film MXene/CNT/GO demonstrated EMI SE of 36.19 dB and 99.97% efficiency meeting the requirements in military applications. It was found that SE_A is twice of SE_R, confirming that the EM interference shielding is dominated by absorption thereby preventing the reflection of EM waves [26]. Chen et al. [28] prepared MXene-based aerogel using MXene and PVA. The dispersions of the MXene/PVA mixture were transferred to a mould with the bottom immersed in liquid N_2. The oriented aerogel (OMP) was acquired through the above process and a randomly distributed product of aerogel (RMP) was obtained when the dispersion was frozen under refrigeration. The aerogel OMP and RMP were then soaked for cross-linking in an aqueous solution containing sodium sulfate, glutaraldehyde, and H_2SO_4 to get GOMP and GRMP respectively. The EMI shielding performance of the cross-linked oriented MXene-poly-vinyl alcohol with varying mass fractions of MXene 20% (GOMP-20), 40% (GOMP-40), 60% (GOMP-40) were tested along radial orientation. The EMI SE_T (Avg) value of GOMP-20 is 11.5 dB. Raising the mass fraction of MXene to 40 and 60%, the EMI SE_T value also increases to 32.3 dB and 32.6 dB respectively. GRMP-40 has an EMI SE_T value of 24% lower compared to GOMP-40. This indicates that the vertical alignment in the structure of composites contributes much to EM wave attenuation. The aerogel samples also demonstrate higher SE_A compared to SE_R [28].

Graphene-Based Shielding Material

The electromagnetic (EM) frequency utilized in defence and other communication systems lies between 8.2 and 12.4 GHz. Hence, the materials designed for EMI shielding should perform in this range [29]. The conventional EMI shielding materials are metals like copper, nickel, silver, tin, brass, and steel which are employed as screens, sheets or foam. The setbacks in the conventional EMI shielding materials are cost, less chemical stability, and processing difficulty [29]. Moreover, the EM shielding in the case of metals, sharply declines with an increase in frequency. Therefore, there is a high requirement for materials with lightweight, high surface area, inexpensive, eco-friendly, scalability with good EM absorption of broad MW frequency bandwidth [29–31]. The conventional materials are restricted as EMI shields owing to their reflection rather than absorption. Layered graphene systems possess high surface area and are porous in nature. The porous nature of graphene increases the dielectric loss, thereby providing effective absorption capacity towards microwaves [29, 32–34]. MXene has a potential shielding effect in terahertz EMI. Its applicability is highly hindered by the high cost and vulnerability to oxidation. This challenge is overcome by spray deposition strategy, by coating PET material with MXene/graphene which exhibits an outstanding shielding performance. The fine layer of graphene serves as a protecting layer. This method helps to design the MXene surface for electromagnetic shielding thereby addressing the scalability and cost issues [35]. In certain cases, however, the conductivity(electrical) of graphene-based systems is not comparable with metals. Therefore, metal nanoparticles or dopants could be suitable additives to graphene, to improve their shielding performance comparable to metal [36].

Wearable products with EMI shielding are prone to several mechanical impacts including compression, and wear away thereby damaging the EMI layer. The existing challenge with EMI materials is their mechanical robustness to prevent cracks or any damage caused by usage. Moreover, a single crack produced on EMI shielding consequently decreases its performance and finally loses its purpose. Hence, apart from the light weight of the EMI shield layers, flexibility is another essential requirement for EMI shields [37].

Self-healing and flexible graphene oxide/Ag nanowire (GOAgNw) printable films were fabricated by Sim et al. The GOAgNw film with 18 μm thickness exhibited 92 dB EMI shielding effectiveness. The composite and the fabric coated with it were forbearing the damage caused either due to rupture or scratch. The crack created damage in the film, consequently reducing the electrical conductance and the mechanical property, decreasing the EMI shielding effectiveness by 16 dB at 8.2 GHz. After healing, it recovered back to 71 dB with the restoration of the mechanical property. The textile was highly flexible with good mechanical stability and no alteration in the activity was noted even after 1000 cycles of bending [37]. The military field has demands on materials with EMI shielding along with infrared (IR) thermal camouflage properties [21, 38].

CNT as Shielding Material

Nanotechnology has provided radical improvement for RAMs. CNTs are one among the promising materials for Radar Absorbing material and EMI shielding material. The molecular architecture of CNT with sp^2 structure possesses mechanical, chemical properties, and electromagnetic properties [16]. The problem associated with graphenes is the difficulty in scaling up owing to predominant wet chemical processing. Hence, CNT and CNT-derived materials are more impressive for the preparation of RAM. Shielding effectiveness was found to be the same for content with 10 wt % of carbon black, and 5 wt% of CNFs, but with a low quantity of 1 wt % of MWCNTs [39]. CNTs possess good and tuneable electrical behaviour via doping or chirality with unique physicochemical behaviour having good thermal conductivity or excellent mechanical characteristics, essential for stealth technology. Less density and ultrathin CNT layers decrease the density of RAMs additionally weakening the skin effect.

The following characteristic features of CNTs make it suitable for attenuation of microwave:

(i) Tunable morphology
(ii) Tunable chirality
(iii) High specific area
(iv) High aspect ratio

Vinyl trimethoxy silane (Pre-polymerized) integrated MWCNT – PVTMS@MWCNT nano-aerogel material was produced through radical polymerization, then sol–gel transition followed by drying using supercritical CO_2. The pore size of the nanogel was found to be in the range of 30–40 nm, with a specific surface area of 559 m^2g^{-1} and a high void fraction of 91.7%. The nanopore efficiently blocks thermal conduction and is beneficial for IR stealth. The heterogenous nature of the interface reflects the IR and MWCNT aids EM wave attenuation [40]. CNT with carbonyl iron powder (CIp) integrated nanohybrid particles further embedded in the polymer matrix is employed as EMI shielding. The EMI SE enhances as CNT@CIp. The shielding effect is further increased with improved alignment of the hybrid nanoparticles induced by magnetization curing [41] (Table 13.1).

Thermal Camouflage in Defense

One of the best methods to ensure the safety and viability of defence targets against IR detection is via thermal camouflage. Due to the technological advances in infrared sensing strategies, thermal camouflage (TC) of military weapons and objects is tedious [63]. Also, the recent detection poses a threat to military instruments and personnel [64]. Hence there is a rising urgent need for effective materials with thermal camouflage property to cope with the requirements of warfare. The purpose of TC

Table 13.1 Nanomaterials and their EMI shielding properties

Material	Thickness (μm)	Shielding effectiveness (dB)	SSE/t (dB cm^2 g^{-1})	Additional feature	References
Graphene oxide/Ag nanowire	18	92	-	Self-healing and flexible	[37]
rGO/Gallium nanocomposite	33 67	80 100	-	Anti-icing/ de-icing	[42]
Carbon nanofiber (CNF) and functionalised CNF (FCNF)/Graphene nanoplates (GNP) incorporated polyurethane (PU) matrix	300	60.55	-	Insignificant change noticed in SE property despite bath-sonicating	[43]
3D-$Ti_3C_2T_x$MXene/ sodium alginate/ CNTs composite	-	48 75 wt % CNTs	-	Ultrahigh conductivity (1630 S m^{-1})	[44]
Biomimetic $Ti_3C_2T_x$ nanocomposite films	4.8	49.4	-	Possess thermal camouflage properties	[21]
Janus Aramid Nanofiber (A mg)/MXene (M mg) Nanocomposite Films A10M6 A10M8 A10M10	0.015 0.015 0.015	15.1 19.9 24.4	20,293.2 26,671.8 32,475.6	Possess Joule heating Provides thermal camouflage	[45]
CNT@graphene in pyrocarbon system	2.5 mm	45.3	172.4	Ultralow density (1.05 g/cm^3)	[46]
Carbon fiber paper @rGO/ Co/C composite 7-layer assembled material	-	38.46 85.83	12	Enhanced conductivity Attenuated nearly 100% of the EM wave	[47]
Carbon/carbon composites/ SiC nanowires	2.7	35.6	77.4	Lightweight	[48]
Hafnium carbide nanowires modified carbon/carbon nanocomposites	5	38	47.8	Relatively low density Good oxidation resistance	[49]
Graphene@MXene	1.67	47	-	Graphene acts as protective layer on MXene and enhanced the shelf life	[35]

(continued)

Table 13.1 (continued)

Material	Thickness (μm)	Shielding effectiveness (dB)	SSE/t (dB cm^2 g^{-1})	Additional feature	References
MXene/CNT/GO	3	36.19	12,063.33 (dB·mm − 1)	Voltage-driven Joule heating Fire resistant	[26]
Co@CNTs/Carbonised wood foams	-	42	-	Voltage-driven Joule heating	[50]
GOMP-60	2	32.6	-	Good thermal insulation property Thermal camouflage Light weight	[28]
MXene-Ni-carbon fiber tri-core–shell composites	2.4 mm	72.4	-	Electrical heating properties Shielding effectiveness almost nearly 100%	[51]
Polyimide(PI)/ MXene-polyhedral oligomeric silsesquioxane (20 wt% MXene)	-	12.7	-	Water resistance Flame retardance	[52]
Cellulose nanofibers/ MXene/FeCo@rGO	-	45.2	-	Good thermal conductivity Good mechanical properties	[53]
Aramid nanofiber (ANF)/ graphene (GN) composite aerogels	-	31.55	3878.78	Ultralow density (0.04 g/cm^3)	[54]
Composites with silver nanoparticles and aramid nanofiber	-	107.9	-	Good mechanical strength Excellent flexibility Good electrical conductivity	[55]
Fe_3O_4 -NH_2 functionalized MWCNT	500 μm	60.7	-		[56]
CNF/Ti3C2Tx/h-BN	47.6 μm	37.29	-	Good conductivity	[57]
Fe_3O_4 nanospheres (HFO)/ MXene	-	33.28	-	-	[58]

(continued)

Table 13.1 (continued)

Material	Thickness (μm)	Shielding effectiveness (dB)	SSE/t (dB cm^2 g^{-1})	Additional feature	References
CNF/MWCNT/ $Ti_3C_2T_x$ MXene	-	45.1	-	Outstanding mechanical stability	[59]
5 vol% of cobalt/carbon dots	-	–54.3	-	Highly flexible	[60]
Kenaf fiber-reinforced high-density polyethylene/ MWCNT 16% (v/v) Kenaf fiber content with 5% (wt.) MWCNTs	-	30	-	Used in unmanned aerial vehicles	[61]
MXene based composite foam using carboxylated nitrile rubber	-	–30.9	441.4	Excellent low density of 0.7 $g.cm^{-3}$	[62]

is to synchronize the object's radiation temperature with its surroundings to counter infrared sensing. Thus, thermal camouflage effectively enhances the difficulty in recognition of defence objects and protects the IR weapons including missiles, tanks, fighters, etc. from striking [63]. Many two-dimensional nanomaterials like graphene [65], MXenes [66], boron nitride [67], and black phosphorus [68] were ideal candidate materials used for the fabrication of thermal camouflage materials.

Principle

Energy from all the substances is radiated outward and travels like EM waves. The emissivity of thermal radiation (TR) is created by the stochastic fluctuation in the quantized internal energy of the emitting material. The total energy radiated is higher when the object is at a higher temperature. TR propagates in the lower frequency region of the visible and infrared spectrum. The TR that falls in the range of the infrared spectrum will be referred to as IR radiation [69].

Depending upon the wavelength, the TR released by a material has various extents of attenuation in the atmospheric transmission. The region of wavelength with small attenuation in the atmosphere shall be the atmospheric window (Fig. 13.5).

Outside of these three bands, the IR radiation is highly opaque. Hence, the infrared detectors operate in these regions. At present, there is point source detection of infrared radiation and image detection, which permits the visual monitoring of the target. The IR band 8–14 μm is utilized by IR imagers and IR cameras. Therefore, materials for thermal camouflage are researched in this band region [63, 70, 71].

Stefan Boltzmann's law, relating the radiation strength of a material in every plausible direction and wavelength is given by [69]

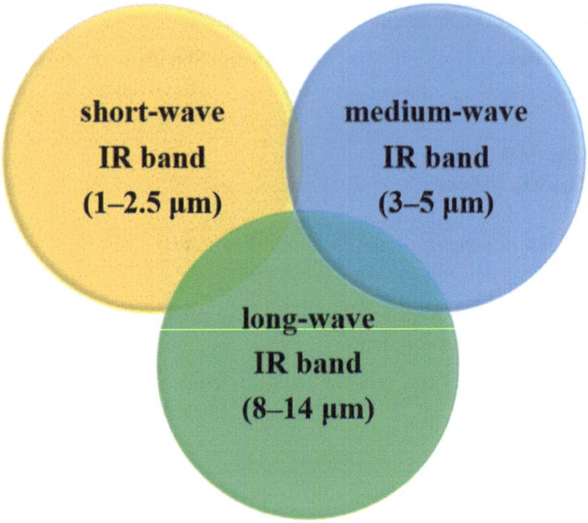

Fig. 13.5 The IR bands present in the atmospheric window

$E = \varepsilon \sigma T^4$

ε denotes the emissivity of the material,

σ is referred to Stepan-Boltzmann constant.

T is the temperature of a material (thermodynamic temperature).

The infrared radiation emitted from the material is related positively with two parameters of the material's surface:(i) Emissivity and (ii) temperature.

Thermal camouflage strategies towards material design are given by (Fig. 13.6).

- Low-emissivity
- Temperature control
- Dynamic temperature

Among the various 2D nanomaterials, the most effective TC materials are graphene and MXene, owing to their dynamic emissivity adjustment. This adjustment allows the object quickly to adapt to the changing temperature of the environment and enhance the camouflage ability. Graphene, a single layer, consisting of hexagonally arranged sp^2 carbons, is a semiconductor with zero band gap. It possesses exceptional conductivity. The structure of graphene permits it to manage the Fermi energy and threshold in optical absorption [63, 72]. This property in graphene was exploited first to design thermal camouflage film [73].

MXenes were discovered in the year 2011 [24], which display fascinating properties owing to their unique structural features and surface groups, possessing excellent electrical conductivity with tunability, outstanding hydrophilicity, and EM shielding performance [63, 74, 75]. Attractively they demonstrate low infrared emissivity (0.19), suitable for thermal camouflage applications [76].

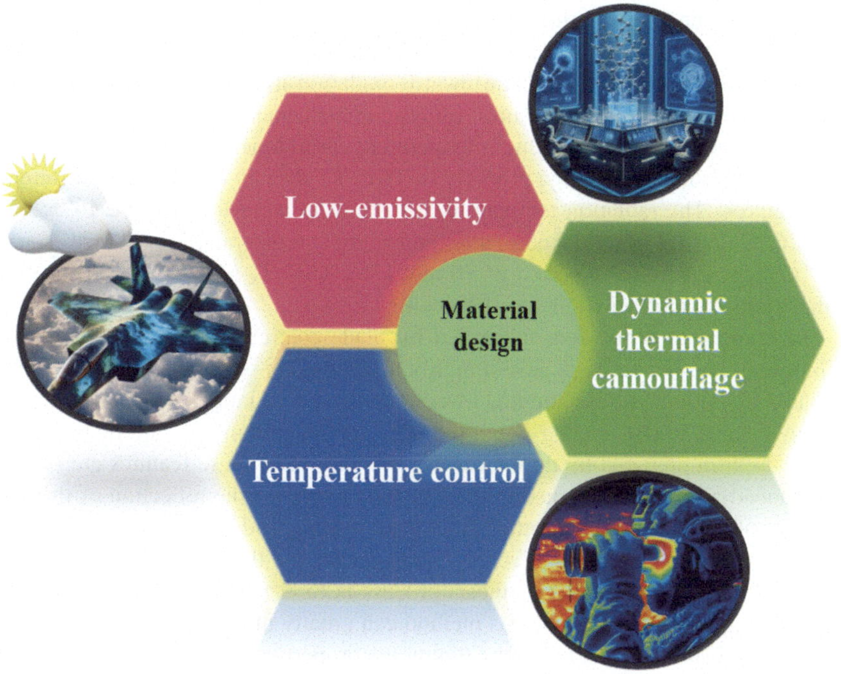

Fig. 13.6 Strategies for tailoring suitable thermal camouflage materials

Tailoring Low Emissive Material

The energy corresponding to TR from a material is reduced by decreasing the emissivity at its surface. Hence, low-emissive materials are preferred for TC applications. $Ti_3C_2T_x$ satisfies this requirement, owing to its low infrared emissivity, similar to metal particles, whereas graphene possesses high emissivity (~0.6) [76, 77].

$Ti_3C_2T_x$ is lightweight and absorbs microwave higher than metal particles [63, 78]. It is used as a camouflage material to protect targets under IT detection and radar [63]. $Ti_3C_2T_x$ MXene was added to polydopamine to produce p-MXene ink of varying emissivity. The emissivity of polydomaine was reduced to 21.5% from 92.2% by enhancing the concentration of $Ti_3C_2T_x$ [79]. Films of $Ti_3C_2T_x$ MXene prepared via vacuum filtration were employed for thermal camouflage at high temperatures by Li et al. [76]. The emissivity being low for $Ti_3C_2T_x$ MXene than GO, montmorillonite and comparable with that of stainless steel, this is much explored. The effect of the thickness of $Ti_3C_2T_x$ MXene (TM) film on its thermal camouflage property is explored by covering an object with different TM films of varying thickness at 510.9 °C radiation temperature and checking the cooling capability under an IR thermal imaging camera. $Ti_3C_2T_x$ films of thickness 29 μm reduced the temperature by 314 °C. Despite the film thickness being 1 μm, the temperature was reduced to 217 °C [76].

Generally, the stable nature of the materials restricts their application towards high- temperature camouflage applications. This is illustrated by the susceptibility of materials like metals and graphene towards oxidation at such elevated temperatures or even continuous exposure to air, leading to enhancement in emissivity. $Ti_3C_2T_x$ exhibited its durability even in high-temperature environments. This was shown by testing the thermal camouflage property of $Ti_3C_2T_x$ after air annelation at different temperatures. The radiation temperatures of the film were found to be 197, 215, and 226 °C (29 μm film thickness) after annealing at temperatures of 200, 400, and 500 °C respectively for a period of 30 min [76]. $Ti_3C_2T_x$ MXene composite film with cellulose nanofiber (CNF) for IR stealth.

The composite films were fabricated by blade-coating (BC) and drop-casting techniques. The BC strategy gave a highly aligned composite structure. The films prepared by both methods exhibited low emissivity (band range 7–14 μm) with the emissivity of 0.56 (BC strategy) and 0.6 (DC method). Similarly, the IR temperature of the film prepared by the BC strategy is very low. These features are owing to enhanced alignment obtained in the film that improves the reflection of the TR than that of DC-film. The structure, functional moiety, and surface morphology of 2-dimensional MXenes are the factors that influence the thermal camouflage capability and their stability. The synthesis process is to be considered to enhance its camouflaging ability. The synthesis method can greatly modify the size, functional moieties, and surface defect, thereby tuning the properties [80] (Table 13.2).

Material for Temperature Control

The control of surface temperature in stationary targets is acquired using thermal camouflage. Materials including porous foams, microcapsules, and aerogels were fabricated to enhance the uptake of heat, thereby merging the IR radiation from the target with the background. Only less information on MXene controls the surface temperature of the target. Graphene exhibits good thermal camouflage properties by exerting control of the surface temperature [63].

Materials like microcapsules possessing hollow structures are commonly employed to control surface temperature. Microcapsules possess a core–shell structure with a core containing wax like octadecane, octacosane or eicosane and GO/PDVB co-shell which stabilizes the core. The wax core is a promising phase change material that provides, at constant temperature, higher energy storage densities. Owing to the good thermal conductivity, GO plays a role in transferring the heat to the wax to enhance thermal storage efficiency. The composite material based on wax exhibited good infrared false detection [83] (Fig. 13.7).

Table 13.2 Camouflage temperature and infrared emissivity of various nanomaterials

Material	Preparation method	Infrared emissivity/ Infrared wavelength range	IR band (μm)	Thickness (μm)	Camouflage temperature (°C)	References
$Ti_3C_2T_x$ nanocomposite films	-	< 0.2	2.5–15.4	5	100 °C	[21]
$Ti_3C_2T_x$ MXene films	Vacuum filtration	0.19	7–14	29	510.9	[76]
Black $Ti_3C_2T_x$ MXene film	Liquid-phase delamination of Ti 3AlC2 powder	0.17	3–16.7	15	100 °C	[81]
MXene/cellulose nanofiber composite film	Blade-coating strategy	0.56	7–14	-	114 °C to 41.9 °C	[82]
Janus films based on ANF and $Ti_3C_2T_x$	-	-	-	-	72 °C for materials with temperatures over 163 °C	[45]

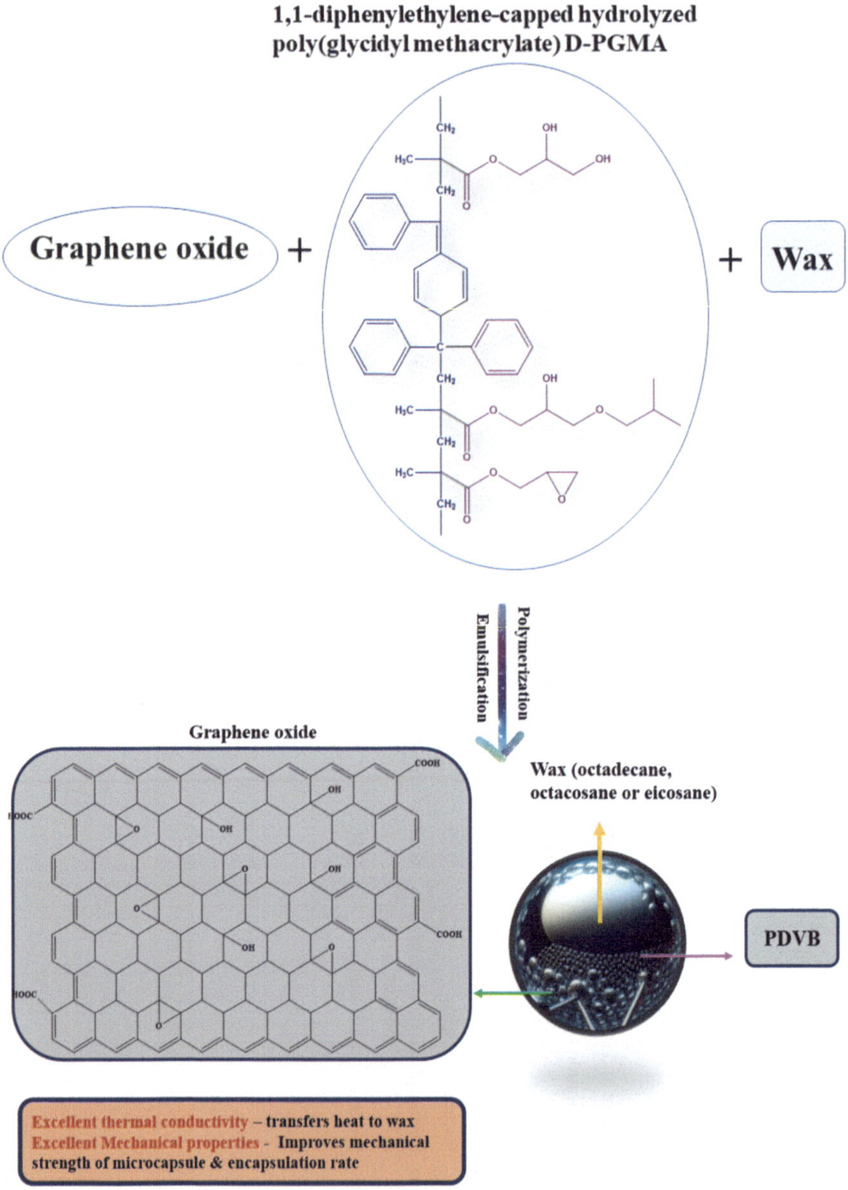

Fig. 13.7 Graphene oxide-based microcapsule phase change material wax/PDVB/GO

Dynamic Thermal Camouflage

The infrared property of the material and their background are the outcome of the interaction between them. Infrared TC aims to match the IR signals of the material unswerving with its surroundings via regulating the emissivity and temperature and thus TC is fixed. In the case of a dynamic background, the target's IR signature becomes different from the background, hence the TC performance diminishes significantly. Chameleons are known for their camouflage performance. Their skin colour is changed dynamically to adapt to their environment. Emissivity modulation is the requirement for dynamic TC in the presence of external stimuli. This is achieved via three types of modulations: (i) electric (ii) phase change (iii) strain.

Visible Range Camouflage

One of the biggest challenges in the military is the attainment of optical invisibility. One of the camouflage methods dates back to 1940, which employs pattern painting. Development of an "Invisibility cloak" to be fabricated via redirecting the incident light. Extending the concept to visible radiation and employing it for practical application is challenging, especially in surroundings that support the propagation of ballistic light [84]. Active camouflage adapts to the environment quickly to become "effectively invisible".

Metal plasmonic nanostructures, by changing the composition and geometries are used to tune the absorption characteristics and their optical reflection in UV to near-IR region. They have produced more saturated colours than the standard technologies. This is illustrated by Wang et al. [84] Au/Ag nano domes were fabricated by electrochemically controlling their structures thereby producing highly ordered structures with dynamic plasmonic properties. These cells were then integrated into a mechanical chameleon capable of blending with background colours [84]. A trilayer aerogel based on MXene/rGO hybrid and *n*-octadecane as materials with phase-changing ability at the lower surface and thermochromic coat on the upper surface was fabricated by Li et al. This composite is integrated with the following for visible and IR camouflage:

(i) Thermal insulating property
(ii) Temperature absorption
(iii) Solar thermal electric conversion
(iv) Thermochromic effect

In the composite, the suppression of heat transfer by the thermal insulating property of aerogel and the temperature absorption characteristics possessed by phase change layer, *n*-octadecane operates synergistically. Therefore, the composite acts as a cloak and hides the signature radiation of the material from IR images of its surrounding in the jungle in the daytime and aids to escape visual surveillance. This helps the military personnel to escape from being tracked through IR images.

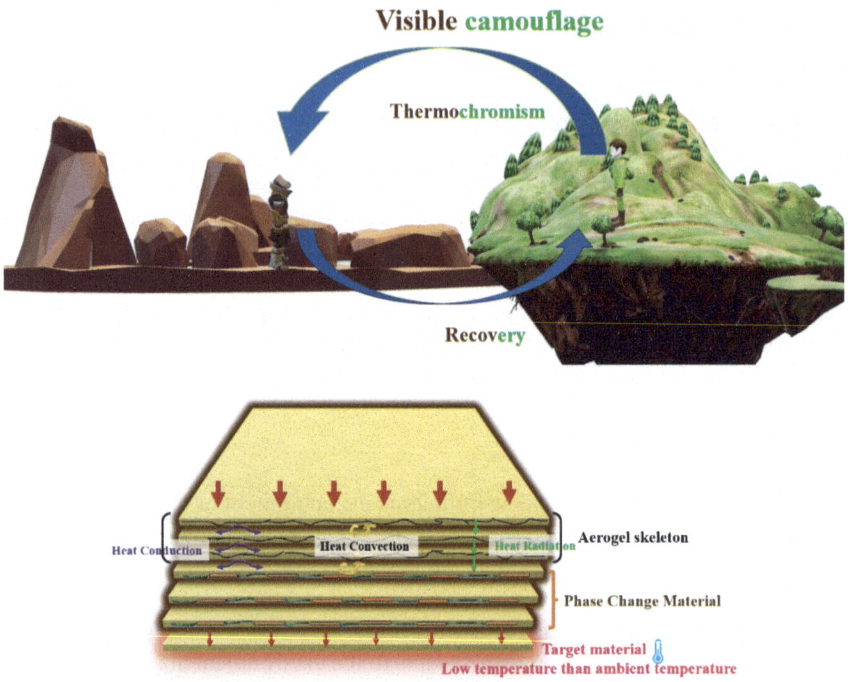

Fig. 13.8 Schematic representation of thermal insulation of composite material

The composites via the solar-thermal conversion can enhance the surface temperature and merge the IR signal of the material with the desert environment. The color of surface changes from green (original) into yellow, so that the target disappears visually from the desert and hilly surroundings [85] (Fig. 13.8).

Nano Tourniquet in Defence

Nearly 5.8 million of the population die annually due to severe trauma, accounting for the death of nearly 16,000 each day [86]. Uncontrollable haemorrhage is responsible for 40% of deaths happening annually resulting from situations of trauma. This has led to death in people aged between 1 and 46 in several decades. In military combat, haemorrhage is the chief cause of death, among which 91% of deaths are preventable. Death owing to haemorrhage could be prevented in a prehospital setting when suitable medical intervention within 1 h is provided, called the "golden hour" in trauma. Present hemostatic dressings and tourniquets do not apply to all hemorrhagic types. They are not efficient for bleeding which originates from internal organs and blood vessels, which is called noncompressible haemorrhage [87]. Commercially available hemostatic agents result in unnecessary secondary effects and responses produced from exothermic reactions and toxicity arising from degraded polymer products. Thus, developing biocompatible materials for hemostasis is essential to safeguard military personnel from haemorrhage and mortality [86].

World Health Organisation in a survey found that the liver is frequently affected internal organs during traumatic cases occurring in warfare, road accidents, and homicides contributing to nearly 10% of the fatality [86]. Rampant and enormous bleeding in the liver, a key post-traumatic condition, yet an avertible reason for fatalities. Despite the timely rescue, the person would suffer from coagulopathy. The core principle in liver trauma during battlefield rescue is quick and efficient hemostasis. The disability of the body to rapidly arrest the bleeding during trauma creates severe blood loss. Thus, haemostatic agents are required to arrest bleeding liver in emergencies is essential [86].

Self-assembled peptide-based biocompatible hydrogels are given much consideration. Self-assembled peptides like EAK16 [88], RADA16 [89], and SLac [90] have the ability to gel at the bleeding spots, potentially decreasing the loss of blood [91]. Yan et al. have fabricated a nano band-aid to arrest the loss of blood rapidly and effectively via the self-assembly of peptides imparted with suitable functional moieties. CRP and CBP co-assembled to produce nanofibers, which were woven into a network by Ca^{2+} ions. The nano band-aid is a viscid fluid until it contacts the blood. The coagulation factor XIIIa activates the nano band-aid at the injury when the band-aid undergoes self-assembly to form a dense compact barrier to arrest and check the loss of blood. The nano band-aid is clinically tested to arrest the bleeding in the liver of rats, where it efficiently decreases the inflammation of the wound and promotes healing. Especially the nano band-aid exhibits good hemostatic activity in case of unsaturated haemorrhage, where the rat suffers from 1/3 of the defect in the liver and noncompressible haemorrhage owing to penetrable wound [86] (Fig. 13.9).

Detonation of explosive devices leads to blast lung injury (PBLI) that causes death owing to internal haemorrhage. Explosions cause injuries up to 79% of combat-based injuries. Hence treatment is required to reduce the death caused by lung trauma [92].

Dexamethasone nanoparticles (DNp) aid to lessen the internal bleeding. The hemostatic DNp is made of block copolymers which are poly(lactic-co-glycolic acid)-poly(l-lys)-poly(ethylene glycol) that are adjoined to suitable peptides. They reduce physiological deprivation produced by the blast and thereby alleviate lung injury [92]. Hemostasis is reached through coagulator cascades with the hemostatic plug being formed by activated platelets, which is then proceeded by activating multiple coagulation factors via intrinsic or extrinsic enzyme pathways which could reinforce the platelet plug. Nanoparticles with positive charge have potential nano hemostats by aggregating platelets and fibrinogen. Assembly of cholic acids (CA) and polyethyleneimines leads to well-assembled positive Nps. These nanoparticles induce the aggregation of platelets and activate them in whole blood and plasma rich in platelets. Polyethyleneimines, itself staunch bleeding in animal models with liver injury. The assembled Nps had a potential haemostatic effect. I.V. injection with nanoparticles efficiently reduces the bleeding time in rats with femoral artery injury by preferentially accumulating at vascular sites [93]. Porous chitosan-based silica composite (PCSC) microspheres were prepared. The hemostatic potential of

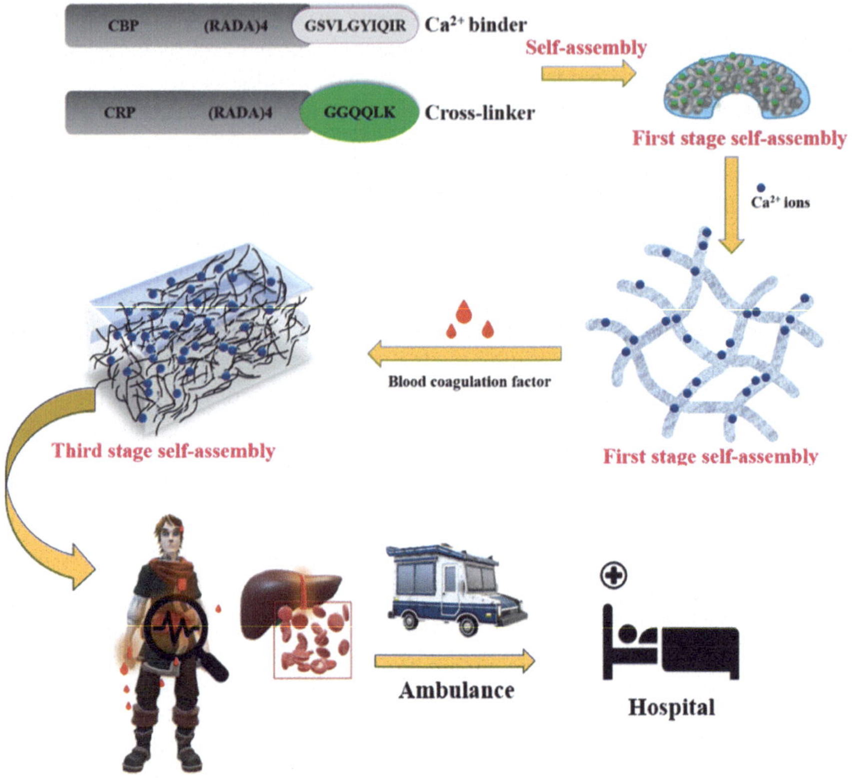

Fig. 13.9 Self-assembly strategy for fabrication of nano band-aid used for treating liver bleeding

the microsphere has improved due to the synergistic hemostatic mechanism. PCSC microsphere formed larger clots exhibiting a haemostatic time of 97 s in rat liver [94].

Smart Textiles in Defence

One of the important elements of military equipment includes textiles which serve various purposes includes

- Uniforms
- Tents
- Parachutes
- Body armour
- Sandbags
- Bomb suit

Fig. 13.10 Timeline showing the different generations of electronic textiles

- Fire retardant material
- Footwear
- Camouflage material [5]

Military personnel should be protected against toxic chemicals, bio-agents with biological threats and nuclear activity. Hence sensing these toxic substances and protection against these agents are mandatory and could be achieved by textiles [95]. Smart nanomaterial-incorporated textiles safeguard the military person from toxic fumes, toxic substances, and infectious agents. Smart textiles, mentioned as electronic textiles, are highly intelligent materials that can detect and respond to external stimuli from the environment. Mechanical, thermal, magnetic, and electric can be the various stimuli and responses. They are of three types based on the generation and the intelligence [96]. Currently, smart textiles are based on the amalgamation of traditional materials and nanomaterials [97] (Fig. 13.10).

i. **1st generation smart textiles:** They are otherwise known as passive smart textiles. Changes in the environment could be well sensed by these materials. They offer additional functions, irrespective of environmental changes. Antimicrobial, anti-odor, and bulletproof materials are typical examples. Silver nanoparticles incorporated with cotton fabrics provide an anti-microbial finish to the fabrics of military personnel and nanoparticles from oxides of metals offer infrared and UV light resistance for the crew [96, 97]. Moreover, this generation of suits which protects the military personnel from toxic chemicals includes polyurethane foam forming the inside coating with fibers of activated carbon. The outside lining is oil with a hydrophobic nature. Currently, these are replaced by customized nanofibers and materials with superhydrophobic properties [95].

ii. **2nd generation smart textiles:** Second-generation textiles are commonly referred to as active smart textiles [97]. These textiles were advanced intelligently to detect and respond according to external stimuli. This is illustrated by sensors, electrically heated fabrics, vapour-absorbing fabrics [96], and thermochromic and shape memory fabrics [97].
iii. **3rd generation smart textiles:** They are super smart textiles with activating capabilities that respond and adapt to environmental conditions [96, 97]. Possessing soft electronics with capabilities of smartly sensing threats and pollutants that the military personnel is exposed to along with nano-generators and offers support to them as an energy reserve [97].

The following aspects are some of the military textile requirements:

- Hydrophobic material
- Hydrophilic material
- Thermal protection
- Strength of the material is high
- Flame-retardant
- good durability
- Safeguarding from ballistic injury
- Withstand chemical threats
- Combats biological threats

Military personnel are involved in tasks in the desert where the tanks and the armored vehicles are subjected to sunstrokes making the environment inefficient to perform operations inside the vehicles for long durations. They may be also affected by heat-related illness, reduced vigilance, and surrounding consciousness. In contrast to this, the crew associated with missions in cold regions are subject to snow blindness, and hypoxemia and suffer from low body temperature. These conditions lead the military personnel to a diseased or deceased situation [5]. Hence, military textile has a crucial role in protecting the troops and making them perform better.

Development in the field of textiles including firefighter uniforms, medical coating, and military garments is increasing. Cotton owing to its softness, ability to absorb moisture, and breathability, is a widely used material. The disadvantage associated is the cost of the chemical finishing. However, the flammability and the infections associated with untreated cotton diminish their utility. Hence, there is a strong requirement for cotton fabric finishing to cater to military and medical garments with multifunctional materials [98]. Additionally, electrically conductive smart textiles offer current flow for electrical purposes retaining the intrinsic properties of textiles like elasticity, flexible nature, and elongation. Incorporating electrical conductivity into textiles may enhance the performance of the fabric and open new avenues for sensors, electromagnetic shielding, camouflage, stealth applications, antibacterial clothing, and many others. Moreover, these applications demand good electrical conductivity with sheet resistance in the range of 0.1 to 100 kΩ/sq [99]. Many nanomaterials are employed to enhance the electrical conductivity of textiles. Therefore, textile materials suitably supported by nanotechnology and the electronic component play an important role in developing military uniforms. Intelligent textile material integrated into electronics helps to enhance the performance of combat soldier through sensing, responding to combat needs adapting themselves to situations and keeping the progress of their mission [100]. Nanotechnology involves the manipulation of materials including textiles leading to products imparted with the required effects. Functionalized products could be fabricated with less amount of nanomaterial. Functionalized textiles with nanomaterials exhibit some of the following qualities: antimicrobial, fire-proof textiles, hydrophobic, and oleophobic, providing finishes of required functionality to the textiles [101].

Chitosan-Based Nanomaterials in Smart Textiles

Chitosan nanoparticles (ChNp) could be well entwined as textile fibres and surface coated onto fabrics to offer an antimicrobial finish suitable for military applications. These nanoparticles are well blended with smart fabrics and incorporated with medications. Wound dressing and drug delivery *via* wearable devices are the potential aspects. CNT and silver nanoparticles impart the textiles fabricated by chitosan nanoparticles with high conductivity. Chitosan nanoparticles are loaded with a variety of active substances like colour-changing dyes, pigments, and phase-change materials, tailoring them to suit various applications [102]. Rehan et al. [98] adopted a green strategy to design nanocomposites based on chitosan. Silver nanoparticles and clay were incorporated into the chitosan to impart multifunctionalities onto the cotton fabrics. Ch/Ag Nps/Clay nanocomposites enhanced the strength, thermal stability, and water absorption with good antimicrobial activity, UV protection, and flame-retardant properties. The nanocomposite-imparted cotton fabrics were stable after 20 washes [98].

Novel nano coating has been designed with multifaceted properties including flame prevention, antimicrobial activity, and the quelling of hazardous fumes. This was illustrated by Attia et al., who prepared casein nanoparticles (CsNp) and ChNp. They designed nanocoating for textile fabrics including cotton and polyester as layers sequentially alternating with CsNp and ChNp. The fire-retardant property of the materials is owing to the combined effect of the nanoparticles, creating a defensive limit on the exterior of the fabric by the char and ingesting the produced gases during the fire which then turns off the fire. The char also reduces the release of hazardous gases [103]. This may guard the soldiers during war from the flame by stopping the spread of the flame as well as from the emitted toxic gases when they get stuck in fire from the combustion in the war zone. Additionally, the fabric also offers antimicrobial activity thereby safeguarding the soldiers in all aspects. The flammability characteristics for various chitosan nanoparticle-based materials are given in Table 13.3.

Table 13.3 Flammability characteristics of chitosan-based nanomaterials

S.No	Material	Distance (mm)	Time (s)	Rate of burning mm/ min	Limiting oxygen index (%)	Additional feature	References
1	CsNp and ChNp nanocoating in cotton fabric (10 multilayer)	0	0	0	34	Antibacterial activity Highly durable	[103]
2	Carboxylmethyl ChNp/G with polypyrrole silver nanocomposite	85	165	30.9	22.8	Antibacterial activity Highly conductive fabrics	[104]
3	PVA/ChNp/ melamine phosphate	-	-	-	57.9%	antibacterial properties Robust even after several washings	[105]
4	Chitosan modified GO nanocomposite	-	-	-	51.2	Improved tensile strength Exhibited fire retardancy even after washing	[106]
5	Polyamide textiles modified $Ch/TiO_2/SiO_2$ Np hybrid	-	-	-	23.5	Decreased wetting time Improved hydrophilicity	[107]
6	Nanocoating of Ch/ phytic acid/ oxidized sodium alginate on polyamide 66 fabrics	-	-	-	21.9	Environmentally friendly material	[108]

Nanocellulose for Smart Textiles

The innovative prospect of nanomaterial aids to establish versatile applications in the textile industry. Incorporation of nanomaterial within the polymer or as surface coating to the textile fabric improves the quality of the fabric suiting the demands in defense purpose. Nanocellulose is acquired from plants in nature. It is known for its high rigidity and robust nature with a high surface area. Moreover, owing to the possibility in surface modification of nanocellulose *via* -OH groups, the textiles can be functionalized. Additionally, the nanocellulose fibers are lightweight, cheap and possess electrical conductivity catering to the wider needs of mankind [109]. Nanocellulose are of the following types: (A) cellulose nanocrystal is a distinctive nanomaterial derived from cellulose (B) Cellulose nanofibrils, an impressive nanocellulose with lightweight (C) bacterial cellulose derived from microbial cellulose [110]. Lightweight yet robust military fabrics aid the soldiers in enhancing their performance by feeling comfortable. Cellulose nanofibers incorporated laminate composites exhibited lightweight and offered better strength [111]. Nanocellulose-derived superhydrophobic materials possess wide application in various fields owing to their unique characteristic features. The uniforms of military personnel become damp and soiled during the execution of their operation. Hence contaminants to be taken off the clothes of soldiers to protect them. This could be achieved *via* surface modification. Suitable architecture with robust nature and minimum surface energy of nanocellulose makes it suitable for superhydrophobic surfaces [112]. E-textiles help to detect vital signs of a soldier, locate him, to communicate with another group of military personnel [113]. Conductivity is one among the property of nanocellulose making them suitable for military applications where electrical signal is essential. Functionalized nanocellulose is employed as an acoustic matter to spot any new vehicles, foes or foes aeroplanes. Cellulose nanocrystals (CNC) possess characteristic features of smart textiles [114]. Incorporation of them into textile fabrics or fibres leads to the improvement of mechanical properties like toughness, elasticity, and strength. Electrically conductive smart textiles are fabricated by incorporating CNTs, and graphene along with CNC, creating wearable sensors and energy harvesting devices [102] (Fig. 13.11).

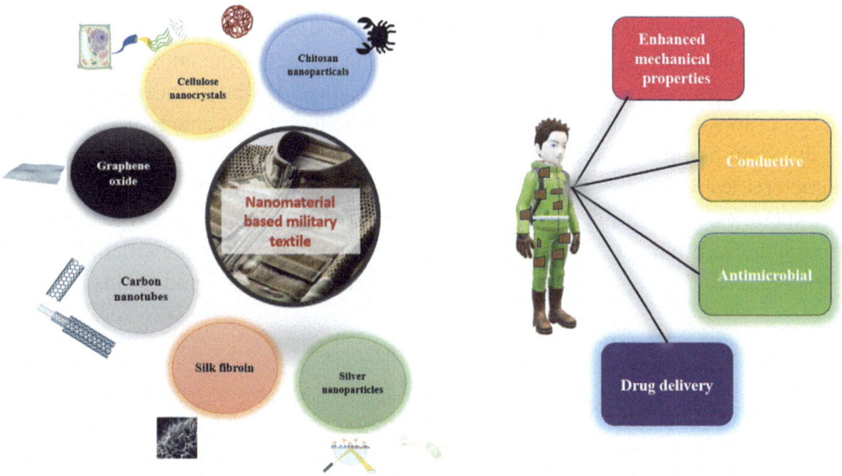

Fig. 13.11 Nanomaterials employed for the fabrication of smart textile material with their properties and applications

Graphene Oxide-Based Smart Textiles

The soldiers face several threats including chemical threats and bio-threats in warfare. It is essential to alert the soldiers about hazardous vapour threats. Many sensors are available that alert the troops of soldiers rather than independent soldiers. Hence smart sensors are required for the soldiers to safeguard them. Additionally, the suits of soldiers in the military suffer from disadvantages like heavy suits with big structures and unreliable lifespans. This problem is overcome with the development of nanomaterials like graphene with robust and lightweight characteristics. Therefore graphene-based smart fabric for protecting the soldiers has been designed. The incorporation of graphene along with cotton fabric could be employed as a sensor material. The resistance of the fabricated graphene-based material was studied using soman (generally used as toxic nerve substance). The electrical resistance of graphene is lowered after the addition of soman, which may be due to the behavior of soman as a n-type dopant in graphene [115]. GO and reduced GO was coated onto a polyester cotton blend in camouflage pattern [116]. GO and reduced GO-coated fabrics decreased the human body's temperature and matching them with the surrounding temperature. Increased concentration of GO and rGO on fabrics enhanced the camouflage property of the textile, even after many washing cycles [116].

Carbon Nanotubes in Smart Textiles

The impact response of carbon fibers-based epoxy textile composites reinforced with CNT was studied by El Moumen et al. [117]. The impact of the amount of CNT on the dynamic properties and the damage were studied. 1% addition of CNT to the textile composites enhanced its ability to assimilate the impact energy and enhances the dynamic nature of the composite panels to nearly 13%. Reduction in dynamic characteristics was noticed with the addition of 4% CNT. This was attributed to the agglomeration of CNTs. The stiffness and rigidity of the composite was increased through the addition of CNTs. The microscopic view and the camera with high speed revealed that the damaged area in panels has decreased with minor additions of CNTs [117].

Silver Nanoparticles in Smart Textiles

Novel multicoloured aramid fabrics with antibacterial activity were fabricated using silver nanoparticles without toxic reagents. Chitosan serves as a reducing and stabilizing agent. The fabric exhibited enhanced thermal resistance and colour with good strength and fastness. The nanoparticles incorporated fabric demonstrated good antibacterial activity after several washings [118]. Firefighting clothes are essential to provide safe personal protection from flame and burn injuries. Additionally, there is a challenge in designing attire with safety towards the flame, signalling the military personnel well ahead of malfunctioning in extreme cases. Self-powered wearable fire alarming e-textile was fabricated using fibres of MXene-based Ag nanowires/aramid nanofibers aerogel. The e-textile integrated clothes sense the temperature in the range of 100–400 °C. The connection established between silver nanowires and nanosheets of $Ti_3C_2T_x$ allows the 30 wt% of textiles to exhibit rapid with repeated alarm at fire in 1.6 s, during burn. Additionally, the synergy in carbonization between aramid nanofiber and MXene, the e-textile demonstrates good flame retardancy [119].

Carbon Quantum Dots in Smart Textiles

Carbon Quantum dots (CQDs) can be well incorporated into the textile matrix. CQDs nucleated out of pyrimidine-derived heterocyclic compound by Emam et al. [120]. CQDs were immobilized separately with native and modified cotton fabrics (cationized cotton fabrics). The UV blocking property and microbicide action of CQDs incorporated into cationized cotton fabrics were better compared to CQDs@cotton. The microbial inhibitions against bacterial strains like *Escherichia coli* and *Staphylococcus aureus* were 63% and 68% respectively and the fungal strain Candida albicans was 67% even after 10 wash cycles. Hence, they could be utilized as superior garments for military purposes [120].

Smart fabrics are associated with suitable programmes for locating the person in safekeeping, security officers and defence personnel. Fabrics can also be linked with AI tools for the safe conveyance of data in military zones. As military uniforms become built with many functionalities and smart devices, power supply becomes essential to make the devices work. Integrating electrical smart devices in fabrics is a challenge. Additionally, the sustenance of e-textiles during the performance of military personnel and washing. The incorporation of many smart components in protective clothing may affect the soldier's consciousness and the ability to recognize dangers. This may create more peril than the advantages of smart textiles. Therefore, great wariness to be considered while designing the e-textiles so that they don't hamper the soldier's ability.

Superthermits in Defence

Energetic compounds (EC) are chemicals including explosives and propelling material, that find application in defence and commercial aspects like mining. Propellants can be.

(i) explosive component in projectiles
(ii) fuel and an oxidizer in rockets
(iii) gas at high pressure, discharging the components immediately upon pressure release

Military propellants vary in the following facets compared to commercial propellants:

(i) Fuel rich
(ii) Increased burning rate
(iii) High energy
(iv) Anti-high overload Insensitive
(v) Minimum characteristic signal [5].

The power of the weapons is enhanced by the addition of other materials called superthermal. These combine nano aluminium with oxides of metal like iron oxide. Following are the advantages of super thermites [121]:

(i) Energy is released at a faster rate
(ii) Rapid chemical reaction releasing high amount of energy
(iii) Different size nano-aluminium powder for varying energy release rate
(iv) Employed for underwater explosive
(v) Nano-aluminum can be employed in torpedoes

Materials based on polydopamine have been researched incorporating in energetic nanocomposites owing to its excellent adherence. Polydopamine-based aluminium/CuO energetic nanocomposites containing aluminium and CuO nanoparticles uniformly distributed. The nanothermits released more heat than the corresponding physical mixture [122].

Nanocellulose is employed to evolve high-energy military explosives compared to the existing explosives, owing to their functionalization ability [5]. Cellulose nanofiber obtained from wood cellulose via TEMPO method of oxidation was employed as nanofillers in solid propellants (double-based). The good aspect ratio, the high suspension ability, and the hydrophilicity of the material endow the material to perform remarkably well than traditional propellants with double base. The material possesses good tensile strength with improved elongation at break by approximately 45%. Additionally, the burning rate was enhanced by 27.5%, and the pressure coefficient was dropped by 20% [123]. The challenge associated with nanocellulose is its semi-rigid property, owing to which it exhibits softening at elevated temperatures and embrittlement when the temperature drops. Striking a balance between energy and mechanical strength is challenging yet attainable. The dispersion of nanocellulose and its compatibility with polymer matrix is challenging yet could be overcome by suitable functionalization. There is a need to impart superhydrophobic properties in nanocellulose- based propellants, as moisture curbs the storage and retards its performance as propellants [5].

Nitrocellulose is derived from natural cellulose, where the –OH groups are replaced by $-NO_2$ groups in cellulose. They are imparted with outstanding physicochemical properties. The concentration of nitro group influences the energy level and the nature of the application. When the content (wt%) of nitrogen in nitrocellulose is less (10.7–12.3%), it suits civilian applications including printing inks. As the content is higher than 12.5 wt%, the compound possesses excellent energy level and mechanical effect, hence employed as an energetic compound and forms a mechanical skeleton in gun propellants. The addition of deterring agents (DA) to nitrocellulose-derived guns and ablation inhibitors (AIn) can enhance the progression of combustion and the life span of barrel weapons. Usually, the DA are inert and have less oxygen content than the prerequisite. The AIs are strenuous to deteriorate leading to problems in ignition worsening the emission of smoke, and reducing the life span of the weapon on the battlefield. Li et al. studied the alliance of denitration of nitrocellulose and ZnO nanoparticles, where ZnO enhances the thermal conductivity ameliorates ablation inhibition and lowers the requirement of an ablation agent. Thus, the gun propellant is imparted with quick ignition, reduced ablation, and minimization of harmful emission with good progression in combustion [124].

Hexanitrohexaazaisowurtzitane, known as CL-20, energy explosive which is in use at present. The primary drawback is they lack safety owing to the molecular structure and its crystallinity. Compared to RDX and β-HMX, CL-20 can easily flame up via unpredicted mechanical external stimulus. To solve the issue of energetic and sensitive property, the following strategies were adopted:

i. Optimize the quality of the crystal
ii. Cocrystals were designed
iii. Coating by nanomaterials

Coating technology is superior to other methods as it can be practically implemented, safe, and economical [125]. Song et al., coupled rGO with CL-20 via polymerization to desensitize CL-20 [126]. The surface coated with rGO enhanced the stability of the material thermally and reduced the sensitive nature of the explosive. The graphene frame can be co-crystallized along with the recrystallized CL-20. Li et al. found that the CL-20 along with the graphene frame maintained the explosion performance and outstanding sensitivity towards mechanical stimuli [127].

Graphene-based materials can act as additives to alter the thermal sensitivity of CL-20. GO/polydopamine was employed to prepare the composite with the explosive CL-20 imparting the composite stability towards mechanical force and stability towards heat [128].

Nanotechnology in Social Justice

Nanotechnology has produced significant social impacts in the area of defense which has a great impact on society. It has a great societal impact in the areas of health, medicine, new drug delivery systems, diagnostic devices, and medical implants. New inventions and research in nanotechnology enhance the lives of people. It reduces the cost in the field of medicine and made available for even economically deprived society. Initially, the therapies involving medicine are highly expensive and nanotechnology serves for the betterment of society. Nanotechnology provides safe drinking water, smart textiles, and healthy food which are the basic needs of society. The evolution of novel materials based on nanotech for defence purposes has pros and cons for society. Modern governments utilize Nanotechnology to collect data about the public. The corporates like banks, insurance companies, and other finance-related corporations already maintain large databases of consumers. The "Radio frequency identification devices" (RFIDs) are nanotechnology-embedded technologies that are used to uniquely identify and read at a distance. The devices become cheaper and more affordable for common people in society just because of the technological advancement in nanotechnology. Protection against human harms, international threats, and safety against non-intentional damage or failure could be well benefited from nanotechnology developments. In quantum computing communication, nanotechnology-incorporated devices pose tantalizing promise on unbreakable encoded and resistant towards eavesdropping in communications. Complex nanostructures can easily and rapidly encrypt and decrypt information in an unbreakable way. Integrated photonics is yet another fast-growing area that highly relies on nanotechnology [129].

Opportunities and Risks

The growth of nanotechnology that emerged in the field of defense in this century is rapid and expansive. All the developed and developing countries are investing huge funds to develop novel materials based on nanotechnology. The advances in nanotechnology have led to the evolution of nano weapons including explosives and machines based on robots. Many opportunities are there in producing energy-absorbing lightweight materials protecting from blasts, sensors to detect toxic chemicals, nano-devices for medical issues, fibre-reinforced nano-materials for missile casings, microchip-based cameras to surveil, and performance monitors, communication devices for remote tags, acoustic materials, GPS receivers, and wire-free communication.

Even though nanotechnology has a wide scope in the field of defence, it has dominant risks associated. Nanotechnology is predominantly involved in the economic growth of a country however it involves an increased threat to the security and safety of a nation. The engineered material used for warfare applications has a lot of health risks as shown in Fig. 13.12. The exposure of mice to single-walled carbon nanotubes SWCNTs causes transitory pulmonary inflammation decreased pulmonary function and reduced bacterial clearance. Even inhalation of 0.5–2 mg/Kg of SWCNTs causes inflammation. The CNTs may enter the human system *via* air inhalation to the lungs, then to blood where it is distributed to the essential organs causing cancer and other diseases. Exposure of skin to nanomaterials occurs *via* handling of products with nanomaterials. Based on studies in mice, it was found that exposure of humans to SMCNT for 20 days may cause budding lesions in the lungs. MWCNT produced fewer inflammations. Nanometals offer an increased energy concentration with minimum raw materials, thereby reducing the cost of weapons. Nanopowders utilized in larger volumes in smaller nano weapons and explosive devices resulted in toxic materials being expelled into the environment. Hence, nanotechnology drastically may change the weaponry in the military by reducing its cost. Mini-nuke

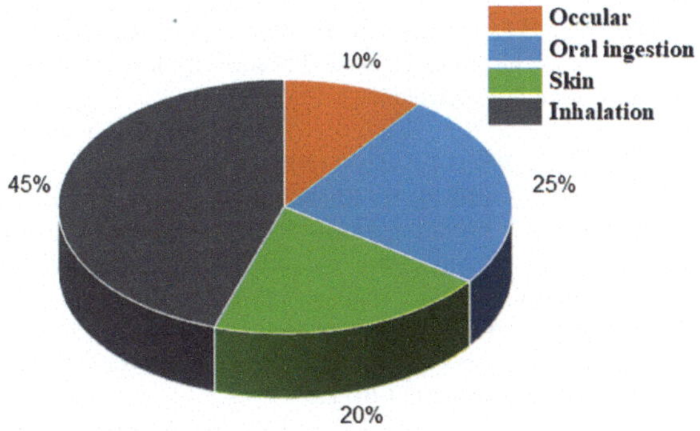

Fig. 13.12 Exposure to engineered nanomaterial and the percentage of risk associated (Modified from source [130])

devices with nanotechnology have the potential to create very tiny nuclear detonators. These devices fit even in a briefcase and could destroy a building. These devices could be dangerous if they move into the hands of terrorists (Fig. 13.12).

Summary

This chapter has presented the role played by nanomaterials in military applications offering great benefits to soldiers in terms of protection and medical treatment. Nanomaterials offer strong, protective lightweight stable supportive products to survive in hazardous environments. Safety and ethical issues are to be explored further. Investigation on the level of toxicity of nanomaterials along with well-being and safety issues are to be considered. Potential risks can be minimized through proper training, adopting health and safety policies with appropriate protection. Nanotechnology has a great role in the economic growth of the country. In contrast, it increases the threat to the safety and security of a nation. No regulations are available to date for the issues arising due to nanotechnology. The rapid development of nanomaterials hinders the construction of relevant regulations. Productive health and safety regulation on the usage of nanomaterials ensures public and environmental protection and concern without restricting the application of nanotechnology.

References

1. Md. Syduzzaman, A. Hassan, Anik, H.R., Akter, M., Md. Islam R.: Nanotechnology for high-performance textiles: a promising frontier for innovation. ChemNanoMat **9**(9) (2023). https://doi.org/10.1002/cnma.202300205
2. World military expenditure grows to $1.8 trillion in 2018 (2018)
3. Hu, Q., Shen, X., Qian, X., Huang, G., Yuan, M.: The personal protective equipment (PPE) based on individual combat: a systematic review and trend analysis. Def. Technol. **28**, 195–221 (2023). https://doi.org/10.1016/j.dt.2022.12.007

4. Doganay, M., Demiraslan, H.: Refugees of the Syrian Civil war: impact on Reemerging infections, health services, and biosecurity in Turkey. Health Secur **14**(4), 220–225 (2016). https://doi.org/10.1089/hs.2016.0054
5. Norrrahim, M.N.F., et al.: Nanocellulose: the next super versatile material for the military. Mater Adv **2**(5), 1485–1506 (2021). https://doi.org/10.1039/D0MA01011A
6. Biradar, A., Arulvel, S., Kandasamy, J.: Significance of ballistic parameters and nanohybridization in the development of textile-based body armor: a review. Int. J. Impact Eng **180**, 104700 (2023). https://doi.org/10.1016/j.ijimpeng.2023.104700
7. U.A. Shakil, Bin Abu Hassan, S., Yahya, Mujiyono, M.Y., Nurhadiyanto, D.: A review of properties and fabrication techniques of fiber reinforced polymer nanocomposites subjected to simulated accidental ballistic impact. Thin-Walled Struct. **158**, 107150 (2021). https://doi.org/10.1016/j.tws.2020.107150
8. Mawkhlieng, U., Majumdar, A., Laha, A.: A review of fibrous materials for soft body armour applications. RSC Adv. **10**(2), 1066–1086 (2020). https://doi.org/10.1039/C9RA06447H
9. Crouch, I.G.: Body armour – new materials, new systems. Def. Technol. **15**(3), 241–253 (2019). https://doi.org/10.1016/j.dt.2019.02.002
10. Bhat, A., Naveen, J., Jawaid, M., Norrrahim, M.N.F., Rashedi, A., Khan, A.: Advancement in fiber reinforced polymer, metal alloys and multi-layered armour systems for ballistic applications – a review. J. Market. Res. **15**, 1300–1317 (2021). https://doi.org/10.1016/j.jmrt.2021.08.150
11. Wang, Z., Zhang, H., Dong, Y., Zhou, H., Huang, G.: Ballistic performance and protection mechanism of aramid fabric modified with polyethylene and graphene. Int. J. Mech. Sci. **237**, 107772 (2023). https://doi.org/10.1016/j.ijmecsci.2022.107772
12. Tsirogiannis, ECh., Daskalakis, E., Hassan, M.H., Omar, A.M., Bartolo, P.: Ballistic design and testing of a composite armour reinforced by CNTs suitable for armoured vehicles. Def. Technol. (2023). https://doi.org/10.1016/j.dt.2023.04.013
13. da Silva, A.O., et al.: Effect of graphene oxide coating on the ballistic performance of aramid fabric. J. Market. Res. **9**(2), 2267–2278 (2020). https://doi.org/10.1016/j.jmrt.2019.12.058
14. Zhang, W., Zhang, Y., Zhao, S., Zhang, Y., Yang, J.: Ballistic impact induced wave propagation and dislocation of three-dimensional graphene origami/copper nanocomposites. Compos. Commun. **45**, 101790 (2024). https://doi.org/10.1016/j.coco.2023.101790
15. Pol, M.H., Liaghat, G., Hajiarazi, F.: Effect of nanoclay on ballistic behavior of woven fabric composites: experimental investigation. J. Compos. Mater. **47**(13), 1563–1573 (2013). https://doi.org/10.1177/0021998312449768
16. Kolanowska, A., Janas, D., Herman, A.P., Jędrysiak, R.G., Giżewski, T., Boncel, S.: From blackness to invisibility – carbon nanotubes role in the attenuation of and shielding from radio waves for stealth technology. Carbon N Y **126**, 31–52 (2018). https://doi.org/10.1016/j.carbon.2017.09.078
17. Zhao, B., Zhao, C., Li, R., Hamidinejad, S.M., Park, C.B.: Flexible, Ultrathin, and high-efficiency electromagnetic shielding properties of poly(Vinylidene Fluoride)/carbon composite films. ACS Appl. Mater. Interf. **9**(24), 20873–20884 (2017). https://doi.org/10.1021/acsami.7b04935
18. Maestre, C.V., Santos, G.N.: Effect of bismuth oxide nanoparticle on the electromagnetic interference shielding and thermal stability of industrial waste based-geopolymer composites. Sci. Rep. **13**(1), 1787 (2023). https://doi.org/10.1038/s41598-023-27623-2
19. Jia, Z., Zhang, M., Liu, B., Wang, F., Wei, G., Su, Z.: Graphene foams for electromagnetic interference shielding: a review. ACS Appl. Nano Mater. **3**(7), 6140–6155 (2020). https://doi.org/10.1021/acsanm.0c00835
20. Wan, Y.-J., Li, X.-M., Zhu, P.-L., Sun, R., Wong, C.-P., Liao, W.-H.: Lightweight, flexible MXene/polymer film with simultaneously excellent mechanical property and high-performance electromagnetic interference shielding. Compos. Part A Appl. Sci. Manuf. **130**, 105764 (2020). https://doi.org/10.1016/j.compositesa.2020.105764

21. Liu, Z., Zhang, G., Chen, W., Wang, J., Zhang, B., Zhang, Q.: Robust biomimetic Ti3C2Tx nanocomposite films enhanced by mussel-inspired polymer for highly efficient electromagnetic shielding and thermal camouflage. Carbon N Y **196**, 410–421 (2022). https://doi.org/10.1016/j.carbon.2022.05.004
22. Orasugh, J.T., Ray, S.S.: Functional and structural facts of effective electromagnetic interference shielding materials: a review. ACS Omega **8**(9), 8134–8158 (2023). https://doi.org/10.1021/acsomega.2c05815
23. Zhang, M., Cao, M.-S., Shu, J.-C., Cao, W.-Q., Li, L., Yuan, J.: Electromagnetic absorber converting radiation for multifunction. Mater. Sci. Eng. R. Rep. **145**, 100627 (2021). https://doi.org/10.1016/j.mser.2021.100627
24. Naguib, M., et al.: Two-dimensional nanocrystals produced by exfoliation of Ti_3AlC_2. Adv. Mater. **23**(37), 4248–4253 (2011). https://doi.org/10.1002/adma.201102306
25. Mercante, L.A., Andre, R.S., Facure, M.H.M., Correa, D.S., Mattoso, L.H.C.: Recent progress in conductive electrospun materials for flexible electronics: energy, sensing, and electromagnetic shielding applications. Chem. Eng. J. **465**, 142847 (2023). https://doi.org/10.1016/j.cej.2023.142847
26. Chang, R., Hao, P., Qu, H., Xu, J., Ma, J.: A fire resistant MXene-based flexible film with excellent Joule heating and electromagnetic interference shielding performance. J. Colloid Interf. Sci. **654**, 437–445 (2024). https://doi.org/10.1016/j.jcis.2023.10.061
27. Li et al.: Graphene oxide-assisted multiple cross-linking of MXene for large-area, high-strength, oxidation-resistant, and multifunctional films. Adv. Funct. Mater **33**(11) (2023). https://doi.org/10.1002/adfm.202213357
28. Chen, X., Yi, Y., Duan, R., Lu, C., Zhou, Z.: Dual cross-linked and vertically oriented MXene/polyvinyl alcohol aerogel for ultrabroadband electromagnetic interference shielding and thermal camouflage. Polymer (Guildf) **291**, 126604 (2024). https://doi.org/10.1016/j.polymer.2023.126604
29. Naghdi, S., et al.: Graphene family, and their hybrid structures for electromagnetic interference shielding applications: Recent trends and prospects. J. Alloys Compd. **900**, 163176 (2022). https://doi.org/10.1016/j.jallcom.2021.163176
30. Gupta, T.K., et al.: MnO2 decorated graphene nanoribbons with superior permittivity and excellent microwave shielding properties. J. Mater. Chem. A. Mater. **2**(12), 4256 (2014). https://doi.org/10.1039/c3ta14854h
31. Singh, V.K., et al.: Microwave absorbing properties of a thermally reduced graphene oxide/nitrile butadiene rubber composite. Carbon N Y **50**(6), 2202–2208 (2012). https://doi.org/10.1016/j.carbon.2012.01.033
32. Wang, J., et al.: Hollow Ni/C microsphere@graphene foam with dual-spatial and porous structure on the microwave absorbing performance. J. Alloys Compd. **873**, 159811 (2021). https://doi.org/10.1016/j.jallcom.2021.159811
33. Shen, B., Li, Y., Yi, D., Zhai, W., Wei, X., Zheng, W.: Microcellular graphene foam for improved broadband electromagnetic interference shielding. Carbon N Y **102**, 154–160 (2016). https://doi.org/10.1016/j.carbon.2016.02.040
34. Xi, J., et al.: Graphene aerogel films with expansion enhancement effect of high-performance electromagnetic interference shielding. Carbon N Y **135**, 44–51 (2018). https://doi.org/10.1016/j.carbon.2018.04.041
35. Danish, M., et al.: Spray deposited Graphene@MXene on a flexible polymer substrate for Terahertz Shielding. Mater. Chem. Phys. **311**, 128573 (2024). https://doi.org/10.1016/j.matchemphys.2023.128573
36. Yu, C., et al.: Fe nanoparticles and CNTs co-decorated porous carbon/graphene foam composite for excellent electromagnetic interference shielding performance. J. Alloys Compd. **820**, 153108 (2020). https://doi.org/10.1016/j.jallcom.2019.153108
37. Sim, H.J., et al.: Self-healing graphene oxide-based composite for electromagnetic interference shielding. Carbon N Y **155**, 499–505 (2019). https://doi.org/10.1016/j.carbon.2019.08.073

38. Liu, H., et al.: Moisture assisted photo-engineered textiles for visible and self-adaptive infrared dual camouflage. Nano Energy **93**, 106855 (2022). https://doi.org/10.1016/j.nanoen.2021.106855
39. Al-Saleh, M.H., Saadeh, W.H., Sundararaj, U.: EMI shielding effectiveness of carbon based nanostructured polymeric materials: a comparative study. Carbon N Y **60**, 146–156 (2013). https://doi.org/10.1016/j.carbon.2013.04.008
40. Ma, H., et al.: Efficient electromagnetic wave absorption and thermal infrared stealth in PVTMS@MWCNT nano-aerogel via abundant nano-sized cavities and attenuation interfaces. Nanomicro Lett **16**(1), 20 (2024). https://doi.org/10.1007/s40820-023-01218-y
41. Jang, D., et al.: Improved electromagnetic interference shielding performances of carbon nanotube and carbonyl iron powder (CNT@CIP)-embedded polymeric composites. J. Market. Res. **18**, 1256–1266 (2022). https://doi.org/10.1016/j.jmrt.2022.02.134
42. Sun, Y., et al.: Slippery graphene-bridging liquid metal layered Heterostructure nanocomposite for stable high-performance electromagnetic interference shielding. ACS Nano **17**(13), 12616–12628 (2023). https://doi.org/10.1021/acsnano.3c02975
43. Tahalyani, J., Akhtar, M.J., Kar, K.K.: Flexible, stretchable, and thin films based on functionalized carbon nanofiber/graphene nanostructures for electromagnetic interference shielding. ACS Appl. Nano Mater. **6**(10), 8178–8191 (2023). https://doi.org/10.1021/acsanm.3c00215
44. Wang, W., et al.: High-efficiency electromagnetic interference shielding from highly aligned MXene porous composites via controlled directional freezing. ACS Appl. Mater. Interf. **15**(40), 47566–47576 (2023). https://doi.org/10.1021/acsami.3c10599
45. Zhao, Y., et al.: One-step method for fabricating janus aramid Nanofiber/MXene nanocomposite films with improved joule heating and thermal camouflage properties. ACS Appl. Mater. Interf. **15**(47), 55150–55162 (2023). https://doi.org/10.1021/acsami.3c13722
46. Feng, L., et al.: Development of light cellular carbon nanotube@graphene/carbon nanocomposites with effective mechanical and EMI shielding performance. Carbon N Y **168**, 719–731 (2020). https://doi.org/10.1016/j.carbon.2020.07.032
47. Liu, Z., et al.: Multi-layer carbon fiber paper @reduced graphene oxide/Co/C composite with adjustable electromagnetic interference shielding properties. Carbon N Y **217**, 118655 (2024). https://doi.org/10.1016/j.carbon.2023.118655
48. Shen, Q., Li, H., Lin, H., Li, L., Li, W., Song, Q.: Simultaneously improving the mechanical strength and electromagnetic interference shielding of carbon/carbon composites by electrophoretic deposition of SiC nanowires. J. Mater. Chem. C. Mater. **6**(22), 5888–5899 (2018). https://doi.org/10.1039/C8TC01313F
49. Tian, S., et al.: 2.5 D carbon/carbon composites modified by in situ grown hafnium carbide nanowires for enhanced electromagnetic shielding properties and oxidation resistance. Carbon N Y **161**, 331–340 (2020). https://doi.org/10.1016/j.carbon.2020.01.080
50. Feng, H., et al.: Enhanced polarization *via* Joule heating in wood-derived carbon materials for absorption-dominated EMI shielding. Mater. Horiz. (2024). https://doi.org/10.1039/D3MH01332D
51. Hu, Y. et al.: Interfacial engineering of hybrid MXene-Ni-CF tri-core-shell composites for electromagnetic interference shielding and E-heating applications. Compos Part A. Appl. Sci. Manuf. 107990 (2023). https://doi.org/10.1016/j.compositesa.2023.107990
52. Yan, Y.-Z., Chen, K., Moon, H.R., Park, S.S., Ha, C.-S.: Flexible and hydrophobic polyimide/MXene-POSS nanocomposite films for electromagnetic interference shielding. Eur. Polym. J., **202**, 112655 (2024). https://doi.org/10.1016/j.eurpolymj.2023.112655
53. Ma, M., et al.: Structural design of asymmetric gradient alternating multilayered CNF/MXene/FeCo@rGO composite film for efficient and enhanced absorbing electromagnetic interference shielding. J. Mater. Chem. A. Mater. (2024). https://doi.org/10.1039/D3TA06467K
54. Luo, N., et al.: Electromagnetic interference shielding performance of lightweight aramid Nanofiber/graphene composite aerogels. J. Mater. Chem. A. Mater. (2024). https://doi.org/10.1039/D3TA07473K
55. Sun, Y.-X., et al.: Metal-grade laminated nanofiber films with outstanding EMI shielding performances and high-temperature resistance. Colloids Surf A Physicochem Eng Asp **680**, 132701 (2024). https://doi.org/10.1016/j.colsurfa.2023.132701

56. Govindasamy, T., Mathew, N.K., Asapu, V.K., Subramanian, V., Subramanian, B.: Evaluating the microwave absorbing performance of polymer-free thin Fe3O4−MWCNT NCs in X-band region. Surf. Interf. **44**, 103716 (2024). https://doi.org/10.1016/j.surfin.2023.103716
57. Qian, K., et al.: Highly thermally conductive Ti3C2Tx/h-BN hybrid films via coulombic assembly for electromagnetic interference shielding. J. Colloid Interf. Sci. **613**, 488–498 (2022). https://doi.org/10.1016/j.jcis.2022.01.060
58. Zheng, X., Tang, J., Cheng, L., Yang, H., Zou, L., Li, C.: Superhydrophobic hollow magnetized Fe3O4 nanospheres/MXene fabrics for electromagnetic interference shielding. J. Alloys Compd. **934**, 167964 (2023). https://doi.org/10.1016/j.jallcom.2022.167964
59. Qian, K., et al.: Yarn-ball-shaped CNF/MWCNT microspheres intercalating Ti3C2Tx MXene for electromagnetic interference shielding films. Carbohydr. Polym. **254**, 117325 (2021). https://doi.org/10.1016/j.carbpol.2020.117325
60. Alshahrani, H., Arun Prakash, V.R.: Development of highly flexible electromagnetic interference shielding composites for electronic applications using Cobalt/ Hevea brasiliensis seed husk carbon dots/Bamboo microfibre-polyvinyl alcohol. Ind. Crops Prod. **191**, 115967 (2023). https://doi.org/10.1016/j.indcrop.2022.115967
61. Singh, A.P., Zafar, S., Suman, S., Pathak, H.: Mechanical and electromagnetic interference shielding properties of natural fiber reinforced polymer composite with carbon nanotubes addition. Polym. Compos. (2024). https://doi.org/10.1002/pc.28075
62. Das, P., et al.: Stretchable and lightweight 2D MXene-based elastomeric composite foam for suppressing electromagnetic interference. J. Alloys Compd. **976**, 173011 (2024). https://doi.org/10.1016/j.jallcom.2023.173011
63. Fan, X., Li, S., Zhang, W., Xu, W.: Recent progress in two-dimensional nanomaterials of graphene and MXenes for thermal camouflage. Ceram. Int. **49**(4), 5559–5572 (2023). https://doi.org/10.1016/j.ceramint.2022.12.034
64. Ding, D., He, X., Liang, S., Wei, W., Ding, S.: Porous nanostructured composite film for visible-to-infrared camouflage with thermal management. ACS Appl. Mater. Interf. **14**(21), 24690–24696 (2022). https://doi.org/10.1021/acsami.2c03509
65. Li, H., Li, Y., Aljarb, A., Shi, Y., Li, L.-J.: Epitaxial growth of two-dimensional layered transition-metal dichalcogenides: growth mechanism, controllability, and scalability. Chem. Rev. **118**(13), 6134–6150 (2018). https://doi.org/10.1021/acs.chemrev.7b00212
66. Hantanasirisakul, K., Gogotsi, Y.: Electronic and optical properties of 2D transition metal carbides and nitrides (MXenes). Adv. Mater. **30**(52) (2018). https://doi.org/10.1002/adma.201804779
67. Wang, J., Ma, F., Sun, M.: Graphene, hexagonal boron nitride, and their heterostructures: properties and applications. RSC Adv. **7**(27), 16801–16822 (2017). https://doi.org/10.1039/C7RA00260B
68. Li, L., et al.: Black phosphorus field-effect transistors. Nat. Nanotechnol. **9**(5), 372–377 (2014). https://doi.org/10.1038/nnano.2014.35
69. J.R., M.M.P., D.K., Howell, S.R.: Thermal radiation heat transfer (2020)
70. Rogalski, A.: Infrared detectors: an overview. Infrared Phys. Technol. **43**(3–5), 187–210 (2002). https://doi.org/10.1016/S1350-4495(02)00140-8
71. Salisbury, J.W., D'Aria, D.M.: Emissivity of terrestrial materials in the 8–14 μm atmospheric window. Remote Sens. Environ. **42**(2), 83–106 (1992). https://doi.org/10.1016/0034-4257(92)90092-X
72. Khan, K., et al.: Recent developments in emerging two-dimensional materials and their applications. J. Mater. Chem. C. Mater. **8**(2), 387–440 (2020). https://doi.org/10.1039/C9TC04187G
73. Salihoglu, O., et al.: Graphene-based adaptive thermal camouflage. Nano Lett. **18**(7), 4541–4548 (2018). https://doi.org/10.1021/acs.nanolett.8b01746
74. Zheng, Z., et al.: In situ reduced MXene/AuNPs composite toward enhanced charging/discharging and specific capacitance. J. Adv. Ceram. **10**(5), 1061–1071 (2021). https://doi.org/10.1007/s40145-021-0491-0

75. Xue, Y., Zheng, Y., Wang, E., Yang, T., Wang, H., Hou, X.: $Ti_3C_2T_x$ (MXene)/Pt nanoparticle electrode for the accurate detection of DA coexisting with AA and UA. Dalton Trans. **51**(11), 4549–4559 (2022). https://doi.org/10.1039/D2DT00110A
76. Li, L. et al.: Ultrathin titanium carbide (MXene) films for high-temperature thermal camouflage. Adv. Funct. Mater **31**(35) (2021). https://doi.org/10.1002/adfm.202101381
77. Li, J., et al.: Multifunctional antimony tin oxide/reduced graphene oxide aerogels with wide-band microwave absorption and low infrared emissivity. Compos. B Eng. **231**, 109565 (2022). https://doi.org/10.1016/j.compositesb.2021.109565
78. Xu, W., Li, S., Zhang, W., Ouyang, B., Yu, W., Zhou, Y.: Nitrogen-Doped $Ti_3C_2T_x$ MXene Induced by plasma treatment with enhanced microwave absorption properties. ACS Appl. Mater. Interf. **13**(41), 49242–49253 (2021). https://doi.org/10.1021/acsami.1c17015
79. Deng, Z., et al.: Controllable surface-grafted MXene Inks for electromagnetic wave modulation and infrared anti-counterfeiting applications. ACS Nano **16**(10), 16976–16986 (2022). https://doi.org/10.1021/acsnano.2c07084
80. Zheng, Z., et al.: The oxidation and thermal stability of two-dimensional transition metal carbides and/or carbonitrides (MXenes) and the improvement based on their surface state. Inorg. Chem. Front. **8**(9), 2164–2182 (2021). https://doi.org/10.1039/D1QI00041A
81. Li, Y. et al.: 2D $Ti_3C_2T_x$ MXenes: visible black but infrared white materials. Adv. Mater. **33**(41) (2021). https://doi.org/10.1002/adma.202103054
82. Feng, S., Yi, Y., Chen, B., Deng, P., Zhou, Z., Lu, C.: Rheology-guided assembly of a highly aligned MXene/cellulose nanofiber composite film for high-performance electromagnetic interference shielding and infrared stealth. ACS Appl. Mater. Interf. **14**(31), 36060–36070 (2022). https://doi.org/10.1021/acsami.2c11292
83. Liu, J., et al.: Preparation of a PCM microcapsule with a graphene oxide platelet-patched shell and its thermal camouflage applications. Ind. Eng. Chem. Res. **58**(41), 19090–19099 (2019). https://doi.org/10.1021/acs.iecr.9b03530
84. Wang, G., Chen, X., Liu, S., Wong, C., Chu, S.: Mechanical chameleon through dynamic real-time plasmonic tuning. ACS Nano **10**(2), 1788–1794 (2016). https://doi.org/10.1021/acsnano.5b07472
85. Li, B.-X., et al.: Adaptive and adjustable MXene/reduced graphene oxide hybrid aerogel composites integrated with phase-change material and thermochromic coating for synchronous visible/infrared camouflages. ACS Nano **17**(7), 6875–6885 (2023). https://doi.org/10.1021/acsnano.3c00573
86. Yan, J., et al.: A bionic nano-band-aid constructed by the three-stage self-assembly of peptides for rapid liver hemostasis. Nano Lett. **21**(17), 7166–7174 (2021). https://doi.org/10.1021/acs.nanolett.1c01800
87. Cau, M.F., Strilchuk, A.W., Kastrup, C.J.: Nanomedicines for hemorrhage control. J. Thromb. Haemost. **19**(4), 887–891 (2021). https://doi.org/10.1111/jth.15211
88. Luo, Z., Wang, S., Zhang, S.: Fabrication of self-assembling d-form peptide nanofiber scaffold d-EAK16 for rapid hemostasis. Biomaterials **32**(8), 2013–2020 (2011). https://doi.org/10.1016/j.biomaterials.2010.11.049
89. Bagrov, D., et al.: Morphology and aggregation of RADA-16-I peptide Studied by AFM, NMR and molecular dynamics simulations. Pept. Sci. **106**(1), 72–81 (2016). https://doi.org/10.1002/bip.22755
90. Kumar, V.A., Wickremasinghe, N.C., Shi, S., Hartgerink, J.D.: Nanofibrous snake venom Hemostat. ACS Biomater. Sci. Eng. **1**(12), 1300–1305 (2015). https://doi.org/10.1021/acsbiomaterials.5b00356
91. Hao, R., et al.: Rapid hemostasis resulting from the synergism of self-assembling short peptide and *O* -Carboxymethyl chitosan. ACS Appl. Mater. Interfaces **12**(50), 55574–55583 (2020). https://doi.org/10.1021/acsami.0c15480
92. Hubbard, W.B., Lashof-Sullivan, M.M., Lavik, E.B., VandeVord, P.J.: Steroid-loaded hemostatic nanoparticles combat lung injury after blast trauma. ACS Macro Lett. **4**(4), 387–391 (2015). https://doi.org/10.1021/acsmacrolett.5b00061

93. Cheng, J., et al.: Facile assembly of cost-effective and locally applicable or injectable nanohemostats for hemorrhage control. ACS Nano **10**(11), 9957–9973 (2016). https://doi.org/10.1021/acsnano.6b04124
94. Sun, X., Fang, Y., Tang, Z., Wang, Z., Liu, X., Liu, H.: Mesoporous silica nanoparticles carried on chitosan microspheres for traumatic bleeding control. Int. J. Biol. Macromol. **127**, 311–319 (2019). https://doi.org/10.1016/j.ijbiomac.2019.01.039
95. Turaga, U., Kendall, R.J., Singh, V., Lalagiri, M., Ramkumar, S.S.: Advances in materials for chemical, biological, radiological and nuclear (CBRN) protective clothing. in Advances in Military Textiles and Personal Equipment, Elsevier, pp. 260–287 (2012). https://doi.org/10.1533/9780857095572.2.260.
96. Alhashmi Alamer, F., Almalki, G.A.: Fabrication of conductive fabrics based on SWCNTs, MWCNTs and graphene and their applications: a review. Polymers (Basel) **14**(24), 5376 (2022). https://doi.org/10.3390/polym14245376
97. Shah, M.A., Pirzada, B.M., Price, G., Shibiru, A.L., Qurashi, A.: Applications of nanotechnology in smart textile industry: a critical review. J. Adv. Res. **38**, 55–75 (2022). https://doi.org/10.1016/j.jare.2022.01.008
98. Rehan, M., El-Naggar, M.E., Mashaly, H.M., Wilken, R.: Nanocomposites based on chitosan/silver/clay for durable multi-functional properties of cotton fabrics. Carbohydr. Polym. **182**, 29–41 (2018). https://doi.org/10.1016/j.carbpol.2017.11.007
99. Rehman, A., Houshyar, S., Reineck, P., Padhye, R., Wang, X.: Multifunctional smart fabrics through nanodiamond-polyaniline nanocomposites. ACS Appl. Polym. Mater. **2**(11), 4848–4855 (2020). https://doi.org/10.1021/acsapm.0c00789
100. Thangakameshwaran, N., Santhoskumar, A.U.: Cotton fabric dipped in carbon nano tube ink for smart textile applications. Int. J. Polym. Mater. Polym. Biomater. **63**(11), 557–562 (2014). https://doi.org/10.1080/00914037.2013.854227
101. Ali, S.W., Basak, S., Shukla, A.: Nanoparticles: a potential alternative to classical fire retardants for textile substrates. in *Handbook of Nanomaterials for Manufacturing Applications*, Elsevier , pp. 265–278 (2020). https://doi.org/10.1016/B978-0-12-821381-0.00011-9
102. Popescu, M., Ungureanu, C.: Green nanomaterials for smart textiles dedicated to environmental and biomedical applications. Materials **16**(11), 4075 (2023). https://doi.org/10.3390/ma16114075
103. Attia, N.F., Mohamed, A., Hussein, A., El-Demerdash, A.-G.M., Kandil, S.H.: Greener bio-based spherical nanoparticles for efficient multilayer textile fabrics nanocoating with outstanding fire retardancy, toxic gases suppression, reinforcement and antibacterial properties. Surf. Interf. **36**, 102595 (2023). https://doi.org/10.1016/j.surfin.2022.102595
104. Goda, E.S., Abu Elella, M.H., Hong, S.E., Pandit, B., Yoon, K.R., Gamal, H.: Smart flame retardant coating containing carboxymethyl chitosan nanoparticles decorated graphene for obtaining multifunctional textiles. Cellulose **28**(8), 5087–5105 (2021). https://doi.org/10.1007/s10570-021-03833-7
105. Makhlouf, G., Abdelkhalik, A., Ameen, H.: Preparation of highly efficient chitosan-based flame retardant coatings with good antibacterial properties for cotton fabrics. Prog. Org. Coat. **163**, 106627 (2022). https://doi.org/10.1016/j.porgcoat.2021.106627
106. Nakhate, A.V., Pethsangave, D.A., Yadav, G.D., Some, S., Tekade, P.V.: Phosphorus grafted chitosan functionalized graphene oxide-based nanocomposite as a novel flame-retardant material for textile and wood. React. Chem. Eng. **8**(4), 804–814 (2023). https://doi.org/10.1039/D2RE00436D
107. Kundu, C.K., Song, L., Hu, Y.: Chitosan-metal oxide nanoparticle hybrids in developing bi-functional polyamide 66 textiles with enhanced flame retardancy and wettability. Appl. Surf. Sci. Adv. **7**, 100202 (2022). https://doi.org/10.1016/j.apsadv.2021.100202
108. Kumar Kundu, C. et al.: A green approach to constructing multilayered nanocoating for flame retardant treatment of polyamide 66 fabric from chitosan and sodium alginate. Carbohyd. Polym. **166**, 131–138 (2017). https://doi.org/10.1016/j.carbpol.2017.02.084
109. Sreeraj, P.R., Kr. Mishra, S., Singh, P.K.: Characteristic features and functions of nanocellulose for its feasible application in textile industry. in *Nanocellulose Materials*, Elsevier, pp. 105–122 (2022). https://doi.org/10.1016/B978-0-12-823963-6.00013-2

110. Lin, N., Dufresne, A.: Nanocellulose in biomedicine: current status and future prospect. Eur. Polym. J. **59**, 302–325 (2014). https://doi.org/10.1016/j.eurpolymj.2014.07.025
111. Toha, N. et al.: Preliminary development of laminated nanocomposite from Nanocellulose-Kevlar for military application. Intern. J. Curr. Res. Sci. Eng. Technol. **1**(Spl-1), 566 (2018). https://doi.org/10.30967/ijcrset.1.S1.2018.566-570
112. Wei, D.W., Wei, H., Gauthier, A.C., Song, J., Jin, Y., Xiao, H.: Superhydrophobic modification of cellulose and cotton textiles: Methodologies and applications. J. Bioresou. Bioprod. **5**(1), 1–15 (2020). https://doi.org/10.1016/j.jobab.2020.03.001
113. Electronic Textiles. Elsevier (2015). https://doi.org/10.1016/C2014-0-03020-1
114. Shojaeiarani, J., Bajwa, D.S., Chanda, S.: Cellulose nanocrystal based composites: a review. Compos. Part C: Open Access **5**, 100164 (2021). https://doi.org/10.1016/j.jcomc.2021.100164
115. Jin, Y., et al.: Fabrication of graphene based durable intelligent personal protective clothing for conventional and non-conventional chemical threats. Nanomaterials **11**(4), 940 (2021). https://doi.org/10.3390/nano11040940
116. Tariq, S., Tariq, Z., Malik, M.H., Anwar, F., Abbas, M., Khan, A.: Development of thermal camouflage polyester-cotton blended fabric for defense security personnel via coating with graphene oxide and reduced graphene oxide. J. Nat. Fibers **19**(16), 14222–14234 (2022). https://doi.org/10.1080/15440478.2022.2117759
117. El Moumen, A., Tarfaoui, M., Lafdi, K., Benyahia, H.: Dynamic properties of carbon nanotubes reinforced carbon fibers/epoxy textile composites under low velocity impact. Compos. B Eng. **125**, 1–8 (2017). https://doi.org/10.1016/j.compositesb.2017.05.065
118. Hasan, K.M.F., Wang, H., Mahmud, S., Genyang, C.: Coloration of aramid fabric via in-situ biosynthesis of silver nanoparticles with enhanced antibacterial effect. Inorg. Chem. Commun. **119**, 108115 (2020). https://doi.org/10.1016/j.inoche.2020.108115
119. He, H., et al.: A wearable self-powered fire warning e-textile enabled by aramid nanofibers/MXene/silver nanowires aerogel fiber for fire protection used in firefighting clothing. Chem. Eng. J. **460**, 141661 (2023). https://doi.org/10.1016/j.cej.2023.141661
120. Emam, H.E., El-Shahat, M., Hasanin, M.S., Ahmed, H.B.: Potential military cotton textiles composed of carbon quantum dots clustered from 4–(2,4–dichlorophenyl)–6–oxo–2–thioxohexahydropyrimidine–5–carbonitrile. Cellulose **28**(15), 9991–10011 (2021). https://doi.org/10.1007/s10570-021-04147-4
121. Kharat, D.K., Muthurajan, H., Praveenkumar, B.: Present and futuristic military applications of nanodevices. Synth. React. Inorg. Met.-Org. Nano-Met. Chem. **36**(2), 231–235 (2006). https://doi.org/10.1080/15533170500524801
122. Song, Z., Liu, W., Xian, M., Jin, M.: Facile fabrication of energetic nanocomposite materials by polydopamine. Int. J. Mol. Sci. **24**(22), 16199 (2023). https://doi.org/10.3390/ijms242216199
123. Zhang-Yunhua, et al.: Preparation and characteristic of modified double-base propellant modified with cellulose nanofibers. Integr. Ferroelectr. **171**(1), 115–123 (2016). https://doi.org/10.1080/10584587.2016.1174561
124. Li, S., et al.: Denitration and nano-ZnO loading strategy to enhance thermal conductivity and ablation inhibition of nitrocellulose. Cellulose **30**(17), 10749–10766 (2023). https://doi.org/10.1007/s10570-023-05556-3
125. Bao, P., et al.: Design and preparation of a novel nano-composite coating for desensitization of CL-20. Diam. Relat. Mater. **141**, 110578 (2024). https://doi.org/10.1016/j.diamond.2023.110578
126. Song, X., Huang, Q., Jin, B., Peng, R.: Preparation of desensitizing CL-20/rGO composites by in-situ reduction. Propell. Explos. Pyrotech. **45**(8), 1293–1299 (2020). https://doi.org/10.1002/prep.202000010
127. Li, Z., Wang, Y., Zhang, Y., Liu, L., Zhang, S.: CL-20 hosted in graphene foam as a high energy material with low sensitivity. RSC Adv. **5**(120), 98925–98928 (2015). https://doi.org/10.1039/C5RA16228A
128. Huang, B., et al.: Stabilization of ε-CL-20 crystals by a minor interfacial doping of polydopamine-coated graphene oxide. Appl. Surf. Sci. **510**, 145454 (2020). https://doi.org/10.1016/j.apsusc.2020.145454

129. Pokrajac, L., et al.: Nanotechnology for a sustainable future: addressing global challenges with the international network4sustainable nanotechnology. ACS Nano **15**(12), 18608–18623 (2021). https://doi.org/10.1021/acsnano.1c10919
130. Kanchi, S., Ahmed, S., Sabela, M.I., Hussain, C.M. (Eds.).: Nanomaterials: biomedical, environmental, and engineering applications. Wiley (2018). https://doi.org/10.1002/9781119370383

Dr. S. K. Suja is working as an Associate Professor in the Department of Chemistry, Lady Doak College, Madurai. Her area of specializations are Biosensors, Electro-catalysis, Carbohydrate polymers, MOFs, Supramolecular gels, and Outcome-Based Education. She has guided 8 M. Phil scholars and 4 Ph. D scholars. She is the life member of Society for Biological Chemists, Society for Advancement of Electrochemical Science and Technology and Indian Society for Electroanalytical Chemistry. She is the recognized reviewer of several journals in Elsevier, RSC, and Springer publishers. She was the recipient of C V Raman Prize - 2022 for outstanding Professional and Research achievements in Chemistry. She has received funded projects from DAE, UGC, and DST. She was the resource person for 11 National and International conferences, workshops, and FDP and published 16 research articles in reputed peer-reviewed journals, 1 review and 4 book chapters. She has h-index 9 and i10-index 9 with the cite score of 213. She has organized 8 workshops and conferences.

Mrs. S. Mathiya is currently employed as Associate Professor, Department of Chemistry, Lady Doak College. She is pursuing her Ph.D under the guidance of Dr. S. K. Suja, Lady Doak College, Madurai. Her research interest includes fabrication of supramolecular gels for treating environmental pollutants. She has been selected twice as the summer research fellow in the Science Academies Summer Research Fellowship Program for students and Teachers. She has published 2 research articles, 1 review and 2 book chapters. She has organized various co-curricular programs.

Chapter 14
Nanotechnology and Governance: Regulatory Framework for Responsible Innovation

Chinnaiah Amutha, Ardhra Gopan, Indhiran Pushbalatatha, Mariappan Ragavi, and Joseph Antony Reneese

Introduction

Nanotechnology is the scientific and engineering field that focuses on via nanoscale manipulation of atoms and molecules, characterized by dimensions of approximately 100 nm or smaller [1]. The major focus remains on the size of particles, as the nanometer (nm) significantly affects material properties. This specific scale allows for the emergence of distinctive phenomena and innovative applications. This involves manufacturing and utilizing materials and devices within the range of a few nanometres to submicron sizes and incorporating these nanostructures into larger systems. Nanotechnology applications involve modelling, imaging, measuring, and manipulating materials at a specific length scale. Nanotechnology refers to revolutionary nanotechnologies, derived from new nanostructured materials, quantum dots' electronic properties, or nanodevice architecture, computation, information storage, and transmission [2]. By using nanostructured structures, these technologies can improve current technologies and offer innovative attributes with significant technical, economic, and societal consequences [3].

Nanoscience is a result of collaboration across disciplines like physics, chemistry, biology, material sciences, and engineering. Its widespread applications include the fields of food, energy, housing, communication, health, mobility, and labor. The

C. Amutha (✉)
Department of Animal Behaviour and Physiology, School of Biological Sciences, Madurai Kamaraj University, Madurai 625021, India
e-mail: amuthaji@gmail.com

A. Gopan · I. Pushbalatatha · M. Ragavi · J. A. Reneese
School of Biological Sciences, Madurai Kamaraj University, Madurai, India
e-mail: i.pushbalatha@gmail.com

J. A. Reneese
e-mail: renesse04@gmail.com

S. C. George and B. Tawiah (eds.), *Nanotechnology in Societal Development*, Advanced Technologies and Societal Change, https://doi.org/10.1007/978-981-97-6184-5_14

field offers numerous potentials for the scientific, engineering, and industrial sectors [4]. Nanotechnology has made significant advancements in recent years, including the production of carbon nanomaterials from plant waste, activated carbon for gold ion electro-reduction, and graphene-based membranes for desalination. It has also led to the development of perovskite photocatalysts for hydrogen production and biocompatible membranes. Nanotechnology has the potential to transform food and agriculture, promoting sustainable farming practices and improving food security, quality, and safety [5].

Nanomaterials, including solid lipids, carbon black, and flexible nanoliposomes, are used in various industries, including farming, shipping, processing, packing, storage, monitoring, and consumption. They offer stability, bioavailability, and UV radiation protection [6]. Nanomaterials have shown promise in addressing cancer, COVID-19, and HIV-related illnesses. Gold nanoparticles can accumulate in tumour tissues or target specific cancer cells. Researchers are exploring therapeutic applications of nanocarriers, such as dendrimers, functionalized with quantum dots [7]. Nanotechnology-based formulations have successfully delivered phages, demonstrating potential in treating eye diseases caused by resistant bacteria. Nanoparticles modified on their surface have various uses in healthcare and everyday life [8].

Governance in nanotechnology is crucial for understanding risk attributes, ensuring public health and environmental protection, and incorporating physical and social concerns. Responsible management involves democratizing governance processes, resolving regulatory capacity, and addressing information asymmetry. Risk governance focuses on including rules, legislation, and public participation, influencing the development and mitigation of risks associated with nanotechnology [9]. Innovation governance directs technological decisions towards socially agreed-upon goals, benefits, and priorities, such as competitiveness, promote health, welfare, social equity or environmental sustainability [10].

The governance of nanomaterials is shifting from risk control to incorporating regulatory and promotional elements, requiring efficient collaboration and trust among parties. The public's perception of nanomaterials is uncertain, necessitating international cooperation and harmonization for efficient regulatory frameworks. Risk governance issues in passive nanostructures include a limited understanding of toxicity, bioaccumulation, and exposure rates. Anticipatory governance faces challenges in size, structure, stakeholder involvement, and unexpected innovation processes [11]. Governance faces challenges due to divergent norms, ideals, and conflicts in different environments. Synchronization of laws and regulations across jurisdictions hinders global multilevel governance. Framing, presentation, and discussion of technology, significantly impact public perception and can lead to disagreement [12]. The aim is to incorporate life cycle and multi-criteria decision analyses into risk assessment and management techniques in developing technology ecosystems. Policymakers must consider nanotechnologies' variables and international governance changes, establishing multiple frameworks and coordinating stakeholders' use. The global nature of nanotechnology necessitates harmonizing rules across countries and regions. Effective governance requires collaboration among governments, regulatory organizations, industry, and civil society. The main

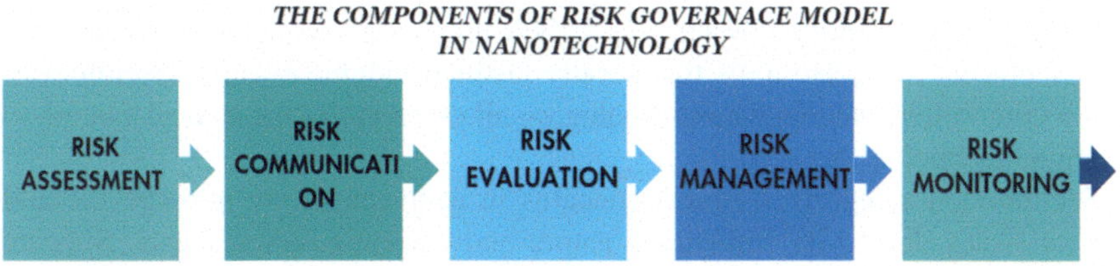

Fig. 14.1 The components of risks governance model in nanotechnology

components of the risk governance model in nanotechnology are listed in Fig. 14.1 [12]

International governance in nanotechnology aims to achieve regulatory uniformity, promote cooperation, and ensure accountable decision-making. It addresses global issues like environmental impact, health, safety, and ethical considerations, encouraging investment and technology transfer [13]. International governance collaboration enhances knowledge exchange, resource pooling, policy formulation, and influence. The organization is partnering with global organizations like ISO and OECD to establish unified frameworks for nanomaterials regulation. It promotes informed decisions, and collective global regulation, and aims to ensure society's security, welfare, human rights, and environmental preservation through statutory and voluntary standards. It is also promoting global collaboration in research and development to address knowledge gaps and evaluate risk promoting transparency and cooperation among governments, industry, academia, and civil society to address nanomaterial regulation challenges [14].

Key Principles and Ethical Considerations in Nanotechnology

The long-term social challenges of nanotechnology are uncertain but expected to be significant. It has the potential to transform healthcare by introducing new drug administration methods, developing diagnostic tools, and creating medical implants. However, it is crucial to assess the unequal allocation of these benefits and potential increased healthcare costs. The ethical dilemmas in nanotechnology emerge from multiple causes, such as the distinct characteristics of nanomaterials at the nanoscale, the social environment within specific laboratories, the fluctuating employment market over time, government goals and funding levels, and the systems of rewards and punishments.

Ethics are crucial in the responsible development of nanotechnology, defining justice, human flourishing, and sustainability. It helps identify opportunities, anticipate challenges, develop standards for assessing potential nanotechnologies, provide ethical capacity for informed decision-making, and identify limits on pursuing goals.

Ethics also help identify obstacles, establish limitations on achieving nanotechnology objectives, create tools for morally justified judgments, and develop evaluation criteria for potential nanotechnologies, all essential for societal adaptation to emerging technologies.

Schulte [15] categorizes ethical concerns in nanotechnology into specific and general issues due to its ambiguous nature and current overstatements. Specific ethical concerns may stem from distinct research procedures, products, and uses, while general ethical concerns may arise from the start, control, and administration of nanotechnology programs, as well as their interpretation in the broader scientific and cultural context. Sandler [16] and Schulte [15] emphasize the importance of ethics in nanotechnology for defining lifestyle enhancement, fairness, and sustainability. They draw attention to the dangers that nanoparticles pose to human and environmental health, which can be harmful if not considered in size and shape-dependent characteristics. As nanotechnology advances, the potential for encountering nanoparticles increases, necessitating a balance between industry benefits and environmental and human health risks.

Nanotechnology, despite its scientific and technological advancements, poses ethical and societal challenges. Privacy and security risks from surveillance devices threaten personal freedom, necessitating the creation of ethical frameworks and safeguards. The significant environmental impact of nanotechnology, particularly nanoparticles accumulating in ecosystems, necessitates comprehensive exploration and mitigation strategies for responsible development and deployment of nanotechnological applications [17]. Nanoparticles pose health and safety risks, requiring rigorous assessments and regulatory measures. Balancing technological progress with ethical principles is crucial, including ensuring harm avoidance and individual freedoms, and achieving a balance for societal benefits [15]. Achieving a balance between technological advancement and ethical principles is crucial for societal benefits, including the equitable distribution of nanotechnology benefits, harm avoidance, and individual freedoms.

The ethical considerations of nanotechnology in the workplace involve identifying and addressing potential risks, avoiding harm, and respecting individual autonomy in decision-making processes [18]. The field of responsible innovation in this rapidly growing field necessitates a comprehensive, interdisciplinary approach involving policymakers, scientists, ethicists, and the public [15].

To ensure the benefits of nanotechnology for humanity, it is crucial to address ethical challenges and address deficiencies in national and global policies. Promoting research that provides insights for policymaking, educates individuals on technology's benefits and limitations, and fosters engagement in technology governance can help mitigate conflicts.

EU member states are implementing programs to raise public awareness about nanotechnology. The **DEEPEN** (Deepening Ethical Engagement and Participation with Emerging Nanotechnologies) initiative in the UK aims to promote the social acceptability of nanotechnology advancement. The **Nanopodium** program in the Netherlands is a prominent social discussion project. Belgium's **Nanosoc** program aims to foster dialogue among researchers, businesses, and society. Germany's

Nanologue project emphasizes the advantages and repercussions of nanotechnology, elucidating its ethical, social, and legal implications. These initiatives aim to foster public discourse and promote understanding of nanotechnology [19].

In 2015 Council of Europe bioethics committee conducted a comprehensive examination of emerging technologies and their impact on human rights. The committee incorporated nanotechnologies into discussions, revealing a lack of awareness among academics and policymakers about the full scope of nanotechnology applications. The working group expressed concerns about the longevity of the human species, as self-replicating artificial intelligence could replace it. They analyzed the convergence of different technologies, and their combined effects on everyday life, societal norms, political determination, and overall quality of life, considering the European Charter for Human Rights. The committee's work aims to address the ethical concerns surrounding the planet's preservation and the societal impact of these technologies.

According to Baren, [19], the ethical aspects of responsible development of nanotechnology encompass several key roles, including:

- ***Clarifying the concepts of justice, human flourishing, and sustainability***: In nanotechnology, ethical discourse can clarify justice, human flourishing, and sustainability.
- ***Identifying opportunities for nanotechnology to achieve these objectives and anticipating potential obstacles:*** Ethical analysis and social science research can help identify nanotechnology's potential applications in addressing societal and environmental issues. However, non-technological obstacles like insufficient research facilities, lack of motivation, intellectual property rights limitations, inefficient distribution networks, compatibility with beneficiaries' lifestyles, and inadequate regulatory capabilities exist.
- ***Establishing standards for evaluating prospective nanotechnologies:*** Ethics are crucial in evaluating the ethical profiles of developing nanotechnologies, which can be equitable, environmentally friendly, empathetic, irresponsible, myopic, non-sustainable, or redundant. However, the feasibility of research initiatives like synthetic biology and memory chips with carbon nanotube technology remains uncertain. Individual evaluations are essential to address ethical concerns, ensure informed resource allocation, and policy design, and avoid public or regulatory responses hindering advancement or commercialization.
- ***Offering ethical tools and resources to individuals and organizations, enabling them to make informed decisions in adapting to emerging nanotechnologies***: Society's adaptability to future nanotechnologies relies heavily on its ethical capabilities, including professional rules, ethical frameworks, case studies, historical antecedents, and competence in recognizing, examining, and resolving ethical dilemmas. Governmental and societal capabilities are crucial for nanotechnology advancement, with government responsiveness often guided by ethical and social capacity [19, 20].
- ***Recognizing boundaries and limitations in pursuing these objectives***.

Ethics dictate the boundaries for pursuing objectives, with certain methods considered unacceptable despite potential benefits, and medical research involving human subjects requires regulation to ensure good intentions and noble objectives [19].

Overview of Existing Regulations and Policies in Nanotechnology

Research shows nanotechnologies pose complex and uncertain risks, making understanding and managing their environmental, health, and societal impacts a global and cross-border responsibility. This requires innovative risk evaluation and control methods within a governance framework, addressing the lack of clear cause-and-effect correlations [21].

Anticipatory governance is emphasized as essential, to direct innovation, preventing inappropriate uses, and promoting sustainable growth. These governance approaches consider the social and ethical consequences of nanotechnology. They aim to create rules and regulations that encourage the appropriate application of nanotechnology and foster public confidence and acceptance of this fast-progressing field. The implementation of nanotechnology governance systems is essential for handling the potential ethical and societal issues that may arise from nanotechnology. These approaches are formulated to predict and control the effects of nanotechnology in diverse domains, such as health and medicine the distribution of authority between citizens and corporations and the distribution of authority between citizens and governments. The prerequisites for successful governance encompass the ability to make well-informed decisions, the presence of empowered institutions, and the significance of education and training. Trust and confidence among stakeholders, especially the public, are crucial for the adoption and further advancement of nanotechnology [22].

The governance approaches strive to provide equitable distribution of the advantages of nanotechnology, thorough consideration and resolution of potential risks and unforeseen repercussions, and active involvement of the public in the decision-making process. In addition, the implementation of nanotechnology governance strategies is crucial for promoting responsible advancement and utilization of nanotechnology, while also addressing concerns related to privacy, security, and environmental consequences. These governance approaches consider the social and ethical consequences of nanotechnology. They aim to create rules and regulations that encourage the appropriate application of nanotechnology and foster public confidence and acceptance of this fast-progressing field.

Risk governance systems involve pre-assessment, appraisal, evaluation, characterization, management, and monitoring, focusing on transparency, involvement, responsibility, efficiency, consistency, and appropriateness. The process of risk assessment involves framing hazards, preparing early warnings, defining risks through risk appraisal, comparing outcomes through evaluation, gathering evidence, and designing actions through risk characterization and management. Continuous monitoring enhances the risk governance process [23].

Various frameworks now exist for the governance and regulation of nanotechnologies, encompassing both mandatory extensions of existing regulatory frameworks and voluntary schemes. The rapid advancement of nanotechnologies and the utilization of nano-enabled items have raised significant apprehension regarding potential hazards to the environment, health, and safety (EHS), as well as broader ethical, legal, and social implications problems (ELSIs) linked to their implementation and utilization or misapplication [24].

Research shows nanotechnologies pose complex and uncertain risks, making understanding and managing their environmental, health, and societal impacts a global and cross-border responsibility. This requires innovative risk evaluation and control methods within a governance framework, addressing the lack of clear cause-and-effect correlations [25].

Analysis of Regulatory Frameworks at the National and International Levels (USA, Europe, Japan, *Etc.*)

Various governance frameworks have been utilized in the field of nanotechnology, such as the International Risk Governance Council (IRGC) Risk Governance Framework, the FramingNano Governance Platform, and the Responsible Care Global Charter. The International Risk Governance Council (IRGC) Framework prioritizes the evaluation and assessment of risks in advance, considering societal values. It also highlights the importance of effective communication, considering the context of stakeholders, and actively involving the public [26, 27]. There is a lack of proof that Nano Risk Framework, CENARIOS, and AssuredNano, among others, effectively involve stakeholders or provide foresight in their risk management processes. The EU FP7 MARINA, SUN, and NANoREG Projects advocate for risk governance that prioritizes prevention and is based on the concept of safety by design. These projects aim to migrate towards innovation governance with NanoReg 2 while ensuring compliance with regulations [28]. The iNTeg-Risk Project, funded by the European Union, encompasses horizon scanning, risk pre-assessment, risk assessment, and risk management procedures, with a specific emphasis on nanotechnologies [29]. The primary objective of the EU H2020 caLIBRAte Project is to streamline the process of evaluating and controlling potential risks associated with both established and new nanotechnologies. This involves the testing and fine-tuning of tools used for risk assessment and management [30].

ISO, or International Organization for Standardization, is responsible for creating standards for nanotechnology, including ISO/TS 80,004–1, ISO 31000:2009 Framework, OECD protocols, and ISO Technical Committee 229. ISO/TS 80,004–1 establishes a uniform language for nanomaterials, [31] ISO 31000:2009 Framework provides risk management principles, [32] and ISO Technical Committee 229 is responsible for creating international standards for nanomaterials characterization, safety testing, and terminology. (Nanotechnologies). These standards help organizations of all sizes and sectors assess and treat potential hazards and dangers related to nanoparticles.

The European Union (EU) and the US have established regulatory frameworks for nanomaterials, including the REACH (Registration, Evaluation, Authorization, and Restriction of Chemicals legislation and the Nanotechnology Research and Development Act, and the National Nanotechnology Initiative (NNI). The REACH Regulation mandates registration and safety assessment of certain chemicals, including nanomaterials, before market placement. The United States (US) also has standards and laws, such as the IEHS (International Nanotechnology Environmental and Health Safety) Forum, aiming to promote safe and responsible advancement and utilization of nanomaterials. Global partnerships and initiatives aim to exchange information and align research efforts [33]. The Significant New Use Rule (SNUR) is being developed by the U.S. Environmental Protection Agency (EPA) to guarantee that nanoscale compounds undergo comprehensive regulatory evaluation. The rule requires individuals to submit an SNUN to the EPA at least 90 days before producing, importing, or the Toxic Substances Control Act (TSCA) Inventory includes hazardous chemicals used in the production of novel nanoscale materials [34].

In the Asian region in China, the State Food and Drug Administration ('SFDA') replaced an existing regulation, classifying medical devices made with nanometer biological materials as Class III medical devices, with the National Centre for Nanoscience and Technology setting standards [35]. Japan's Ministry of Economy, Trade, and Industry (METI) established the Committee on Safety Management for Nanomaterials to improve understanding of nanotechnology risk management [36]. The Republic of South Korea established the inter-ministerial 'National Nano-safety Strategic Plan (2012–2016)' [87]. Taiwan: The Industrial Technology Research Institute (ITRI) has been using the Nanomark certification system since 2004. The National Nanotechnology Center (Nanotec) manages an industrial standards certification system known as NanoQ [37, 38].

Committees and panels play a crucial role in setting voluntary standards for consistency, repeatability, accuracy, and safety in nanotechnology research, development, and commercialization. They are:

- ANSI-Nanotechnology Standards Panel in the U.S.
- ASTM Committee E56 British Standards Committee for Nanotechnologies (NTI/1)
- European Committee for Standardization (CEN/TC 352)
- IEC group nanotechnology standardization for electrical and electronic products and systems (TC 113)
- The Technical Committee on Standardization TC 441 'Nanotechnologies and Nanomaterials' RUSSIA [39].

The Code of Conduct for Responsible Nanosciences and Nanotechnologies Research by the European Commission [40]. The Code, sometimes known as NanoCode, was enforced throughout the European Union in 2008 after a phase of public participation and deliberation. The primary objective of the Code is to encourage secure and advantageous advancement in the realm of nanotechnologies by Member States, pertinent research funding entities (both public and commercial), research institutions, the academic research community, and standardization bodies [34, 40].

The ANSI Nanotechnology Standards Panel (ANSI-NSP) creates voluntary consensus standards for nanotechnology, considering health, safety, and environmental aspects [41]. The (Organisation for Economic Co-operation and Development) OECD Working Party on Nanotechnology develops guidance documents on nanomaterial risk assessment and safe innovation [42]. The FAO/WHO Codex Alimentarius sets international food safety standards, including for nanomaterials in food [43]. The International Council on Nanotechnology (ICNT) promotes responsible development of nanotechnologies through collaboration and knowledge sharing [44].

Challenges and Gaps in Nanotechnology Governance.

Nanotechnology has the potential to revolutionize fields like medicine, electronics, and energy, but it also poses environmental, health, and ethical risks. The use of nanoparticles in consumer goods raises privacy, security, and misuse concerns. Governance in nanotechnology involves voluntary codes of conduct, reporting schemes, and standardization efforts, addressing ethical, legal, and social issues. However, governance challenges persist, and the growing field outpaces current regulations, leaving uncertainty about long-term risks and environmental impact. The lack of clear regulations on nanotechnology advancements could lead to a lack of accountability and transparency, affecting public safety [21]. The diverse materials and uses of nanotechnology pose challenges in creating a comprehensive regulatory framework, and the rapid pace of advancements may not match the rapid pace of technical advancements [45].

The complexity of risk evaluation due to a limited understanding of nanoparticles is exacerbated by insufficient data on exposure rates and regulatory mechanisms. Current monitoring systems, categorization methods, and risk assessment methodologies may not be suitable for nanoparticles. Improved communication is needed to establish confidence in potential dangers and advantages [45].

Nanotechnology management is a complex issue with numerous challenges, including regulatory deficiencies, moral dilemmas, and ambiguities. It requires flexibility, transparency, and a commitment to comprehensive procedures. The lack of understanding of Environment, Health, and Safety (EHS) and sustainability, fragmented institutional structures, and legal support for nanotechnology risk governance further complicate matters. Regulatory agencies often focus on individual incidents rather than the entire subject's lifespan. Addressing these issues directly can ensure nanotechnology's potential for a future prioritizing safety, ethics, and societal welfare (Renn and Roco's [45]).

Transparent communication and understanding of scientific topics are crucial for building trust with the public. Involving stakeholders like NGOs, industry representatives, and local communities in decision-making processes is essential for responsible development in nanotechnology research. This involves ethical behaviour, considering both positive and negative consequences, and active public participation. Implementing public participation is challenging but essential for responsible development, as it ensures fair progress and avoids unfair progress [46].

The challenges in nanotechnology governance include rapid advancements, diverse materials and applications, uncertainties in knowledge, ethical, legal, and social issues, current protocols' sufficiency, and global collaboration [45]. These issues are exacerbated by Regulatory capacity, cooperation across agencies, information asymmetry, and overlapping mandates and functions [21].

Ethical Challenges

- arise from the potential hazards and benefits of nanotechnology, including concerns of justice, safety, and risks. The integration of nanotechnology with domains like AI and gene editing raises ethical issues, necessitating proactive ethical frameworks to ensure appropriate utilization and mitigate potential misuse. The discipline faces uncertainties, including inherent hazards and advantages, and lack of understanding about the long-term consequences of nanoparticle exposure. A delicate balance between innovation, progress, public safety, and ethical accountability is essential [47]. Anticipating ethical dilemmas related to active nanostructures and synthetic biology is essential yet challenging.
- ***Effective management*** of nanomaterials and environmental assessments are crucial. The absence of synchronization among stakeholders and political agencies, coupled with deficiencies in global and national regulatory frameworks, is a significant challenge in nanotechnology. Regulatory ambiguities hinder industrial innovation, especially for small organizations, and companies face difficulties in accurately assessing their risk profiles [48]. Cognitive biases and deficiencies in global governance also impact decision-making in nanotechnology matters [3].
- ***Regulatory gaps*** in nanotechnology refer to the lack of clear directives for its advancement and usage, leading to a lack of responsibility and transparency. Obtaining evidence for regulatory changes can be time-consuming and costly. The advancement of nanotechnology has exceeded the need for a regulatory framework, and the scarcity of information about potential risks poses challenges in evaluating and controlling these risks. Governments' inertia and interconnected institutional interests also hinder nanotechnology governance. The convergence of academic research and business incentives also blurs the line between independent research and commercialization [49].

Governance and Risk Assessment in Nanotechnology

Role of Government Agencies, Stakeholders, and Experts in Governing Nanotechnology

The successful governance of nanotechnology relies on the cooperation and involvement of government agencies, stakeholders, and experts. Government agencies are tasked with developing regulatory frameworks to mitigate risks and protect public health and the environment, funding research to understand nanotechnology's effects, and promoting international collaboration (Fig. 14.2) The scientific community researches to inform policy decisions and enhance nanotechnology's understanding. Industry participates in nanotech product design, production, and marketing, adhering to regulations and promoting responsible innovation. Civil Society Organizations advocate for transparency, public participation, and equitable distribution of nanotechnology benefits [50]. Experts emphasize the importance of technical expertise in shaping policy and decision-making processes, communicating the latest scientific findings to policymakers and industry, and fostering interdisciplinary collaboration to address nanotechnology's full spectrum of implications [9].

The aim is to incorporate life cycle and multi-criteria decision analyses into risk assessment and management techniques in developing technology ecosystems. Policymakers must consider nanotechnologies' variables and international governance changes, establishing multiple frameworks and coordinating stakeholders' use. The global nature of nanotechnology necessitates harmonizing rules across countries and regions. Effective governance requires collaboration among governments, regulatory organizations, industry, and civil society [12].

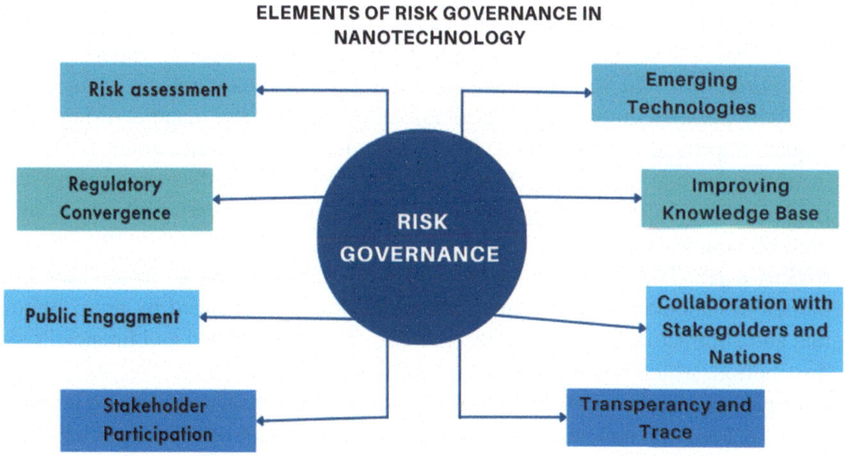

Fig. 14.2 Elements of risk governance in nanotechnology

The government plays a crucial role in addressing these concerns and ensuring the genuine benefit of nanotechnology to humanity. Risk management emphasizes risk management throughout the life cycle of products containing nanomaterials, including research, development, manufacturing, use, disposal, and recycling. Strengthening technical expertise and institutional infrastructure within regulatory bodies enhances their effectiveness in monitoring and enforcing regulations related to nanotechnologies. Public engagement builds support for policies and aligns with societal values and expectations [9].

(i) Risk assessment methods and practices in nanotechnology

Enhanced risk governance mechanisms require improved foresight and supplementary risk governance mechanisms, such as those based in the insurance sector, to fill gaps in regulatory and risk management frameworks. Predictive methods in nanotechnology governance offer numerous benefits, including enabling informed decision-making, regulatory compliance, best practices, and risk management choices. They can establish a universal standard for nanomaterial regulation at national and international levels. Advanced tools and strategies, such as risk mitigation, risk banding, risk transfer considering user capacities and needs, Decision Support Systems (DSS), risk screening and ranking schemes, Quantitative Multicriteria Decision Analysis (MCDA), and in silico methods, can enhance risk decision-making. These techniques can mitigate risks, enhance exposure controls, streamline testing processes, and reduce time and expense. They also provide relevant information about occupational, consumer, and environmental contexts, including various stages of the nanomaterial's life cycle. Overall, predictive methods can help mitigate risks, streamline testing processes, and reduce costs associated with nanotechnology.

- ***Risk banding:*** Following the precautionary principle, the risk assessment method identifies potential threats to human and environmental health and the environment through the following steps: problem formulation, exposure assessment, hazard assessment, and risk and uncertainty characterization. It then focuses on the most serious possible scenarios to determine the likelihood of negative impacts [51].
- ***Risk mitigation:*** The strategy aims to minimize the risks associated with nanomaterials by implementing technological measures such as safer manufacturing techniques, improved handling procedures, and enhanced exposure controls. These measures include reducing technical risks, improving manufacturing processes, implementing safer handling methods, and enhancing exposure controls like high-efficiency filters [52].
- ***Risk transfer:*** The technique involves creating legally binding agreements among parties in nanomaterials life cycle and value chains, influenced by noncontractual and contractual responsibilities, risk insurability, and legal and financial risk distribution, which significantly impacts decision-making [53].
- ***User capacities and needs:*** The tool assesses user needs and abilities to provide practical advice on nanomaterials decision-making throughout the value chain and life cycle. Customization of outcomes is crucial for the governance framework's value to stakeholders, ensuring understanding and proficiency [54].

- ***Decision support systems (DSS)***: Many FP7-funded initiatives, including SUN and GUIDEnano, have created decision support systems (DSSs) and prototypes for risk assessment and management of nanomaterials. These systems integrate information from risk assessment data, regulatory requirements, and economic considerations to help decision-makers make informed choices about nanotechnology development and application [55].
- ***Risk screening***: Hristozov [56] came up with a novel way to find nanomaterial health risks by using multicriteria decision analysis (MCDA). This approach quantifies uncertainty from hazard and exposure evaluations, prioritizing exposure scenarios and ranking risks. The methodology evaluates data quality, considering adequacy, reliability, statistical significance, and toxicological significance.
- **In silico *methods:*** Methods such as nano-QSAR can be employed to complete data gaps in groups of comparable nanomaterials, thereby diminishing the time and expense of testing [57].
- Evaluation of the effectiveness of governance mechanisms and risk assessment in ensuring responsible innovation

Risk analysis (RATA) is a scientific method used to evaluate the health effects of chemicals, based on toxicology, statistics, ecology, and chemistry It includes exposure, life cycle, and technology assessments. Technology Assessment (TA) evaluates the potential effects and futures of technology in society, focusing on public perception, stakeholder perspectives, changes in responsibilities, and liabilities among actors. TA is rooted in social sciences and involves participatory methods to address social impacts. The combination of RATA and TA provides a scientific basis for assessing risks associated with new technological developments, contributing to responsible innovation [58]. Risk analysis and technology assessment were integrated into one RATA theme, while the Technology Assessment (TA) program studied nanotechnology developments' dynamics, societal impacts, ethics, social equity, norm protection, public perception, and governance [59]. Several evaluation techniques, including technology evaluation, risk analysis, and life cycle analysis, are intended to be combined in the TranSTEP process [58].

Life Cycle Assessment (LCA) tools determine how nanoparticles affect both human and environmental health from extraction to disposal, identifying potential risks and promoting sustainable practices. They help identify hotspots and develop more sustainable practices by assessing the entire lifecycle of a product or process. The SUNDS Tool for International Nanotechnology Governance integrates technical criteria and stakeholder values for sustainable nanotechnology decisions. It uses multi-criteria decision analysis and LICARA NanoSCAN framework for SMEs to evaluate life-cycle benefits and risks, with potential for expansion [60]. Control Banding tools categorize nanomaterials based on hazard and exposure potential, offering a framework for risk management measures like Stoffenmanager, Nano, and NanoSafer [61].

To ensure responsible and sustainable use, manage potential hazards and environmental repercussions, and contribute to effective governance in this field, these tools guide the development and regulation of nanotechnologies.

Public Perception and Awareness of Nanotechnology and Its Governance Issues

The public's involvement in nanotechnology regulatory decisions is crucial for responsible governance. Research emphasizes understanding public perception, potential hazards and benefits, and regulatory frameworks. Efficient regulatory decision-making requires a comprehensive understanding of public views and stakeholder confidence. Current discussions focus on implementing a coordinated regulatory strategy and creating a Nanotechnology Regulatory Coordination Agency for centralized supervision. Public perception plays a crucial role in determining regulatory decisions and governance frameworks, promoting responsible and beneficial advancements in nanotechnology. Therefore, incorporating public opinions into regulatory decision-making is essential for responsible governance. Transforming toxicological methods and creating nano-specific recommendations are needed.

Nanotechnology exhibits promise in environmental preservation, sustainable energy, healthcare, and electronics. Nanotechnology understanding is low, with country variances, according to research. The public's view of nanotechnology is influenced by factors such as media portrayal, faith in regulatory agencies, knowledge of toxicity, environmental hazards, authenticity, psychometric variables, cultural prejudice, religious affiliation, socioeconomic status, and level of education, as stated by Satterfield [62]. The public views it as remote, unknown, and intangible. Nanotechnology's numerous applications and risk profiles make communication difficult [63]. Government agencies emphasize concerns over the toxicity of nanomaterials and the absence of effective mechanisms to enforce legislation about environmental, health, and safety issues.

However, it is important to be cautious about relying solely on education, as people may resist change based on their values. Religion negatively affects nanotechnology support, but it has mixed effects on perceived risk [64]. Gender-based mixed attitudes towards nanotechnology are influenced by mass media, with women showing less stability, and opinions can shift rapidly, with information not always changing opinions [15].

Most people in the world are unaware of the potential hazards and benefits of nanotechnology, despite its generally favorable public sentiment. Public trust in science authority affects media coverage of scientific ambiguity [65]. Enhancing public awareness, conducting scientific research, and ensuring honest communication of risks and benefits are crucial for regulatory decision-making. According to surveys, academics and non-researchers are increasingly concerned about the toxicity of poorly understood nanomaterials [66]. Researchers, communicators, and politicians should adapt their strategies by engaging in open discourse and considering varied perspectives to foster trust and promote responsible development and adoption of this emerging technology. To tackle issues concerning the public, it is necessary to integrate nanotechnology into science education, establish programs that promote understanding of risks, and utilize evidence maps to simplify complex information [67].

Nanotechnology's nature, benefits, hazards, and safety are unclear, making it hard to express hopes and anxieties [68]. The European Commission and other international agencies are studying nanomaterial diffusion's toxicology [69]. Nanotechnology dangers must be communicated clearly. Proper labeling of nanotechnology products is a crucial strategy for communicating risks, and further research is required to comprehend the impact of labeling on public perception [70]. To establish confidence in nanotechnology, it is important to prioritize openness, cultivate public confidence in government and regulatory bodies, acknowledge the influence of reputable institutions and scientists in disseminating information, enhance trust in scientists through seminars and active involvement, and explore proactive and transparent communication strategies [71, 72]. Policy frameworks are difficult to construct without a scientific basis, and health and environmental implications require urgent investigation and testing. In silico and mathematical models can forecast nanoproduct toxicity.

The below paragraph emphasizes the importance of public participation in policy processes, promoting open discussion, raising awareness, addressing safety concerns, and promoting responsible development. It also emphasizes ethical standards, strict rules, independent investigation, and involving experts.

Potential strategies to shift public perception toward a more balanced and informed way to Enhance public awareness and comprehension are:

- ***Effective communication and outreach initiatives***. They are crucial for understanding nanotechnology. Break down complex concepts using relatable examples and visual aids. Organize workshops, demonstrations, and public events to provide hands-on experiences and encourage the use of educational resources for diverse age groups and backgrounds.
- ***Promote independent investigation and risk evaluation:*** The text emphasizes the importance of transparency and open dialogue in addressing potential hazards and uncertainties in nanotechnology. It emphasizes the role of trust in science in shaping public views but acknowledges the complexity and ethical dilemmas in the field despite growing information and limited personal experience [73]. Advocate for autonomous research on nanomaterial safety and ecological consequences, promote data availability, strict rules, and public participation in regulatory procedures.
- ***Emphasize responsible development and ethical considerations:*** Honor ethical nanotechnology developers and researchers. Increase voices for safe and beneficial applications. These tactics can improve public awareness of nanotechnology and generate confidence and support for its responsible development and use for society. To promote ethical nanotechnology development and use, support the EC's Code of Conduct for Responsible Nanoscience and Nanotechnologies Research.
- Society faces challenges like mass media dominance, fragmented media representations, regulatory fragmentation, and definition disagreements in nanotechnologies. Limited reporting, unclarity in media coverage, regulatory inconsistency, and communication barriers hinder progress and influence popular opinion [74].

Nanotechnology has low dread risk, making it hard to grab public attention and communicate risk scenarios [75].

- ***Promote social and environmental responsibility:*** Show how nanotechnologies can sustainably and fairly address climate change, energy, and healthcare.
- ***Make decisions with stakeholders:*** Discuss nanotechnology development and its effects with residents, community groups, and other stakeholders.
- ***Display nanotechnology applications:*** Show how nanoparticles are helping medicine, energy, and environmental cleanup. Share responsible development and positive effect successes
- ***Community engagement and need-addressing***: Support local nanotechnology projects that address community issues. Show how nanotechnology may improve local development.

International Cooperation and Harmonization of Regulatory Frameworks Across Different Countries.

International collaboration is crucial for coordinating rules, exchanging expertise, and facilitating capacity development in underdeveloped nations. Regulatory frameworks must be flexible, and adaptable, and maintain safety and ethical norms. Open communication and public involvement in decision-making foster trust. Investing in research and development helps understand long-term risks and inform governance actions. The policy and regulatory framework for nanotechnology is disjointed and inadequate, both domestically and globally. A well-coordinated international governance system is needed to address gaps in regulation, particularly in environmental, health, and safety concerns. International governing agencies are adapting to oversee nanotechnologies' advantages and potential hazards [76].

International collaborations in nanotechnology risk assessment and management have resulted in mutual recognition agreements, simplified regulatory compliance, and facilitated international trade and innovation. They have also fostered transparent communication strategies, public trust, and capacity building in developing countries. Collaboration has also enabled collective foresight and scenario planning, enabling proactive approaches to address risks and opportunities posed by new nanotechnology applications. Overall, these collaborations have contributed to responsible governance and inclusivity in the field [26].

International collaboration is vital in addressing nanotechnology risks by sharing knowledge, harmonizing data formats, and developing risk assessment frameworks. This ensures responsible development and application of nanotechnology, filling knowledge gaps and accelerating global development. This helps establish standards for reliable risk assessment and comparison across national boundaries, avoiding trade barriers and enabling data comparison across national boundaries [77].

(i) Overview of international collaborations and initiatives in regulating nanotechnology

The International Risk Governance Council (IRGC) has identified a governance gap between the legislative and regulatory framework and the unique properties of nanotechnology, which has led to effective international cooperation in risk assessment and management. The IRGC has established a framework for nanotechnology risk governance, addressing environmental, health, and safety concerns, developing common regulatory principles, and mutual recognition of testing data and regulatory decisions [76, 77].

The EU H2020 caLIBRAte project is developing the Emerging Risk Management Framework (ERMF) to improve risk assessment and management in nanotechnologies. EU-funded iNTeg-Risk combined with (Emerging Risk Management Framework) ERMF, which includes horizon scanning, risk pre-assessment, risk assessment, risk management, monitoring and review, and risk communication activities, is a significant step in international collaboration in developing risk governance frameworks for nanotechnologies. The 'Work health and safety assessment tool for handling engineered nanomaterials' is an international tool developed to assist stakeholders in managing risks associated with nanomaterials, guiding risk assessment, management, and communication [76, 78].

Public–private partnerships in nanotechnology aim to increase collaboration, study biological interactions of nanoscale materials, and investigate prospects for innovation in generating safe and effective products under regulatory control. The Nanotechnology Characterization Laboratory (NCL) is a public–private partnership between the National Cancer Institute (NCI), the U.S. Food and Drug Administration (FDA), and NIST (National Cancer Institute the National Institute of Standards and Technology) offering preclinical characterization and testing services for nanoparticles for cancer therapies. The Nanotechnology Innovation Center of Kansas State (NICKS), a public–private partnership between Kansas State University and the Kansas Bioscience Authority, offers research and development services for nanotechnology-based products. National Research Council Canada (NRC) is a public–private partnership between the University of Missouri and the Missouri Technology Corporation. The NRC provides research and development services for nanotechnology-based products. The Nanotechnology Research and Development Partnership (NRDP), a public–private collaboration between the University of California, Berkeley, and Dow Chemical Company, focuses on developing nanotechnology-based water purification solutions. The Nanotechnology Research Partnership (NRP), a public–private collaboration between the Los Angeles, the California NanoSystems Institute, and University of California, focuses on developing nanotechnology-based energy storage solutions [79].

The Nanotechnology Research Partnership (NRP), a public–private collaboration between the University of California, Los Angeles, and the California NanoSystems Institute, focuses on developing nanotechnology-based solutions. MIT, Stanford, and Berkeley have established nanotechnology research centers and programs, collaborating with industry and NGOs to advance the field. NGOs like the Project on Emerging Nanotechnologies, Environmental Defense Fund, and Nanotechnology Consumer Education and Advocacy Project are promoting public awareness, responsible nanotechnology development, and governance issues [79].

The International Risk Governance Council (IRGC) in 2006 identified the following roles of international collaboration in nanotechnology governance:

- ***Risk Assessment and management***: Collaborative efforts in nanomaterials risk identification, assessment, and management can establish consistent approaches, avoid duplication, and resolve uncertainties and knowledge gaps. This approach enables experts from different disciplines to share knowledge and address uncertainties.
- ***Regulatory Convergence and Trade Facilitation***: Collaboration between nations can establish nanomaterial regulation principles, simplify compliance, and promote international trade and innovation by reducing testing requirements and promoting mutual recognition agreements.
- ***Public Engagement and Transparency***: International collaboration fosters transparency in communication about nanotechnology risks and benefits, fostering public trust and capacity building in developing countries for risk assessment, regulation, and public engagement.
- ***Adapting to Emerging Technologies and Applications***: Collaboration in nanotechnology allows for proactive risk and opportunity management, enabling countries to design adaptable governance frameworks that can evolve with rapid innovation.

Thus, International collaboration is crucial for responsible nanotechnology development and application. It involves sharing knowledge, addressing risks, facilitating trade, building public trust, and adapting to the evolving landscape, ultimately enabling a sustainable and prosperous future [76].

Emerging Technologies and Future Challenges

As nanotechnology continues to evolve and expand, governance becomes increasingly vital to managing its risk and realizing its potential. The integration of nanotechnology with other disciplines is revolutionizing the interaction with the world. Nanoparticles coated with antibodies can be used for precise drug delivery in cancer therapy, while nanoelectronics merges nanotechnology and electronics to produce smaller, quicker, and more effective devices. Quantum dots improve display quality and lower power usage in display technology. Nanomanufacturing allows precise control at atomic and molecular scales, while 3D printing using nanomaterials provides personalized implants for medical purposes [79, 80]. Nano informatics combines nanotechnology and information science to handle large volumes of data produced in nanoscience research. Nanoenergy is merging with energy systems to provide advanced energy storage and conversion technologies. Carbon-based nanomaterials, such as carbon dots, graphene quantum dots, and carbon nanotubes, are leading advancements in electronics, health, and energy due to their unique characteristics and wide-ranging uses. Artificial Intelligence (AI) is essential for evaluating large datasets in nanotechnology research and development, enhancing data collection using inexpensive, unobtrusive sensing devices, facilitating precision medicine by identifying intricate patterns in biomedical data, and improving the effectiveness of nanotechnological procedures using functional recognition imaging (FR-SPM) [81]. AI models are used to optimize medication compounds by testing numerous iterations against biological data. Green nanotechnology addresses environmental, health, and safety concerns in nanoscale materials for sustainable development. It includes technologies for hazardous waste cleanup, water purification, food nutrition improvement, and pollutant identification. The focus is on ecologically friendly production using green chemistry concepts. Key aspects of nanotechnology governance include Green Engineering lifespan, comprehensive assessment, sustainable design, and environmentally friendly production. These technologies can be developed using renewable elements or nanoparticles to avoid water dissolution [82].

The governance of nanotechnology requires improvement, with predictive tools and international collaboration being crucial for effective governance. Regulatory frameworks are crucial in regulating nanotechnology development and commercialization, ensuring safety and minimizing risks through risk assessment, data generation, and establishment of regulatory bodies. Governance in nanotechnology aims to establish regulations, promote collaboration, and ensure responsible decision-making [82]. Governments should address concerns about nanomaterial toxicity and lack of enforcement for safety, health, and environmental issues. The public perception of nanotechnology is still developing, and integrating social and ethical studies into development is essential. Public participation, open communication about its advantages and risks, and risk assessment are crucial for building public trust in developing nations. Proactive risk management must adapt to changing technology and applications, risk banding, risk reduction, and risk transfer [9]. This remarkable advancement has significantly enhanced the world and will continue to do so in the future. International cooperation must play important roles in managing risks, improving regulatory compliance, international trade, innovation, transparency, public trust, and capacity building in developing nations. Public–private partnerships in nanotechnology foster cooperation and must improve nanoscale biological interactions, and explore safe and efficient products within the regulatory preview. Ethical and social implications of nanotechnology need to be addressed through social scientific and economic research. Policies for responsible nanotechnology development are crucial, as the societal impacts are hard to measure or predict, but are likely to impact military applications, intellectual property issues, and labour market inequalities.

Acknowledgements The authors extend gratitude to the Department of Animal Behaviour and Physiology, School of Biological Sciences, Madurai Kamaraj University for the resources made available via Rashtriya Uchchatar Shiksha Abhiyan (RUSA) and UGC NRCBS.

References

1. Bhushan, B. (Ed.).: Springer handbook of nanotechnology. Springer (2017)
2. Wang, G.: Nanotechnology: the new features (2018). *arXiv preprint* arXiv:1812.04939
3. Roco, M.C., Hersam, M.C., Mirkin, C.A., Roco, M.C., Harthorn, B., Guston, D., Shapira, P.: Innovative and responsible governance of nanotechnology for societal development. *Nanotechnology Research Directions for Societal Needs in 2020: Retrospective and Outlook*, pp. 561–617 (2011)
4. Jopp, K., Jopp, K.: Inspiration für Chemie und neue Materialien. *Nanotechnologie—Aufbruch ins Reich der Zwerge*, pp. 97–138 (2003)
5. Ashraf, S.A., Siddiqui, A.J., Elkhalifa, A.E.O., Khan, M.I., Patel, M., Alreshidi, M., Moin, A., Singh, R., Snoussi, M., Adnan, M.: Innovations in nanoscience for the sustainable development of food and agriculture with implications on health and environment. Sci. Total Environ. **768**, 144990 (2021). https://doi.org/10.1016/j.scitotenv.2021.144990
6. Gupta, V., Mohapatra, S., Mishra, H., Farooq, U., Kumar, K., Ansari, M.J., Iqbal, Z.: Nanotechnology in cosmetics and cosmeceuticals—a review of latest advancements. Gels **8**(3), 173 (2022)
7. Paul, S., Mukherjee, S., Banerjee, P.: Recent advancement in the Nanomaterials encapsulated drug delivery vehicles towards combating of Cancer, COVID-19 and HIV like chronic diseases. Mater. Adv. (2023)
8. Patil, S.S., Shedbalkar, U.U., Truskewycz, A., Chopade, B.A., Ball, A.S.: Nanoparticles for environmental clean-up: a review of potential risks and emerging solutions. Environ. Technol. Innov. **5**, 10–21 (2016). https://doi.org/10.1016/j.eti.2015.11.001
9. Malakar, Y., Lacey, J., Bertsch, P.M.: Towards responsible science and technology: how nanotechnology research and development is shaping risk governance practices in Australia. Human. Soc. Sci. Commun. **9**(1), 17 (2022)
10. Rafols, I., van Zwanenberg, P., Morgan, M., Nightingale, P., Smith, A.: Missing links in nanomaterials governance: bringing industrial dynamics and downstream policies into view. J. Technol. Transf. **36**, 624–639 (2011)
11. Read, S.A., Kass, G.S., Sutcliffe, H.R., Hankin, S.M.: Foresight study on the risk governance of new technologies: the case of nanotechnology. Risk Anal. **36**(5), 1006–1024 (2016)
12. Fadel, T.R., Steevens, J.A., Thomas, T.A., Linkov, I.: The challenges of nanotechnology risk management. Nano Today **10**(1), 6–10 (2015)
13. Auplat, C.A.: The challenges of nanotechnology policy making part 1. Discussing mandatory frameworks. Glob. Policy **3**(4), 492–500 (2012)
14. Li, W., Yigitcanlar, T., Browne, W., Nili, A.: The making of responsible innovation and technology: an overview and framework. Smart Cities **6**(4), 1996-2034 (2023)
15. Schulte, P.A., Salamanca-Buentello, F.: Ethical and scientific issues of nanotechnology in the workplace. Environ. Health Perspect. **115**(1), 5–12 (2007)
16. Sandler, R.: Nanotechnology: the social and ethical issues (2009)
17. Khan, A. (2014). Ethical and social implications of nanotechnology. *QScience Proceedings, 2015*(4),
18. Bainbridge, W.S.: Social and ethical implications of nanotechnology. *Springer Handbook of Nanotechnology*, 1135 (2004)
19. Baran, A.: Nanotechnology: legal and ethical issues. Ekonomia i Zarządzanie, **8**(1) (2016)
20. Malsch, I., Subramanian, V., Semenzin, E., Hristozov, D., Marcomini, A.: Supporting decision-making for sustainable nanotechnology. Environ. Syst. Decis. **35**(1), 54–75 (2015)
21. Hankin, S.M., Read, S.A.K.: Governance of nanotechnology: context, principles and challenges. In: Murphy, F., McAlea, E., Mullins, M. (eds.), Managing Risk in Nanotechnology. Innovation, Technology, and Knowledge Management. Springer, Cham (2016). https://doi.org/10.1007/978-3-319-32392-3_3
22. Harding, R., Fisher, E.: Perspectives on the precautionary principle, pp. 2–3. Federation Press, Sydney (1999)

23. Ho, S.S., Scheufele, D.A., Corley, E.A.: Making sense of policy choices: understanding the roles of value predispositions, mass media, and cognitive processing in public attitudes toward nanotechnology. J. Nanopart. Res. **12**, 2703–2715 (2010)
24. Shatkin, J.A., Abbott, L.C., Bradley, A.E., Canady, R.A., Guidotti, T., Kulinowski, K.M., Löfstedt, R.E., Louis, G., MacDonell, M., Maynard, A.D., Paoli, G., Sheremeta, L., Walker, N., White, R., Williams, R.: Nano risk analysis: ad vancing the science for nanomaterials risk manage ment. Risk Anal. **30**(11), 1680–1687 (2010). https://doi.org/10.1111/j.1539-6924.2010.01493.x
25. Roig, A.: Nanotechnology governance: from risk regulation to informal platforms. Nano Ethics **12**(2), 115–121 (2018)
26. Renn, O., Roco, M.: White paper on nanotechnology risk governance. International Risk Governance Council, Geneva (2006)
27. Florin, M.V., Bürkler, M.T.: Introduction to the IRGC risk governance framework (No. REP_WORK). EPFL (2017)
28. Noorlander, C., Sips, A., Hock, J., Honeher, K., Lehmann, H.C.: NANoREG safe-by-design (SbD) concept. *RIVM/TEMAS* (2016)
29. Isigonis, P., Hristozov, D., Benighaus, C., Giubilato, E., Grieger, K., Pizzol, L., Marcomini, A.: Risk governance of nanomaterials: review of criteria and tools for risk communication, evaluation, and mitigation. Nanomaterials **9**(5), 696 (2019)
30. Cooper, R.G.: Stage-gate systems: a new tool for managing new products. Bus. Horiz. **33**(3), 44–54 (1990)
31. ISO, I.: TS 80004–1: Nanotechnologies—Vocabulary—Part 1: Core Terms. *ISO (the International Organization for Standardization): Geneva, Switzerland* (2015)
32. Hutchins, G.: ISO 31000: 2018 enterprise risk management. Greg Hutchins (2018)
33. Schierow, L.J.: The toxic substances control act (TSCA): implementation and new challenges. Congressional Research Service, Library of Congress (2008)
34. Bergeson, L.L., Engler, R.E.: Chemical innovation and new TSCA: the good, the bad, and the evolving. *IRCL*, 157 (2019)
35. Tan, Y., Xie, Y., Qiao, X., Bai, S.: The china food and drug administration (Cfda). *Approaching China's Pharmaceutical Market: A Fundamental Guide to Clinical Drug Development*, pp. 239–284 (2015)
36. DEMİRBAŞ, K., ÇEVİK, S.: Regulatory policies for safety of nanomaterials. Open J. Nano **5**(1), 1-16 (2020)
37. NanoMalaysia: NanoVerify Program (2018). http://www.nanomalaysia.com.my/img/nanoverify/NANOVerify-Brochure.pdf
38. TANIDA: Nanomark (2018). Retrieved from Taiwan Nanotechnology Industry Development Association http://www.tanida.org.tw/nanomark_e.php?mn=nanomark_e&nm=markIntroductione
39. Soltani, A.M., Pouypouy, H.: Standardization and regulations of nanotechnology and recent government policies across the world on nanomaterials. Advances in phytonanotechnology (pp. 419–446). Academic Press (2019)
40. Ruggiu, D.: Responsibilisation phenomena: the EC code of conduct for responsible nanosciences and nanotechnologies research. Europ. J. Law Technol. **5**(3) (2014)
41. Kalpin, M.C., Hoffer, M., Hale, W.C.P., Dorr, L.L.P.: Nanotechnology and the Environment: Will Emerging Environmental Regulations Stifle the Promise?. Wilmer Cutler Pickering Hale and Dorr LLP (2005)
42. Gottardo, S., Mech, A., Drbohlavová, J., Małyska, A., Bøwadt, S., Sintes, J.R., Rauscher, H.: Towards safe and sustainable innovation in nanotechnology: state-of-play for smart nanomaterials. NanoImpact **21**, 100297 (2021)
43. World Health Organization.: FAO/WHO expert meeting on the application of nanotechnologies in the food and agriculture sectors: potential food safety implications: meeting report. World Health Organization (2010)
44. Guston, D.H. (ed.).: Encyclopedia of nanoscience and society (Vol. 2). Sage (36) (2010)

45. Roco, M., Renn, O., Jäger, A.: Nanotechnology risk governance. In: Global risk governance: concept and practice using the IRGC framework, pp. 301–327. Springer, Netherlands, Dordrecht (2008)
46. Lin, P.: Nanotechnology bound: evaluating the case for more regulation. Nano Ethics **1**(2), 105–122 (2007)
47. Boholm, Å., Larsson, S.: What is the problem? a literature review on challenges facing the communication of nanotechnology to the public. J. Nanopart. Res. **21**, 1–21 (2019)
48. Porter, A.L., Youtie, J., Shapira, P., Schoeneck, D.J.: Refining search terms for nanotechnology. J. Nanopart. Res. **10**, 715–728 (2008)
49. Whitman, J.: The governance of nanotechnology. Sci. Public Policy **34**(4), 273–283 (2007)
50. Coussens, C., Goldman, L.: (Eds.).: Implications of nanotechnology for environmental health research (2005)
51. van Leeuwen, C.J., Vermeire, T.G. (eds.): Risk assessment of chemicals: an introduction (Vol. 94). Springer, Dordrecht (2007)
52. Marchant, G.E.: 'Soft law'mechanisms for nanotechnology: liability and insurance drivers. J. Risk Res. **17**(6), 709–719 (2014)
53. Murphy, F., Mullins, M., Hester, K., Gelwick, A., Scott-Fordsmand, J.J., Maynard, T.: Insuring nanotech requires effective risk communication. Nat. Nanotechnol. **12**(8), 717–719 (2017)
54. Scheufele, D.A., Corley, E.A., Dunwoody, S., Shih, T.J., Hillback, E., Guston, D.H.: Scientists worry about some risks more than the public. Nat. Nanotechnol. **2**(12), 732–734 (2007)
55. Stone, V., Führ, M., Feindt, P.H., Bouwmeester, H., Linkov, I., Sabella, S., Poortvliet, P.M.: The essential elements of a risk governance framework for current and future nanotechnologies. Risk Anal. **38**(7), 1321–1331 (2018)
56. Hristozov, D.R., Zabeo, A., Foran, C., Isigonis, P., Critto, A., Marcomini, A., Linkov, I.: A weight of evidence approach for hazard screening of engineered nanomaterials. Nanotoxicology **8**(1), 72–87 (2014)
57. Puzyn, T., Puzyn, T., Jeliazkova, N., Sarimveis, H., Marchese Robinson, R.L., Lobaskin, V., Rallo, R., Richarz, A.N., Gajewicz, A., Papadopulos, M.G., Hastings, J., Cronin, M.T.D., Benfenati, E., Fernandez, A.: Perspectives from the nanosafety modelling cluster on the validation criteria for (Q) SAR models used in nanotechnology
58. van Wezel, A.P., van Lente, H., van de Sandt, J.J., Bouwmeester, H., Vandeberg, R.L., Sips, A.J.: Risk analysis and technology assessment in support of technology development: putting responsible innovation in practice in a case study for nanotechnology. Integr. Environ. Assess. Manag. **14**(1), 9–16 (2018)
59. Walhout, B., Konrad, K.: Practicing responsible innovation in NanoNextNL. Bowman, D.; Dijkstra, A.; Fautz, C (2015)
60. van Harmelen, T., Zondervan-van den Beuken, E.K., Brouwer, D.H., Kuijpers, E., Fransman, W., Buist, H.B., Som, C.: LICARA nanoSCAN-A tool for the self-assessment of benefits and risks of nanoproducts. Environ. Int. **91**, 150–160 (2016)
61. Liguori, B., Hansen, S.F., Baun, A., Jensen, K.A.: Control banding tools for occupational exposure assessent of nanomaterials—Ready for use in a regulatory context? NanoImpact **2**, 1–17 (2016)
62. Satterfield, T., Kandlikar, M., Beaudrie, C.E.H., Conti, J., Herr Harthorn, B.: Anticipating the perceived risk of nanotechnologies. Nat. Nanotechnol. **4**(11), 752–758 (2009. https://doi.org/10.1038/nnano.2009.265
63. Shatkin, J.A., Abbott, L.C., Bradley, A.E., Canady, R.A., Guidotti, T., Kulinowski, K.M., Williams, R.: Nano risk analysis: advancing the science for nanomaterials risk management. Risk Analy. Intern. J. **30**(11), 1680–1687 (2010)
64. Cacciatore, M.A., Scheufele, D.A., Corley, E.A.: From enabling technology to applications: the evolution of risk perceptions about nanotechnology. Public Underst. Sci. **20**(3), 385–404 (2011)
65. Binder, A.R., Hillback, E.D., Brossard, D.: Conflict or caveats? Effects of media portrayals of scientific uncertainty on audience perceptions of new technologies. Risk Anal. **36**(4), 831–846 (2016)

66. Bottini, M., Gloria, F., Adanti, S., Corradino, N., Bergamaschi, A., Magrini, A.: Public optimism towards nanomedicine. Intern. J. Nanomed. **3473** (2011). https://doi.org/10.2147/ijn.s26340
67. Gardner, G., Jones, G., Taylor, A., Forrester, J., Robertson, L.: Students' risk perceptions of nanotechnology applications: implications for science education. Int. J. Sci. Educ. **32**(14), 1951–1969 (2010)
68. Bostrom, A., Löfstedt, R.E.: Nanotechnology risk communication past and prologue. Risk Analy. Intern. J. **30**(11), 1645–1662 (2010)
69. European Commission.: Intern. J. Health Care Qual. Assur. **17**(1) (2004). https://doi.org/10.1108/ijhcqa.2004.06217aab.004
70. Berube, D.M., Cummings, C.L., Frith, J.H., Binder, A.R., Oldendick, R.: Comparing nanoparticle risk perceptions to other known EHS risks. J. Nanopart. Res. **13**(8), 3089–3099 (2011). https://doi.org/10.1007/s11051-011-0325-z
71. Simons, J., Zimmer, R., Vierboom, C., Härlen, I., Hertel, R., Böl, G.F.: The slings and arrows of communication on nanotechnology. J. Nanopart. Res. **11**, 1555–1571 (2009)
72. Capon, A., Gillespie, J., Rolfe, M., Smith, W.: Comparative analysis of the labelling of nanotechnologies across four stakeholder groups. J. Nanopart. Res. **17**, 1–13 (2015)
73. Retzbach, A., Marschall, J., Rahnke, M., Otto, L., Maier, M.: Public understanding of science and the perception of nanotechnology: the roles of interest in science, methodological knowledge, epistemological beliefs, and beliefs about science. J. Nanopart. Res. **13**, 6231–6244 (2011)
74. Allan, J., Belz, S., Hoeveler, A., Hugas, M., Okuda, H., Patri, A., Anklam, E.: Regulatory landscape of nanotechnology and nanoplastics from a global perspective. Regul. Toxicol. Pharmacol. **122**, 104885 (2021)
75. Renn, O., Benighaus, C.: Perception of technological risk: insights from research and lessons for risk communication and management. J. Risk Res. **16**(3–4), 293–313 (2013)
76. IRGC: White paper on nanotechnology risk governance. International Risk Governance Council, Geneva (2006)
77. Owen, R., Baxter, D., Maynard, T., Depledge, M.: Beyond regulation: risk pricing and responsible innovation (2009)
78. Zingg, R., Fischer, M.: The rise of private–public collaboration in nanotechnology. Nano Today **25**, 7–9 (2019)
79. Paddock, L.: An integrated approach to nanotechnology governance. UCLA J. Envtl. L. Pol'y **28**, 251 (2010)
80. Chia, H.N., Wu, B.M.: Recent advances in 3D printing of biomaterials. J. Biol. Eng. **9**(1), 1–14 (2015)
81. Isigonis, P., Afantitis, A., Antunes, D., Bartonova, A., Beitollahi, A., Bohmer, N., Dusinska, M.: Risk governance of emerging technologies demonstrated in terms of its applicability to nanomaterials. Small **16**(36), 2003303 (2020)
82. Bergeson, L.L.: Sustainable nanomaterials: emerging governance systems (2013)

Dr. C. Amutha is an Assistant Professor in the Department of Animal Behaviour and Physiology at Madurai Kamaraj University. She holds a Ph.D in Zoology With 14 years of teaching experience and 20 years in research, Dr. Amutha specializes in endocrine disruption, enzyme defense mechanisms, pollution monitoring, and aquarium management. Her research focuses on the impacts of endocrine disruption on aquatic organisms and its bioremediation using nanoparticles for pollutants like PAHs, PCBs, PBPs, and heavy metals. She also conducts epigenomic studies on herbicides like Glyphosate and atrazine-affected aquatic species. Dr. C. Amutha, a renowned academic, has made significant contributions to research through numerous publications, authored four books, and filed four patents.

Chapter 15
Considerations in the Development and Deployment of Nanotechnology

Vernica Verma, Priya Gupta, Peramjeet Singh, and Narendra Kumar Pandey

Introduction: Nanotechnology's Impact on Diverse Domains

In the vast landscape of scientific exploration, few frontiers have sparked as much intrigue and promise as nanotechnology. Originating as a theoretical concept, nanotechnology has evolved into a tangible and revolutionary force, profoundly impacting diverse scientific and technical domains, and thus human life [1].

Nanotechnology, as a concept, emerged from the realm of physics and chemistry, where scientists envisioned manipulating matter at the nanoscale dimensions of the order of nanometers, or one billionth of a meter. The transition from theoretical musings to practical applications marks a significant milestone in the evolution of nanotechnology [2]. The promise of nanotechnology lies in its ability to engineer materials and devices at the atomic and molecular levels, granting unprecedented control over their properties. This newfound mastery over the nanoscale has catalyzed breakthroughs across a myriad of scientific and technical disciplines [3].

One of the primary domains profoundly impacted by nanotechnology is the field of miniaturized electronic components. The relentless pursuit of smaller, more efficient electronic devices has led to the development of nanoscale transistors, capacitors, and other components. This downsizing not only enhances the performance of electronic devices but also opens avenues for innovative applications, such as wearable technology and flexible electronics [4].

In the realm of sensors and imaging systems, nanotechnology has ushered in a new era of precision and sensitivity. Nanosensors, equipped with the ability to detect minute changes in their environment, find applications in fields as

V. Verma · P. Singh · N. K. Pandey (✉)
Department of Physics, University of Lucknow, Lucknow, UP 206007, India
e-mail: nkp1371965@gmail.com

P. Gupta
Electronics Department, Manyawar Kanshiram Government Polytechnic, Tirwa, Kannauj, UP 209732, India

S. C. George and B. Tawiah (eds.), *Nanotechnology in Societal Development*, Advanced Technologies and Societal Change, https://doi.org/10.1007/978-981-97-6184-5_15

diverse as healthcare, environmental monitoring, and security. Nanoscale imaging systems, employing advanced techniques like scanning tunnelling microscopy, enable researchers to visualize and manipulate matter at the atomic level, revolutionizing our understanding of the microscopic world [5].

Nano-mechanics, a branch of nanotechnology exploring the mechanical properties of nanomaterials, has had a profound impact on materials science and engineering. The manipulation of materials at the nanoscale grants them unique mechanical properties, paving the way for the development of stronger, lighter, and more resilient materials with applications in aerospace, construction, and manufacturing.

Nano-computers represent another frontier where nanotechnology has left an indelible mark. The miniaturization of computing components to the nanoscale allows for the creation of more powerful and energy-efficient computers. Quantum dots, for instance, show promise in revolutionizing computing by harnessing the principles of quantum mechanics for unprecedented processing capabilities.

Energy harvesting, a critical concern in an era grappling with sustainability, has found a powerful ally in nanotechnology. Nanomaterials, such as nanostructured solar cells and nanogenerators, hold the potential to transform ambient energy sources into usable electrical power. This innovation not only contributes to renewable energy solutions but also addresses the growing demand for efficient energy utilization.

In the realm of propulsion systems, nanotechnology has implications for advanced space exploration. Nanoscale materials and devices offer opportunities for enhancing propulsion efficiency, reducing spacecraft mass, and enabling unprecedented capabilities in interplanetary travel.

As we explore further in this chapter, it becomes evident that this transformation is not without its challenges. The increasing complexity of nanotechnology applications raises ethical concerns, necessitating a nuanced understanding and proactive approach to address potential pitfalls. From synthetic biology to the potential for military applications, the ethical landscape of nanotechnology is as vast and intricate as the nanoscale itself.

In this chapter, we are navigating through the ethical considerations inherent in the development and deployment of nanotechnology. The aim is not only to celebrate the remarkable advancements made possible by nanotechnology but also to engage in a thoughtful exploration of the responsibilities that come with wielding such transformative power at the nanoscale. The evolution of nanotechnology represents not just a scientific and technological triumph but a call to action for responsible innovation that considers the broader societal and ethical implications of our endeavours in this exciting frontier.

Evolution of Nanotechnology: From Promise to Reality

Nanotechnology, once a realm of scientific promise, has evolved from theoretical concepts to tangible reality, revolutionizing diverse fields. Initially conceptualized in the 1950s by physicist Richard Feynman, the journey from promise to reality has been marked by relentless innovations. Breakthroughs in material science and engineering have enabled the manipulation of matter at the nanoscale, bringing about transformative applications in medicine, electronics, and materials. Nanotechnology's precision at the atomic and molecular levels has paved the way for advancements like targeted drug delivery, efficient energy storage, and ultra-miniaturized electronic components. As we witness the seamless integration of nanotech into everyday life, it becomes evident that its evolution has indeed transcended promises, shaping a future defined by microscopic wonders.

The relentless march of technological progress has ushered in a transformational era, significantly impacting scientific and technical domains. Breakthroughs in artificial intelligence, biotechnology, and quantum computing have not merely shaped industries but have redefined the very fabric of research and development. Enhanced data processing capabilities, precision engineering at the nanoscale, and unprecedented insights into genetic codes are catalyzing discoveries at an unprecedented pace. The synergy of disciplines, facilitated by cutting-edge technology, fosters interdisciplinary collaborations, propelling us towards novel solutions and unforeseen innovations. This transformation is more than a paradigm shift; it's a dynamic reconfiguration of the scientific landscape, heralding a future where possibilities are boundless.

Contributions Across Domains

Miniaturized Electronic Components and Their Implications

Nanotechnology's profound contribution to miniaturized electronic components has revolutionized the landscape of modern electronics, offering unparalleled advancements and implications. At the heart of this evolution lies the ability to manipulate materials at the nanoscale, allowing for the creation of smaller, more efficient, and powerful electronic devices. Nanoscale transistors, for instance, enable the production of microprocessors with increased computing power while reducing physical size. This not only enhances the performance of gadgets but also facilitates the development of compact and wearable technologies. The implications extend beyond consumer electronics, influencing fields like medical devices, aerospace, and communication technologies. Smaller, faster, and more energy-efficient electronic components are pivotal in shaping a future where technology seamlessly integrates into every facet of our lives, fostering innovation and efficiency.

Advancements in Sensors, Imaging Systems, and Their Real-World Applications

Nanotechnology has emerged as a cornerstone in advancing sensors and imaging systems, revolutionizing their capabilities and real-world applications. At the nanoscale, materials exhibit unique properties that can be harnessed for precise sensing and imaging. Quantum dots, for instance, enable highly sensitive and selective detection, crucial in medical diagnostics and environmental monitoring. Nanoscale structures enhance the resolution of imaging systems, providing unprecedented clarity in medical imaging and surveillance technologies. The miniaturization enabled by nanotech has led to the development of wearable sensors, transforming healthcare with real-time monitoring. In environmental science, nanosensors detect pollutants at ultra-low concentrations, ensuring timely interventions. Nanotechnology's contribution transcends theoretical realms, actively shaping a tangible and transformative future in sensing and imaging technologies.

Nano-Mechanics and Its Role in Materials Science and Engineering

Nanotechnology has made remarkable contributions to nano-mechanics, playing a pivotal role in materials science and engineering. At the nanoscale, materials exhibit unique mechanical properties that differ significantly from their macroscopic counterparts. Nanotechnology allows precise manipulation of structures at this level, enabling the design of materials with tailored mechanical characteristics. Nanomechanics investigates the behaviour of materials at the atomic and molecular levels, providing insights crucial for developing advanced materials [6].

In materials science and engineering, nanotechnology has facilitated the creation of novel materials with enhanced strength, flexibility, and conductivity. Carbon nanotubes, for instance, showcase exceptional mechanical properties, making them ideal reinforcements in composite materials. This not only contributes to lighter and stronger materials but also opens avenues for innovation in aerospace, automotive, and structural engineering.

Furthermore, nanotechnology has revolutionized the understanding of material fatigue, wear, and fracture at the nanoscale, leading to the development of more durable and reliable materials. Nanostructured materials, engineered with precision using nanotechnology, exhibit improved mechanical performance, making them invaluable in applications ranging from medical devices to high-performance electronics.

In essence, nanotechnology's influence on nano-mechanics has not only deepened our comprehension of material behaviour but has also empowered engineers and scientists to craft materials with unprecedented properties. This symbiotic relationship between nanotechnology, nano-mechanics, and materials science is propelling the field towards groundbreaking innovations and reshaping the landscape of modern engineering.

Nano-Computers and Their Potential for Computing Revolution

Nanotechnology has emerged as a game-changer in the realm of computing, promising a revolution through the development of nano-computers. At the forefront of this technological frontier is the ability to manipulate materials and processes at the nanoscale. Nano-computers leverage the unique properties of nanomaterials to enable faster, more efficient, and smaller computing devices. The potential for a computing revolution lies in their capacity to surpass the limitations of traditional silicon-based technology, offering unprecedented speed and energy efficiency. Quantum dots, carbon nanotubes, and other nanostructures hold the key to building ultra-compact, high-performance computing systems that could redefine the future of information processing. As nanotechnology continues to advance, the prospect of nano-computers stands as a testament to its transformative impact on the computing landscape.

Energy Harvesting and Sustainability

Nanotechnology has emerged as a pivotal player in the quest for sustainable energy solutions, making significant contributions to energy harvesting and environmental preservation. At the forefront, nanomaterials exhibit unique properties that enhance energy conversion and storage efficiency. Nanoscale structures, such as quantum dots and nanowires, optimize solar cell performance, boosting solar energy harvesting. Moreover, nanotechnology facilitates the development of advanced energy storage devices, enhancing the capacity and lifespan of batteries crucial for renewable energy integration.

In the realm of sustainability, nanomaterials play a key role in water purification and pollution remediation. Nano-enabled filtration systems ensure efficient removal of contaminants, mitigating environmental impact. The application of nanotechnology in catalysis also aids in the development of greener processes for energy production, reducing reliance on conventional, pollutant-intensive methods.

The marriage of nanotechnology with energy harvesting not only enhances efficiency but also fosters a more sustainable future. As research continues, nanotechnology promises to be a cornerstone in addressing global energy challenges and advancing the cause of environmental stewardship.

Advanced Propulsion Systems and Implications for Space Exploration

Nanotechnology has emerged as a game-changer in the realm of advanced propulsion systems, significantly influencing the future of space exploration. In the quest for efficient and sustainable propulsion, nanomaterials play a pivotal role. Lightweight yet robust nanocomposites enhance structural integrity while minimizing overall spacecraft weight. Nanoscale sensors provide real-time monitoring, optimizing fuel consumption and performance. Moreover, nanotechnology facilitates the development of advanced propulsion concepts, such as nano-electromagnetic thrusters and nanoscale propulsion systems, enabling unprecedented precision and control. These advancements not only enhance spacecraft manoeuvrability but also reduce fuel requirements, extending mission durations and reaching distant celestial bodies.

Nanotechnology's contribution to advanced propulsion systems is pivotal, unlocking new frontiers in the cosmos and reshaping the future of space exploration.

Potential Risks and Concerns of Nanotechnology Deployment

Assessment of the Potential Negative Impact of Nanotechnology on Human Health and the Environment

Although nanotechnology holds immense potential from revolutionizing medicine to transforming materials, this exciting frontier poses profound ethical dilemmas that demand careful consideration before we plunge headlong into the nanoscale world. Nanotechnology involves the manipulation of matter at the molecular scale and holds immense potential to revolutionize major fields from medicine and energy to materials science and computing. Yet, apart from these lucrative possibilities lie equally important questions about the social and ethical implications of this nascent field. A major concern involves the potential impact of nanomaterials on human health and environment. The minuscule size of nanoparticles allows them to penetrate biological barriers and travel long distances, raising concerns about their toxicity and long-term effects on the ecosystem [7]. One of the challenges is the potential for nanoparticles to accumulate in the environment. The small size of nanoparticles enables them to infiltrate air, water, and soil systems, potentially causing unintended ecological implications, imagine tiny, invisible particles wreaking havoc on delicate ecosystems, disrupting food chains, and contaminating water sources and marine life, this chilling prospect demanding careful consideration. These unintended and unforeseen consequences of nanotechnological applications, such as toxicity or environmental impacts, necessitate a thorough examination of ethical guidelines and regulations. Researchers must carefully study the toxicity of nanoparticles before they can be used in humans. Additionally, the uncertain consequences arising from the unique properties of nanoparticles at the cellular level necessitates careful consideration of potential harm as the minute size of nanoparticles facilitates easy penetration into cells and tissues, leading to undesired outcomes, few reports also suggest a high-risk of severe diseases like breast cancer, pulmonary fibrosis, DNA damage due to nanoparticles [8]. The long-term effects on human health are still unknown, raising concerns about respiratory issues, cellular damage, and unforeseen health disruptions.

Hence, it presents a dilemma to prioritize thorough safety testing or prioritize innovation and risk the potentially irreversible consequences of unforeseen harm. Human augmentation advancements like genetic engineering and brain-computer interfaces raise profound threats to the human race. Concepts of nanotechnology interfaced with the Internet of Things (IoT) or Artificial Intelligence (AI) could technologically enhance human bodies and minds raising grave questions about mortality or humanoids taking over humanity at the cost of our identity, privacy, and autonomy.

Hence, obtaining informed consent over the use of nanoparticles on the uncertainties becomes a challenging issue and requires urgent actions towards the ethical development of AI powered by nanotech [9]. Hence, striking a balance between fostering scientific progress and ensuring ethical boundaries is paramount, requiring interdisciplinary collaboration and on-going dialogue among scientists, policymakers, and ethicists to navigate the intricate ethical terrain accompanying nanotechnological advancements [10].

Safety Issues and Environmental Pollution Caused by Nanomaterials

One of the primary safety concerns regarding nanomaterials revolves around their potential toxicity. Due to their small size and increased surface area, nanomaterials can exhibit different chemical, physical, and biological properties compared to bulk materials, which might lead to unforeseen health hazards. For instance, studies have shown that certain nanoparticles can penetrate biological barriers, such as the blood–brain barrier, posing risks of systemic toxicity and neurological disorders [11]. Moreover, the unique properties of nanomaterials, such as high reactivity and catalytic activity, raise concerns about their long-term impact on human health. Inhalation of nanoparticles, for example, could lead to respiratory issues and lung damage, while dermal exposure might result in skin irritation and inflammation [12]. Additionally, the potential for nanomaterials to accumulate in vital organs raises questions about their bioaccumulative properties and chronic health effects. Furthermore, the lack of standardized testing protocols and regulatory frameworks for nanomaterials complicates the assessment of their safety. Traditional toxicity testing methods may not be suitable for evaluating nanomaterials due to their complex interactions with biological systems [13]. Therefore, there is a pressing need for robust safety assessment strategies tailored specifically to nanomaterials to ensure consumer and environmental protection. Figure 15.1 shows how nanomaterials affect the environment, human exposure, and health.

In addition to human health concerns, nanomaterials also pose environmental risks, primarily through pollution of air, water, and soil [14]. The increasing use of nanomaterials in consumer products and industrial processes raises the likelihood of their release into the environment during the manufacturing, usage, and disposal stages. Once released, nanomaterials can interact with various environmental components and organisms, potentially leading to adverse effects on ecosystems and biodiversity. Air pollution represents a significant pathway for the dispersion of nanomaterials into the environment. Aerosolized nanoparticles from industrial processes or consumer products can remain suspended in the air for extended periods, facilitating their inhalation by humans and wildlife. Moreover, nanomaterials deposited on land and water surfaces can undergo re-suspension due to wind or water currents, further contributing to air pollution. Water pollution is another pressing concern associated

Fig. 15.1 Effect of nanomaterials on environment, human exposure pathway, and health

with nanomaterials, particularly due to their widespread use in wastewater treatment, contaminant remediation, and antimicrobial agents. While nanomaterials hold promise for improving water quality, their unintended release into aquatic ecosystems can have detrimental effects on aquatic organisms and disrupt ecosystem dynamics. Furthermore, the potential for nanomaterials to adsorb and transport other pollutants raises concerns about their role in facilitating the spread of contaminants in water bodies. Soil pollution from nanomaterials primarily occurs through the application of nanomaterial-based agrochemicals and soil remediation techniques. Nanoparticles introduced into soil environments can interact with soil microorganisms, plants, and invertebrates, affecting nutrient cycling, plant growth, and soil fertility [15] recover, the persistence of nanomaterials in soil matrices raises concerns about their long-term impact on terrestrial ecosystems and food safety.

Challenges in the Regulation and Responsible Use of Nanotechnology

Despite the rapid progress in nanotechnology that has altogether brought a new era of scientific innovation, it also has rendered intricate ethical dilemmas that demand careful consideration and discourse. As researchers delve into the microscopic realms of materials and devices, questions about the responsible use of these advancements become increasingly complex and a major cause of concern around potential health and environmental risks associated with these novel nanomaterials. The rapid advancement of this technology not only brings incredible opportunities but also throws up complex ethical dilemmas. In harnessing the complete potential of nanotechnology, robust governance, and effective regulations are essential to ensure its benefits reach all areas of society without sacrificing safety, equity, and ethical considerations. However, the exciting future of nanotechnology comes intertwined with significant challenges regarding responsible use and effective regulation [16]. For the proper implementation of the ethical framework, it becomes crucial to recognize and address these challenges. The rapid evolution of nanotechnology due to its dynamic nature presents a challenge for regulators; new applications emerge quickly, making it difficult to keep pace with evolving risks and potential impacts. Hence, this technology brings constant scrutiny that would in turn require a constant observation of its developments. Further, the unique properties of nanoparticles such as increased reactivity and the ability to cross biological barriers, present challenges in assessing their safety and environmental impact compared to traditional materials. In addition, these offer the unpredictability of the future of such technology [17]. Therefore it becomes important to define and standardize the constituents of "nanomaterial", which is crucial for effective regulation, but achieving consensus on a definition remains a complex issue. The globalized nature of research and development necessitates international collaboration and harmonization of regulations to avoid inconsistencies and unintended consequences. Furthermore, the capacity for enhanced surveillance, privacy infringement, and the creation of powerful, yet potentially uncontrollable technologies raise ethical challenges regarding the responsible development and deployment of nanotechnological solutions. From the potential biases of AI algorithms to the privacy concerns surrounding facial recognition, these challenges demand careful consideration and nuanced solutions [18]. Existing regulatory frameworks for approving similar versions of established drugs (generics) might not be suitable for complex nanomedicines. The concept of "pharmaceutical equivalence", a cornerstone of generic approval, becomes problematic due to the difficulty in isolating and identifying the "active ingredient" in some nanomedicines. This lack of clarity hinders the development and approval of safe and effective nanomedicines.

Moreover, regulatory agencies often lack the necessary scientific expertise and clear guidelines specifically tailored for nanomedicines. This knowledge gap makes it challenging to formulate adequate regulations that effectively ensure patient safety and regulate the use of nanomedicines in clinical settings. While infrastructure like national characterization labs exist, translating their findings and recommendations into actionable regulatory frameworks remains a hurdle.

The global spread of nanotechnology poses regulatory challenges due to its widespread impact and transnational applications. Diverse legal systems, cultural norms, and economic contexts complicate efforts to establish consistent standards. Variation in regulatory approaches across regions creates confusion and barriers to international collaboration and trade. Inconsistent regulations may hinder innovation and fragment markets, raising concerns about safety and environmental protections. To address these issues, cooperation among regulatory bodies is essential. International organizations like ISO, OECD, and WHO facilitate dialogue and standard development. Bilateral and multilateral agreements promote harmonization, reducing trade barriers and fostering global diffusion of nanotechnology. Overall, collaborative efforts are crucial to bridge regulatory gaps, ensure consistency, and maximize the benefits of nanotechnology worldwide. Figure 15.2 is the pictorial depiction of certain challenges faced in the regulation and responsible use of nanotechnology.

In navigating this ethical maze, on-going dialogue and collaboration are not just important, they are essential. Hence, scientists and philosophers from across the globe should come forth to brainstorm the ethical implications of emerging technologies before they are fit to be released into the world. Policy makers and community leaders

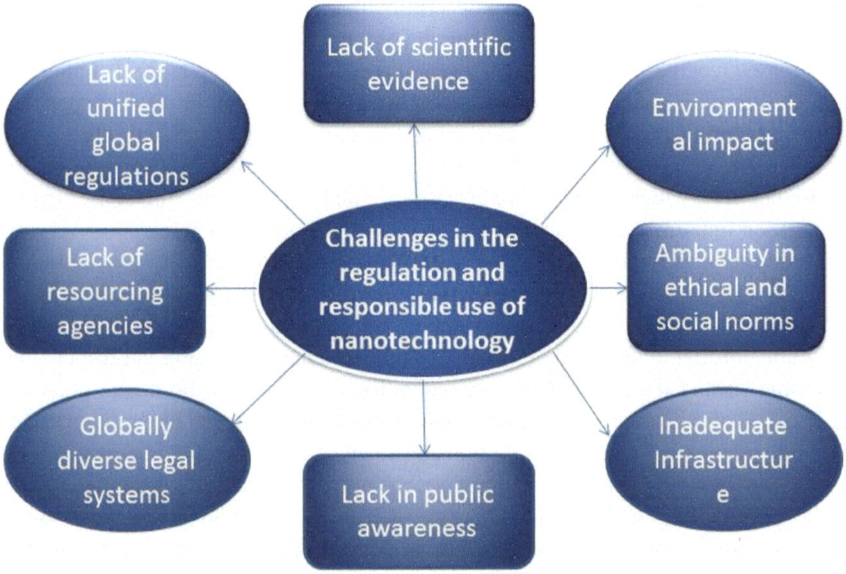

Fig. 15.2 Challenges in the regulation and responsible use of nanotechnology

are expected to work together to develop regulations that reflect diverse perspectives and protect vulnerable populations. Additionally, the developers of the technology and civil society organizations should collaborate to design technologies that are not only powerful but also fair and inclusive.

These interconnected challenges significantly restrict the potential of nanotechnology. Addressing these issues necessitates a collaborative effort involving various stakeholders, including researchers, regulatory bodies, and industry experts. By developing comprehensive and adaptable regulatory frameworks, fostering knowledge exchange and constantly refining testing methodologies, the path towards safe and responsible utilization of nanotechnology is certainly possible.

Ethical Issues in Research and Development

Ethical Implications in Experimental Design and Testing

Ethical considerations in experimental design and testing are paramount in ensuring the integrity of research outcomes and safeguarding the well-being of participants. The process involves a delicate balance between the pursuit of scientific knowledge and the protection of human or animal subjects involved. This chapter explores the ethical implications inherent in experimental design and testing, highlighting key principles, common dilemmas, and strategies for ethical decision-making. Ethical considerations play a crucial role in both the design and testing phases of experiments, ensuring the integrity of research outcomes and the protection of participants' rights and well-being. Guided by principles, such as respect for autonomy, beneficence, non-maleficence, justice, and integrity, researchers strive to conduct studies that uphold ethical standards. Informed consent, risk–benefit assessment, confidentiality, and respect for vulnerable populations are key elements in ethical experimental design. Similarly, in testing procedures, fairness, validity, reliability, minimization of harm, and confidentiality are paramount. Ethical dilemmas often arise, including the balancing act between scientific rigor and participant welfare, conflicts of interest, dual-use research, and post-study obligations. To address these challenges, researchers employ strategies, such as seeking ethical review, consultation, ethical reflection, transparency, and ongoing education. By adhering to ethical principles and implementing sound ethical practices, researchers can ensure that their work contributes to the advancement of knowledge while upholding the rights and dignity of all involved [19].

Importance of Informed Consent and Protection of Research Participants

In the realm of research and development (R&D), the cornerstone principles of informed consent and protection of research participants stand as imperatives, essential for upholding ethical standards and ensuring the integrity of scientific inquiry. These principles serve as the bedrock upon which the trust between researchers and participants is built, fostering a relationship founded on transparency, respect, and mutual understanding. At its essence, informed consent embodies the concept of autonomy, affirming an individual's right to make informed decisions regarding their participation in research endeavours. It requires that participants are provided with comprehensive information regarding the nature, purpose, risks, and potential benefits of the study, empowering them to make voluntary and rational choices about whether to engage in the research [20]. This process extends beyond mere acquiescence to participation; rather, it necessitates genuine comprehension and comprehension of the information conveyed, ensuring that individuals are cognizant of what their involvement entails. Through informed consent, participants become active partners in the research process, contributing their consent with full awareness and agency. Moreover, informed consent serves as a protective mechanism, shielding participants from potential harm and exploitation. By elucidating the risks associated with participation and outlining the measures put in place to mitigate them, researchers uphold their duty to prioritize the well-being of participants above all else. This commitment to safeguarding individuals is particularly critical in R&D endeavours where the experimental nature of the work may entail unforeseen risks or uncertainties. In such contexts, the informed consent process serves as a vital conduit for open dialogue and risk communication, enabling researchers to address any concerns or uncertainties voiced by participants and to adapt protocols accordingly.

Furthermore, informed consent fosters transparency and accountability within the research enterprise, establishing a framework of ethical conduct that is rooted in integrity and honesty. By adhering to rigorous standards of informed consent, researchers not only uphold their professional obligations but also honour the implicit social contract between science and society. This transparency cultivates public trust in the scientific community, reinforcing the credibility and reliability of research findings. In an era marked by increasing scrutiny of scientific practices and growing public scepticism, the preservation of trust through transparent and ethical conduct is paramount. Concomitant with the principle of informed consent is the imperative to protect the rights and welfare of research participants. This entails a multifaceted approach encompassing both procedural safeguards and substantive protections to ensure that individuals are treated with dignity, respect, and fairness throughout the research process. Central to this mandate is the concept of beneficence, which obligates researchers to maximize potential benefits and minimize potential harms to participants [21, 22]. To this end, regulatory frameworks and ethical guidelines mandate the implementation of rigorous protocols for the recruitment, informed consent, and on-going monitoring of participants in research studies. Institutional review boards (IRBs) play a pivotal role in evaluating the ethical soundness of research protocols, scrutinizing the potential risks and benefits, and ensuring that adequate safeguards are in place to protect participants' rights and well-being. Through the oversight of IRBs, research institutions demonstrate their commitment to upholding ethical standards and prioritizing the interests of participants above all else [23].

Furthermore, the principle of justice underscores the importance of equitable distribution of the burdens and benefits of research participation, guarding against exploitation and discrimination. It necessitates that the selection of research participants is fair and inclusive, reflective of the diversity of the population being studied, and that the benefits derived from research are fairly distributed among all stakeholders. In addition to these procedural safeguards, researchers have a moral obligation to cultivate a culture of respect, empathy, and inclusivity within the research community. This entails fostering open communication, soliciting feedback from participants, and creating an environment where individuals feel empowered to voice their concerns and advocate for their rights. By prioritizing the perspectives and experiences of participants, researchers can enrich the quality and relevance of their research endeavours, ultimately contributing to more robust and impactful scientific outcomes [24].

In conclusion, the principles of informed consent and protection of research participants are indispensable pillars of ethical research and development, underpinning the trust, integrity, and credibility of the scientific enterprise. By upholding these principles with unwavering dedication and diligence, researchers reaffirm their commitment to the highest standards of ethical conduct and ensure that the pursuit of knowledge remains grounded in respect for human dignity and welfare.

Power Dynamics and Conflicts of Interest in Nanotechnology Research

In the dynamic landscape of nanotechnology research, power dynamics and conflicts of interest emerge as significant considerations that can shape the trajectory and outcomes of scientific inquiry. At the heart of these issues lie complex interplays between various stakeholders, including researchers, funding agencies, industry collaborates, policymakers, and the broader public. Understanding and navigating these dynamics are crucial for upholding the integrity, transparency, and societal benefit of nanotechnology research endeavours [25]. One prominent aspect of power dynamics in nanotechnology research revolves around the allocation and distribution of resources, both financial and intellectual. As nanotechnology represents a multidisciplinary field encompassing physics, chemistry, engineering, and biology, collaborations and partnerships between academic institutions, industry players, and governmental agencies are common. However, these collaborations can give rise to asymmetries in power, with industry partners wielding significant influence over research agendas, priorities, and outcomes. Industry funding may come with implicit or explicit expectations regarding the direction of research, potentially leading to conflicts of interest and compromising the autonomy and objectivity of researchers [26]. Moreover, the pursuit of commercialization and intellectual property rights in nanotechnology research introduces additional complexities to the power dynamics at play. Researchers and institutions may be incentivized to prioritize research avenues that promise commercial viability or patentability, potentially overlooking other areas of inquiry that may hold greater scientific or societal value. This emphasis on commercialization can create tensions between the pursuit of scientific knowledge for its own sake and the imperative to generate tangible economic returns, raising questions about the ethical conduct and motivations underlying research activities.

Conflicts of interest further exacerbate these power dynamics, posing ethical challenges to the integrity and impartiality of nanotechnology research. Conflicts of interest can arise in various forms, including financial interests, professional affiliations, and personal relationships, all of which have the potential to bias research outcomes and decision-making processes. For instance, researchers may have financial ties to industry partners or hold positions on advisory boards, creating conflicts between their fiduciary responsibilities and their obligations to scientific rigor and public welfare. Similarly, institutional pressures to secure funding or achieve commercial success can create conflicts that compromise the independence and objectivity of research endeavours [27]. Addressing power dynamics and conflicts of interest in nanotechnology research requires a multifaceted approach that encompasses both institutional safeguards and cultural norms within the scientific community. Transparency and disclosure mechanisms play a crucial role in mitigating conflicts of interest, enabling researchers to openly declare financial interests, affiliations, and potential biases that may influence their work. Institutional policies mandating the disclosure of conflicts of interest, as well as the establishment of independent oversight committees, such as ethics review boards or conflict of interest committees, can help uphold research integrity and mitigate the influence of undue external pressures [28].

Furthermore, fostering a culture of scientific integrity, accountability, and ethical conduct is essential for navigating the complex terrain of power dynamics in nanotechnology research. Education and training programs can raise awareness about the ethical implications of conflicts of interest and empower researchers to recognize and address potential biases in their work. Peer review processes, scientific conferences, and professional associations provide forums for dialogue and critique, enabling researchers to engage in constructive discourse and uphold the highest standards of ethical conduct [29]. Ultimately, the responsible stewardship of nanotechnology research necessitates a commitment to transparency, accountability, and the equitable distribution of benefits and risks. By acknowledging and addressing power dynamics and conflicts of interest head-on, stakeholders can foster an environment conducive to scientific excellence, innovation, and societal impact, ensuring that nanotechnology research serves the collective good and advances human well-being.

Social Inequality and Access to Nanotechnology

Potential Disparities in Access to Nanotechnology Across Different Socio-Economic Groups

In harnessing the complete potential of nanotechnology, robust governance, and effective regulations are essential to ensure its benefits reach all areas of society without sacrificing safety, equity, and ethical considerations. We must prioritize safety across various socioeconomic groups and address potential risks before widespread deployment. The precautionary principle should guide our approach, demanding rigorous research, transparent communication, and open public dialogue on potential environmental, health, and societal impacts. The rapid nanoscale revolution transcending borders calls for harmonized international regulations that are crucial to prevent harmful applications, ensure safety standards, and promote responsible development across the globe. Responsible governance necessitates incorporating diverse perspectives. As nanotechnology is a global accomplishment, international cooperation is vital to ensure harmonized regulations and prevent any war or exploitation. Scientists, engineers, ethicists, policymakers, industry leaders, and the public from across the globe must have firm notions in shaping the holistic future of nanotechnology [30].

Building a strong ethical framework for nanotechnology requires responsible governance by establishing a devoted commission for developing, regulating, and reinforcing ethical laws in nanotechnology under the supervision of experts from various fields like science, engineering, ethics, law, and public policy. Moreover, the agencies providing research funding should consider ethical implications alongside scientific merit when evaluating nanotechnology proposals. This can encourage responsible research aligned with ethical principles like sustainability, inclusivity, and risk mitigation. To foster transparency and build trust regular public forums and consultations must be taken up to gather feedback and address public concerns on potential ethical dilemmas in nanotechnology developments.

Regulatory frameworks including implementation of a mandatory risk assessment system for new nanomaterials before they can enter the market would ensure comprehensive and transparent assessment considering potential environmental, health, and safety risks. Further, a classification and labelling system must be developed to classify nanomaterials based on their risk level and require clear labelling on products containing nanomaterials. This information empowers consumers to make informed choices. This system must also be backed by strict liability for environmental and health concerns, holding manufacturers and users of nanomaterials accountable for any harm caused by their products. Encouragement of global cooperation in developing and implementing common ethical and regulatory frameworks for nanotechnology should be implemented to avoid market fragmentation, ensure responsible global development, and facilitate knowledge sharing.

Ethical Considerations in the Distribution and Affordability of Nanotechnology Products and Services

The inequality arising due to nanotechnological advancement is deeply concerning as the developed countries with greater research infrastructure and financial resources are likely to reap the initial benefits of this technology. Meanwhile, developing countries may struggle to access the necessary training, equipment, and infrastructure, which would potentially widen the gap between rich and poor nations. Ensuring equitable access to its benefits requires proactive measures like targeted interventions. The rapid pace of development, fuelled by technological advancements and societal changes, often brings about unintended consequences and challenges that demand careful consideration. As innovations swiftly reshape industries and lifestyles, unforeseen impacts may emerge, ranging from environmental degradation to societal inequalities. For instance, the adoption of new technologies may inadvertently contribute to job displacement, creating economic disparities and unemployment challenges. Additionally, the speed of technological progress can outstrip regulatory frameworks, raising concerns about the ethical use of emerging technologies and the protection of individual rights. Environmental consequences, such as increased waste and resource depletion, may also accompany rapid development. Balancing the benefits of progress with the potential negative outcomes requires proactive measures, adaptive policies, and a heightened awareness of the intricate web of consequences that can arise from the accelerated pace of change. Addressing these challenges necessitates a holistic and collaborative approach from governments, industries, and societies to ensure that development is not only rapid but also sustainable and inclusive [31].

Furthermore, issues of equity and access arise. Nanotechnology-based treatments might be expensive, potentially excluding low-income populations from this potentially life-saving technology. Moreover, the questions surrounding the distribution of benefits and risks, as certain populations might be disproportionately exposed to potential harms during research or early adoption stages. Balancing the potential benefits of nanomedicine with these ethical concerns requires a multi-faceted approach. Open discussions and on-going research are crucial to ensure transparency and understanding. Additionally, robust ethical frameworks and regulations are needed to guide responsible development and prevent misuse. Ultimately, we must strive to harness the power of nanomedicine for good, ensuring equitable access and protecting human dignity while safeguarding against potential risks.

Privacy and Security Concerns

Ethical Implications of Data Privacy and Security in Nanotechnology

The intersection of nanotechnology and data privacy/security raises complex ethical considerations that encompass a range of issues, from the collection and storage of sensitive data to the potential risks of misuse or unauthorized access [32]. As nanotechnology continues to advance, enabling the development of increasingly sophisticated sensors, devices, and systems at the nanoscale, the volume and diversity of data generated present both opportunities and challenges for safeguarding privacy and ensuring security. Exploring the ethical implications of data privacy and security in nanotechnology requires a nuanced examination of the multifaceted issues at stake, as well as the development of ethical frameworks and regulatory mechanisms to address them.

One of the primary ethical concerns surrounding data privacy and security in nanotechnology pertains to the collection and utilization of personal or sensitive information. Nanotechnology-enabled sensors and devices can gather vast amounts of data, including biometric data, health information, and behavioural patterns, often without individuals' explicit consent or awareness. This raises questions about the ethical implications of data collection practices, particularly regarding the protection of privacy, autonomy, and individual rights. The indiscriminate collection and use of personal data without adequate safeguards can undermine trust, erode privacy rights, and potentially lead to unintended consequences, such as discrimination or stigmatization based on sensitive information.

Furthermore, the storage and management of data generated by nanotechnology applications pose significant challenges to ensuring security and preventing unauthorized access or exploitation. As data storage capacities continue to expand and nanoscale materials are increasingly integrated into computing and communication systems, the risk of data breaches, cyber-attacks, and privacy violations escalates. Vulnerabilities in nanotechnology-based data storage and processing systems could expose individuals to identity theft, financial fraud, or surveillance, raising profound ethical concerns about the protection of sensitive information and the preservation of privacy rights in the digital age.

Moreover, the globalization of nanotechnology research and development introduces additional complexities to the ethical landscape of data privacy and security. Collaborative research initiatives and multinational partnerships may involve the exchange of data across borders, each governed by its own regulatory frameworks and cultural norms regarding privacy and data protection. Discrepancies in regulatory standards and enforcement mechanisms can create legal ambiguities and ethical dilemmas, particularly in cases where conflicting interests or values intersect. Harmonizing international standards for data privacy and security in nanotechnology research is essential for promoting consistency, coherence, and accountability across diverse contexts and jurisdictions [33].

Addressing the ethical implications of data privacy and security in nanotechnology necessitates a proactive and interdisciplinary approach that integrates insights from ethics, law, technology, and public policy. Ethical frameworks, such as the principles of fairness, transparency, and accountability, can guide the development and implementation of responsible data governance practices, ensuring that the collection, use, and dissemination of data are conducted in a manner consistent with ethical values and societal norms. Additionally, regulatory mechanisms, such as data protection laws, privacy regulations, and cyber-security standards, play a crucial role in establishing legal safeguards and enforcement mechanisms to protect individuals' privacy rights and mitigate risks associated with data breaches or misuse [34].

Furthermore, fostering a culture of ethical awareness and responsibility within the nanotechnology community is essential for promoting ethical conduct and mitigating potential risks. Education and training programs can raise awareness about the ethical implications of data privacy and security in nanotechnology research, equipping researchers, policymakers, and industry stakeholders with the knowledge and skills to navigate complex ethical dilemmas and make informed decisions [35]. Additionally, interdisciplinary collaboration and stakeholder engagement provide opportunities for dialogue, consensus-building, and collective action, enabling stakeholders to collaboratively develop ethical guidelines, best practices, and governance mechanisms that reflect diverse perspectives and values.

In conclusion, the ethical implications of data privacy and security in nanotechnology underscore the need for proactive and interdisciplinary approaches to address complex challenges and safeguard individuals' rights and well-being [36]. By integrating ethical principles, regulatory mechanisms, and stakeholder engagement into the fabric of nanotechnology research and development, stakeholders can foster a culture of responsible innovation, uphold privacy rights, and promote the ethical use of data for the benefit of society.

Protection of Personal Information in the Context of Nanotechnology Applications

In the realm of nanotechnological applications, the protection of personal information is paramount due to the increasing integration of nanoscale sensors, devices, and systems into various facets of daily life. These applications, ranging from biomedical sensors to environmental monitoring devices, can collect and process sensitive data, including biometric information, health records, and behavioural patterns. As nanotechnology continues to advance, the volume and diversity of data generated pose significant challenges to safeguarding privacy and ensuring the security of personal information. Ensuring the protection of personal information in the context of nanotechnology applications requires robust regulatory frameworks, technological safeguards, and ethical considerations to mitigate risks and uphold individuals' privacy rights.

Regulatory frameworks play a crucial role in establishing legal safeguards and enforcement mechanisms to protect personal information in nanotechnology applications. Data protection laws, privacy regulations, and industry standards govern the collection, use, and dissemination of personal data, imposing obligations on data controllers to implement measures to safeguard data privacy and security. For example, the General Data Protection Regulation (GDPR) in Europe sets stringent requirements for the processing of personal data, including provisions for informed consent, data minimization, and data breach notification [37]. Similarly, regulatory agencies, such as the U.S. Food and Drug Administration (FDA) may impose data security requirements for medical devices incorporating nanotechnology, ensuring that personal health information is protected from unauthorized access or disclosure [38]. Technological safeguards also play a critical role in protecting personal information in nanotechnology applications. Encryption, authentication mechanisms, and access controls can help prevent unauthorized access to sensitive data, reducing the risk of data breaches or cyber-attacks. Additionally, anonymization and pseudonymization techniques can be employed to de-identify personal information, mitigating privacy risks while still enabling data analysis and research. Furthermore, secure data storage and transmission protocols, such as blockchain technology, offer decentralized and tamper-resistant solutions for safeguarding personal information, enhancing data integrity and confidentiality in nanotechnology applications.

Ethical considerations are essential for guiding the responsible development and deployment of nanotechnology applications while upholding individuals' privacy rights. Ethical frameworks, such as principles of autonomy, beneficence, and justice, provide guidelines for ethical decision-making and responsible innovation in nanotechnology. Researchers, policymakers, and industry stakeholders must balance the potential benefits of nanotechnology applications with the ethical imperative to protect individuals' privacy and autonomy. Transparency, accountability, and stakeholder engagement are crucial for fostering trust and ensuring that ethical considerations are integrated into all stages of the research and development process [39].

In conclusion, protecting personal information in the context of nanotechnology applications requires a multifaceted approach that encompasses regulatory, technological, and ethical considerations. By implementing robust regulatory frameworks, leveraging technological safeguards, and adhering to ethical principles, stakeholders can mitigate risks, uphold privacy rights, and promote responsible innovation in nanotechnology. Ultimately, safeguarding personal information is essential for fostering public trust, advancing scientific progress, and maximizing the societal benefits of nanotechnology applications [40].

Ethical Considerations in Nanomedicine

Ethical Implications in the Use of Nanotechnology in Medicine

Nanotechnology can undoubtedly revolutionize the advancements in health care through designing, producing, and applying systems, structures, and devices at a tiny scale. Nanotechnology enhances materials used in medicines, making them more effective, such as improved drug delivery systems. Microfluidic devices as if miniaturized blood analysis tools demonstrate the potential for faster, more accurate diagnostics. Nano-biosensors can detect specific proteins linked to diseases like cancer or even track the effectiveness of personalized asthma treatments. Furthermore, nano-functionalized robots promise to be beneficial for the detection of early disease and the delivery of therapeutic agents. These nanobots may be able to recognize targets within the body (eg. a cancer cell) and perform specialized and targeted treatment. Such treatment supports delivery by releasing a drug in a localized area that minimizes the potential systemic side effects of convectional drug therapy.

The examination of ethical implications in the use of nanotechnology in medicine is a critical aspect of contemporary healthcare discourse. As nanotechnology continues to revolutionize medical treatments, diagnostics, and drug delivery systems, ethical considerations become increasingly significant. Nanomedicine raises questions about patient privacy, informed consent, and equitable access to innovative treatments. The potential risks associated with nanomaterials and nanoparticles, such as toxicity and environmental impact, also demand careful ethical scrutiny. Ethicists, policymakers, healthcare professionals, and researchers must collaborate to address these challenges and ensure that the benefits of nanotechnology in medicine are balanced with ethical principles and societal values, ultimately advancing patient care and well-being.

The simultaneous surge in advanced technologies like machine learning, neural networks, and artificial intelligence offers opportunities for personalized healthcare at the nanoscopic level. Machine learning advancements in the diagnosis of any disease enable faster cross-platform communication and interpretations that are more accurate. However, integrating nanomedical devices with the Internet of Things raises legal and ethical concerns. Innovations like nano-sensors in medical devices facilitate health monitoring yet they also raise questions about privacy and security. Despite the potential of nanotechnology to revolutionize healthcare, ethical and legal challenges persist.

Nanotechnology, with its potential to revolutionize medicine, also raises significant ethical concerns. One key issue is informed consent. As this technology is still evolving, the long-term effects and risks of using nanoparticles for medical use are not explored completely. A significant concern highlighted pertains to the potential risks associated with the use of nanomedical technology, primarily due to the heightened reactivity of nanomaterials compared to their bulk counterparts. A few reports also suggest a high risk of severe diseases like breast cancer, pulmonary fibrosis, and DNA damage due to nanoparticles [41]. Hence, obtaining informed consent from patients who may not fully grasp these uncertainties becomes a challenging issue. Additionally, the potential for unintended consequences arising from the unique properties of nanoparticles at the cellular level necessitates careful consideration of potential harm.

Promising medical technologies like somatic stem cell therapy, tissue engineering, and therapeutic cloning raise concern over the source of the stem cells, as they are obtained either from the umbilical cord or from the spinal cord of humans. Hence, ethical issues including informed consent and risk analysis surround the use of stem cells in such technologies and raise the question of the principle of justice as to who can be offered the treatment irrespective of their socio-economic status. Furthermore, such therapies could encourage malpractices of organ harvesting and feticides. Therefore, the central ethical question revolves around determining the status of the human embryonic stem cells that deserve protection asserting the moral dignity that should not suffer at the hand of scientific research.

These methods offer potential avenues for stem cell research without involving the moral ambiguity associated with stem cell use.

Finally, the potential for misuse of nanotechnology in medicine cannot be ignored. The ability to manipulate matter at the atomic level raises concerns about the development of undetectable performance-enhancing drugs. Hence, strict ethical frameworks and regulations need to be developed to ensure the responsible and safe application of this powerful technology in healthcare.

Balancing the Potential Benefits of Nanomedicine with Ethical Concerns

As nanomedicine transitions from research to real-world applications, a complex set of ethical questions confronts healthcare professionals. While these interventions hold great promise for improving health, they also introduce unfamiliar territory for doctors in terms of prevention, diagnosis, and treatment. Therefore, balancing responsible development with economic potential is a complicated discussion to face. The ethical analysis also examines value conflicts between the personal freedom to use economic resources to obtain advanced nano-medicinal treatments and what society at large considers desirable or ethically acceptable. Concern for economic competitiveness and other economic values (economic growth) may come into conflict with respect for human dignity. The unrestricted freedom of some may endanger the health and safety of others. Therefore, a balance has to be struck between values that are all legitimate in our culture. Unlike a single technology, nanomedicine encompasses diverse tools applied to devices, materials, procedures, and treatment methods. Two main categories exist: diagnostic and therapeutic. Diagnostic nanomedicine involves using nanoparticles for imaging, detecting tumours, or employing microchips for various diagnostic purposes. Therapeutic nanomedicine, on the other hand, utilizes nanoparticles for drug delivery, gene therapy, or other treatment modalities. As nanomedicine progresses, safeguarding patient privacy and confidentiality becomes paramount. This emerging field promises to collect vast amounts of data at the cellular and subcellular levels, raising ethical concerns about potential misuse and individual autonomy. Nanotechnology allows for implantable devices miniaturized to function at the subcellular level, monitoring and collecting data on cellular activity and biochemical events within individual organs, tissues, or even cells. These devices might even be remotely accessible for data transmission. Companies offer implantable microchips that store unique identification numbers, accessible for various purposes, including accessing medical information or authentication. Only by carefully balancing the promise and pitfalls of nanomedicine can we ensure its safe and ethical advancement for the benefit of all.

Another emerging question of the current trends in this discipline of ethics qualifies for nanotechnology application. Are the too restrictive ethical considerations barriers to the rapid integration of nanotechnological developments to the betterment of human life? Take an example, neuro-technology, a rapidly growing field in the intersection between nano and biotechnology, has the potential to affect the mental and moral capacities of the brain, suggesting that ethics itself may be affected by nanotechnology. When it comes to life-saving medications and medicines being developed via nano-biotechnology the ethical considerations have got to be sidelined in the interest of humanity. These advancements raise critical questions like; control over patients' collected data, and their control over their health information, ensuring transparency and access can be challenging. Another aspect is data security and the potential for misuse of sensitive health data poses significant risks, requiring robust security measures. A rapidly emerging opinion that suggests rather than artificially putting barriers in the progress of nano-bio-medical applications by advocating ethics, society muster the collective consciousness and wisdom in terms of logistic, financial, and moral courage to allow such extraordinarily powerful medicines to be deployed for human betterment, of course integrating it with essential ethical considerations.

International Perspectives on Nanotechnology Ethics

Comparison of Different Countries' Approaches and Regulations Regarding Ethical Considerations in Nanotechnology

Analysing different countries' approaches and regulations concerning ethical considerations in nanotechnology provides valuable insights into the diverse perspectives, priorities, and challenges shaping the governance of this rapidly advancing field. Nanotechnology, with its potential to revolutionize various industries and domains, raises complex ethical issues related to safety, privacy, equity, and societal impact [40]. By examining the approaches of key countries, such as the United States, the

European Union, China, and Japan, we can gain a comprehensive understanding of the regulatory landscape and the strategies employed to address ethical concerns in nanotechnology research and development.

The United States has been a pioneer in nanotechnology research and innovation, with a strong emphasis on promoting scientific progress and technological advancement. Ethical considerations in nanotechnology are primarily addressed through a combination of federal regulations, voluntary guidelines, and public–private partnerships. The National Nanotechnology Initiative (NNI), launched in 2000, coordinates nanotechnology research efforts across various federal agencies and provides guidance on ethical, legal, and societal implications (ELSI) through its Nanotechnology Signature Initiatives. Additionally, regulatory agencies, such as the Food and Drug Administration (FDA) and the Environmental Protection Agency (EPA) oversee the safety and environmental impact of nanotechnology products and applications, ensuring compliance with existing laws and regulations. Despite these efforts, critiques of the U.S. approach highlight gaps in oversight, fragmented regulation, and limited public engagement, necessitating greater coordination and transparency to address emerging ethical challenges effectively.

In contrast, the European Union (EU) has adopted a more precautionary approach to nanotechnology governance, prioritizing risk assessment, public participation, and stakeholder engagement. The EU's regulatory framework for nanotechnology encompasses a range of directives, regulations, and initiatives aimed at ensuring the safe and responsible development and use of nanomaterials. The European Commission's Nanotechnology Observatory and the European NanoSafety Cluster facilitate information sharing, risk assessment, and policy development, while the European Group on Ethics in Science and New Technologies advises policymakers on ethical issues. Moreover, the EU's Horizon 2020 program funds research projects focusing on ethical, legal, and societal aspects of nanotechnology, promoting interdisciplinary collaboration and responsible innovation [42]. Despite these efforts, challenges persist in harmonizing regulations across member states, addressing ethical concerns proactively, and integrating societal values into decision-making processes.

China, as a global leader in nanotechnology research and investment, has prioritized economic development and technological innovation, often at the expense of ethical considerations and environmental concerns. While China has established regulatory frameworks for nanotechnology, enforcement and oversight remain inadequate, leading to lax safety standards, insufficient risk assessment, and limited public engagement. The Chinese government's Nanotechnology Strategic Action Plan and funding initiatives support research and development in nanotechnology, with an emphasis on industrial applications and commercialization. However, concerns have been raised regarding the protection of workers' health and safety, environmental pollution, and the potential misuse of nanotechnology for surveillance and control purposes. Despite recent efforts to strengthen regulations and enhance transparency, China's approach to nanotechnology governance continues to face scrutiny over its ethical implications and societal impact [43]. Japan, renowned for its leadership in technological innovation and industrial competitiveness, has adopted a balanced approach to nanotechnology governance, integrating ethical considerations with scientific progress and economic growth [44]. Japan's regulatory framework for nanotechnology encompasses laws, guidelines, and industry standards aimed at ensuring safety, quality, and ethical conduct. The Japanese government's Council for Science, Technology, and Innovation advises policymakers on ethical, legal, and societal issues related to nanotechnology, fostering dialogue among stakeholders and promoting responsible innovation. Additionally, Japan's Nanotechnology Innovation Act encourages collaboration between academia, industry, and government to address ethical challenges, enhance public trust, and maximize the societal benefits of nanotechnology. Despite these efforts, Japan faces on-going challenges in addressing public concerns, fostering transparency, and balancing economic imperatives with ethical considerations in nanotechnology research and development [45].

In conclusion, a comparative analysis of different countries' approaches and regulations regarding ethical considerations in nanotechnology reveals a diverse landscape shaped by varying cultural, political, and socioeconomic factors. While each country has made strides in addressing ethical challenges and promoting responsible innovation in nanotechnology, significant gaps and shortcomings remain, necessitating greater international collaboration, information sharing, and dialogue. By learning from each other's experiences, best practices, and lessons learned, stakeholders can work towards establishing a robust and inclusive framework for governing nanotechnology that upholds ethical values, safeguards human health and the environment, and promotes the common good.

International Collaborations and Efforts to Address Ethical Concerns

International collaborations and efforts to address ethical concerns in research and development (R&D) represent crucial endeavours aimed at promoting responsible innovation, fostering global cooperation, and addressing shared challenges. In an increasingly interconnected world, where scientific advancements transcend national borders and impact diverse communities, international collaboration is essential for addressing complex ethical issues that arise in the pursuit of scientific progress. By examining key initiatives, partnerships, and frameworks at the international level, we can gain insights into the strategies employed to navigate ethical dilemmas, promote ethical conduct, and maximize the societal benefits of R&D.

One prominent example of international collaboration in addressing ethical concerns in R&D is the Global Research Council (GRC), an alliance of research funding agencies from around the world committed to promoting the quality, integrity, and societal impact of research. The GRC's Statement of Principles for Scientific Merit Review emphasizes the importance of ethical conduct, transparency, and accountability in research, providing guidance to funding agencies on best practices for peer review and research evaluation. Additionally, the GRC's annual meetings and working groups facilitate dialogue, information sharing, and capacity building among member countries, enabling them to address ethical challenges collaboratively and harmonize policies and practices [46].

Another notable initiative is the Responsible Research and Innovation (RRI) framework, promoted by the European Union (EU) and adopted by various international organizations and research consortia. RRI emphasizes the integration of ethical, societal, and environmental considerations into the entire research and innovation process, from the conception of research projects to their implementation and dissemination. Through initiatives, such as the Horizon 2020 program and the European Research Area (ERA), the EU supports RRI by funding interdisciplinary research projects, promoting stakeholder engagement, and fostering a culture of openness, inclusivity, and accountability in R&D. Moreover, the EU's Global Approach to Research and Innovation (GARI) promotes RRI principles in international collaborations, ensuring that ethical concerns are addressed comprehensively across borders and disciplines [47].

Furthermore, international organizations, such as the United Nations Educational, Scientific and Cultural Organization (UNESCO) and the World Health Organization (WHO) play vital roles in addressing ethical concerns in R&D at the global level [48]. UNESCO's Universal Declaration on Bioethics and Human Rights provides ethical principles and guidelines for biomedical research, emphasizing respect for human dignity, autonomy, and justice. Similarly, the WHO's Global Observatory on Health Research and Development promotes the ethical conduct of health research worldwide, supporting countries in strengthening their research governance systems, capacity-building initiatives, and ethical review processes. Through these initiatives, international organizations contribute to the development of ethical norms, standards, and guidelines that guide R&D practices and protect the rights and well-being of individuals and communities.

Moreover, industry-led initiatives and partnerships play a significant role in addressing ethical concerns in R&D, particularly in sectors, such as biotechnology, pharmaceuticals, and information technology. Collaborative initiatives, such as the Pharmaceutical Research and Manufacturers of America's (PhRMA) Principles on Conduct of Clinical Trials and Communication of Clinical Trial Results and the Information Technology Industry Council's (ITI) Framework for Responsible and Ethical AI emphasize industry's commitment to ethical conduct, transparency, and accountability in R&D. These initiatives guide companies on best practices for ethical decision-making, stakeholder engagement, and risk management, while also fostering collaboration with academia, government, and civil society to address shared ethical challenges [49–51].

Despite these efforts, several challenges persist in addressing ethical concerns in international R&D collaborations. One challenge is the lack of harmonization and consistency in ethical standards and regulatory frameworks across countries and regions, which can lead to confusion, inconsistency, and regulatory burden for researchers and institutions operating in multiple jurisdictions. Additionally, cultural, social, and political differences may influence perceptions of ethical issues and the acceptability of certain research practices, posing challenges to effective collaboration and communication among diverse stakeholders. Moreover, the rapid pace of technological innovation and the emergence of novel research areas, such as gene editing and artificial intelligence, raise new ethical questions and dilemmas that require ongoing dialogue, reflection, and adaptation of existing ethical frameworks and guidelines.

In conclusion, international collaborations and efforts to address ethical concerns in R&D play a crucial role in promoting responsible innovation, fostering global cooperation, and maximizing the societal benefits of scientific progress. By leveraging partnerships, frameworks, and initiatives at the international level, stakeholders can address complex ethical dilemmas, harmonize standards and practices, and ensure that research advances are conducted ethically, transparently, and with respect for human dignity and rights. Moving forward, sustained commitment to ethical conduct, interdisciplinary collaboration, and stakeholder engagement will be essential for navigating the evolving landscape of R&D and addressing emerging ethical challenges in a rapidly changing world.

Ethical Frameworks and Guidelines

Existing Ethical Frameworks and Guidelines for the Development and Deployment of Nanotechnology

Rapid technical advancements often pose unprecedented ethical dilemmas, compelling the evaluation and re-establishment of equipped tools to keep a check on the ever-evolving nanotechnological developments. Ethics are the driving force, to handle responsibly the complexities of technological advancements, through establishing and implementing feasible frameworks to social and judicial norms.

The existing framework in nanotechnology encompasses a multifaceted landscape of scientific, regulatory, and industrial frameworks designed to guide the development, application, and governance of nanoscale materials and technologies. At its core, this framework integrates interdisciplinary research efforts spanning physics, chemistry, biology, and engineering to understand and manipulate matter at the nanoscale. Key regulatory bodies worldwide have established guidelines to ensure the safe production, use, and disposal of nanomaterials, addressing concerns related to environmental impact, health hazards, and ethical considerations. Moreover, industry standards and best practices are continually evolving to promote innovation while mitigating potential risks associated with nanotechnology applications across various sectors, including healthcare, electronics, energy, and environmental remediation. Collaborative efforts between academia, government agencies, and industry stakeholders play a pivotal role in shaping this framework, fostering transparency, accountability, and responsible stewardship of nanotechnology advancements in today's rapidly evolving scientific landscape. As research progresses and technologies mature, the existing framework in nanotechnology will continue to adapt and evolve to meet the dynamic challenges and opportunities inherent in this transformative field. Recognizing the international nature of nanotechnology research and development, international organizations like the OECD (Organization for Economic Co-operation and Development) play a crucial role in developing best practices and promoting information sharing. Organizations like ISO (International Organization for Standardization) are developing standards for characterizing nanomaterials, assessing their safety and environmental impact, and ensuring responsible production and life cycle management. Many countries and organizations issue guidance documents to provide clarity on how existing regulations apply to nanotechnology and to offer recommendations for responsible research and development practices.

Rapid technical advancements often pose unprecedented ethical dilemmas, compelling the evaluation and re-establishment of equipped tools to keep a check on the ever-evolving nanotechnological developments. Ethics are the driving force, to handle responsibly the complexities of technological advancements, through establishing and implementing feasible frameworks, on social and judicial norms. Drafting of these frameworks involves consideration of various aspects of the brimming technology and its unprecedented consequences. These ethical considerations ought to safeguard and ensure responsible development prioritizing overall human well-being and environmental sustainability. Implementation of such novel frameworks also requires awareness and trust of the society which can be achieved through proper transparency and accountability. This, in turn, encourages broader societal acceptance and maximizes the positive impact of technology.

Recognizing the critical role of ethics is the first step towards responsible innovation and integration of nanotechnology globally. By integrating ethical considerations into the core of nanotechnological endeavours, we can harness the power of technology for good, building a future that is not only technologically advanced but also just, equitable, and sustainable. The future we create is ultimately a collective responsibility, and ethical considerations are the compass guiding us towards a brighter tomorrow.

The conception stage of the ethical frameworks should ensure a holistic approach that anticipates and mitigates the underlying potential, technical evolutions, and associated harm. Ethical frameworks must be shaped through open dialogues on public platforms ensuring transparency and better understanding and through collaborations, encompassing the voices of not only scientists and engineers but also ethicists, policymakers, and the communities affected by such technological advancements. Hence, these ever-evolving advancements demand acceptable and adaptable working frameworks only possible through continuous revisiting and refinements to make them robust, relevant, acceptable, and effective in addressing the dynamical challenges. Embedding ethical frameworks equips us with the tools to navigate these dilemmas responsibly, making informed decisions that align with our collective aspirations [52].

Of course, this kind of collaboration is not always easy. It requires us to step outside our comfort zones, listen to opposing viewpoints, and be willing to compromise. Nevertheless, the potential rewards are immense. By working together, we can ensure that technology is used for good, making the world a more just and equitable place for everyone. So, let us break down the misconceptions about this brimming concept by open dialogue, and start navigating the ethical maze of nanotechnology together. Remember, we are all in this boat, and we are all better off when we row in the same direction.

Importance of Ethical Decision-Making and Responsibility in Nanotechnology

Ethical complexities are often multifaceted, no single individual or group has all the answers, by collective opinions and perspectives, a more comprehensive understanding of the challenges and potential solutions are expected to be addressed, that are more likely to support new technologies and regulations. Collaboration fosters trust and keeps the dialogue open and feasible for their deployment across diversified scenarios across borders. As technology evolves, so too must our ethical considerations and regulations help us identify new challenges and adapt our approaches accordingly thereby, creating safe spaces where people can express their concerns without fear of judgment which is essential for fostering open and honest dialogue. Enough transparency, educating the population, and convincing them to follow these regulations is required. Honest reviews on the emerging technology, feasible flexibility of the regulations, and unbiased frameworks.

In the realm of science and technology, few fields hold as much promise and potential as nanotechnology. With the ability to manipulate matter at the nanoscale, scientists and engineers have unlocked a plethora of opportunities for innovation across various industries, from healthcare to electronics. However, amidst the excitement and progress, it is imperative to underscore the critical importance of ethical decision-making and responsibility in the development and application of nanotechnology. Nanotechnology offers groundbreaking solutions to some of the most pressing global challenges. It has the potential to revolutionize medicine through targeted drug delivery systems, enhance energy efficiency with advanced materials, and even contribute to environmental sustainability through pollution remediation technologies. However, the sheer power of nanotechnology necessitates a careful examination of its ethical implications.

One of the primary concerns surrounding nanotechnology pertains to environmental and health risks. As researchers delve deeper into the manipulation of materials at the nanoscale, the potential for unintended consequences increases. Nanoparticles, for instance, may exhibit unique properties that could pose unforeseen risks to ecosystems and human health. Therefore, ethical decision-making demands thorough risk assessments and mitigation strategies to ensure the responsible development and deployment of nanotechnological solutions. Moreover, the equitable distribution of the benefits of nanotechnology is another ethical consideration. While nanotechnology holds immense promise, there is a risk that it may exacerbate existing social and economic disparities if access to its benefits is unequal. Therefore, efforts must be made to ensure that the fruits of nanotechnological innovation are accessible to all segments of society, particularly those in underserved communities.

Ethical considerations also extend to issues of privacy and security in the age of nanotechnology. As the boundaries between the physical and digital worlds blur, the potential for intrusive surveillance and data breaches becomes more pronounced. Nanoscale sensors and devices embedded in everyday objects could collect vast amounts of personal data, raising concerns about individual privacy and autonomy. Ethical frameworks must be established to govern the responsible use and safeguarding of sensitive information in the nanotechnology era. Furthermore, the responsible conduct of research and innovation in nanotechnology requires adherence to established ethical principles and guidelines. Scientists and engineers must conduct their work with integrity, transparency, and respect for human rights. This includes accurately reporting research findings, obtaining informed consent from research participants, and considering the broader societal implications of their work.

Institutions and regulatory bodies also play a crucial role in ensuring ethical standards are upheld in nanotechnology research and development. Robust oversight mechanisms, ethical review boards, and regulatory frameworks are essential for safeguarding against potential harm and promoting the responsible advancement of nanotechnology. Education and public engagement are key components of fostering ethical awareness and responsibility in nanotechnology. By promoting dialogue and awareness about the ethical implications of nanotechnology, we can empower stakeholders to make informed decisions and advocate for ethical practices in research, development, and deployment.

The ethical dimensions of nanotechnology cannot be overstated. As we continue to unlock the vast potential of nanotechnology, it is imperative that we do so with a deep sense of ethical responsibility. By integrating ethical considerations into every stage of the nanotechnology lifecycle, we can ensure that the benefits of this transformative field are realized in a manner that upholds human dignity, promotes social justice, and protects the well-being of present and future generations.

Public Engagement and Education

Importance of Public Awareness and Understanding of Nanotechnology and Its Ethical Implications

The rapid advancement of technology not only brings incredible opportunities but also throws up complex ethical dilemmas. From the potential biases of AI algorithms to the privacy concerns surrounding facial recognition, these challenges demand careful consideration and nuanced solutions. In navigating this ethical maze, public awareness, open dialogue, and collaboration are not just important but are essential. Hence, scientists and philosophers from across the globe should come forth to brainstorm the ethical implications of emerging technologies before they are fit to be

Table 15.1 Assessment of public awareness towards nanotechnology [54]

Classification	Strongly agree (%)	Agree (%)	Neutral (%)	Disagree (%)	Strongly disagree (%)
Transparent information about nanotechnology is required	41.5	42.3	14.5	1.2	0.4
Appropriate communication about the risks of nanotechnology	39.5	37.1	21.4	0.8	1.2
Promotion and education on nanotechnology are necessary	36.7	42.7	19.0	0.8	0.8

released into the world. Policymakers and community leaders are expected to work together to understand and develop regulations that reflect diverse perspectives and protect vulnerable populations. Additionally, the developers of the technology and civil society organizations should collaborate to design technologies that are not only powerful but also fair and inclusive [53].

To foster a holistic understanding and acceptance of nanotechnology to the very root of humanity it becomes very necessary to reach out and engage the public through educational programs, interactive exhibits, participation, and initiatives that can demystify nanotechnology and spark informed discussions. Moreover, scientists, policymakers, and industry leaders must communicate explicitly on open forums about nanotechnology's potential risks and benefits, using accessible language and engaging formats. Open forums, citizen panels, and public consultations ensure diverse voices are heard and inform decision-making processes. Ethical complexities are often multifaceted, no single individual or group has all the answers, by collective opinions and perspectives, a more comprehensive understanding of the challenges and potential solutions are expected to be addressed, and they are more likely to support new technologies and regulations. Collaboration fosters trust and keeps the dialogue open and feasible for their deployment across the diversified scenarios across the borders. As technology evolves, so too must our ethical considerations and regulations help us identify new challenges and adapt our approaches accordingly thereby, creating safe spaces where people can express their concerns without fear of judgment which is essential for fostering open and honest dialogue Table 15.1. Summarizes the assessment of public awareness towards nanotechnology.

Strategies for Effective Communication and Education About Nanotechnology Ethics

Nanotechnology raises profound questions about humanity, applications like genetic engineering and brain-computer interfaces blur the lines between the natural and artificial. These ethical quandaries demand open and inclusive dialogue, ensuring technological progress that aligns with our shared values and aspirations. Addressing these concerns is not simply a matter of caution; but is imperative for responsible innovation. By integrating social and ethical considerations into the very fabric of nanotechnology research and development, we can ensure that this powerful technology serves as a force for good, shaping a future that is not only technologically advanced but also just, sustainable and equitable [55]. Researchers and governments all over the world are engrossed with concerns about the fast-growing integration of human life and artificial entities being developed through nanotechnology. A need is being felt to encourage trans-disciplinary understanding by a multidisciplinary group of researchers from fields like nanotechnology, biochemistry, medical science, artificial intelligence, robotics, anthropology, philosophy, psychology, and other cognitive sciences. The common agenda for the specialized group should be to identify nanotech-ethical issues and analyse the concerns regarding the integration of human lives and fast-developing dynamic artificial entities. Continued research is essential to understand the long-term risks and benefits of nanotechnology, inform policy decisions, and develop safer materials and applications. These specialized agendas can be implemented feasibly and successfully through proper investment in educating and training citizens around the globe. The nations should take the initiative to equip their people with the skills and knowledge necessary to participate in the nanotechnology revolution. This includes supporting education, providing resources, and infrastructures, training programs in nanotechnology applications, and fostering knowledge-sharing initiatives. Open dialogue engaging stakeholders from diverse backgrounds, including scientists, policymakers, ethicists, and the public, in open and transparent dialogues is crucial for developing responsible and effective regulations and addressing ethical concerns.

Additionally, adequate access to the internet and digital tools is crucial for participating in the era of nanotechnology. Investments in infrastructure, affordable technology, and digital literacy programs are essential to bridge the digital divide and ensure that everyone has the opportunity to benefit from this technology. Encouraging and supporting local research and development efforts in developing countries will ensure that nanotechnology solutions are tailored to their specific needs and contexts. This could involve funding local research initiatives, creating technology hubs, and promoting collaboration between local and international researchers for innovations in nanotechnology. Fostering international collaboration and harmonization of regulations is critical to ensure the responsible development and use of nanotechnology while minimizing potential risks and maximizing its benefits for all. Finally, by promoting open access and knowledge sharing, continuous research trials and technological advancements freely available will accelerate the development and diffusion of nanotechnology in developing countries. This could involve open-access publishing platforms, technology transfer initiatives, and collaborative research projects. Raising public awareness about the potential of nanotechnology, alongside its risks and ethical considerations, empowers individuals to engage in informed discussions and participate in decision-making processes. By acknowledging these challenges and actively engaging in multi-faceted solutions, we can navigate the complex landscape of nanotechnology and harness its potential for a better future while ensuring responsible and ethical development. Remember, the journey towards a brighter nanoscope requires a collective effort built on transparency, open dialogue, and a commitment to responsible innovation.

Summary of Key Ethical Considerations in the Development and Deployment of Nanotechnology

Nanotechnology, since its commencement in the 1980s, has made a tremendous journey of both dazzling possibilities and ethical landmines. From its nascent discovery, this manipulation of matter at the atomic and molecular level has blossomed into a field brimming with promises of revolutionizing everything from medicine to materials science. However, amidst the excitement, questions about the safety, equity, and even the very definition of humanity have shadowed its progress, early concerns focused on the potential environmental havoc due to nanomaterials, because of their minuscule size that could potentially cause unforeseen harm to ecosystems and human health. As the field matured, however, this ethical landscape broadened, and issues like the accessibility of this powerful technology, the potential for misuse in autonomous systems, and the blurring lines between humans and machines through genetic engineering and brain-computer interfaces became pressing concerns. The development and implementation of a consolidated ethical framework beyond borders, across disciplines, and through generations would ensure a safe future because the choices of today will determine the legacy of nanotechnology. An unwavering commitment to ethics and sustainability paves a path towards the future where the ultimate power of this technology serves the well-being of future generations. A holistic implementation of a compatible ethical framework secures a future where nanotechnology reflects our shared values and aspirations. Let this be not a footnote in history, but a testament to collective responsibility. NOW is the perfect time for ACTION. The journey of nanotechnology is by no means over. Hence, it will be a continuous effort to harness immense potential and ensure a responsible development of nanotechnology. With careful consideration and unwavering ethical vigilance, this microscopic revolution can illuminate a brighter future for all. This is not the end of the conversation, but the beginning. As we explore the nanoscale world, let us do so with open minds, ethical hearts, and a shared commitment to building a future where nanotechnology illuminates a brighter path for all.

Future Challenges and Directions in Nanotechnology Ethics Research and Implementation

Future challenges and directions in nanotechnology ethics research and implementation are crucial as this field continues to burgeon with innovative applications. As nanotechnology advances, ethical considerations become increasingly pertinent.

One of the primary challenges lies in ensuring the responsible development and deployment of nanotechnology across diverse domains, such as medicine, environment, and consumer products. Key ethical concerns include issues related to privacy, security, equity, and informed consent, particularly concerning the use of nanomaterials in medical interventions and surveillance technologies. Moreover, the potential for unintended consequences and unforeseen risks necessitates robust ethical frameworks and risk assessment methodologies. As nanotechnology becomes more integrated into society, there is a growing need for interdisciplinary collaboration among scientists, ethicists, policymakers, and stakeholders to address these complex ethical dilemmas. Additionally, enhancing public awareness and engagement regarding the ethical dimensions of nanotechnology is essential for fostering trust and accountability. Moving forward, prioritizing research into ethical implications, developing guidelines for responsible innovation, and implementing mechanisms for ethical oversight will be paramount in navigating the evolving landscape of nanotechnology ethically and responsibly. By proactively addressing these challenges, we can harness the transformative potential of nanotechnology while upholding ethical principles and societal values.

References

1. Karn, B., Wong, S.S.: Ten years of green nanotechnology. Sustainable nanotechnology and the environment: advances and achievements, ACS Publication, pp. 1–10 (2013)
2. Societal impact of Nanomaterials, Narendra Kumar Pandey and Priya Gupta, Book: Versatile Solicitations of Materials Science ISBN: 978–1–53619–763–1 Nova Science Publishers, Inc. (2021)
3. Chen, M., He, Y., Ye, Q., Wang, X., Hu, Y.: Shape-dependent solar thermal conversion properties of plasmonic Au nanoparticles under different light filter conditions. Sol. Energy **182**, 340–347 (2019)
4. Danesfahani, F., Tabatabaie, M.R.R.: Nanotechnology in electronics: a survey, by, Nanoscale Research Letters (2017)
5. Khan, I., Saeed, K., Khan, I.: Nanoparticles: properties, applications and toxicities. Arab. J. Chem. **12**(7), 908–931 (2019)
6. Guo, D., Xie, G., Luo, J.: Mechanical properties of nanoparticles: basics and applications. J. Phys. D Appl. Phys. **47**, 13001 (2014)
7. Schummer, J.: Societal and ethical implications of nanotechnology: meanings, interest groups and social dynamics. Nanotechnol. Chall. **2006**, 413–449 (2006)
8. Song, Y., Li, X., Du X.: Eur. Respir. J. **34**, 559 567. Song, Exposure to nanoparticles ıs related to pleural effusion, pulmonary fibrosis and granuloma, Eu J Resp. (2009). 10.118309031936.00178308
9. Saravanan, A. et al.: Environmental and health effects of nanomaterials. Nanomater. Applic. Biof. Bioene. Product. Syst. 701–711 (2021)
10. Dreher, K.L.: Health and environmental impact of nanotechnology: toxicological assessment of manufactured nanoparticles. Toxicol. Sci. **77**(1), 3–5 (2004)
11. Oberdörster, G., Oberdörster, E., Oberdörster, J.: Nanotoxicology: an emerging discipline evolving from studies of ultrafine particles. Environ. Health Perspect. **113**(7), 823–839 (2005)
12. Nel, A., Xia, T., Mädler, L., Li, N.: Toxic potential of materials at the nanolevel. Science **311**(5761), 622–627 (2006)

13. Handy, R.D., Owen, R., Valsami-Jones, E.: The ecotoxicology of nanoparticles and nanomaterials: current status, knowledge gaps, challenges, and future needs. Ecotoxicology **17**(5), 315–325 (2008)
14. Gottschalk, F., Sonderer, T., Scholz, R.W., Nowack, B.: Modeled environmental concentrations of engineered nanomaterials (TiO2, ZnO, Ag, CNT, fullerenes) for different regions. Environ. Sci. Technol. **43**(24), 9216–9222 (2009)
15. Kah, M., Tufenkji, N., White, J.C.: Nano-enabled strategies to enhance crop nutrition and protection. Nat. Nanotechnol. **5**(7), 573–577 (2009)
16. Linsinger, T. et al.: Requirements on measurements for the implementation of the european commission definition of the term "Nanomaterial." Publications Office of the European Union (2012)
17. Nel, A. et al.: Badesha. Nanotechnology environmental, health, and safety issues. in *Nanotechnology research directions for societal needs in 2020* (2011)
18. Price, N.: Artificial intelligence in health care: applications and legal issues. SciTech Lawyer **14**(1), 10–13 (2017)
19. Beauchamp, Childress (2019), Emanuel et al. (2008), the National Institutes of Health (2018), and the World Medical Association (2013)
20. Emanuel, E.J.: What makes clinical research ethical? JAMA **283**(20), 2701–2711 (2000)
21. National Institutes of Health: Protecting Human Research Participants (PHRP) - NIH Office of Intramural Training & Education (2018)
22. World Medical Association: World medical association declaration of Helsinki: ethical principles for medical research involving human subjects. JAMA **310**(20), 2191–2194 (2013)
23. Beauchamp, T.L., Childress, J.F.: Principles of biomedical ethics (8th ed.). Oxford University Press (2019)
24. Resnik, D.B.: What is ethics in research & why is it important? National Institute of Environmental Health Sciences (2015)
25. Kulinowski, K.M., Westervelt, R.M. (Eds.).: Ethics in Nanotechnology (1st ed.). John Wiley & Sons (2011)
26. Macnaghten, P., Kearnes, M.B., Wynne, B.: Nanotechnology, governance, and public deliberation: what role for the social sciences? Sci. Commun. **27**(2), 268–291 (2005)
27. Salamanca-Buentello, F., Persad, D.L., Court, E.B., Martin, D.K., Daar, A.S., Singer, P.A.: Nanotechnology and the developing World. PLoS Med. **2**(5), e97 (2005)
28. National Nanotechnology Initiative: Environmental, health, and safety research needs for engineered nanoscale materials (2011)
29. Resnik, D.B., Tinkle, S.S., Kissling, G.E.: Research ethics in the Era of nanotechnology. J. Nanopart. Res. **12**(3), 835–853 (2010)
30. Buentello, F.S., Daar, A.: Nanotechnology, equity and global health. Nat. Nanotechnol. **16**(4), 1–4 (2021)
31. Del Monte, I.A.: Nano weapons; a growing threat to humanity. Potomac Books; An imprint of University of Nabraska, Lincoln (2017)
32. Lin, P., Abney, K.: Robot Ethics 2.0: From Autonomous Cars to Artificial Intelligence. Oxford University Press (2016)
33. Moor, J.H.: Why we need better ethics for emerging technologies. Ethics Inf. Technol. **7**(3), 111–119 (2005)
34. Roco, M.C.W.S. (Eds.).: Converging technologies for improving human performance: nanotechnology, biotechnology, information technology and cognitive science. Springer (2013)
35. Scherer, M.U.: Regulation of nanotechnology: principles and legal frameworks. John Wiley & Sons (2016)
36. Tavani, H.T.: Ethics and technology: controversies, questions, and strategies for ethical computing. John Wiley & Sons (2015)
37. European Union: General data protection regulation (GDPR) (2016). https://eur-lex.europa.eu/eli/reg/2016/679/oj

38. U.S. Food and Drug Administration: Considerations for the Inclusion of Women in Clinical Trials and Analysis of Sex Differences (2019). https://www.fda.gov/regulatory-information/search-fda-guidance-documents/considerations-inclusion-women-clinical-trials-and-analysis-sex-differences
39. Culnan, M.J., Williams, C.C.: How ethics can enhance organizational privacy: lessons from the choicepoint and TJX data breaches. MIS Q. **33**(4), 673–687 (2009)
40. European Commission: Ethics Guidelines for Trustworthy AI (2021). https://ec.europa.eu/digital-single-market/en/news/ethics-guidelines-trustworthy-ai
41. Zheng, Y., Sanche, L.: Gold nanoparticles enhance DNA damage induced by anti-cancer drugs and radiation. Radiat. Res. **172**, 114–119 (2009)
42. National Nanotechnology Initiative: Ethical, legal, and societal implications (ELSI) of nanotechnology (2020). https://www.nano.gov/elsi
43. Nanotechnology Industries Association: Nanotechnology regulations in China (2020). https://www.nanotechia.org/our-activities/nanotechnology-regulations-in-china
44. Council for Science, Technology and Innovation (CSTI): White paper on science and technology (2020). https://www8.cao.go.jp/cstp/english/annualreport/2020.html
45. U.S. Food and Drug Administration: Nanotechnology task force report (2021). https://www.fda.gov/science-research/science-and-research-special-topics/nanotechnology-task-force-report
46. Global Research Council: About GRC (2021). https://www.globalresearchcouncil.org/about-grc
47. European Commission: Responsible research and innovation (2021). https://ec.europa.eu/programmes/horizon2020/en/h2020-section/responsible-research-innovation
48. United Nations Educational, Scientific and Cultural Organization (UNESCO): Universal Declaration on Bioethics and Human Rights (2005). https://unesdoc.unesco.org/ark:/48223/pf0000146055
49. World Health Organization (WHO): Research for universal health coverage: World Health Report 2013 (2012). https://www.who.int/whr/2013/report/en
50. Pharmaceutical Research and Manufacturers of America (PhRMA): Principles on conduct of clinical trials and communication of clinical trial results (2021). https://www.phrma.org/-/media/Project/PhRMA/PhRMA-Org/PhRMA-Org/PDF/Clinical-Trials/PhRMA_Principles_Clinical_Trials_and_Clinical_Trial_Results.pdf
51. Information Technology Industry Council (ITI): Framework for responsible and ethical AI (2018). https://www.itic.org/policy/ethical-ai-pledge
52. Bhatia, P., Chugh, A.: A multilevel governance framework for regulation of nanomedicine in India. Nanotechnol. Rev. **6**(4) (2016)
53. Jayanthi, A.P.: Nanotechnology: 'risk governance' in India. Econ. Polit. Weekly **47**(4), 34–40 (2012)
54. Araybeyyat, Z.H., et al.: Public awareness of nanotechnology and its implications for health in Jordan. Sustainability **14**(10), 5786 (2022). https://doi.org/10.3390/su14105786
55. Daar, A.S. et al.: Harnessing nanotechnology to improve global equity. Nat. Acad. Sci. Eng. Med. **21**(4) (2005)

Dr. Vernica Verma is former assistant professor at the Department of Physics, BBDNIIT, Lucknow, and completed her Ph.D. in Physics from the Department of Physics, University of Lucknow. She did her B.Sc. and M.Sc. in Physics from the University of Lucknow in 2005 and 2007, respectively. Her specialization in research is fabrication, characterization like XRD, SEM, AFM, UV-VIS, FT-IR, Raman, gamma irradiation, etc., and humidity and gas sensors.

Dr. Priya Gupta has been working as a lecturer at Manyawar Kanshiram Government Polytechnic, Tirwa, Kannauj, since January 2022. She has completed a Ph.D. from Integral University, Lucknow, in collaboration with Lucknow University, Lucknow, in the year 2021. She did her

M. Tech. from Madan Mohan Malviya Engineering College, Gorakhpur, in 2014, and her B.Tech. from Integral University, Lucknow.

Dr. Peramjeet Singh has completed his Ph.D. from the Department of Physics, University of Lucknow, Uttar Pradesh, India, in the field of materials science. He did his M.Sc. from Barkatullah Vishwavidyalaya, Bhopal, and his B.Sc. from GDC, Udhampur, University of Jammu, India, in 2017 and 2015, respectively. His area of research is nanotechnology, gas sensors, and the characterization of materials.

Prof. Narendra Kumar Pandey received his degree (hons.) in Physics from St. Xavier's College, Ranchi, India, in 1985, and the M.Sc. degree in Physics and the Ph.D. degree from the Indian Institute of Technology Delhi, India, in 1987 and 1995, respectively. He is currently professor and head of department with the Department of Physics, University of Lucknow, Lucknow, India. His areas of interest include humidity and gas sensing, nanotechnology, solid state and optical fibre sensors.

SPRINGER NATURE

GPSR Compliance

The European Union's (EU) General Product Safety Regulation (GPSR) is a set of rules that requires consumer products to be safe and our obligations to ensure this.

If you have any concerns about our products, you can contact us on ProductSafety@springernature.com

In case Publisher is established outside the EU, the EU authorized representative is:

Springer Nature Customer Service Center GmbH
Europaplatz 3
69115 Heidelberg, Germany

The manufacturer's authorised representative in the EU is Springer Nature Customer Service Centre GmbH, Europaplatz 3, 69115 Heidelberg, Germany. If you have any concerns regarding our products, please contact ProductSafety@springernature.com

Printed and bound by CPI Group (UK) Ltd, Croydon, CR0 4YY
01/10/2026
02225857-0016